Methods in Cell Biology

VOLUME 77

The Zebrafish: Genetics, Genomics, and Informatics

Series Editors

Leslie Wilson

Department of Molecular, Cellular and Developmental Biology
University of California
Santa Barbara, California

Paul Matsudaira

Whitehead Institute for Biomedical Research
Department of Biology
Division of Biological Engineering
Massachusetts Institute of Technology
Cambridge, Massachusetts

Methods in Cell Biology

VOLUME 77

The Zebrafish: Genetics, Genomics, and Informatics

Edited by

H. William Detrich, III

Department of Biology
Northeastern University
Boston, Massachusetts

Monte Westerfield

Department of Biology
University of Oregon
Eugene, Oregon

Leonard I. Zon

Division of Hematology/Oncology
Children's Hospital Boston
Boston, Massachusetts

ELSEVIER
ACADEMIC
PRESS

AMSTERDAM • BOSTON • HEIDELBERG • LONDON
NEW YORK • OXFORD • PARIS • SAN DIEGO
SAN FRANCISCO • SINGAPORE • SYDNEY • TOKYO

Elsevier Academic Press
525 B Street, Suite 1900, San Diego, California 92101-4495, USA
84 Theobald's Road, London WC1X 8RR, UK

This book is printed on acid-free paper. ∞

For all information on all Academic Press publications
visit our Web site at www.academicpress.com

ISBN: 0-12-564172-9

PRINTED IN THE UNITED STATES OF AMERICA
04 05 06 07 08 9 8 7 6 5 4 3 2 1

CONTENTS

4. Target-Selected Gene Inactivation in Zebrafish

Erno Wienholds and Ronald H. A. Plasterk

5. A High-Throughput Method for Identifying *N*-Ethyl-*N*-Nitrosourea (ENU)-Induced Point Mutations in Zebrafish

Bruce W. Draper, Claire M. McCallum, Jennifer L. Stout, Ann J. Slade, and Cecilia B. Moens

6. Production of Zebrafish Germline Chimeras by Using Cultured Embryonic Stem (ES) Cells

Lianchun Fan, Jennifer Crodian, and Paul Collodi

PART II   The Zebrafish Genome and Mapping Technologies

PART III   Transgenesis

PART IV   Informatics and Comparative Genomics

PART V   Infrastructure

CONTRIBUTORS

Numbers in parentheses indicate the pages on which the authors' contributions begin.

Violaine Alunni (505), Institut de Génétique et de Biologie Moléculaire et Cellulaire, UMR 7104 CNRS/INSERM/ULP, BP 10142, CU de Strasbourg, 67404 Illkirch Cedex, France

Chris T. Amemiya (545), Molecular Genetics Program, Benaroya Research Institute, Seattle, Washington 98101

Adam Amsterdam (3), Center for Cancer Research and Department of Biology, Massachusetts Institute for Technology, Cambridge, Massachusetts 01239

Hideki Ando (159), Laboratory for Developmental Gene Regulation, Brain Science Institute, RIKEN (The Institute of Physical and Chemical Research), Saitama 351-0198, Japan and CREST (Core Research for Evolutional Science and Technology), Japan Science and Technology Corporation (JST), Chuo-ku, Tokyo 103-0027, Japan

Shuichi Asakawa (173), Department of Molecular Biology, Keio University School of Medicine, Tokyo 160-8582, Japan

Nathan Bahary (305), Howard Hughes Medical Institute, Harvard Medical School, Boston, Massachusetts 02115

Herwig Baier (49), Department of Physiology, University of California, San Francisco, San Francisco, California 94143

Darius Balciunas (349), The Arnold and Mabel Beckman Center for Transposon Research, Department of Genetics, Cell Biology and Development, University of Minnesota Medical School, Minneapolis, Minnesota 55455

Bruce Barut (305), Howard Hughes Medical Institute, Harvard Medical School, Boston, Massachusetts 02115

Stéphane Berghmans (645), Department of Pediatric Oncology, Dana–Farber Cancer Institute, Harvard Medical School, Boston, Massachusetts 02115

M. A. Black (521), Department of Statistics, University of Auckland, 1001 Auckland, New Zealand

Michael J. Carvan (255), Great Lakes WATER Institute, University of Wisconsin-Milwaukee, Milwaukee, Wisconsin 5320

Peter Cattin (593), Department of Molecular Medicine and Pathology, School of Medical Sciences, The University of Auckland, Auckland, New Zealand

Michael Choob (137), Active Motif, Carlsbad, California 92008

Paul Collodi (113), Department of Animal Sciences, Purdue University, West Lafayette, Indiana 47907

Mark S. Cooper (439), Department of Biology and Center for Developmental Biology, University of Washington, Seattle, Washington 98195

Bryan D. Crawford (439), Department of Biology, University of Alberta, Edmonton, Alberta, Canada T6G 2E9

Jennifer Crodian (113), Department of Animal Sciences, Purdue University, West Lafayette, Indiana 47907

Phil Crosier (593), Department of Molecular Medicine and Pathology, School of Medical Sciences, The University of Auckland, Auckland, New Zealand

Laura M. Cross (137), Zygogen, Atlanta, Georgia 30303

Alan Davidson (305), Howard Hughes Medical Institute, Division of Hematology/ Oncology, Children's Hospital, Dana-Farber Cancer Institute, and Harvard Medical School, Boston, Massachusetts 02115

Ann E. Davidson (349), The Arnold and Mabel Beckman Center for Transposon Research, Department of Genetics, Cell Biology and Development, University of Minnesota Medical School, Minneapolis, Minnesota 55455

Agnès Degrave (505), Institut de Génétique et de Biologie Moléculaire et Cellulaire, UMR 7104 CNRS/INSERM/ULP, BP 10142, CU de Strasbourg, 67404 Illkirch Cedex, France

H. William Detrich, III (475), Department of Biology, Northeastern University, Boston, Massachusetts 02115

Jie Dong (363), Department of Life Sciences, Indiana State University, Terre Haute, Indiana 47809

Bruce W. Draper (91), Howard Hughes Medical Institute and Division of Basic Science, Fred Hutchinson Cancer Research Institute, Seattle, Washington 98109

Stephen C. Ekker (121, 349), The Arnold and Mabel Beckman Center for Transposon Research, Department of Genetics, Cell Biology and Development, University of Minnesota Medical School, Minneapolis, Minnesota 55455

Lianchun Fan (113), Department of Animal Sciences, Purdue University, West Lafayette, Indiana 47907

Steven A. Farber (137), Department of Microbiology & Immunology and Kimmel Cancer Center, Thomas Jefferson University, Philadelphia, Pennsylvania 19107

Allan Force (545), Molecular Genetics Program, Benaroya Research Institute, Seattle, Washington 98101

Toshiaki Furuta (159), Department of Biomolecular Science, Toho University, Funabashi Chiba 274-8510, Japan and PREST (Precursory Research for Embryonic Science and Technology), Japan Science and Technology Corporation (JST), Tokyo 103-0072, Japan

Makoto Furutani-Seiki (173), ERATO, 14 Yoshida-kawaramachi, Sakyo-ku, Kyoto 606-8305, Japan

Ethan Gahtan (49), Department of Physiology, University of California, San Francisco, San Francisco, California 94143 and Department of Biology, Rennselaer Polytechnic Institute, Troy, New York 12180

Robert Geisler (295), Max-Planck-Institut für Entwicklungsbiologie, D-72076 Tübingen, Germany

Clemens Grabher (381), Developmental Biology Program, European Molecular Biology Laboratory (EMBL), 69117-Heidelberg, Germany

Thorsten Henrich (173), Developmental Biology Programme, EMBL-Heidelberg, D-69012 Heidelberg, Germany

Spencer Hermanson (349), The Arnold and Mabel Beckman Center for Transposon Research, Department of Genetics, Cell Biology and Development, University of Minnesota Medical School, Minneapolis, Minnesota 55455

Vincent Heyer (505), Institut de Génétique et de Biologie Moléculaire et Cellulaire, UMR 7104 CNRS/INSERM/ULP, BP 10142, CU de Strasbourg, 67404 Illkirch Cedex, France

Heinz Himmelbauer (173), Max-Planck-Institute of Molecular Genetics, D-14195 Berlin, Germany

Yunhan Hong (173), Department of Biological Sciences, National University of Singapore, Singapore 117543

Nancy Hopkins (1), Center for Cancer Research and Department of Biology, Massachusetts Institute for Technology, Cambridge, Massachusetts 01239

Karl Hsu (333), Department of Pediatric Oncology, Dana-Farber Cancer Institute and Division of Hematology/Oncology, Beth Israel Deaconess Medical Center, Harvard Medical School, Boston, Massachusetts 02115

Haigen Huang (403), Department of Molecular, Cellular, and Developmental Biology, University of California Los Angeles, Los Angeles, California 90095

Sean Humphray (295), Wellcome Trust Sanger Institute, Wellcome Trust Genome Campus, Hinxton, Cambridge CB10 1SA, United Kingdom

Kerstin Jekosch (225), Wellcome Trust Sanger Institute, Cambridge CB10 1SA, United Kingdom

Jean-Stephane Joly (381), INRA, Institute de Neurobiologie A. Fessard, CNRS, 91198 Gif-Sur-Yvette, France

Bensheng Ju (403), Department of Molecular, Cellular, and Developmental Biology, University of California Los Angeles, Los Angeles, California 90095

John P. Kanki (333, 645), Department of Pediatric Oncology, Dana-Farber Cancer Institute, Harvard Medical School, Boston, Massachusetts 02115

Koichi Kawakami (201), Division of Molecular and Developmental Biology, National Institute of Genetics, Mishima, Shizuoka 411-8540, Japan

Johanne Kirchner (505), Institut de Génétique et de Biologie Moléculaire et Cellulaire, UMR 7104 CNRS/INSERM/ULP, BP 10142, CU de Strasbourg, 67404 Illkirch Cedex, France

Romke Koch (295), Hubrecht Laboratory, Uppsalalaan 8, 3584 CT Utrecht, The Netherlands

Akihiko Koga (173), Division of Biological Sciences, Graduate School of Science, Nagoya University, Nagoya 464-8602, Japan

Mariko Kondo (173), Biocenter, University of Wuerzburg, Am Hubland, D-97074 Wuerzburg, Germany

Hisato Kondoh (173), ERATO, 14 Yoshida-kawaramachi, Sakyo-ku, Kyoto 606-8305, Japan

Charles Lee (241), Department of Pathology, Brigham and Women's Hospital, and Harvard Medical School, Boston, Massachusetts 02115

Ki-Young Lee (403), Department of Molecular, Cellular, and Developmental Biology, University of California Los Angeles, Los Angeles, California 90095

Shuo Lin (403), Department of Molecular, Cellular, and Developmental Biology, University of California Los Angeles, Los Angeles, California 90095

A. Thomas Look (333, 645), Department of Pediatric Oncology, Dana-Farber Cancer Institute, Harvard Medical School, Boston, Massachusetts 02115

D. R. Love (521), Molecular Genetics and Development Group, School of Biological Sciences, University of Auckland, 1001 Auckland, New Zealand

Aline Lux (505), Institut de Génétique et de Biologie Moléculaire et Cellulaire, UMR 7104 CNRS/INSERM/ULP, BP 10142, CU de Strasbourg, 67404 Illkirch Cedex, France

Jennifer L. Matthews (617), The Zebrafish International Resource Center, University of Oregon, Eugene, Oregon 97403

Claire M. McCallum (91), Anawah, Inc., Seattle, Washington 98104

Matthew B. McCarthy (439), Department of Biology and Center for Developmental Biology, University of Washington, Seattle, Washington 98195

Hiroshi Mitani (173), Department of Integrated Biosciences, Graduate School of Frontier Sciences, The University of Tokyo, Kashiwa City, Chiba 277-8562, Japan

Cecilia B. Moens (91), Howard Hughes Medical Institute and Division of Basic Science, Fred Hutchinson Cancer Research Institute, Seattle, Washington 98109

John P. Morris IV (645), Department of Pediatric Oncology, Dana-Farber Cancer Institute, Harvard Medical School, Boston, Massachusetts 02115

Mary C. Mullins (21), University of Pennsylvania, Department of Cell and Developmental Biology, Philadelphia, Pennsylvania 19104

Akira Muto (49), Department of Physiology, University of California, San Francisco, San Francisco, California 94143

Indrajit Nanda (173), Biocenter, University of Wuerzburg, Am Hubland, D-97074 Wuerzburg, Germany

Masaru Nonaka (173), Department of Biological Sciences, Graduate School of Sciences, The University of Tokyo, Bunkyo-ku, Tokyo 113-0033, Japan

Hitoshi Okamoto (159), Laboratory for Developmental Gene Regulation, Brain Science Institute, RIKEN (The Institute of Physical and Chemical Research), Saitama 351-0198, Japan and CREST (Core Research for Evolutional Science and Technology), Japan Science and Technology Corporation (JST), Tokyo 103-0027, Japan

Michael B. Orger (49), Department of Physiology, University of California, San Francisco, San Francisco, California 94143

Patrick Page-McCaw (49), Department of Physiology, University of California, San Francisco, San Francisco, California 94143 and Department of Biology, Rennselaer Polytechnic Institute, Troy, New York 12180

Jean-Paul Parkhill (505), Institut de Génétique et de Biologie Moléculaire et Cellulaire, UMR 7104 CNRS/INSERM/ULP, BP 10142, CU de Strasbourg, 67404 Illkirch Cedex, France

Francisco Pelegri (21), University of Wisconsin—Madison, Laboratory of Genetics, Madison, Wisconsin 53706

Carey Phillips (439), Department of Biology, Bowdoin College, Brunswick, Maine 04011

F. B. Pichler (521), Molecular Genetics and Development Group, School of Biological Sciences, University of Auckland, 1001 Auckland, New Zealand

Ronald H. A. Plasterk (69, 295), Hubrecht Laboratory, Center for Biomedical Genetics, Uppsalalaan 8, 3584 CT Utrecht, The Netherlands

Cara T. Poage (439), Department of Biology and Center for Developmental Biology, University of Washington Seattle, Washington 98195

John Postlethwait (255), Institute of Neuroscience, University of Oregon, Eugene, Oregon 97403

David Ransom (305), Division of Hematology/Oncology, Children's Hospital, Dana-Farber Cancer Institute, and Harvard Medical School, Boston, Massachusetts 02115

Gerd-Jörg Rauch (295), Max-Planck-Institut für Entwicklungsbiologie, Tübingen, Germany

Barrie Robison (599), Department of Biological Sciences, University of Idaho, Moscow, Idaho 83844

Amy Rubinstein (137), Zygogen, Atlanta, Georgia 30303

Victor Ruotti (255), Human & Molecular Genetics Center, Medical College of Wisconsin, Milwaukee, Wisconsin 53226

Takashi Sasaki (173), Department of Molecular Biology, Keio University School of Medicine, Tokyo 160-8582, Japan

Manfred Schartl (173), Biocenter, University of Wuerzburg, Am Hubland, D-97074 Wuerzburg, Germany

Michael Schmid (173), Biocenter, University of Wuerzburg, Am Hubland, D-97074 Wuerzburg, Germany

Lynn M. Schriml (415), National Center for Biotechnology Information (NCBI), National Institutes of Health, Bethesda, Maryland 20894

Iban Seiliez (505), Institut de Génétique et de Biologie Moléculaire et Cellulaire, UMR 7104 CNRS/INSERM/ULP, BP 10142, CU de Strasbourg, 67404 Illkirch Cedex, France

Cooduvalli Shashikant (545), College of Agricultural Sciences, The Pennsylvania State University, University Park, Pennsylvania 16802

Jennifer Shepard (305), Division of Hematology/Oncology, Children's Hospital, Dana-Farber Cancer Institute, and Harvard Medical School, Boston, Massachusetts 02115

Akihiro Shima (173), Department of Integrated Biosciences, Graduate School of Frontier Sciences, The University of Tokyo, Kashiwa City, Chiba 277-8562, Japan

Nobuyoshi Shimizu (173), Department of Molecular Biology, Keio University School of Medicine, Tokyo 160-8582, Japan

Sridhar Sivasubbu (349), The Arnold and Mabel Beckman Center for Transposon Research, Department of Genetics, Cell Biology and Development, University of Minnesota Medical School, Minneapolis, Minnesota 55455

Ann J. Slade (91), Anawah, Inc., Seattle, Washington 98104

Matthew C. Smear (49), Department of Physiology, University of California, San Francisco, San Francisco, California 94143

Amanda Smith (241), Department of Pathology, Brigham and Women's Hospital, Boston, Massachusetts 02115

Greg Sommers-Herivel (439), Department of Biology and Center for Developmental Biology, University of Washington Seattle, Washington 98195

Anhua Song (459), Children's Hospital Boston and Harvard Medical School, Boston, Massachusetts 02115

Judy Sprague (415), Zebrafish Information Network (ZFIN), University of Oregon, Eugene, Oregon 97403

Peter Stadler (545), Bioinformatics, Department of Computer Science, University of Leipzig, D-04103 Leipzig, Germany

Howard Stern (305), Division of Hematology/Oncology, Children's Hospital, Dana-Farber Cancer Institute, and Harvard Medical School, Boston, Massachusetts 02115

Nitzan Sternheim (137), Department of Developmental Biology, Stanford University, Stanford, California 94305

Jennifer L. Stout (91), Howard Hughes Medical Institute and Division of Basic Science, Fred Hutchinson Cancer Research Institute, Seattle, Washington 98109

Gary W. Stuart (363), Department of Life Sciences, Indiana State University, Terre Haute, Indiana 47809

Bernard Thisse (505), Institut de Génétique et de Biologie Moléculaire et Cellulaire, UMR 7104 CNRS/INSERM/ULP, BP 10142, CU de Strasbourg, 67404 Illkirch Cedex, France

Christine Thisse (505), Institut de Génétique et de Biologie Moléculaire et Cellulaire, UMR 7104 CNRS/INSERM/ULP, BP 10142, CU de Strasbourg, 67404 Illkirch Cedex, France

Xiaobing Tian (137), Department of Biochemistry & Molecular Pharmacology, Thomas Jefferson University, Philadelphia, Pennsylvania 19107

Peter J. Tonellato (255), Human & Molecular Genetics Center, Medical College of Wisconsin, Milwaukee, Wisconsin 53226

Nikolaus Trede (305), Division of Hematology/Oncology, Children's Hospital, Dana-Farber Cancer Institute, and Harvard Medical School, Boston, Massachusetts 02115

Bill Trevarrow (565, 599), Institute of Neuroscience, University of Oregon, Eugene, Oregon 97403

Karen A. Urtishak (137), Department of Microbiology & Immunology and Kimmel Cancer Center, Thomas Jefferson University, Philadelphia, Pennsylvania 19107

Eric Wickstrom (137), Department of Biochemistry & Molecular Pharmacology, Department of Microbiology & Immunology, and Kimmel Cancer Center, Thomas Jefferson University, Philadelphia, Pennsylvania 19107

Erno Wienholds (69), Hubrecht Laboratory, Center for Biomedical Genetics, Uppsalalaan 8, 3584 CT Utrecht, The Netherlands

L. C. Williams (521), Molecular Genetics and Development Group, School of Biological Sciences, University of Auckland, 1001 Auckland, New Zealand

Joachim Wittbrodt (173, 381), Developmental Biology Program, European Molecular Biology Laboratory (EMBL), D–69012, Germany

Donald A. Yergeau (475), Department of Biology, Northeastern University, Boston, Massachusetts 02115

Yi Zhou (273, 305, 459), Division of Hematology/Oncology, Children's Hospital, Dana-Farber Cancer Institute, and Harvard Medical School, Boston, Massachusetts 02115

Leonard I. Zon (305), Howard Hughes Medical Institute, Division of Hematology/Oncology, Children's Hospital, Dana-Farber Cancer Institute, and Harvard Medical School, Boston, Massachusetts 02115

PREFACE

Monte, Len, and I welcome you to two new volumes of *Methods in Cell Biology* devoted to *The Zebrafish: Cellular and Developmental Biology* and *Genetics, Genomics, and Informatics*. In the five years since publication of the first pair of volumes, *The Zebrafish: Biology* (Vol. 59) and *The Zebrafish: Genetics and Genomics* (Vol. 60), revolutionary advances in techniques have greatly increased the versatility of this system. At the Fifth Conference on *Zebrafish Development and Genetics*, held at the University of Wisconsin in 2003, it was clear that many new and compelling methods were maturing and justified the creation of the present volumes. The zebrafish community responded enthusiastically to our request for contributions, and we thank them for their tremendous efforts.

The new volumes present the post-2000 advances in molecular, cellular, and embryological techniques (Vol. 76) and in genetic, genomic, and bioinformatic methods (Vol. 77) for the zebrafish, *Danio rerio*. The latter volume also contains a section devoted to critical infrastructure issues. Overlap with the prior volumes has been minimized intentionally.

The first volume, *Cellular and Developmental Biology*, is divided into three sections: Cell Biology, Developmental and Neural Biology, and Disease Models. The first section focuses on microscopy and cell culture methodologies. New microscopic modalities and fluorescent reporters are described, the cell cycle and lipid metabolism in embryos are discussed, apoptosis assays are outlined, and the isolation and culture of stem cells are presented. The second section covers development of the nervous system, techniques for analysis of behavior and for screening for behavioral mutants, and methods applicable to the study of major organ systems. The volume concludes with a section on use of the zebrafish as a model for several diseases.

The second volume, *Genetics, Genomics, and Informatics*, contains five sections: Forward and Reverse Genetics, The Zebrafish Genome and Mapping Technologies, Transgenesis, Informatics and Comparative Genomics, and Infrastructure. In the first, forward-genetic (insertional mutagenesis, maternal-effects screening), reverse-genetic (antisense morpholino oligonucleotide and peptide nucleic acid gene knockdown strategies, photoactivation of caged mRNAs), and hybrid (target-selected screening for ENU-induced point mutations) technologies are described. Genetic applications of transposon-mediated transgenesis of zebrafish are presented, and the status of the genetics and genomics of *Medaka*, the honorary zebrafish, is updated. Section 2 covers the zebrafish genome project, the cytogenetics of zebrafish chromosomes, several methods for mapping zebrafish genes and mutations, and the recovery of mutated genes via positional cloning.

The third section presents multiple methods for transgenesis in zebrafish and describes the application of nuclear transfer for cloning of zebrafish. Section 4 describes bioinformatic analysis of the zebrafish genome and of microarray data, and emphasizes the importance of comparative analysis of genomes in gene discovery and in the elucidation of gene regulatory elements. The final section provides important, but difficult to find, information on small- and large-scale infrastructure available to the zebrafish biologist.

The attentive reader will have noticed that this Preface was drafted by the first editor, Bill Detrich, while he (I) was at sea leading the sub-Antarctic ICEFISH Cruise (International Collaborative Expedition to collect and study Fish Indigenous to Sub-antarctic Habitats; visit www.icefish.neu.edu). Wearing my second biological hat, I study the adaptational biology of Antarctic fish and use them as a system for comparative discovery of erythropoietic genes. Antarctic fish embryos generally hatch after six months of development, and they reach sexual maturity only after several years. Imagine attempting genetic studies on these organisms! My point is that the zebrafish system and its many advantages greatly inform my research on Antarctic fish, while at the same time I can move genes discovered by study of the naturally evolved, but very unusual, phenotypes of Antarctic fish into the zebrafish for functional analysis. We the editors emphasize that comparative strategies applied to multiple organisms, including the diverse fish taxa, are destined to play an increasing role in our understanding of vertebrate development.

We wish to express our gratitude to the series editors, Leslie Wilson and Paul Matsudaira, and the staff of Elsevier/Academic Press, especially Kristi Savino, for their diligent help, great patience, and strong encouragement as we developed these volumes.

H. William Detrich, III
Monte Westerfield
Leonard I. Zon

This volume is dedicated to Jose Campos-Ortega and Nigel Holder,
departed colleagues whose wisdom and friendship will be missed
by the zebra fish community

PART I

Forward and Reverse Genetics

CHAPTER 1

Retroviral-Mediated Insertional Mutagenesis in Zebrafish

Adam Amsterdam and Nancy Hopkins

Center for Cancer Research and Department of Biology
Massachusetts Institute for Technology
Cambridge, Massachusetts 01239

I. Introduction

Large-scale chemical mutagenesis screens have resulted in the isolation of thousands of mutations in hundreds of genes that affect zebrafish embryonic development (Driever *et al.*, 1996; Haffter *et al.*, 1996). These screens have used an alkylating agent, ethyl nitrosourea (ENU), to induce mutations, primarily by causing base pair substitutions. Approximately 100 of the genes disrupted by these mutations have been isolated till March 2004, primarily through a candidate gene approach and less frequently by pure positional cloning (Postlethwait and Talbot, 1997), and many other chapters in this volume are devoted to describing this. However, positional cloning remains arduous.

Insertional mutagenesis is an alternative to chemical mutagenesis in which exogenous DNA is used as the mutagen. Although insertional mutagenesis is usually less efficient than ENU, insertions serve as a molecular tag to aid in the isolation of the mutated genes. Several methods can be employed to insert DNA into the zebrafish genome, including DNA microinjection (Culp *et al.*, 1991; Stuart *et al.*, 1988), or microinjection of DNA aided by retroviral integrases (Ivics *et al.*, 1993) or a transposable element's transposase (Davidson *et al.*, 2003; Kawakami *et al.*, 2000; Raz *et al.*, 1997); reviews and updates on these methods can be found in other chapters in this volume. However, to date, by far the most efficient way to make a large number of insertions in the zebrafish genome is to use a pseudotyped retrovirus.

Retroviruses have an RNA genome and, on infection of a cell, reverse transcribe their genome to a DNA molecule, the provirus. The provirus integrates into a host cell chromosome, where it remains stable and is thus inherited by all the descendants of that cell. Replication-defective retroviral vectors, unlike nondefective retroviruses, are infectious agents that can integrate into host DNA, but whose genetic material lacks the coding sequences for the proteins required to make progeny virions. Retroviral vectors are made in split-genome packaging cells, in which the genome of the retroviral vector is expressed from one integrated set of viral sequences, whereas the retroviral genes required for packaging, infection, reverse transcription, and integration are expressed from another locus. The most widely used retroviral vectors have been derived from a murine retrovirus, the Moloney murine leukemia virus (MoMLV), resulting in replication-defective viruses that can be produced at very high titers. Initially, these retroviruses were only capable of infecting mammalian cells, but their host range can be expanded as described later.

Retroviruses have a host range, or tropism, which is frequently determined by their envelope protein, which recognizes and binds to some specific component, usually a protein, on the surface of the cell to be infected. Cell types that have an appropriate receptor can be infected by the retrovirus, whereas those that do not are refractory to infection. The host range of a virus can be changed by pseudotyping, a process in which virions acquire the genome and core proteins of one virus but the envelope protein of another. One way to enable this situation in split-genome packaging cells is to simply substitute the gene encoding the alternative envelope protein for the usual one. Although there is some specificity as to which envelope proteins can be pseudotyped with which viral genomes, one such combination that is particularly useful allows the MoMLV viral genomes and core proteins to be pseudotyped with the envelope glycoprotein (G-protein) of the vesicular stomatitis virus (VSV; Weiss *et al.*, 1974). VSV is a rhabdovirus that is apparently pantropic; it can infect cells of species as diverse as insects and mammals (Wagner, 1972). MoMLV vectors pseudotyped with VSV-G possess two qualities essential for their use in high-frequency germline transgenesis in zebrafish: the extended host range allows for the infection of fish cells, and the VSV-G-pseudotyped virions are unusually stable, allowing viruses to be concentrated 1000-fold by centrifugation (Burns *et al.*, 1993).

When pseudotyped retroviral vectors are injected into zebrafish blastulae, many of the cells become independently infected, producing a mosaic organism in which different cells harbor proviral insertions at different chromosomal sites. When cells destined to give rise to the germ-line are infected, some proportion of the progeny of the injected fish will contain one or more insertions (Lin *et al.*, 1994). When a sufficiently high-titer virus is used, one can infect a very high proportion of the germline of injected fish (Gaiano *et al.*, 1996a). With very high-titer virus, on average, about 25–30 independent insertions can be inherited from a single founder, though any given insertion will only be present in about 3–20% of the offspring (Chen *et al.*, 2002; A. Amsterdam, unpublished data). However, the progeny are nonmosaic for the insertions and transmits them in a Mendelian fashion to 50% of then progeny. Furthermore, because more than one virus can infect a single cell, some germ cells contain multiple insertions, and thus offspring can be born with as many as 10–15 independently segregating insertions (Amsterdam *et al.*, 1999; Chen *et al.*, 2002; Gaiano *et al.*, 1996a). This remarkable transgenesis rate has made it possible to conduct an insertional mutagenesis screen, which has allowed isolation of hundreds of insertional mutants and rapid cloning of the mutated genes (Amsterdam *et al.*, 1999, 2004a; Golling *et al.*, 2002).

II. Mutagenesis

To establish the frequency of mutagenesis with retroviral vectors in the zebrafish, we carried out a pilot screen in which we inbred more than 500 individual proviral insertions, one at a time, and screened for recessive phenotypes that could be visually scored in the first 5 days of embryonic development. We found six recessive embryonic lethal mutations, a frequency of about one mutation per 80–100 insertions (Allende *et al.*, 1996; Becker *et al.*, 1998; Gaiano *et al.*, 1996b; Young *et al.*, 2002). We also found one viable dominant insertional mutation (Kawakami *et al.*, 2000). Although this rate was too inefficient to conduct a large-scale screen by breeding one insertion at a time, by using the ability of founders to transmit multiple insertions to individual F1 progeny an average of about 12 inserts can be screened per family, allowing the recovery of about one insertional mutation per seven families screened (Amsterdam *et al.*, 1999; Amsterdam, unpublished data). This is only 7- to 10-fold lower than the frequency observed in analogously performed (three-generation diploid) ENU screens (Driever *et al.*, 1996; Haffter *et al.*, 1996; Mullins *et al.*, 1994; Solnica-Krezel *et al.*, 1994). The strategy to produce, select, and breed the fish for such an insertional mutagenesis screen is outlined in this section.

A. Making Founder Fish That Transmit Proviral Inserts at High Frequency to Their F1 Progeny

Founders are produced by injecting late-blastula-stage (512–2000 cells) embryos. Virus must be injected into the space between the cells, and blastula-stage embryos ideally accommodate the injected fluid. At this time, there are four primordial germ cells, and these cells divide two or three times over the course of the infection (Yoon et al., 1997). Thus, the injected embryos grow up to be founder fish (F0) with mosaic germ lines. With very good viral stocks, individual founders can contain 25–30 different insertions in their germ lines, with any given insertion present in 3–20% of the gametes (Amsterdam, unpublished data; Chen et al., 2002). Individual F1 fish can inherit up to 10 different insertions, and when founders are bred to each other F1 fish can be found with up to 20 different insertions. F1 fish are not mosaic and transmit all their insertions in a Mendelian fashion.

Because the efficiency of the screen relies on the generation of F1 fish with a high number of inserts, it is essential to perform quality control assays on the viral stocks and founder injections before raising and breeding the founders. For every batch of injected embryos, several embryos are sacrificed for DNA preparation at 48 h postinjection and subjected to quantitative Southern analysis or real-time polymerase chain reaction (PCR) to determine the average number of insertions per cell in the entire infected embryo. This number is called the embryo assay value (EAV; Amsterdam et al., 1999). In our experience, if the average EAV is above 15 and does not vary much among the individual analyzed embryos, the rest of the founders from that injection session transmit inserts at the rates mentioned previously. Batches with average EAV below 15 transmit somewhat fewer inserts, and usually have greater founder-to-founder variation, and those with average EAV below 5 are quite inconsistent in transmitting multiple inserts to their progeny.

B. Breeding and Screening for Mutations

The breeding scheme for a diploid F3 insertional mutagenesis screen is outlined in Fig. 1. In essence, the goal is to create families with a large number of independent insertions that can be screened simultaneously. This is achieved by selecting and breeding F1 fish that inherit the most inserts from the mosaic founders.

Founder fish can be bred to each other or outcrossed to nontransgenic fish. For reasons that remain unclear, a majority of injected fish grow up to be males; thus, it is most efficient to outcross the best male F0 fish (those from batches with the highest EAV) and inbreed the rest. F1 families of 30 fish are raised, and at 6 weeks of age the fish are fin clipped for DNA preparation and analysis by quantitative Southern or real-time PCR to determine which fish harbor the most insertions. Keeping up to the three top fish per family with at least five inserts strikes a balance between throwing away too many inserts (if fewer fish were kept) and keeping too many 'repeat inserts' (i.e., the same insert inherited by sibling F1 fish). The repeat insert rate is quite low if only three fish are kept, as the average mosaicism

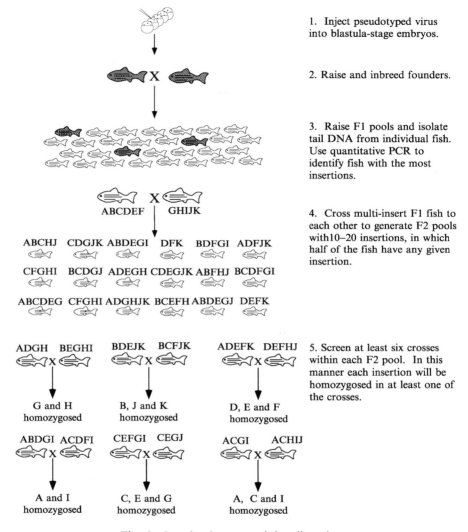

1. Inject pseudotyped virus into blastula-stage embryos.

2. Raise and inbreed founders.

3. Raise F1 pools and isolate tail DNA from individual fish. Use quantitative PCR to identify fish with the most insertions.

4. Cross multi-insert F1 fish to each other to generate F2 pools with 10–20 insertions, in which half of the fish have any given insertion.

5. Screen at least six crosses within each F2 pool. In this manner each insertion will be homozygosed in at least one of the crosses.

Fig. 1 Insertional mutagenesis breeding scheme.

(i.e., proportion of F1 inheriting a given insert) is about 8%. In our screen, only 3% of the recovered mutations have been caused by reisolating such repeat inserts.

The selected multiinsert F1 fish are pooled together and eventually bred to make F2 families that harbor at least 10 different independently segregating inserts, and in which each insert is present in half the fish. Multiple sibcrosses are then conducted between the F2 fish; because half the fish have any given insert, including one causing a mutation, should be homozygosed in one quarter of the crosses. On average, six crosses will homozygose 83% of the inserts in the family and ten crosses will screen 95% of them. Every F3 clutch from each F2 family is screened for a phenotype in one quarter of the embryos. In our screen, embryos were scored for any morphological defect visible in a dissecting microscope at 1, 2, and 5 days post fertilization (dpf), as well as for defects in motility and touch response. One aid to screening is that more than 98% of these mutants fail to inflate their swim bladders by 5 dpf; because this is such a highly visible structure, a quick screen for clutches in which one quarter of the embryos fail to inflate their swim bladder often signals the presence of a mutation.

III. Cloning the Mutated Genes

A. Identifying the Mutagenic Insert

The great advantage to using insertional mutagenesis over chemical mutagenesis is that the mutagenic insertion provides a molecular tag that can be used to identify the disrupted gene. However, because the mutagenesis screen described uses multiple insertions to increase the rate of recovery of mutations, the first step after identifying a mutation is to determine which (if any) insertion appears to be responsible for the mutation. DNA is prepared from the tails of the parents of all of the crosses from the F2 family, and, using Southern analysis to distinguish the different insertions, one looks for an insertion that segregates with the phenotype (Fig. 2A). A linked insert (represented by a band of a specific size) will be shared by both parents of every cross that had the phenotype and be in only one or neither of the parents of every cross that lacked the phenotype. In addition, DNA prepared from the mutant embryos is subjected to the same analysis; an unlinked insert would be in only three quarters of the mutant embryos, but a linked insert must be present in all of them (Fig. 2B).

It is possible that no insert appears linked to the phenotype; in our screen, we found that about one quarter of the mutants recovered were not linked to a detectable insertion. In addition, it is important to note that the identification of an insertion initially linked to the phenotype is not proof that the identified insert is *tightly* linked to the mutation; it is merely a way to either identify the insertion that is a *candidate* for causing the mutation or to conclude that the mutation is not linked to any insert if no insert meets the criteria. This is because recombination rates in the male germline are much lower than in the female germline (Johnson, personal communication; Amsterdam, unpublished data); thus, an insert inherited from an F1 male that is merely on the same chromosome as a noninsertional mutation will often meet the previously mentioned criteria. The mutation and the insert will not have segregated in the F2 generation, and because the mutant F3 embryos must inherit the mutant locus from both parents, even if there is recombination in the female germ line all the mutant embryos will receive the insert from their father. Thus, additional linkage experiments that can distinguish

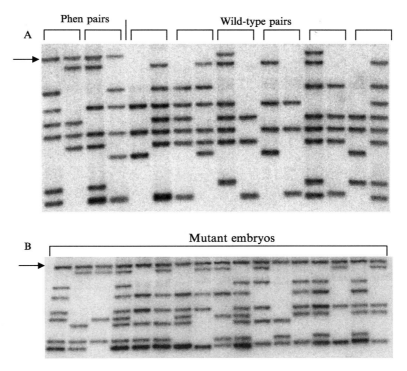

Fig. 2 Identification of the mutagenic insert. (A) Southern analysis of DNA prepared from tail fins of F2 fish: the arrow indicates an insert that is homozygosed in phenotypic pairs but not any of the wild-type pairs. (B) Southern analysis of DNA from individual mutant embryos from the second phenotypic pair in (A) the arrow indicates that the same insert is also present in all the mutant embryos.

heterozygosity from homozygosity for the insert are required, but it is not possible to perform these until genomic DNA flanking the candidate insert is cloned.

Sometimes more than one insert meets the previously mentioned criteria, and thus more than one are candidates to have caused the mutant phenotype. This can be for one of several reasons. First, if more than one insert in the family is on the same chromosome, for the reasons described previously they might fail to segregate from each other. Often this can be resolved by outcrossing a female carrier and repeating the analysis in the next generation, either by further random sib crosses followed by molecular analysis or by using Southern analysis first to identify fish with one or the other insertion and then performing test crosses. Another possibility is that multiple copies of the virus have integrated in tandem, which happens about 3–4% of the time. Usually when this happens, there is a higher-intensity provirus-sized band (if the enzyme used cuts the insert only once) in addition to the true junction fragment band. Finally, there might be too many inserts in the family to accurately distinguish all the inserts (more than 15–20), and this might complicate the analysis. In these cases a female carrier fish, preferably already shown to have relatively few inserts, must be outcrossed and the analysis repeated in the next generation. It is essential not to focus on a single insertion as the cause of a mutation unless it is very clear that no other insertion could also be linked to the mutant phenotype.

B. Cloning the Flanking Genomic DNA

After identifying the candidate mutagenic insertion, inverse PCR or linker-mediated PCR can be used to clone genomic DNA flanking one or both sides of the mutagenic provirus (Fig. 3A). Because all the inserts have the same sequence, to clone the correct junction fragment, one must know the size expected for a given enzyme used. Often it is necessary to analyze the DNA samples by Southern analysis with several enzymes in order to identify which enzyme will be best for obtaining the desired insert (Fig. 3B). After cloning and sequencing the putative junction fragment, one can design a PCR primer in the sequence that points back at the provirus and use PCR with this and a viral primer on DNA isolated from tail fin samples of fish known to be carriers or noncarriers to confirm that the correct junction was cloned.

After cloning the genomic DNA flanking one side of the virus, it is necessary (for reasons explained later) to obtain sequence on the other side of the insertion as well; it might also be desirable to obtain additional sequence extending further from the virus on the side originally cloned. One way to do this is to use the cloned sequence as an anchor for additional inverse PCR or linker-mediated PCR. However, as the zebrafish genome assembly becomes increasingly complete, this step is becoming increasingly dispensable; often even a small amount of sequence adjacent to the virus is sufficient to place the insertion site on a large contig of known sequence.

One of the uses of the cloned sequence is that it allows one to perform an assay to distinguish transgenic and nontransgenic chromosomes in a codominant fashion. Such an assay is essential to demonstrate that the insertion is tightly linked to the mutation and is thus most likely its cause. One method is to use the junction fragment as a probe on a Southern blot, as the transgenic and nontransgenic chromosomes will each produce hybridizing fragments of a different size (Fig. 4A). Alternatively, PCR can be conducted with three primers, one on each side of the insert and one pointing out of the insert, such that different-sized products are amplified by insert-bearing and non-insert-bearing chromosomes (Fig. 4B). In either case, the assay is used to demonstrate that mutant embryos are invariably homozygous for the insertion, whereas wild-type embryos never are. Every mutant analyzed is the equivalent of observing one meiotic event (only counting the female germline); every wild type analyzed is the equivalent of observing one third of a meiosis. (Only one in three recombination events between a mutation and a marker in a dihybrid cross will lead to a wild-type embryo that is homozygous for the marker; thus scoring for wild types that are homozygous for a marker detects only one third of the recombination events between these

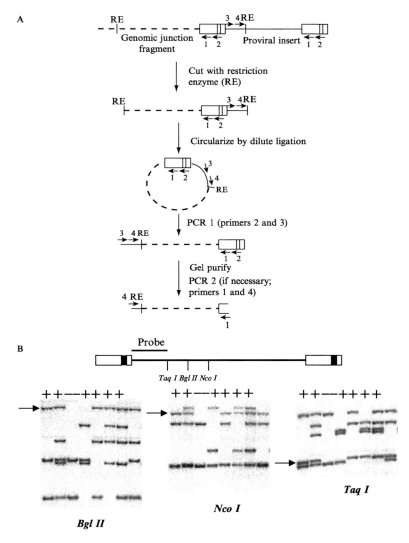

Fig. 3 Inverse PCR. (A) Schematic of the inverse PCR process. (B) Selection of the correct enzyme to use. For an insert already identified as in Fig. 2, DNA from several tails already known to be positive or negative for that insert is analyzed by Southern after digestion with different restriction enzymes. In this example, the junction fragments with *Bgl II* and *Nco I* are too big to successfully amplify by inverse PCR, but the *Taq I* junction fragment should amplify easily.

loci.) If any recombinants are observed between the mutation and the insertion, the insertion cannot be the cause of the mutation. We standardly analyze 50–100 meioses in this fashion; although not absolute proof that the insert is the cause of

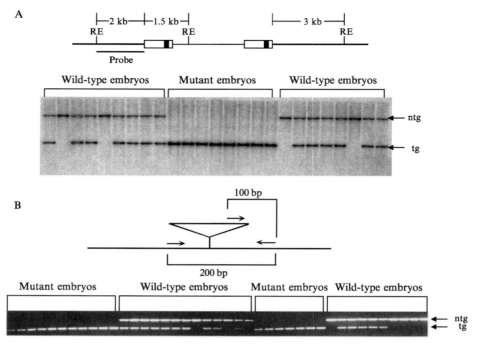

Fig. 4 Tight linkage assays. In either assay, mutant embryos should always be homozygous for the insert whereas wild-type embryos should never be. (A) Southern analysis of DNA prepared from individual wild-type or mutant embryos. The sequence of the junction fragment on one side of the virus is used as the probe. In this example, insert-bearing chromosomes (tg) will give a 3.5-kb band, whereas noninsert chromosomes (ntg) will give a 5-kb band. Thus, each embryo can be genotyped as homozygous for the insertion (smaller band only), heterozygous (both bands), or homozygous noninsertion (larger band only). (B) PCR analysis of DNA prepared from individual wild-type or mutant embryos. The PCR reaction is run with three primers such that (as with the Southern method) the presence of either chromosome is indicated by a unique-sized band.

the mutation, given the size of the genome, the relative rates of spontaneous and insertional mutations, and the average number of inserts in each family, less than 0.5% of mutations which meet this criteria should have a cause other than the insertion. Establishing tight linkage with more observed meioses can linearly decrease the likelihood that the mutation is not caused by the insertion, but linkage alone cannot reduce this likelihood to zero.

One exception to the requirement for absolute linkage is in cases in which there is incomplete penetrance of the phenotype; thus by definition the phenotype and genotype do not always match. This is evident when consistently less than 25% of the embryos are phenotypic. In these cases, although all the mutant embryos must still be homozygous for the insert, some of the phenotypically wild-type embryos will also be homozygous (Amsterdam *et al.*, 1999; Golling *et al.*, 2002).

C. Gene Identification

The sequence of the junction fragment is useful for allowing genotypic identification of carriers and is required for the tight linkage experiments described previously, but its greatest utility is in the ability to identify the mutated gene. Given up to a few kilobases of sequence on either side of the insertion, in more than 80% of the cases exonic sequence can be found by BLAST (Altschul et al., 1997) based either on nucleotide identity to a zebrafish cDNA or expressed sequence tag (EST) or on amino acid homology to known or predicted proteins from other organisms when translated. RT-PCR or 3′ and 5′ rapid amplification of cDNA ends (RACE) can then be used to complete the cDNA if necessary. As the zebrafish genome assembly and annotation become more complete, merely blasting to the genome will be sufficient to identify the gene into which the virus has inserted.

It is important that the virus actually is in—as opposed to just near—the gene identified, or it is possible to identify the wrong gene. Zebrafish genes are sometimes very near each other, with as few as half a kilobase separating them. Thus, an insert could be less than 1 kb from an easily recognized or annotated gene, but actually disrupt another gene not found in the BLAST search. Zebrafish genes often have first exons that are entirely 5′ untranslated or include the coding sequence for only a few amino acids, and such initial exons could easily be missed by a BLAST search of the genomic DNA sequence. In several of our insertional mutants, the gene originally recognized in the flanking sequence either began or ended about a kilobase from the provirus; only on more careful inspection was it found that another gene began between the originally identified gene and the virus. Analysis of mRNA expression in wild-type and mutant embryos in two of these cases demonstrated that only the proximal gene's expression was affected by the insertion. On the other hand, in the case of several insertional mutants, the insert is outside the affected gene, presumably in the promoter region, and affects the transcription of the gene. Thus, the finding that the mutagenic insert lies outside of a gene does not necessarily mean that the nearest identified gene is *not* the gene of interest, but rather that further analysis is required.

Cloning a gene that is proximal to a mutagenic insertion is not absolute proof that the correct gene has been identified. First, it is always possible—if very unlikely—that there is a noninsertional mutation very near the insert. Second, it is possible that the expression of a neighboring gene has been affected as well. Although this has not been observed to date in our studies, we have investigated it in very few cases. Similarly, even when a provirus lies between two exons of a gene and disrupts its expression, it is still possible that another gene (e.g., one lying within an intron of the first gene) might be the gene responsible for the mutant. Thus, to be absolutely certain that mutation of the identified gene is responsible for the phenotype, the gene identity needs to be confirmed by independent means. The finding of a second insertional allele, or noncomplementation to a chemical allele with a demonstrated point mutation, would make the likelihood of a nearby noninsertional mutation exceedingly low. The ideal proof is rescue—whether reintroduction of the gene in *trans* into mutant embryos can rescue the phenotype. Rescue is not easily accomplished in stable transgenics, but can be done transiently for some mutations. Alternatively, phenocopy of the phenotype by morpholino injection (Nasevicius and Ekker, 2000) can independently verify that mutation of this gene leads to this phenotype.

D. Phenotypic Consequence of Insertions

Insertion of a provirus into a gene can affect that gene in a number of different ways. Unlike chemical mutagenesis, which by causing point mutations has the potential to create hypomorphic or neomorphic alleles by amino acid substitution, insertional mutagenesis generally works by more broadly knocking down or out gene expression, although there are exceptions. Although only about one third of our mutagenic inserts actually interupt exons, which would obviously impair gene expression, nearly half land in the first intron and the rest in downstream introns (Fig. 5A). We have used Northern analysis or quantitative RT-PCR, or both, to analyze gene expression in mutants in many of these cases, and the most common effect that we have seen is a reduction or elimination of mRNA, anywhere from five-fold to undetectable levels (Golling *et al.*, 2002). Thus, although some insertional mutants are nulls, some are hypomorphs, as expression has not been completely abrogated.

In addition, some insertional mutants can affect the nature of the message (and the protein it produces) rather than merely the level of expression. First, some mutations cause exon skips instead of downregulation of expression. For example, we have three insertional alleles of the *vhnF1* gene, two in the first or second exon, which appear to be nulls, and one in the fifth intron, which leads to two different splice variants, skipping either the fourth or third and fourth exons (Sun and Hopkins, 2001). This allele is predicted to make a truncated protein and in fact has a less severe phenotype. Second, the virus used to make most of our insertional mutations contains a splice-in, splice-out frameshift-producing gene trap cassette (Chen *et al.*, 2002). When the virus inserts in an intron in the correct orientation, it is possible for the preceding splice donor to splice to this exon in the provirus and splice out to the next endogenous exon, thus creating a frameshift and presumably a truncated protein (Fig. 5B). If this happens, analysis of mRNA by RT-PCR will show the presence of an increased-sized band, indicating the inclusion of the trapped exon. This is likely to lead to a truncated protein and could thus act as a hypomorph or neomorph. We have not found this to be a common mechanism of mutation, but it has occurred in some of our mutants.

IV. Future Directions

The current insertional mutant collection in our laboratory includes more than 500 mutants, which include mutations in nearly 400 genes; so far we have identified more than 300 of these genes. We believe that this represents approximately

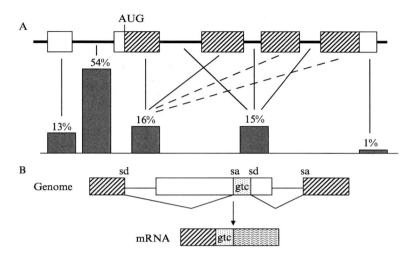

Fig. 5 Mutagenic insertion sites. (A) Distribution of mutagenic insertions from 413 insertional mutants in 298 different genes. White boxes indicate 5′ and 3′ untranslated regions (UTRs), striped boxes indicate coding exons, and lines between boxes indicate introns. The sample gene here has an untranslated first intron (the first white box), but in about half the cases the first exon contains the initiation codon. The percentage of mutagenic insertions that lie in the 5′ UTR or promoter, the first intron, coding exons, downstream introns, or the 3′ UTRs, is shown below the gene. (B) Consequence of gene trap event. If the insertion is in an intron in the correct orientation, the splice donor from the previous exon can splice to the gene trap cassette and then splice out to the next endogenous exon, thus creating a frameshift. Striped boxes, exons; large white box, provirus; stipled box (gtc), gene trap cassette; sd, splice donor; sa, splice acceptor; wavy-lined box, exon with frameshift mutation.

25% of the genes that can be mutated to a visible (and usually lethal) phenotype in the fish (Amsterdam *et al.*, 2004a). Although the characterization of the phenotypes is somewhat rudimentary at present, numerous shelf screens of this collection are being conducted, including staining with various antibodies, *in situ* hybridization markers, and other reagents that illuminate the patterning and development of specific tissues. Thus, a substantial portion of the genes required for the proper formation of all these structures will be identified.

The existing collection can also be used to monitor the long-term effects of heterozygosity of these genes in adults, as mutations that have recessive embryonic phenotypes might predispose adults to disease in the heterozygous state. The genes mutated in several of our mutants are known to be autosomal-dominant disease genes in humans. For example, mutations in the zebrafish *vhnF1* gene affect both kidney and pancreas development (Sun and Hopkins, 2001). *vhnF1* mutations in the heterozygous state, while not yet investigated in fish, can lead to either kidney disease or diabetes in humans (Horikawa *et al.*, 1997; Nishigori *et al.*, 1998). Similarly, the *jellyfish* mutation, which results in defective cartilage differentiation and morphogenesis, is caused by disruption of the *sox9a* gene (Yan *et al.*, 2002), whereas heterozygosity for *sox9* mutations in humans leads to campomelic dysplasia (Foster *et al.*, 1994). Thus, it might not be surprising if heterozygotes of insertional mutants are predisposed to a variety of diseases that might be screenable.

Similarly, many genes whose mutation in the heterozygous state is known to predispose mammals to cancer (tumor suppressor genes) cause prenatal death in mice in the homozygous state (Jacks, 1996). An insertional mutation of the *NF2a* gene, an ortholog of a known human tumor suppressor gene (Ruttledge *et al.*, 1994; Trofatter *et al.*, 1993), is embryonic lethal in zebrafish. Heterozygosity for this *NF2a* mutation in zebrafish predisposes them to development of tumors of the nervous system (Amsterdam *et al.*, 2004b), as it does in mammals. Among a large collection of recessive embryonic lethal mutations one might expect to find other mutations in which heterozygotes are more prone to develop cancer, and in fact a number of the insertional mutations have increased rates of tumirogenesis as heterozygous adults (Amsterdam *et al.*, 2004b).

The screen we have conducted has netted hundreds of embryonic lethal mutations, but it is also possible that the retroviral technology could be used in new screens targeted for specific phenotypes in more efficient ways. The screen described was quite labor intensive and required a lot of tanks for the several generations of breeding involved. There are several ways in which the mosaicism of the founders' germline might be used for more efficient screening, although these approaches would require the ability to recognize the mutants when present at far less than the usual 25% of the clutch. It might be difficult to conduct a general morphological screen in such a way, but this might work to screen for individual structures or patterns, for example, by antibody staining or *in situ* hybridization. Alternatively, such screens might be conducted in strains that allow the easy detection of mutations for certain processes, such as GFP transgenic lines that illuminate structures of interest.

One method would be to sequentially backcross individual F1 fish to their founder parent. Because the average germline clone size for the insertions is small, the phenotype would have to be scored where only 3–10% of the embryos are mutant. Nonetheless, this would likely allow for the screening of more insertions per F0 fish than the three-generation screen that we conducted, and mutations would be identified one generation earlier. Phenotypes of interest could then be recovered by outcrossing the carrier F1 fish and reidentifying the mutation in the F2 generation, where the mutagenic insertion could be identified and its adjacent DNA cloned.

Alternatively, one could conduct gynogenetic screens, using either haploids or early pressure diploids (Beattie *et al.*, 1999; Walker, 1999). One possibility would be to screen the progeny of F0 fish, thus screening dozens of insertions per clutch, without any of the generations of breeding or selection of high-insert-number fish. Again, because of the germline mosaicism, only a small percentage of the embryos will inherit any given insertion; therefore, the screen would have to be sensitive enough to detect only one or a few mutants per clutch, without too high a false positive rate. In a pilot screen in which the haploid progeny of about 300 mosaic F0 fish were screened for both brain pattern formation by *in situ* hybridization at tailbud stage and brain morphology at 32 hpf, six insertional mutants were recovered (Wiellette, Grinblat, Austen, and Sive, personal communication). Alternatively, if one required a higher proportion of mutant embryos to be sure of the phenotype, one could screen the progeny of multiinsert F1 fish, selected as in the three-generation screen. This would add the work and resources to raise all F1 families and select the high-insert-number F1 fish, but one could still screen around 10 inserts per clutch and it would still be one fewer generation to breed and many fewer clutches to screen than the diploid F3 screen. Either of these methods has the potential to approach saturation for a given phenotype.

There is also room for newer vectors or other methods of transgenesis to improve insertional mutagenesis, many of which are described in other chapters. Vectors employing a gene trap with a visible marker such as GFP could preselect for insertions in genes with expression patterns of interest (Kawakami, 2004; Kawakami *et al.*, 2004). Such insertions could be selected in F1 fish after passage through the germline, or gene trap events could be selected *in vitro*, as is often done in ES cells in mice, and then cloned by nuclear transfer (Ju *et al.*, 2004; Lee *et al.*, 2002). Alternatively, it has been suggested that vectors including transcriptional regulatory elements, such as the tetracycline-responsive promoter, could create inducible gain-of-function mutations (Chen *et al.*, 2002). Retroviruses need not be the only possible tool. Several transposons have been shown to be able to integrate into the zebrafish genome (Davidson *et al.*, 2003; Kawakami *et al.*, 2000; Raz *et al.*, 1997); if they could be mobilized in a controlled fashion similar to P elements in flies (Cooley *et al.*, 1988; Spradling *et al.*, 1999), they could prove to be a very effective mutagenesis tool. Recent advances with transposons such as *Sleeping Beauty* are described in this volume (Kawakami, 2004; Hermanson *et al.*, 2004).

Finally, insertional technologies, be they retroviral or transposon based, could also be used for reverse genetics by generating a library of insertions, as has become popular in other model organisms. Thousands of P-element fly lines have been cataloged with their junction sequences and thus chromosomal location now that the genome is complete (FlyBase Consortium, 2002). Similarly, a consortium of several labs is producing thousands of murine ES cell clones with gene trap insertions and identifying the trapped genes by inverse PCR or RACE (Stanford *et al.*, 2001; Wiles *et al.*, 2000). Also, hundreds of *Sleeping Beauty* insertions in live mouse lines are being mapped (Roberg-Perez *et al.*, 2003). In *C. elegans*, hundreds of thousands of Tc1 insertion lines have been isolated and their genomic DNA has been arrayed to provide a PCR-screenable panel for insertions in any gene of interest (Zwaal *et al.*, 1993). This model might be the one most applicable to zebrafish. Sperm samples could be taken from male founders, each sample representing about 30 insertions. Half of each sample would be frozen down, and the rest used to prepare DNA for the PCR screening array. The sperm from 30,000 founders should include nearly 1×10^6 insertions, likely to disrupt every gene at least once. One could then screen the panel by PCR for an insertion in a gene of interest and thaw out the correct sperm sample for use in *in vitro* fertilization. Alternatively, one could clone and sequence the flanking sequences the flanking sequences of all or most of the inserts from each founder and determine what gene disruptions might be found in each sperm sample. One concern might be that because *in vitro* fertilization with frozen sperm sometimes results in low fertilization rates, it might not be possible to recover the desired insertion if it were present in only a few percent of the sperm. This would certainly have to be tested. If this proved infeasible, sperm could still be frozen from multiinsert F1 fish, although this would require sperm and DNA from about three times as many fish. Either way, screening such a library, once it were made, would provide an alternative to screening chemically mutagenized sperm libraries that currently exist and have to be screened one or a few samples at a time by direct sequencing or TILLING (Draper *et al.*, 2004; Stemple, 2004; Wienholds *et al.*, 2002; Weinholds and Plasterk 2004).

Acknowledgments

We thank past and present members of the Hopkins laboratory for their contribution to the insertiontional mutagenesis screen, and AMGEN and the NIH for funding support.

References

Allende, M. L., Amsterdam, A., Becker, T., Kawakami, K., Gaiano, N., and Hopkins, N. (1996). Insertional mutagenesis in zebrafish identifies two novel genes, *pescadillo* and *dead eye*, essential for embryonic development. *Genes Dev.* **10**, 3141–3155.

Altschul, S. F., Gish, W., Miller, W., Myers, E. W., and Lipman, D. J. (1997). Gapped BLAST and PSI-BLAST: A new generation of protein database search programs. *Nucleic Acids Res.* **25**, 3389–3402.

Amsterdam, A., Burgess, S., Golling, G., Chen, W., Sun, Z., Townsend, K., Farrington, S., Haldi, M., and Hopkins, N. (1999). A large-scale insertional mutagenesis screen in zebrafish. *Genes Dev.* **13**, 2713–2724.

Amsterdam, A., Nissen, R. M., Sun, Z., Swindell, E. C., Farrington, S., and Hopkins, N. (2004a). Identification of 315 genes essential for early zebrafish development. *Proc. Natl. Acad. Sci. USA* **101**, 12792–12797.

Amsterdam, A., Sadler, K. C., Lai, K., Farrington, S., Bronson, R. T., Lees, J. A., and Hopkins, N. (2004b). Many ribosomal protein genes are cancer genes in zebrafish. *PloS Bio.* **2**, 690–698.

Beattie, C. E., Raible, D. W., Henion, P. D., and Eisen, J. S. (1999). Early pressure screens. *Methods Cell Biol.* **60**, 71–86.

Becker, T. S., Burgess, S. M., Amsterdam, A. H., Allende, M. L., and Hopkins, N. (1998). *Not really finished* is crucial for development of the zebrafish outer retina and encodes a transcription factor highly homologous to human Nuclear Respiratory Factor-1 and avian Initiation Binding Repressor. *Development* **125**, 4369–4378.

Burns, J. C., Friedmann, T., Driever, W., Burrascano, M., and Yee, J.-K. (1993). Vesicular stomatitis virus G glycoprotein pseudotyped retroviral vectors: Concentration to very high titer and efficient gene transfer into mammalian and non-mammalian cells. *Proc. Natl. Acad. Sci. USA* **90**, 8033–8037.

Chen, W., Burgess, S., Golling, G., Amsterdam, A., and Hopkins, N. (2002). High-throughput selection of retrovirus producer cell lines leads to markedly improved efficiency of germ line-transmissible insertions in zebra fish. *J. Virol.* **76**, 2192–2198.

Cooley, L., Kelley, R., and Spradling, A. (1988). Insertional mutagenesis of the *Drosophila* genome with single P elements. *Science* **239**, 1121–1128.

Culp, P., Nusslein-Volhard, C., and Hopkins, N. (1991). High-frequency germ-line transmission of plasmid DNA sequences injected into fertilized zebrafish eggs. *Proc. Natl. Acad. Sci. USA* **88**, 7953–7957.

Davidson, A. E., Balciunas, D., Mohn, D., Shaffer, J., Hermanson, S., Sivasubbu, S., Cliff, M. P., Hackett, P. B., and Ekker, S. C. (2003). Efficient gene delivery and gene expression in zebrafish using the *Sleeping Beauty* transposon. *Dev. Biol.* **263**, 191–202.

Draper, B. W., McCallum, C. M., Stout, J. L., Slade, A. J., and Moens, C. B. (2004). A highthroughput method for indentifying ENU-induced point mutations in zebrafish. *Methods Cell Biol.* **77**, 91–112.

Driever, W., Solnica-Krezel, L., Schier, A. F., Neuhauss, S. C. F., Malicki, J., Stemple, D. L., Stainier, D. Y. R., Zwartkruis, F., Abdelilah, S., Rangini, Z., Belak, J., and Boggs, C. (1996). A genetic screen for mutations affecting embryogenesis in zebrafish. *Development* **123**, 37–46.

FlyBase Consortium(2002). The FlyBase database of the *Drosophila* genome projects and community literature. *Nucleic Acids Res.* **30**, 106–108.

Foster, J. W., Dominguez-Steglich, M. A., Guioli, S., Kwok, C., Weller, P. A., Stevanovic, M., Weissenbach, J., Mansour, S., Young, I. D., Goodfellow, P. N., Brook, J. D., and Schafer, A. J. (1994). Campomelic dysplasia and autosomal sex reversal caused by mutations in an SRY-related gene. *Nature* **372**, 525–530.

Gaiano, N., Allende, M., Amsterdam, A., Kawakami, K., and Hopkins, N. (1996a). Highly efficient germ-line transmission of proviral insertions in zebrafish. *Proc. Natl. Acad. Sci. USA* **93**, 7777–7782.

Gaiano, N., Amsterdam, A., Kawakami, K., Allende, M., Becker, T., and Hopkins, N. (1996b). Insertional mutagenesis and rapid cloning of essential genes in zebrafish. *Nature* **383**, 829–832.

Golling, G., Amsterdam, A., Sun, Z., Antonelli, M., Maldonado, E., Chen, W., Burgess, S., Haldi, M., Artzt, K., Farrington, S., Lin, S.-Y., Nissen, R. M., and Hopkins, N. (2002). Insertional mutagenesis in zebrafish rapidly identifies genes essential for early vertebrate development. *Nat. Genet.* **31**, 135–140.

Haffter, P., Granato, M., Brand, M., Mullins, M. C., Hammerschmidt, M., Kane, D. A., Odenthal, J., van Eeden, F. J. M., Jiang, Y.-J., Heisenberg, C.-P., Kelsh, R. N., Furutani-Seiki, M., Vogelsang, E., Beuchle, D., Schach, U., Fabian, C., and Nusslein-Volhard, C. (1996). The identification of genes with unique and essential functions in the development of the zebrafish, *Danio rerio*. *Development* **123**, 1–36.

Hermanson, S., Davidson, A. E., Sivasubba, S., Balciunas, D., and Ekker, S. C. (2004). *Sleeping Beauty* transposon for efficient gene delivery. *Methods Cell Biol.* **77**, 357–364.

Horikawa, Y., Iwasaki, N., Hara, M., Furuta, H., Hinokio, Y., Cockburn, B. N., Linder, T., Yamagata, K., Ogata, M., Tomonaga, O., Kuroki, H., Kasahara, T., Iwamoto, Y., and Bell, G. I. (1997). Mutation in hepatocyte nuclear factor-1 beta gene (TCF2) associated with MODY. *Nat. Genet.* **17**, 384–385.

Ivics, Z., Izsvak, Z., and Hackett, P. B. (1993). Enhanced incorporation of transgenic DNA into zebrafish chromosomes by a retroviral integration protein. *Mol. Mar. Biol. Biotech.* **2**, 162–173.

Jacks, T. (1996). Tumor suppressor gene mutations in mice. *Annu. Rev. Genet.* **30**, 603–636.

Ju, B., Huang, H., Lee, K-Y., and Lin, S. (2004). Cloning of zebrafish by nuclear transfer. *Methods Cell Biol.* **77**, 405–413.

Lin, S., Kawakami, K., Shima, A., and Kawakami, N. (2000). Identification of a functional transposase of the Tol2 element, an Ac-like element from the Japanese medaka fish, and its transposition in the zebrafish germ lineage. *Proc. Natl. Acad. Sci. USA* **97**, 11403–11408.

Kawakami, K., (2004). The transgenesis and the gene trap methods in zebrafish using the *Tol2* transposable element. *Methods Cell Biol.* **77**, 201–222.

Kawakami, K., Takeda, H., Kawakami, N., Kobayashi, M., Matsuda, N., and Mishina, M. (2004). A transposon-mediated gene trap approach identifies developmentally regulated genes in zebrafish. *Dev. Cell* **7**, 133–144.

Lee, K. Y., Huang, H., Ju, B., Yang, Z., and Lin, S. (2002). Cloned zebrafish by nuclear transfer from long-term-cultured cells. *Nat. Biotechnol.* **20**, 795–799.

Lin, S., Gaiano, N., Culp, P., Burns, J. C., Friedmann, T., Yee, J.-K., and Hopkins, N. (1994). Integration and germ-line transmission of a pseudotyped retroviral vector in zebrafish. *Science* **265**, 666–669.

Mullins, M. C., Hammerschmidt, M., Haffter, P., and Nusslein-Volhard, C. (1994). Large-scale mutagenesis in the zebrafish: In search of genes controlling development in a vertebrate. *Curr. Biol.* **4**, 189–202.

Nasevicius, A., and Ekker, S. C. (2000). Effective targeted gene 'knockdown' in zebrafish. *Nat. Genet.* **26**, 216–220.

Nishigori, H., Yamada, S., Kohama, T., Tomura, H., Sho, K., Horikawa, Y., Bell, G. I., Takeuchi, T., and Takeda, J. (1998). Frameshift mutation, A263fsinsGG, in the hepatocyte nuclear factor-1beta gene associated with diabetes and renal dysfunction. *Diabetes* **47**, 1354–1355.

Postlethwait, J. H., and Talbot, W. S. (1997). Zebrafish genomics: From mutants to genes. *Trends Genet.* **13,** 183–190.

Raz, E., van Luenen, H. G. A. M., Schaerringer, B., Plasterk, R. H. A., and Driever, W. (1997). Transposition of the nematode *Caenorhabditis elegans Tc3* element in the zebrafish *Danio rerio.* *Curr. Biol.* **8,** 82–88.

Roberg-Perez, K., Carlson, C. M., and Largaespada, D. A. (2003). MTID: A database of Sleeping Beauty transposon insertions in mice. *Nucleic Acids Res.* **31,** 78–81.

Ruttledge, M. H., Sarrazin, J., Rangarantnam, S., Phelan, C. M., Twist, E., *et al.* (1994). Evidence for the complete inactivation of the NF2 gene in the majority of sporadic meningiomas. *Nat. Genet.* **6,** 180–184.

Solnica-Krezel, L., Schier, A. F., and Driever, W. (1994). Efficient recovery of ENU-induced mutations from the zebrafish germline. *Genetics* **136,** 1401–1420.

Spradling, A. C., Stern, D., Beaton, A., Rhem, E. J., Laverty, T., Mozden, N., Misra, S., and Rubin, G. M. (1999). The Berkeley *Drosophila* Genome Project gene disruption project: Single P-element insertions mutating 25% of vital *Drosophila* genes. *Genetics* **153,** 135–177.

Stanford, W. L., Cohn, J. B., and Cordes, S. P. (2001). Gene-trap mutagenesis: Past, present and beyond. *Nat. Rev. Genet.* **2,** 756–768.

Stemple, D. L. (2004). TILLING—a high-throughput harvest for functional genomics. *Nat. Rev. Genet.* **5,** 145–150.

Stuart, G. W., McMurray, J. V., and Westerfield, M. (1988). Replication, integration and stable germ-line transmission of foreign sequences injected into early zebrafish embryos. *Development* **103,** 403–412.

Sun, Z., and Hopkins, N. (2001). vhnf1, the MODY5 and familial GCKD-associated gene, regulates regional specification of the zebrafish gut, pronephros, and hindbrain. *Genes Dev.* **15,** 3217–3229.

Trofatter, J. A., MacCollin, M. M., Rutter, J. L., Murrell, J. R., Duyao, M. P., Parry, D. M., Eldridge, R., Kley, N., Menon, A. G., Pulaski, K., *et al.* (1993). A novel moesin-, ezrin-, radixin-like gene is a candidate for the neurofibromatosis 2 tumor suppressor. *Cell* **72,** 791–800.

Wagner, R. R. (1972). Rhabdoviridae and their replication. *In* Fundamental Virology (B. N. Fields and D. M. Knipe, eds.), pp. 489–503. Raven Press, New York.

Walker, C. (1999). Haploid screens and gamma-ray mutagenesis. *Methods Cell Biol.* **60,** 43–70.

Weiss, R. A., Boettiger, D., and Love, D. N. (1974). Phenotypic mixing between vesicular stomatitis virus and avian RNA tumor viruses. *Cold Spring Harbor Symp. Quant. Biol.* **39,** 913–918.

Wienholds, E., Schulte-Merker, S., Walderich, B., and Plasterk, R. H. (2002). Target-selected inactivation of the zebrafish rag1 gene. *Science* **297,** 99–102.

Wienholds, E., and Plasterk, R. H. (2004). Target-selected gene activation in zebrafish. *Methods Cell Biol.* **77,** 69–90.

Wiles, M. V., Vauti, F., Otte, J., Fuchtbauer, E. M., Ruiz, P., Fuchtbauer, A., Arnold, H. H., Lehrach, H., Metz, T., von Melchner, H., and Wurst, W. (2000). Establishment of a gene-trap sequence tag library to generate mutant mice from embryonic stem cells. *Nat. Genet.* **24,** 13–14.

Yan, Y. L., Miller, C. T., Nissen, R. M., Singer, A., Liu, D., Kirn, A., Draper, B., Willoughby, J., Morcos, P. A., Amsterdam, A., Chung, B. C., Westerfield, M., Haffter, P., Hopkins, N., Kimmel, C., and Postlethwait, J. H. (2002). A zebrafish sox9 gene required for cartilage morphogenesis. *Development* **129,** 5065–5079.

Yoon, C., Kawakami, K., and Hopkins, N. (1997). Zebrafish vasa homologue RNA is localized to the cleavage planes of 2- and 4-cell-stage embryos and is expressed in the primordial germ cells. *Development* **124,** 3157–3165.

Young, R. M., Marty, S., Nakano, Y., Wang, H., Yamamoto, D., Lin, S., and Allende, M. L. (2002). Zebrafish yolk-specific *not really started (nrs)* gene is a vertebrate homolog of the *Drosophila spinster* gene and is essential for embryogenesis. *Dev. Dyn.* **223,** 298–305.

Zwaal, R. R., Broeks, A., van Meurs, J., Groenen, J. T., and Plasterk, R. H. (1993). Target-selected gene inactivation in *Caenorhabditis elegans* by using a frozen transposon insertion mutant bank. *Proc. Natl. Acad. Sci. USA* **90,** 7431–7435.

CHAPTER 2

Genetic Screens for Maternal–Effect Mutations

Francisco Pelegri★ and Mary C. Mullins[†]

★University of Wisconsin–Madison
Laboratory of Genetics
Madison, Wisconsin 53706

[†]University of Pennsylvania
Department of Cell and Developmental Biology
Philadelphia, Pennsylvania 19104

I. Introduction

In all animals, development from fertilization to the activation of the zygotic genome at the midblastula transition (MBT; Newport and Kirschner, 1982a,b; Signoret, 1971) depends on maternal factors made during oogenesis and activated upon fertilization. By necessity, all cellular and developmental processes that occur during this time window are carried out solely by such maternal factors. Although the activation of zygotic gene expression at the MBT marks the beginning of zygotic gene control during development, it does not imply an absolute shift between the use of maternal or zygotic products. Rather, in many instances, perduring maternal products interact with newly expressed zygotic products to control developmental processes even after the activation of the zygotic genome. Moreover, maternal products establish both the dorsal-ventral and animal-vegetal axes. Formation of the dorsal-ventral axis occurs during the early cleavage stages prior to the MBT, whereas the animal-vegetal axis is established during oogenesis and marks the prospective anterior-posterior axis of the embryo.

Genetic analysis in invertebrate model organisms, such as *Drosophila* and *Caenorhabditis elegans*, has revealed networks of maternal factors involved in basic cellular functions, establishment of egg polarity, and regulation of cell fates (Kemphues and Stome, 1997; Schnabel and Priess, 1997; St. Johnston and Nüsslein, 1992). Studies in teleost fish, including the zebrafish, have begun to address the requirement of maternally driven genes in early development. Such maternal processes span basic cellular functions such as fertilization, egg activation, and early cellular and nuclear divisions, as well as induction of embryonic cell fates and execution of morphogenetic movements during gastrulation (reviewed in Pelegri, 2003; see also Dekens *et al.*, 2003; Dosch *et al.*, 2004; Kishimoto *et al.*, 2004; Pelegri *et al.*, 1999, 2004; Wagner *et al.*, 2004). However, our knowledge of maternal gene functions in zebrafish early development remains superficial and disconnected. Much of this work reflects knowledge initially acquired in other vertebrate model organisms such as *Xenopus*.

The genetic attributes of zebrafish allow the powerful method of forward genetics, so effective in *Drosophila* and *C. elegans*, to be applied in a vertebrate to identify and study the functions of maternal genes through loss-of-function analysis. Such an approach can identify in a systematic and unbiased manner a large majority of genes essential for maternal processes. Therefore, it is well suited to establish pathways of genes or fill gaps in our knowledge in particular maternal processes and to provide insights in unpredicted directions. The first forward genetic screens to isolate recessive maternal-effect mutations have recently been performed in zebrafish (Dekens *et al.*, 2003; Dosch *et al.*, 2004; Kishimoto *et al.*, 2004; Pelegri and Schulte-Merker, 1999; Pelegri *et al.*, 1999, 2004; Wagner *et al.*, 2004). Here, we describe approaches and methodologies to carry out such genetic screens for maternal-effect mutations.

II. Strategies for Maternal–Effect Screens

Two approaches have been used to identify *de novo* recessive maternal-effect mutations in zebrafish: (1) an F_4 screen based solely on natural crosses; and (2) an F_3 screen based on gynogenesis, specifically the technique of Early Pressure (EP). These alternatives differ in various important ways, which are summarized in Table I and described throughout this chapter. Here, we discuss these alternatives and provide detailed protocols to implement them.

A. F_4 Screen Based on Natural Crosses That Integrates a Mapping Strategy

One method to produce recessive, homozygous maternal-effect mutants is through a four-generation inbreeding strategy. Although at first glance this method seems to occupy an enormous amount of tank space, consolidating the F_3 generation into a single tank composed of an F_3-extended family makes this approach considerably more practical. This scheme begins similarly to F_3 zygotic screens (Driever *et al.*, 1996; Haffter *et al.*, 1996; Mullins *et al.*, 1994). As shown in

Table I

Summary of an F_4 Screen Based Solely on Natural Crosses and an F_3 Screen Based on Gynogenesis (Early Phase, EP)

	Genetic approach	
	F_4 based on natural crosses (F3-extended family)	F_3 based on parthenogenesis
Background strain used in screen	Lethal and sterile free	Lethal and sterile free Amenable to IVF and EP-based gynogenesis Needs to produce females under gynogenetic conditions
Basic methodology	Sibling pair matings	Induction of gynogenetic clutches using EP
Number of generations needed	Three, plus maternal-effect test	Two, plus maternal-effect test
Amount of space needed	Large to moderate	Moderate to small
Fraction of females within a family expected to exhibit maternal-effect phenotype	1/16	Variable, from 50% to 0%, depending on distance from the locus to the centromere
Incorporation of mapping scheme	Feasible	Difficult, because of low number of surviving gynogenotes
Identification of maternal-zygotic and paternal-effect mutations	Feasible	Feasible, but more difficult because of reduced fertility of gynogenotes

Fig. 1A, G_0 male fish are mutagenized with 3 or 3.3 mM ENU as described (Mullins *et al.*, 1994), crossed to wild-type females, and their F_1 progeny raised. Each F_1 fish carries a different set of mutagenized genes derived from the independently mutagenized spermatogonial cells of their fathers. F_1 fish are interbred and the F_2 progeny, referred to as an F_2 family, are raised to adulthood. Each F_2 family contains two mutagenized genomes, one from its mother and one from its father. In a zygotic screen, the fish of a given F_2 family are intercrossed, and the F_3-embryos are screened for recessive mutant defects. To identify maternal-effect mutations, the F_3 progeny are instead raised to adulthood. As with a zygotic mutation, for a maternal-effect mutation, one quarter of the F_3 families from an F_2 family (the F_2 intercrosses) will yield 25% maternal-effect mutant females.

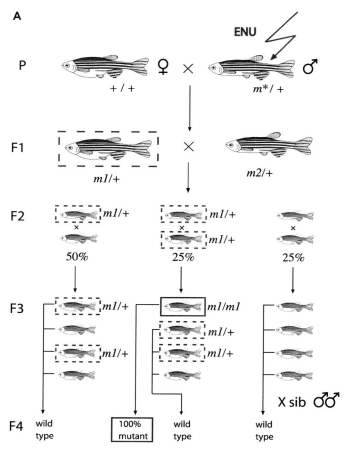

Test for maternal-effect embryonic phenotypes

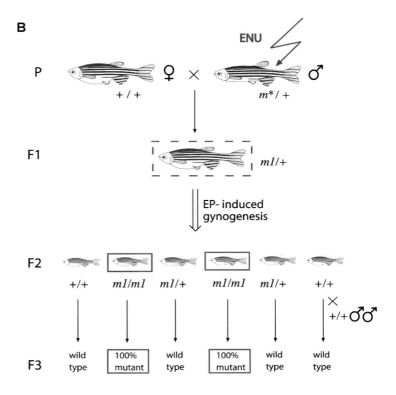

Test for maternal-effect embryonic phenotypes

Fig. 1 (A) F_4 natural crosses screen strategy. Males of the parental generation (P) are mutagenized with ethylnitrosourea (ENU) to induce new mutations (m^*) and crossed to wild-type females. F_1 fish are raised, each of which carries a different set of mutagenized genes. Two mutations are shown, $m1$ and $m2$, each carried by one of the two F_1 fish. Only mutation $m1$ is followed in subsequent generations for simplicity. Two F_1 fish are intercrossed and an F_2 family raised. Half the individuals of the F_2 family are heterozygous for $m1$. F_2 fish are intercrossed to make an F_3-extended family (see text), composed of equal numbers of F_3 fish from each of the F_2 intercrosses of one family. One quarter of the F_2 intercrosses are between $m1$ heterozygotes (boxed with hatched lines), producing $m1$ homozygotes (boxed with solid lines) in 25% of their F_3 progeny. F_3 females are tested for maternal effects in the F_4 generation. (B) Early Pressure (EP)-based screen strategy. F_1 heterozygous females carrying newly induced mutations [boxed with hatched lines; m^* and $m1$ as in (A)] are treated to induce gynogenetic F_2 clutches, which can contain homozygotes for maternal-effect mutations (boxed with solid lines). F_2 females are tested for maternal effects in the F_3 generation. A fraction of EP-derived progeny will be heterozygous for the mutation. Hypothetical results are shown by using a gene with an average centromere-locus distance (see text).

Similar to a zygotic mutation, if a recessive maternal-effect mutation exists in an F_2 family, the probability of identifying it directly depends on the number of F_3 families generated from F_2 intercrosses and the number of F_3 females screened from each F_3 family. (See Section II.A.3 for formulas.) To obtain a

90% probability of making a mutation homozygous, eight F_2 intercrosses are required, corresponding to eight F_3 families raised from each F_2 family. If these eight F_3 families each occupy a separate tank, then an eightfold increase in tank space is required over that needed for an F_3 zygotic screen. To make such a maternal-effect screen practical for the moderately sized fish facility, multiple F_3 crosses from a single F_2 family are pooled into an F_3-extended family, composed of siblings and cousins, and raised in a single tank. Specifically, if eight crosses from each F_2 family are desired, then equal numbers of each cross are pooled and raised together. If it is necessary to set up the F_2 family more than once to obtain the desired eight crosses, then separate pools are generated on different days, which are then pooled at a later point.

In an F_3-extended family, recessive maternal-effect mutants represent one sixteenth of the total females rather than one quarter, because F_2 intercrosses generating mutant and nonmutant progeny are pooled. The advantage is that only one tank is occupied and screened rather than eight, which is a significant saving in space. However, the same number of females are screened, regardless of whether they are in eight separate tanks or one consolidated tank. To obtain an 80% probability of identifying a mutant if it exists in an F_3-extended family (see also Section II.A.3), 25 F_3 females must be screened. F_3 females are screened for maternal-effect phenotypes by crossing them to sibling or wild-type males and examining their F_4 progeny for defects. Far fewer F_4 embryos are examined for defects in a maternal-effect screen than in a zygotic screen, because all or nearly all embryos are affected, in contrast to 25% for a zygotic mutant phenotype.

1. An F_3-Extended Family Approach with Integrated Mapping

A mapping cross can be integrated into an F_4 natural crosses approach, which allows one to map the maternal-effect mutation to a chromosomal position in the F_3 generation. Maternal-effect mutations are difficult to propagate because they typically produce all nonviable progeny. Thus, it is necessary to identify heterozygous females and males or homozygous males. This can be greatly facilitated by mapping the mutation to a chromosomal position, which has the additional value of initiating the molecular isolation of the mutated gene. The mapping strategy is discussed further in Section V.A.

2. Identification of Maternal–Zygotic, Male–Sterile, and Paternal–Effect Mutations

An advantage to screening the F_3 females in crosses to F_3 sibling/cousin males is that maternal-zygotic, zygotic, as well as male-sterile and paternal-effect mutations can also be isolated. F_3 intercrosses provide a $\sim$50% probability of detecting a mutant that requires loss of both maternal and zygotic gene activity, which is not possible if the F_3 female is crossed to a wild-type male. All zygotic mutations in the F_2 family are still present in the F_3 family, and therefore the F_4 embryos from F_3 intercrosses can be screened not only for maternal-zygotic but also for zygotic mutants if desired. In F_3 sibling/cousin crosses, paternal-effect and male-sterile mutations will also be revealed.

If a mutant is identified in F_4 embryos, the type of mutation induced can be distinguished by crossing the F_3 female and male parental fish separately to wild-type fish and examining the progeny for defects. If the mutation is a maternal or paternal effect or a female- or male-sterile mutation, then the defect is evident in the F_4 embryos, even when the F_3 fish is crossed to wild type. For a maternal effect or female-sterile mutation, the F_3 female will be the cause of the defective embryos, whereas for a paternal-effect or male-sterile mutation, the F_3 male parent will be the sole cause of the defect. For recessive-zygotic and maternal-zygotic mutations, the F_4 embryonic defect will depend on both F_3 parents. A different fraction of affected F_4 embryos is expected for fully penetrant maternal-zygotic versus zygotic mutations. For a maternal-zygotic mutant, the F_3 female is homozygous and the F_3 male heterozygous for the mutation, resulting in 50% mutant progeny, whereas a zygotic mutation yields 25% F_4 mutants.

3. Assessment of the F_4 Natural Crosses Screen

Typical values for several parameters of an F_4 natural crosses screen are shown in Table II. The results of an F_4 natural crosses strategy, using an F_3-extended family and incorporating a mapping cross, have recently been published (Dosch *et al.*, 2004; Wagner *et al.*, 2004). In such an F_4 screen, the number of mutagenized genomes (G) screened contributed by a given F_3-extended family is determined with the following formula: $G = (1 - 0.9375^n) \times 2 \times (1 - 0.75^m)$. The term $(1 - 0.9375^n)$ is the probability of identifying an F_3 recessive maternal-effect mutant female present in an F_3-extended family if n females are screened within that family. The factor 2 represents the two mutagenized genomes derived from the

Table II
Statistics in an F_4 Natural Crosses Screen

% F_2 families used to generate F_3 families[a]	80%
No. of F_2 intercrosses generating an F_3-extended family	≥8
Fraction of F_1 mutagenized genomes homozygous in F_3-extended family	≥90%
No. of F_3 females screened/F_3-extended family	24
Fraction of genomes screened in F_3 family	79%
No. of haploid genomes screened/F_3-extended family[b]	1.4
Maternal-effect mutants identified/genome screened	0.11
Fraction of candidate mutations recovered	95%

[a]Because of the female bias in hybrid strains that are needed to make the mapping cross, a fraction of the F_2 families have three or less males. These families are difficult to work with and are therefore discarded.

[b]Because F_1 fish are interbred to make the F_2 generation, there are two mutagenized genomes present in the F_2 family and therefore more than one mutagenized genome is ultimately screened in each F_3-extended family.

two F_1 fish. The term $(1 - 0.75^m)$ is the fraction of the two mutagenized genomes expected to be homozygous in the F_3 generation, where m is the number of F_2 crosses that comprise an F_3-extended family.

B. F_3 Screen Based on Early Pressure (EP)–Induced Gynogenesis

Artificially induced gynogenesis in zebrafish involves diploidization of the maternal haploid genome, producing viable offspring with solely a maternal genetic contribution (Streisinger et al., 1981). Incorporation of gynogenesis into a genetic scheme for maternal-effect mutations allows the direct production of homozygotes for induced mutations from a single heterozygous F_1 carrier, by-passing one generation in comparison to a scheme based solely on natural crosses (compare Fig. 1A and B). Because of the large number of chromosomes present in zebrafish, genetic screens in this organism involve the whole genome and are essentially blind, screening all mutagenized chromosomes simultaneously rather than individually, as done in the fly and worm. Therefore, each generation in a screen generates an exponentially increasing number of crosses. Thus, by-passing one generation through the use of EP allows a significant reduction of the time and space required to carry out a maternal-effect screen.

In a basic gynogenesis-based scheme (Fig. 1B), mutations are induced in the germline of parental (P) males by exposing them to the point-mutagen N-ethyl-N-nitroso-urea (ENU; Mullins et al., 1994; Solnica-Krezel et al., 1994; van Eeden, 1999). P males are then crossed to produce F_1 progeny heterozygous for induced mutations. Eggs are stripped from F_1 females and gynogenesis is induced. This allows newly induced mutations to become homozygous in up to 50% of the gynogenetic F_2 generation (see Section II.B.1). Adult F_2 females are screened for maternal effects by testing their F_3 progeny for embryonic phenotypes. In the EP-based screen, the production of F_3 clutches is best achieved by in vitro fertilization (IVF) using wild-type sperm. This is because sib-sib crosses mate inefficiently because of the semisterility of sibling males caused by both the female-rich genetic background (see Section III.B.2) and EP-induced inbreeding. On the other hand, IVF is facilitated by females from the background line being easily stripped of eggs. The use of IVF precludes the possibility of identifying maternal-zygotic mutations (see Section II.A.2), although the use of highly selected lines might allow this in the future. However, IVF has the advantage that it allows the production of F_3 clutches that can be immediately observed and followed synchronously, thus facilitating the identification of early phenotypes (see Section II.C). In addition, because all clutches are fertilized by the same batch of sperm solution, IVF allows the rapid identification of maternal mutations that affect the ability of the egg to become fertilized (which would be obscured by unfertilized clutches sporadically observed in natural matings from wild-type parents).

Because of the relatively small number of surviving individuals in the F_2 gynogenetic clutches, it is not currently practical to incorporate a mapping strategy within an EP-based genetic screen as described for F_4 screens based solely on natural crosses (see Sections II.A.1 and V.A), although this might be possible in the future through the use of selected polymorphic lines (see Section III.B). Rather, in an EP-based screen, mapping crosses are currently initiated after recovery of the mutation (see Section V.B).

Although incorporation of gynogenesis can simplify a maternal-effect genetic screen, gynogenesis itself is only efficiently induced under specific conditions. In a scheme for a gynogenesis-based maternal-effect screen, the main goal is the efficient production of fertile gynogenetic F_2 females that are homozygous for newly induced mutations. A number of variables need to be optimized to carry out this procedure, which we discuss next. First, a suitable method of gynogenesis needs to be selected. Second, an appropriate mutagenesis dosage needs to be chosen to induce a reasonably high rate of mutations while allowing the production of viable homozygous adult mutants. In addition, lines amenable to gynogenetic procedures need to be selected (see Section III.B).

1. Choice of Gynogenetic Method

There are two main techniques for the artificial induction of gynogenesis in zebrafish: Early Pressure (EP) and Heat Shock (HS; Streisinger *et al.*, 1981; see also Fig. 3 in Pelegri and Schulte-Merker, 1999). In both methods, eggs are first artificially fertilized with sperm whose genetic material has been inactivated by UV irradiation. In the absence of further treatment, these eggs would develop into haploid embryos that are inviable. Both EP and HS lead to the diploidization of the genetic content of the egg, thus producing viable diploid embryos.

In EP, diploidization is induced by the application of hydrostatic pressure between Minutes 1.33 and 6 after egg activation (see Section VI.C.6). This treatment inhibits completion of the second meiotic division and the expulsion of the second polar body, resulting in a diploid egg. HS, on the other hand, inhibits cytokinesis of the first mitotic division of haploid embryos by applying a heat pulse during Minutes 13 to 15 after egg activation (see Section VI.C.5), transforming haploid embryos into diploid ones. Hydrostatic pressure, applied late, has also been used as an alternative method to inhibit the first mitosis, although it has been found to be less effective and more cumbersome than HS (Streisinger *et al.*, 1981).

In theory, HS is more efficient than EP in the direct induction of homozygosity and therefore might be the technique of choice in a maternal-effect screen. This is because HS-derived progeny are homozygous at every single locus, and therefore 50% of HS-derived F_2 progeny are homozygous for a mutation present in heterozygous form in the F_1 mother. EP, on the other hand, because of recombination during meiosis, leads to a variable degree of homozygosity ranging from 50% towards 0%, depending, respectively, on whether loci are linked to the centromere or are distally located. Thus, HS would in principle provide the highest possible yield of homozygous mutant adults for all loci regardless of their chromosomal location. Moreover, the expectation of a fixed percentage of homozygous mutant females would aid in the assessment of newly identified phenotypes.

In spite of these obvious theoretical advantages of HS over EP, in practice EP is superior to HS as a gynogenetic method for a number of reasons. First, HS is about twofold less efficient than EP in inducing viable diploid gynogenotes (Table III; see also Streisinger *et al.*, 1981), presumably because of a greater intrinsic ease of inhibiting the extrusion of the polar body during meiosis rather than cytokinesis during the first mitosis. Moreover, EP-derived adults, probably because of their higher heterozygosity, show viability and fertility rates that combined are about fourfold higher than those in HS-derived clutches (Table III). Thus, the final yield of fertile adult gynogenotes derived from EP is about eightfold higher than that derived from HS.

Higher levels of heterozygosity in EP-derived gynogenotes are beneficial for additional reasons. First, under mutagenic conditions, the yield of HS-gynogenetic clutches is expected to be further reduced by a factor of 0.5 per induced zygotic recessive lethal mutation, whereas EP-derived clutches are expected to be reduced by a factor of only 0.23 (see Section II.B.2). Second, the increased heterozygosity of EP gynogenotes improves the odds of recovering newly identified mutations. This occurs because the overall fraction of *fertile* siblings that carry a given mutation, because of a decrease in the fraction of the (sterile) homozygous mutant females, is higher in EP-derived clutches than in HS-derived ones (see Section IV.A.2).

The main drawback of the higher heterozygosity of EP gynogenotes is that it leads to an intrinsic bias against the identification of distally located mutations. However, measurements of the frequency of homozygosity (F_m) of random zygotic mutations after EP-induced diploidization range from 0.50 to 0.04, with an average value of 0.23 (16 loci; Streisinger *et al.*, 1986; Neuhauss, 1996). With the assumption that maternal genes are similarly distributed throughout the chromosomes, these data suggest that the majority of these genes are sufficiently close to a centromere to be identified through an EP-based screen. For these reasons, we chose EP over HS as a gynogenetic method for our screen, although HS might become applicable in the future with the use of highly selected lines.

Table III
Comparison of Heat Shock- and Early Pressure-Induced Gynogenesis (Gol-Mix Line)

	Heat shock	Early pressure
Viability at Day 5 (viable/fertilized eggs)[a]	0.09 ($n = 3590$)	0.21 ($n = 4368$)
Fraction clutches with >6 viable Day 5 fish	0.41 ($n = 29$)	0.93 ($n = 29$)
Clutch size (viable Day 5 fish/clutch)	10 ($n = 29$)	37 ($n = 29$)
Adult viability (viable at 3 months/Day 5 viable)	0.53 ($n = 324$)	0.66 ($n = 218$)
Fertility (fertile adults/total adults)	0.23 ($n = 13$)	0.65[b] ($n = 226$)

[a]Viable at Day 5: fish that can inflate their swim bladders.
[b]Value from F_2 descendants of P males mutagenized with 2 mM ENU.

2. Mutagenesis Dose

In the F_2 gynogenetic generation, homozygosity for mutations in essential zygotic genes will lead to a decreased survival of gynogenotes. For example, the mutagenic dosage used in large-scale zygotic screens (3 × 1 h treatments with 3 mM ENU) is expected to induce about one embryonic lethal and one larval lethal per haploid genome (Mullins *et al.*, 1994; Haffter *et al.*, 1996; Solnica-Krezel *et al.*, 1994). This implies that under this mutagenic condition, only 59% of what would be otherwise viable EP-derived gynogenotes (25% using HS) would survive to adulthood. Thus, we reduced the ENU dosage in our maternal-effect screen experiments. Similar reductions in the strength of mutagenic treatments were adopted for maternal-effect screens in *Drosophila* and *C. elegans* (see, e.g., Lehmann and Nüsslein-Volhard, 1986, and Kemphues *et al.*, 1988). We have observed that a mutagenic dosage of 3 × 1 h 2 mM ENU treatments begins to have a mild effect on the viability of F_2 gynogenetic clutches (not shown). These conditions lead to a mutagenic rate, as assayed by the frequency of newly induced *albino* alleles, estimated to be about one third of the rate induced by the standard (3 mM ENU) treatment (data not shown; Mullins *et al.*, 1994; Solnica-Krezel *et al.*, 1994), or about 0.3–0.4 embryonic lethal mutations per haploid genome. We chose for our screen the ENU concentration of 2 mM as a compromise between a moderate mutagenic rate and a practical level of viability.

3. Assessment of EP-Based Screens

The results from genetic screens using an EP-based method have been described elsewhere (Pelegri and Schulte-Merker, 1999; Pelegri *et al.*, 2004, see also Dekens *et al.*, 2003; Pelegri *et al.*, 1999). Typical survival and yield values are presented in Table IV. To estimate the number of genomes screened using an EP-based method, one needs to keep in mind that the number of genomes screened depends on the level of EP-induced homozygosity (F_m), which, in turn, is inversely related to the centromere-to-locus distance. Thus, the number of genomes screened will differ according to the position of genes with respect to the centromere. To estimate such values, one can first estimate the critical number of F_2 females that need to be screened in a given family to result in a 90% probability of detecting a newly induced mutation present in that family. For example, with an average F_m value of 0.23 (see Section II.B.1), nine F_2 tested females per clutch would be needed to detect a newly induced mutation with a 90% probability. This estimate corresponds to an average locus and varies greatly, depending on the centromere–locus frequency: to reach a similar frequency of detection for centromere-linked (F_m 0.5) and distal (F_m 0.05) loci, the critical number of F_2 females tested per clutch is 4 and 44, respectively. For specific F_m values, each family in which the number of tested F_2 females is equal to or higher than the critical number of F_2 females contributes one screened haploid genome. In cases in which the number of F_2 females tested per family is less than the critical value for a given F_m, the

Table IV
Statistics in an F_3 Gynogenesis–Based Screen

% F_2 clutches grown to adulthood[a]	45%
% F_2 clutches with fertile adult females[b]	20%
No. of screened haploid genomes/No. of F_2 clutches with fertile females[c]	0.45 (proximal) – 0.27 (average)
Maternal-effect mutants identified/No. of haploid genomes screened[d]	0.11–0.19
% Candidate mutations recovered[e]	44%

[a] EP-derived F_2 clutches with at least six viable fish on Day 6 of development.

[b] Fertile females are defined as those that produce normal eggs, which on activation exhibit the wild-type translucent appearance and can be fertilized to exhibit either a normal or a characteristic abnormality in the early cleavage pattern.

[c] The number of genomes screened depends not only on the number of females tested but also on the average distance of the loci to the centromere (Section II.B.1). The values presented are derived from the number of females tested for each family, so that mutations present in the family have a 0.97 chance of being identified, assuming F_m values of 0.50 and 0.23, respectively, for centromere-linked loci and loci at an average distance to the centromere. F_2 families are considered to carry a candidate mutation when they contain females that produce a phenotype in 100% of the F_3 offspring and the phenotype appears in more than one independent F_3 clutch.

[d] The range given is estimated by assuming that the isolated mutations are all either proximal (0.11) or at an average distance to the centromere (0.19).

[e] Mutations in some lines are not recovered because of a variety of reasons: false positives in the original tests, inability to recover the line due to insufficient fish to perform recovery crosses, and variability in the penetrance of the mutation.

number of tested F_2 females can be pooled in order to find a combined number of screened genomes contributed by that pool. The latter value is calculated by dividing the pooled number of tested F_2 females by the critical value needed to screen one haploid genome at a 90% certainty. The total number of haploid genomes screened is the sum of all families with more than the critical number of tested F_2 females and the combined number of genomes calculated from the pooled number of tested F_2 females.

C. Screening Embryos

Once females that might be homozygous for a maternal-effect mutation are produced, either in the F_3 generation in a screen with solely natural crosses or the F_2 generation in an EP-based screen, embryos from those females are screened for potential defects. Ideally, embryos are collected as early as possible, at most within 2 h after fertilization. As mentioned in Section II.B, in EP-based strategies, clutches can be derived by IVF. This allows one to both observe and synchronize their development immediately after fertilization. On the other hand, if the screened embryos are produced through natural crosses, early observation and synchronization of the clutches can be approached by taking advantage of the propensity of zebrafish to lay eggs during the early hours of their daylight cycle. In practice, this is done by setting up crosses toward the end of the light cycle and collecting embryos during the early hours of the following light cycle. The early collection of egg clutches allows one to discard clutches with eggs that have undergone aberrant ovulation, which normally occurs in a fraction of clutches from wild-type females and which, if undetected, would provide false positives in the screen. This early observation also allows one to determine whether the cleavage and cellularization pattern characteristic of wild-type embryos is normal. Because unfertilized embryos also exhibit a pattern of irregular cleavages (pseudocleavages; Kane and Kimmel, 1993), it is also important that such early embryos be carefully observed to reveal potential differences between pseudocleavage formation and an abnormal early cellular pattern. Once the regular pattern of cellular cleavage characteristic of normal fertilized embryos is detected, fertilized embryos are sorted and transferred to a clean plate at low densities (40 embryos per 10.5 cm diameter plate). Embryos are subsequently screened for deviations in the wild-type developmental pattern (Kimmel *et al.*, 1995; see also van Eeden *et al.*, 1999, for a sample scoring chart).

This screening strategy relies on the incorporation of the sperm into the embryo, which is necessary for patterns of cleavage distinct from those that occur in unfertilized eggs. Maternal-effect mutations acting at earlier steps in oogenesis and egg maturation, for example, those affecting the animal-vegetal axis of the egg (Fig. 2), can also be identified by looking at clutches shortly after fertilization. In addition, it is possible to identify mutations affecting oogenesis by selecting for nonlaying females and screening for defects in the ovary by sectioning or dissection of the ovary (Bauer and Goetz, 2001; T. Gupta, F. Marlow, MCM, unpublished).

III. Selection of Lines for Genetic Screens

A. Selection for Lethal-/Sterile-Free Background Lines

An important characteristic desired in a genetic background is the absence of preexisting mutations, either maternal or zygotic. In any kind of screen, whether based on gynogenetic techniques or natural crosses, the use of lines free of preexisting mutations is important for two reasons. First, lines free of lethal mutations diminish unwanted background lethality, which reduces brood sizes and can preclude the isolation of new mutations closely linked to the background mutation. In addition, the use of lines free of preexisiting mutations eliminates the possibility of isolating multiple copies of a mutant allele already present in the genetic background. Absence of lethal or sterile mutations can be selected for in two ways.

1. Continuous Inbreeding

Wild-type stocks free of zygotic lethal and sterile mutations can be obtained by inbreeding individuals for two generations and essentially screening the F_2 generation for lethality and sterility phenotypes. In a stock maintained by mixed breeding of many individuals (to maintain genetic diversity and prevent unhealthy, highly

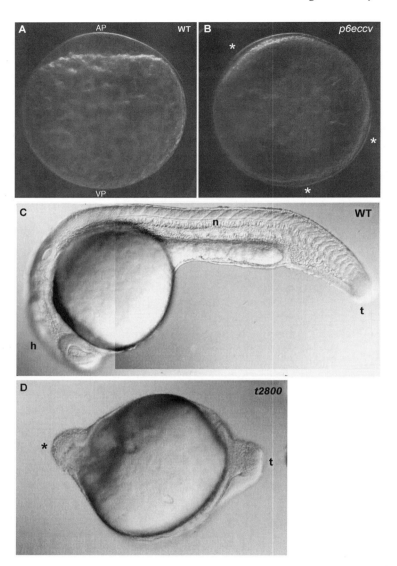

Fig. 2 Two mutants identified and recovered in systematic screens for maternal-effect mutants. (A, B) The animal-vegetal polarity mutant *p6eccy* was identified in an F$_4$ natural crosses screen (Dosch *et al.*, 2004). (A) A wild-type egg shortly after fertilization displaying the blastodisc prominently at the animal pole. (B) In contrast, in the *p6eccv* mutant the cytoplasm segregates to multiple locations around the circumference of the egg (asterisks). AP and VP are the animal and vegetal poles, respectively (A,B photo courtesy of Florence Marlow). (C, D) The mutation *t2800*, recovered in an EP-based screen (Pelegri *et al.*, 2004), results in defects in the induction of dorsoanterior cell fates. (C) A wild-type embryo 24 hrs after fertilization shows the normal body plan, including the head (h) and the notochord (n), a dorsal mesoderm derivative. (D) An embryo from *t2800* mutant mothers lacks anterodorsal structures and is radially symmetric. (t) indicates the tail region in both wild-type and mutant embryos, which is less extended and contains multiple folds in the mutant. The asterisk indicates a group of cells that accumulates at the anterior of the embryo.

inbred stocks), two generations of inbreeding of several pairs of fish can reduce the likelihood of background mutations being present in the parental generation. Individual pairs of wild-type fish are intercrossed and their respective F1 progeny raised in separate tanks. The F_1 progeny are then intercrossed and screened for zygotic lethal mutations in the F_2 generation. A reliable indicator of zygotic lethal mutations is the lack of swim bladder inflation 5 days postfertilization, in addition to obvious defects at earlier stages in 25% of the brood. By examining F_2 embryos from at least 12 intercrosses from one F_1 family, a >95% probability exists that a mutation will be detected if it exists in that particular family.

If lethality is not observed in any of the 12 crosses, then the F_2 fish from the 12 individual F_1 intercrosses are raised in a separate tank and screened for late lethal mutations, as well as maternal and paternal effects and female and male sterile mutations. The total number of F_2 adults is counted and compared to the number of larvae initially raised. If a late lethal mutation exists, then 25% of the larvae will not survive to adulthood. Several control crosses can be raised between unrelated individuals to control for nongenetic lethality associated with normal fish raising. Only lethality significantly beyond that of normal baby raising is then considered as a potential late lethal mutation. From each F_2 family 12 males and females are intercrossed and their embryos examined. If the F_3 embryos are normal, then the probability is >95% that such maternal and paternal effects and sterile mutations do not exist in those F_2 fish. If two such F_2 lines are established from different F_1 fish, then the males can be mutagenized and then interbred to females from another F_2 line in the parental generation of the screen to prevent further inbreeding.

2. Whole Genome Homozygosity Through Heat-Shock (HS)-Induced Gynogenesis

The gynogenetic method of HS induces homozygosity at every single locus (see Section II.B.1) and is thus particularly effective at selecting, in one single generation, for fish that lack *any* background mutations. After growing a large number of HS-derived gynogenetic clutches from our substrate line gol-mix, we selected for adult fish free of lethal or sterile mutations. From our starting gol-mix population, we generated two lines, golFL-1 and golFL-2, from four different HS-derived individuals. These two lines were combined to create golFL-3. (golFL-2 was 100% male and could not be propagated as a pure stock.) Similar strategies have been used previously to select for such lethal-/sterile-free strains, which can be further propagated through EP to generate clonal lines (Streisinger *et al.*, 1981). Selection of lines through HS and EP can also lead to stocks of higher viability under gynogenetic conditions, presumably by the reduction of background detrimental alleles (Streisinger *et al.*, 1981).

B. Specific Requirements for Lines in EP–Based Screens

The majority of lines we have examined, including lines recently derived from the wild, tend to produce low yields of fertile gynogenotes (unpublished observations). Selection of appropriate lines is therefore very important for an efficient gynogenetic-based maternal screen.

1. Selection of Lines That Produce a High Yield of Gynogenotes

Experimental induction of gynogenesis relies on the manipulation of *in vitro* fertilized eggs at very early stages. Therefore, it is necessary that females should, as a first requirement, readily yield eggs when manually stripped. Different fish strains differ greatly in their ability to be manually stripped of eggs (not shown). The capacity to be stripped of eggs is distinct from being fertile and to successfully mate under standard natural crosses in the laboratory (Eaton and Farley, 1974; our observations). This might be related to the fact that, under natural conditions, release of mature oocytes from their follicles into the ovarian lumen requires hormonal stimulation (Selman *et al.*, 1994), which might normally be triggered by vigorous chasing by the males (Eaton and Farley, 1974). Lines that can be most easily stripped of eggs appear to be those that have been propagated by artificial fertilization methods, which also involve stripping of eggs, such as those derived from the AB Oregon line (Streisinger *et al.*, 1981). In contrast, lines from the wild or laboratory lines that have been propagated mostly by natural crosses tend not to be easily stripped of eggs. We found that the gol-mix line, a hybrid line with both AB and Tübingen genetic backgrounds, is robust, and that its females can be easily stripped of eggs, as well as produce a high yield of gynogenotes. Thus, we chose to continue our selections and genetic screen schemes with this starting population. In addition, because this line is marked with the recessive pigment marker *golden*, it allows the detection of unwanted products of incompletely inactivated sperm (isolated from *golden*[+] males) after the EP procedure.

2. Selection for Favorable Sex Ratios Under Gynogenetic Conditions

Sex determination in fishes varies from organisms with sex-determining chromosomes to multifactor autosomal ones, and in some cases sex has been shown to be influenced by external factors (reviewed in Chan and Yeung, 1983). The mechanism of sex determination in zebrafish, although poorly understood, appears to fall in the latter category, lacking a single sex chromosome and being sensitive to growing conditions. Most gynogenetic clutches after grown to adulthood exhibit sex ratios that are strongly biased toward maleness (86–88% males; see Fig. 4 in Pelegri and Schulte-Merker, 1999). The phenomenon of sex bias in gynogenetic clutches is likely related to the tendency of zebrafish and other teleosts to develop into males under suboptimal conditions, for example, in overcrowded conditions or in subviable genetic backgrounds (see Chan and Yeung, 1983; our observations). Presumably, gynogenetic clutches, because of their high degree of inbreeding, also have a suboptimal genetic background that under normal circumstances produces males. Nevertheless, a small fraction of the gynogenetic clutches (5–10%) are composed of at least 50% females.

The observation of rare gynogenetic clutches with a high female-to-male ratio suggests that it is possible to select for genetic backgrounds that produce a high proportion of females even under gynogenetic conditions. In fact, the gynogenetic procedure itself might act as a selection for female-rich genetic backgrounds, as exemplified by the fact that one out of two lethal-/sterile-free lines that we derived from HS-derived adult gynogenotes consists of mostly females (92% females). In this line, go1FL-1, there exists a small fraction of males, which can mate with wild-type females, but tend to produce unfertilized eggs. Nevertheless, treatment of this line with testosterone for the first 14 days of development leads to the production of larger percentages of fertile males (72% males), and thus allows the production of males both for mutagenesis and for the propagation of the line through natural crosses.

C. A Hybrid/Inbred Approach

Selection of lines can increase the frequency of certain desired traits, but such selection also leads to inbreeding, which often causes a reduction in overall robustness and fertility (Thorgard, 1983). Thus, the best lines for gynogenetic-based maternal screens might be hybrids between gynogenetically selected lethal-/sterile-free lines. This approach is essential for genetic screens that incorporate a simultaneous mapping strategy. In this case, both polymorphic lines can be selected independently for the characteristics desired in the screen.

IV. Recovery and Maintenance of Maternal–Effect Mutations

Once females are identified as exhibiting a maternal-effect phenotype, the mutation needs to be recovered. The observed maternal-effect phenotypes are expected to be caused by maternal homozygosity for recessive mutations, because dominant mutations are unlikely to be propagated through generations that occur prior to screening. Because homozygosity for recessive maternal-effect mutations in females leads to the inviability of their progeny, a genetic scheme has to allow the recovery of the mutations through genetically related individuals.

A. General Methods for the Recovery of Mutations

A mutation can be recovered by three general means: (1) through known heterozygous carriers, (2) through siblings of homozygous mutant females, and (3) through survivor progeny derived from homozygous mutant females.

1. Recovery Through Known Heterozygous Carriers

Individuals that produce homozygous maternal-effect mutant females (F_2 parents in an F_4 natural crosses screen and F_1 females in a F_3 EP screen) are heterozygous carriers of the mutations. Thus, such fish can be stored separately until their progeny reach adulthood and are tested for maternal-effect phenotypes. After the 3 to 5 months that are required to grow up and test their progeny, we find that the majority of the separated individual fish are still alive and fertile, and thus can be used to recover the mutation.

2. Recovery Through Siblings of Homozygous Mutant Females

In some instances, a known carrier for a mutation is not available or is not fertile. In these cases, maternal-effect mutations can be recovered by performing crosses between siblings of homozygous mutant females (F_3 siblings in an F_4 natural crosses screen and F_2 siblings in an F_3 EP-based screen), a fraction of which are carriers of the mutation. If the mutation is mapped (see Section V.A), siblings can be selected that are either homozygous or heterozygous for the mutation by genotyping.

If the mutation has not yet been mapped, propagation is ensured by raising the progeny of multiple intercrosses of siblings. In an F_4 natural crosses screen using an F_3-extended family (composed of siblings and cousins), ~50% of the sibling males and females are heterozygous carriers and 6.25% of the males are homozygous carriers. Thus, 25% of F_3 sibling/cousin intercrosses are between heterozygotes of the mutation and yield F_4 mutant females. If the F_3 (or subsequent) generation is made between two heterozygous carriers, then 75% of all F_3 males are carriers (50% heterozygous and 25% homozygous for the mutation), and 67% of the sibling females are heterozygous carriers for the mutation. In this case, 50% of the intercrosses should yield F_4 homozygous mutants.

In EP-based screens, it is preferable to generate outcrosses rather than incrosses for the recovery of mutations. This is because EP-derived fish do not mate as efficiently as wild-type fish. In addition, the background used in an EP-screen results in female-rich tanks, and such abnormal ratios interfere with subsequent propagation of the mutation. We have found that outcrossing to a line such as *Leopard Long Fin* (also known as TLF), which tends to have a slight bias toward maleness, results in hybrid stocks that have normal sex ratios in subsequent generations. Outcrossing also improves the general robustness and fertility of the line.

Outcrossing is preferably carried out through sibling F_2 males rather than sibling females, because females homozygous for a maternal-effect mutation are sterile whereas homozygous males should be fertile unless the mutated gene also affects male fertility. This is particularly true in the case of centromere-linked loci (F_m towards 0.5), when most *fertile* females are expected to be homozygous for the wild-type allele (see Section II.B.1 and later) and therefore cannot transmit the mutation.

In EP-derived clutches, the frequency of heterozygotes and homozygotes for a given mutation varies depending on the centromere–locus distance. For centromere-linked loci (F_m close to 0.5), 50% of the siblings are homozygous for the mutation. As the centromere–locus distance increases, the fraction of homozygous siblings (F_m) decreases, but the fraction of heterozygous siblings increases two times as rapidly. For a distal mutation of $F_m = 0.05$, for example, 5% of the F_2 siblings are homozygous mutants whereas 90% are heterozygous carriers. Thus, the overall frequency of F_2 carrier siblings (homozygous or heterozygous), varies from 50% for centromere-linked loci to percentages approaching 100% for distal loci. Therefore, the recovery of mutations through F_2 siblings can also be an efficient strategy in EP screens. In large F_2 EP-derived clutches, F_m, and therefore the fraction of siblings heterozygous or homozygous carriers for the mutation, can be estimated by the proportion of F_2 females that exhibit the maternal-effect mutant phenotype.

3. Recovery Through Rare Survivors of the Maternal Effect

In cases of incompletely penetrant phenotypes, mutations can also be recovered through rare survivors from clutches that exhibit the maternal-effect phenotype (e.g., F_4 clutches in natural crosses screens and F_3 clutches in EP screens). The presence of such escapers might be due to variability in the phenotype caused by residual function of a hypomorphic allele or some degree of redundancy in the affected pathways. Escapers are expected to be heterozygous carriers for the mutation, and the mutation can be propagated by incrossing fish derived from them. Of the given options, the schemes in IV.A.1 and IV.A.3 are the most efficient, because they use individuals that are known carriers of the mutation.

B. Maintenance of Maternal–Effect Mutations

Whether a mutation has been mapped or not, it is tempting to maintain stocks carrying maternal-effect mutations by either repeated inbreeding or through escaper embryos (see Section IV.A.3). However, repeated inbreeding eventually generates inbred stocks that are weak and have aberrant sex ratios that typically lead toward maleness (see Section III.B.2), thus interfering with the identification of homozygous mutant females. Moreover, maintenance of the mutation by repeated propagation through escaper embryos might select for genetic constellations that gradually weaken the mutant phenotype. Lastly, in an F_4 natural crosses approach, the high ENU dose used leads to induction of multiple lethal mutations, typically unlinked to the maternal-effect mutation, but which can nevertheless reduce the size of intercross families and are best crossed out of the maternal-effect mutant background. To address these issues, mutations can be routinely propagated through cycles of crosses to a wild-type stock (outcrosses), followed by crosses between siblings (incrosses). Typically, an outcross that is known to carry a mutation can be kept for a period of time and additional incrosses performed from the outcross fish to produce new families containing homozygous mutant females. It works well to perform a cycle of one outcross, which can generate several incrosses over a period of a year or more, and then initiate a new cycle by carrying out an outcross from one of the more recent incrosses.

1. Maintenance of Mapped Mutations

If mutations are mapped to a chromosomal position, genotyping is used to identify males homozygous for the mutation from the siblings of homozygous mutant females (25% of the male siblings expected to be homozygotes). Outcrosses are then initiated from such homozygous males to wild-type females carrying alleles, for example SSLP markers, flanking the mutation that are polymorphic to those of the homozygous male. The progeny of this cross are all heterozygous carriers and can be interbred to produce a family that contains 25% homozygous mutant females. This strategy allows the unambiguous identification of heterozygous and homozygous carriers in this generation through the use of polymorphic markers flanking the mutation.

2. Maintenance of Unmapped Mutations

If the mutations are not mapped, a similar strategy is followed, except that it is through multiple, random crosses. As when recovering the mutations (see Section IV.A.2), it is more efficient to initiate the crosses through males, 75% of which are expected to be either homozygous or heterozygous carriers for the mutation. Thus, 75% of outcrosses derived from males that are siblings of homozygous mutant females consist of families of carrier individuals. Multiple crosses ensure the propagation of the mutation. For example, five outcrosses from such sibling males of homozygous mutant females ensure a 99.9% probability of transmission of the allele to at least one of the outcrosses. Within such outcrossed-derived families, the percent of heterozygotes is expected to be 100% or 50%, depending, respectively, on whether the original outcrossed male is homozygous or heterozygous for the mutation. Multiple incrosses from such families allow the recovery of homozygous mutant females in the next generation. For example, eight incrosses from a tank that consists of 50% heterozygous carriers lead to a 90% probability of finding homozygous females (at a 25% frequency) in at least one of the incrossed families.

A variation of this approach is to outcross identified heterozygous or homozygous carrier males. These males are identified as carriers by interbreeding them with sibling females, raising the progeny, and determining whether their offspring yield mutant females. Parental males yielding mutant female progeny are then outcrossed. These outcross progeny are then inbred to produce a new generation of homozygous mutant females, as discussed previously. In this modified approach, two blind generations of intercrosses are avoided by first identifying the males prior to outcrossing them.

During the maintenance of mutations, individual carriers should be outcrossed to fish of the same strain. This avoids increasing the degree of polymorphism in the carrier line, which in turn facilitates the subsequent process of mapping (see Section V).

V. Mapping Maternal–Effect Mutations

Mapping a mutation to a chromosomal position can be carried out either simultaneously with an F_4 screen using natural crosses (see Section II.A) or at any time after the identification and recovery of the mutation. Specific details on mapping protocols have been previously described (Geisler, 2002; Talbot and Schier, 1999). Here we describe the modification of this approach for mapping maternal-effect mutations. Briefly, the approach consists of outcrossing a carrier for the mutation to a polymorphic wild-type stock to yield F_1 hybrid families. Incrosses from the F_1 family in turn allow the production of F_2 adult females, which can be tested for homozygosity of the maternal-effect mutation and analyzed for linkage to DNA markers throughout the genome.

A. Mapping Concomitant with F_4 Genetic Screens

A mapping cross can be integrated into an F_4 natural crosses screen strategy. Two strains that are polymorphic to each other, for example, TU and AB, are mutagenized. The mutagenized males are crossed to females of their respective strain to produce an F_1 generation. F_1 fish are then interbred between the two strains to make a hybrid F_2 generation. F_2 fish are intercrossed to make the mutations homozygous in the F_3 generation and F_3 mutant females can be used to map the mutation. The F_1 grandparent DNA is crucial in examining linkage by using bulk segregant analysis. Thus, the F_1 fish are frozen and kept for mapping purposes, should a mutation be found that one wants to map.

Intercrossing strains to make a map cross gives rise to very robust stocks, through so called hybrid vigor. As a consequence of interbreeding F_1 fish of different strains, we find the F_2 hybrid generation to be particularly healthy and prolific, with increased reproductive longevity, compared with either independent strain. This is advantageous in regenerating the maternal-effect mutation to produce additional females for mapping (see later). However, a drawback is that hybrid vigor leads to an increased propensity to produce females in the F_2 hybrid generation. Thus, typically 20% of F_2 families yield three or less males. We assess the sex ratio at about 2 months of age and discard those with less than four males. In the F_3 generation, the sex ratio is not distorted and we rarely find such sex-biased families. It is possible that future lines could be developed that do not exhibit the sex bias in the F_2 hybrid generation.

As with zygotic mutations (Geisler, 2002; Talbot and Schier, 1999), bulk segregant analysis is used to map maternal-effect mutations. However, it is considerably more difficult to generate maternal-effect mutant individuals than zygotic mutant embryos for this purpose. We have found that 12 mutant females are sufficient to map efficiently a mutation. In using an F_3-extended family (see Section II.A), mutant females represent 1/16 of the total, producing 1–5 mutant females in a tank of 60 fish. Even five females is insufficient to map efficiently a mutation. Thus, in performing such a screen, we keep up to 10 pairs of fish of the

F_2 generation in a small 2-liter tank. If we are interested in a mutation, we return to the F_2 fish and regenerate the mutant through F_2 pairwise crosses. Each F_3 family is raised separately and the F_2 fish stored individually. Regenerated F_3 families are screened for mutant females (as described in Section IV.B.2); the F_2 parents of those that yield mutants can then be used to produce more mutant females. The F_3 mutant and nonmutant sibling females are then used for mapping the mutation. We routinely use as few as 12 mutant females to map a mutation and have mapped mutations with just 9 mutant individuals. In the latter cases, considerably more false-positive linkages are detected. We have reliably regenerated >20 maternal-effect mutants and efficiently mapped most of the mutations using this strategy (Dosch *et al.*, 2004; Wagner *et al.*, 2004).

B. Mapping After Identification and Recovery of Mutations

Genetic mapping can be initiated by a process similar to that carried out for the maintenance of unmapped mutations (see Section IV.B.2) by performing multiple outcrosses from sibling (P) males (from families containing homozygous mutant females) to females from polymorphic strains, such as WIK. F_1 hybrid individuals from such outcrosses are incrossed at random to generate F_2 crosses. Outcrosses from homozygous mutant sibling males, 25% of which are expected to be homozygous for the mutation, yield F_1 families in which all individuals are heterozygous carriers for the mutation, so that all random F_2 incrosses contain homozygous mutant females. However, outcrosses from heterozygous carrier males, expected at a 50% frequency among the siblings of homozygous mutant females, produce F_1 families in which 50% of individuals are heterozygous carriers, so that only 25% of random F_2 incrosses yield homozygous mutant F_2 females. Thus, it is more efficient to initiate the mapping strategy using homozygous males, because a much larger fraction of incrosses from the F_1 hybrids (100% compared with 25%) yields homozygous F_2 females. Parental males can be identified in advance as homozygotes and then the mapping cross initiated with such a male. Alternatively, the mapping strategy can be initiated with eight unidentified sibling males outcrossed to polymorphic females. This relatively large number of outcrosses increases the probability that at least one of the outcrosses originates from a homozygous male. After the F_1 hybrid fish are incrossed to make an F_2 generation, they are kept separately. Multiple incrosses from F_1 hybrids and testing F_2 females for maternal effects allows inferring whether the original P male was homozygous for the mutation, if it was not identified in advance, and which pairs of F_1 hybrids are heterozygous carriers. Such pairs, now identified, can be crossed repeatedly to produce more mutant F_2 females for mapping.

F_2 females are separated into two phenotypic classes, the maternal-effect mutant females that yield mutant embryos (i.e., females homozygous mutant for the mutation) and their wild-type siblings that produce wild-type embryos (i.e., females either heterozygous or homozygous for the wild-type allele). Identified mutant females can be tested a second time to ensure that they produce a phenotype at high penetrance and to check for potential errors in handling.

After their classification into phenotypic classes, F_2 females are anesthesized, their tail fin clipped, and the tail fin DNA isolated. (The remaining part of the body is also frozen and serves as a backup in case additional DNA needs to be isolated.) Individual tail fin DNA from 20 females of each phenotypic class is used to make two DNA pools. The pools are used to carry out a first-pass mapping (see Section II.A; Geisler, 2002; Talbot and Schier, 1999) to identify linkage. Once linkage has been found, DNA from single fish is analyzed separately with respect to polymorphisms to markers within the linked region (Geisler, 2002; Talbot and Schier, 1999).

C. Efficient Fine Mapping of Maternal–Effect Mutations

Fine mapping a maternal-effect mutation can be performed much more efficiently than the initial mapping. Narrowing down the location of a mutation through fine mapping is necessary to identify the molecular nature of the mutated gene through either candidate gene or positional cloning approaches. Once a maternal-effect mutation is mapped, homozygous male and heterozygous female carriers can then be identified by polymorphic markers flanking the mutation. Such F_0 fish are intercrossed to map finely the position of the mutation in the F_1 progeny. In a cross of a heterozygous female to a homozygous male, mapping is performed with recombinants generated only through meiosis in the F_0 female, because the male is homozygous for the mutation. Thus, each F_1 fish represents a single meiosis rather than two meioses for progeny from heterozygote intercrosses. However, the loss of recombination events from the homozygous male is offset by meiotic recombination being suppressed in males compared to females so that the vast majority of all recombinants generated in intercrosses of heterozygotes are from female and not male meioses (S. Johnson, personal communication; Mullins, unpublished; Singer *et al.*, 2002). Thus, little is lost in crossing heterozygous females to homozygous males, and there is considerable gain in using this strategy, as discussed below.

Fine mapping of a mutation through crosses between heterozygous females and homozygous males is similar to haploid mapping of zygotic mutations (Postlethwait and Talbot, 1997). In both cases, all fish can be examined for re-combinants. For a maternal-effect mutation, both the phenotypically mutant and wild-type F_1 females are examined for recombination between, respectively, the mutation and a wild-type-linked flanking marker or the wild-type allele of the gene and a mutant-linked flanking marker. Thus, all the female progeny from a cross between a heterozygous female and a homozygous male are informative, in contrast to only one fourth of the female progeny in intercrosses of heterozygotes. Because it is significantly more effort to generate adult individuals to map mater-nal-effect mutations than to generate embryos or young larvae to map finely zygotic mutations, this strategy saves considerable effort and tank space.

Fine mapping is made even more efficient in a map cross between a heterozygous female and homozygous male by genotyping all F_1 female progeny rather than phenotyping them in crosses through the examination of their progeny. All F_1 females (and males, see later) are genotyped with the closest polymorphic markers flanking the mutation to determine whether they are nonrecombinant mutants or heterozygotes or recombinants within the interval of the flanking markers. Only the small subset of recombinants is phenotyped to determine whether they are mutant or wild-type females and thereby establish where the recombination occurred relative to the mutation. As the critical interval is narrowed and closer polymorphic markers are defined, fewer recombinants are identified and consequently fewer females phenotyped in crosses. We typically genotype individual fish at 2 months of age and only maintain the small fraction of recombinants until breeding age to determine their phenotype and the position of the recombination (the recombination break point) relative to the mutation. The total number of F1 females genotyped is compared to the number of recombinants to determine the genetic distance between the mutation and the flanking markers.

Once the interval of the maternal-effect mutation is narrowed to a ~0.5-centi-Morgan region, we also find it worthwhile genotyping F_1 males from the intercross of the heterozygous female and homozygous male for recombination within the critical interval. The disadvantage of males is that a test cross must be performed between the recombinant male and a heterozygous female to determine whether the male is homozygous or heterozygous for the mutation. The female progeny of this cross must be tested to determine whether 25% or 50% are mutant to assess the genotype of the male. Although this is considerable effort, identifying a recombinant that narrows down the interval can be so valuable that we have found it worth the effort.

VI. Solutions, Materials, and Protocols

A. Solutions

- *MESAB stock solution*: 0.2% ethyl-*m*-aminobenzoate methanesulfonate. Adjust to pH 7.0 with 1 M Tris pH 9.0. Keep at 4 °C.
- *MESAB working solution*: 7 ml stock solution per 100 ml fish water.
- *Hank's solutions*: Stock solutions 1, 2, 4, and 5 and premix can be stored at 4°C. Stock solution 6 is prepared fresh and added to the premix to form the final Hank's solution.
  - *Solution 1*: 8.0 g NaCl, 0.4 g KCl-in 100 ml double-distilled (dd) H_2O.
  - *Solution 2*: 0.358 g Na_2HPO_4 anhydrous, 0.60 g KH_2PO_4, in 100 ml ddH_2O.
  - *Solution 4*: 0.72 g $CaCl_2$ in 50 ml ddH_2O.

- *Solution 5*: 1.23 g $MgSO_4 \cdot 7H_2O$ in 50 ml ddH_2O.
- *Hank's premix*: Combine the following in order
  - 10.0 ml Solution 1
  - 1.0 ml Solution 2
  - 1.0 ml Solution 4
  - 86.0 ml ddH_2O
  - 1.0 ml Solution 5
- *Solution 6 (prepare fresh)*: 0.35 g $NaHCO_3$ in 10 ml ddH_2O
- Hank's (final): 990 μl Hank's premix
    10 μl Solution 6
- *E2 saline (used specially during testosterone treatment because of its higher buffering properties)*: 15 mM NaCl, 0.5 mM KCl, 1 mM $CaCl_2$, 1 mM $MgSO_4$, 0.15 mM KH_2PO_4, 0.05 mM Na_2HPO_4, 0.7 mM $NaHCO_3$.
- *E3 saline (a simpler version of E2 used for routine embryo raising)*: 5 mM NaCl, 0.17 mM KCl, 0.33 mM $CaCl_2$, 0.33 mM $MgSO_4$, 10^{-5}% methylene blue.
- *Testosterone stock*: 150 mg testosterone in 50 ml absolute ethanol. Store in aliquots at $-20\,°C$.
- *Testosterone working solution*: While stirring, add 10 μl of stock solution per 600 ml of (a) E2 saline for babies before Day 16 or (b) fish water supplemented with 3 g/l Red Sea salt (Read Sea Fish pHarm, Israel) for larvae between Day 6 and Day 15. In our hands, E2 (instead of E3) and Red Sea salt in the fish water improve the survival of testosterone-treated larvae. Stir for 10 min.

B. Other Materials

- *UV lamp*: Sylvania 18-in., 15-W germicidal lamp.
- French Press Cell, 40 ml (SLM-Aminco)
- French Pressure cell press (SLM-Aminco) or Hydraulic Laboratory Press (Fisher)
- *Heat shock baskets*: These can be made by cutting off the bottom of Beckman Ultraclear centrifuge tubes and heat-sealing a fine wire mesh to the bottom edge of the tube.
- *EP vials*: Disposable glass scintillation vials, with plastic caps (3.2-cm height and 2.2-cm diameter, Wheaton) or similar vials. The plastic caps are perforated several times with a needle to better allow exposure to the hydrostatic pressure. Only two vials of these type can fit at one time in a pressure cell. To fit four vials in one cell, we have custom-built shorter plastic vials (1.8-cm height, including cap, 2.5-cm diameter, 0.3-mm wall thickness), which fit the plastic caps from the scintillation vials.

C. Protocols

1. Sperm Collection (Adapted from Ransom)

A sperm solution can be made with testes dissected from 10 males for each 1 ml of Hank's solution. Keep the isolated testes and Hank's solution on ice. Shear the testes with a small spatula and by pipetting up and down with a 1000-μl pipetteman. Allow debris to settle and transfer supernatant to a new tube. Sperm solution on ice is effective for about 2 h. Sperm still remaining inside the sheared, settled testes can be further collected by adding 300 μl of fresh Hank's solution and letting the mixture rest for 30 min or longer. For more details see Ransom and Zon, this issue.

2. UV Inactivation of Sperm

Transfer the sperm solution to a watch glass. Avoid pieces of debris, as they might shield sperm from the UV light. Place the watchglass on ice at a distance of 38 cm (15 in.) directly under the UV lamp. Irradiate for 2 min with gentle stirring every 30 sec. Transfer to a new Eppendorf tube with a clean pipette tip. UV-treated sperm solution on ice is effective for about 2 h.

3. Stripping of Eggs

• Our observations suggest that females are more amenable to manual stripping if removed from their tank and placed in a clean tank (1–10 females per 2-l tank) the evening before stripping. Best stripping and egg clutch quality are obtained during the first 4 h after the start of the light cycle on the first day after separation of the females. The presence of males together with the separated females does not significantly affect the ability of gol-mix females to be stripped (our observations), although it might have an effect when working with other fish lines (Eaton and Farley, 1974).

• Anesthetize females in MESAB working solution until they reduce their gill movements (2–4 min, MESAB solution might have to be boosted through time with more stock solution in 0.5–1 ml increments). Overexposure to MESAB will impede recovery of the female, and fish should be placed in fresh water if they are not going to be used within 1 or 2 min after they stop their movements.

• With the aid of a spoon, rinse a female in fish water and place her on several paper towels to remove excess moisture.

• Place the female on the bottom half of a petri plate. With a soft tissue, dry further the anal fin area. Excess water can prematurely activate the eggs.

• Slightly moisten the index fingers of both hands. (Dry hands will stick to the skin of the fish.) With one finger support the back of the female and with the other gently press her belly. Females that can be stripped will release their eggs on gentle pressure. Healthy eggs have a translucent, yellowish appearance. Separate the eggs from the female with a small, dry spatula. Females can be placed separately in boxes, and identifying tags can be attached to the box with the female and the corresponding egg clutch. If necessary, clutches can wait for several minutes before being activated. In this case, we cover the clutches with the petri plate lid to reduce drying of the clutch. Fertilization can occur after even longer delays (in our hands, up to 6 min), although not in a consistent manner. Egg activation can be delayed for periods of 1.5 h or more with ovarian fluid from the rainbow trout or coho salmon (Corley-Smith *et al.*, 1995), or with Hank's saline buffer supplemented with 0.5% BSA (Sakai *et al.*, 1997), although we have not tested these methods in combination with gynogenesis.

4. *In Vitro* Fertilization

Add 25 μl of untreated or UV-irradiated sperm to the egg clutch. Mix the sperm and eggs by moving the pipette tip without lifting it from the petri plate (to minimize damage to the eggs). If desired, proceed at this point to heat shock or early pressure protocols. If not, add 1 ml of E3 saline to activate the eggs, and after 1 min fill the petri plate with E3. Incubate at 27–29 °C.

5. Heat Shock

- After IVF with UV-treated sperm, add 1 ml of E3 saline to activate the eggs and start the timer.
- Add more E3 after 30 sec. Transfer the eggs to a heat shock basket. Immerse the basket in a water bath with stirring and E3 saline at 28.5 °C.
- At 13.0 min, blot briefly the bottom of the basket onto a stack of paper towels and transfer the basket to a water bath with stirring and E3 saline at 41.4 °C.
- At 15.0 min, blot briefly the bottom of the basket and transfer the basket back to the 28.5 °C E3 bath.
- Allow the embryos to rest for about 45 min and transfer to a petri plate. Allow embryos to develop in a 27–29 °C incubator. (See note under "Early Pressure.")

6. Early Pressure

To maximize the number of clutches produced, we work on cycles in which we include up to four clutches in separate vials within the pressure cell. For this, we typically anesthetize 6–12 females. Once four healthy-looking clutches are obtained, the females that have not yet been stripped of eggs are transferred to fresh fish water until they completely recuperate. It works well to begin to anesthetize females for the next EP cycle at around Minute 4 within a current cycle.

- After mixing eggs with UV-treated sperm (see IVF), activate up to four clutches simultaneously by adding 1 ml of E3 saline to each clutch and start the timer. (At least two people are required to timely manipulate four clutches.)
- After 12 sec, add more E3. A squirt to the side of the petri plate will make the fertilized eggs collect in the middle of the plate.

• With a plastic pipette, transfer the fertilized eggs to an EP vial. Fill the vial with E3 and cap it with the perforated plastic lid. Avoid large air bubbles. Place the vials inside the pressure cell, ensuring that no air remains trapped inside it. Record the relative position of the clutch within the pressure cell by placing the tags in the corresponding order on a dry surface. Fill the pressure cell with E3 and close it allowing excess E3 to be released from the side valve. Close the side valve without overtightening. Insert entire assembly on the French Press apparatus and apply pressure to 8000 lb/sp. in. by time 1 min 20 sec after activation. For different strains and/or presses, different pressure values might be optimal (see Gestl *et al.*, 1997).

• At 6.0 min, release the pressure and remove the pressure cell from the French Press apparatus. Maintaining the relative order of the vials, remove the vials from the pressure cell, dry them with a towel, and label them with their corresponding number tags. Place the vial in a 27–29 °C incubator.

• After all EP cycles have been completed, allow the embryos to rest in the vial for at least 45 min but no more than 4 h. Transfer embryos with their corresponding tags to petri plates. Let embryos develop in a 27–29 °C incubator.

Note: Because for the large amount of embryonic lethality induced by the HS and EP procedures, we incubate the embryos at a low density of 80 embryos maximum per 94-mm petri plate. (This is particularly important for the first 24 h of development.)

7. Testosterone Treatment

• Before embryos reach 24 h of development, remove the chorions from the embryos, remove as much E3 as possible, and replace with testosterone/E2 working solution.

• Each consecutive day, replace half the testosterone/E2 with fresh testosterone/E2.

• On Day 6, transfer the embryos to mouse cages with 1 l of testosterone solution in fish water supplemented with 3 g/l Coral Reef salt. Start feeding as normally. Continue replacing half the solution every day by carefully aspirating the solution and refilling with fresh testosterone solution.

• On Day 15, remove the testosterone by aspirating most of the solution and refilling with fresh fish water. Rinse again by repeating this procedure. Embryos can now be connected to the water system.

VII. Conclusions

We describe methodologies to identify, recover, maintain, and map maternal-effect mutations. Two main genetic screening strategies are described: an F_4 screen based solely on natural crosses and an F_3 EP-based screen. Each of these strategies has advantages and disadvantages. F_4 screens based solely on natural crosses are technically relatively simple and allow simultaneous mapping as well as identification of maternal-zygotic mutations. However, such an approach requires larger amounts of space, generation time, and labor. On the other hand, EP screens require substantial selection of specialized lines amenable to the procedure but can be carried out using less generation time and space and are more amenable to the observation after IVF of events immediately after fertilization. Both these methods, however, have allowed the unbiased identification of many maternal-effect mutants (Dosch *et al.*, 2004; Pelegri and Schulte-Merker, 1999; Pelegri *et al.*, 2004; Wagner *et al.*, 2004). These efforts should lead to the identification of additional mutations in maternal-effect genes in the zebrafish and eventually the genetic analysis of early development in this vertebrate species.

Acknowledgments

F.P. is grateful to Stefan Schulte-Merker, Marcus Dekens, Hans-Martin Maischein, and Catrin Weiler for their participation in the original EP screen, as well as Christiane Nüsslein-Volhard for support and advice at the Max-Planck Institut für Entwicklungsbiologie, Tübingen. Current support in the laboratory of F.P at the University of Wisconsin—Madison is provided by a March of Dimes Birth Defects Foundation Grant #5-FY00-597 and an NIH grant RO1 GM65303. M.C.M thanks Daniel Wagner, Roland Dosch, and Keith Mintzer for their contributions in performing the F4 natural crosses screen described here and numerous discussions; Florence Marlow for contributing Fig. 1(A) and (B) M.C.M was supported in part by research Grant No. 1-FY02–24 from the March of Dimes Birth Defects Foundation, NIH Grant ES11248, and by an award from the American Heart Association. Stefan Schulte-Merker was a co-author in a previous version of this chapter.

References

Bauer, M. P., and Goetz, F. W. (2001). Isolation of gonadal mutations in adult zebrafish from a chemical mutagenesis screen. *Biol. Reprod.* **64,** 548–554.

Chan, S. T. H., and Yeung, W. S. B. (1983). Sex control and sex reversal in fish under natural conditions. *In* "Fish Physiology" (W. S. Hoar, D. J. Randall, and E. M. Donaldson, eds.), Vol IX, Part B, pp. 171–222. Academic Press, New York.

Corley-Smith, G. E., Lim, C. J., and Brandhorst, B. P. (1995). Delayed *in vitro* fertilization using coho salmon ovarian fluid. *In* "The Zebrafish Book: A Guide for the Laboratory Use of Zebrafish (*Daniorerio*)" online version. (M. Westerfield, ed.), pp. 722–725. University of Oregon Press, Eugene, OR.

Dekens, M. P., Pelegri, F. J., Maischein, H. M., and Nüsslein-Volhard, C. (2003). The maternal-effect gene futile cycle is essential for pronuclear congression and mitotic spindle assembly in the zebrafish zygote. *Development* **130,** 3907–3916.

Dosch, R., Wagner, D. S., Mintzer, K. A., Runke, G., Wiemelt, A. P., and Mullins, M. C. (2004). Maternal control of vertebrate development before the midblastula transition: Mutants from the zebrafish I. *Dev. Cell* **6,** 771–780.

Driever, W., Solnica-Krezel, L., Schier, A. F., Neuhauss, S. C. F., Malicki, J., Stemple, D. L., Stainier, D. Y. R., Zwartkruis, F., Abdelilah, S., Rangini, Z., Belak, J., and Boggs, C. (1996). A genetic screen for mutations affecting embryogenesis in zebrafish. *Development* **123,** 37–46.

Eaton, R. C., and Farley, R. D. (1974). Spawning cycle and egg production of zebrafish, *Brachydanio rerio*, in the laboratory. *Copeia* **1,** 195–204.

Geisler, R. (2002). Mapping and cloning. *In* "Zebrafish" (C. Nüsslein-Volhard and R. Dahm, eds.), Practical Approach Series, Vol. 261, pp. 175–212. Oxford University Press, Oxford.

Gestl, E. E., Kauffman, E. J., Moore, J. L., and Cheng, K. C. (1997). New conditions for generation of gynogenetic half-tetrad embryos in the zebrafish (*Danio rerio*). *J. Hered.* **88,** 76–79.

Haffter, P., Granato, M., Brand, M., Mullins, M. C., Hamerschmidt, M., Kane, D. A., Odenthal, J., van Eeden, F. J. M., Jiang, Y.-J., Heisenberg, C.-P., Kelsh, R. N., Furutani-Seiki, M., Vogelsang, E., Beuchle, D., Schach, U., Fabian, C., and Nüsslein-Volhard, C. (1996). The identification of genes with unique and essential functions in the development of the zebrafish *Danio rerio*. *Development* **123,** 1–36.

Kane, D. A., and Kimmel, C. B. (1993). The zebrafish midblastula transition. *Development* **119,** 447–456.

Kemphues, K. J., Kusch, M., and Wolf, N. (1988). Maternal-effect lethal mutations on linkage group II of *Caenorhabditis elegans*. *Genetics* **120,** 977–986.

Kemphues, K. J., and Strome, S. (1997). Fertilization and establishment of polarity in the embryo. *In* "C. elegans II" (D. L. Riddle, T. Blumenthal, B. J. Meyer, and J. R. Priess, eds.), pp. 335–359. Cold Spring Harbor Laboratory Press, Cold Spring Harbor, New York.

Kimmel, C., Ballard, W. W., Kimmel, S. R., Ullmann, B., and Schilling, T. F. (1995). Stages of embryonic development in the zebrafish. *Dev. Dyn.* **203,** 253–310.

Kishimoto, Y., Koshida, S., Furutani-Seiki, M., and Kondoh, H. (2004). Zebrafish maternal-effect mutations causing cytokinesis defects without affecting mitosis or equatorial vasa deposition. *Mech. Dev.* **121,** 79–89.

Lehmann, R., and Nüsslein-Volhard, C. (1986). Abdominal segmentation, pole cell formation, and embryonic polarity require the localized activity of *oskar*, a maternal gene in *Drosophila*. *Cell* **47,** 141–152.

Mullins, M. C., Hammerschmidt, M., Haffter, P., and Nüsslein-Volhard, C. (1994). Large-scale mutagenesis in the zebrafish: In search of genes controlling development in a vertebrate. *Curr. Biol.* **4,** 189–202.

Neuhauss, S. (1996). Craniofacial development in zebrafish (*Danio rerio*): Mutational analysis, genetic characterization, and genomic mapping. Ph.D. Thesis, Fakultät für Biologie, Eberhard-Karl-Universität Tübingen.

Newport, J., and Kirschner, M. (1982a). A major developmental transition in early *Xenopus* embryos. I. Characterization and timing of cellular changes at the midblastula stage. *Cell* **30,** 675–686.

Newport, J., and Kirschner, M. (1982b). A major developmental transition in early *Xenopus* embryos. II. Control of the onset of transcription. *Cell* **30,** 687–696.

Pelegri, F. (2003). Maternal factors in zebrafish development. *Dev. Dyn.* **228,** 535–554.

Pelegri F., Dekens M. P. S., Schulte-Merker S., Maischein H.-M., Weiler C., Nüsslein-Volhard C. (2004). Identification of recessive maternal-effect mutations in the zebrafish using a gynogenesis-based method. *Dev. Dyn.* **231,** 325–334.

Pelegri, F., Knaut, H., Maischein, H. M., Schulte-Merker, S., and Nüsslein-Volhard, C. (1999). A mutation in the zebrafish maternal-effect gene nebel affects furrow formation and vasa RNA localization. *Curr. Biol.* **9,** 1431–1440.

Pelegri, F., and Schulte-Merker, S. (1999). A gynogenesis-based screen for maternal-effect genes in the zebrafish, *Danio rerio*. *In* "The Zebrafish: Genetics and Genomics" (W. Detrich, L. I. Zon, and M. Westerfield, eds.), Methods in Cell Biology, Vol. 60, pp. 1–20. Academic Press, San Diego.

Postlethwait, J. H., and Talbot, W. S. (1997). Zebrafish genomics: From mutants to genes. *Trends Genet.* **13,** 183–190.

Ransom, D. G., and Zon, L. I. (1999). Collection, storage, and use of zebrafish sperm. *In* "The Zebrafish: Genetics and Genomes" (W. Detrich, L. I. Zon, and M. Westerfields, eds.), Methods in cell Biology, Vol. 60, pp. 1–20. Academic Press, San Diego.

Sakai, N., Burgess, S., and Hopkins, N. (1997). Delayed *in vitro* fertilization of zebrafish eggs in Hank's saline containing bovine serum albumin. *Mol. Mar. Biotechnol.* **6,** 84–87.

Schnabel, R., and Priess, J. R. (1997). Specification of cell fates in the early embryo. *In* "*C. elegans* II" (D. L. Riddle, T. Blumenthal, B. J. Meyer, and J. R. Priess, eds.), pp. 361–382. Cold Spring Harbor Laboratory Press, Cold Spring Harbor, New York.

Selman, K., Petrino, T. R., and Wallace, R. A. (1994). Experimental conditions for oocyte maturation in the zebrafish, *Brachydanio rerio. J. Exp. Zool.* **269,** 538–550.

Signoret, J., and Lefresne, J. (1971). Contribution a l'etude de la segmentation de l'oeuf d'axolotl: I. Definition de la transition blastuleenne. *Ann. Embryol. Morphog.* **4,** 113.

Singer, A., Perlman, H., Yan, Y., Walker, C., Corley-Smith, G., Brandhorst, B., and Postlethwait, J. (2002). Sex-specific recombination rates in zebrafish (*Danio rerio*). *Genetics* **160,** 649–657.

Solnica-Krezel, L., Schier, A. F., and Driever, W. (1994). Efficient recovery of ENU-induced mutations from the zebrafish germline. *Genetics* **136,** 1401–1420.

St. Johnston, D., and Nüsslein-Volhard, C. (1992). The origin of pattern and polarity in the *Drosophila* embryo. *Cell* **68,** 201–219.

Streisinger, G., Singer, F., Walker, C., Knauber, D., and Dower, N. (1986). Segregation analyses and gene-centromere distances in zebrafish. *Genetics* **112,** 311–319.

Streisinger, G., Walker, C., Dower, N., Knauber, D., and Singer, F. (1981). Production of clones of homozygous diploid zebra fish (*Brachydanio reriol*). *Nature* **291,** 293–296.

Talbot, S., and Schier, A. F. (1999). Positional cloning of mutated zebrafish genes. *In* "The Zebrafish: Genetics and Genomics" (H. W. Detrich, M. M. Westerfield, and L. I. Zon, eds.), Methods in Cell Biology,Vol. 60, pp. 259–286. Academic Press, San Diego.

Thorgard, G. H. (1983). Chromosome set manipulation and sex control in fish. *In* "Fish Physiology" (W. S. Hoar, D. J. Randall, and E. M. Donaldson, eds.), Vol. IX, Part B, pp. 405–434. Academic Press, New York.

van Eeden, F. J. M., Granato, M., Odenthal, J., and Haffter, P. (1999). Developmental mutant screens in the zebrafish. *In* "The Zebrafish: Genetics and Genomics" (H. W. Detrich, M. Westerfield, and L. I. Zon, eds.), Methods in Cell Biology, Vol. 60, pp. 21–41. Academic Press, San Diego.

Wagner D. S., Dosch R., Mintzer K. A., Wiemelt A. P., Mullins M. C. (2004). Maternal control of development at the mid-blastula transition and beyond: Mutants from the zebrafish II, *Dev. Cell.* **6,** 781–790.

CHAPTER 3

Behavioral Screening Assays in Zebrafish

Michael B. Orger,* **Ethan Gahtan,***,‡ **Akira Muto,***
Patrick Page-McCaw,*,† **Matthew C. Smear,*** **and**
Herwig Baier*

*Program in Neuroscience
Department of Physiology
University of California, San Francisco
San Francisco, California 94143

†Department of Biology
Rennselaer Polytechnic Institute
Troy, New York 12180

‡Department of Psychology
University of Massachusetts, Amherst
Amherst, Massachusetts 01003

I. Introduction

All animals show innate behaviors, which depend on the correct development and function of their nervous systems. The genes each individual inherits specify how the brain develops and operates and the way in which experience affects these processes. The search for the genetic underpinnings of behavior is rapidly expanding. Systematic screens for behavioral mutants provide an unbiased method to find the underlying genes (Benzer, 1973). In this review, we describe some of the behavioral screening assays we have devised in our laboratory to isolate mutations affecting the zebrafish nervous system. Zebrafish are ideally suited for a behavioral genetic approach (Guo, 2004; Li, 2001; Neuhauss, 2003). The larvae show a wide range of interesting behaviors, yet are small and can be produced in large numbers. Importantly, larvae do not need to be fed to survive until 8 days postfertilization (dpf.), and all the assays described in this review can be carried out in this period. Therefore, millions of fish can be tested in a large-scale screen while requiring relatively little maintenance. Adult behaviors can also be used for screens, on a smaller scale. Dozens of larval and adult behaviors have been described. Table I shows a small selection for which screens have been carried out or proposed in zebrafish.

What can we discover by using a behavioral genetic approach? First, such an approach is uniquely suited to discover the genes involved in the proper execution of behavioral programs, that is, in the acute function of the nervous system. Second, behavior provides a sensitive readout of developmental disruptions, many of which might be too subtle to be picked up by anatomical or histological screens. Thus, we can expect to identify factors required for neural cell fate decisions, differentiation, axon guidance, and synapse formation. Third, mutants obtained in a behavioral screen provide an alternative lesioning technique, complementary to surgical ablations but with different temporal and spatial resolution (Gahtan and Baier, 2004). Therefore, mutations provide unique insights into the function of neural circuits by identifying their essential components.

Well-designed behavioral screens are focused to find mutations specific to a particular neural system. Yet behavior is the endpoint of neural processing often involving tens or hundreds of cell types in a neural circuit or pathway, and therefore mutations might exert their effects at many places in the brain. This breadth should be considered an advantage of a behavioral genetic approach, because it allows the researcher to investigate a neural system in its entirety. This tight functional focus combined with a systemwide view can reveal rare but highly significant links between brain and behavior that would probably escape discovery by any other means.

II. General Considerations

Behavioral screens require responses to be reliably evoked in the laboratory. Thus, not all behaviors are equally conducive to this approach. As a rule, when looking for recessive mutations, the wild-type response probability in a given screening assay should be higher than 90% to be significantly distinct from the 75% expected from Mendelian ratios in a mutant clutch. Otherwise, a large number of false positives will be identified. In our experience, screening out false positives is

Table I
Zebrafish Behaviors Used in Genetic Screens

Behavior	Age	Selected references	Screen references
Swimming/motility	>4 dpf	Budick and O'Malley, 2000; Liu and Westerfield, 1988; Saint-Amant and Drapeau, 1998	Granato et al., 1996
Photoentrainment and circadian regulation of activity; sleep	>4 dpf	Cahill et al., 1998; Hurd et al., 1998; Zhdanova et al., 2001	
Touch-evoked twitching/fast start	>27 hpf	Eaton et al., 1977; Gahtan et al., 2002; Liu and Fetcho, 1999; Liu et al., 2003; Lorent et al., 2001; O'Malley et al., 1996; Ribera and Nüsslein-Volhard, 1998; Saint-Amant and Drapeau, 1998	Granato et al., 1996
Auditory fast start (acoustical/vibrational startle)	>5 dpf	Eaton et al., 1977; Kimmel et al., 1974	Bang et al., 2002; Nicolson et al., 1998
Optokinetic response (eye movements pursuing visual motion)	>73 hpf	Easter and Nicola, 1996; Kainz et al., 2003; Rick et al., 2000; Roeser and Baier, 2003	Brockerhoff et al., 1995, 1997; Neuhauss et al., 1999
Optomotor response (whole-body movements following visual motion)	>4 dpf	Bilotta, 2000; Krauss and Neumeyer, 2003; Maaswinkel and Li, 2003; Orger et al., 2000, Roeser and Baier, 2003	Neuhauss et al., 1999
Visually mediated background adaptation (dispersal and aggregation of pigment granules in the skin)	>5 dpf	Kay et al., 2001	Kelsh et al., 1996; Neuhauss et al., 1999
Visually mediated escape	Adult	Dill, 1972, 1974; Li and Dowling, 2000a,b	Li and Dowling, 1997
Prey capture	>5 dpf	Borla et al., 2002; Budick and O'Malley, 2000	
Addiction; drug responses	Adult	Gerlai et al., 2000	Darland and Dowling, 2001; Lockwood et al., 2004

the most time-consuming part of a behavioral screen, even for our most robust assays. It is therefore essential to optimize the screening assay, ideally by using wild-type fish in a small-scale mock screen. Time allotted to perfecting the assay under "screen" conditions will certainly be recouped as time saved in the screen.

Choosing the time of day to run a particular assay is very important. Circadian rhythms can affect many aspects of behavior, including visual sensitivity (Li and Dowling, 1998). The robustness of both the optomotor and optokinetic responses is significantly affected by circadian rhythms. A reduction in wild-type responsiveness of just a few percent can seriously impair a screening assay.

Once the screen has begun, retesting of putants (putative mutants), by setting up the same pair of carriers and reevaluating the behavioral phenotype of their offspring, is essential. In our hands, with the screening assays presented here, we find that three trials with consistent results are sufficient to weed out false positives and inconsistent phenotypes. The recovery rate in the subsequent generation is higher than 50% for most assays, a rate comparable to morphological screens.

Innate behaviors are generally more robust than learned responses and therefore better suited for genetic approaches. None of the associative learning paradigms known to us fulfills the 90% criterion stated previously. However, we found that nonassociative learning (habituation and sensitization) is genetically tractable in zebrafish, as shown later. A systematic genetic dissection of learning and nervous system plasticity will be an important research program for the future, and it is likely that zebrafish will be an attractive system for this approach.

III. Behavioral Assays

A. The Optomotor Response

Larval zebrafish show a rich repertoire of visual behaviors (Clark, 1981; for reviews see Li, 2001 and Neuhauss, 2003). These include the optomotor response (OMR), in which a whole-field moving stimulus evokes swimming in the direction of stimulus motion. This response allows the fish to eliminate unwanted self-motion and avoid being swept away by water currents (Rock and Smith, 1986). When presented with a strong stimulus, 6–7 dpf. larvae respond more than 90% of the time. When fish perform an OMR appropriately, they physically separate themselves from those that do not, making this an excellent assay for an automated large-scale genetic screen.

We have designed a setup for automated testing of the OMR (Fig. 1A) (Orger *et al.*, 2004). Visual stimuli are displayed on a flat-screen CRT monitor that faces upward. Larvae are placed in custom-built long and narrow plexiglass tanks, or racetracks. These essentially restrict the larvae to swimming in one dimension, along the length of the racetrack. Twelve racetracks can be placed side by side on the monitor, and each can hold a clutch of up to 50 larvae. The stimuli, which consist of moving sinusoidal gratings, are generated in Matlab, using the Psychophysics Toolbox extensions (Brainard, 1997; Pelli, 1997), which can be downloaded for free (http://psychtoolbox.org). The gamma function of the CRT is measured by using a Minolta LS-100 light meter and corrected using Matlab, so that pixel brightness is linearly related to pixel value in our movies.

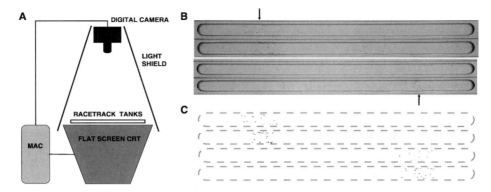

Fig. 1 The optomotor assay. (A) Schematic of the optomotor assay. Six- to seven-day-old fish larvae swim in shallow, elongated racetrack tanks on an upturned flat screen CRT monitor. A computer (Apple) controls stimulus presentation and a digital still camera (Nikon). External light is excluded. (B) Racetrack tanks pictured from above following stimuli consisting of converging sinusoidal gratings. In the upper image the grating converges at a point under the left half of the tanks (upper arrow). In the lower image, the convergence point is under the right half of the tanks (lower arrow). (C) Results of subtracting the images in (B) from each other and thresholding the resulting images. The outlines of the tanks disappear, and their position is indicated by the dotted line. The fish can be clearly seen clustered around the stimulus convergence points.

A digital camera (Nikon CoolPix) suspended above the monitor captures an image of the racetracks before and after each stimulus (Fig. 1B). Matlab can trigger the camera remotely by using a set of serial commands. These images are downloaded from the camera offline and analyzed by using custom macros in Object-Image (http://simon.bio.uva.nl/object-image.html). After subtracting two consecutive images to remove the background, the position of each fish is determined by using the "analyze particles" function of Object-Image (Fig. 1C). The average position of the fish in each tank before a stimulus is then subtracted from the average position after the stimulus. This gives the average distance swum by all the fish, which we call the optomotor index (OMI). Using this method, it is possible to screen thousands of clutches for recessive mutations affecting the OMR. However, because individual fish are not being tracked, it does not identify individual mutant fish. To sort mutants for further characterization and mapping, we use the following simple method. We play a movie in the leftward direction and move nonresponders into a new racetrack with a transfer pipette. After 30 sec the movie reverses, and the process is repeated. Even mutants with subtle defects can be efficiently sorted by using repeated iterations of this technique.

What mutants can we find with the OMR assay? First and most obviously, blind mutants are unable to see the stimulus. For example, mutants lacking photoreceptors or retinal ganglion cells have no OMR (Neuhauss *et al.*, 1999). A second class of mutants has motor deficits that impede the mutants' swimming ability. Finally, segregation of function is a common organizational feature of visual systems (e.g., Simpson, 1984), and so the most interesting class of mutations could specifically affect the OMR while leaving other visually mediated behaviors intact. Finding such a mutation would help elucidate the neural circuit that mediates the OMR and the genes necessary for its development and function.

Computer-generated stimuli provide versatility because they can easily and rapidly be varied in several parameters, allowing us to ask more specific questions about the visual system. For example, we use cone-isolating stimuli in a motion-nulling paradigm to study color vision. Another study has focused on the acuity of the optomotor response by systematically varying the spatial frequency of the stimulus. By alternating stimuli, we can screen for mutants that respond to low but not high spatial frequencies.Acuity mutants can help us understand the limiting factors for spatial resolution in the visual system, such as the organization and grain of photoreceptor mosaics or the accuracy of retinotopy.

B. The Optokinetic Response

The optokinetic response (OKR) is a reflexive response, in which the eyes move to follow a large-field motion stimulus. In zebrafish, the OKR develops a few days after fertilization (Easter and Nicola, 1996). In our experiments, we use 6–7 dpf. larvae, which exhibit a robust OKR.

For optokinetic stimulation, a drum with black-and-white vertical stripes on its internal wall can be rotated mechanically around the zebrafish larvae (Brockerhoff *et al.*, 1995; Neuhauss *et al.*, 1999). An alternative method, which we use, is to project a computer-generated visual stimulus onto the internal wall of a drum (Roeser *et al.*, 2003). The advantage of this method is that we can change the color, spatial frequency, contrast, speed, or any other condition of the stimulus at any time during the recording to study a specific aspect of vision.

We use a public domain program, Image-J (http://www.rsb.info.nih.gov/ij/), for both stimulus generation and processing of the captured image. Image-J is a Java version of the NIH Image application, programmed by the same author, Wayne Rasband.

The setup for the OKR assay is shown in Fig. 2A, B. An animation of a windmill pattern of sine-wave gratings is generated by a computer. It is projected from below onto a white paper wall inside a drum, using an LCD projector (InFocus LP550), where the windmill pattern becomes vertical stripes. A robust OKR is elicited by using sine-wave gratings with a spatial frequency of 20 degrees/cycle moving at 10 degrees/sec. To focus the image, a wide-angle conversion lens (Kenko VC-050Hi, Japan) equipped with a close-up lens (King CU+1, Japan) and a neutral density filter (Hoya, ND4, Japan) is placed in front of the projector. At the center of the drum (6 cm height, 5.6 cm inner diameter), the zebrafish larvae are immobilized in 2.5% methylcellulose in egg water (0.3 g Instant Ocean/liter water) with their dorsal side up in the inverted lid of a 3.5-cm-diameter petri dish, which is placed on a glass table. To shield the fish from the direct beam of the projector and also to obtain a high-contrast image of the larvae, a white diffuser is

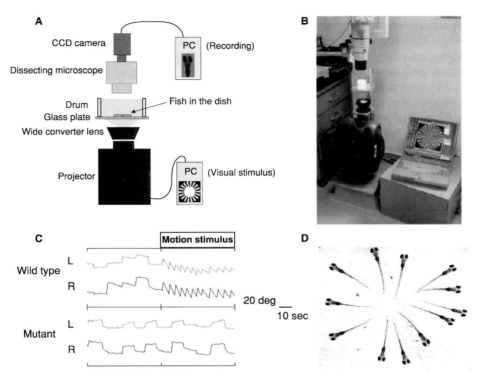

Fig. 2 The optokinetic assay. (A) A schematic of the optokinetic response (OKR) recording setup. An LCD projector projects computer-generated movies onto the inside of a drum, which surrounds a dish containing fish larvae. A CCD camera records movies of the larvae onto a computer, which analyzes their eye positions. (B) A picture of the OKR recording setup. A windmill grating stimulus can be seen on the computer screen. (C) OKR traces from a wild-type fish and a blind mutant fish at 7 dpf. Changes in eye positions of a wild-type larva and a blind mutant larva were plotted over time. Motion stimulus was presented during the second half of the recording period. L: left eye, R: right eye. Wild-type fish respond to the motion stimulus with alternating smooth tracking movements and fast reset saccades. (D) Image of 12 larvae in the recording chamber. Using such an arrangement, many larvae can be tested simultaneously.

placed below the dish. The fish are imaged by using a dissecting microscope (Nikon SMZ-800) with an arm stand and a CCD camera (Cohu MOD8215-1300).

Images are captured by a second computer through an LG-3 video capture board (Scion Corp.) at 2 frames/sec with Scion Java Package 1.0 for Image-J Windows (http://scioncorp.com). A custom Image-J plug-in identifies the eyes based on their dark pigmentation and then calculates their angle by using a standard algorithm (e.g., Seul *et al.*, 2002). The output is a plot of eye angle versus time. Figure 2C shows an example of eye position records of a wild-type and a blind mutant fish with and without an optokinetic stimulus.

Arranging larvae so that they are facing outward in a clock-face formation, as shown in Fig. 2D, enables us to record the eye movements of many fish simultaneously, allowing for high-throughput screening. We typically use batches of 12 fish. Wild-type response probability is close to 100%, and therefore screening 12 fish from a single clutch gives a high probability of detecting a recessive mutation.

C. Spontaneous Activity

An assay of larval spontaneous swimming activity can be used as both a primary screen assay and as an important control for other behavioral tests. Spontaneous locomotor movements in zebrafish develop through stereotyped stages, beginning at approximately 1 dpf (Budick and O'Malley, 2000; Saint-Amant and Drapeau, 1998). Variability among clutch mates is low enough to allow even small differences in activity, such as those caused by drugs (Zhdanova *et al.*, 2001), circadian cycles (Cahill *et al.*, 1998), or mutations (unpublished observations from our laboratory), to be detected. As a primary screening assay, spontaneous swimming activity is a blunt tool, because many different physiological defects could influence swimming. Therefore, we have found spontaneous activity most useful as a secondary test of locomotor function. For example, the OMR and prey capture assays are designed to test vision, but both depend on swimming ability, so a secondary spontaneous swimming test can help distinguish visual from motor defects.

The spontaneous activity assay described here can be run after larvae hatch and are freely swimming. We generally test only 7- to 8-day-old larvae with inflated swim bladders. Fish are tested in groups, with up to 6 fish in each rectangular well (3 cm × 7.5 cm) of a four-well clear acrylic plate (12.8 cm × 7.7 cm; Nunc, Roskilde, Denmark; Fig. 3A). The four-well plates are placed on a glass pane and imaged from below with a digital camcorder (Sony TRV-9). A light-diffusing white acrylic sheet is placed on top of the plates to produce a uniform light background in recorded images. Images are captured directly to the computer at a rate of 0.5 Hz, using the "time lapse" capture feature in Adobe Premiere. We generally record for 20 min, but longer durations or multiple sessions are possible. Recorded movies are analyzed by using Image-J. To be opened correctly in Image-J, the movies must first be saved without compression.

Image processing begins by subtracting each frame from the previous frame through the entire movie (Fig. 3B). The subtracted movie reveals only pixels that changed in value from one frame to the next; if a fish does not move in the 2-sec interframe interval, it will not be visible (dashed circle in Fig. 3A and B). The subtracted movie is thresholded at a level that includes the high-contrast fish but eliminates the background noise. A maximum projection of this movie provides a way to visualize the level of activity in each well at a glance (Fig. 3C). For a quantitative analysis, the "analyze particles" feature in Image-J is then used to count all fish in each well on each frame. The software is configured to count particles that are within a predetermined size range matching the size of the fish.

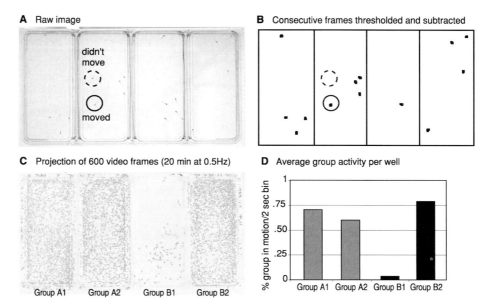

Fig. 3 The spontaneous activity assay. (A) Raw video frame. There are five fish in each well, which are visible as dark dots on the light background (e.g., inside the two circles). (B) An example of the successive frame subtraction procedure. This is the same image as in (A) after the previous image was subtracted from it. In this case, for clarity, the fish in the subtracted image were highlighted using the "Pixel Dilate" function in Image J. The dashed circle indicates a fish, visible in (A), which did not move and was subtracted out, and the solid circle indicates another fish that did move. (C) A projection of all 600 video frames after successive subtraction. Group B1 shows the fewest movement episodes. (D) Plot of the average number of movements per well across the 20-min observation period, divided by the number of fish. If every fish moved between every video frame, the value would be 1.

By keeping track of the number of fish in each well, we can determine the percentage of fish that have moved between any two frames. We quantify spontaneous activity as the average number of movements across all frames divided by number of fish in the well. Figure 3D shows the results of a single trial of two different OMR mutants and their siblings, whose projected tracks are shown in Fig. 3C. We test multiple fish per well, because we have observed in this assay and other behavioral assays that individually housed larvae are less active and responsive. To establish a statistically significant difference between two classes of fish, experiments need to be repeated multiple times.

D. Prey Capture

We have developed a prey capture assay as a test of fine visuomotor control (Fig. 4A). We have shown in control experiments in which fish feed in the light or the dark that feeding on paramecia is, to a large extent, visually mediated (Fig. 4B).

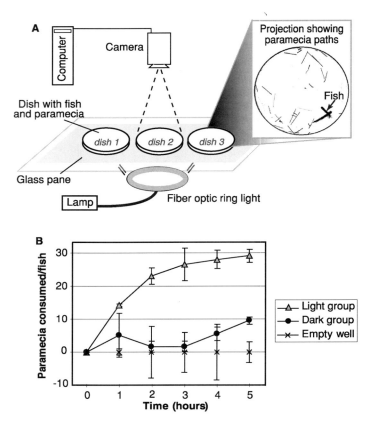

Fig. 4 The prey capture assay. (A) Schematic of the prey capture apparatus. Two hundred video frames are recorded for each dish (60 frames/sec), and paramecia are visualized as dark streaks on the projected image (inset). Sliding the glass pane facilitates recording multiple dishes in succession. (B) Fish kept in the dark during testing are impaired, suggesting a strong visual contribution to prey detection.

Our assay consists of tracking the number of paramecia in a dish over time as the larvae are feeding.

Live paramecia cultures and protozoa food pellets are obtained from Carolina Biological Supply Company (Burlington, NC). Cultures are grown in 500-ml plastic flasks at 28.5 °C to a density of approximately 100 paramecia per milliliter, and new cultures are started every 2–4 weeks. Larvae are tested at 7–8 dpf., either individually in 3.5-cm petri dishes or in groups of up to four larvae in 5.5-cm petri dishes. Between 0.5 ml and 1.5 ml of paramecia culture is added to larvae dishes to achieve a ratio of about 50 paramecia per fish. Paramecia are also added to a dish containing E3 medium but no fish to determine the viability of paramecia over the 5-h assay period.

The equipment used is shown as a schematic in Fig. 4A. Dishes are placed on a glass pane illuminated from below with a fiber optic ring light to provide even illumination across the circular petri dish. Video images of the dish are recorded with a high-speed digital camera (Redlake Motionscope PCI) positioned 30 cm above the dish. Two hundred video frames, captured at 60 Hz, are recorded for each dish, and dishes are recorded successively. Recording multiple dishes is facilitated by using a moveable glass base. An imaging-based counting method is used to determine the number of paramecia immediately after they are introduced (Time 0) and again hourly for 5 h. (The time course was determined by pilot studies.) Imaging-based counting is necessary because individual paramecia cannot reliably be seen in single still images from these recordings, but projecting multiple video frames allows each paramecium to be visualized as a dark streak across the background (Fig. 4A; paramecia can easily be distinguished from fish in the projected image).

Images are processed by using Image-J. First, the movies are projected by using the standard deviation z-projection method, which highlights changes in pixel values caused by the movement of paramecia or fish. Paramecia are marked and counted manually from the projected image, which is saved, whereas the larger movie file is deleted. Results are expressed as the number of paramecia consumed per fish. This number is then corrected for the spontaneous decline in paramecia over time, as determined by counts in the empty wells. In addition to spontaneous decline in paramecia number, there is also some variability in counting precision. In pilot experiments in which individual empty wells were counted repeatedly, we determined that the counting error was less than 5%. Analysis of variance can be used to determine whether treatment groups differ in prey capture performance. As in the spontaneous activity analysis, each treatment group should be run multiple times to determine trial-to-trial variability.

E. Startle Plasticity

Startle is a relatively simple reflex behavior that develops early, is homologous across species, and activates a motor response intended to facilitate escape from a threatening stimulus (Landis and Hunt, 1939). Startle can be elicited by multiple stimuli in different sensory modalities: in experimental systems, acoustic, visual, and tactile cues are most commonly used. The startle response is regulated by associative learning (e.g., fear conditioning), prepulse inhibition, and nonassociative learning (sensitization and habituation; Koch, 1999). Sensitization is the increase in response or response likelihood, and habituation is the decrease in response or response likelihood to a repeated stimulus. Defects in the regulation of startle have been observed in human diseases such as schizophrenia and Tourette's syndrome (Geyer et al., 2001). Genetics are postulated to contribute to the etiologies of these defects. A screen for mutations that affect the regulation of startle will identify the genes, and ultimately the molecular and circuit mechanisms, that are required for experience-dependent plasticity in this system. In fish, startling stimuli elicit a specialized, highly stereotyped startle behavior called the fast start (Eaton and Hackett, 1984). The fast start consists of a large turn away from the stimulus followed by rapid swimming. The response latency (the time

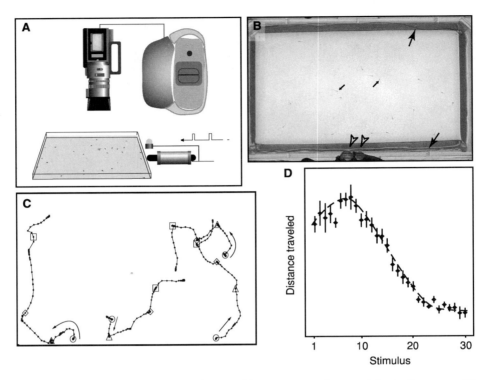

Fig. 5 The startle plasticity assay. (A) Diagram of the apparatus used to record the startle response. (B) One frame from a movie showing 7-day-old zebrafish larvae in the dish. Approximately 20 fish (small arrows) are placed into a rectangular dish illuminated from below. The dish-tap stimulus is presented by using a solenoid, and electrical shock stimuli ($\sim$1 VDC/cm) are presented by using a stainless steel electrode mesh running along the long sides of the dish (large arrows). All stimulus trains are controlled and timed by a G3 Macintosh using a custom AppleScript macro communicating over a serial bus to a custom-built stimulus controller. The stimulus controller also illuminates light-emitting diodes (arrowheads) in synchrony with the stimulus. Video of the behavior is captured by using a video camera, which streams the video data directly to a Macintosh computer for later processing. (C) Results of the tracking algorithm. Four fish are shown tracked across four stimulus responses. The position of the fish at each time point is indicated by the symbols, and the position of the fish at the time of the stimulus is indicated by the large symbols (circle, first stimulus; triangle, second stimulus; diamond, third stimulus; square, fourth stimulus; arrows indicate direction of motion). (D) Startle responses as a function of stimulus number as measured by mean distance traveled in response to a stimulus ($\pm$SEM, seven dishes of 20 fish). The response increases over the first few stimuli (sensitization) and then decreases (habituation).

between stimulus and response) is very short; it is among the fastest known motor responses. The turn, the rapid acceleration, and the short latency are all stereotyped.

High-speed videography (1000 frames/sec) of the fast start has been used for the detailed examination of the body kinematics during this behavior. (For a review, see Domenici and Blake, 1997.) The high temporal and spatial resolution of the data in these experiments come at the cost of allowing the study of only one startle response in one animal per experiment. However, if the interest is in behavioral output rather than in kinematics, a much lower sampling rate and spatial resolution can be used. Using conventional video, we record the behavior of 20 fish simultaneously over 25 sec, extracting such startle parameters as the change in the animals heading, the velocity and acceleration during the response, the distance traveled and the duration of response.

Determining these parameters of a startle response for small groups of fish consists of four steps (Fig. 5A–D):

1. *Stimulus presentation*: The stimulus train is generated by a computer, allowing the presentation of multiple stimuli at precisely timed intervals. We typically use two different stimuli to elicit the startle response: a dish-tap and a mild electrical shock. The dish-tap stimulus is produced by a solenoid mounted so that the solenoid core taps the dish when activated. The electrical-shock stimulus is produced by stainless steel mesh electrodes running the length of the dish. Stimulus presentation at 2 Hz produces good nonassociative learning while keeping the video data relatively short (and the video files manageably small).

2. *Video capture*: To facilitate analysis of the motor behavior, video of the behavior is recorded from a Sony video camera directly to a computer hard drive. We determined that the camera's frame rate (deinterlaced to yield about 60 frames/sec) is sufficient to quantify startle behavior. The camera is set to record with short exposure times ($\leq 1/6000$ sec) to ensure that the image is not blurred during high-speed swimming. Video capture is controlled through a VideoScript macro (http://www.videoscript.com).

3. *Movie processing*: The resulting videos of the startle responses are processed to remove the image background, allowing identification of each fish in each video frame. To remove background from the video images, an image of the dish with no fish in it is subtracted from each movie frame. Each frame is deinterlaced to extract the full temporal resolution of the movie. Background subtraction and deinterlacing are performed by macros written in VideoScript. Fish are then detected by using the "analyze particles" function of Object-Image; the x and y coordinates of each fish in each frame are recorded.

4. *Tracking*: Fish are tracked using the (x,y) positions supplied by Object-Image. Tracking allows the assignment of startle parameters to individual fish. The tracking algorithm attempts to find the nearest fish position in frame $i + 1$ to the track end found in frame i. Tracking and subsequent calculation of velocity, acceleration, heading, latency, and duration of response are made using macros written in Object-Image and MatLab (www.mathworks.com).

We use this assay to study nonassociative learning of the startle response. The startle-plasticity assay can be used to identify mutants that affect the parametric measures of startle, such as duration of response or distance traveled. We find that the distance traveled in response to a stimulus provides a reliable measure of the stimulus response. This assay can be used to find mutations that increase or decrease this behavioral parameter, but because decreases in startle response might result from mutations that affect sensory or motor systems as well as integrative (i.e., learning) functions, we focus on mutants that have wild-type initial responses and show larger than wild-type response to later stimuli, because habituation is delayed. A screen for mutants with delayed habituation selects for the presence, rather than the absence, of a response, filtering out mutants with general sensory or motor deficits.

IV. Conclusions

We have described a set of assays of larval behavior in zebrafish. The setups are inexpensive, automated, and very adaptable, making them ideal as primary screening assays or for testing existing mutants in shelf screens. Although these assays cover a diverse set of behaviors, they do not come close to exhausting the range of behaviors shown by larvae. Moreover, by altering the stimuli, different types of screens can be conducted by using one assay. For example, we have described screening assays for mutants with altered stimulus specificity (Section III-A) and response plasticity (Section III-E). With so many possibilities, we can expect zebrafish to be a preferred model for behavioral screens in the years to come.

Acknowledgments

H. B. was supported by the grants from the following organizations: NIH RO1–EY12406, Packard Foundation, Klingenstein Foundation, and Sloan Foundation. Additional support was provided by NRSA (E.G.), NSF, UC Chancellor's Award, American Heart Association (M.C.S.), NARSAD (P.P-M.) Uehara, Naito (A.M.), and HHMI (M.B.O.). Thanks to Megan Carey for helpful comments on the manuscript.

References

Bang, P., Yelick, P., Malicki, J., and Sewell, W. (2002). High-throughput behavioral screening method for detecting auditory response defects in zebrafish. *J. Neurosci. Meth.* **118**(2), 177–187.

Benzer, S. (1973). Genetic dissection of behavior. *Sci. Am.* **229**(6), 24–37.

Bilotta, J. (2000). Effects of abnormal lighting on the development of zebrafish visual behavior. *Behav. Brain Res.* **116**(1), 81–87.

Borla, M. A., Palecek, B., Budick, S., and O'Malley, D. M. (2002). Prey capture by larval zebrafish: Evidence for fine axial motor control. *Brain Behav. Evol.* **60**(4), 207–229.

Brainard, D. H. (1997). The psychophysics toolbox, *Spatial Vision.* **10**, 433–436.

Brockerhoff, S. E., Hurley, J. B., Janssen-Bienhold, U., Neuhauss, S. C., Driever, W., and Dowling, J. E. (1995). A behavioral screen for isolating zebrafish mutants with visual system defects. *Proc. Natl. Acad. Sci. USA* **92**(23), 10545–10549.

Brockerhoff, S. E., Hurley, J. B., Niemi, G. A., and Dowling, J. E. (1997). A new form of inherited red-blindness identified in zebrafish. *J. Neurosci.* **17**(11), 4236–4242.

Budick, S. A., and O'Malley, D. M. (2000). Locomotor repertoire of the larval zebrafish: Swimming, turning and prey capture. *J. Exp. Biol.* **203**(17), 2569–2579.

Cahill, G. M., Hurd, M. W., and Batchelor, M. M. (1998). Circadian rhythmicity in the locomotor activity of larval zebrafish. *Neuroreport* **9**, 3445–3449.

Clark, D. T. (1981). "Visual Responses in Developing Zebrafish." University of Oregon Press, Eugene, OR.

Darland, T., and Dowling, J. E. (2001). Behavioral screening for cocaine sensitivity in mutagenized zebrafish. *Proc. Natl. Acad. Sci. USA* **98**(20), 11691–11696.

Dill, L. M. (1972). Visual mechanism determining flight distance in zebra danios (*Brachydanio rerio*, Pisces). *Nature* **236**, 30–32.

Dill, L. M. (1974). The escape response of the zebra danio (*Brachydanio rerio*). II. The effect of experience. *Anim. Behav.* **22**, 723–730.

Domenici, P., and Blake, R. W. (1997). The kinematics and performance of fish fast-start swimming. *J. Exp. Biol.* **200**, 1165–1178.

Easter, S. S. Jr., and Nicola, G. N. (1996). The development of vision in the zebrafish (*Danio rerio*). *Dev. Biol.* **180**(2), 646–663.

Eaton, R. C., Farley, R. D., Kimmel, C. B., and Schabtach, E. (1977). Functional development in the Mauthner cell system of embryos and larvae of the zebrafish. *J. Neurobiol.* **8**, 151–172.

Eaton, R. C., and Hackett, J. T. (1984). The role of the Mauthner cell in fast-starts involving escape in teleost fishes. *In* "Neural Mechanisms of Startle Behavior" (R. C. Eaton, ed.), pp. 213–262. Plenum Press, New York.

Gahtan E., and Baier H. (2004). Of lasers, mutants, and see-through brains: Assigning cells and circuits to behaviors in zebrafish. *J. Neurobiol.* **59**, 147–161.

Gahtan, E., Sankrithi, N., Campos, J. B., and O'Malley, D. M. (2002). Evidence for a widespread brain stem escape network in larval zebrafish. *J. Neurophysiol.* **87**(1), 608–614.

Gerlai, R., Lahav, M., Guo, S., and Rosenthal, A. (2000). Drinks like a fish: Zebra fish (*Danio rerio*) as a behavior genetic model to study alcohol effects. *Pharmacol. Biochem. Behav.* **67**(4), 773–782.

Geyer, M. A., Krebs-Thomson, K., Braff, D. L, and Swerdlow, N. R. (2001). Pharmacological studies of prepulse inhibition models of sensorimotor gating deficits in schizophrenia: A decade in review. *Psychopharmacology* **156**, 117–154.

Granato, M., van Eeden, F. J., Schach, U., Trowe, T., Brand, M., Furutani-Seiki, M., Haffter, P., Hammerschmidt, M., Heisenberg, C. P., Jiang, Y. J., Kane, D. A., Kelsh, R. N., Mullins, M. C., Odenthal, J., and Nüsslein-Volhard, C. (1996). Genes controlling and mediating locomotion behavior of the zebrafish embryo and larva. *Development* **123**, 399–413.

Guo, S. (2004). Linking genes to brain, behavior, and neurological diseases: What can we learn from Zebrafish? *Genes Brain Behav.* **3**, 63–74.

Hurd, M. W., Debruyne, J., Straume, M., and Cahill, G. M. (1998). Circadian rhythms of locomotor activity in zebrafish. *Physiol. Behav.* **65**, 465–472.

Kainz, P. M., Adolph, A. R., Wong, K. Y., and Dowling, J. E. (2003). Lazy eyes zebrafish mutation affects Müller glial cells, compromising photoreceptor function and causing partial blindness. *J. Comp. Neurol.* **463**(3), 265–280.

Kay, J. N., Finger-Baier, K. C., Roeser, T., Staub, W., and Baier, H. (2001). Retinal ganglion cell genesis requires lakritz, a zebrafish atonal homolog. *Neuron* **30**(3), 725–736.

Kelsh, R. N., Brand, M., Jiang, Y. J., Heisenberg, C. P., Lin, S., Haffter, P., Odenthal, J., Mullins, M. C., van Eeden, F. J., Furutani-Seiki, M., Granato, M., Hammerschmidt, M., Kane, D. A., Warga, R. M., Beuchle, D., Vogelsang, L., and Nüsslein-Volhard, C. (1996). Zebrafish pigmentation mutations and the processes of neural crest development. *Development* **123**, 369–389.

Kimmel, C. B., Patterson, J., and Kimmel, R. J. (1974). The development and behavioral characteristics of the startle response in the zebra fish. *Dev. Psychobiol.* **7**, 47–60.

Koch, M. (1999). The neurobiology of startle. *Prog. Neurobiol.* **59**(2), 107–128.

Krauss, A., and Neumeyer, C. (2003). Wavelength dependence of the optomotor response in zebrafish (*Danio rerio*). *Vision Res.* **43**(11), 1275–1284.

Landis, C., and Hunt, W. A. (1939). "The Startle Pattern." Farrah and Rinehart, New York.

Li, L. (2001). Zebrafish mutants: Behavioral genetic studies of visual system defects. *Dev. Dyn.* **221**(4), 365–372.

Li, L., and Dowling, J. E. (1997). A dominant form of inherited retinal degeneration caused by a non-photoreceptor cell-specific mutation. *Proc. Natl. Acad. Sci. USA* **94**, 11645–11650.

Li, L., and Dowling, J. E. (1998). Zebrafish visual sensitivity is regulated by a circadian clock. *Vis. Neurosci.* **15**(5), 851–857.

Li, L., and Dowling, J. E. (2000a). Disruption of the olfactoretinal centrifugal pathway may relate to the visual system defect in night blindness b mutant zebrafish. *J. Neurosci.* **20**(5), 1883–1892.

Li, L., and Dowling, J. E. (2000b). Effects of dopamine depletion on visual sensitivity of zebrafish. *J. Neurosci.* **20**(5), 1893–1903.

Liu, K. S., and Fetcho, J. R. (1999). Laser ablations reveal functional relationships of segmental hindbrain neurons in zebrafish. *Neuron* **23**(2), 325–335.

Liu, D. W., and Westerfield, M. (1988). Function of identified motoneurones and co-ordination of primary and secondary motor systems during zebra fish swimming. *J. Physiol. Lond.* **403**, 73–89.

Liu, K. S., Gray, M., Otto, S. J., Fetcho, J. R., and Beattie, C. E. (2003). Mutations in deadly seven/notchla reveal developmental plasticity in the escape response circuit. *J. Neurosci.* **23**(22), 8159–8166.

Lockwood, B., Bjerke, S., Kobayashi, K., and Guo, S. (2004). Acute effects of a alcohol on larval zebrafish: A genetic system for large-scale screening. *Pharmacol. Biochem. Behav.* **77**, 647–654.

Lorent, K., Liu, K. S., Fetcho, J. R., and Granato, M. (2001). The zebrafish space cadet gene controls axonal pathfinding of neurons that modulate fast turning movements. *Development* **128**(11), 2131–2142.

Maaswinkel, H., and Li, L. (2003). Spatio-temporal frequency characteristics of the optomotor response in zebrafish. *Vision Res.* **43**(1), 21–30.

Neuhauss, S. C. (2003). Behavioral genetic approaches to visual system development and function in zebrafish. *J. Neurobiol.* **54**(1), 148–160.

Neuhauss, S. C., Biehlmaier, O., Seeliger, M. W., Das, T., Kohler, K., Harris, W. A., and Baier, H. (1999). Genetic disorders of vision revealed by a behavioral screen of 400 essential loci in zebrafish. *J. Neurosci.* **19**, 8603–8615.

Nicolson, T., Rüsch, A., Friedrich, R. W., Granato, M., Ruppersberg, J. P., and Nüsslein-Volhard, C. (1998). Genetic analysis of vertebrate sensory hair cell mechanosensation: The zebrafish circler mutants. *Neuron* **20**, 271–283.

O'Malley, D. M., Kao, Y. H., and Fetcho, J. R. (1996). Imaging the functional organization of zebrafish hindbrain segments during escape behaviors. *Neuron* **17**(6), 1145–1155.

Orger, M. B., Smear, M. C., Anstis, S. M., and Baier, H. (2000). Perception of Fourier and non-Fourier motion by larval zebrafish. *Nat. Neurosci.* **3**(11), 1128–1133.

Pelli, D. G. (1997). The VideoToolbox software for visual psychophysics: Transforming numbers into movies. *Spatial Vision* **10**, 437–442.

Ribera, A. B., and Nüsslein-Volhard, C. (1998). Zebrafish touch-insensitive mutants reveal an essential role for the developmental regulation of sodium current. *J. Neurosci.* **18**, 9181–9191.

Rick, J. M., Horschke, I., and Neuhauss, S. C. (2000). Optokinetic behavior is reversed in achiasmatic mutant zebrafish larvae. *Curr. Biol.* **10**(10), 595–598.

Rock, I., and Smith, D. (1986). The optomotor response and induced motion of the self. *Perception* **15**(4), 497–502.

Roeser, T., and Baier, H. (2003). Visuomotor behaviors in larval zebrafish after GFP-guided laser ablation of the optic tectum. *J. Neurosci.* **23**(9), 3726–3734.

Saint-Amant, L., and Drapeau, P. (1998). Time course of the development of motor behaviors in the zebrafish embryo. *J. Neurobiol.* **37**(4), 622–632.

Seul, M., O'Gorman, L., and Michael, J. S. (2002). "Practical Algorithms for Image Analysis: Descriptions, Examples, and Code." Cambridge University Press, Cambridge.

Simpson, J. L. (1984). The accessory optic system. *Annu. Rev. Neurosci.* **7**, 13–41.

Zhdanova, I. V., Wang, S., Leclair, O. U., and Danilova, N. (2001). Melatonin promotes sleep-like behavior in zebrafish. *Brain Res.* **903**(12), 263–268.

CHAPTER 4

Target-Selected Gene Inactivation in Zebrafish

Erno Wienholds and Ronald H. A. Plasterk

Hubrecht Laboratory
Center for Biomedical Genetics
Uppsalalaan 8
3584 CT Utrecht
The Netherlands

I. Introduction

A. Reverse Genetics in a "Forward" Genetic Model Organism

The zebrafish was developed as a vertebrate genetic model organism (Kimmel, 1989; Streisinger *et al.*, 1981) because of properties that allow forward genetic screens: small size, relative short generation time, and many offspring. The logistics of breeding allow mutant screens of such size that most genes whose loss results in a mutant phenotype can be hit multiple times (Mullins *et al.*, 1994; Solnica-Krezel *et al.*, 1994). This can produce a virtually complete, and saturated inventory of genes specifically involved in different stages of embryonic development. Large-scale forward mutagenesis screens have identified thousands of mutants defective in a variety of embryological processes (Driever *et al.*, 1996; Haffter *et al.*, 1996) and more sophisticated forward mutagenesis screens, such as modifier screens, are being conducted to find more genes involved in embryonic development.

An alternative way to link gene and function is by reverse genetics. Here, a specific targeted gene is inactivated and the consequences are studied. The vertebrate model at present commonly used for reverse genetics is the mouse. However, there is also a growing need for reverse genetics in the fish for a variety of reasons: (1) With the genome being sequenced, researchers recognize zebrafish orthologs of interesting genes from other species and want to obtain mutants to study the functions. (2) Genes initially resulting from a forward mutant hunt in the zebrafish might deserve more detailed attention; for example, the mutant originally isolated might be a missense mutant, and it can be of interest to obtain a known null mutant. (3) Large-scale studies in zebrafish on gene expression (using microarrays or *in situ* hybridizations) or screens for morpholino effects might identify genes of interest, and again the question is what the mutant phenotype will be.

B. Knockouts and Knockdowns

Strategies to reduce or completely knock out gene function fall into two classes: those that aim to knock down gene expression (RNAi and morpholinos) and those that aim at isolation of a mutation in the gene. Both approaches have advantages. In zebrafish, knockdowns are mostly done by morpholinos (Nasevicius *et al.*, 2000); given the success of this approach, RNAi with short interfering RNAs (siRNAs; Elbashir *et al.*, 2001) has not yet been thoroughly tested in zebrafish. Such knockdowns have great advantages: they are quick and relatively easy, can be done at a large scale, and (except for the costs of morpholinos) are cheap. Nevertheless, there are also advantages in having genuine mutations. They are permanent, they can be combined with other mutations, and mutant fish can be used as starting point for subsequent screens for modifier genes. Also, in principle, one can obtain mutations that are guaranteed nulls, whereas with knockdowns one can never be sure. In practice, it is often useful to apply both morpholino and mutation studies to the same gene. This chapter further addresses the generation of zebrafish that have mutations in a gene of choice.

II. Gene Targeting Strategies

A. Homologous Disruption

The method of choice for gene targeting in mouse is homologous disruption by recombination in embryonic stem (ES) cells (Capecchi, 1989; Thompson *et al.*, 1989). Although this approach is very efficient in mouse, it has not yet successfully been applied to any other model organism, primarily because of the lack of isolation or cultivation of pluripotent ES cells. In zebrafish, ES-like cells have been isolated (Sun *et al.*, 1995). After cultivation *in vitro* and transplantation into the developing embryo, these cells have been shown to be able to contribute to the germline (Ma *et al.*, 2001). However, this has not yet resulted in a targeted knockout. An alternative to the use of ES cells is to perform homologous recombination in somatic cell lines followed by nuclear transfer into oocytes. In sheep and pigs, this has resulted in gene targeting (Dai *et al.*, 2002; McCreath *et al.*, 2000). In zebrafish, it is now also possible to perform nuclear transfer of genetically modified cultured cells (Lee *et al.*, 2002), but a knockout has not yet been reported.

In *Drosophila*, gene targeting by homologous recombination has been performed successfully *in vivo*. Extrachromosomal linear DNA fragments recombined with the target gene (Rong *et al.*, 2000, 2001). Similarly, in *Arabidopsis*, homologous disruption also occurred on transformation (Miao *et al.*, 1995). In principle, one can envisage that microinjection of DNA fragments into zebrafish oocytes or early embryos might provide a similar method for gene targeting. Presumably, the limiting factor is not so much the frequency of such homologous recombination events (which is also often low in mouse ES cells) as the absolute numbers of injected eggs. Coinjection of proteins involved in homologous recombination (Cui *et al.*, 2003) or genetic backgrounds that enhance homologous recombination (de Wind *et al.*, 1995; Hanada *et al.*, 1997) might reduce the number of embryos considerably. Although the methodologies described next might mean that homologous disruption is not required for simple gene knockouts, it would still be a valuable method to develop (e.g., for specific knockins or gene fusions).

B. Target-Selected Mutagenesis

Other approaches currently available for making gene knockouts are, strictly speaking, not gene targeting, because lesions in the genes of interest are not generated in a targeted manner but randomly throughout the genome. Therefore, it is more precise to describe them as 'target-selected' gene inactivations: the mutagenesis per se is random, but mutations are sought in a targeted manner in a gene of interest by analysis of the genomic DNA (Fig. 1). Target-selected mutagenesis has successfully been used in various organisms.

The first class of target-selected gene inactivation is insertional mutagenesis. Here, the mutagen is a transposon or virus, which is inserted randomly throughout the genome, thereby disrupting genes. A (pooled) library of mutagenized animals is screened for insertions in target genes. By PCR with insert-specific primers and gene-specific primers, insertions in target genes can be discovered in a (pooled) library of mutagenized animals. This relatively easy method has been successful in *Drosophila* (Ballinger *et al.*, 1989; Kaiser *et al.*, 1990), *C. elegans* (Zwaal *et al.*, 1993), and plants (Das *et al.*, 1995; Koes *et al.*, 1995; Meissner *et al.*, 1999). Alternatively, the insert is used as a starting point for adaptor-mediated or inverse PCR. The sequences flanking the inserts are determined and mapped to the genome and (predicted) genes by, for example, BLAST (Altschul *et al.*, 1990). By determining the flanking sequences of a comprehensive number of insertions, disruptions can be identified in virtually any of the (predicted) target genes. This has been proven feasible in *C. elegans* (Korswagen *et al.*, 1996; Martin *et al.*, 2002) and mouse (Zambrowicz *et al.*, 1998) and highly efficient in *Drosophila* (Spradling *et al.*, 1999) and *Arabidopsis* (Alonso *et al.*, 2003; Parinov *et al.*, 1999; Sessions *et al.*, 2002; Tissier *et al.*, 1999).

In zebrafish, a large-scale insertional mutagenesis screen has been performed by using retroviral vectors as mutagen (Amsterdam *et al.*, 1999; Golling *et al.*, 2002). Here, the goal is a forward screen followed by rapid cloning of the affected genes. However, these same founder fish might also be used for the target-selected mutagenesis approaches described previously. The founder fish harbor approximately 1,000,000 insertions and 1 embryonic lethal mutation is expected per 70–100 insertions (Amsterdam *et al.*, 1999). Thus, in this collection, at least 10,000 insertions are expected to disrupt genes essential for embryonic development (five times the number of genes expected to be essential for embryonic development). In principle, insertions in these and other (nonessential) genes can be found by the methods described previously. However, the mutagenesis efficiency is such that high numbers of fish (30,000–100,000; Amsterdam, 2003) have to be archived to find at least one insertion per gene. It remains to be seen whether this is feasible in practice. An alternative insertion mutagen is a transposon (Davidson *et al.*, 2003; Raz *et al.*, 1998), although insertion frequencies are not optimal.

The second class of target-selected mutagenesis is chemical mutagenesis. Chemicals can introduce various kinds of mutations. In *C. elegans*, chemical mutagenesis is performed using *N*-ethyl-*N*-nitrosourea (ENU), ethylmethanesulfonate (EMS), or trimethylpsoralen (TMP) in order to generate small deletions. These deletions can be detected by a simple PCR method (Jansen *et al.*, 1997; Liu *et al.*, 1999). Individuals carrying deletions in any gene can then be recovered from a pooled library of millions of animals. In zebrafish, ENU is also widely used as mutagen in forward screens. Screening these animals for small deletions is possible

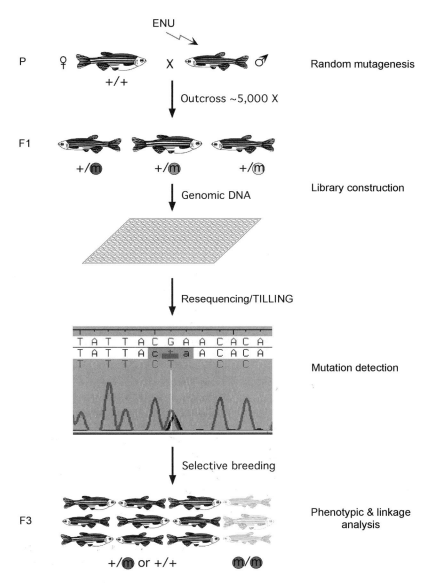

Fig. 1 Target-selected mutagenesis in zebrafish. Approximately 100 adult male zebrafish are randomly mutagenized with *N*-Ethyl-*N*-nitrosourea (ENU) and outcrossed against wild-type females. A library of approximately 5000 F1 animals is constructed, in principle having independent mutations. Genomic DNA of these F1 animals is isolated and arrayed in 384-well PCR plates, suitable for robotic handling. The DNA is screened for mutations in target genes by resequencing or TILLING. Animals with interesting mutations are recovered from the library [reidentified from a pool of living F1 fish or recovered by *in vitro fertilization* (IVF) with frozen sperm] and outcrossed against wild-type fish or incrossed with other mutants. Finally, the mutations are homozygosed and animals are analyzed for phenotypes and linkage to the mutation. (See Color Insert.)

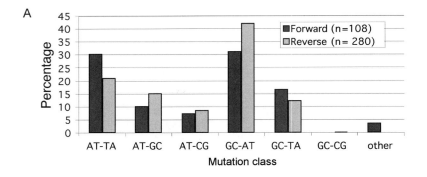

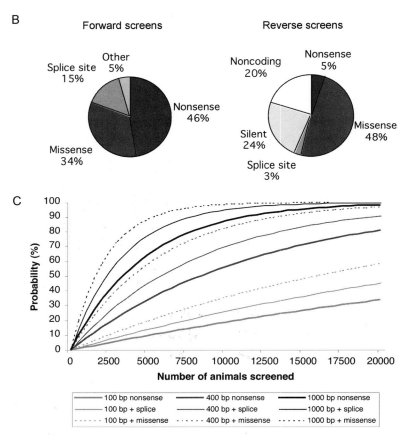

Fig. 2 ENU-induced mutations in zebrafish. (A) Mutations spectra of ENU-induced mutations in forward and reverse genetic screens. (B) Consequence of mutations at coding level. Other mutations in forward screens are insertions and deletions. The ratio of mutations, strictly in the coding regions in reverse screens, is 6.9% nonsense, 62.2% missense, and 30.9% silent. (C) Probabilities for finding at least one potential loss-of-function mutation in ENU-mutagenized F1 zebrafish. Binomial distribution

in theory (Lekven *et al.*, 2000), but the number of animals needed to hit any gene at least once is probably unfeasable (at least 100,000 if 10% of the mutations are small deletions as in *C. elegans*; De Stasio *et al.*, 1997). In addition, small deletions have seldom been recovered after ENU mutagenesis of mouse spermatogonial germ cells and ES cells (Chen *et al.*, 2000; Douglas *et al.*, 1995) and zebrafish premeiotic germ cells in forward screens (Fig. 2B), indicating that using ENU for creation of small deletions is not practical. Alternatively, one could use EMS or TMP, but mutagenesis frequencies are too low and only large deletions have been recovered so far (Lekven *et al.*, 2000; Mullins *et al.*, 1994; Solnica-Krezel *et al.*, 1994). The most powerful and widely used mutagens for mutagenesis screens are ENU and EMS, which cause predominantly point mutations. Point mutations are more difficult to detect than insertions and small deletions, but the mutagenesis efficiency is so much higher that only a limited number of animals are needed to reach saturation (typically a few thousand). For organisms such as *C. elegans* and *Drosophila*, the ability to raise a larger number of animals might not be a limiting factor, but for vertebrates it definitely is. Recent methods for the identification of point mutations have improved considerably in speed and cost. One can now extensively screen a limited number of animals and still recover knockout alleles of genes of interest. Therefore target-selected mutagenesis using mutagens causing point mutations, such as ENU and EMS, is likely to be the method of choice for creating knockouts. Various target-selected mutation detection methods have already been successfully applied to *Arabidopsis* (McCallum *et al.*, 2000), lotus (Perry *et al.*, 2003), *Drosophila* (Bentley *et al.*, 2000), *C. elegans* (R. Plasterk and E. Cuppen, unpublished data), mouse (Beier, 2000; Coghill *et al.*, 2002), rat (Smits *et al.*, 2004; Zan *et al.*, 2003), and zebrafish (Wienholds *et al.*, 2002, 2003b).

III. Target-Selected Mutagenesis in Zebrafish

At present, the only working method for making knockouts in zebrafish is chemical mutagenesis followed by targeted screening for point mutations (Fig. 1). The first gene in zebrafish to be knocked was *rag1* (Wienholds *et al.*, 2002). In this case, and in other approaches described hereafter, the mutagen was the standard chemical mutagen ENU. Because ENU is the same mutagen that is being used for forward genetic screens, one can use one mutagenesis regime for forward as well as reverse genetic goals. For finding mutations in the *rag1* gene, we screened the F1 fish from the Tübingen 2000 mutagenesis screen and found a

---

probabilities are calculated for three different coding sequence fragment sizes (100, 400, and 1000 bp) with a mutagenesis efficiency of 1 in 250,000 bp (0.00004 mutations per base). Nonsense, splice site, and missense mutations that potentially result a phenotypic change represent approximately 5%, and 2.5%, and 3.5% of the mutations, respectively. The last two are accumulated to the class of nonsense mutations (nonsense and splice site mutations; nonsense, splice site and missense mutations).

premature stop codon by straightforward DNA resequencing of amplicons containing most of the coding regions of the gene. This turned out to be a genuine null allele. Since then, a faster and cheaper method has been developed to prescreen fish for the presence of mutations within an amplicon and then sequence only those amplicons known to contain a mutation. The method is called TILLING, and was initially developed by the laboratory of Steve Henikoff for *Arabidopsis* (Colbert *et al.*, 2001) and applied to the zebrafish (Fig. 4) by us and others (Wienholds *et al.*, 2003b). Noted that there are also potential alternative approaches for this prescreen step, such as a method based on phage Mu transposition described recently (Yanagihara *et al.*, 2002), and possibly other methods. To date, the only methods that have generated the target-selected zebrafish mutants are resequencing and TILLING, and the latter is described here in more detail.

A. Mutagenesis

The success rate of ENU mutagenesis screens depends on several factors. First, the mutagenesis must be optimal. The better the mutagenesis, the more chance to obtain mutants in a given set of animals. Typically, young adult male zebrafish are mutagenized with 4–6 consecutive treatments of ENU (van Eeden *et al.*, 1999), and after a few weeks they are crossed with wild-type females to obtain an F1 generation of fish carrying nonmosaic mutations (Fig. 1). Increasing ENU dose or the number of treatments might result in higher mutation frequencies, but also in decreased survival (Solnica-Krezel *et al.*, 1994), reduced fertility, and increased chance of clonal mutants. Prior to outcrossing, the mutagenesis efficiency is efficiently monitored with a specific locus test. Mutagenized founders are crossed with tester females carrying known homozygous mutations that can be scored easily, for example, pigmentation mutants *albino* or *sparse*. Depending on the marker gene assayed, a good mutagenesis typically gives a specific locus rate of 0.2–0.3% (Mullins *et al.*, 1994; Solnica-Krezel *et al.*, 1994). The second factor influencing the possibility of recovering mutants is the number of animals screened. Screening more animals will lead to better chance of finding at least one mutant allele. The third factor influencing the possibility of finding mutant alleles is the size of the target gene. The larger the gene, the more chance of finding potential loss-of-function alleles. In forward screens, the mean mutation frequency per gene is approximately 1 in 1000 genomes. Thus, for an average gene one would have to screen about 1000 animals to get a potential loss-of-function allele. Screening smaller genes requires more animals.

To date, more than a hundred ENU-induced mutations have been cloned in forward genetic screens. The spectrum of these mutations (Fig. 2A) is biased, because each of these mutations has a phenotype, mostly loss-of-function. About half (46%) introduce a nonsense codon; 15% alter a splice site, resulting in insertions and deletions in the mRNA; and 34% are missense mutations (Fig. 2B). The ENU spectrum of 280 mutations identified in reverse genetics screens is unbiased.

However, it is similar to that of forward screens (Fig. 2A). Of the mutations found in noncoding regions, only the ones in splice sites (2.5%, Fig. 2B) will, with great certainty, alter protein function. The influence of the mutations found in coding regions can be calculated from the codon usage in zebrafish (Nakamura *et al.*, 2000, http://www.kazusa.or.jp/codon/) and the observed spectrum of ENU-induced mutations in unbiased (thus reverse) genetic screens. Of these, 29% are expected to be silent, 66% missense, and 5.0% nonsense. (The last is predicted to be 5.6% if the ENU spectrum from forward genetic screens is used.) The number of nonsense mutations found in reverse screens thus far (Fig. 2B) is even somewhat higher (6.9%), which could reflect statistical variation or local codon usage differences. Splice-site and nonsense mutations are the most likely candidates to be loss of function. However, the pool of missense mutations is also likely to contain functional changes. In forward screens, the ratio between missense and nonsense mutations is 0.7 (37/52). Extrapolation to reverse screens suggests that 3.5% (0.7 times 5%) of the missense mutations found in coding regions also give rise to a phenotypic chance. The challenge is to predict which missense mutations might give rise to a functional change of the protein (Chasman *et al.*, 2001; Ng *et al.*, 2001; Sunyaev *et al.*, 2001).

Together, up to 11% of the mutations (nonsense, splice site, and missense) are expected to have a (partial) loss-of-function phenotype or other influence on protein function. The probability of finding at least one such allele in different-sized target fragments in a library of mutagenized animals, with a per-base mutation frequency of approximately 1 in 250,000 bp (Wienholds *et al.*, 2003b), is shown in Fig. 2C. For example, to have an 80% chance of finding at least one nonsense or splice site mutant in a 1-kb gene, one needs to screen approximately 6000 animals.

B. Knockout Libraries

There are two ways to store the library of mutagenized fish. For inactivation of the *rag1* gene and screening of several other genes, we have frozen the sperm of 2679 mutagenized F1 males (Brand *et al.*, 2002; Wienholds *et al.*, 2002). Keeping the library as a frozen stock results in a permanent library that can be screened many times for many different genes over an unlimited period of time. In addition, it saves a lot of space, compared to an aquarium facility. Some drawbacks are that the cryopreservation of sperm samples is quite laborious and mutants need to be recovered by *in vitro* fertilization (IVF). For recovery of the *rag1* loss-of-function allele this worked well, but we were unable to recover several other mutants by IVF (unpublished observation). However, recent advances in sperm freezing methods (Morris *et al.*, 2003) and IVF indicate that the creation of a reliable, permanently frozen library of sperm is feasible. A fast alternative to freezing sperm is to keep the mutagenized animals alive during the screening process. Such a library is relatively easy to construct. As soon as the outcrossed mutagenized F1 animals are old enough, pieces of the fins are removed for DNA isolations (fin-clipped). The fish are then pooled to minimize the aquarium space required. For the second knockout library, we fin-clipped 4608 fish and grouped them into 384 pools of 12 (Wienholds *et al.*, 2003b). A consequence of pooling is that after screening the complete library, the fish of a positive pool have to be fin-clipped and genotyped again. An advantage of such a living library is that both males and females can be screened. If different interesting mutations are found in the same gene, but in opposite sexes, these can be crossed at once, generating transheterozygous fish. In addition to all background mutations then being heterozygous, this saves a full generation time for analysis. A drawback of living libraries is that they can be screened for only a limited period of time: from the point the fish are fin-clipped until they lose fertility (approximately 1.5 years). If aquarium space is limiting, one can construct several small libraries (e.g., 384 animals) that are extensively screened for many genes (e.g., 100 genes), as is being done for rats (Smits *et al.*, 2004). Of course, a combination of a living and a cryopreserved library is possible. One can start constructing and screening a living library, remove the males, and cryopreserve their sperm.

C. Mutation Detection by Resequencing

The most straightforward and reliable method for point mutation detection is DNA sequencing. During the past couple of years, DNA sequencing techniques have improved considerably in throughput and cost. This has resulted in the human genome being sequenced in about 1 year (Lander *et al.*, 2001; Venter *et al.*, 2001).

The first library used for target-selected inactivation of the *rag 1* gene required only 2679 animals. To maximize the chance of finding a guaranteed loss-of-function allele in this limited set of animals, we have chosen to screen them by resequencing in a one-to-one manner, minimizing the number of false negatives. The pipeline for this resequencing is simple. First, we PCR-amplified most of the coding regions of the *rag 1* gene. Because this resulted in variations in yield, we used a small volume of this PCR as template in a nested PCR with internal primers. After the nested PCR, all samples had equal yield. Next, the PCR products were either sequenced directly or were first purified and then sequenced with internal primers designed to cover approximately 600 bp per read. Sequence samples were purified and analyzed on a 96-capillary sequencer (ABI3700 DNA analyzer). Although initially only the liquid handling steps were performed by robots, we have now automated the entire process, including all the PCR and purification steps.

Because we screened F1 animals, the mutations were heterozygous. Therefore, we could not do a simple alignment to detect these mutations and filter out the differences. We detected heterozygous positions by parsing the trace files in batches of 96 through Phred (Ewing *et al.*, 1998), Phrap (Gordon *et al.*, 2001), and Polyphred (Nickerson *et al.*, 1997) and visualized them with Consed (Gordon *et al.*, 1998). The batches of 96 samples were inspected simultaneously for heterozygous mutations. New mutations could be discriminated from natural-ly occurring single nucleotide polymorphisms (SNPs), because these were present in multiple samples whereas the genuine mutations were present only in one sample.

To find a knockout of the *rag 1* gene, we analyzed almost 12,500 sequence reactions covering most of the coding regions of the gene. This yielded nine amino acid changes and one premature stop. This nonsense mutation was in the middle of the catalytic domain of the recombinase and therefore expected to be a null. Indeed, fish homozygous for this stop allele were defective in V(D)J recombination and immunodeficient, indicating that this allele was a true loss of function. In addition to this *rag 1* loss-of-function allele, we found dominant-negative muta-tions in *p53* (S. Berghmans, R. Murphey *et al.*, manuscript in preparation) and several potential loss-of-function mutations in *tie2* (A. M. Kuechler *et al.*, manu-script in preparation). Screening these genes took us approximately 1 month for each gene and costs are estimated around U.S. $12,500. More recently, the method of choice for finding mutations is TILLING (Section III-D). However, new technological advances in sequencing techniques and automation will probably reduce both time and cost for finding mutants by direct resequencing significantly, so that in the future resequencing might again be the method of choice.

D. Mutation Detection by Targeting Induced Local Lesions in Genomes (TILLING)

TILLING (targeting induced local lesions in genomes) was originally developed for *Arabidopsis* and used denaturing high-performance liquid chromatography (DHPLC) to detect EMS-induced mutations (McCallum *et al.*, 2000). The imple-mentation of enzyme-mediated mismatch recognition (Oleykowski *et al.*, 1998) allowed a high throughput detection of mutations (Colbert *et al.*, 2001). This led to the *Arabidopsis* TILLING Project (ATP) (Till *et al.*, 2003), which has isolated more than 100 potential knockouts within 1 year. Because TILLING is approxi-mately 10 times faster and cheaper than resequencing, we implemented it to prescreen for ENU-induced mutations in zebrafish (Wienholds *et al.*, 2003b).

The setup for TILLING in zebrafish is similar to that of resequencing. It involves nested PCR followed by mutation detection in the amplified fragments (Fig. 3). First the target is amplified by PCR with gene-specific primers. The maximum length of the fragments that can be analyzed, with good sensitivity, is approximately 1000 bp. The continuous stretches of coding regions or exons in zebrafish (and most other vertebrates) are usually smaller (around 100 bp). There-fore, amplicons are designed to contain several of these smaller exons or, occa-sionally, single exons. In addition, exons in the 5' region of the gene are favored for screening, because a nonsense or splice-site mutation might lead to a complete removal of downstream part of the protein. In nested PCR, the concentrations of the amplified fragments are equalized and the fragments are labeled with fluorescent dyes. These labels can be incorporated in two ways. The nested set of gene-specific primers can be directly labeled or the nested set of primers can be

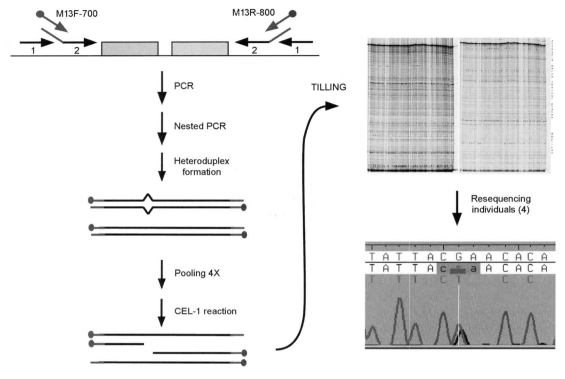

Fig. 3   Mutation detection by TILLING. See text for detailed description of all the steps involved. (See Color Insert.)

tailed with universal adaptor sequences, for which the corresponding labeled universal primers are mixed into the PCR reaction. A major advantage of the latter method is that the same fluorescent labeled primers can be used to label any fragment, thereby reducing costs considerably: unmodified primers are cheaper than fluorescent primers and the two universal fluorescent primers are ordered in large batches. After PCR, heteroduplexes between wild-type and mutant PCR fragments are formed by denaturing and reannealing the PCR fragments. This is necessary to form mismatches, which can be recognized by the mismatch recognition enzyme CEL-1 isolated from celery (Oleykowski *et al.*, 1998; Yang *et al.*, 2000). To increase the throughput, samples are pooled four times. Additional pooling is expected to result in a decreased sensitivity because the complex nature of the vertebrate genome. SNPs present in the target fragments might mask the discovery of new mutations. Pooled fragments are treated with the CEL-1 enzyme. Fragments are purified using G50 minicolumns or isopropanol precipitation. After purification, the labeled digested fragments are separated and visualized on slab gel sequencers (e.g., see Fig. 4). Each gel can be used to analyze

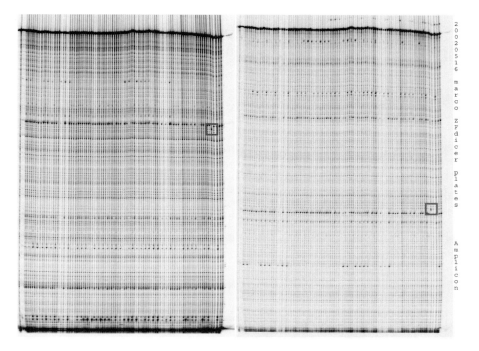

Fig. 4 Example TILLING gel. Each of the 96 lanes contains pooled CEL-1-digested PCR fragments of the *dicer1* gene of four different animals. The IR Dye 700 and IR Dye 800 channels are shown on the left and right, respectively. A potential mutation is boxed in both the 700 and 800 IR Dye channels. After sequencing, this mutation turned out to be nonsense and causes a loss-of-function phenotype.

96 samples at one time. Because the samples are pooled four times, these gels represent 384 animals. The gels are manually processed and inspected with the Photoshop software. In principle, a mutation should give a strong signal in both lanes at reciprocal heights (Fig. 4), but we found that this is not an absolute prerequisite to confirm a mutation; often any signal that is different in one lane is confirmed successfully. The mutation can be confirmed by resequencing the individual four samples of the pooled lane. First, the target amplicon is reamplified from the first PCR to confirm the TILLING results. Next it is amplified from the original genomic DNA again to exclude any PCR artifacts or other mistakes.

The TILLING procedure described can be performed at different scales. We use a sophisticated robotic setup, with a 96-channel pipette and eight integrated 384-well PCR machines and six slab gel sequencers to screen for mutations in a large number of genes. With this setup, a single person can prescreen a library of 9216 animals by TILLING in 1 to 2 days. The same protocol can be used at a smaller scale if only a small number of genes need to be mutated. All the steps can be done manually, using multichannel pipettes. With a few PCR machines and

Table I
Summary of Genes Screened and Mutations Found by TILLING in Zebrafish

Gene	Name	Amplicon Length/cds (bp)[a]	Total	Nonsense	Splice	Missense	Silent	Noncoding
gene1	A	569/264	6	—	—	—	1	5
gene2	A	777/381	15	—	1	6	2	6
gene3	A	639/327	18	1	2	7	3	5
gene4	A	911/879	14	1	—	8	5	—
gene5	A	812/327	18	2	1	2	3	10
gene6	A	541/436	6	1	—	2	3	—
gene7	A	590/264	6	—	1	2	1	2
gene8	A	442/270	6	—	1	2	2	1
gene9	A	841/279	20	—	—	7	2	11
	B	448/230	7	—	1	2	2	2
gene10	B	664/395	4	1	—	3	—	—
gene11	A	720/720	11	—	—	6	5	—
gene12	A	496/469	6	—	—	3	3	—
gene13	A	816/443	9	1	—	4	1	3
apc[c]	A	953/953	16	—	—	14	2	—
	B	873/873	13	1	—	7	5	—
	C	813/813	10	—	—	5	5	—
	G	921/881	11	1	—	6	4	—
gene15	A	955/955	18	2	—	12	4	—
dicer1[d]	A	798/775	17	2	—	9	6	—
	C	752/418	16	1	—	8	2	5
	E	736/526	8	—	—	4	2	2
Average		730/540	11.6	0.6	0.3	5.4	2.9	2.4
Total		16,067/11,878	255	14	7	119	63	52

[a]Length of nested PCR amplicon, excluding primer sequences; length of coding sequence (cds) in amplicon.
[b]Potential mutations were found by TILLING and confirmed by resequencing.
[c]Hurlstone *et al.* (2003).
[d]Wienholds *et al.* (2003a).
Source: Adapted from Wienholds, E. *et al.* (2003b). Efficient target selected mutagenesis in zebrafish. *Genome Res.* **13,** 2700–2707, with permission.

a single slab gel sequencer, one person should be able to screen the same number of animals for one gene within a week. Recently, we screened a library of 4608 mutagenized F1 animals for mutations in 16 different genes by TILLING (Table I, Wienholds *et al.*, 2003b). We found 458 potential mutations, of which 255 could be confirmed by resequencing. These included 119 missense, 14 nonsense, and 7 splice site mutations, the last two classes found in 13 genes and expected to be loss of function. Thus, we potentially knocked out 13 different genes by using the TILLING setup for mutation detection. One of the genes we screened for

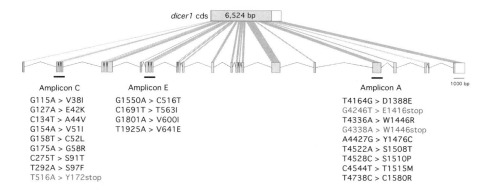

Fig. 5 Target-selected inactivation of the *dicer1* gene. *Dicer1* mRNA and genomic organization are schematically drawn. Exons in the genomic structure are indicated as boxes. The three fragments screened by TILLING are underlined. Nucleotide and amino acid positions are given with regard to the predicted start codon. On homozygosing, the three nonsense mutations were all loss of function (Wienholds *et al.*, 2003a).

mutations is the *dicer 1* gene (Fig. 5). Three different amplicons were designed to cover as much of the coding region as possible. In these three fragments, 78 potential mutations were identified. Approximately half (41) of them could be confirmed by resequencing (Table I). Of these, 21 were missense mutations and 3 were nonsense mutations (Fig. 5). All three nonsense mutations displayed similar phenotypes as homozygotes: developmental arrest and failure in micro-RNA processing (Wienholds *et al.*, 2003a). The influence of the missense mutations on the protein function is currently being investigated.

IV. Discussion

A. Linking Genotypes and Phenotypes

There is something counterintuitive about this approach: to target a gene one mutagenizes the entire genome. The expected number of genes per haploid genome that is knocked out by a standard ENU mutagenesis is of the order of 10 to 20 (which fits with the expected numbers of genes in a vertebrate genome of about 30,000, and the chance for an average gene to be knocked out in 1 in 1000–2000 animals). So if one recognizes by TILLING that an F1 fish has a mutation in a gene of interest, it will most likely contain multiple other mutations. How are these filtered? How do you know the mutation causes the observed phenotype? These are questions most commonly raised when this approach is presented.

First, the collateral damage is equally big in any forward genetic protocol (in which it does not seem to raise nearly as many eyebrows). In practice, it is hardly ever a problem, for the following reasons:

1. Even in the very first backcross to homozygose a mutation observed in an F1 fish, one recognizes linkage between being homozygous for the mutation and having a given phenotype. It is common that in such backcrosses multiple additional phenotypes are seen, many with Mendelian segregation patterns, but genotyping the offsping will quickly sort out the unlinked mutations.

2. As with any mutagenesis, it is wise to outcross a mutant once or twice more before too much analysis is done. Then only tightly linked mutations might blur the analysis.

3. Fortunately, one often encounters more than one reduction-of-function- or even loss-of-function allele in one screen. These can then be used to address the specificity question in two ways: if two independent mutations in the same gene have the same phenotype, this strongly argues for a causal effect, because it would be highly unlikely that in both cases a genetically linked second mutation causes the same phenotype. We have also used two alleles to further prove causation: one can create heteroallelic animals, in which one allele has one mutation and the homologous allele has the other. Then this fish is heterozygous for all other possibly linked mutations, and if it still has the same phenotype as fish homozygous for each individual mutation, one can be basically sure of a causal effect.

4. Ideally, one could rescue a mutation by transgenesis with the wild-type gene. In practice, this is not needed to prove causality, if the first two or three points are followed.

5. A final argument to consider, although with caution, is that one might have good reasons to expect a very specific phenotype. For the *rag1* gene, required for V(D)J joining in mice and humans, it was not unreasonable to expect that the mutant would fail to show V(D)J joining and be immunodeficient; when it had those phenotypes, a causal relation seemed likely. With more common phenotypes (such as lethality), this reasoning does not apply.

B. Null Alleles, Weak Alleles, or Silent Alleles

The first goal of most gene targeting work is to obtain a guaranteed null or loss-of-function allele. We prefer to focus on stop mutations, or (less preferred) splice site mutations. With stops the only concern is that an alternative splice removes the part of the exon that contains the stop. Therefore, an extra safeguard might be to focus initial mutant searches on exon domains encoding a conserved portion of a protein, so that any alternative splice that by-passes the mutation will not encode a functional protein. In some cases, knowledge of protein structure might help predict that a missense mutation can reasonably be expected to be a null or strong reduction of function. Again, as mentioned previously; having two different stop mutations with the same phenotype is a strong argument that both are null.

In practice, approximately 1 in 20 mutations in coding regions is a stop; the missense mutations can be highly valuable as well. To enhance possible weak phenotypic effects of such mutations, one can cross them to a null allele and analyze the phenotype of the transheterozygote (it will have half the gene function of the homozygous missense mutant). In theory, one might be able to sort a series of missense mutants into an allelic series of weak to strong mutants. If the null allele is lethal or sterile, and therefore in some cases of limited use for some experimental studies, weaker alleles might be quite useful.

V. Materials and Methods

A. Zebrafish Mutagenesis and Libraries

Zebrafish are raised under standard conditions. Adult wild-type (TL) male zebrafish (4 months old) are mutagenized by six consecutive treatments with 3.0 mM ENU as described (van Eeden *et al.*, 1999). For the specific locus test, the surviving and fertile fish are outcrossed with *albino* (*alb*) or *sparse* (*spa*) females. To generate F1 progeny for the library, the mutagenized males are outcrossed against wild-type (TL) females. To prevent clonal mutants, it is recommended to raise a maximum of a few hundred progeny per mutagenized F1 male. For a living library, all healthy looking adult fish are fin-clipped and stored in pools of 12. This can be more if the aquarium facility is limiting. To construct a permanently frozen library, the sperm of the mutagenized males is cryopreserved according to the protocol described by Morris *et al.* (2003).

B. Genomic DNA Isolation

For a living library, genomic DNA is isolated from fin-clips in 96-well deep-well plates (1.0 ml capacity per well). Freshly cut fin-clips are directly transferred to plates kept on dry ice. Fins are lysed by incubation in 400 μl prewarmed lysis buffer (100 mM Tris-HCl pH 8–8.5, 200 mM NaCl, 0.2% SDS, 5 mM EDTA, and 100 μg/ml proteinase K) at 55 °C for at least 3 h with occasional vortexing. DNA is precipitated by adding 300 μl isopropanol and centrifugation at $>6000 \times g$, followed by washing with 70% ethanol. Finally, pellets are dissolved in 500 μl TE. For screening, the DNA is diluted 10 times and 5-μl aliquots are arrayed in 384-well PCR plates using a 96-channel pipette (HYDRA-96, Robbins Scientific). PCR plates are covered with aluminum foil tape (3 M) and stored at -20 °C. DNA from a permanently frozen library can be isolated similarly from complete tails (approximately 1.0 cm), except that tails are lysed in 1.0 ml lysis buffer in 2.0 ml 96-well deep-well plates and precipitated with 700 μl isopropanol. DNA is dissolved in 200 μl TE and diluted 50-fold (in water) prior to screening.

C. Screening by Resequencing

Gene-specific primers are designed to amplify most of the coding regions of target genes by a nested PCR approach. In the first PCR, target genes are amplified in 384-well PCR plates with a standard PCR program (94 °C for 60 sec; 30 cycles of

92 °C for 20 sec, 58 °C for 30, and 72 °C for 60 sec, and an additional extension step of 72 °C for 180 sec; GeneAmp9700, Applied Biosystems). PCR samples contain 5 μl genomic DNA, 0.2 μM forward (f1) and 0.2 μM reverse (r1) primer, 200 μM of each dNTP, 20 mM Tris-HCl (pH 8.4), 50 mM KCl, 1.5 mM MgCl$_2$, and 0.2 U Taq DNA polymerase in a total volume of 10 μl. For the nested PCR, a small volume is transferred by 384-well replicators to a new 5 μl PCR mixture with internal gene-specific primers (f2 and r2). The nested PCR is empirically optimized to have maximal yield and have minimal residual primers, dNTPs, and Taq DNA polymerase left after cycling. (Typically this is 0.1 μM of each primer, 50 μM dNTPs, and 0.1 U Taq DNA polymerse.) The cycling program is identical to that of the first PCR. Next, PCR fragments are diluted 10-fold with water and from this 1 μl is used as template in a sequencing reaction. Alternatively, PCR fragments can first be purified with DNA-binding filterplates (Itoh et al., 1997) according to manufacturer instructions (Whatman). Sequencing reactions contain 0.25–0.35 μl BigDye Terminator (Applied Biosystems), 3.75–3.65 μl dilution buffer (200 mM Tris-HCl, pH 9.0, and 10 mM MgCl$_2$), 5% DMSO, and 0.5 μM of one of the nested primers (f2 or r2) in a total volume of 10 μl. Cycling conditions are as recommended by the manufacturer. Sequencing products are purified using Sephadex G50 (superfine coarse) minicolumns and analyzed on a 96-capillary 3700 DNA analyzer (Applied Biosystems) for which the running time is adjusted to the fragment lengths. Mutations are found by parsing the trace files through Phred (Ewing et al., 1998), Phrap (Gordon et al., 2001), and Polyphred (Nickerson et al., 1997) and are visualized with Consed (Gordon et al., 1998).

D. Screening by TILLING

The CEL-I enzyme is isolated from celery according to Oleykowski et al. (1998) and Yang et al. (2000), with minor modifications (for a detailed protocol see http://www.niob.knaw.nl/researchpages/cuppen/cel1.html). ENU-induced mutations are screened by using CEL-I-mediated heteroduplex cleavage, as described for *Arabidopsis* (Colbert et al., 2001), but with several adaptations described next. All pipetting steps are done on a Genesis Workstation 200 (Tecan) and Microlab 2200 (Hamilton) or using multichannel pipets. Target genes are amplified by a nested PCR approach in 384-well plates. In the first PCR with gene-specific primers, a touchdown cycling program is used (94 °C for 60 sec; 30 cycles of 94 °C for 20 sec, 65 °C for 30 sec with a decrement of 0.5 °C per cycle, and 72 °C for 60 sec; followed by 10 cycles of 94 °C for 20 sec, 58 °C for 30 sec and 72 °C for 60 sec and an additional extension step of 72 °C for 180 sec; GeneAmp9700, Applied Biosystems). PCR samples contain 5 μl genomic DNA, 0.2 μM forward (f1) and 0.2 μM reverse (r1) primer, 200 μM of each dNTP, 25 mM Tricine, 7.0% glycerol (m/v), 1.6% DMSO (m/v), 2 mM MgCl$_2$, 85 mM ammonium acetate pH 8.7 and 0.2 U Taq DNA polymerase in a total volume of 10 μl.

After the first PCR reactions, the samples are diluted with 20 μl water and 1 μl is used as template for the second nested PCR reaction. This reaction contains a mixture of gene-specific forward (M13F-f2, 0.08 μM) and reverse (M13R-r2, 0.04 μM) primers that contain universal M13 adaptor sequences at their 5' end, and the two corresponding universal M13F (5'-TGTAAAACGACGGCCAGT, 0.12 μM) and M13R (5'AGGAAACAGCTATGACCAT, 0.16 μM) primers labeled with fluorescent dyes (IR Dye 700 and IR Dye 800, respectively) for detection. In addition, the PCR samples contain 200 μM of each dNTP, 20 mM Tris-HCl, pH 8.4, 50 mM KCl, 1.5 mM MgCl$_2$, and 0.1 U Taq DNA polymerase in a total volume of 5 μl. Standard cycling conditions are used for the nested PCR reactions (30 cycles of 94 °C for 20 sec, 58 °C for 30 sec, and 72 °C for 60 sec, followed by an additional extension step of 72 °C for 180 sec).

Directly following the nested PCR, heteroduplex formation is done by incubation at 99 °C for 10 min and 70 cycles of 70 °C for 20 sec with a decrement of 0.3 °C per cycle. Next, 1.25 μl aliquots of four individual PCR reactions are pooled (total volume 5 μl) and incubated with 0.01 μl CEL-I enzyme solution in a total volume of 15 μl (buffered in 10 mM Hepes, pH 7.0, 10 mM MgSO$_4$, 10 mM KCl, 0.002% Triton X-100, 0.2 μg/ml BSA) at 45 °C for 15 min. CEL-I reactions are stopped by adding 5 μl of 75 mM EDTA. Fragments are purified by using Sephadex G50 (medium coarse) minicolumns in 96-wells filter plates (Multiscreen HV, Millipore) and eluted into plates prefilled with 5 μl formamide loading buffer [37% (v/v) deionized formamide, 4 mM EDTA, pH 8.0, 90 μg/ml bromophenol blue] per well or purified by isopropanol precipitation. Samples are concentrated to about 1 μl by heating at 85 °C for 45–60 min without cover. A 0.4 μl sample is applied to a 96-lane membrane comb (The Gel Company) and loaded on 25 cm denaturing 6% polyacrylamide gels on LI-COR 4200 DNA analyzers. Raw TIFF images produced by the analyzers are manipulated by using Adobe Photoshop and potential mutations are detected and scored manually.

Acknowledgments

We thank Ewart de Bruijn for help with collecting ENU data from forward genetic screens, Elizabeth Greene for advice on expectations after ENU mutagenesis, our collaborators for sharing (unpublished) sequences, and Edwin Cuppen and Robin May for useful comments on this manuscript. This work is sponsored by a NWO Genomics grant.

References

Alonso, J. M., *et al.* (2003). Genome-wide insertional mutagenesis of *Arabidopsis thaliana*. *Science* **301**, 653–657.

Altschul, S. F., *et al.* (1990). Basic local alignment search tool. *J. Mol. Biol.* **215**, 403–410.

Amsterdam, A. (2003). Insertional mutagenesis in zebrafish. *Dev. Dyn.* **228**, 523–554.

Amsterdam, A., *et al.* (1999). A large-scale insertional mutagenesis screen in zebrafish. *Genes Dev.* **13**, 2713–2724.

Ballinger, D. G., and Benzer, S. (1989). Targeted gene mutations in *Drosophila*. *Proc. Natl. Acad. Sci. USA* **86**, 9402–9406.

Beier, D. R. (2000). Sequence-based analysis of mutagenized mice. *Mamm. Genome* **11**, 594–597.

Bentley, A., MacLennan, B., Calvo, J., and Dearolf, C. R. (2000). Targeted recovery of mutations in *Drosophila. Genetics* **156,** 1169–1173.

Brand, M., Granato, M., and Nusslein-Volhard, C. (2002). Keeping and raising zebrafish. *In* "Zebrafish" (C. Nusslein-Volhard and D. Ralf, eds.), Vol. 261, pp. 7–37. Oxford University Press, Oxford.

Capecchi, M. R. (1989). Altering the genome by homologous recombination. *Science* **244,** 1288–1292.

Chasman, D., and Adams, R. M. (2001). Predicting the functional consequences of non-synonymous single nucleotide polymorphisms: Structure-based assessment of amino acid variation. *J. Mol. Biol.* **307,** 683–706.

Chen, Y., *et al.* (2000). Genotype-based screen for ENU-induced mutations in mouse embryonic stem cells. *Nat. Genet.* **24,** 314–317.

Coghill, E. L., *et al.* (2002). A gene-driven approach to the identification of ENU mutants in the mouse. *Nat. Genet.* **30,** 255–256.

Colbert, T., *et al.* (2001). High-throughput screening for induced point mutations. *Plant Physiol.* **126,** 480–484.

Cui, Z., *et al.* (2003). RecA-mediated, targeted mutagenesis in zebrafish. *Mar. Biotechnol. (NY)* **5,** 174–184.

Dai, Y., *et al.* (2002). Targeted disruption of the alpha 1,3-galactosyltransferase gene in cloned pigs. *Nat. Biotechnol.* **20,** 251–255.

Das, L., and Martienssen, R. (1995). Site-selected transposon mutagenesis at the hcf106 locus in maize. *Plant Cell* **7,** 287–294.

Davidson, A. E., *et al.* (2003). Efficient gene delivery and gene expression in zebrafish using the Sleeping Beauty transposon. *Dev. Biol.* **263,** 191–202.

De Stasio, E., *et al.* (1997). Characterization of revertants of unc-93(e1500) in *Caenorhabditis elegans* induced by *N*-ethyl-*N*-nitrosourea. *Genetics* **147,** 597–608.

de Wind, N., *et al.* (1995). Inactivation of the mouse Msh2 gene results in mismatch repair deficiency, methylation tolerance, hyperrecombination, and predisposition to cancer. *Cell* **82,** 321–330.

Douglas, G. R., *et al.* (1995). Temporal and molecular characteristics of mutations induced by ethylnitrosourea in germ cells isolated from seminiferous tubules and in spermatozoa of lacZ transgenic mice. *Proc. Natl. Acad. Sci. USA* **92,** 7485–7489.

Driever, W., *et al.* (1996). A genetic screen for mutations affecting embryogenesis in zebrafish. *Development* **123,** 37–46.

Elbashir, S. M., *et al.* (2001). Duplexes of 21-nucleotide RNAs mediate RNA interference in cultured mammalian cells. *Nature* **411,** 494–498.

Ewing, B., and Green, P. (1998). Base-calling of automated sequencer traces using phred. II. Error probabilities. *Genome Res.* **8,** 186–194.

Golling, G., *et al.* (2002). Insertional mutagenesis in zebrafish rapidly identifies genes essential for early vertebrate development. *Nat. Genet.* **31,** 135–140.

Gordon, D., Abajian, C., and Green, P. (1998). Consed: A graphical tool for sequence finishing. *Genome Res.* **8,** 195–202.

Gordon, D., Desmarais, C., and Green, P. (2001). Automated finishing with autofinish. *Genome Res.* **11,** 614–625.

Haffter, P., *et al.* (1996). The identification of genes with unique and essential functions in the development of the zebrafish, *Danio rerio. Development* **123,** 1–36.

Hanada, K., *et al.* (1997). RecQ DNA helicase is a suppressor of illegitimate recombination in *Escherichia coli. Proc. Natl. Acad. Sci. USA* **94,** 3860–3865.

Hurlstone, A. T., *et al.* (2003). The W$_{NT}$/beta-catenin pathway regulates cardiac value formation. *Nature* **425,** 633–637.

Itoh, M., *et al.* (2003). W$_{NT}$/beta-catenin pathway regulates cardiac value formation. *Nature* **425,** 633–637.

Jansen, G., Hazendonk, E., Thijssen, K. L., and Plasterk, R. H. (1997). Reverse genetics by chemical mutagenesis in *Caenorhabditis elegans. Nat. Genet.* **17,** 119–121.

Kaiser, K., and Goodwin, S. F. (1990). "Site-selected" transposon mutagenesis of *Drosophila*. *Proc. Natl. Acad. Sci. USA* **87**, 1686–1690.

Kimmel, C. B. (1989). Genetics and early development of zebrafish. *Trends Genet.* **5**, 283–288.

Koes, R., *et al.* (1995). Targeted gene inactivation in petunia by PCR-based selection of transposon insertion mutants. *Proc. Natl. Acad. Sci. USA* **92**, 8149–8153.

Korswagen, H. C., Durbin, R. M., Smits, M. T., and Plasterk, R. H. (1996). Transposon Tc1-derived, sequence-tagged sites in *Caenorhabditis elegans* as markers for gene mapping. *Proc. Natl. Acad. Sci. USA* **93**, 14680–14685.

Lander, E. S., *et al.* (2001). Initial sequencing and analysis of the human genome. *Nature* **409**, 860–921.

Lee, K. Y., *et al.* (2002). Cloned zebrafish by nuclear transfer from long-term-cultured cells. *Nat. Biotechnol.* **20**, 795–799.

Lekven, A. C., *et al.* (2000). Reverse genetics in zebrafish. *Physiol. Genom.* **2**, 37–48.

Liu, L. X., *et al.* (1999). High-throughput isolation of *Caenorhabditis elegans* deletion mutants. *Genome Res.* **9**, 859–867.

Ma, C., *et al.* (2001). Production of zebrafish germ-line chimeras from embryo cell cultures. *Proc. Natl. Acad. Sci. USA* **98**, 2461–2466.

Martin, E., *et al.* (2002). Identification of 1088 new transposon insertions of *Caenorhabditis elegans*: A pilot study toward large-scale screens. *Genetics* **162**, 521–524.

McCallum, C. M., Comai, L., Greene, E. A., and Henikoff, S. (2000). Targeted screening for induced mutations. *Nat. Biotechnol.* **18**, 455–457.

McCreath, K. J., *et al.* (2000). Production of gene-targeted sheep by nuclear transfer from cultured somatic cells. *Nature* **405**, 1066–1069.

Meissner, R. C., *et al.* (1999). Function search in a large transcription factor gene family in *Arabidopsis*: Assessing the potential of reverse genetics to identify insertional mutations in R2R3 MYB genes. *Plant Cell* **11**, 1827–1840.

Miao, Z. H., and Lam, E. (1995). Targeted disruption of the TGA3 locus in *Arabidopsis thaliana*. *Plant J.* **7**, 359–365.

Morris, J. P. T., *et al.* (2003). Zebrafish sperm cryopreservation with *N,N*-dimethylacetamide. *Biotechniques* **35**, 956–958, 960, 962 passim.

Mullins, M. C., Hammerschmidt, M., Haffter, P., and Nusslein-Volhard, C. (1994). Large-scale mutagenesis in the zebrafish: In search of genes controlling development in a vertebrate. *Curr. Biol.* **4**, 189–202.

Nakamura, Y., Gojobori, T., and Ikemura, T. (2000). Codon usage tabulated from international DNA sequence databases: Status for the year 2000. *Nucleic Acids Res.* **28**, 292.

Nasevicius, A., and Ekker, S. C. (2000). Effective targeted gene 'knockdown' in zebrafish. *Nat. Genet.* **26**, 216–220.

Ng, P. C., and Henikoff, S. (2001). Predicting deleterious amino acid substitutions. *Genome Res.* **11**, 863–874.

Nickerson, D. A., Tobe, V. O., and Taylor, S. L. (1997). Polyphred: Automating the detection and genotyping of single nucleotide substitutions using fluorescence-based resequencing. *Nucleic Acids Res.* **25**, 2745–2751.

Oleykowski, C. A., Bronson Mullins, C. R., Godwin, A. K., and Yeung, A. T. (1998). Mutation detection using a novel plant endonuclease. *Nucleic Acids Res.* **26**, 4597–4602.

Parinov, S., *et al.* (1999). Analysis of flanking sequences from dissociation insertion lines: A database for reverse genetics in *Arabidopsis*. *Plant Cell* **11**, 2263–2270.

Perry, J. A., *et al.* (2003). A TILLING reverse genetics tool and a web-accessible collection of mutants of the legume *Lotus japonicus*. *Plant Physiol.* **131**, 866–871.

Raz, E., *et al.* (1998). Transposition of the nematode *Caenorhabditis elegans* Tc3 element in the zebrafish *Danio rerio*. *Curr. Biol.* **8**, 82–88.

Rong, Y. S., and Golic, K. G. (2000). Gene targeting by homologous recombination in *Drosophila*. *Science* **288**, 2013–2018.

Rong, Y. S., and Golic, K. G. (2001). A targeted gene knockout in *Drosophila*. *Genetics* **157**, 1307–1312.

Sessions, A., *et al.* (2002). A high-throughput *Arabidopsis* reverse genetics system. *Plant Cell* **14**, 2985–2994.

Smits, B. M., Mudde, J., Plasterk, R. H., and Cuppen, E. (2004). Target-selected mutagenesis of the rat. *Genomics* **83**, 332–334.

Solnica-Krezel, L., Schier, A. F., and Driever, W. (1994). Efficient recovery of ENU-induced mutations from the zebrafish germline. *Genetics* **136**, 1401–1420.

Spradling, A. C., *et al.* (1999). The Berkeley Drosophila Genome Project gene disruption project: Single P-element insertions mutating 25% of vital *Drosophila* genes. *Genetics* **153**, 135–177.

Streisinger, G., *et al.* (1981). Production of clones of homozygous diploid zebra fish (*Brachydanio rerio*). *Nature* **291**, 293–296.

Sun, L., *et al.* (1995). ES-like cell cultures derived from early zebrafish embryos. *Mol. Mar. Biol. Biotechnol.* **4**, 193–199.

Sunyaev, S., *et al.* (2001). Prediction of deleterious human alleles. *Hum. Mol. Genet.* **10**, 591–597.

Thompson, S., *et al.* (1989). Germ line transmission and expression of a corrected HPRT gene produced by gene targeting in embryonic stem cells. *Cell* **56**, 313–321.

Till, B. J., *et al.* (2003). Large-scale discovery of induced point mutations with high-throughput TILLING. *Genome Res.* **13**, 524–530.

Tissier, A. F., *et al.* (1999). Multiple independent defective suppressor-mutator transposon insertions in *Arabidopsis*: A tool for functional genomics. *Plant Cell* **11**, 1841–1852.

van Eeden, F. J., Granato, M., Odenthal, J., and Haffter, P. (1999). Developmental mutant screens in the zebrafish. *Methods Cell Biol.* **60**, 21–41.

Venter, J. C., *et al.* (2001). The sequence of the human genome. *Science* **291**, 1304–1351.

Wienholds, E., *et al.* (2003a). The microRNA-producing enzyme Dicer1 is essential for zebrafish development. *Nat. Genet.* **35**, 217–218.

Wienholds, E., Schulte-Merker, S., Walderich, B., and Plasterk, R. H. (2002). Target-selected inactivation of the zebrafish rag1 gene. *Science* **297**, 99–102.

Wienholds, E., *et al.* (2003b). Efficient target-selected mutagenesis in zebrafish. *Genome Res.* **13**, 2700–2707.

Yanagihara, K., and Mizuuchi, K. (2002). Mismatch-targeted transposition of Mu: A new strategy to map genetic polymorphism. *Proc. Natl. Acad. Sci. USA* **99**, 11317–11321.

Yang, B., *et al.* (2000). Purification, cloning, and characterization of the CEL I nuclease. *Biochemistry* **39**, 3533–3541.

Zambrowicz, B. P., *et al.* (1998). Disruption and sequence identification of 2,000 genes in mouse embryonic stem cells. *Nature* **392**, 608–611.

Zan, Y., *et al.* (2003). Production of knockout rats using ENU mutagenesis and a yeast-based screening assay. *Nat. Biotechnol.* **21**, 645–651.

Zwaal, R. R., *et al.* (1993). Target-selected gene inactivation in *Caenorhabditis elegans* by using a frozen transposon insertion mutant bank. *Proc. Natl. Acad. Sci. USA* **90**, 7431–7435.

CHAPTER 5

A High-Throughput Method for Identifying *N*-Ethyl-*N*-Nitrosourea (ENU)-Induced Point Mutations in Zebrafish

Bruce W. Draper,⋆,† **Claire M. McCallum,**‡ **Jennifer L. Stout,**⋆,† **Ann J. Slade,**‡ **and Cecilia B. Moens**⋆,†

⋆Howard Hughes Medical Institute
Fred Hutchinson Cancer Research Institute
Seattle, Washington 98109

†Division of Basic Science
Fred Hutchinson Cancer Research Institute
Seattle, Washington 98109

‡Anawah, Inc.
Seattle, Washington 98104

I. Introduction

The zebrafish has become an important model system for vertebrate biology. Although forward genetic screens have uncovered the functions of many zebrafish genes, until recently no reliable, inexpensive, and high-throughput technology for targeted gene disruptions was developed. Here we outline an approach for identifying mutations in any gene of interest by TILLING (Targeting Induced Local Lesions in Genomes), a sensitive method for detecting single nucleotide polymorphisms (SNPs) in mutagenized genomes (Colbert *et al.*, 2001; McCallum *et al.*, 2000a,b). This method uses the CEL1 assay to detect mutations in DNA isolated from *N*-ethyl-*N*-nitrosourea (ENU)-mutagenized F_1 individuals, which are heterozygous for randomly induced mutations (Fig. 1A). The CEL1 endonuclease specifically cleaves DNA 3′ to any single base pair mismatches present in heteroduplexes between wild-type and mutant DNA (Oleykowski *et al.*, 1998). Genomic DNA isolated from mutagenized individuals is used as template for PCR amplification with gene-specific, fluorescently labeled (IRDye) primers. Because both wild-type and mutant alleles are amplified from heterozygous fish, heteroduplex PCR fragments are formed by denaturing and slowly reannealing the fragments. After CEL1 digestion, cleavage products are separated on a high-resolution polyacrylamide sequencing gel to reveal the presence and approximate location of induced mutations in the target sequence. This method allows the detection of rare ENU-induced mutations in the background of preexisting polymorphisms that are present even in inbred zebrafish strains.

To date, we have screened 25,303 kb from 5050 mutagenized genomes, using the approach described in this chapter, and have identified 48 new mutations (1 mutation per 527 kb screened). We anticipate, based on our data and data of others (Till *et al.*, 2003; Wienholds *et al.*, 2003) that about 5% of the mutations in coding DNA identified by TILLING will result in loss-of-function alleles. We

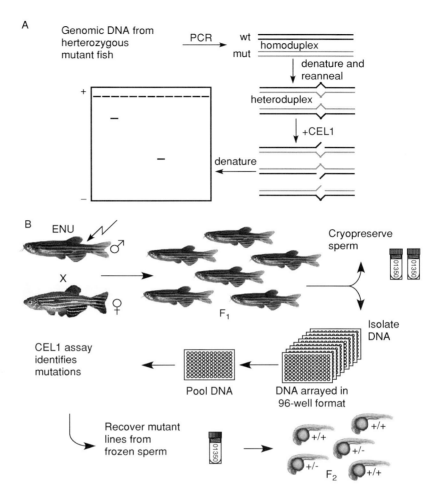

Fig. 1 The CEL1 assay and a pipeline for zebrafish TILLING. The CEL1 endonuclease cleaves DNA 3′ to single base-pair mismatches present in heterodupexed DNA formed between wild-type and mutant PCR fragments as outlined in (A). First, genomic DNA from fish heterozygous for induced mutations is used as template for PCR amplification with gene-specific primers. Heteroduplexes are then formed by denaturing and slowly reannealing the PCR fragments. After CEL1 treatment, the digestion products are denatured and run on a LI-COR polyacrylamide sequencing gel. Unique bands indicate the presence and approximate location of induced point mutations within the analyzed fragment. (B) Pipeline for TILLING in zebrafish. ENU-mutagenized males are mated with wild-type females to produce F_1 progeny heterozygous for induced mutations. Sperm from adult F_1 males are cryopreserved and DNA is prepared from euthanized sperm donors in a 96-well format for screening. DNA templates are then pooled four- to eightfold prior to screening. The CEL1 assay identifies ENU-induced mutations and sequence analysis determines whether these mutations are likely to be deleterious. Finally, lines of fish carrying interesting mutations are recovered from cryopreserved sperm by *in vitro* fertilization.

therefore project that with DNA and frozen sperm from 10,000 individuals it should be possible to generate an allelic series that consists of on average, 20 mutations, including at least one loss-of-function and several hypomorphic alleles, provided the gene has an open reading frame that is ≥ 1 kb in size. We estimate that a library prepared as outlined later can be screened at least 50,000 times before the initial DNA resource is depleted, at which time more DNA can be isolated from reserve tissue.

The screening approach is outlined in Fig. 1B. Briefly, sperm collected from ENU-mutagenized F_1 adult males is cryopreserved in liquid nitrogen, using an efficient and rapid sperm cryopreservation protocol that archieves two sperm samples per F_1 male. Importantly, our sperm cryopreservation protocol allows the recovery of an average of 109 ± 84 ($n = 46$) viable F_2 progeny (or $28\% \pm 18\%$ fertility) when one of the two samples is used for *in vitro* fertilization (see later). Genomic DNA is then purified from euthanized sperm donors, using a 96-well DNA isolation format. Next, mutants are identified with the CEL1 assay in a two-step screening process. For the primary screen, template DNA is pooled from 4 to 8 individuals, such that 384–768 mutagenized genomes are analyzed per 96-well plate of pooled DNA. PCR using IRDye-labeled primers and partial CEL1 digestion is performed in a 96-well format. Because the forward and reverse primers are labeled with unique fluorescent tags, it is possible to confirm mutations identified in one channel with the presence of the corresponding cleavage product in the second channel (Fig. 2). After identifying positive pools, mutant

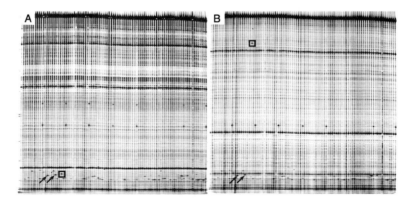

Fig. 2 Mutation detection by CEL1 in zebrafish. The CEL1 assay identifies an induced point mutations in a zebrafish gene. Typical images of a single LI-COR gel as visualized in the IRDye 700 (A) and IRDye 800 (B) channels. This gel represents a screen of 384 F_1 individuals (96 fourfold pools) for induced mutations in this fragment. Low-molecular-weight markers are included in every 12th lane to aid in determining lane numbers. A unique band in Lane 22 (square) indicates the presence of a ENU-induced polymorphism detected following CEL1 digestion. Note that both digestion products can be visualized by viewing separately the IRD700 and IRD800 gel images. Bands that appear in multiple lanes result from preexisting polymorphism in the population, whereas unique bands that appear in the same position on both gel images are PCR artifacts. (Arrows point to two examples.)

individuals are identified in a secondary CEL1 assay, and the nature of the mutation is determined by sequencing. Finally, fish lines carrying mutations that are likely to be deleterious are recovered by using the cognate cryopreserved sperm sample to fertilize eggs isolated from wild-type females (Fig. 1). The presence of the mutation in the F_2 generation is confirmed by PCR of tail-clipped DNA, using allele-specific primers or restriction fragment length polymorphism (RFLP) detection.

Our method differs in several ways from that of Wienholds *et al.* (2003):

1. F_1 males are preserved as frozen sperm rather than as a living library. This eliminates the need for maintaining several thousand live fish during the screening process and provides a long-term (i.e., many-year) resource for mutation detection.

2. Genomic DNA is normalized and pooled prior to PCR amplification rather than after PCR. This approach has allowed us to detect mutations efficiently in eightfold pooled samples rather than in fourfold pooled samples and thus decrease the expense of the mutation detection process (Draper, Moens, Till, Comai, and Henikoff, unpublished).

3. A single-step PCR approach that uses gene-specific labeled primers is used rather than a nested PCR approach that uses universal labeled primers. Although this is more expensive in terms of primer cost, it requires less liquid handling capacity and therefore might be better suited to the smaller laboratory.

An example of the data produced by the CEL1 assay is presented in Fig. 2. In this example, a 443-bp fragment was screened for induced mutations in fourfold pools of template DNA. Following PCR amplification with IRDye-labeled primers, heteroduplexes were formed, digested with the CEL1 endonuclease, and separated on a LI-COR sequencing gel. The gel images generated for the IRD700 and IRD800 primer channels are shown in Fig. 2A and 2B, respectively. Bands present in multiple lanes identify preexisting polymorphisms in the target sequence, whereas unique bands present in one channel and having the corresponding cleavage product in the second channel (boxed in Fig. 2) indicate an induced mutation, one of which is visible on this gel.

II. Rationale for Reverse Genetics in Zebrafish

Zebrafish forward genetic screens have been—and continue to be—exceptionally productive. However, as the content of the zebrafish genome becomes available in the form of primary sequence information, it becomes increasingly evident that many essential genes have not been identified by this approach. There are several possible explanations for why mutations in certain zebrafish genes have so far not been identified. First, mutant phenotypes might be subtle or even undetectable in forward genetic screens because of the nature of the screen. For example, most genetic screens performed to date have focused on identifying phenotypes during the embryonic period while the embryo is still transparent and have therefore been easy to screen for morphological defects in the light microscope or following staining with tissue-specific markers. In contrast, genes that function primarily in larvae and adults have remained largely inaccessible to genetic analysis. Furthermore, though existing screening tools might allow for an assessment of organ differentiation and shape, they rarely allow assessment of organ physiology or function.

Second, mutant phenotypes might also be subtle or undetectable because the functions of some genes are compensated for by gene duplicates or redundant pathways. Vertebrate genomes have undergone whole genome duplication events (Furlong and Holland, 2002), and it is common for multiple copies of individual ancestral genes to be present and have overlapping expression patterns. Thus, mutations in single members of a duplicate pair can often lead to subtle phenotypes, whereas loss of function of both family members causes more severe phenotypes, as has been shown in numerous cases in the zebrafish (e.g., Draper *et al.*, 2003; Feldman *et al.*, 1998; Waskiewicz *et al.*, 2002). The problem of redundancy is compounded in the teleost lineage, in which an additional genome duplication is thought to have taken place since its divergence from the tetrapod lineage (Amores *et al.*, 1998; Prince, 2002). Therefore, a reverse genetic approach that allows the identification of mutations in gene duplicates independent of phenotype will be beneficial, even necessary, to assess gene function in zebrafish in a comprehensive manner. By identifying mutations independent of phenotype, a reverse genetic approach will provide access to genes and biological processes that have thus far been beyond the reach of zebrafish genetics.

III. Rationale for Using the CEL1 Assay to Detect Induced Mutations

There are currently several technologies available that allow the identification of SNP, the most common class of mutations induced by ENU in zebrafish spermatogonia (Imai *et al.*, 2000). These include denaturing high-performance liquid chromatography (dHPLC), temperature gradient capillary electrophoresis (TGCE), and the gel-based CEL1 endonuclease assay. Among these different technologies, the gel-based CEL1 assay has several advantages that make it the ideal choice for reverse genetics in zebrafish. First, the CEL1 nuclease assay can detect induced SNPs in the context of a highly polymorphic genome such as that of the zebrafish. The CEL-1 TILLING methodology was originally developed for *Arabidopsis* (Colbert *et al.*, 2001), an organism that is highly inbred and therefore does not have a high degree of heterozygosity for preexisting SNPs. In contrast, commonly used zebrafish lines, such as AB and Tübingen, are much less inbred and thus have a high degree of heterozygosity for preexisting SNPs (Nechiporuk *et al.*, 1999). We have found an average of one to three preexisting SNPs per 500-bp fragment analyzed. However, as exemplified in Fig. 3, the protocol outlined

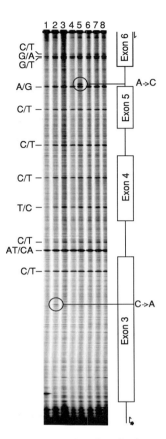

Fig. 3 CEL1 can detect ENU-induced mutations in a background of preexisting poymorphisms. Circles indicate ENU-induced mutations in this 1032-bp fragment that are unique to single F1 fish Preexisting polymorphisms appear as a band present in all lanes and were confirmed to be bona fide polymorphisms by sequencing this fragment from multiple F1 fish (sequence shown on left). Note that because the 3′ primer is labeled (i.e., at the bottom of the gel), the A → C change in Lane 5 is detected in spite of the presence of eight preexisting polymorphisms between it and the labeled primer.

here allows the identification of induced SNPs in fragments containing as many as 11 preexisting SNPs. Thus, in contrast to an induced SNP detection method that uses dHPLC or TGCE (McCallum and Slade, unpublished; McCallum *et al.*, 2000a,b), the presence of multiple preexisting SNPs appears to have little or no effect on the ability of the gel-based CEL1 assay to detect induced mutations.

A second advantage of the gel-based CEL1 assay is that it is high throughput. The CEL1 assay has been used to identify induced mutations in PCR fragments that range from 400 to 1500 bp in length and that are amplified from four- to eightfold pools of genomic DNA (our results; Draper, Moens Till, Comai, and

Henikoff, unpublished). Thus in a single 96-well assay, it is possible to screen over 1000 kb of sequence for induced mutations. With a mutation frequency of 1 in 500 kb, as we have observed in our library, we can identify one to two mutations per gel for a 1-kb fragment or two mutations per gel for a 1.5-kb fragment. Therefore, a relatively small laboratory with only two gel apparatuses can reasonably expect to identify 10–20 induced mutations per week. Because approximately 5% of mutations in coding DNA are expected to be deleterious (Till *et al.*, 2003; Wienholds *et al.*, 2003; our unpublished data), it is possible to identify useful mutations every 1–2 weeks of screening.

A final advantage of the CEL1 assay is that the primary screen not only identifies positive pools but also indicates the position of the induced point mutation relative to the fluorescently labeled primers. The resolution of the LI-COR gel is such that mutations can be localized to within 10 bp in the primary screen. Thus, for primers that amplify both exon and intron sequences, as exemplified in Fig. 3, only induced mutations that localize to coding DNA or sufficiently close to the canonical splice donor/acceptor sites need to be analyzed further.

IV. Rationale for Generating a Cryopreserved Mutant Library

Prior to screening for ENU-induced mutations in zebrafish, we chose to generate a library consisting of cryopreserved sperm isolated from the F_1 progeny of ENU-mutagenized males. Although it is possible to screen for mutations in fish that are kept alive, a cryopreserved library has several advantages. First, for some genes it might be necessary to screen as many as 10,000 mutagenized genomes to identify a useful mutation. For example, we anticipate that 5% of all induced mutations will be deleterious to gene function. Thus, with a mutation frequency of 1 in 500 kb, it will be necessary to screen 1 kb of coding DNA in 10,000 mutagenized genomes to identify a single deleterious mutation. For small- and medium-sized fish facilities, this could be a prohibitive number of live fish to maintain at any one time. Second, under even the best conditions, zebrafish are only fecund for 1.5–2 years of age. Thus, it is necessary to generate a new living library on a yearly basis. In contrast, a cryopreserved library can be large enough to ensure the identification of mutations in almost any gene and once made is stable indefinitely and can therefore be a resource for many years.

V. Method of *N*–Ethyl–*N*–Nitrosourea (ENU) Mutagenesis and Rearing of F_1 Founder Fish

A mutant library can efficiently be produced by randomly mutagenizing adult zebrafish spermatogonia with ENU following a standard protocol (Solnica-Krezel *et al.*, 1994). We have added the following modifications that ensure a consistent mutagenesis efficiency and maximize the production of F_1 offspring with the minimum amount of effort.

A. Determining ENU Concentration

After resuspension of approximately 1 g of ENU (Sigma) in 85 ml of ENU dilution buffer (0.03% Instant Ocean, 10 mM $NaPO_4$, pH 6.5), determine the concentration by measuring the OD_{398} of a 1:20 dilution. A 1 mg ENU/ml solution has an OD_{398} of 0.72 (Justice *et al.*, 2000).

B. Producing F_1 Mutagenized Founder Fish and Testing Specific Allele Frequency by *In Vitro* Fertilization

One month after the final mutagenesis, F_1 founder fish are most efficiently produced by *in vitro* fertilization, using sperm squeezed from mutagenized males (Westerfield, 1995). Resuspend squeezed sperm in 100 μl Hank's saline (see Section VI.B.2), and use 10 μl of this solution to fertilize eggs isolated from wild-type females (for more on *in vitro* fertilization, see Section IX-A or Westerfield, 1995). Using this strategy, 1500–2000 F_1 progeny per mutagenized male can routinely be produced in a single day. To prevent the isolation of multiple mutations that result from a single mutagenic event, it is important to keep track of all F_1 founders that are the progeny of a single ENU-mutagenized male. To minimize this possibility, we limit to 1000 the number of F_1 progeny a single mutagenized male can contribute to the mutant library.

An additional advantage of the *in vitro* fertilization strategy for generating F_1 founders is that a specific allele frequency can simultaneously be obtained by using the excess sperm. For example, we routinely use excess sperm to fertilize eggs squeezed from females homozygous for the *nacre* mutation. *nacre* is an ideal tester locus because it is homozygous viable and mutant embryos lack body pigmentation, a phenotype easily scored 2 days post-fertilization. In addition, the *nacre* gene is of average size (1.2 kb) and is thus a representative target (Lister *et al.*, 1999). We routinely identify new *nacre* alleles by noncomplementation at a frequency that ranges from 1 in 500 to 1 in 1200 haploid genomes screened.

VI. Generating a Cryopreserved Mutant Library

The sperm cryopreservation method we use to generate a cryopreserved mutant library is an adaptation of the Harvey method (Harvey *et al.*, 1982; Westerfield, 1995) that both streamlines the procedure and increases sample uniformity. First, after sperm is isolated from individual males, the volume is normalized by using freezing medium that does not contain cryoprotectant, prior to adding freezing medium containing cryoprotectant. Second, cryopreserved sperm are stored in screw cap cryovials instead of capillary tubes. With these simple modifications, teams of two people can collect and cryopreserve the sperm from 100 males in 2 h. Importantly, by this method, we routinely achieve an average of 28% $\pm$ 18% fertility, recovering 109 $\pm$ 84 ($n =$ 46) viable F_2 progeny following *in vitro* fertilization.

A. Equipment

1. 10-μl disposable pipettes (Fisherbrand Cat. No. 22–358697).
2. 250-ml beakers with fish water containing Tricaine.
3. Watch glasses (Pyrex Cat. No. 9985–75).
4. P20 pipettman (or equivalent) and tips.
5. Sponge with slit cut in top to hold male fish while squeezing.
6. Plastic spoon.
7. Dissecting microscope with above-stage lighting.
8. 2.0-ml screw cap cryogenic vials (Corning Cat. No. 430488).
9. 2.0-ml microcentrifuge tubes for freezing and storing male fish after sperm isolation.
10. Cryogenic freezer [e.g., Taylor-Warton (Theodore, AL) 10K].
11. 10 $\times$ 10 Cryoboxes for storing sperm in liquid nitrogen (Nalgene Cat. No. 03–337–7AA).
12. 15-ml conical tubes (Falcon No. 352099).
13. Large styrofoam cooler (8 in. $\times$ 12 in. internal dimensions) filled with at least 6-in.-deep finely pulverized dry ice for freezing sperm.
14. Ice bucket filled with dry ice for storing males.
15. Dry ice crusher (e.g., Clawson Ice Crusher, Model RE-2).
16. Large Dewar flask (e.g., Nalgene 101, No. 4150-9000) containing liquid nitrogen.
17. Long forceps (e.g., CMS Fisher Health Care Cat. No. 10-316B).
18. Cryogloves.

B. Reagents and Buffers

1. *Tricaine anesthetic*: (3-aminobenzoic acid ethyl ester) (Sigma No. A5040) 4.2 ml tricaine solution mixed with 100 ml fish water.
   a. *Tricaine solution*: 400 mg tricaine dissolved in 97.9 ml double-distilled (dd) H_2O. Adjust to pH 7.0 with approximately 2.1 ml 1 M Tris (pH 9).
2. *Hank's saline*: 0.137 M NaCl, 5.4 mM KCl, 0.25 mM Na_2HPO_4, 0.44 mM KH_2PO_4, 1.3 mM $CaCl_2$, 1.0 mM $MgSO_4$, 4.2 mM $NaHCO_3$.
   a. *Hank's final*: 9.9 ml Hank's premix, 0.1 ml Stock 6.
   b. *Hank's premix*: Combine the following solutions in order: 10 ml Stock1, 1.0 ml Stock 2, 1.0 ml Stock 4, 86.0 ml ddH_2O, 1.0 ml Stock 5.

c. *Hank's stock solutions:*

Stock 1	Stock 2	Stock 4	Stock 5	Stock 6
8.0 g NaCl, 0.4 g KCl in 100 ml ddH$_2$O	0.358 g Na$_2$HPO$_4$ anhydrous, 0.60 g K$_2$H$_2$PO$_4$ in 100 ml ddH$_2$O	0.7 g CaCl$_2$ in 50 ml ddH$_2$O	1.23 g MgSO$_4$ · 7H$_2$O in 50 ml ddH$_2$O	0.35 g NaHCO$_3$ in 10 ml ddH$_2$O

3. *Ginsburg's Fish Ringers*: For 500 ml final volume add in order: 450 ml ddH$_2$O, 3.25 g NaCl, 0.125 g KCl, 0.175 g CaCl$_2$ · H$_2$O, and 0.10 g NaHCO$_3$. Adjust final volume to 500 ml with ddH$_2$O and store at 4 °C. This solution should be made fresh every 3 days.

4. *Freezing medium without methanol*: For 10 ml final volume, add in order: 9 ml Ginsburg's Fish Ringers (room temperature) and 1.5 g powdered skim milk. Adjust final volume to 10 ml with Ginsburg's Fish Ringers.

5. *Freezing medium with methanol*: For 10 ml final volume, add in order 8 ml Ginsburg's Fish Ringers (room temperature), 1 ml methanol, and 1.5 g powdered skim milk. Adjust final volume to 10 ml with Ginsburg's Fish Ringers. After assembling freezing media, mix well for 20 min on orbital shaker or rocker and aliquot into 1-ml microcentrifuge tubes.

C. Method

1. *Mark capillary tubes*: Prior to beginning, use a lab pen to place a mark 16.5 mm from the bottom of a 10-μl capillary tube. This mark indicates the target sperm volume of 3.3 μl.

2. *Anesthetize male*: Place male(s) in a 250-ml beaker containing 100 ml of tricaine anesthetic.

3. *Dry fish*: Once anesthetized, remove male from beaker with plastic spoon and blot dry by gently rolling fish on paper towel, paying special attention to dry ventral side. Water activates sperm, therefore, it is important to dry thoroughly around the urogenital pore. It is important that no pressure be applied to the torso while drying as this can result in premature expulsion of milt.

4. *Position fish on sponge holder*: Position the male on a sponge holder, ventral side up, and place on the dissecting microscope stage.

5. *Collect sperm*: Expose the urogenital pore by carefully spreading apart the anal fins, using the end of the capillary tube. Expel sperm by gently squeezing the sides of the fish between your index finger and thumb, massaging in an anterior to posterior direction. Collect sperm in capillary tube as it is expelled, using gentle suction, avoiding feces that might be expelled with sperm.

6. *Freeze male on dry ice*: While the sperm donor is still anesthetized, place the donor in a labeled 2.0-ml microcentrifuge tube and freeze on dry ice. Store males at −80 °C until time to isolate genomic DNA (see DNA isolation method).

7. *Normalize sperm volume to 3.3 µl*: If the volume of sperm isolated reaches or exceeds the pen mark on the capillary tube (i.e., 3.3 µl or greater), then proceed to Step 8. If sperm volume is less than 3.3 µl, then normalize to 3.3 µl using room-temperature freeze medium WITHOUT methanol. The minimum amounts of sperm acceptable varies with the quality of sperm. Good sperm is white and opaque; poor sperm looks watery. In general, we accept as minimums 1 µl good sperm and 2 µl poor sperm.

8. *Add cryoprotectant to sperm*: Gently aspirate room-temperature freeze medium with methanol to the orange band on the capillary tube (total volume now 20 µl). Expel sperm and cryoprotectant mixture onto clean area of a watch glass, paying special attention not to introduce bubbles. Gently mix by pipetting.

9. *Aliquot 10 µl Sperm into each of two cryovials*: Pipette 10 µl of the sperm solution into the bottom of two separate cryovials labeled with relevant information. Cap vials and drop them into the bottom of room-temperature 15-ml Falcon tubes with one cryovial/tube. Cap Falcon tube.

10. *Freeze sperm for 20 min on dry ice*: Immediately insert the pair of cryovial-containing Falcon tubes into crushed dry ice. The tubes should be inserted into dry ice deep enough that only their caps show. To keep track of tubes in dry ice, number pairs of tubes from 1 to 20 (on caps) and record the time they go into the dry ice. (*Note: Speed is important*; Steps 6–9 should take no more than 30 sec.)

11. *Place cryovials into liquid nitrogen*: After 20 min, transfer cryovials to a liquid-nitrogen-containing Dewar flask. Store vials here until time to place into the liquid nitrogen freezer. When placing in the freezer boxes, place freezer box in a bath of liquid nitrogen to maintain temperature. Use long metal forceps to recover vials from the Dewar flask and handle with cryogloves. Store cryovials long term in a cryogenic liquid nitrogen freezer. To maintain the viability of sperm, it is important to store vials immersed in liquid nitrogen and not in the vapor phase.

VII. Isolating Genomic DNA

High-quality genomic DNA is an essential reagent for the TILLING methodology. To assure the highest quality, it is necessary to store fish tissue at −80 °C until time to isolate DNA. We use a 96-well-format DNA isolation system to minimize sample handling and increase throughput. Although there are several equivalent options for preparing high-quality DNA, our protocol uses the QIA-GEN 96-well-format DNeasy Tissue kit (Qiagen, Valencia, CA). However, with appropriate modifications, other kits can be substituted. Using this method, we routinely recover 20–40 µg of high-molecular-weight genomic DNA per sample, enough to screen over 50,000 times.

A. Equipment

1. High-speed tabletop centrifuge (e.g., Qiagen 4-15C) equipped with a 96-well-plate compatible rotor (e.g., Qiagen plate rotor 2×96).
2. Dog nail clippers (scissor style; available at most pet stores).
3. Multichannel pipettes (8 or 12 channels).
4. Three medium-sized styrofoam boxes filled with finely pulverized dry ice.
5. 96-Position microcentrifuge tube rack (e.g., Fisher Cat. No. 05-541-29).
6. Additional 96 × 1.4 ml plates [Micronic Systems (McMurray, PA) Cat. No. M42000].

B. Reagents and Kits

DNeasy 96-tissue kit (Qiagen Cat. No. 69582).

C. Method

1. Prior to tissue isolation, array the fish-containing microcentrifuge tubes into the appropriate positions of a 96-position microcentrifuge tube rack. To prevent tissue from prematurely thawing, the rack should be partially buried in crushed dry ice.

2. Remove fish from tube, and, while still frozen, use dog nail clippers to clip off the head just behind the gills. Transfer the head to the appropriate DNA preparation tube (e.g., a 2-ml, 96-well-format tube included with the DNA isolation kit). These tubes should also be partially buried in crushed dry ice. Place the remaining fish carcass back into its original microcentrifuge tube and place on dry ice. This remaining tissue is stored long term at $-80\,^{\circ}$C, and, if necessary, can be used for isolating additional DNA.

3. Isolate the DNA from the fish tissue following the kit manufacturer's recommendation, with the following modification: because a large amount of insoluble material is present following tissue lysis (e.g., bone, cartilage, and scales), it is necessary to clear this debris from the lysis by centrifugation and to transfer the cleared lysate to a new set of 2-ml, 96-well-format tubes. Following the manufacturer's recommendation, RNase-treat the DNA samples at this stage prior to proceeding with DNA isolation.

4. Once the DNA has been isolated, it is necessary to normalize its concentrations. The concentration of the DNA samples can be determined by number of methods. For example, DNA concentrations can be rapidly determined in a 96-well format by using the PicoGreen (MolecularProbes, Eugene, OR) fluorescent assay (Singer *et al.*, 1997) and a flourescence microplate reader. Alternatively, DNA concentrations can be estimated by comparing the intensity of genomic

DNA bands to lambda DNA standards following brief (20 min at 20 V/cm) electrophoresis through a 1% agarose gel (Till, Comai, and Henikoff, personal communication). Once the individual concentrations have been determined, all DNA samples are normalized to a standard concentration (e.g., 40 ng/μl) by adding the appropriate amount of TE buffer (10 mM Tris, pH 8.0, 1 mM EDTA). Prior to CEL 1 analysis, DNA samples are pooled either four or eightfold and diluted to a final concentration of 0.8ng total DNA/μl in TE. To detect all possible mutations in pooled samples, it is essential that the individual DNA concentrations be accurately normalized prior to pooling.

VIII. Choosing Fragments to Screen

The ability to identify useful mutations in any gene of interest depends on three factors: mutation frequency, target size, and size of library. PCR fragments between 0.4 and 1.5 kb can be screened by using the CEL1 assay (our results; Draper, Moens, Till, Comai, and Henikoff, unpublished), and in some cases it might be possible to screen 1.0 kb of protein-coding sequence in a single assay. If the frequency of mutations in the library is reasonably high (e.g., one induced mutation/500 kb screened), then it should be possible to identify 15–20 mutations in a 1.0-kb fragment by screening 7500–10,000 mutagenized genomes. However, most genes in zebrafish do not consist of large exons, but rather are composed of many small exons separated by large introns. Thus, for the majority of genes, it will be necessary to screen multiple smaller amplicons to assay a sufficient amount of protein-coding DNA to assure the identification of useful mutations.

For genes that have several possible fragments that can be screened, gene structure/function models can be used to determine the optimal gene fragment(s) by considering the following three criteria: (1) the likelihood that a mutation within a particular gene fragment would disrupt a conserved functional domain, (2) the statistical likelihood that an ENU-induced mutation within a particular gene fragment would create a premature stop codon, and (3) exon size. To simplify this analysis, we employ the CODDLE Web-based analysis program (http://www.proweb.org/coddle/; McCallum *et al.*, 2000b). CODDLE builds gene structure/function models and identifies exons that have the highest proportion of codons that could mutate to either a nonconservative amino acid substitution or a premature stop codon, given the spectrum of mutations induced by ENU in zebrafish.

IX. CEL1 Endonuclease Assay

Screening for ENU-induced point mutations with the CEL 1 assay consists of two basic steps: mutation detection by using pooled DNA templates and mutation confirmation by using individual DNA templates, as outlined next. It is possible to detect unique SNP in 12-fold pools of individual template DNA, but the signal from four- to eightfold pools is more robust.

A. Equipment

1. Thermal cycler with a 96-well block.
2. LI-COR Global IR^2 gel scanner (LI-COR, Lincoln, NE).
3. Computer with Internet access.
4. Table-top centrifuge with 96-well-plate compatible rotor.
5. 96-Well PCR plates (e.g., ABgene, Rochester NY).
6. Multichannel pipettors (8 or 12 channels).
7. 100-Tooth membrane combs (The Gel Company, San Francisco, CA).
8. Loading tray for membrane combs (The Gel Company, San Francisco, CA), or comb loading robot (MWG, Biotech).
9. IRDye700 50-700 sizing standard (LI-COR Cat. Vo. 4200-60)
10. Image analysis software [e.g., Photoshop (Adobe, San Jose, CA) or equivalent].
11. Apricot 96-channel pipettor (Perkin Elmer, Boston, MA; optinal).
12. 96-well heat blocks (optional).

B. Reagents, Buffers, and Kits

1. *Ex Taq polymerase (Takara Mirus Bio, Kyoto, Japan)*.

2. *Primer mix*: 45 μM IRD700-labeled forward primer, 5 μM unlabeled forward primer, 45 μM IRD800-labled reverse primer, 5 μM unlabeled reverse primer. IRD-labeled primers are light sensitive, and therefore care should be taken to minimize exposure to light during the assay.

3. *CEL 1 digestion solution*: 10 mM HEPES (pH 7.5), 10 mM MgSO$_4$, 0.002% (w/v) Triton X-100, 20 ng/ml bovine serum albumin, and 1:300–1:10,000 dilution of CEL 1. Because the activity of CEL 1 varies from batch to batch, the exact concentration necessary to achieve optimum digestion must be determined empirically.

4. *CEL 1*: This enzyme is isolated from celery (Oleykowski *et al.*, 1998) and can be obtained commercially from Transgenomic Ltd. (Omaha, NE) under the trade name Surveyor.

5. *100-bp IRD-labeled lane marker*: To make it easier to determine lane numbers following electrophoresis, we add a 100-bp IRD-labeled marker in every 12th lane. This marker can easily be made by PCR, using dye-labeled primers designed to amplify any 100-bp product. The exact amount to add per lane must be determined empirically.

6. 6.5% KB^{Plus} polyacrylamide (LI-COR, Lincoln, NE).

7. *Formamide load solution*: 1 mM EDTA (pH 8) and 200 μg/ml bromophenol blue in 33% deionized formamide.

8. *1% Ficoll solution*: 1% Ficoll (w/v) in dH$_2$O.

9. *Stop solution*: 75 mM EDTA (pH 8), 2.5 M NaCl.

C. Method

All PCR reactions are performed in 96-well plates with a final reaction volume of 15 μl. Because the same template DNAs are used repeatedly, it is convenient to aliquot the appropriate amount of pooled template DNA into many 96-well plates, seal, and store at -20 °C until needed. Primers are designed with the aid of Primer3 (http://www.broad.mit.edu/cgi-bin/primer/primer3.cgi/primer3_www.cgi; Rozen and Skaletsky, 2000) and to have a melting temperature between 65 and 72 °C and a length between 25 and 30 nucleotides. IRD700 and IRD800 dye-labeled primers can be obtained from MWG Biotech (Ebersberg, Germany). Resuspend all primers to 100 μM in TE and store at -20 °C. Because IRD-labeled primers are photosensitive, care should be taken to minimize light exposure when handling primer-containing solutions.

1. *Standard PCR reaction*: Aliquot 10 μl of the PCR master mix into each well of a 96-well plate that contains 5 μl of pooled template DNA (4 ng total DNA/well). For the PCR master mix, mix the following on ice:

124 μl	10 × PCR buffer
73 μl	25 mM MgCl$_2$
198 μl	2.5 mM (each) dNTP mix
6.6 μl	Primer mix
11 μl	Ex Taq DNA polymerase
687 μl	ddH$_2$O
Total Volume 1100 μl	

2. *Standard PCR cycling profile*: Our PCR cycling is carried out in an MJ Research (Waltham, MA) DNA Engine thermal cycler, using the following parameters:

Step	Temp.	Time
1.	95 °C	2 min
2.	94 °C	20 sec
3.	$T_m + 3$ °C	30 sec (decrease 1 °C/cycle to $T_m - 4$ °C) Annealing
4.	Ramp 0.5 °C/sec to 72 °C	
5.	72 °C	1 min Extension
6.	Repeat Steps	2–4 seven more times (eight cycles total)

7.	94 °C	20 sec
8.	$T_m - 5$ °C	30 sec Annealing
9.	Ramp 0.5 °C/sec to 72 °C	
10.	72 °C	1 min (for 600–1000 base amplicon) Extension
11.	Repeat steps	6–8 29–39 more times (30–40 cycles total)
12.	72 °C	5 min
13.	98 °C	8 min Denaturation
14.	80 °C	20 sec
15.	80 °C	7 sec (decrease 0.3 °C/cycle) Reannealing
16.	Repeat Step	15 69 more times (70 cycles total)
17.	8 °C	Hold

3. *CEL1 digestions*

   a. Place PCR plate on ice and add 30 μl CEL 1 digestion solution.

   b. Incubate plate at 45 °C for 30 min in PCR machine.

   c. Stop digestion by adding 10 μl stop solution.

   d. Add appropriate amount of 100-bp lane marker to wells that will be loaded every 12th lane on the gel (e.g., add to Column 12 of the 96-well plate).

   e. Precipitate digested DNA by adding 80 μl isopropanol per well, seal plate with plastic cover, and incubate at room temperature overnight, protected from light.

   f. Pellet DNA by centrifugation at 3220 RCF for 30 min in a table-top centrifuge.

   g. Resuspend pellet in 8 μl formamide load solution and incubate in a thermal cycler at 80 °C for 7 min, followed by 95 °C for 2 min.

4. *Running LI-COR gel*

   a. Membrane combs are loaded as follows: approximately 0.3 μl of each sample is spotted onto the middle 96 teeth of a 100-tooth membrane comb with an automated comb-loading robot (MWG, Biotech). Alternatively, combs can be loaded manually with the aid of a comb-loading tray (The Gel Company, San Francisco, CA) and a mutichannel pipettor that allows the tips to be variably spaced (e.g., Matrix Equalizer, Matrix Technologies, Hudson, NH).

   b. After the samples have been loaded onto the comb, spot 0.3 μl of the IRDye 700 50-700 sizing standard (LI-COR) onto the empty combs that flank the samples.

    c. Prepare a polyacrylamide gel using LI-COR 6.5% KB^{Plus} poly-acrylamide in 1 × TBE, following manufacturer's recommendations (LI-COR).

    d. Once the gel is positioned in the LI-COR apparatus, fill both upper and lower buffer chambers with 0.8 × TBE and perform a prerun focusing step for 20 min at 1500-V, 40-W, and 40-mA limits at 50 °C.

    e. After the prerun, but prior to inserting the comb, remove as much buffer in the upper buffer chamber as necessary to expose the comb well. Then, with the aid of a syringe and strips of Whatman paper, remove as much buffer from the comb well as possible. Next, use a syringe to refill the comb well with a 1% Ficoll in dH$_2$O solution, and carefully insert the comb until the teeth just touch the surface of the gel. Comb insertion is most easily accomplished by holding the comb at a 45° vertical angle to the gel. After the comb has been inserted, slowly and carefully refill the upper buffer chamber with 0.8 × TBE.

    f. Run the gel for 3–4 h at 1500-V, 40-W, and 40-mA limits at 50 °C. Gel images are stored on the LI-COR machine as TIFF files and can be retrieved and viewed by using any computer with Internet access and appropriate software (e.g., Photoshop) following manufacturer's instructions.

  5. *Secondary screen identifies individual mutant*: After the primary CEL1 assay has identified DNA pools that contain mutations, rescreen the individual DNA samples that are contained within each positive pool, using the same parameters outlined previously. Finally, determine the nature of the mutation by sequencing the PCR fragment amplified from the positive individual, using only unlabeled primers.

D. Analysis of Mutations

Mutations that introduce premature stop codons 5′ to the conserved protein-coding domains are predicted to be null alleles because they cause protein truncations. However, it is less straightforward to predict the effect of missense mutations on protein function. To aid in assessing the likelihood that a particular missense mutation could have deleterious effects on gene function, we use the PARSESNP (Project Aligned Related Sequences and Evaluate SNPs) Web-based program (Taylor and Greene, 2003; http://www.proweb.org/parsesnp/). PARSESNP builds intron/exon gene models of target fragments by comparing genomic and coding sequence information and then automatically identifies conserved domains. Once the mutation information is entered, PARSESNP uses sequence homology in the conserved domains to predict whether an amino acid substitution at a particular protein position will be tolerated or deleterious to protein function by calculating a position-specific scoring matrix (PSSM) difference score for each altered residue. PSSM difference scores range from 0 to 30, and missense mutations that score above 10 are predicted to be deleterious to protein function.

X. Recovery of Mutations from Cryopreserved Sperm

Live fish lines heterozygous for interesting mutations are recovered from cryo-preserved sperm stocks by *in vitro* fertilization. Because this is a very important step, extreme care should be taken to fertilize only high-quality eggs that have a uniform, yellowish appearance.

A. Equipment

1. 33 °C water bath.
2. Dewar flask containing liquid nitrogen.

B. Reagents and Buffers

1. *Tricane anesthetic*: See Section VI-B-1.
2. *Hank's saline*: See Section VI-B-2.

C. *In Vitro* Fertilization with Cryopreserved Sperm

1. Isolate eggs from anesthetized females by gentle squeezing and collect in a 35-mm plastic culture dish (Westerfield, 1995). To maximize recovery of fertile eggs, it is possible to combine clutches of eggs isolated from multiple females prior to fertilization.

2. Thaw cryopreserved sperm by removing the cryovial cap and immersing the cryovial half way into a 33 °C water bath for 8–10 sec.

3. Add 70 μl of room-temperature Hank's solution to cryovial and gently mix with sperm.

4. Quickly add the resuspended sperm solution to the eggs and gently mix with the pipette tip.

5. Activate eggs and sperm by adding 750 μl of room-temperature fish water.

6. Following incubation at room temperature for 5 min, add an additional 5 ml of water to the dish and place in a 28 °C incubator.

7. After incubating for 4 h at 28 °C, sort fertile eggs into 100-mm culture dishes (70 embryos/dish) containing fish water and raise fry by using standard conditions.

D. Tracking Recovered Mutations

Once fish carrying interesting mutations have been recovered, it is useful to develop a genotyping assay that is both easy and reliable. An ideal method is to design a PCR-based assay to screen DNA isolated from caudal fin-clips. For cases in which the identified mutation either creates or eliminates a restriction enzyme cleavage site in the genomic DNA, PCR primers can be designed that amplify the polymorphic sequence and a simple restriction digest of the PCR fragment will reveal the genotype. However, for mutations that do not affect a restriction site, it is possible to design mismatched PCR primers that will create an allele-specific restriction site in the amplified fragment by a technique called dCAPS (derived cleavage amplified polymorphic sequence; Neff *et al.*, 1998). dCAPS primers can be designed for nearly any sequence, and primer design can be facilitated by using the Web-based program dCAPS Finder 2.0 (http://helix.wustl.edu/dcaps/dcaps.html; Neff *et al.*, 2002).

XI. Materials Cost Estimate

We have presented an efficient method for identifying ENU-induced mutations in specific genes of interest in zebrafish. We have observed a mutation detection frequency of approximately one mutation per 500 kb screened. Because only about 5% of ENU-induced mutations in protein-coding DNA are expected to be deleterious, it is necessary to screen, on average, 10,000 kb of coding target gene sequence to identify one or two deleterious mutations. Thus with DNA and frozen sperm from 10,000 F_1 individuals, it will be possible to generate an allelic series including loss-of-function mutations in any target ≥ 1 kb in size. The estimated cost of TILLING is summarized in Table I and depends on the target to be screened: targets consisting of a single, long, contiguous open reading frames can be screened as a single fragment and thus will be less expensive than targets

Table I
Estimated Cost ($) of Materials for TILLING

	One fragment	Two fragments	Three fragments	Four fragments
Primary screen	780	1560	2340	3120
Primers	260	520	780	1040
Secondary screen	120	120	120	120
Sequencing	140	140	140	140
Total	1300	2340	3380	4420

Note: Estimated cost of screening 10,000 mutagenized genomes for mutations in 1.0 kb of coding DNA. For some genes, 1.0 kb of coding DNA can be screened in a single fragment, whereas other genes require analysis of 2–4 fragments. These estimates assume that eight fold pools of DNA are analyzed in the primary screen.

that must be screened as multiple small fragments. We estimate that the per-plate materials cost of TILLING is $60.00, including enzymes, plasticware, ladder, and chemicals. IRDye-labeled primers cost approximately $130.00 each. Thus, the cost of screening a single 1-kb fragment in 10,000 eightfold-pooled individuals is $60.00 × 13 + $260.00 = $1040.00. The primary screen is expected to identify, on average, 20 mutations; therefore, a secondary screen of a further two 96-well plates will add $120.00 to the total cost. Finally, sequencing 20 mutations will add an additional $140.00. Thus, the total cost of TILLING a single 1-kb fragment in 10,000 individuals is approximately $1300.00; larger fragments will cost less because they can be screened in fewer individuals, while multiple smaller fragments will cost more (Table I). Thus, the CEL1 assay is an efficient and cost-effective method for reverse genetics in zebrafish.

Acknowledgments

We thank Brad Till and the University of Washington Seattle Tilling Project for sharing information on mutation detection frequencies, cost estimates of reagents, and for critical reading of this manuscript. C. B. M. is an assistant investigator with the Howard Hughes Medical Institute.

References

Amores, A., Force, A., Yan, Y. L., Joly, L., Amemiya, C., Fritz, A., Ho, R. K., Langeland, J., Prince, V., Wang, Y. L., Westerfield, M., Ekker, M., and Postlethwait, J. H. (1998). Zebrafish hox clusters and vertebrate genome evolution. *Science* **282**, 1711–1714.

Colbert, T., Till, B. J., Tompa, R., Reynolds, S., Steine, M. N., Yeung, A. T., McCallum, C. M., Comai, L., and Henikoff, S. (2001). High-throughput screening for induced point mutations. *Plant Physiol.* **126**, 480–484.

Draper, B. W., Stock, D. W., and Kimmel, C. B. (2003). Zebrafish fgf24 functions with fgf8 to promote posterior mesoderm development. *Development* **130**, 4639–4654.

Feldman, B., Gates, M. A., Egan, E. S., Dougan, S. T., Rennebeck, G., Sirotkin, H. I., Schier, A. F., and Talbot, W. S. (1998). Zebrafish organizer development and germ-layer formation require nodal-related signals. *Nature* **395**, 181–185.

Furlong, R. F., and Holland, P. W. (2002). Were vertebrates octoploid? *Philos. Trans. R. Soc. Lond. B. Biol. Sci.* **357**, 531–544.

Harvey, B., Kelley, R. N., and Ashwood-Smith, M. J. (1982). Cryopreservation of zebra fish spermatozoa using methanol. *Can. J. Zool.* **60**, 1867–1870.

Imai, Y., Feldman, B., Schier, A. F., and Talbot, W. S. (2000). Analysis of chromosomal rearrangements induced by postmeiotic mutagenesis with ethylnitrosourea in zebrafish. *Genetics* **155**, 261–272.

Justice, M. J., Carpenter, D. A., Favor, J., Neuhauser-Klaus, A., Hrabe de Angelis, M., Soewarto, D., Moser, A., Cordes, S., Miller, D., Chapman, V., Weber, J. S., Rinchik, E. M., Hunsicker, P. R., Russell, W. L., and Bode, V. C. (2000). Effects of ENU dosage on mouse strains. *Mamm. Genome.* **11**, 484–488.

Lister, J. A., Robertson, C. P., Lepage, T., Johnson, S. L., and Raible, D. W. (1999). Nacre enconds a zebrafish microphthalmia-related protein that regulates neural-crest-derived pigment cell fate. *Development* **126**, 3757–3767.

McCallum, C. M., Comai, L., Green, E. A., and Henikoff, S. (2000a). Targeted screening for induced mutations. *Nat. Biotech.* **18**, 455–457.

McCallum, C. M., Comai, L., and Henikoff, S. (2000b). Targeting induced local lesions IN genomes (TILLING) for plant functional genomics. *Plant Physiol.* **123,** 439–442.

Nechiporuk, A., Finney, J. E., Keating, M. T., and Johnson, S. L. (1999). Assessment of polymorphism in zebrafish mapping strains. *Genome Res.* **9,** 1231–1238.

Neff, M. M., Neff, J. D., Chory, J., and Pepper, A. E. (1998). dCAPS, a simple technique for the genetic analysis of single nucleotide polymorphisms: Experimental applications in *Arabidopsis thaliana* genetics. *Plant J.* **14,** 387–392.

Neff, M. M., Turk, E., and Kalishman, M. (2002). Web-based primer design for single nucleotide polymorphism analysis. *Trends Gen.* **18,** 613–615.

Oleykowski, C. A., Bronson Mullins, C. R., Godwin, A. K., and Yeung, A. T. (1998). Mutation detection using a novel plant endonuclease. *Nucleic Acids Res.* **26,** 4597–4602.

Prince, V. E. (2002). The hox paradox: more complex(es) than imagined. *Dev. Biol.* **249,** 1–15.

Rozen, S., and Skaletsky, H. J. (2000). Primer3 on the WWW for general users and for biologist programmers. *In* "Bioinformatics Methods and Protocols: Methods in Molecular Biology" (S. Krawetz and S. Misener, eds.), pp. 365–386.

Singer, V. L., Jones, L. J., Yue, S. T., and Haugland, R. P. (1997). Characterization of PicoGreen reagent and development of a fluorescence-based solution assay for double stranded DNA quantitation. *Anal. Biochem.* **249,** 228–238.

Solnica-Krezel, L., Schier, A. F., and Driever, W. (1994). Efficient recovery of ENU-induced mutations from the zebrafish germline. *Genetics* **136,** 1401–1420.

Taylor, N. E., and Greene, E. A. (2003). PARSESNP: A tool for the analysis of nucleotide polymorphisms. *Nucleic Acids Res.* **31,** 3808–3811.

Till, B. J., Reynolds, S. H., Greene, E. A., Codomo, C. A., Enns, L. C., Johnson, J. E., Burtner, C., Odden, A. R., Young, K., Taylor, N. E., Henikoff, J. G., Comai, L., and Henikoff, S. (2003). Large-scale discovery of induced point mutations with high-throughput TILLING. *Genome Res.* **13,** 524–530.

Waskiewicz, A. J., Rikhof, H. A., and Moens, C. B. (2002). Eliminating zebrafish pbx proteins reveals a hindbrain ground state. *Dev. Cell* **3,** 723–733.

Westerfield, M. (1995). "The Zebrafish Book." University of Oregon Press, Eugene, OR.

Wienholds, E., van Eeden, F., Kosters, M., Mudde, J., Plasterk, R. H., and Cuppen, E. (2003). Efficient target-selected mutagenesis in zebrafish. *Genome Res.* **13,** 2700–2707.

CHAPTER 6

Production of Zebrafish Germline Chimeras by Using Cultured Embryonic Stem (ES) Cells

Lianchun Fan, Jennifer Crodian, and Paul Collodi

Department of Animal Sciences
Purdue University
West Lafayette, Indiana 47907

I. Introduction

The development of an embryonic stem (ES)-cell-mediated approach to gene targeting in zebrafish will require that methods be available for the production of germline chimeras. To generate a knockout, ES cells carrying the targeted mutation are selected *in vitro* and introduced into recipient embryos to produce germline chimeras, which are then bred to establish the mutant line (Capecchi, 1989; Doetschman *et al.*, 1987). We have produced zebrafish germline chimeras, using ES cells maintained for up to 6 weeks (6 passages) in culture (Fan *et al.*, 2004a,b). The capacity of ES cells to contribute to the germ cell lineage of a host embryo is maintained when the ES cells are cultured on a feeder layer of growth-arrested rainbow trout spleen cells (RTS34st; Ganassin and Bols, 1999; Fan *et al.*, 2004c; Ma *et al.*, 2001). In addition to the germline, ES cells contribute to multiple tissues of the chimeric host (Fan *et al.*, 2004b).

By using the microinjection techniques described in this chapter, cultured ES cells can be introduced into a large number of zebrafish embryos (300 embryos/h), with a recipient survival rate of up to 50% one week after injection. These methods are based on the microinjection techniques first developed by Dr. Wilber Long (Lin *et al.*, 1992). Germline contribution of ES cells is confirmed by demonstrating that F1 individuals, produced by the sexually mature chimeras, possess a pigmentation pattern and marker gene donated by ES cells (Fig. 1). Although the frequency of germline chimera production is low (approximately 2–4%), the ability to conveniently inject a large number of embryos with ES cells makes it feasible to produce a sufficient number of germline chimeras to establish a transgenic or knockout line of fish.

To facilitate the identification of germline chimeras, a zebrafish ES cell line was established from transgenic embryos that constitutively express the green fluorescent protein (GFP; Fan *et al.*, 2004b; Higashijima *et al.*, 1997). One week following ES cell injection, the potential germline chimeras can be identified by the presence of GFP+ cells in the region of the developing gonad (Fig. 2; Fan *et al.*, 2004b). Approximately 40% of the injected embryos identified in this manner were later confirmed to be germline chimeras by transmission of GFP to the F1 generation.

II. Methods

A. Preparation of Recipient Embryos for Embryonic Stem (ES) Cell Injection

1. Collect zebrafish embryos and incubate them at 28 °C until they reach the midblastula stage of development (approximately 4 h; Westerfield, 1995). To facilitate germline chimera identification, embryos from the GASSI strain of zebrafish that lack melanocyte pigmentation on the body can be used as recipients (Gibbs and Schmale, 2000). When ES cells derived from wild-type embryos are injected into GASSI recipients, the germline chimeras are identified by the presence of F1 individuals that inherit a wild-type pigmentation pattern donated by ES cells (Fig. 1). Germline chimeras have also been generated by using wild-type host embryos injected with ES cells carrying a marker gene.

2. Before injection, dechorionate the embryos by incubating them in pronase solution in a 60-mm petri dish. Release the embryos from the partially digested chorions by gently swirling them in the dish. Remove the pronase solution with the suspended empty chorions and rinse the dechorionated embryos two times with egg water. Leave the embryos in the egg water until they are needed for ES cell injection.

3. Before performing the injections, transfer a group of approximately 50 embryos to a shallow depression made in agarose (1.5%) contained in a 60-mm petri dish filled with egg water. The depression will prevent the embryos from moving during the injection procedure.

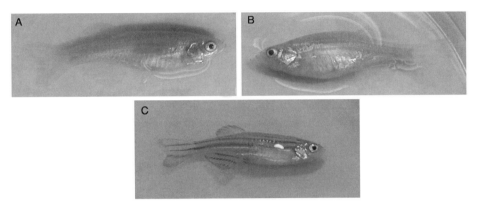

Fig. 1 A zebrafish germline chimera (A) was produced by injecting a passage 6 wild-type embryonic stem (ES) cell culture into a GASSI host embryo. Germline contribution of ES cells was confirmed by breeding the chimera with a GASSI mate (B) to produce the pigmented F1 individual (C).

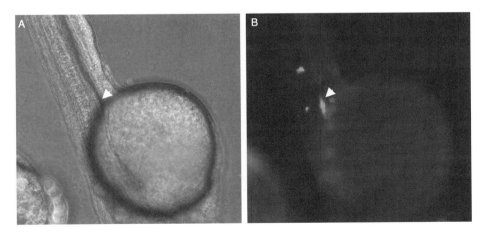

Fig. 2 Identification of germline chimeras is facilitated by the use of ES cells that express green fluorescent protein (GFP). One week after ES cell injection, the potential germline chimeras are identified by the presence of GFP-positive cells in the region of the gonad. The same embryo is shown under bright field (A) and UV (B) and the region containing the GFP positive cells is indicated (arrow).

B. Preparation of ES Cells for Injection into Host Embryos

To prepare for the injection, harvest a zebrafish ES cell culture (Fan *et al.*, 2004c) by trypsinization. A single confluent ES cell culture contained in a 35-mm dish will yield a sufficient number of cells to inject several hundred embryos. ES cells should be derived from wild-type embryos or embryos obtained from a transgenic line of fish that expresses a marker gene such as GFP (Higashijima *et al.*, 1997). To harvest the cells, add 1 ml of trypsin/EDTA solution to the dish and incubate approximately 30 sec or until ES cells begin to round up and detach from the feeder layer. Transfer the trypsin/EDTA solution containing the suspended ES cells to a 15-ml plastic centrifuge tube and add 100 μl of FBS to stop the action of the trypsin. Gently rinse the dish one time with LDF culture medium to remove the loosely attached ES cells and add to the centrifuge tube. Most of the feeder cells will be left behind in the dish. Collect the harvested ES cells by centrifugation (500 g, 5 min) and suspend the pellet in LDF medium at a density of 2–3 $\times$ 10^6 cells/ml.

C. Procedure for Injecting ES Cells into Host Embryos

1. Draw approximately 1–3 μl of the ES cell suspension into the tip of a needle formed from a drawn-out Pasteur pipette connected to a Pipet-Aid pipettor (VWR) with Tygon tubing (Fig. 3A). The Pasteur pipette is drawn out over a flame and the end is broken to produce a sharp opening with a width of approximately 20 μm.

2. Inject ES cells into host embryos that are at the blastula stage of development. Insert the needle into the center of the cell mass of the blastula so that the end of the needle reaches to a depth that is just above the yolk interface (Fig. 3B). Turn the wheel on the Pipet-Aid to release approximately 100 cells into the embryo and then gently remove the needle. Because ES cells tend to adhere to each other, it is necessary to occasionally clear the cell aggregates that block the end of the needle. This is done by turning the wheel on the Pipet-Aid until the blockage is cleared and then immediately releasing the pressure in the needle by using the valve to prevent loss of additional cells.

3. Allow the embryos to recover for 24 h before transferring them from the agarose to a petri dish (100 mm) containing egg water and four drops of methylene blue solution (0.01%). Incubate the embryos (28 °C) for 7 days, replacing approximately 50% of the egg water daily.

4. Transfer the embryos to a beaker (200 ml) containing egg water and incubate (28 °C) for 10 days before transferring them to a 2.5-gallon tank.

D. Identification of Germline Chimeras

If ES cells derived from wild-type embryos are injected into GASSI recipients, all the surviving recipient embryos are raised to sexual maturity and the fish are bred with noninjected GASSI mates. Germline chimeras are identified by screening the F1 generation for the presence of pigmented individuals (Fig. 1C). If a large

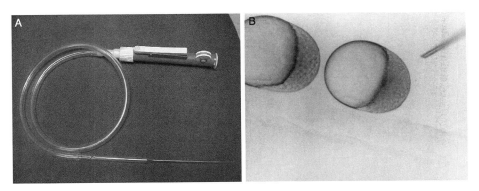

Fig. 3 (A) Injector used to introduce ES cells into host embryos. (B) The cells are injected into recipients embryos at the blastula stage of development.

number of fish are to be tested, the initial screen can be conducted by breeding groups of 25–30 injected fish and screening the F1 embryos visually for the presence of pigmented individuals or by PCR for the presence of a marker gene donated by ES cells. Once a group of fish containing a germline chimera is identified, each member of the group is bred with a single mate and the F1 embryos examined for the presence pigment or the marker gene to identify the chimera.

The use of ES cells expressing a marker gene such as GFP greatly facilitates the screening process by eliminating the need to raise all the injected embryos to sexual maturity. The potential germline chimeras are identified several days after ES cell injection by screening the larvae by fluorescence microscopy for the presence of GFP$^+$ cells in the region of the developing gonad (Fig. 2). Only the identified individuals are saved and raised to sexual maturity. Germline contribution of ES cells is confirmed by breeding the chimeras and examining F1 individuals for GFP expression.

III. Reagents

1. *Cell culture media*: Leibowitz's L-15 (Cat. No. 41300-039), Ham's F12 (Cat. No. 21700-075), and Dulbecco's modified Eagle's media (Cat. No. 12100-046) are available from GIBCO-BRL, Grand Island, NY. One liter of each medium is prepared separately by dissolving the powder in ddH$_2$O and adding HEPES buffer (final concentration 15 mM, pH 7.2), penicillin G (120 μg/ml), ampicillin (25 μg/ml), and streptomycin sulfate (200 μg/ml). LDF medium is prepared by combining Leibowitz's L-15, Dulbecco's modified Eagles, and Ham's F12 media (50:35:15) and supplementing with sodium bicarbonate (0.180 g/l) and sodium selenite (10^{-8} M). The medium is filter-sterilized before use.

2. *Trypsin/EDTA solution*: 2 mg/ml trypsin, 1 mM EDTA is prepared in PBS. The solution is filter-sterilized before use. Trypsin (Cat. No. T-7409) and EDTA (Cat. No. E-6511) are available from Sigma, St. Louis, MO.

3. Pronase (Cat. No. P6911) is available from Sigma and is prepared at 0.5 mg/ml in Hanks solution.

4. *Egg water*: 60 μg/ml aquarium salt.

IV. Future Directions

Several strategies are currently being pursued to improve the efficiency of germ-line chimera production from zebrafish ES cell cultures. Various strains of fish are being investigated both as a source of ES cells and as recipient embryos for chimera production. Also, *in vitro* markers of ES cell pluripotency and germline competency are being identified for use in optimizing the ES cell culture system for germline chimera production. Finally, methods to inhibit germ cell formation in the recipient embryo are being used to improve the efficiency of donor cell contribution to the germ-cell lineage of the host (Ciruna *et al.*, 2002).

Acknowledgments

This work was supported by grants from the U.S. Department of Agriculture NRI 01-3242, Illinois-Indiana SeaGrant R/A-03-01, and the National Institutes of Health R01-GM069384-01.

References

Capecchi, M. (1989). Altering the genome by homologous recombination. *Science* **244**, 1288–1292.

Ciruna, B., Weidinger, G., Knaut, H., Thisse, B., Thisse, C., Raz, E., and Schier, A. (2002). Production of maternal-zygotic mutant zebrafish by germ-line replacement. *Proc. Natl. Acad. Sci. USA* **99**, 14919–14924.

Doetschman, T., Gregg, R. G., Maeda, N., Hooper, M. L., Melton, D. W., Thompson, S., and Smithies, O. (1987). Targeted correction of a mutant HPRT gene in mouse embryonic stem cells. *Nature* **330**, 576–578.

Fan, L., Alestrom, A., Alestrom, P., and Collodi, P. (2004a). Development of zebrafish cell cultures with competency for contributing to the germ line. *Crit. Rev. Euk. Gene Expr.* **14**, 43–51.

Fan, L., Crodian, J., Liu, X., Alestrom, A., Alestrom, P. and Collodi, P. (2004b). Zebrafish ES cells remain pluripotent and germ-line competent for multiple passages in culture. *Zebrafish* **1**, 21–26.

Fan, L., Crodian, J., and Collodi, P. (2004c). Culture of embryonic stem cell lines from zebrafish. *Methods Cell Biology* **76**, 149–158.

Ganassin, R., and Bols, N. C. (1999). A stromal cell line from rainbow trout spleen, RTS34st, that supports the growth of rainbow trout macrophages and produces conditioned medium with mitogenic effects on leukocytes. *In Vitro Cell Dev. Biol. Animal* **35**, 80–86.

Gibbs, P. D., and Schmale, M. C. (2000). GFP as a genetic marker scorable throughout the life cycle of transgenic zebrafish. *Mar. Biotech.* **2,** 107–125.

Higashijima, S., Okamoto, H., Ueno, N., Hotta, Y., and Eguchi, G. (1997). High-frequency generation of transgenic zebrafish which reliably express GFP in whole muscles or the whole body by using promoters of zebrafish origin. *Dev. Biol.* **192,** 289–299.

Lin, S., Long, W., Chen, J., and Hopkins, N. (1992). Production of germ line chimeras in zebrafish by cell transplants from genetically pigmented to albino embryos. *Proc. Natl. Acad. Sci. USA* **89,** 4519–4523.

Ma, C., Fan, L., Ganassin, R., Bols, N., and Collodi, P. (2001). Production of zebrafish germ-line chimeras from embryo cell cultures. *Proc. Natl. Acad. Sci. USA* **98,** 2461–2466.

Westerfield, M. (1995). The Zebrafish Book. 3rd ed., University of Oregon Press, Eugene, OR.

CHAPTER 7

Nonconventional Antisense in Zebrafish for Functional Genomics Applications

Stephen C. Ekker

University of Minnesota
Arnold and Mabel Beckman Center for Transposon Research
Department of Genetics, Cell Biology and Development
Minneapolis, Minnesota 55455

I. Introduction

The ability of the various genome projects to acquire gene sequence data has far outpaced our ability to ascribe biological functions to these new genes. How can we determine which of the increasing number of these uncharacterized gene products are involved in a given biological or disease process? This challenge has led to the concept of a scientific field called functional genomics, which can be defined as the attempt to match biological function with gene sequence on a genome scale. For the many biological processes that are well conserved in evolution, model systems with rapid genetic tools such as *D. melanogaster* have opened the door to functional genomics. In this paradigm, genes with specific biological roles are identified first in the model organism and genome databases are subsequently used to identify human homologs. Many biological and biochemical pathways are not conserved between flies and vertebrates, and, consequently, studies directly using biologically more complex model systems such as the mouse and the fish are warranted. Examples include neural crest formation, most organogenesis pathways, and signaling cascades such as those induced by vascular endothelial growth factor.

Genetic approaches have been developed to approach this problem, but each are a significant cost in terms of animal husbandry needs and time for large-scale analyses in vertebrates. Classical chemical mutagenesis screens are efficient at generating altered genetic loci but require extensive housing for the study of the necessary families, and the molecular characterization of the genetic locus is slow and costly. In mice, ES cell technology is expensive ($30,000 for the knockout cell line alone for a single gene without functional analysis) and requires extensive time for the subsequent necessary F3 mouse work. In zebrafish, an insertional mutagenesis screen is ongoing from the Hopkins lab, with the goal of assigning function to ~250 genes (Golling et al., 2002). This latter collection includes a variety of morphologically complex phenotypes, with an estimated 40% (~100) displaying phenotypic effects specific to a particular biological problem or pathway (Golling et al., 2002). The high expense in labor and steep initial capital cost of this method make the replication of this screen prohibitive in most other research laboratories. In addition, phenotypes obtained from ENU or retroviral insertional approaches might be due to a gene whose function has already been well characterized from previous work, resulting in potential significant redundancy in effort as our knowledge base associated with core vertebrate genes grows.

Alternatives to genetic approaches for sequence-driven screens have begun to be developed for functional genomics applications, even for systems without the high animal costs and significant time and infrastructure commitments associated with vertebrate model systems. RNAi-based screening in the nematode (Barstead, 2001) and in fly tissue culture cells (Lum et al., 2003) has explored the use of this knockdown strategy for sequence-specific annotation. In each of these examples, the extensive development of the genomic infrastructure has significantly augmented the ability to make these screens comprehensive for a specific subset of the genome. siRNA approaches in mammalian systems have made practical similar approaches in tissue culture models (McManus and Sharp, 2002) but are impractical for large-scale in vivo work. A sequence-based loss-of-function technology using a high-throughput F0 whole animal vertebrate assay system would be the method of choice for vertebrate functional genomics applications.

We reported the development of morpholinos (MOs) as effective sequence-specific translational inhibition agents in zebrafish (Nasevicius and Ekker, 2000). Morpholinos are chemically modified oligonucleotides with similar base-stacking abilities as natural genetic material but a morpholine moiety instead of a riboside (Summerton, 1999; Summerton and Weller, 1997). In addition, a phosphorodiamidate linkage is used, resulting in a neutral charge backbone. These two modifications form a modified and highly soluble polymer capable of hybridizing single-stranded nucleic acid sequences with high affinity and little cellular toxicity and free of most or all antisense-related side effects (Summerton, 1999; Summerton and Weller, 1997). Indeed, MOs are not subject to any known endogenous enzymatic degradation activity. MOs have been shown to bind to and block translation of mRNA in both cell-free assays and tissue culture cells (Summerton, 1999; Summerton and Weller, 1997). With RNAase H-mediated approaches, nonspecific interactions can result in significant message destruction of nontargeted mRNAs, thus limiting the levels of oligonucleotides that can be delivered. Alternatively, the MO antisense approach relies on a steric interaction that leaves nontargeted messages intact, allows the delivery of larger amounts of oligonucleotides, and significantly reduces nonspecific effects. Results obtained over the past 3 years have demonstrated efficacy of this tool *in vivo* in a variety of therapeutic and genomics applications (Arora *et al.*, 2000; Heasman *et al.*, 2000; Nasevicius and Ekker, 2000; Qin *et al.*, 2000) in a large number of model systems, including frog, fly, nematode, mouse, snail, and leech (see Heasman 2002, for review; see also Chen *et al.*, in press) in addition to their extensive use in zebrafish. (See Sumanas and Larson (2002) for a compilation of the first ~50 works published that used this tool.) The zebrafish represents a system well suited to MO targeting because of the ease of delivery and rapid embryonic development. This tool offers the opportunity to pursue sequence-specific gene targeting studies in a whole animal without the necessity of laborious, time consuming, and expensive F3 vertebrate genetic testing or the limitations associated with studying functions by using *in vitro* biological models. MO and other knockdown approaches also offer the opportunity for the generation of animals with the equivalent of a genetic allelic series through regulation of the specific inhibitory dose, generating phenotypes that might be masked because of severe effects from a null genetic phenotype. In addition, these partial loss-of-function animals are more likely to mimic such human genetic disease states because of only a partial reduction in gene function. The use of MO knockdowns in screening approaches have now been published by three distinct research groups (Chen *et al.*, 2004; Doitsidou *et al.*, 2002; Sumanas *et al.*, 2003), demonstrating the feasibility of using MO for the study of gene collections for functional genomics applications.

II. General Use of Nonconventional Antisense Tools in Zebrafish

Although MOs have received maximum attention by zebrafish researchers, these oligonucleotides represent only one of many potential chemistries for nonconventional antisense applications. Indeed, the biology of the early zebrafish embryo greatly facilitates the use of this model organism for the simultaneous testing of efficacy, specificity, and toxicity of a diverse array of oligonucleotide chemistries (Fig. 1; Nasevicius and Ekker, 2001). Recently, a derivative of the peptide nucleic-acid-based backbone chemistry (gripNA) was shown to also work well in zebrafish on early acting genes (Urtishak *et al.*, 2003). GripNA oligonucleotides appear to be more sensitive to mismatch in the hybridization sequence

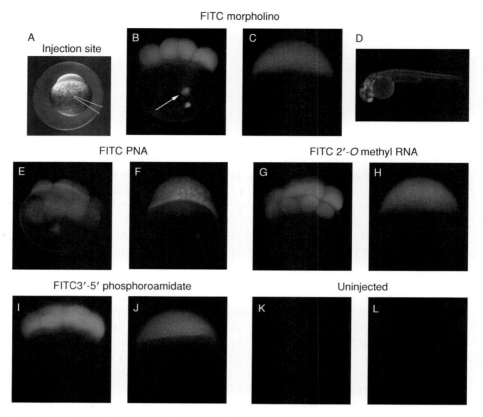

Fig. 1 Uniform distribution of fluorescein isothiocyanate (FITC)-modified oligonucleotides in zebrafish embryos after microinjection. FITC-labeled modified oligonucleotides of various chemistries were injected at the one- to two-cell stage as described (Nasevicius and Ekker, 2000). (A–D) Morpholino (MO)-injected embryos. (E, F) Peptide nucleic acid (PNA)-injected embryos. (G, H) 2′-O methyl RNA-injected embryos. (I, J) 3–5′ Phosphoroamidate oligonucleotide-injected embryos. (K, L) Uninjected embryos. Fluorescence was assayed by using a modified FITC filter set. Fluorescence indicates the injected oligonucleotide localization. The compounds are completely translocated to blastomeres as early as the eight-cell stage (<1 h after injection, Panels B, E, G, and I compared with Panel K). Later in development oligonucleotides remain uniformly distributed among blastomeres (midblastula, Panels C, F, H, and J compared with Panel L) and at 30 h of development (panel D). From Nasevicius and Ekker, (2001). The zebrafish as a novel system for functional genomics and therapeutic development applications. *Curr. Opin. Mol. Ther.* **3,** 224–228, with permission. (See Color Insert.)

(Urtishak *et al.*, 2003), but they also appear to have a reduced perdurance when compared to MOs. For example, gripNAs do not appear to effectively target the function of the *nacre* gene (Pickart and Ekker, unpublished observations), whose function is required later in embryonic development for pigment cell development and which can be effectively targeted by MOs (Nasevicius and Ekker, 2000).

III. Use of Morpholinos (MOs) as Antisense Tools in Zebrafish

A. Reagent Preparation

1. Storage

Chemically, MOs are inherently very stable molecules. A highly pure preparation maintains its activity over several years (Ekker, unpublished observations). For greatest preservation of activity, lyophilization is effective but not very convenient. Aqueous solutions can be stored at $-80\,°C$ for long-term storage where only limited access is required. Storage of aqueous solutions for ready access is more problematic, however. Multiple freeze–thaw cycles of aliquots stored at $-20°C$ can reduce efficacy, presumably through the formation of complex interoligonucleotide interactions. Sometimes, this effect can be partially reversed by heating the affected solution. To reduce this effect, storage of solutions at $4\,°C$ is also viable, but this method requires airtight containers and careful mixing of the solution prior to each use to avoid unexpected changes in activity, such as a concentration from the loss of aqueous solvent.

In one instance, we have identified a reproducible loss of activity on storage by one specific sequence oligonucleotide. The wnt5 mutation *pipetail* can be effectively phenocopied by using a MO targeted to the *wnt5* gene (Lele *et al.*, 2001). In two independent syntheses, the same sequence oligonucleotide irreversibly lost activity in two separate laboratories (Kim, Hammerschmidt, and Ekker, unpublished). The origin of this activity loss is currently unknown.

2. Resuspension Solutions

Lyophilized MO oligonucleotide preparations can be resuspended in a variety of solutions. The use of high-purity water, pretested for lack of toxicity on injection in zebrafish embryos, is, however, convenient for MO use in zebrafish embryos as it allows subsequent analysis of the solution for concentration. One useful protocol for measuring concentration of MOs uses the absorbance peaks of the bases at 265 nm in 0.1 M HCl, with the following polymer subunit estimates: adenine (mass = 339.290, molar A_{265} = 12660), cytosine (mass = 315.268, molar A_{265} = 8880), guanine (mass = 355.298, molar A_{265} = 10080), and thymine (mass = 330.278, molar A_{265} = 9830) (Morcos and Summerton, personal communication).

3. Injection Buffers

As is the case with mRNA delivery studies, a variety of injection buffers have been used for MO applications in zebrafish to reduce the osmotic shock of pure water injections. Danieau solutions of $0.3 \times$ to 1×58 mM NaCl, 0.7 mM KCl,

$0.4\,mM$ $MgSO_4$, $0.6\,mM$ Ca $(NO_3)_2$, $5\,mM$ HEPES pH 7.6 have been used as modestly buffered solutions that have no overt effects on injection of up to 15 nl per embryo. The buffering capacity can be supplemented with $10\,mM$ Tris-HCl without adverse effect on the embryo.

4. Delivery by Microinjection

One essential ingredient for successful MO use in any system is a proficient delivery methodology. In zebrafish, the method of microinjection is extremely potent (Fig. 1), and an accomplished investigator can inject a thousand one- to eight-cell-stage embryos in a day. The activity of MOs is dose dependent (see more later), and calibration of the delivered volume is essential for reproducibility between experiments. One such apparatus suitable for MO injections and associated protocols has been previously described (Hyatt and Ekker, 1999). It is worth noting that standard practice in zebrafish laboratories involves the regulation of embryonic development by altering the thermal environment of these cold-blooded embryos. This practice of shifting temperature to facilitate or reduce the rate of embryonic development for the convenience of the investigator has one potential complication of altering the hybridization kinetics of MO injected embryos.

B. Sequence Design

1. General Considerations

MOs derive their activity from the binding of RNA *in vivo*. The core sequence design rules developed for nucleic acid hybridization should be followed regardless of the targeting strategy used. The target sequence should ideally be unique, have high complexity, and should not be subject to likely internal interactions such as potential hair-pins or other likely complementary subsequences. In practice, antisense MO oligonucleotides are typically selected with a 40–60% GC content to maintain sequence complexity and for consistency in binding activity. Solubility and synthesis issues empirically restrict MOs to a less than 37% G content and a lack of any consecutive tri- or tetra-G nucleotide sequences (Morcos, personal communication). In addition, intra- and intersequence homology between one or two selected oligonucleotide targets is selected to minimize self or pair sequence homology. A semiautomated algorithm for assisting in MO design has been established and will be published separately (Pickart, Klee, Bllis, and Ekker, unpublished).

2. Translational Inhibitors

The ability of MOs to serve as translational inhibitors of gene function was pioneered by Summerton and colleagues, using a cell-free translational assay system. [See Summerton (1999) for a review]. In these studies, the most reproducible and effective gene inhibition was noted for oligonucleotides that bound to the untranslated leader sequence and those sequences around and including the initiating methionine. These effects are well reflected by *in vivo* targeting in zebrafish (Fig. 2). For example, during the process of positional cloning of the *parachute* mutation, Lele *et al.* (2002) developed MOs against several potential initiating methionines in the mRNA. The MO that bound to the 5′-most putative initiating methionine demonstrated the greatest efficacy. Note that one sequence designed against an internal coding sequence also demonstrated detectable activity, as was noted for some sequences in *in vitro* studies (Summerton, 1999). Thus, although the leader sequence is not absolutely required for MO efficacy, this information does appear to be essential for high-throughput protocols in which multioligonucleotide design and testing is cost or time prohibitive. One major distinction of translational targeting is the potential to inhibit both zygotic and maternal functions in some instances (Nasevicius and Ekker, 2000), opening the door to the analysis of functions not normally accessible to standard (zygotic-centric) genetic studies in zebrafish.

3. Transcriptional Targeting

An approach first demonstrated by Kole and colleagues (Schmajuk *et al.*, 1999), the use of MOs for altering pre-mRNA splicing was recently added to the zebrafish genetic toolbox (Draper *et al.*, 2001). One advantage of this targeting approach is the ability to use RT-PCR as a rapid efficacy assay. In addition, the

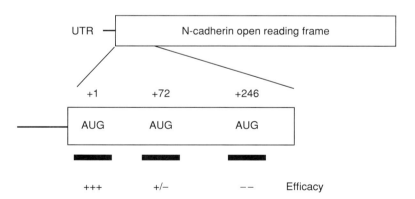

Fig. 2 Example of MO targeting effectiveness as translational inhibitors of the zebrafish *parachute*/N-cadherin gene. Three MOs were designed against three distinct in-frame methionine codons encoded within the N-cadherin presumptive open reading frame. Efficacy at phenocopying the characterized mutation is shown below, with +++ indicating a high penetrance and a strong phenocopy was observed. In contrast, the two MO-targeting sequences within the downstream open reading frame had little (+/−) or no (−,−) detectable efficacy. Data from Lele *et al.* (2002). *parachute*/N-cadherin is required for morphogenesis and maintained integrity of the zebrafish neural tube. *Development* **129**, 3281–3294 and Hammerschmidt, personal communication, with permission.

advent of the zebrafish whole genome sequence means there will be no lack of available target site sequences for MO use. The difficulty in using this approach is the predicted effects of splice-site targeting results in the requirement of evaluating each of the resulting altered splice forms in addition to monitoring the loss of wild-type mRNA (see example in Chen *et al.*, in press). In addition, translational inhibition can block maternal and zygotic function, whereas splice-site-blocking MOs target zygotic gene function only. This latter distinction indicates that the range of gene functions amenable to study by using this approach is similar to that for classical genetic studies.

4. Length

Oligonucleotides that are 25 bases long have been commonly used in the field to date due in part to strong recommendations by the manufacturer. However, work with gripNAs (Urtichak *et al.*, 2003) and PNAs has led some to try shorter-length MO oligos. An 18-mer *chordin* targeted oligonucleotide is a very effective agent, whether the backbone chemistry is a gripNA (Urtishak *et al.*, 2003) or MO (Pickart, Farber, and Ekker, unpublished data). In addition, an 18-mer MO was found to be very effective against the *syn-2* gene (Chen and Ekker, unpublished). The rules for effective length of MO design are still to be determined.

C. Efficacy Testing

1. Mutant Phenocopy Dataset

Estimates of the average rate of success of single MO sequences can perhaps best be determined by using MO-targeting genes whose loss-of-function phenotypes are already known. One such examination of a gene set of well-characterized mutations was summarized in Ekker (2000), and an expanded analysis is presented in Table I. Data from this and related studies indicate that the overwhelming majority ($\geq$80%) of genes with high-quality bioinformatics and sequence analysis can be targeted with a single oligonucleotide to reduce the targeted protein levels by 90% or more. The primary limitation is a sequence-dependent toxicity in $\sim$20% of synthesized oligonucleotides whose effect precludes the generation of tissue for examination (see later). This can often be overcome by the synthesis of an additional MO of independent sequence designed against the same gene target or through the use of a gripNA oligo (Urtishak *et al.*, 2003).

2. Dose Dependency of Phenotype

Table I lists a series of characterized MOs as a function of two delivered doses of oligonucleotide: low to moderate ($\leq$5 ng/embryo) and high (6–9 ng/embryo). The overwhelming majority of MOs elicit a robust and specific effect that is dose dependent in either strength of phenotype or penetrance in these dose ranges. In practice, few MOs that give specific phenotypes fail to demonstrate at least a

Table I
Dose-Dependent Estimates of Efficacy, Penetrance, and Mistargeting Rates for Morpholinos (MOs) With Known Efficacy Rates

Gene	Moderate-dose injections (~5 ng)			High-dose injections (~6–9 ng)			Ref.
	Efficacy[a]	Penetrance[b]	Mistargeting	Efficacy[a]	Penetrance[b]	Mistargeting	
twhh	++	+++	−	+++	+++	−	Nasevicius and Ekker, 2000
shh-1	+	+++	−	++	+++	−	Nasevicius and Ekker, 2000
shh-2	+++	+++	+/−	+++	+++	++	Ekker and Larson, 2001
oep-1	+++	+	−	+++	+	+	Nasevicius and Ekker, 2000
oep-2	+++	+++	−	+++	+++	−	Ekker, unpublished
ntl	+++	+++	−	+++	+++	−	Nasevicius and Ekker, 2000
smo-1	+	+	+	ND[c]	ND	++	Ekker, unpublished
smo-2	+/−	+/−	+	ND	ND	++	Ekker, unpublished
GFP	++	+++	−	+++	+++	+	Nasevicius and Ekker, 2000
chd	+++	+++	−	+++	+++	−	Nasevicius and Ekker, 2000
nacre	++	+++	−	+++	+++	−	Nasevicius and Ekker, 2000
sparse	++	+++	−	+++	+++	−	Nasevicius and Ekker, 2000
urod	+++	+++	−	+++	+++	−	Nasevicius and Ekker, 2000
boz	+	++	+	++	++	++	Nasevicius and Ekker, 2000
pax2.1	−	−	++	ND	ND	+++	Nasevicius and Ekker, 2000; Ekker and Larson, 2001

[a] +++: Null or near-null efficacy.
[b] +++: 80–100% penetrance; +: 25–50% penetrance; +/−: <25% penetrance.
[c] ND: Not determined because extensive toxicity at this dose.

moderate effect in over half the injected embryos at 5 ng or lower doses. Higher doses are sometimes required to elicit more extreme effects, but this strategy raises the risk of observing nonspecific effects.

Two outliers from these general rules are worth noting. First, even at extreme doses (up to 18 ng/embryo), the shh-1 MO does not yield a null phenocopy of the *sonic-you* mutation (Nasevius and Ekker, 2000). A second MO against the same gene, shh-2, does, indicating that the *sonic hedgehog* gene is capable of being quantitatively inhibited by MO oligonucleotides. At higher shh-2 MO doses, however, noticeable mistargeting can be detected (Table I). A second example is represented by the oep-1 MO: even at the highest doses, at which many affected embryos resemble a full mutant phenotype, a maximum penetrance of 50% was noted (Nasevicius and Ekker, 2000). An additional MO (oep-2) was tested, and this oligonucleotide displayed full penetrance and efficacy in phenocopying the *oep* phenotype (Table I). Thus, the use of a second MO of independent sequence can often overcome even these unusual efficacy or penetrance activities occasionally observed for individual oligonucleotide sequences.

3. Perdurance of MO Action

How long do MOs act after injection? The fish embryo can develop for about 10 days without adding biomass if left unfed. Any reduction in concentration *in vivo* is thus because of dilution of the MO through a differential expansion of a particular cellular lineage or because of some other cellular filtration or sequestration process removing the active molecule from the cell. A MO that translationally inhibits the transcription factor encoded by the *nacre* locus is fully penetrant beyond 50 h of development, demonstrating that MOs can function to inhibit translation through all embryogenesis (Nasevicius and Ekker, 2000). A recent study by Smart *et al.* (2004) used Western analyses to show effective MO translational inhibition in a subset of injected embryos through day 5 of development. The duration of transcriptional inhibition has also been measured for a variety of MOs. *Sox9a* MOs showed effective splice-site targeting through day 3 of development, but this effect was considerably attenuated by day 4 (Yan *et al.*, 2002). Another study generated a MO phenocopy of *endothelin* and demonstrated effective transcriptional targeting through day 5 of development (Kimmel *et al.*, 2003). Together, these studies indicate that the perdurance of MO action appears to be comparable for both gene knockdown strategies regularly employed in zebrafish.

4. Efficacy Measurements for Genes of Unknown Function

An assessment of the degree of knockdown is often a critical measure when studying genes of unknown function. A comparison of some common methods is listed in Table II. For translational blocking approaches, direct measurements of gene knockdown are usually difficult unless an antibody is available. Consequently, several indirect methods have been developed by using *in vitro* or

Table II

Comparison of Major Efficacy Testing Strategies for MO Use in Zebrafish

Translational blocking	Advantages	Disadvantages
Western	Detects endogenous gene product	Requires protein-specific antibody
Cell-free translation	Quick, no embryos or cells required	Cell free
Reporter fusion	*In vivo* efficacy test	Indirect measure of gene product inhibition
Transcriptional targeting		
RT-PCR	Detects endogenous gene product	Detailed effects on splicing must be determined for each oligo for each transcript
Western	Detects endogenous gene product	Requires protein-specific antibody

in vivo translational assays. For transcriptional targeting applications, efficacy measurements are in principle more straightforward because the RNA gene product can be readily measured by quantitative RT-PCR or other methods, but direct estimates of protein levels also require an antibody. Transcriptional targeting can, however, result in a variety of currently unpredictable effects on mRNA splicing. Effects ranging from heterogenous exon skipping (Draper *et al.*, 2001) to intron inclusion (Chen *et al.*, in press) have been noted. Thus, efficacy testing of transcriptional modifying MOs requires the unique assessment of the effect of each oligonecleotide on a particular message.

5. Toxicity and Other Effects

Table I gives a summary of the observed effectiveness as well as unexpected phenotypes from the mutant phenocopy dataset. In these studies, mistargeting is defined as effects on development that cannot be attributed to the specific loss of function of the targeted gene. In some cases, such mistargeting effects can overlap the efficacy curve (such was the case for the *boz* MO), or be so extreme so as to preclude the use of single MOs (such as the *smo* MOs) or block any observable efficacy (*pax2.1*). To minimize these effects, a reduced dosing scheme for individual oligonucleotides, the use of two independent oligonucleotides at doses at which no observable effect can be seen alone (such as *smo*, which work well when used at 1 ng each as a cocktail), or the simple testing of several oligonucleotide of distinct sequence are strategies that can be employed.

Some toxic effects can be recognized by their regular appearance in multiple oligonucleotide preparations. For example, an initially localized (at lower doses) and more extensive (at higher doses) neural cell death common to some MOs has been described (Ekker and Larson, 2001). In contrast, a variety of other highly heterogeneous nonspecific effects have been observed in a subset of MOs. For example, the *oep-1* MO at high concentrations results in a total failure in eye development, a phenotype not noted in even maternal/zygotic *one-eyed pinhead* mutant embryos (Gritsman *et al.*, 1999). In contrast, high-dose injections of a GFP MO cause a localized neural death phenotype (Davidson and Ekker, unpublished) indistinguishable from the *curly-up* subclass of embryonic mutations (Brand *et al.*, 1996). The nonoverlapping nature of these effects and the observations that related but distinct sequence oligonucleotides (such as a four-base mismatch MO) fail to elicit these effects have led to the conclusion that these are sequence-specific mistargeting events. One mechanism would be for these oligos to bind to and inhibit a gene of related sequence during embryogenesis, an effect likely to occur at higher frequency as the MO dose increases. The ability to distinguish the specific phenotypes from the sequence-specific toxic effects of MOs is essential and requires subsequent specificity testing studies.

D. Specificity Testing

Table III summarizes some of the more common strategies for MO specificity testing currently employed in zebrafish. Key elements to RNA and DNA rescue approaches include reintroduction of the targeted gene without MO binding sites in the resulting RNA at sufficient levels at the right time in development to

Table III
Comparisons of Major Specificity Testing Strategies for MO Use in Zebrafish

Approach	Advantages	Major disadvantages
Mutation comparison	Genetic test	Mutation might not be characterized or readily available
		Most mutations reveal zygotic gene function only
		Many mutations are not a confirmed null
mRNA rescue	RNA can be uniformly delivered by microinjection	Difficult to make synthetic mRNA for large genes
		Ectopic overexpression effects can confound analyses
		Protein is synthesized very early in development and can be toxic.
DNA rescue	Easy reagent preparation	Mosaic and nonuniform delivery of DNA upon microinjection
	Delivery of protein is delayed compared to RNA injection	Protein delivery delay
Multiple, independent sequence oligos	Easy reagent preparation if sequence is available	Does not directly assay protein function
	Works well on large genes	
	Can also conduct synergy tests	
4/5 Base mismatch oligo	Demonstrates requirement of selected sequence in target	Does not exclude targeting of a second gene with related sequence

ameliorate the MO-induced effects. Such temporal and spatial constraints might make it impossible to effect rescue in some instances, however. In addition, the technical constraints on *in vitro* mRNA synthesis will usually preclude this approach for large transcripts.

For genes in which rescue is not a viable option, the use of multiple independent sequence oligonucleotides against the same transcripts can be used for specificity testing approaches. Overlapping phenotypes not noted in other oligonucleotide injections are likely to be specific to the targeted gene of unknown function. In addition, two nonoverlapping sequence MOs usually will have a more than additive effect on gene function (Ekker and Larson, 2001). Finally, candidate specific activity should be significantly attenuated by a four or five-base mismatch oligonucleotide.

IV. Comparison of MO-Based Screening Success to Mutational Methods

How do the results of MO-based gene inhibition to date compare to other genotype–phenotype assignment methods? Chemical mutagenesis is a very powerful method for the identification of genes required for specific biological processes (Driever *et al.*, 1996; Haffter *et al.*, 1996). This work has established the rate of visible effects due to a single gene mutation in zebrafish. Saturation estimates using visible morphological phenotyping criteria suggest that 2400 total genes of unique function can be identified using that approach (Driever *et al.*, 1996; Golling *et al.*, 2002). Assuming that the zebrafish genome includes 36,000 genes (numbers extrapolated from the human and fugu genome projects; Aparicio *et al.*, 2002; Venter *et al.*, 2001), this suggests that ~1 in 15 genes when mutated yields a detectable phenotype visible during the first 5 days of development. Of those, only 40% result in specific defects (Driever *et al.*, 1996; Haffter *et al.*, 1996), suggesting that the rate of identifying overt, biologically specific phenotypes from removing function from a random gene set is ~2–3%. Thus, a biologically specific phenotype is likely to be observed for only a small minority of genes of unknown function within the genome.

Data from a pilot MO (25-gene) screen and an initial (50 EST) screen suggest a significantly higher potential phenotype detection rate by using MOs (16%; 12/75; Pickart, Klee, Ellis, Farber, Hammerschmidt, and Ekker, unpublished). These numbers represent confirmed, specific phenotypes that had passed one or more standard specificity tests (see previously). We attribute this higher detection rate to several factors. First, some of the noted phenotypes would not have been detected by standard morphological criteria, including observed defects in lipid metabolism and vascular function. Second, we examined conserved genes in these studies, and we believe that this type of gene set is enriched for proteins with essential gene functions and will be more likely to elicit phenotypes with regional or specific defects. Third, translational blocking MOs are able to target both maternal and zygotic messages (Nasevicius and Ekker, 2000), suggesting that some functions can be uncovered by using MOs that would not have been detected using standard mutagenesis approaches. Finally, the ability of MOs to elicit a full range of phenotypes because of altered dosing might identify hypomorphic-like phenotypes that would be too difficult to analyze from a strong, near-null allele. We consider the current estimate of 16% to be a lower estimate of observable specific phenotypes from MO screening, as further analysis will examine these MOs using a variety of novel assays for developmental aspects not readily visible by morphological criteria. MO screening thus represents a potent functional genomics methodology amenable to zebrafish and offers the first opportunity for a comprehensive analysis of these processes by using as template an entire vertebrate genome.

Acknowledgments

The author thanks P. Morcos, J. Summerton, M. Hammerschmidt, M. Pickart, A. Nasevicius and E. Chen for helpful comments and for sharing unpublished data. This work was supported by the National Institutes of Health (GM63904).

References

Aparicio, S., Chapman, J., Stupka, E., Putnam, N., Chia, J. M., Dehal, P., Christoffels, A., Rash, S., Hoon, S., Smit, A., Gelpke, M. D., Roach, J., Oh, T., Ho, I. Y., Wong, M., Detter, C., Verhoef, F., Predki, P., Tay, A., Lucas, S., Richardson, P., Smith, S. F., Clark, M. S., Edwards, Y. J., Doggett, N., Zharkikh, A., Tavtigian, S. V., Pruss, D., Barnstead, M., Evans, C., Baden, H., Powell, J., Glusman, G., Rowen, L., Hood, L., Tan, Y. H., Elgar, G., Hawkins, T., Venkatesh, B., Rokhsar, D., and Brenner, S. (2002). Whole-genome shotgun assembly and analysis of the genome of *Fugu rubripes*. *Science* **297,** 1301–1310.

Arora, V., Knapp, D. C., Smith, B. L., Statdfield, M. L., Stein, D. A., Reddy, M. T., Weller, D. D., and Iversen, P. L. (2000). c-Myc antisense limits rat liver regeneration and indicates role for c-Myc in regulating cytochrome P-450 3A activity. *J. Pharmacol. Exp. Ther.* **292,** 921–928.

Barstead, R. (2001). Genome-wide RNAi. *Curr. Opin. Chem. Biol.* **5,** 63–66.

Brand, M., Heisenberg, C. P., Warga, R. M., Pelegri, F., Karlstrom, R. O., Beuchle, D., Picker, A., Jiang, Y. J., Furutani-Seiki, M., van Eeden, F. J., *et al.* (1996). Mutations affecting development of the midline and general body shape during zebrafish embryogenesis. *Development* **123,** 129–142.

Chen, E., Hackett, P. B., and Ekker, S. C. (in press). Gene knockdown approaches using unconventional antisense oligonucleotides. *In* "Molecular Aspects of Fish and Marine Biology Vol. 2: Fish Developmental Biology: Zebrafish and Medaka" (Z. Gong and V. Korzh, eds.) World Scientific.

Chen, E., Hermanson, S., and Ekker, S. C. (2004). Syndecan-2 is essential for angiogenic sprouting during zebrafish development. *Blood* **103,** 1710–1719.

Doitsidou, M., Reichman-Fried, M., Stebler, J., Koprunner, M., Dorries, J., Meyer, D., Esguerra, C. V., Leung, T., and Raz, E. (2002). Guidance of primordial germ cell migration by the chemokine SDF-1. *Cell* **111,** 647–659.

Draper, B. W., Morcos, P. A., and Kimmel, C. B. (2001). Inhibition of zebrafish fgf8 pre-mRNA splicing with morpholino oligos: A quantifiable method for gene knockdown. *Genesis* **30,** 154–156.

Driever, W., Solnica-Krezel, L., Schier, A. F., Neuhauss, S. C., Malicki, J., Stemple, D. L., Stainier, D. Y., Zwartkruis, F., Abdelilah, S., Rangini, Z., Belak, J., and Boggs, C. (1996). A genetic screen for mutations affecting embryogenesis in zebrafish. *Development* **123,** 37–46.

Ekker, S. C. (2000). Morphants: A new systematic vertebrate functional genomics approach. *Yeast* **17**, 302–306.

Ekker, S. C., and Larson, J. D. (2001). Morphant technology in model developmental systems. *Genesis* **30**, 89–93.

Golling, G., Amsterdam, A., Sun, Z., Antonelli, M., Maldonado, E., Chen, W., Burgess, S., Haldi, M., Artzt, K., Farrington, S., Lin, S. Y., Nissen, R. M., and Hopkins, N. (2002). Insertional mutagenesis in zebrafish rapidly identifies genes essential for early vertebrate development. *Nat. Genet.* **31**, 135–140.

Gritsman, K., Zhang, J., Cheng, S., Heckscher, E., Talbot, W. S., and Schier, A. F. (1999). The EGF-CFC protein one-eyed pinhead is essential for nodal signalling. *Cell* **97**, 121–132.

Haffter, P., Granato, M., Brand, M., Mullins, M. C., Hammerschmidt, M., Kane, D. A., Odenthal, J., van Eeden, F. J., Jiang, Y. J., Heisenberg, C. P., Kelsh, R. N., Furutani-Seiki, M., Vogelsang, E., Beuchle, D., Schach, U., Fabian, C., and Nusslein-Volhard, C. (1996). The identification of genes with unique and essential functions in the development of the zebrafish. *Danio rerio. Development* **123**, 1–36.

Heasman, J. (2002). Morpholino oligos: Making sense of antisense? *Dev. Biol.* **243**, 209–214.

Heasman, J., Kofron, M., and Wylie, C. (2000). β-catenin signaling activity dissected in the early *Xenopus* embryo: A novel antisense approach. *Dev. Biol.* **222**, 124–134.

Hyatt, T. M., and Ekker, S. C. (1999). Vectors and techniques for ectopic gene expression in zebrafish. *In* "Methods Cell Biology" (W. Detrich, ed.), Vol. 59, pp. 117–126. Academic Press, San Diego.

Kimmel, C. B., Ullmann, B., Walker, M., Miller, C. T., and Crump, J. G. (2003). Endothelin 1-mediated regulation of pharyngeal bone development in zebrafish. *Development* **130**, 1339–1351.

Lele, Z., Bakkers, J., and Hammerschmidt, M. (2001). Morpholino phenocopies of the swirl, snailhouse, somitabun, minifin, silberblick, and pipetail mutations. *Genesis* **30**, 190–194.

Lele, Z., Folchert, A., Concha, M., Rauch, G. J., Geisler, R., Rosa, F., Wilson, S. W., Hammerschmidt, M., and Bally-Cuif, L. (2002). parachute/n-cadherin is required for morphogenesis and maintained integrity of the zebrafish neural tube. *Development* **129**, 3281–3294.

Lum, L., Yao, S., Mozer, B., Rovescalli, A., Von Kessler, D., Nirenberg, M., and Beachy, P. A. (2003). Identification of Hedgehog pathway components by RNAi in *Drosophila* cultured cells. *Science* **299**, 2039–2045.

McManus, M. T., and Sharp, P. A. (2002). Gene silencing in mammals by small interfering RNAs. *Nat. Rev. Genet.* **3**, 737–747.

Nasevicius, A., and Ekker, S. C. (2000). Effective targeted gene 'knockdown' in zebrafish. *Nat. Genet.* **26**, 216–220.

Nasevicius, A., and Ekker, S. C. (2001). The zebrafish as a novel system for functional genomics and therapeutic development applications. *Curr. Opin. Mol. Ther.* **3**, 224–228.

Qin, G., Taylor, M., Ning, Y. Y., Iversen, P., and Kobzik, L. (2000). *In Vivo* evaluation of a morpholino antisense oligomer directed against tumor necrosis factor-alpha. *Antisense Nucleic Acid Drug Dev.* **10**, 11–16.

Schmajuk, G., Sierakowska, H., and Kole, R. (1999). Antisense oligonucleotides with different backbones. Modification of splicing pathways and efficacy of uptake. *J. Biol. Chem.* **274**, 21783–21789.

Smart, E. J., De Rose, R. A., and Farber, S. A. (2004). Annexin 2-caveolin 1 complex is a target of ezetimibe and regulates intestinal cholesterol transport. *Proc. Natl. Acad. Sci. USA* **101**, 3450–3455.

Sumanas, S., and Larson, J. (2002). Morpholino oligonucleotides in zebrafish: A recipe for functional genomics? *Brief. Func. Genom. Proteom.* **1**, 239–256.

Sumanas, S., Larson, J. D., and Miller Bever, M. (2003). Zebrafish chaperone protein GP96 is required for otolith formation during ear development. *Dev. Biol.* **261**, 443–455.

Summerton, J. (1999). Morpholino antisense oligomers: The case for an RNase H-independent structural type. *Biochim. Biophys. Acta* **1489**, 141–158.

Summerton, J., and Weller, D. (1997). Morpholino antisense oligomers: Design, preparation, and properties. *Antisense Nucleic Acid Drug Dev.* **7**, 187–195.

Urtishak, K. A., Choob, M., Tian, X., Sternheim, N., Talbot, W. S., Wickstrom, E., and Farber, S. A. (2003). Targeted gene knockdown in zebrafish using negatively charged peptide nucleic acid mimics. *Dev. Dynam.* **228,** 405–413.

Venter, J. C., Adams, M. D., Myers, E. W., Li, P. W., Mural, R. J., Sutton, G. G., Smith, H. O., Yandell, M., Evans, C. A., Holt, R. A., Gocayne, J. D., Amanatides, P., Ballew, R. M., Huson, D. H., Wortman, J. R., Zhang, Q., Kodira, C. D., Zheng, X. H., Chen, L., Skupski, M., Subramanian, G., Thomas, P. D., Zhang, J., Gabor Miklos, G. L., Nelson, C., Broder, S., Clark, A. G., Nadeau, J., McKusick, V. A., Zinder, N., Levine, A. J., Roberts, R. J., Simon, M., Slayman, C., Hunkapiller, M., Bolanos, R., Delcher, A., Dew, I., Fasulo, D., Flanigan, M., Florea, L., Halpern, A., Hannenhalli, S., Kravitz, S., Levy, S., Mobarry, C., Reinert, K., Remington, K., Abu-Threideh, J., Beasley, E., Biddick, K., Bonazzi, V., Brandon, R., Cargill, M., Chandramouliswaran, I., Charlab, R., Chaturvedi, K., Deng, Z., Di Francesco, V., Dunn, P., Eilbeck, K., Evangelista, C., Gabrielian, A. E., Gan, W., Ge, W., Gong, F., Gu, Z., Guan, P., Heiman, T. J., Higgins, M. E., Ji, R. R., Ke, Z., Ketchum, K. A., Lai, Z., Lei, Y., Li, Z., Li, J., Liang, Y., Lin, X., Lu, F., Merkulov, G. V., Milshina, N., Moore, H. M., Naik, A. K., Narayan, V. A., Neelam, B., Nusskern, D., Rusch, D. B., Salzberg, S., Shao, W., Shue, B., Sun, J., Wang, Z., Wang, A., Wang, X., Wang, J., Wei, M., Wides, R., Xiao, C., Yan, C., *et al.* (2001). The sequence of the human genome. *Science* **291,** 1304–1351.

Yan, Y. L., Miller, C. T., Nissen, R. M., Singer, A., Liu, D., Kirn, A., Draper, B., Willoughby, J., Morcos, P. A., Amsterdam, A., Chung, B. C., Westerfield, M., Haffter, P., Hopkins, N., Kimmel, C., Postlethwait, J. H., and Nissen, R. (2002). A zebrafish sox9 gene required for cartilage morphogenesis. *Development* **129,** 5065–5079.

CHAPTER 8

Downregulation of Gene Expression with Negatively Charged Peptide Nucleic Acids (PNAs) in Zebrafish Embryos

Eric Wickstrom,[*,†,‡] **Karen A. Urtishak,**[†,‡] **Michael Choob,**[§]
Xiaobing Tian,[*] **Nitzan Sternheim,**[‖] **Laura M. Cross,**[¶]
Amy Rubinstein,[¶] **and Steven A. Farber**[†,‡]

[*]Department of Biochemistry & Molecular Pharmacology
Thomas Jefferson University
Philadelphia, Pennsylvania 19107

[†]Department of Microbiology & Immunology
Thomas Jefferson University
Philadelphia, Pennsylvania 19107

[‡]Kimmel Cancer Center
Thomas Jefferson University
Philadelphia, Pennsylvania 19107

[§]Active Motif
Carlsbad, California 92008

[‖]Department of Developmental Biology
Stanford University
Stanford, California 94305

[¶]Zygogen
Atlanta, Georgia 30303

METHODS IN CELL BIOLOGY, VOL. 77
Copyright 2004, Elsevier Inc. All rights reserved.
0091-679X/04 $35.00

I. Introduction

A. Zebrafish Genetics and Vertebrate Physiology

Genetic analyses in the zebrafish *Danio rerio* have proved quite useful for identifying genes that direct vertebrate development (Brownlie *et al.*, 1998; Driever *et al.*, 1996; Haffter *et al.*, 1996; Nusslein-Volhard, 1994). Since the completion of the original large-scale chemical mutagenesis screens in 1997, the phenotypic and molecular characterizations of many mutations have confirmed that *D. rerio* is an important model system for functional studies of vertebrate genes (Ackermann and Paw, 2003; Brownlie *et al.*, 1998; Detrich *et al.*, 1999). Molecular studies are further enhanced by the publically available draft of the zebrafish genome (http://www.sanger.ac.uk/Projects/D_rerio/), along with over 200,000 expressed sequence tag (EST) sequences (http://www.genetics.wustl.edu/fish_lab/frank/cgi-bin/fish/).

B. Zebrafish Developmental Genes

Several genes necessary for normal embryonic development have been characterized by visual characterization of morphological phenotypes. For example, the *chordin* gene encodes an antagonist of ventralizing BMP signals, and *chordin* mutants have a ventralized phenotype characterized by a reduction of anterior neural structures and an expansion of ventral tail structures and blood (Fisher *et al.*, 1997; Haffter *et al.*, 1996). Another gene expressed early in zebrafish development is *no tail* (*ntl*), which encodes a T-box transcription factor; *ntl* mutants lack a notochord and tail (Halpern *et al.*, 1993; Schulte-Merker *et al.*, 1994).

Uroporphyrinogen decarboxylase, an enzyme essential for heme biosynthesis, is also essential for embryonic development. Mutants for *uroD* exhibit hepatoerythropoietic porphyria (HEP), which is characterized by excessive accumulation of porphyrin compounds, and, as a result, autofluorescent, photosensitive erythrocytes (Nasevicius and Ekker, 2000; Wang *et al.*, 1998). Another gene vital for correct morphological development is *dharma*, which encodes a homeodomain transcription factor acting in specification of dorsal fates. *Dharma* mutant embryos exhibit a variable phenotype, ranging from loss of eyes, telencephalon, and all axial mesendoderm to cyclopia and moderate notochord defects to wild type, depending on genetic background and maternal age (Ekker and Larson, 2001; Fekany, 1999; Sirotkin *et al.*, 2000).

C. Oligonucleotide Knockdown

The ability to turn off individual genes at will in growing cells provides a powerful tool for elucidating the role of a particular gene, for diagnosis, and for therapeutic intervention. Knockdown oligonucleotides (Fig. 1) were first conceived as alkylating complementary oligonucleotides directed against naturally occurring nucleic acids (Belikova *et al.*, 1967) and first successfully used against Rous sarcoma virus (Zamecnik and Stephenson, 1978). Since those proofs of principle, antisense DNA derivatives have been used to inhibit the expression of a wide variety of target genes, in viral, bacterial, plant, and animal systems, in cells (Wickstrom, 1991), in animals (Agrawal, 1996b), and in humans (Wickstrom, 1998).

Novel oligonucleotide analogs (Fig. 2) have been synthesized to act as knockdown agents to improve the biological stability, solubility, cellular uptake, and ease of synthesis (Wickstrom, 1992). The simplest oligodeoxynucleotide modification involves blocking the 3' terminus to prevent attack by 3' exonucleases, the predominant extracellular degradative mechanism for oligodeoxynucleotides (Zendegui *et al.*, 1992). Other modifications focus on protecting the internucleoside linkage by changing the phosphodiester linkages to phosphorothioates (Stec *et al.*, 1991), methylphosphonates (Miller, 1991), or boranophosphates (Shaw *et al.*, 2000). Each of these structural changes affects not only nuclease susceptibility but also cellular uptake, cellular trafficking, and RNase H activation (Wickstrom, 1992). Among the derivatives described, only phosphodiester, phosphorothioate, and boranophosphate DNAs direct RNase H degradation of hybridized RNA.

Phosphorothioate oligonucleotides are the only derivatives that have been administered so far to humans. Despite their efficacy, however, phosphorothioate DNAs exhibit less sequence specificity in their effects than do phosphodiesters or methylphosphonates (Ho *et al.*, 1991; Wickstrom, 1991), because of significant binding to a spectrum of plasma and cellular proteins (Agrawal, 1996a). Although these modifications increase the *in vivo* half-life of oligonucleotides, they also weaken hybridization to the RNA target sites because of the creation of chiral phosphorus diastereomers (Lebedev and Wickstrom, 1996).

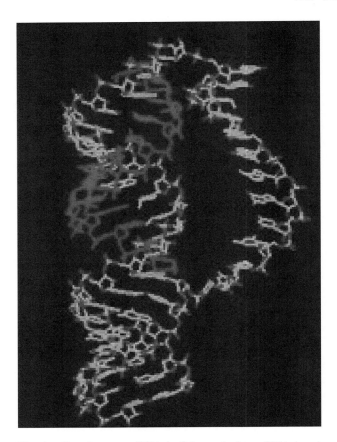

Fig. 1 Complementary DNA (red) basepaired to mRNA target.

The deoxyribose can be modified to 2′-*O*-alkyl RNAs, such as 2′-*O*-methyl, strengthening hybridization and resisting nuclease attack (Iribarren *et al.*, 1990). Encouraging results have been obtained recently which suggest that greater potency and specificity might be possible with 2′-*O*-alkyl RNA/DNA/2′-*O*-alkyl RNA phosphorothioate chimeras (Agrawal *et al.*, 1997; Monia *et al.*, 1993) or peptide–DNA conjugates (Basu and Wickstrom, 1995; Hughes *et al.*, 2000). Similar improvements result from preparing 3′-amino phosphoramidates (Gryaznov *et al.*, 1996) or morpholino phosphorodiamidates (Summerton and Weller, 1997).

D. Morpholino Phosphorodiamidates

The development of morpholino-based knockdown technology in zebrafish (Heasman, 2002; Nasevicius and Ekker, 2000) has enabled sequence-based reverse genetic screens. In addition, it can be used to elucidate rapidly the function of any targeted gene in this model vertebrate. These reagents offer the potential to explore

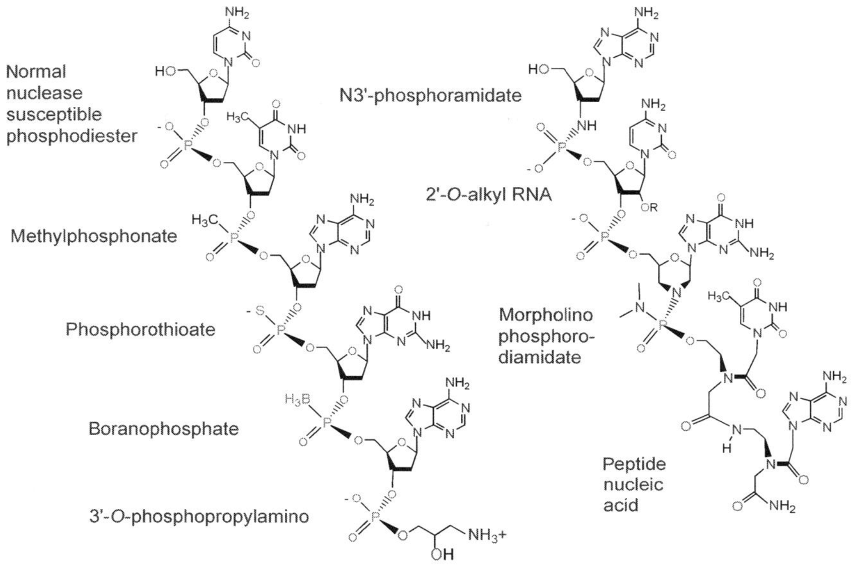

Fig. 2 Oligonucleotide backbone derivatives.

the molecular mechanisms of organ development and physiology by directing antisense oligonucleotide derivatives (Fig. 2) against any gene of interest. Typically, the most time-consuming component of any experiment is waiting for the antisense reagent to be synthesized and shipped (typically 2–3 weeks). Experimentally, an antisense MO or HypNA-pPNA oligomer targeting the initiation codon region of an mRNA of interest is pressure-injected into the yolk of a one- to eight-cell embryo. Numerous studies have successfully phenocopied a number of mutants by using this approach (Nasevicius and Ekker, 2000). MO oligomers (Fig. 2) display good hybridization properties as well as base specificity (Summerton, 1989; Summerton *et al.*, 1997). They are inherently immune to a broad range of cellular and circulating degradative enzymes and exhibit high solubility in water despite their lack of charge, because of their strong polarity (Summerton and Weller, 1997). Despite MO injections having emerged as a powerful approach to perform reverse genetic studies in zebrafish embryos, not all MOs efficiently inhibit translation. Moreover, for unexplained reasons, some MOs exhibit nonspecific mistargeting effects (Ekker and Larson, 2001). It is perplexing that other typical oligonucleotide derivatives, such as phosphorothioate, 2'-*O*-methyl, or peptide nucleic acids (PNAs) have not been successful knockdown agents in zebrafish.

E. Peptide Nucleic Acids (PNAs)

The most radical modifications are found in PNAs (Fig. 2), in which both the phosphodiester linkages and sugars are replaced with a peptide-like backbone of (*N*-2-aminoethyl) glycine units, with the bases directly attached by methylene-carbonyl linkers (Nielsen *et al.*, 1993). Compared with other oligonucleotide derivatives, PNAs display the highest T_ms for duplexes formed with single-stranded DNA or RNA (Egholm *et al.*, 1993). Receptor-specific uptake into cells has been demonstrated for PNA–peptide chimeras (Basu and Wickstrom, 1997). The higher affinity of PNAs toward DNA and RNA, along with more stringent mismatch discrimination and resistance to proteases and nucleases, appears more promising for diagnosis and therapy than MO oligomers, although PNAs are less soluble than MOs (Hyrup and Nielsen, 1996) and have not displayed knockdown activity in zebrafish. Recently, however, a new type of DNA mimic composed of alternating phosphonate PNA analogs and *trans*-4-hydroxy-L-proline PNA analogs (HypNA-pPNA) was synthesized and characterized (Efimov *et al.*, 1999b; Fig. 3). The negatively charged HypNA-pPNAs display excellent hybridization properties toward DNA and RNA while preserving the high single mismatch discrimination and nuclease/protease resistance of PNAs (Efimov *et al.*, 1999a,b; Phelan *et al.*, 2001; Urtishak *et al.*, 2003).

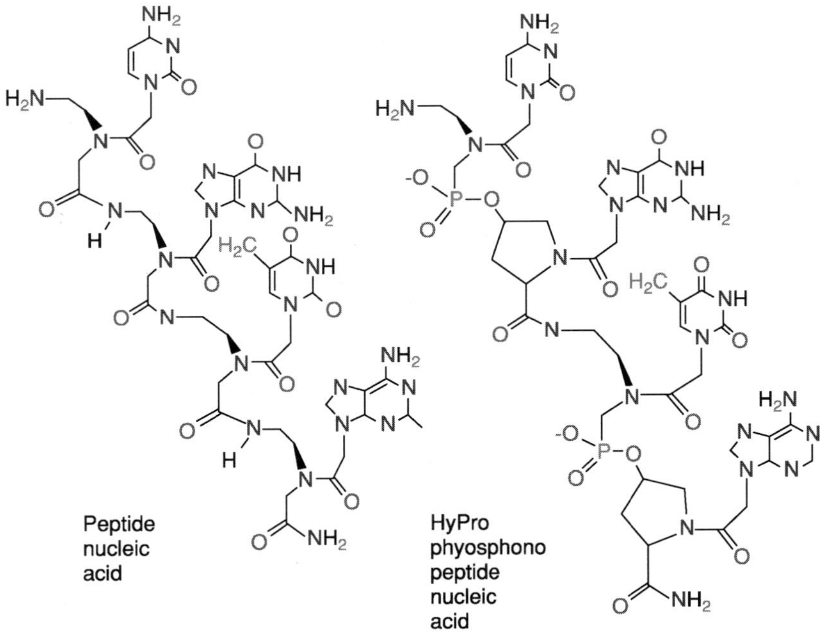

Fig. 3 HypNA-pPNAs are alternating heterooligomers constructed of phosphonate PNA monomers and *trans*-4-hydroxy-L-proline PNA monomers.

II. Materials and Methods

A. Oligonucleotide Synthesis

In published experiments with scrape-loaded cells, MO oligomers >20 nt, bracketing the initiation codon, have been used to reach the highest antisense efficacy (Summerton and Weller, 1997; Summerton *et al.*, 1997). In the case of zebrafish embryos injected with antisense MOs, 25-mers were used (Nasevicius and Ekker, 2000). Morpholino phosphorodiamidate (MO) sequences (Table I) with 3' fluorescein modification were obtained from Gene Tools, Corvallis, OR.

Taking into account the stronger hybridization characteristics of PNAs, a 12-mer, a 13-mer, and an 18-mer bracketing the *chordin* initiation codon (Table I) were synthesized by solid-phase synthesis with Fmoc coupling on an Applied Biosystems 8909 synthesizer as described (Tian and Wickstrom, 2002).

Table I
Antisense and Mismatch Sequences

Derivative	Target	Sequence	T_m (°C)	ΔT_m (°C)
PNA	*chordin*	FITC-N-ATCCACAGCAGC-Lys-C	ND	
PNA	*chordin*	N-CCCTCCATCATCC-Lys-C	ND	
PNA	*chordin*	N-GCAGCCCCTCCATCATCC-Lys-C	ND	
HypNA-pPNA	*chordin*	FITC-N-GCAGCCCCTCCATCATCC-C	87.9 ± 0.5	
HypNA-pPNA	*chordin* 2-mis	FITC-N-GCAGCCgCTCCtTCATCC-C	69.1 ± 0.3	19
HypNA-pPNA	*chordin* 4-mis	FITC-N-GCtGCCgCTCCtTCtTCC-C	69.81 ± 0.05	18
Morpholino	*chordin*	6'-ATCCACAGCAGCCCCTCCATCATCC-3'-FITC	88.0 ± 0.4	
Morpholino	*chordin* 2-mis	6'-ATCCACAcCAGCCCCTCgATCATCC-3'-FITC	73.3 ± 0.5	15
Morpholino	*chordin* 4-mis	6'-ATCCtCAGCcGCCCCaCCATgATCC-3'-FITC	67.7 ± 0.3	20
HypNA-pPNA	*ntl*	N-TGAGGCAGACATATTTCC-C	ND	
HypNA-pPNA	*ntl* 1-mis	N-TGAGGCAGgCATATTTCC-C	ND	
Morpholino	*ntl*	6'-GACTTGAGGCAGACATATTTCCGAT-3'-FITC	78 ± 0.5	
Morpholino	*ntl* 1-mis	6'-GACTTGAGGCAGgCATATTTCCGAT-3'-FITC	76.5 ± 0.5	1
HypNA-pPNA	*uroD*	FITC-N-AACTGTCCTTATCCATCA-C	ND	
HypNA-pPNA	*uroD* 2-mis	FITC-N-AACTGaCCTTtTCCATCA-C	ND	
Morpholino	*uroD*	6'-GAATGAAACTGTCCTTATCCATCA-3'-FITC	ND	
HypNA-pPNA	*dharma*1	N-TGCCATGTTCAAGTGTAG-C	75.82 ± 0.02	
HypNA-pPNA	*dharma*2	N-TCAAGTGTAGGGGTGCC-C	84.4 ± 0.4	
Morpholino	*dharma*	6'-TGCCATGTTCAAGTGTAGGGGTGCC-3'-FITC	87.1 ± 0.4	
HypNA-pPNA	GRCFP	N-GTGCTTGGACTGGGCCAT-C	ND	
Morpholino	GRCFP	6'-TCAGGCCGTGCTTGGACTGGGCCAT-3'	ND	

Note: Duplicate equimolar 2.5 μM HypNA-pPNA/RNA or MO/RNA mixtures were annealed to 90 °C for 3 min in 10 mM Na₂HPO₄, 1.0 M NaCl, 0.5 mM EDTA, pH 7.0, then cooled gradually to room temperature. Samples were then heated at a rate of 1 °C/min from 20 to 95 °C. Changes in A_{260} were recorded and a T_m value was calculated for each duplex. The lowercase base in the sequence represents the mismatch.

PNA oligomers were purified by preparative reversed-phase HPLC on a 10 mm $\times$ 250 mm Alltima C_{18} column eluted with a gradient over 25 min from 5 to 70% CH_3CN in aqueous 0.1% CF_3CO_2H at 1 ml/min and 50 °C, monitored at 260 nm. Purified oligomers were analyzed by SELDI-TOF mass spectroscopy on a Ciphergen mass spectrometer with a 338-nm laser.

For comparable experiments with HypNA-pPNAs, we predicted that 18-mers (Table I) would be long enough to provide sequence uniqueness, and two mismatches in an 18-mer would greatly destabilize HypNA-pPNA/RNA complexes in the zebrafish model. HypNA-pPNA dimer building blocks are synthesized as described (Efimov *et al.*, 1998, 1999b). The solid-phase synthesis of all HypNA-pPNA oligomers was carried out by using derivatized CPG supports on an Applied Biosystems 380B automated DNA synthesizer, with phosphotriester coupling of dimer synthons. Coupling yields of 85–90% are typical (Efimov *et al.*, 1998, 1999b). The phosphonate protecting groups are removed with triethylammonium thiophenolate before the final deprotection by ammonolysis. Some sequences are labeled by fluorescein phosphoramidite (Glen Research, Sterling, VA) coupling to the 5′ terminalamine. After desalting by gel filtration, all oligomers are purified by denaturing electrophoresis on 15% polyacrylamide gels in 7 M urea with 100 mM Tris-H_3BO_3, 1 mM EDTA, pH 8.3. The purity of the HypNA-pPNAs oligomers was determined by analytical gel electrophoresis to be 85–90%.

Oligomer concentrations are determined by ultraviolet absorbance spectra measured at room temperature on a Shimadzu UV-160, assuming molar absorptivities at 260 nm of Ade, 15.4; Gua, 11.7; Thy, 8.8; and Cyt, 7.3 $\times$ 10^3/M · cm in 50 mM Et_3N-H_2CO_3, pH 7.0, at 25 °C. HypNA-pPNAs and MOs are further purified on SepPaks (Waters) to remove any remaining salts that were found to be toxic to the embryos, a requirement when injected at the highest concentrations.

B. Animal Experiments

Zebrafish are the most appropriate species for this analysis because they are small in size, easy to maintain and breed, and produce large numbers of progeny on a daily basis. Their embryos develop rapidly and are optically clear, permitting direct observation of the developing organs.

Zebrafish also contain orthologs for almost all human genes. Many developmental and metabolic genes have already been characterized and placed on the zebrafish genetic map. In addition, most of the zebrafish genome has been sequenced, and many of the genes have been assembled and annotated, enabling rapid identification of many oncogene orthologs. A large number of ESTs are available for detecting expressed messages. Microarray chips covering ~10,000 genes are available at the Thomas Jefferson University to enable analysis of transcription profiles of zebrafish with and without treatment, control vs. antisense injected, or mutant vs. wild type.

C. Animal Care

Zebrafish are housed in a separate facility consisting of approximately 500 tanks units of varying sizes (1, 3.75, and 9 l). Environmental conditions are carefully monitored for disease prevention and to maintain fish in perpetual breeding condition. Male and female fish are reared at a density of no more than eight fish per liter at a constant temperature and light cycle (27–29 °C with light/dark cycle kept at 14/10 h) in pretreated water (heated, charcoal filtered, and UV sterilized). They are fed twice daily with a variety of dried and live foods. At capacity, the facility is expected to house approximately 25,000 adult (3 months to 1–2 years old) and juvenile fish.

The zebrafish facility at Thomas Jefferson University is a certified animal care unit, with attendant veterinary care, as required. The institution has an approved Animal Welfare Assurance on file with the Office for Protection from Research Risks and is accredited by the American Association for the Accreditation of Laboratory Animal Care. The animals are maintained under all current guidelines and legislation governing the use of laboratory animals, including those set forth in the Animal Welfare Act.

D. Injections

AB and pet store zebrafish strains are raised by using standard methods (Westerfield, 2000). Embryos are collected at the one-cell stage <1 h after fertilization and placed in 60 < 15 mm petri dishes with embryo medium (EM, Westerfield, 2000). Embryos are injected using a pressure injector (PLI-100, Harvard Apparatus, Cambridge, MA). Electrodes are pulled using a Flaming/ Brown micropipette puller (Model P-97) and filled with either MO or HypNA-pPNA (0.1–0.4 mM oligomer, ∼1 g/l) dissolved in sterile-filtered, double-deionized water and dyed with phenol red solution (0.2% final concentration) to visualize injections (Nasevicius and Ekker, 2000). For gene knockdown studies, 1–2 nl of each oligomer is injected into the margin between the yolk and nucleus of 30 or more one- to eight-cell-stage embryos.

Bright-field and fluorescent images are captured by using either a Zeiss Axiocam 2 mounted on a Leica MZFL-III stereo microscope or a Nikon Coolpix 995 mounted on a Nikon SMZ1500 stereo microscope. Once injected, the embryos are incubated at 28 °C. Between 2 and 6 hpf, all infertile embryos are removed from the dishes and the number of survivors recorded. The mortality and any phenotypes observed in injected embryos are recorded at 28 h, 52 h, and 3 dpf. Larvae injected with *chordin* oligomers are photographed at 24–32 hpf, *ntl* at 3 dpf, *uroD* at 2 dpf, and *dharma* at 29–34 hpf.

E. Morphology

Embryos were scored according to the following criteria. For *chordin*, embryos exhibiting a shrunken head and a severely ventralized tail with some necrosis were classified as severe. Ventralized embryos with slightly shrunken heads were classified as moderate, and embryos with only slight tail ventralization were classified as mild.

F. Transcription Profiles

Microarray analyses of total RNA extracted from groups of four zebrafish embryos injected with phenol red, MO antisense, or HypNA-pPNA antisense 24 h after injection were carried out in the KCC Microarray Facility. Five Micrograms of RNA was reverse transcribed, labeled, and hybridized to chips spotted with 16,399 5′-amino-C6-oligodeoxynucleotide sequences of 65 nt (Compugen/Sigma-Genosys), using a GeneMachine OmniGridder 100. The sequences represent 16,288 unique gene clusters or ~10,000 genes. The library includes 172 zebrafish beta-actin internal control 65-mers distributed over the entire library (approximately 4/384-well plate) and our own control 65-mers from six different bacterial sequences.

Posthybridization signal detection, chip scanning, and data analysis were carried out by the microarray facility. A single fluorophore is used to label all the chip sequences, because this approach is not susceptible to interference resulting from differing dye incorporation efficiencies, allows a single normalized control chip to be compared with multiple experimental chips, provides increased accuracy because of less variance, and less total RNA is required.

III. Results

A. Antisense PNA Injections

Unmodified PNA oligomers exhibit low solubility at pH 7, are barely soluble at pH 5, and are toxic to embryos (data not shown). Next, we prepared PNAs with a C-terminal L-lysine to increase solubility. Injections of ≈1 pmol of the FITC-12-mer, 13-mer, and 18-mer (Table I) had no discernable effects on zebrafish embryo morphology. At ≈2 pmol of the 18-mer, nonspecific toxicity was apparent.

B. Anionic PNA Analogs

The lack of knockdown activity by PNAs led us to consider the negatively charged HypNA-pPNA analog (Efimov *et al.*, 1999b; Fig. 3). The negatively charged HypNA-pPNAs display excellent hybridization properties toward DNA and RNA while preserving the high single mismatch discrimination and nuclease/protease resistance of PNAs (Efimov *et al.*, 1999a,b; Phelan *et al.*, 2001; Urtishak *et al.*, 2003).

Based on the melting temperatures (T_ms) of a test HypNA-pPNA 17-mer with complementary DNA 16-mers (Urtishak *et al.*, 2003), we predicted that an 18-mer purine-pyrimidine HypNA-pPNA/RNA duplex with GC/AT pair ratios of about 1:1 should have a T_m of 60–65 °C (4–5 μM of each oligomer in 150 mM NaCl,

10 mM MgCl$_2$, 20 mM Tris-HCl, pH 7.5), similar to regular DNA/RNA duplexes. Similarly, we estimated that a single mismatch in a HypNA-pPNA/RNA duplex would lead to a drop in T_m of 10–17 °C, relative to fully complementary duplexes, depending on base mismatch position in an oligomer. In accordance with the HypNA-pPNA/DNA results (Urtishak *et al.*, 2003), we predicted that two mismatches in an 18-mer would greatly destabilize HypNA-pPNA/RNA complexes in the zebrafish model.

Direct measurements of T_ms of HypNA-pPNA knockdown oligonucleotide duplexes with zebrafish mRNA targets (Table I) revealed even higher T_ms than predicted and confirmed the hypothesis of stringent mismatch discrimination. The high T_ms, 76–88 °C, of HypNA-pPNA 18-mer duplexes with RNA 25-mers were equivalent to T_ms of MO 25-mers. The chordin sequence with two mismatches displayed nearly the same T_m as with four mismatches, which might reflect minimal destabilization by T:T mismatches (Peyret *et al.*, 1999). Based on our measurements of the stability and specificity of HypNA-pPNA/RNA duplexes, we hypothesized that injection of HypNA-pPNA 18-mers would allow targeted knockdown of genes in zebrafish.

The first test of knockdown activity in zebrafish embryos with ≈1 pmol of *chordin* FITC-HypNA-pPNA revealed excellent distribution of fluorescein label, except when injected into the yolk, and a strong phenocopy (Fig. 4). Thus, we proceeded to compare the efficacy and specificity of HypNA-pPNA with those of MO oligomers (Urtishak *et al.*, 2003).

C. HypNA-pPNA vs. MO Oligomers Against *chordin*

Embryos injected with *chordin* HypNA-pPNA were severely ventralized and were indistinguishable from larvae injected with *chordin* MO (Fig. 5B and 2D; Urtishak *et al.*, 2003). Embryos injected with *chordin* MO or purified HypNA-pPNA at 0.4 mM yielded strongly ventralized larvae 98% and 80% of the time, respectively. At half the concentration, both MO and HypNA-pPNA injections resulted in a ventralized phenotype of varying degrees approximately 99% of the time (Fig. 5G). To compare the specificity of HypNA-pPNA and MO injections two- and four-base mismatches of HypNA-pPNA and MO were injected. The two- and four-base mismatch *chordin* HypNA-pPNAs had a normal phenotype (Fig. 5C), but on rare instances (≈3%) a mild *chordin*$^{-/-}$ phenotype was observed for both. The two-base mismatch MO morphants had a *chordin*$^{-/-}$ phenotype (96% Fig. 5E) and the four-base mismatch MO morphants had abnormal somites (67%; Fig. 5F).

D. HypNA–pPNA vs. MO Oligomers Against *notail*

To test more rigorously whether HypNA-pPNAs are comparable in efficiency to MOs, another gene expressed early in development, *notail* (*ntl*; Halpern *et al.*, 1993) was tested. Embryos injected with *ntl* antisense oligomers were compared

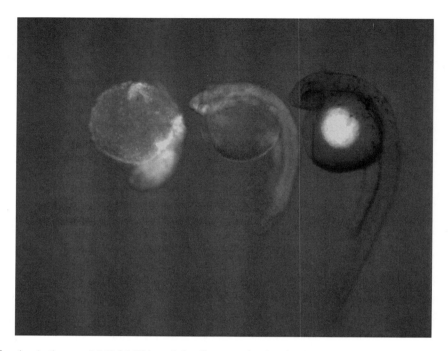

Fig. 4 Antisense gripNA inhibition of chordin expression in zebrafish larvae. Approximately 2 ng of chordin antisense FITC-gripNA was injected into each one-day embryo. One day later, the chordin phenotype and bodywide fluorescence were observed in the left and center larvae. The wild-type larva on the right retained FITC-gripNA in the yolk and displayed no phenotype. 2-mismatch and 4-mismatch gripNAs were inactive.

with the known phenotype of $ntl^{-/-}$. If injected embryos had a short tail and linear somites, they were considered a phenocopy of the known mutant. Originally, the HypNA-pPNA sequence was designed based on the published MO sequence (Nasevicius and Ekker, 2000). However, when this oligomer was injected, the $ntl^{-/-}$ mutant phenotype was observed only 30% of the time (data not shown). To understand why a low $ntl^{-/-}$ frequency was seen with HypNA-pPNA, the MO and HypNA-pPNA sequences were compared to the ntl 5' UTR (GenBank Accession Number NM_131162), which revealed that both the MO and Hyp-NA-pPNA sequences contained a one-base mismatch (Fig. 6B). To resolve this problem, a new HypNA-pPNA sequence was synthesized without the mismatch (Fig. 6B) and injected. Injections with the new ntl HypNA-pPNA sequence yielded the $ntl^{-/-}$ phenotype (Fig. 6A) 96% of the time. Having observed greater specificity with antisense HypNA-pPNA 18-mers than with MO 25-mers, we then examined the potency and specificity of MO 18-mers against *chordin* and ntl, which displayed improved specificity than MO 25-mers, although still less than the HypNA-pPNAs (data not shown).

E. HypNA-pPNA vs. MO Oligomers Against *uroD*

Having observed that the HypNA-pPNA can effectively knock down early acting genes, the later-acting *uroD* was targeted. In the case of *uroD*, embryos were classified according to number of fluorescent blood cells. Embryos were considered to have a bright phenotype if many fluorescent cells were seen and a dim phenotype if only a few fluorescent cells were seen. Both HypNA-pPNA and MO injections resulted in hepatoerythropoietic porphyria (HEP) embryos (Nasevicius and Ekker, 2000; Wang *et al.*, 1998) in high frequency (Fig. 7). Bright fluorescent orange blood cells were observed in ≈99% of embryos injected with both HypNA-pPNA (Fig. 7B) and MO (Fig. 7D). However, at the highest concentration of HypNA-pPNA, body mutations were also induced in 45% of the embryos with the HEP phenotype. The frequency of the HEP phenotype for HypNA-pPNA was observed to decrease as the concentration was lowered; in contrast, the HEP frequency remained constant in larvae injected with antisense MO. To verify the specificity of the HypNA-pPNA, a two-base mismatch was injected, which yielded a normal phenotype in ≈58% of injected larvae (Figs. 7E, 4F, and 4G).

F. HypNA-pPNA vs. MO Oligomers Against *dharma*

Knowing that MOs can cause nonspecific mistargeting effects, we targeted a gene that is poorly inhibited by a known MO at low doses and at higher doses exhibits head and tail necrosis (mistargeting). Embryos injected with *dharma* MO were classified as mild if only a slight tail defect was observed and the somites were U-shaped. With the moderate phenotype, the head and tail size were reduced along with U-shaped somites. At the most severe level, head and tail were severely reduced associated with a loss of eyes. The MO targeting the *dharma* (or *bozozok*) gene exhibited this pattern (Ekker and Larson, 2001). Because the 25-mer MO was significantly longer than the 18-mer HypNA-pPNA, two HypNA-pPNAs were designed to cover the entire sequence targeted by the MO (Fig. 8H). When the MO was injected, only a slight tail mutation was seen at the lowest dose (Fig. 8A). However, as the concentration increased, widespread necrosis was observed throughout the embryo (Fig. 8B, C). Embryos injected with *dharma*2 exhibited little or no necrosis even at high doses and resulted in embryos exhibiting the *dharma* phenotype (Fig. 8E, F). Like the MO, the lowest dose of HypNA-pPNA caused a slight tail mutation (Fig. 8D). Interestingly, when *dharma*1 was injected, a normal phenotype (Fig. 8G) was seen 97% of the time and infrequently embryos resembled the lowest dose of *dharma*2. It is possible that this result might be due to the secondary structure of *dharma* mRNA, which could determine accessibility to antisense agents.

G. HypNA-pPNA vs. MO Persistence over Time

To compare the relative stabilities of HypNA-pPNA vs. MO oligomers, we designed sequences to target the translational start site of the mRNA for green

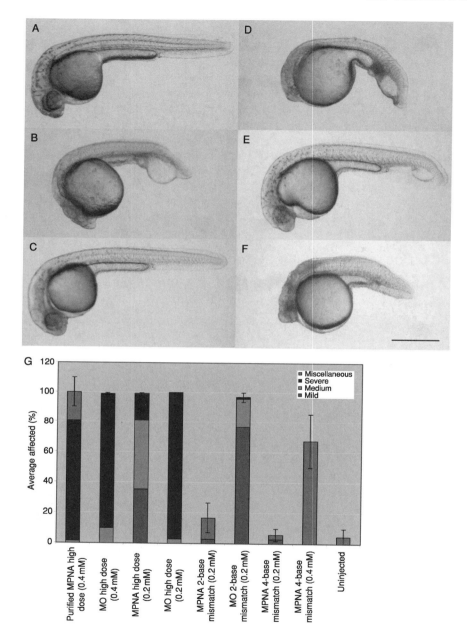

Fig. 5 *Chordin* inhibition. Uninjected embryos displayed wild-type morphology at 24 hpf (A). Embryos injected with *chordin* HypNA-pPNA phenocopy the null *chordin*$^{-/-}$ mutation (B). Embryos injected with *chordin* two-base-mismatch HypNA-pPNA appear wild type (C). Embryos injected with *chordin* MO phenocopy the null *chordin*$^{-/-}$ mutation (D). Chordin MO-injected embryo. Slight

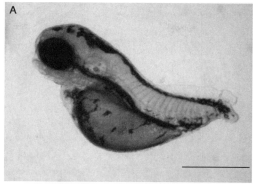

Fig. 6 *Notail* inhibition. *notail* null ($ntl^{-/-}$) mutant was phenocopied. HypNA-pPNA injection (72 hpf, scale bar 500 μm) (A). *notail* 5′ UTR sequence region used (nt 78–95) for HypNA-pPNA targeting (B). From Urtishak, K. A., *et al.* (2003). Targeted gene knockdown in zebrafish using negatively charged peptide nucleic acid mimics. *Dev. Dynam.* **228**, 405–413, with permission.

reef coral fluorescent protein (GRCFP; Matz *et al.*, 1999, Table I). Aliquots of 5 pmol of each oligomer were injected into transgenic zebrafish embryos that express GRCFP specifically in blood vessels (Cross *et al.*, 2003). At 24 hpf, GRCFP was observed in the dorsal aorta, caudal vein, cranial vessels, and intersegmental vessels of mock-injected embryos (Fig. 9A). Fluorescence was almost undetectable in embryos injected with either the HypNA-pPNA or the MO oligomer (Fig. 9B, C). By 48 hs, fluorescence was observed in the blood vessels of HypNA-pPNA-injected embryos, though not at the same level as mock-injected embryos (Fig. 9D, E). MO-injected embryos continued to exhibit almost no fluorescence at this stage (Fig. 9F). At 72 h, when MO-injected embryos began to express GRCFP, the level of fluorescence was reduced compared with the corresponding HypNA-pPNA-injected and mock-injected embryos (Fig. 9G–I).

ventralization was seen in the tail of an embryo injected with a *chordin* two base mismatch MO (E). Embryos injected with a *chordin* four base-mismatch MO displayed mistargeting (F). Scale bar 500 μm. The bar graph presentation (G) depicts the average phenotype rate for a specific dose, along with the range of severities seen. Miscellaneous mutations are those that do not phenocopy the null *chordin*$^{-/-}$ mutation. From Urtishak, K. A., *et al.* (2003). Targeted gene knockdown in zebrafish using negatively charged peptide nucleic acid mimics. *Dev. Dynam.* **228**, 405–413, with permission.

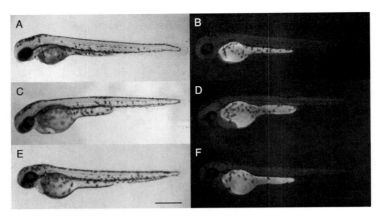

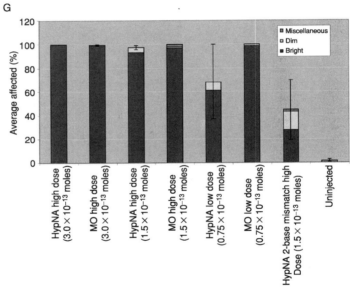

Fig. 7 *UroD* inhibition. Embryos injected with *uroD* HypNA-pPNA (A, B) or MO (C, D) displayed hepatoerythopoietic porphyria (HEP) observed at 48 hpf. Embryos injected with *uroD* HypNA-pPNA 2 base mismatch (E, F) were often wild type. Summary of injections (G): bright or dim refers to the intensity and number of fluoresceinated blood cells and miscellaneous refers to other phenotypes not characteristic of HEP. Scale bar 500 μm. From Urtishak, K. A., *et al.* (2003). Targeted gene knockdown in zebrafish using negatively charged peptide nucleic acid mimics. *Dev. Dynam.* **228**, 405–413, with permission.

H. Lack of HypNA-pPNA Oral Availability

The *uroD* antisense and mismatch FITC-HypNA-pPNAs were added to final concentration of 25 μM and 50 μM in glass tubes with 100 μl of embryo medium containing four 2-day-old embryos. A damp Kimwipe was inserted into the top of each tube, which was then wrapped with Parafilm so that no moisture could

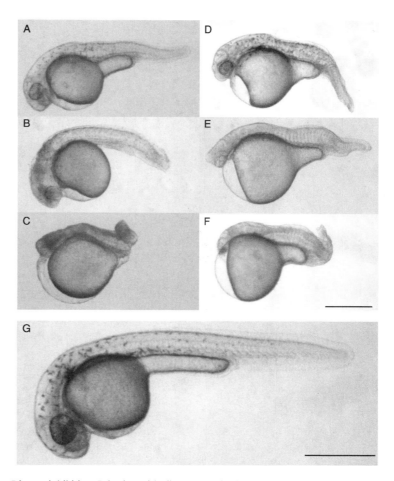

Fig. 8 *Dharma* inhibition. Injection with *dharma* MO leads to mistargeting. In embryos injected with *dharma* MO at the lowest dose only, a tail phenotype was seen (A). As the concentration of *dharma* MO increased, necrosis in the head was observed (B, C). The lowest dose of *dharma*2 HypNA-pPNA induced the tail phenotype (D). As the concentration of *dharma*2 HypNA-pPNA increased, so did the severity of the phenotype (E, F). With *dharma*2 HypNA-pPNA, necrosis was observed less frequently and with less severity than with *dharma* MO. Embryos injected with *dharma*1 HypNA-pPNA were wild-type (G). Embryos were examined ≈33 hpf. Scale bars 500 μm. From Urtishak, K. A., *et al.* (2003). Targeted gene knockdown in zebrafish using negatively charged peptide nucleic acid mimics. *Dev. Dynam.* **228**, 405–413, with permission.

escape. The embryos were allowed to swim freely in the solution for 1 day at 25 μM or 3 days at 50 μM, by which time they had developed into larvae with open mouths. The treated larvae were compared with untreated control larvae for the distribution of green fluorescence reflecting the localization of FITC-HypNA-pPNAs and for orange fluorescence in the blood pool as a consequence of HEP.

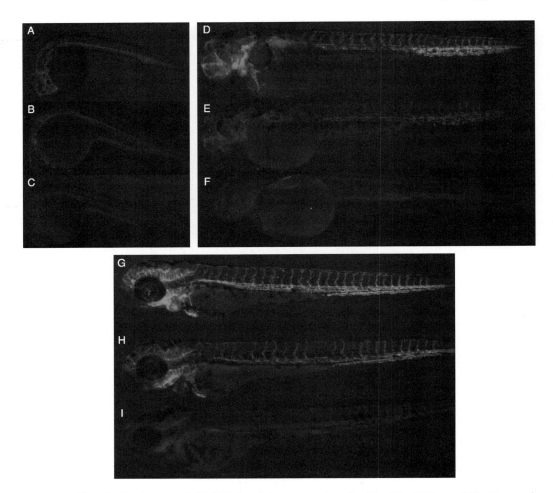

Fig. 9 Persistence of GRCFP knockdown over time for HypNA-pPNA vs. MO. Transgenic zebrafish embryos that express GRCFP in blood vessels were injected with 0.2% phenol red (A, D, G; $n = 115$), 5 pmol GRCFP HypNA-pPNA in 0.2% phenol red (B, E, H; $n = 87$), or 5 pmol GRCFP MO in 0.2% phenol red (C, F, I; $n = 84$). Embryos at 24 hpf (A–C), 48 hpf (D–F), and 72 hpf (G–I).

Green fluorescence was observed along the alimentary canal, but was not dispersed to the tissues (not shown). No orange fluorescence was observed in any larva. Further modification is necessary to enable capillary uptake from the gut and cellular uptake from circulation.

I. Transcription Profile Identification of a Dependent Downstream Gene

Zebrafish embryos were injected with three overlapping MO and HypNA-pPNA antisense sequences against a tumor suppressor gene whose mechanism of action remains unclear. First pass bioinformatics analysis of the transcription profiles from RNA samples extracted from treated and control embryos identified up- and downregulated genes, of which one clear example was p53. Among the three overlapping MO sequences, p53 mRNA increased by 7.1 ± 0.3 fold; for the three overlapping HypNA-pPNA sequences, p53 mRNA increased by 5.5 ± 0.2 fold.

IV. Discussion

On testing our hypothesis with HypNA-pPNA and MO oligomers containing zero, two, or four mismatches with *chordin* antisense, the HypNA-pPNA mismatches displayed no activity, whereas the MO with two mismatches induced frequent *chordin*$^{-/-}$ phenotypes and the MO with four mismatches induced a nonspecific phenotype. The biological results agreed precisely with the physical chemical results.

Similar specificity was observed for *uroD* with two mismatches in the HypNA-pPNA. In the most rigorous attempt to disprove the hypothesis of HypNA-pPNA specificity, a single base mismatch in the *ntl* HypNA-pPNA induced the *ntl*$^{-/-}$ phenotype in only 30% of the injected larvae, whereas the antisense HypNA-pPNA induced the *ntl*$^{-/-}$ phenotype in virtually all larvae injected. These results are all consistent with the model that 18-mer HypNA-pPNA oligomers display high sequence specificity relative to MO 25-mers.

On the other hand, it was apparent that HypNA-pPNA oligomers lost activity in zebrafish embryos sooner than MO oligomers did and were not orally available. Hence, further derivatization to resist metabolism and enable uptake from the gut would be desirable.

To query the functions of oncogene orthologs after Day 5 of development, we will have to design and synthesize novel negatively charged PNA analogs capable of cellular uptake. This might require conjugation of small molecules or peptides that will facilitate internalization, along with fluorophores for detection in later-stage larvae and free-swimming unpigmented fish. Uptake could present a new problem, if there are different selection rules for a phosphono PNA than have been observed for classical PNAs. That is not particularly likely, because it is apparent that basic peptides also facilitate uptake of normal phosphodiester oligonucleotides (Soomets *et al.*, 1999).

Toxicity was an issue for embryos injected with PNA-trifluoroacetate salts, the form in which PNAs emerged from reversed-phase HPLC, and continued to be a problem for HypNA-pPNA until we desalted them on Waters Sep-Paks. Perhaps a new mode of toxicity will become apparent when anionic PNA derivatives capable of cellular uptake are presented to free-swimming fish. Such a result would require a new iteration of our design cycle to reduce toxicity while maintaining cellular uptake and knockdown capabilities.

V. Summary

We found that negatively charged, highly soluble PNA analogs with alternating phosphonates (HypNA-pPNAs) are effective and specific antisense agents in zebrafish embryos, showing comparable potency and greater specificity against *chordin*, *ntl* and *uroD*. In addition, we successfully phenocopied a *dharma* mutant that had not been found susceptible to MO knockdown. Both MO and HypNA-pPNAs against a tumor suppressor gene induced comparable upregulation of p53, illustrating similar effects on transcription profiles. HypNA-pPNAs are therefore a valuable alternative for reverse genetic studies, enabling the targeting of previously inaccessible genes in zebrafish or validating newly identified orthologs, and perhaps for reverse genetic studies in other organisms.

Acknowledgments

We thank Dr. Steven Ekker for helpful discussions and reagents, and Dr. John Archdeacon for his valuable advice and encouragement. S. A. F and E. W. are funded by the NIH, and N. S. is supported by a predoctoral fellowship from HHMI.

References

Ackermann, G. E., and Paw, B. H. (2003). Zebrafish: A genetic model for vertebrate organogenesis and human disorders. *Front Biosci.* **8**, d1227–d1253.

Agrawal, S. (1996a). Antisense oligonucleotides: Towards clinical trials. *Trends Biotechnol.* **14**(10), 376–387.

Agrawal, S. (1996b). "Antisense Therapeutics." Humana Press, Totowa, NJ.

Agrawal, S., Jiang, Z., Zhao, Q., Shaw, D., Cai, Q., Roskey, A., Channavajjala, L., Saxinger, C., and Zhang, R. (1997). Mixed-backbone oligonucleotides as second generation antisense oligonucleotides: *In vitro* and *in vivo* studies. *Proc. Natl. Acad. Sci. USA* **94**(6), 2620–2625.

Altschul, S. F., Madden, T. L., *et al.* (1997). Gapped BLAST and PSI-BLAST: A new generation of protein database search programs. *Nucleic Acids Res.* **25**(17), 3389–3402.

Basu, S., and Wickstrom, E. (1995). Solid phase synthesis of a D-peptide-phosphorothioate oligodeoxynucleotide conjugate from two arms of a polyethylene glycol-polystyrene support. *Tetrahedr. Lett.* **36**, 4943–4946.

Basu, S., and Wickstrom, E. (1997). Synthesis and characterization of a peptide nucleic acid conjugated to a D-peptide analog of insulin-like growth factor 1 for increased cellular uptake. *Bioconj. Chem.* **8**(4), 481–488.

Beier, D. R. (1998). Zebrafish: Genomics on the fast track. *Genome Res.* **8**(1), 9–17.

Belikova, A. M., Zarytova, V. F., and Grineva, N. I. (1967). Synthesis of ribonucleosides and diribonucleoside phosphates containing 2-chloroethylamine and nitrogen mustard residues. *Tetrahedr. Lett.* **37**, 3557–3562.

Bishop, J. M. (1991). Molecular themes in oncogenesis. *Cell* **64**(2), 235–248.

Brownlie, A., Donovan, A., Oates, A. C., Brugnara, C., Witkowska, H. E., Sassa, S., and Zon, L. I. (1998). "Positional cloning of the zebrafish sauternes gene: A model for congenital sideroblastic anaemia. *Nat. Genet.* **20**(3), 244–250.

Cross, L. M., Cook, M. A., Lin, S., Chen, J. N., and Rubinstein, A. L. (2003). Rapid analysis of angiogenesis drugs in a live fluorescent zebrafish assay. *Arterioscler. Thromb. Vasc. Biol.* **23**(5), 911–912.

Detrich, H. W., 3rd, Westerfield, M., Zon, L. I. (eds.) (1999). "The Zebrafish: Genetics and Genomics" *Methods Cell Biol.* Academic Press San Diego, CA.

Driever, W., Solnica-Krezel, L., Schier, A. F., Neuhauss, S. C., Malicki, J., Stemple, D. L., Stainier, D. Y., Zwartkruis, F., Abdelilah, S., Rangini, Z., Belak, J., and Boggs, C. (1996). A genetic screen for mutations affecting embryogenesis in zebrafish. *Development* **123**, 37–46.

Efimov, V. A., Buryakova, A. A., and Chakhmakhcheva, O. G. (1999a). Synthesis of polyacrylamides *N*-substituted with PNA-like oligonucleotide mimics for molecular diagnostic applications. *Nucleic Acids Res.* **27**(22), 4416–4426.

Efimov, V. A., Buryakova, A. A., Choob, M., and Chakhmakhcheva, O. G. (1999b). Peptide nucleic acids and their phosphonate analogues. II. Synthesis and physicochemical properties of hybrids containing serine and 4-hydroxyproline residues. *Bioorganicheskaya Khimia* **25**(8), 611–622.

Efimov, V. A., Choob, M. V., Buryakova, A. A., Kalinkina, A. L., and Chakhmakhcheva, O. G. (1998). Synthesis and evaluation of some properties of chimeric oligomers containing PNA and phosphono-PNA residues. *Nucleic Acids Res.* **26**(2), 566–575.

Egholm, M., Buchardt, O., Christensen, L., Behrens, C., Freier, S. M., Driver, D. A., Berg, R. H., Kim, S. K., Norden, B., and Nielsen, P. E. (1993). PNA hybridizes to complementary oligonucleotides obeying the Watson-Crick hydrogen-bonding rules [see comments]. *Nature* **365**(6446), 566–568.

Ekker, S. C., and Larson, J. D. (2001). Morphant technology in model developmental systems. *Genesis* **30**(3), 89–93.

Fisher, S., Amacher, S. L., and Halpern, M. E. (1997). Loss of cerebum function ventralizes the zebrafish embryo. *Development* **124**(7), 1301–1311.

Gryaznov, S., Skorski, T., Cucco, C., Nieborowska-Skorska, M., Chiu, C. Y., Lloyd, D., Chen, J. K., Koziolkiewicz, M., and Calabretta, B. (1996). Oligonucleotide N3' → P5' phosphoramidates as antisense agents. *Nucleic Acids Res.* **24**(8), 1508–1514.

Haffter, P., Granato, M., Brand, M., Mullins, M. C., Hammerschmidt, M., Kane, D. A., Odenthal, J., van Eeden, F. J., Jiang, Y. J., Heisenberg, C. P., Kelsh, R. N., Furutani-Seiki, M., Vogelsang, E., Beuchle, D., Schach, U., Fabian, C., and Nusslein-Volhard, C. (1996). The identification of genes with unique and essential functions in the development of the zebrafish, *Danio rerio. Development* **123**, 1–36.

Halpern, M. E., Ho, R. K., Walker, C., and Kimmel, C. B. (1993). Induction of muscle pioneers and floor plate is distinguished by the zebrafish no tail mutation. *Cell* **75**(1), 99–111.

Heasman, J. (2002). Morpholino oligos: Making sense of antisense? *Dev. Biol.* **243**(2), 209–214.

Ho, P. T., Ishiguro, K., Wickstrom, E., and Sartorelli, A. C. (1991). Non-sequence-specific inhibition of transferrin receptor expression in HL-60 leukemia cells by phosphorothioate oligodeoxynucleotides. *Antisense Res. Dev.* **1**(4), 329–342.

Hughes, J., Astriab, A., Yoo, H., Alahari, S., Liang, E., Sergueev, D., Shaw, B. R., and Juliano, R. L. (2000). *In vitro* transport and delivery of antisense oligonucleotides [In Process Citation]. *Methods Enzymol.* **313**, 342–358.

Hyrup, B., and Nielsen, P. E. (1996). Peptide nucleic acids (PNA): Synthesis, properties and potential applications. *Bioorg. Med. Chem.* **4**(1), 5–23.

Iribarren, A. M., Sproat, B. S., Neuner, P., Sulston, I., Ryder, U., and Lamond, A. I. (1990). 2'-O-alkyl oligoribonucleotides as antisense probes. *Proc. Natl. Acad. Sci. USA* **87**(19), 7747–7751.

Knudson, A. G., Jr. (1970). Genetics and cancer. *Postgrad. Med.* **48**(5), 70–74.

Lebedev, A. V., and Wickstrom, E. (1996). The chirality problem in P-substituted oligonucleotides. *In* "Perspectives in Drug Discovery and Design" (G. Trainor, ed.), Vol. 4, pp. 17–40. ESCOM Science Publishers, Leiden.

Matz, M. V., Fradkov, A. F., Labas, Y. A., Savitsky, A. P., Zaraisky, A. G., Markelov, M. L., and Lukyanov, S. A. (1999). Fluorescent proteins from nonbioluminescent Anthozoa species. *Nat. Biotechnol.* **17**(10), 969–973.

Miller, P. S. (1991). Oligonucleoside methylphosphonates as antisense reagents. *Biotechnology (NY)* **9**(4), 358–362.

Monia, B. P., Lesnik, E. A., Gonzalez, C., Lima, W. F., McGee, D., Guinosso, C. J., Kawasaki, A. M., Cook, P. D., and Freier, S. M. (1993). Evaluation of 2′-modified oligonucleotides containing 2′ deoxy gaps as antisense inhibitors of gene expression. *J. Biol. Chem.* **268**(19), 14514–14522.

Nasevicius, A., and Ekker, S. C. (2000). Effective targeted gene 'knockdown' in zebrafish. *Nat. Genet.* **26**(2), 216–220.

Nielsen, P. E., Egholm, M., Berg, R. H., and Buchardt, O. (1993). Peptide nucleic acids (PNAs): Potential antisense and anti-gene agents. *Anticancer Drug Des.* **8**(1), 53–63.

Nusslein-Volhard, C. (1994). Of flies and fishes. *Science* **266**(5185), 572–574.

Peyret, N., Seneviratne, P. A., Allawi, H. T., and SantaLucia, J., Jr. (1999). Nearest-neighbor thermodynamics and NMR of DNA sequences with internal A.A, C.C, G.G, and T.T mismatches. *Biochemistry* **38**(12), 3468–3477.

Phelan, D., Hondrop, K., Choob, M., Efimov, V., and Fernandez, J. (2001). Messenger RNA isolation using novel PNA analogues. *Nucleosides Nucleotides Nucleic Acids* **20**(4–7), 1107–1111.

Schulte-Merker, S., van Eeden, F. J., Halpern, M. E., Kimmel, C. B., and Nusslein-Volhard, C. (1994). No tail (ntl) is the zebrafish homologue of the mouse T (Brachyury) gene. *Development* **120**(4), 1009–1015.

Shaw, B. R., Sergueev, D., He, K., Porter, K., Summers, J., Sergueeva, Z., and Rait, V. (2000). Boranophosphate backbone: A mimic of phosphodiesters, phosphorothioates, and methyl phosphonates. *Methods Enzymol.* **313**, 226–257.

Soomets, U., Hallbrink, M., and Langel, U. (1999). Antisense properties of peptide nucleic acids. *Front Biosci.* **4**, D782–D786.

Stec, W. J., Grajkowski, A., Koziolkiewicz, M., and Uznanski, B. (1991). Novel route to oligo(deoxyribonucleoside phosphorothioates). Stereocontrolled synthesis of *P*-chiral oligo(deoxyribonucleoside phosphorothioates). *Nucleic Acids Res.* **19**(21), 5883–5888.

Summerton, J. (1989). *In* "Discoveries in Antisense Nucleic Acids" (C. Brakel, ed.), pp. 71–80. Portfolio Publishing, The Woodlands, TX.

Summerton, J., Stein, D., Huang, S. B., Matthews, P., Weller, D., and Partridge, M. (1997). Morpholino and phosphorothioate antisense oligomers compared in cell-free and in-cell systems. *Antisense Nucleic Acid Drug Dev.* **7**(2), 63–70.

Summerton, J., and Weller, D. (1997). Morpholino antisense oligomers: design, preparation, and properties. *Antisense Nucleic Acid Drug Dev.* **7**(3), 187–195.

Tian, X., and Wickstrom, E. (2002). Continuous solid-phase synthesis and disulfide cyclization of peptide-PNA-peptide chimeras. *Org. Let.* **4**(23), 4013–4016.

Urtishak, K. A., Choob, M., Tian, X., Sternheim, N., Talbot, W. S., Wickstrom, E., and Farber, S. A. (2003). Targeted gene knockdown in zebrafish using negatively charged peptide nucleic acid mimics. *Dev. Dynam.* **228**, 405–413.

Wang, H., Long, Q., Marty, S. D., Sassa, S., and Lin, S. (1998). A zebrafish model for hepatoerythropoietic porphyria. *Nat. Genet.* **20**(3), 239–243.

Westerfield, M. (2000). The Zebrafish Book. A Guide for the Laboratory use of Zebrafish (*Danio rerio.*). University of Oregon Press, Eugene, OR.

Wickstrom, E. (1991). "Prospects for Antisense Nucleic Acid Therapy of Cancer and AIDS." Wiley-Liss, New York.

Wickstrom, E. (1992). Strategies for administering targeted therapeutic oligodeoxynucleotides. *Trends Biotechnol.* **10**(8), 281–287.

Wickstrom, E. (1998). "Clinical Trials of Genetic Therapy with Antisense DNA and DNA Vectors." Marcel Dekker, New York.

Wooster, R. (2000). Cancer classification with DNA microarrays is less more? *Trends Genet.* **16**(8), 327–329.

Zamecnik, P. C., and Stephenson, M. L. (1978). Inhibition of Rous sarcoma virus replication and cell transformation by a specific oligodeoxynucleotide. *Proc. Natl. Acad. Sci. USA* **75**(1), 280–284.

Zendegui, J. G., Vasquez, K. M., Tinsley, J. H., Kessler, D. J., and Hogan, M. E. (1992). *In vivo* stability and kinetics of absorption and disposition of 3′ phosphopropyl amine oligonucleotides. *Nucleic Acids Res.* **20**(2), 307–314.

CHAPTER 9

Photo–Mediated Gene Activation by Using Caged mRNA in Zebrafish Embryos

Hideki Ando,[*] Toshiaki Furuta,[†] and Hitoshi Okamoto[*]

[*]Laboratory for Developmental Gene Regulation
Brain Science Institute, RIKEN (The Institute of Physical and Chemical Research)
Saitama 351-0198, Japan
and
CREST (Core Research for Evolutional Science and Technology)
Japan Science and Technology Corporation (JST)
Tokyo 103-0027, Japan

[†]Department of Biomolecular Science
Toho University
Funabashi Chiba 274-8510, Japan
and
PREST (Precursory Research for Embryonic Science and Technology)
Japan Science and Technology Corporation (JST)
Tokyo 103-0072, Japan

I. Introduction

Recent advances in systematically compiling information on sequences of the whole genome and the expressed sequence tags (ESTs) with expression patterns of individual genes has made identification of novel genes easier than ever (Barbazuk *et al.*, 2000; Lo *et al.*, 2003). Recently, a knockdown technology by injection of antisense morpholino oligonucleotide into zebrafish embryos was developed to facilitate direct assignment of functions to such genes (Nasevicius and Ekker, 2000). Another powerful approach for studying gene functions is a gain-of-function approach by ectopic expression of genes in a temporal- and spatial-specific manner.

One promising approach is RNA caging, in which RNA is inactivated by covalent attachment of a photoremovable protecting group (caging group) and then reactivated by photoillumination with light of a specific wavelength. RNA caging was first achieved by the site-specific modification of the 2′-hydroxyl nucleophile in the substrate RNA of the hammerhead ribozyme with a caging functionality, *O*-(2-nitrobenzyl) caging group (Chaulk and MacMillan, 1998). Susceptibility of the substrate RNA to hammerhead-catalyzed cleavage reaction was abolished by this modification, but recovered rapidly and efficiently after removal of this group by photoillumination (308 nm, $10 \, J/cm^2$) with excimer laser. Another pioneering attempt to achieve spatiotemporal control of gene expression has been made by caging DNA with the 1-(4, 5-dimethoxy-2-nitrophenyl)ethyl (DMNPE) group (Monroe *et al.*, 1999). HeLa cells trasfected with the plasmid, which encoded GFP and was caged with this agent, showed reduced levels (0–25%) of GFP expression compared with cells transfected with the intact plasmid, and exposure of these cells to UV light (365 nm, $0.25–0.5 \, J/cm^2$) doubled the expression level.

To apply the RNA-caging technology for controlling gene expression at spatially and temporally high resolution in rapidly developing zebrafish embryos, it is essential to use a caging agent that is easy to react with mRNA *in vitro* and to be removed *in vivo* by a minimum amount of photoillumination. To meet this requirement, we developed an RNA caging system that uses the caging agent 6-bromo-4-diazomethyl-7-hydroxycoumarin (Bhc-diazo; Ando *et al.*, 2001; Tsien and Furuta, 2000a,b). 6-Bromo-7-hydroxycoumarin-4-yl methyl (Bhc)-caged mRNA loses almost all translational activity, but illumination with 350 through 365 nm ultraviolet (UV) light removed Bhc from caged mRNA, resulting in recovery of translational activity. Bhc can be removed from RNA by photoillumination with a low level of energy ($100 \, mJ/cm^2$). This advantage has enabled the precise control of expression of genes in live zebrafish embryos by photoillumination without giving a critical damage to the tissue.

In this chapter, we describe the basic principle of the design and synthesis of Bhc-diazo and the detailed method for caged mRNA technology based on what we previously described (Ando and Okamoto, 2003; Ando *et al.*, 2001).

II. Synthesis of Bhc–Caged mRNA

A. Design and Synthesis of Bhc–Diazo

Caged compounds that can be activated by photoirradiation should offer an ideal method to control the intracellular concentration of a signaling molecule and will provide an opportunity to control gene expressions with high temporal and spatial resolution (Adams et al., 1993; Pelliccioli et al., 2002). Among the caging groups available, we chose the Bhc group to mask the translational activity of mRNA. Bhc is a newly developed photochemically removable protecting group designed to protect carboxylates, amines, phosphates, and alcohols, and has several advantages over those reported previously: high photosensitivity on one-photon excitation, large cross-sections on two-photon excitation, and improved stability in the dark (Ando et al., 2001; Furuta et al., 1999; Lin et al., 2002; Lu et al., 2003; Montgomery et al., 2002; Robu et al., 2003; Suzuki et al., 2003; Tsien and Furuta, 2000a,b). All these properties are favorable, especially for cell biological applications. For example, Bhc-caged glutamates show 5–10 times higher photosensitivity than CNB-glutamate on one-photon UV irradiation both in cuvette experiments and on brain slices. Therefore, we can reduce the intensity of the uncaging light nearly 10-fold with Bhc-caged compounds to achieve the same magnitude of activation, which leads to the minimization of unfavorable tissue damage.

Synthesis of caged RNA can be achieved in two ways. In the first, a full-length mRNA is used as the starting material and the synthesis involves a chemoselective protection of one of the functional groups in the RNA with a caging agent. In the second, a caged nucleoside monomer is the precursor for a chemical synthesis of the caged full-length mRNA. The latter approach requires the rather complicated synthesis of an appropriately protected caged monomer, and, at present, the chemical synthesis can only be applied to short RNAs. The former is straightforward and easier in access for most biologists if an appropriate caging agent is available. Furthermore, there is no limitation of the length and sequence of the starting nucleotides.

We designed Bhc-diazo as a phosphate-specific caging agent. Bhc-diazo is expected to react with the phosphate moiety of the sugar–phosphate backbone of RNA to yield the 6-bromo-7-hydroxycoumarin-4-ylmethyl ester of the phosphates, because most diazoalkanes are known to react with phosphoric acids without any catalysts or additives to form alkyl esters.

Bhc-diazo was synthesized starting from commercially available 4-bromoresorcinol (A) in five steps, with an overall yield of 50% (Scheme 1). The use of a fresh and finely powdered selenium dioxide is necessary to achieve reproducible product yields in Step 3. In this reaction, the aldehyde (D) was purified by recrystallization from toluene. A trace amount of by-product probably derived from a selenoalkyl intermediate was often detected and must be completely removed during the purification, otherwise Bhc-diazo is contaminated by unidentified impurities, which affect the efficiency and reproducibility of the final caging reaction.

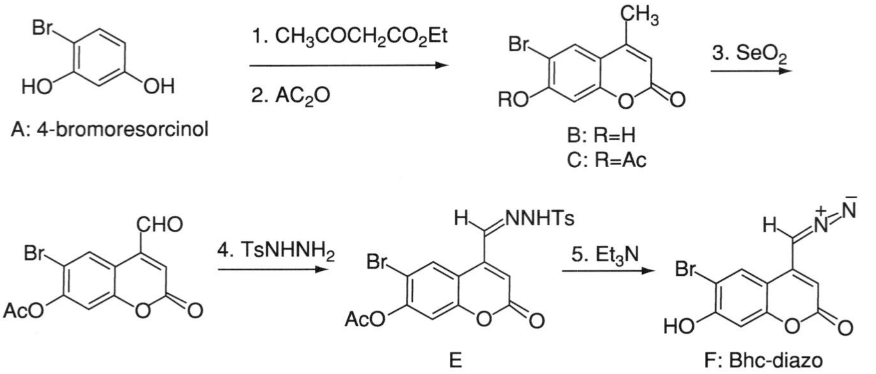

Scheme 1 Synthesis of Bhc-diazo.

B. Chemical Properties of Bhc–Diazo

Bhc-diazo is a stable, nonvolatile light-yellow solid diazo compound and can be stored for at least 6 months at room temperature without any detectable decomposition. The compound is soluble in dimethyl sulfoxide (DMSO), *N,N*-dimethylformamide (DMF), and acetone; slightly soluble in methanol, ethanol, and chloroform; and practically insoluble in water. Scheme 2 illustrates plausible mechanisms of the caging reaction of RNA with Bhc-diazo. A nucleophilic displacement reaction initiated by the protonation of the diazo carbon atom with a free phosphoric acid (nonionized form) can lead to the Bhc ester of the phosphate (Route a). The carbene generated from Bhc-diazo can be an intermediate and inserted into the O-H bond of a free phosphoric acid to form the Bhc ester (Route b). Both mechanisms require the nonionized form of a phosphoric acid.

Caution. In general, volatile organic diazo compounds, especially with low molecular weight (such as diazomethane), can explode. The molecular weight of Bhc-diazo is 281, and it has a polar phenolic hydroxyl group. In addition, the chemical reactivity of Bhc-diazo is modest. Therefore, Bhc-diazo is not volatile and seems to be not explosive. Although we usually scratch the powder of Bhc-diazo to pick up from a glass tube and apply heating (about 80 °C), we have never experienced its explosion since we first synthesized Bhc-diazo in 1998. Nevertheless, note that a possibility of explosion cannot be completely excluded.

C. Synthesis of Bhc–Caged mRNA

Special note. All procedures described next should be performed under ultraviolet-free conditions. All experiments treating Bhc-diazo and caged mRNA are performed under illumination of the fluorescent lamp laminated by UV-shielding film, which is developed for use in the semiconductor industry. Toshiba supplies such a lamp (FLR40S. Y/M.P.NU) that emits light with wavelengths longer than 450 nm. UV-shielding film can also be purchased from the same

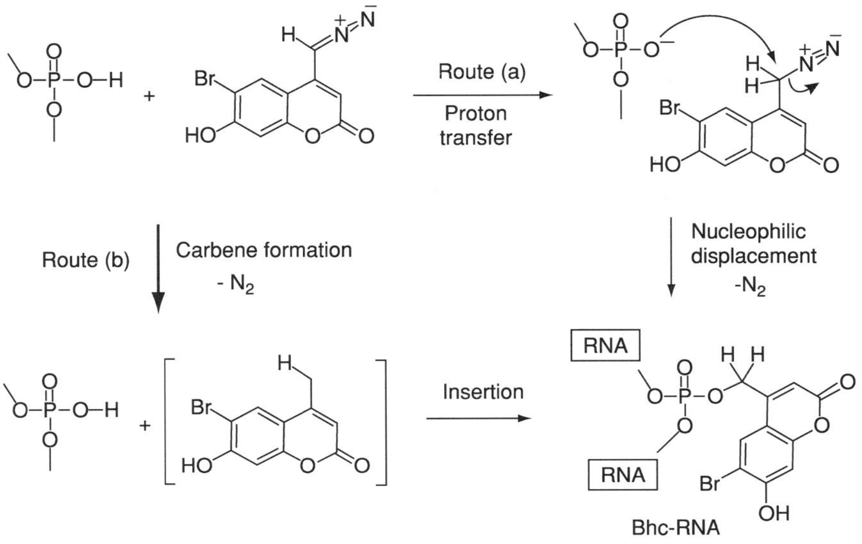

Scheme 2 Plausible reaction mechanisms for RNA caging with Bhc-diazo.

company and be easily used to coat conventional fluorescent lamps by placing the lamp in it and heating the tube with a high-power air blower (PJ208A, Ishizaki Electric MFG).

1. Preparation of *In Vitro* Synthesis of mRNA

mRNAs to be caged with Bhc-diazo were synthesized *in vitro* from full-length cDNAs in the pCS2+ vector, a general expression vector (Rupp *et al.*, 1994; Turner and Weintraub, 1994). We used the mMESSAGE mMACHINE (Ambion, Austin, TX) kit for *in vitro* transcription. Although, depending on length of the mRNA, 10–20 μg of mRNA was obtained in a single procedure. The mRNA pellets were rinsed with 80% ethanol and stored at $-80\,^\circ$C in 80% ethanol.

For the caging reaction, after centrifuging (15k rpm, $4\,^\circ$C, 5 min), ethanol should be carefully removed completely, because the remaining ammonium acetate tends to prevent the following chemical reaction with Bhc-diazo. The mRNA pellet was dried in air. A vacuum dessicator should not be used, because it might excessively dry the mRNA. The overdried pellet cannot be dissolved in DMSO. Just before the pellet became completely dry (at the time the pellet became nearly transparent), it was mixed with 5 μl of DMSO completely by gentle pipetting to resolve mRNA. Pipetting was continued until the solution started to retain the air bubbles for a moment after each pipetting because of the increase in viscosity (approximately 100 times). The concentration of mRNA in this solution was 2–4 μg/μl.

2. Preparation of Bhc–Diazo Solution

Bhc-diazo powder was stored at $-20\,°C$ in a light-shielded glass vial. Before the caging experiment started, the vial was placed at room temperature for at least 10 min. Approximately $900\,\mu g$ of Bhc-diazo was transferred from a vial into a 1.5-ml eppendorf plastic microtube with a small autoclaved stainless steel microspatula (No. 9-866-01, Laboran), and then precisely weighed with an electronic precision balance (AG245, Mettler). The powder was then dissolved in the appropriate volume of DMSO to make a $60–80\,\mu g/\mu l$ solution. (The volume of DMSO used was $15–11\,\mu l$). Because caging efficiency varies with the species of mRNA, we recommend varying the amount of Bhc-diazo from 40 to $80\,\mu g/\mu l$ in several preliminary experiments to identify the best conditions for the individual species of mRNA to be caged. The color of the solution is very important for checking the purity of Bhc-diazo. A solution of pure Bhc-diazo is pale yellowish, but samples with contamination look dark brown. Because contamination of metallic substance significantly prevents the caging reaction, such samples should not be used.

3. Caging Reaction

A $5\,\mu l$ volume of the Bhc-diazo solution was transferred to a microtube containing the mRNA solution to prepare the reaction mixture (final volume $10\,\mu l$). It was mixed well by gently pipetting several times, and the mixture was incubated for 60 mins at room temperature. As the caging reaction proceeds, the solution becames paler than the initial color or almost clear.

4. Removal of Unbound Bhc–Diazo from Caged mRNA

After caging for 60 min, unbound Bhc-diazo was removed by column chromatography. A $500\,\mu l$ volume of Sephadex G50 (Pharmacia) that had been allowed to swell and was washed in DMSO was loaded onto a 1-ml micropipette tip that had been plugged with cotton, autoclaved, and dried in oven at $80\,°C$ for at least 1 h. Care was taken not to incorporate any bubbles in the tip. Immediately after excess DMSO stopped dripping from the tip of the column, the column was washed once with $500\,\mu l$ of DMSO. To prevent the resin from drying, new solution was always loaded immediately after the last drop. After washing, the reaction mixture ($10\,\mu l$) was carefully loaded onto the center of the surface of the resin. After the reaction mixture was loaded, $200\,\mu l$ of DMSO was loaded twice for elution. The eluate was collected in a 1.5-ml microtube and mixed with $500\,\mu l$ of isopropyl alcohol and $50\,\mu l$ of 10 M ammonium acetate. After pipetting up and down 20–30 times with a 1-ml micropipette, the mixture was centrifuged at 15,000 rpm for 10 min at $4\,°C$ and the supernatant completely removed by aspiration with a 1-ml micropipette. The pellet appeared as a tightly packed film with a dark spot at the center of the pellet. The pellet was washed with $150\,\mu l$ of 70% ethanol by

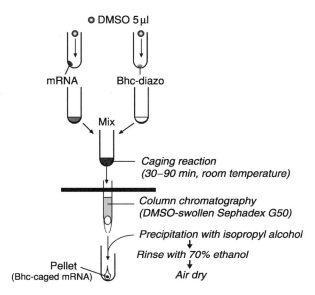

Fig. 1 Schematic illustration of the procedure of synthesis and recovery of Bhc-caged mRNA.

centrifugation at 15,000 rpm for 5 min. After removing the supernatant, the pellet was air-dried.

5. Preparation of the Caged mRNA Solution

Dried pellet was dissolved in 5 μl of nuclease-free water or 10 μl of DMSO. When the pellet was dissolved in water, the pellet was air-dried for 1 h. In contrast, when the pellet was dissolved in DMSO, it was dissolved just before it became completely dry (till it became nearly transparent), as it could not be dissolved in DMSO when it was overdried, as mentioned previously. In both cases, solutions were prepared just before they were microinjected into embryos. The pellet of Bhc-caged mRNA is difficult to dissolve in water or DMSO and tends to float like a thin film in the solution. Pipetting more than 100 times was necessary until the solution became viscous. The procedures for the caging reaction and column chromatography are summarized in Fig. 1.

III. Microinjection of the Bhc–Caged mRNA

The instruments for microinjection of the Bhc-caged mRNA were essentially the same as those described in Westerfield (2000). A glass microcapillary is prepared beforehand by pulling a hollow glass capillary (G-1, Narishige; outer diameter 1 mm and inner diameter 0.6 mm) with a microelectrode puller. A 2 μl

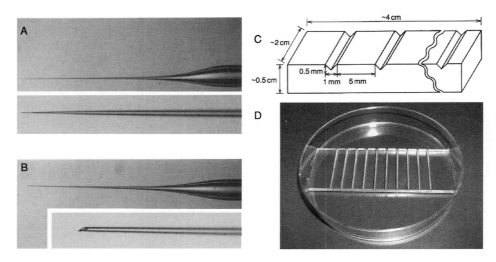

Fig. 2 Tools for injection of caged mRNA into the one-cell-stage zebrafish embryos. (A, B) The microcapillary for injection before (A) and after (B) the tip was broken manually with microforceps. The insets are close-up views of the tips. (C) A schematic diagram of the acrylic plate with the V-shaped grooves for holding the embryos during injection. (D) The acrylic plate for holding the embryos fixed in the plastic culturing dish.

volume of caged mRNA solution was loaded into the capillary with the 200 Original Eppendorf Microloader (Order No. 5242 956.003). Just before microinjection, the tip of microcapillary was broken manually by a fine microforceps while observing it through a 16× dissecting microscope (Stemi DRC, Carl Zeiss, Germany), so that the opening at the tip of the capillary had a diameter of approximately 20 μm (Fig. 2A,B).

Fertilized eggs were collected and aligned on a transparent acrylic plate (custom made by a local workshop) with V-shape grooves, which was attached to the bottom of a plastic petri dish (diameter 90 mm, and depth 15 mm I-90, INA-OPTIKA) containing fish-breeding water (Fig. 2C,D). The water was removed by aspirating almost to the bottom edges of the grooves by a plastic pipette (No. 6583, Nissui Seiyaku) so that the surface tension of the water kept the eggs attached to the grooves.

The plastic dish carrying the aligned eggs was placed on the stage of the dissecting microscope, and the caged mRNA solution was pressure-injected into the yolk of the embryo at approximately 30 psi for 40–50 msec. The pressure and duration of each injection were adjusted so that the size of the injected solution spread to form a drop having a diameter approximately one third of the cytoplasm. The pressure for microinjection (30 psi) was exceptionally high because the caged mRNA solution was very viscous. For this reason, the capillary tended to clog after injection into 20–30 embryos. We usually recovered the residual solution from the microcapillary by using the Microloader and reloaded the solution into a new capillary for continuous injection. The injected embryos were transferred to fish-breeding water and kept in the dark at 28.5 °C.

IV. Uncaging by Illumination of UV Light

For photo-mediated uncaging of mRNA, the injected embryos were aligned again on the grooves of the acrylic plate 3–12 h postfertilization (hpf) and transferred onto the stage of a microscope (BX-51 WI, Olympus, http://microscope.olympus.com/ga/products/Bx51_Bx61_E/) equipped with a 10× objective lens (UplanApo, Olympus) for UV illumination. An electronically controlled shutter (F77, Suruga Seiki) installed between the pinhole plate and a 100-W mercury lamp was used to illuminate the body parts of choice with a spot of UV light (365 nm) for 1 sec (Fig. 3A). A plate with a pinhole of 20 through 400 μm in diameter was

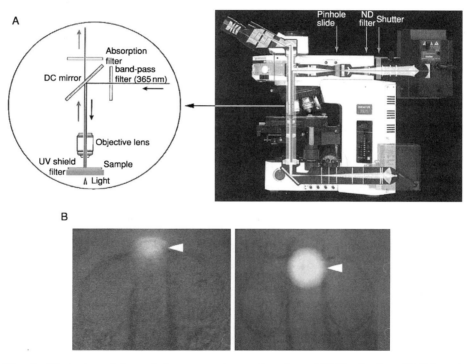

Fig. 3 Illustration of the optical system and examples of spot-UV illumination. (A) Schematic illustration of the UV-light path through the fluorescence microscope equipment with a pinhole slide for a spot-UV illumination (courtesy of Olympus). A pinhole slide was inserted at the confocal plane of the light path for projecting the spot UV light image on the sample. (B) Examples of the embryos being illuminated with the spot UV on the telencephalon (arrowhead).

inserted at the confocal plane of the light path from the UV lamp to illuminate sharp spots having diameters of approximately 4 through 70 μm at a discrete focal plane in the embryos. For example, we used a pinhole 200 μm in diameter to specifically illuminate a spot 36 μm in diameter in the forebrain of the 12-hpf embryos (Fig. 3B). No neutral density filter was inserted in the light path.

V. Titration of Caging Efficiency

The caging efficiency of Bhc-diazo might be affected by its purity and storage conditions. We therefore recommend checking the titer of Bhc-diazo by the simple procedure described next before actually using Bhc-diazo for the mRNA of interest. This titration test takes 1 day and ensures that the batch of Bhc-diazo is active. We recommend use of the β-galactosidase or GFP mRNA because of the simplicity of detecting the translational product (either by histochemical staining or by simple examination with a fluorescence microscope).

A. Synthesis of Caged β-Galactosidase or GFP mRNA

A 300 μg/5 μl solution of Bhc-diazo in DMSO and a 10 μg/5 μl solution of *in vitro* synthesized mRNA in DMSO were mixed together and allowed to react at room temperature for 30 min (for β-galactosidase mRNA) or for 1 h (for GFP mRNA). After purification, the caged mRNA was dissolved in 5 μl of nuclease-free water and injected into the one-cell-stage embryos.

B. Uncaging

At 3 hpf, half the injected embryos showing normal development were randomly selected for UV illumination of the entire blastoderm for 1 sec with a 10× objective lens and 25% neutral density filter installed between the pinhole plate and the UV lamp. Both the uncaged and non-uncaged injected embryos were then kept in the dark at 28.5 °C for an additional 3 h, until the embryos developed to early gastrula stage (shield stage).

C. Histochemical Detection of β-Galactosidase Activity

For the easy detection of β-galactosidase expression, the embryos were fixed at 6 hpf with 0.8% glutaraldehyde in PBS (0.1 M phosphate, pH 7.4) for 1 h at room temperature. After removal of the chorion, the embryos were rinsed twice with PBS. The histochemical reaction was performed according to the standard procedure at room temperature (Sanes *et al.*, 1986). The embryos were immersed in NDP solution (0.02% NP40, 0.01% sodium deoxycholate in 0.1 M phosphate, pH 7.4). After incubation for 10 min at room temperature, they were transferred to the reaction mixture (prepared by adding 20 μl of 200 mM potassium ferrocyanide, 20 μl of

200 mM potassium ferricyanide, $2\,\mu l$ of 1 M magnesium chloride, and $20\,\mu l$ of 20 mg/ml solution of X-gal in DMSO to 1 ml NDP solution). When both the caging and uncaging reactions were successful, the chromogenic reaction product began to appear as a blue color in the entire animal pole region of the uncaged embryos within 30 min, whereas almost no staining appeared in non-uncaged injected embryos. When we used old or contaminated Bhc-diazo, leaky expressions appeared in the non-uncaged embryos within that time (Fig. 4).

For quantitative measurement of β-galactosidase activity, we used the method of Miller (1972). Both uncaged and non-uncaged embryos, 50 each, were homogenized in 0.5 ml of Z buffer (60 mM disodium hydrogenphosphate, 40 mM sodium dihydrogenphosphate, 10 mM potassium chloride, 1 mM magnesium sulfate, and 50 mM β-mercaptoethanol) in a 1.5 ml microtube and centrifuged at 4 °C for 10 mins at 10,000 rpm. Fifty microliters of the supernatant was mixed with $450\,\mu l$ of Z buffer. The mixed solution was incubated for 30 min at room temperature and then with 0.1 ml of 4 mg/ml o-nitrophenyl-β-D-galactoside (ONPG). The reaction mixture was incubated for 30 min at room temperature, and then the chromogenic reaction was stopped by adding 0.25 ml of 1 M sodium carbonate. After a brief centrifugation, supernatants were measured for their optical density at 420 nm. When caging and uncaging succeeds, OD_{420} of the solution derived from uncaged

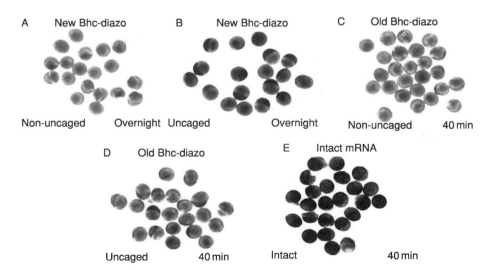

Fig. 4 Comparison of the efficiency of mRNA caging with fresh and old Bhc-diazo and by injection of caged mRNA dissolved in DMSO. The embryos were injected with β-galactosidase mRNA caged with fresh (A, B) or old (C, D) Bhc-diazo, uncaged at 3 hpf, and stained for β-galactosidase activity at 6 hpf. The embryos injected with intact β-galactosidase mRNA were fixed at 6 hpf and stained for β-galactosidase activity for comparison (E). (See Color Insert.)

embryos should be approximately fivefold higher than that of the solution derived from non-uncaged embryos.

VI. Injection of Dimethyl Sulfoxide (DMSO) Solution of Caged mRNA

As described previously, the caging efficiency of Bhc-diazo varies considerably. It is important to suppress leaky expression in non-uncaged embryos. The presence of DMSO in the solution injected might help prevent the hydrolysis of the P-O bonds between Bhc and phosphate by phosphatase (Grzyska *et al.*, 2002). In addition, caged mRNA with Bhc shows higher solubility in DMSO than in water; therefore, we can dissolve caged mRNA completely much more easily.

When the uncaged and non-uncaged embryos were observed at various developmental stages, the embryos injected with caged mRNA dissolved in DMSO had greatly reduced levels of leaky expression. Non-uncaged embryos also showed much lower or no leaky expression. Although approximately half the injected and uncaged embryos showed severe malformation, possibly owing to the toxicity of DMSO, the remaining normally developing embryos showed prominent ectopic expression in the illuminated region.

Acknowledgments

This research was supported by the internal research grant of the Brain Science Institute, RIKEN, a Grant-in-Aid for Scientific Research on Priority Areas (C) "Genome Science" to H. A. and (B) to H. O. and in part by the Special Coordination Fund to H. O. from the Ministry of Education, Science, Technology, Sports and Culture of Japan and by grants to H. O. for Core Research for Evolutionary Science from the Japan Science and Technology Corporation. A part of this chapter is adapted from the authors' previous publication (Andoh and Okamoto, 2003) under the copyright permission of Kluwer Academic Publishers.

References

Adams, S. R., and Tsien, R. Y. (1993). Controlling cell chemistry with caged compounds. *Annu. Rev. Physiol.* **55,** 755–784.

Ando, H., Furuta, T., Tsien, R. Y., and Okamoto, H. (2001). Photo-mediated gene activation using caged RNA/DNA in zebrafish embryos. *Nat. Genet.* **28,** 317–325.

Ando, H., and Okamoto, H. (2003). Practical procedure for ectopic induction of gene expression in zebrafish using Bhc-cged mRNA. *Methods Cell Sci.* **25,** 25–31.

Barbazuk, W. B., Korf, I., Kadavi, C., Heyen, J., Tate, S., Wun, E., Bedell, J. A., McPherson, J. O., and Johnson, S. L (2000). The syntenic relationship of the zebrafish and human genomes. *Genome Res.* **10,** 1351–1358.

Chaulk, S. G., and MacMillan, A. M. (1998). Caged RNA: Photo-control of a ribozyme reaction. *Nucleic Acids Res.* **26,** 3173–3178.

Furuta, T., Wang, S. S., Dantzker, J. L., Dore, T. M., Bybee, W. J., Callaway, E. M., Denk, W., and Tsien, R. Y. (1999). Brominated 7-hydroxycoumarin-4-ylmethyls: Photolabile protecting groups with biologically useful cross-sections for two photon photolysis. *Proc. Natl. Acad. Sci. USA* **96**, 1193–1200.

Grzyska, P. K., Czyryca, P. G., Golighty, J., Small, K., Larsen, P., Hoff, R. H., and Hengge, A. C. (2002). Generality of solvation effects on the hydrolysis rates of phosphate monoesters and their possible relevance to enzymatic catalysis. *J. Org. Chem.* **67**, 1214–1220.

Lin, W., and Lawrence, D. S. (2002). A strategy for the construction of caged diols using a photolabile protecting group. *J. Org. Chem.* **67**, 2723–2726.

Lo, J., Lee, S., Xu, M., Liu, F., Ruan, H., Eun, A., He, Y., Ma, W., Wang, W., Wen, Z., and Peng, J. (2003). 15000 unique zebrafish EST clusters and their future use in microarray for profiling gene expression patterns during embryogenesis. *Genome Res.* **13**, 455–466.

Lu, M., Fedoryak, O. D., Moister, B. R., and Dore, T. M. (2003). Bhc-diol as a photolabile protecting group for aldehydes and ketones. *Org. Lett.* **5**, 2119–2122.

Miller, F. (1972). Glycopeptides of human immunoglobulins. 3. The use and preparation of specific glycosidases. *Immunochemistry* **9**, 217–228.

Monroe, W. T., McQuain, M. M., Chang, M. S., Alexander, J. S., and Haselton, F. R. (1999). Targeting expression with light using caged DNA. *J. Biol. Chem.* **274**, 20895–20900.

Montgomery, H. J., Perdicakis, B., Fishlock, D., Lajoie, G. A., Jervis, E., and Guillemette, J. G. (2002). Photo-control of nitric oxide synthase activity using a caged isoform specific inhibitor. *Bioorg. Med. Chem.* **10**, 1919–1927.

Nasevicius, A., and Ekker, S. C. (2000). Effective targeted gene 'knockdown' in zebrafish. *Nat. Genet.* **26**, 216–220.

Pelliccioli, A. P., and Wirz, J. (2002). Photoremovable protecting groups: Reaction mechanisms and applications. *Photochem. Photobiol. Sci.* **1**, 441–458.

Robu, V. G., Pfeiffer, E. S., Robia, S. L., Balijepali, R. C., Pi, Y., Kamp, T. J., and Walker, J. W. (2003). Localization of functional endothelin receptor signaling complexes in cardiac transverse tubules. *J. Biol. Chem.* **278**, 48154–48164.

Rupp, R. A., Snider, L., and Weintraub, H. (1994). Xenopus embryos regulate the nuclear localization of XMyoD. *Genes Dev.* **11**, 1311–1323.

Sanes, J. R., Rubenstein, J. L., and Nicolas, J. F. (1986). Use of a recombinant retrovirus to study post-implantation cell lineage in mouse embryos. *EMBO J.* **5**, 3133–3142.

Suzuki, A. Z., Watanabe, T., Kawamoto, M., Nishiyama, K., Yamashita, H., Ishii, M., Iwamura, M., and Furuta, T. (2003). Coumarin-4-ylmethoxycarbonyls as phototriggers for alcohols and phenols. *Org. Lett.* **5**, 4867–4870.

Tsien, R. Y., and Furuta, T. (2000a). Protecting groups with increased photosensitivities. *US Patent Application* WO/00/31588.

Tsien, R. Y., and Furuta, T. (2000b). Preparation of halogenated coumarins, quinoline-2-ones, xanthenes, thioxanthenes, selenoxanthenes, and anthracenes as photolabile protecting groups with increased photosensitivities. PCT International Application WO 2000031588.

Turner, D. L., and Weintraub, H. (1994). Expression of achaete-scute homolog 3 in *Xenopus* embryos converts ectodermal cells to a neural fate. *Genes Dev.* **8**, 1434–1447.

Westerfield (2000). "The Zebrafish Book. A guide for the Laboratory Use of Zebrafish (*Danio rerio*)." University of Oregon Press, Eugene, OR.

CHAPTER 10

Current Status of Medaka Genetics and Genomics

The Medaka Genome Initiative (MGI)

Manfred Schartl,[*] **Indrajit Nanda,**[*] **Mariko Kondo,**[*] **Michael Schmid,**[*] **Shuichi Asakawa,**[†] **Takashi Sasaki,**[†] **Nobuyoshi Shimizu,**[†] **Thorsten Henrich,**[‡] **Joachim Wittbrodt,**[‡] **Makoto Furutani-Seiki,**[§] **Hisato Kondoh,**[§] **Heinz Himmelbauer,**[‖] **Yunhan Hong,**[¶] **Akihiko Koga,**[#] **Masaru Nonaka,**[**] **Hiroshi Mitani,**[††] **and Akihiro Shima**[††]

[*]Biocenter, University of Wuerzburg
Am Hubland, D-97074 Wuerzburg, Germany

[†]Department of Molecular Biology
Keio University School of Medicine
Tokyo 160-8582, Japan

[‡]Developmental Biology Programme
EMBL-Heidelberg
D-69012 Heidelberg, Germany

[§]ERATO
14 Yoshida-kawaramachi, Sakyo-ku
Kyoto 606-8305, Japan

[‖]Max-Planck-Institute of Molecular Genetics
D-14195 Berlin, Germany

[¶]Department of Biological Sciences
National University of Singapore
Singapore 117543

[#]Division of Biological Sciences
Graduate School of Science
Nagoya University, Nagoya 464-8602, Japan

[**]Department of Biological Sciences
Graduate School of Sciences
The University of Tokyo
Bunkyo-ku, Tokyo 113-0033, Japan

METHODS IN CELL BIOLOGY, VOL. 77

††Department of Integrated Biosciences 102
Graduate School of Frontier Sciences
The University of Tokyo
Kashiwa City, Chiba 277-8562, Japan

I. Introduction

The egg-laying killifish medaka, *Oryzias latipes*, is an old (100 years) and very well established genetic model system for developmental genetics and many other areas of biological and environmental research. Although zebrafish and its features as an experimental system are well known, to date the publicity of medaka was restricted (with a few exceptions) to its home range, Japan, and some countries in the Far East.

In the early twentieth century, medaka was recruited as a model species for biological and genetic research. The study of the inheritance of body color in medaka by K. Toyama in 1916 (a review in Japanese) first proved that Mendelian laws also apply to fish. Another milestone was the discovery of Y-chromosome-linked inheritance by Aida in 1921. The growing rate of publications on medaka per year within the past decade illustrates the increasing popularity of medaka as a model species.

The key technologies that made zebrafish such a successful model species (discussed in detail in this volume) fully apply to medaka too. Genetics in medaka offers several advantages, for example, the availability of divergent, completely inbred strains and a genome of only 800 Mb, less than half the size of the zebrafish genome. Also, as a practical aspect, the sexes are easily distinguished in medaka, in contrast to zebrafish, facilitating breeding in general and genetic studies in particular.

With the release of the draft versions of the first teleost whole genome sequences of *Fugu rubripes* and the closely related freshwater pufferfish (*Tetraodon nigriviridis*), the postgenomic era has reached fish. As both pufferfish species cannot be bred in captivity, medaka is the closest 'approximation' with fully developed genetics and experimental embryology. Phylogenetic studies indicate that the last common ancestor of medaka and pufferfish lived between 40 and 60 million years ago, less than the evolutionary distance between humans and mouse. Medaka and zebrafish (*Danio rerio*), on the other hand, are more distant cousins that have evolved separately for at least 140 million years. (The same holds true for the distance between zebrafish and fugu). This situation—the availability of two fully developed model systems—the genomes of which should be available within 2004, is unique to vertebrates and will allow researchers to address the mechanisms of genome evolution on a functional level. This will elucidate developmental mechanisms that are employed generally as much as the specifics and peculiarities of the model system worked on.

II. Current Status of Medaka Genetics

A. Germ Cell Mutagenesis

Mutations are permanent alterations that occur in the genetic materials of cells. Unlike germ cell mutations, mutations in somatic cells are not transmitted to the progeny. They might result in mosaics if they occur in early embryos, and are responsible for important biological phenomena, such as the generation of antibody diversity, if they take place in the immune system, or induction of cancer, if induced in cancer-related genes.

Mutations can arise spontaneously (see Section II.E), but can be induced by various agents such as radiation and chemicals. For the use of an organism as a tool for genetic dissection of a biological phenomenon, to target and to induce a mutation in the gene responsible for the phenomenon of interest is most desirable. Alternatively, germ cell mutagenesis with as high a yield of viable mutants as possible by using mutagens, however, affecting the genome at random, can facilitate genetic dissection. The basic protocols for radiation (Shima and Shimada, 1991) and chemical (Shima and Shimada, 1994) germ cell mutagenesis have already been established and are successfully applied to mutagenesis screening in zebrafish (Solnica-Krezel *et al.*, 1994) and medaka (Loosli *et al.*, 2000).

Safeguards against germ cell mutagenesis suggested earlier (Shima and Shimada, 2001) were substantiated as differences in the major countermeasure among male germ cells of different maturation stages against genomic alterations: stem spermatogonia—DNA repair, early differentiating spermatogonia—apoptosis, late-differentiating spermatogonia—acceleration of maturation to yield abortive gametes, and sperm-dominant lethals (Kuwahara *et al.*, 2002, 2003).

In the mouse, it is generally accepted that female gametes are less sensitive to radiation-induced mutagenesis than are male gametes (Alpen, 1998). By using the medaka specific-locus test system, this trend was confirmed (Shima and Shimada, unpublished). Further, it was found that male germ cells of the HNI strain were more susceptible to radiation than those of the Sakura strain, most remarkable in spermatogonia (Shima and Shimada, unpublished). Revealing the mechanisms underlying these sex and strain differences could help in further understanding germ-cell mutagenesis.

An *in vitro* culture system was established that allows analyses of proliferation of spermatogonia and preleptotene spermatocytes and differentiation of spermatocytes to haploid gametes in medaka (Song and Gutzeit, 2003). Combining this culture system with transgenesis technology would facilitate further elucidation of mechanisms of germ cell mutagenesis. For instance, the curious finding that mutant spermatogonia are positively selected before the start of meiosis in mice is worth verifying in other species (Goriely *et al.*, 2003). As regards somatic mutation, medaka is successfully used in carcinogenesis testing (Hawkins *et al.*, 2003).

B. Mutagenesis Screening

A large-scale mutagenesis screen in zebrafish identified many genes and genetic pathways regulating development (Driever *et al.*, 1996; Haffter *et al.*, 1996). However, because in vertebrates genomic functions are mediated by sets of multiple genes and usage of individual gene functions might be divergent to a certain extent from one species to another, mutant screening that uses one species might not suffice to uncover the functions of all genes. Thus, medaka, phylogenetically distant from zebrafish, was employed as the second fish species for large-scale mutant screening.

Like zebrafish, medaka is suitable for forward genetics based on chemically induced mutations and might even be advantageous in certain aspects. The medaka genome is less than half the size of the zebrafish genome and is highly polymorphic in nucleotide sequences without large-scale rearrangements. Useful tools have been established, including inbred medaka fish lines, conditions for chemical mutagenesis (Shima and Shimada, 1991), and a high-resolution genetic map (Naruse *et al.*, 2000). Mutants of medaka identified in pilot screens affecting development of eyes and the nervous system included many showing phenotypes never seen in the zebrafish mutant collection (Ishikawa, 2000; Loosli *et al.*, 2000), corroborating the notion that medaka mutants will complement zebrafish mutants in uncovering genomic functions.

A large-scale systematic screen in medaka for mutations affecting embryonic and larval development was carried out in the Kondoh Differentiation Signaling Project (ERATO; Furutani-Seiki *et al.*, 2004). Male founders were mutagenized with ENU, and induced mutations were screened by homozygous phenotypes at the F3 generation. Exhaustive screening of live embryos for morphology of developing tissues and organs was carried out. Various molecular and cellular markers were also employed to screen mutations by using fixed embryos, for example, immunostaining of acetylated tubulins for axonal trajectories, DiI and DiO labeling of optic nerves, PED 6 staining for lipid metabolism in liver, *rag1* expression detected by *in situ* hybridization for thymus development, and *vasa* expression for migration and differentiation of germ cells (Table I).

By screening 1135 F2 families corresponding to approximately 1900 mutagenized haploid genomes, 2031 embryonic lethal mutations were detected. Of these, 312 mutations were maintained that cause specific embryonic and larval patterning defects in homozygotes, of which 126 mutations have been characterized for homozygous phenotypes in morphology, expression of marker genes, and by complementation testing. The analysis has defined 105 genetic loci, most of them represented by single alleles but some by up to noncomplementing four alleles.

Table I
Assays Used in the Kyoto Medaka Mutant Screen

Tissues and developmental process of interest	Detection of mutant phenotypes	Embryo and larval specimens	Refs.
Embryo morphogenesis and body patterning	Morphology of embryos and their tissues	Live	Elmasri *et al.*, 2004; Furutani-Seiki *et al.*, 2004; Kitagawa *et al.*, 2004; Loosli *et al.*, 2004; Watanabe *et al.*, 2004
Migration and distribution of primordial germ cells	Primordial germ cells detected by *in situ* hybridization for *vasa*	Fixed	Sasado *et al.*, 2004
Gonadal development	Germ cells in gonads detected by *in situ* hybridization for *vasa*	Fixed	Morinaga *et al.*, 2004
Thymus development	Thymocytes detected by *in situ* hybridization for *rag1*	Fixed	Iwanami *et al.*, 2004
Lateral line development	Lateral line nerves visualized by immunostaining of acetylated tubulins	Fixed	Yasuoka *et al.*, 2004
Tectal projection of retinal ganglion cell (RGC) axons	RGC axons stained by retinal injection of DiI or DiO	Fixed	Yoda *et al.*, 2004
Heme metabolism	Color of bile	Live	Watanabe *et al.*, 2004
Lipid metabolism in the hepatic system	Lipid uptake and transport to the liver and gall bladder, detected by fluorogenic conversion of PED6	Live	Watanabe *et al.*, 2004
γ-Ray sensitivity	Recovery from γ-ray irradiation with sublethal doses	Live	Aizawa *et al.*, 2004

Mutations mainly affecting the forebrain either in morphological development or in expression of specific genes are somewhat unique to medaka (Kitagawa *et al.*, 2004). They have been classified into five phenotypic groups (specification, regionalization, morphogenesis, axonal projection, and ventricular formation) and are attributed to 25 genes.

Although the comparison of all mutant phenotypes in medaka and zebrafish is not possible, it is clear that many mutants in medaka display phenotypes previously undescribed in zebrafish. A substantial fraction of the mutants in medaka have phenotypes, which through their similarity might point to possible counterparts in zebrafish (Table II; Furutani-Seiki *et al.*, 2004). Note that this kind of correspondence is not on a gene-to-gene basis. For instance, medaka mutants with a *one-eyed-pinhead (oep)*-like embryonic phenotype are assigned to three different loci, but in zebrafish only a single gene is known to cause the *oep* phenotype. Analogously, mutations of three independent genetic loci give rise to a *parachute*-like phenotype, whereas *parachute* mutants in zebrafish all belong to a single *N-cadherin* gene. The nonidentical mutant phenotypes observed between medaka and zebrafish can be partly ascribed to nonsaturation of the mutagenesis in either of the fish species. However, the identification of multiple medaka genes contributing to a mutant phenotype class defined by a single gene in zebrafish, in spite of more extensive mutant survey in the latter, argues that genetic regulatory cascades involved in the genesis of a tissue are conserved in their framework but can be considerably divergent in their details between medaka and zebrafish. This is corroborated by the recent report that the downstream components of the *rx3* regulatory pathway are considerably different between medaka and zebrafish, although *rx3* mutations cause analogous eyeless phenotypes in both species (Loosli *et al.*, 2003).

Table II
Medaka Mutants That Show Phenotypes Similar to those in Zebrafish

Zebrafish mutants	Medaka mutants	Common phenotypes
one-eyed-pinhead (oep)	*akebono (ake), akatsuki (aka), mochizuki (moc)*	Fused eyes, narrow brain, ventral deficiencies including floor plate
parachute (pac)	*Oobesshimi (oob), samidare (sam), shigure (shi)*	Exfoliating cells in the brain ventricles
cyclops (cyc)	*karakasa (ksa)*	Fused eyes, ventral deficiencies including floor plate
no ithmus (noi)	*zeppeki (zep)*	Absence of isthmus, cerebellum, and tectum
acellebellar (ace)	*kappa (kap)*	Absence of isthmus and cerebellum, tectum enlarged
spadetail (spd)	*shogoin (sho)*	Enlarged tailbud, no trunk somites

Analysis of the battery of chemically induced mutants in medaka, as well as molecular cloning and characterization of their responsible genes, will benefit from the genetic and genomic resources, tools, and information being developed by the Medaka Genome Initiative (MGI) and other groups. Insertional mutagenesis, for example, using transposon vectors (Grabher *et al.*, 2003), provides another means for cloning mutated genes effectively, but genome wide survey of mutations with this approach is still challenging.

In conclusion, the combined use of medaka and zebrafish mutants, as well as comparative studies of genetic regulatory cascades of these two fish species, will contribute greatly to understanding the genetic and molecular mechanisms of vertebrate development.

C. Medaka Expression Pattern and Phenotype Databases

To store and integrate information about gene expression during medaka development, the Medaka Expression Pattern Database (MEPD) was established (Henrich *et al.*, 2003). Expression patterns of cDNAs (cloned genes, unigene clones as well as ESTs) are generated at high throughput (Henrich and Wittbrodt, 2000; Quiring *et al.*, 2004), documented with images and descriptions of staining parameters (e.g., intensity, category, comments) and through a medaka fish ontology (see later). Sequences of stored cDNAs are available and searchable through BLAST. Sequence entries are clustered on entry to the database and have been blasted against public databases. BLAST results that are updated regularly are stored within the database and are fully searchable. A new version of MEPD at EMBL (http://www.embl.de/mepd/) based on its first implementation at http://medaka.dsp.jst.go.jp/MEPD supports a detailed anatomy ontology, thumbnails, and other new features and at present contains more than 7000 entries compared with 800 in the previous version. More than 13,000 new entries from a unigene library are expected in the near future. These expression data will be analyzed by bioinformatics tools and complemented with microarray data.

Genetic Screen Database (GSD) is a software package that allows storing and integrating data from genetic screens (Henrich *et al.*, 2004). GSD originates from a large-scale F3 mutagenesis screen for developmental mutants of medaka fish (see Section II.B; Furutani-Seiki *et al.*, in press; Loosli *et al.*, in press). The original version was subsequently altered to support a wide range of different screens (mutagenesis, RNAi, morpholinos, transgenesis, and others), using different model organisms (medaka, zebrafish, and others).

Data are stored in a relational database and are made accessible by Web interfaces. Screeners can enter data describing phenotypes obtained in a genetic screen. They can keep track of statistics, submit pictures, and describe the occurring phenotypes, using a phenotype classification ontology.

In addition, a list of mutant lines resulting from this screen can be readily organized. These lines (mutant alleles, transgenic lines) are described in the same way as are screened individuals. Raw data from the screen can be integrated to describe these lines. A query module searching this list can be used to publish the screen results on the Internet. A test version is installed at http://www.embl.de/wittbrodt/gsd, and the software can be downloaded from this site.

To enable cross-references of expression patterns, transgenic lines, and mutants between medaka and zebrafish, a medaka ontology was developed in close collaboration with the Zebrafish Information Network (ZFIN; http://zfin.org) and submitted to Open Biological Ontologies (OBO; http://obo.sourceforge.net), from where it can be downloaded in a DAGedit format. During development, great care was taken to use the same terms for corresponding structures in both model systems.

The ontology uses a unique identifier space, Medaka Fish Ontology (MFO), and consists of three parts. The first part has 46 terms representing the *developmental stages* described by Iwamatsu (1994). Definitions for these terms were taken from this publication. The second part has more than 4173 terms describing the *anatomy* of medaka. This ontology is mainly used to describe expression patterns in the MEPD (http://www.embl-heidelberg.de/mepd/), but can also be implemented in GSD, for example, to describe the GFP expression patterns in transgenic lines or a well-characterized phenotype when combined with an ontology such as the Phenotype and Trait Ontology (PATO).

During an initial screen, it is usually not possible to describe an embryo in detail. Researchers typically record approximate descriptions that allow categorization of phenotypes, thus supporting discovery of the functional relatedness of the mutated genes. For this reason, we developed a relatively high-level *phenotype classification* ontology. This ontology comprises 106 structure and 29 modifier terms. A *structure* term can be further specified by a list of appropriate *modifiers*. For example, the structure term *eyes* which *is-part-of* the *sensory system* can be modified by the terms *abnormal, absent, enlarged, reduced, cyclopic*. This ontology was based on the experience from a large-scale medaka mutagenesis screens as well as with the mutants described in ZFIN (Sprague *et al.*, 2003) and the Tuebingen zebrafish screen (Haffter *et al.*, 1996).

The phenotype classification terms are matched with the anatomy terms in order to facilitate cross-referencing between the two databases (GSD and MEPD). The ontology is represented in a directed acyclic graph (DAG), which is similar to a hierarchy, but allows more than one parent for each node. It has been developed using DAGedit and self-written tools and implements the three different relations among the terms *is-part-of*, *is-a*, and *develops-from*.

D. Genetic Approaches to Understand Gene Function

The following sections describe two phenomena—the major histocompatibility complex (MHC) gene organization and sex determination—for which in medaka the genetic analysis has already reached a substantial level and which therefore became paradigms for other approaches.

1. Major Histocompatibility Complex (MHC)

The MHC is one of the most intensively characterized regions of the vertebrate genome, because of the interest in the curious accumulation of immunologically important genes in this region and the question of how this has evolved. The human MHC occupies about 4 Mb of chromosome 6p and harbors genes essential for both adaptive and innate immune responses, such as MHC class I and II genes, genes involved in class I antigen presentation, complement component genes, and TNF gene (The MHC sequencing Consortium, 1999). By convention, the mammalian MHC is divided into three subregions, class I, II and III, and phylogenetic analysis indicates that the linkage among MHC class I, II, and III genes is conserved throughout jawed vertebrate evolution from cartilaginous fish, the most primitive extant vertebrate to possess MHC, to mammals (Flajnik and Kasahara, 2001; Terado *et al.*, 2003).

In all bony fish examined thus far, however, the MHC class I and II genes are mapped to multiple chromosomes (Flajnik and Kasahara, 2001). Moreover, in medaka and zebrafish, the mammalian MHC class III complement genes are linked neither to class I nor Class II genes (Kuroda *et al.*, 2000; Samonte *et al.*, 2002), indicating that extensive chromosomal rearrangements in the bony fish lineage dispersed the MHC genes to several chromosomes. Even in bony fish, however, the MHC class I gene and several MHC genes are linked to each other, defining the bony fish MHC class I region.

Nucleotide sequence analysis of two BAC clones of Hd-rR, one inbred strain derived from the Southern Japan population, which together span about 430 kb of the MHC class I region, identified 22 putative genes and 3 truncated pseudogenes (Matsuo *et al.*, 2002). Except for three genes whose human orthologs are mapped to different chromosomes, 19 of the 22 genes have their orthologs in the human MHC, indicating a high degree of synteny conservation of these genes between mammals and bony fish. Two classical class I alpha chain genes and the six other genes directly involved in class I antigen presentation form an uninterrupted cluster in the medaka MHC class I region, whereas the human orthologs of these genes disperse over the entire human MHC, suggesting that the genes directly involved in class I antigen presentation are the evolutionarily conserved core of the vertebrate MHC. The main evolutionary role of the MHC might be to provide an opportunity for these structurally unrelated and functionally linked genes to coevolve to establish an efficient class I antigen presentation system.

To assess the intraspecies polymorphism of the medaka MHC class I region, a second inbred strain, HNI, derived from the Northern Japan population, was analyzed (K. Tsukamoto *et al.*, unpublished data). The Hd-rR and HNI MHC class I show only about 95% nucleotide identity in the region that can be aligned. Moreover, a region of about 100 kb encompassing two classical class I genes and a PSMB8 gene is so divergent between Hd-rR and HNI that it is impossible to align. Thus, the medaka MHC class I region is one of the most polymorphic regions of the vertebrate genome analyzed thus far and ongoing analysis of this region of wild medaka populations will reveal intriguing facts about the evolution of the bony fish MHC.

2. Sex Determination

In zebrafish and fugu, the other main model fish species, it is not known whether sex is determined genetically or environmentally. In contrast, the mechanism of sex determination has been known for quite some time in medaka. Medaka has an XX–XY system. Sexually mature females and males are distinguished morphologically by apparent secondary sex characters, for example, the shape of the dorsal and anal fins. Sex can be easily reversed by hormone treatment of both genetic sexes and fully fertile fish are obtained (Wittbrodt *et al.*, 2002; Yamamoto, 1975).

Linkage analyses of the sex-linked markers and the sex-determining gene (*y*) produced a high-resolution map of the gonosomes. It revealed that there is a region on the Y chromosome that harbors *y*, and that recombination of the markers in the vicinity of the sex-determining region is highly suppressed (Kondo *et al.*, 2001).

In a candidate gene approach for finding the master sex regulatory gene, medaka orthologs of a transcription factor gene, *DMRT1* (*doublesex* and *mab-3* related transcription factor 1), were cloned and analyzed (Brunner *et al.*, 2001), and two copies of *dmrt1*, *dmrt1a*, and *dmrt1bY*, the last linked to male sex, were isolated. The Y-specific region was found to be about 280 kb, containing many pseudogenes and repetitive sequences, and the only gene that seems to be functional is the *dmrt1bY* gene (Matsuda *et al.*, 2002; Nanda *et al.*, 2002). Simultaneously and independently, using positional cloning the same gene was isolated as the obvious candidate for the male sex-determining gene (Matsuda *et al.*, 2002). It was named *DMY*, a synonym for the name *dmrt1bY*, the latter following the accepted nomenclature for naming duplicated genes in fish. The coding sequence of this gene consists of five exons and covers about 50 kb of genomic sequence. Naturally occurring mutations in this gene caused XY females (Matsuda *et al.*, 2002). Therefore, this gene is the most likely candidate sex-determining gene of medaka.

In the Y-specific region, pseudogenes of two other genes located close to *dmrt1a* in the 5' region as well as a part of *dmrt3*, which is the gene positioned 3' to *dmrt1a* on the autosome, were found (Nanda *et al.*, 2002). Therefore, duplication of the autosomal *dmrt1a* region in LG9 resulted in the formation of the sex-determining gene as well as the Y-specific region, which defines the identity of the Y chromosome.

Although orthologs of *dmrt1bY* were expected to be the sex-determining genes of other related fish species, thus far the duplicated version of the autosomal *dmrt1* gene is found in only medaka and its sister species *Oryzias curvinotus* but not in a more distantly related species, *O. celebensis* (Kondo *et al.*, 2003; Matsuda *et al.*, 2003). Therefore, appearance of the sex-determining gene and the sex chromosome of medaka seems to be a rather recent event in evolution. Identification of the sex-determining gene, the relatively stable sex-determining mechanism of medaka, and the frequent occurrence of XX males in some strains (Nanda *et al.*, 2003) make this species a useful tool to study the molecular processes and the evolution of sex determination.

E. Transposable Elements

Transposable elements are a major source of mutations. They also cause, even when they have lost their transposition activity, chromosomal rearrangements because of their presence as repetitive sequences. Because of this nature, they are thought to be factors contributing to genome evolution. Another feature of transposable elements is that they can serve as tools for various techniques in genetic engineering.

Eukaryotic transposable elements fall into two general classes: RNA-mediated elements and DNA-based elements. The former is further divided into three major groups: viral family elements, long interspersed nuclear elements (LINEs), and short interspersed nuclear elements (SINEs). Elements of all these types have been found in the medaka genome (Koga *et al.*, 2002a; Volff *et al.*, 2001).

It is common among eukaryotes for transposable elements to occupy significant fractions of the genomes, but the majority of the copies are defective. In particular, DNA-based elements appear to have been nearly or completely inactivated in vertebrates. However, medaka is remarkable in that highly active DNA-based elements have been found (Koga *et al.*, 2002b). The *Tol1* element was identified as an extra insertion sequence in a pigmentation gene of a spontaneous body-color mutant, and a reversion mutation due to its excision has recently been observed (S. Hamaguchi, personal communication). *Tol2* is present at 10–20 copies per haploid medaka genome, and virtually all copies are autonomous or potentially autonomous, carrying an intact gene for their transposition (Koga and Hori, 1999). This is the only DNA-based element among vertebrates that carries a functional transposase gene.

Tol2 is capable of movement not only in medaka but also in zebrafish (Kawakami *et al.*, 2000) and mammals (Koga *et al.*, 2003). A gene transfer system using this element has already been established (Koga *et al.*, 2002a). Another prominent element is *Sleeping Beauty*, reconstructed based on sequence information of *Tc1* family elements of salmonids (Ivics *et al.*, 1997). Similarly to its activity in zebrafish and other vertebrates, it exhibits a high transposition rate and provides an efficient genetic tool for gene transfer and enhancer trapping in medaka (Grabher *et al.*, 2003).

The genus *Oryzias* includes 14 known species whose phylogenetic relationships have already been well examined. This is a major advantage of medaka in conducting evolutionary studies on the time scale of speciation. A significant result with respect to transposable elements is that *Tol2* invaded the medaka genome recently (Koga *et al.*, 2000) and has become widespread in the species over a relatively short time span (Koga and Hori, 1999).

F. Medaka Embryonic Stem Cells

The isolation and genetic manipulation of embryonic stem (ES) cells represent one powerful tool in mammalian developmental biology. After blastocyst injection, they are able to give rise to all cell lineages, including the germline after blastocyst injection (Bradley *et al.*, 1984). Mouse ES cells have widely been exploited for gene targeting for the production of knockout animal models for the analysis of gene functions and human disease. ES cells also provide a universal source for induced/directed differentiation in basic research and regeneration medicine.

Fish ES cell cultivation began approximately 10 years ago in zebrafish and medaka. (See Part I.G. for zebrafish ES cells.) Wakamatsu *et al.* (1994) adopted the feeder layer technique and reported the first medaka ES-like cell line, OLES1. In parallel, a feeder-free culture system was developed in which blastula-derived stem cells are cultured on gelatin-coated substrata (Hong and Schartl, 1996). With this protocol, several stable cell lines were obtained. One of these, MES1, was characterized in more detail. MES1 cells are eudiploid pluripotent *in vitro* (Hong *et al.*, 1996) and chimera competent *in vivo* (Hong *et al.*, 1998a) and display pluripotency-specific gene expression as shown by the ability to activate the mouse Oct4 regulatory sequence (Hong *et al.*, 2004a). This feeder-free protocol has also been adopted for several other fish species and has resulted in ES-like cultures from the gilthead seabream (Bejar *et al.*, 2002), red seabream (Chen *et al.*, 2003), and zebrafish (Hong and Schartl, unpublished).

Three variables are critical for ES cell derivation in medaka: culture conditions, strains, and stages. Of the 16 medaka strains, populations, and species tested (Table III), 9 allow for ES derivation. These include SOK, HB32C, and HNI of the medaka and its closely related species *Oryzias minutillus, O. mekongensis*, and *O. curvinotus*, whereas others were impossible to cultivate (Hong *et al.*, 1998b; Y. Hong and M. Schartl, unpublished). Such strain differences in ES cell derivation have been also documented in mice (Kawase *et al.*, 1994).

For initiating ES cell cultures, early to midblastula stages (512–1028 cells) are used, as cells from earlier stages are difficult to culture. Special media have been developed that contain growth factors and a required extract prepared from medaka embryos (Hong and Schartl, 1996).

One major application of ES cells is to provide a system for studying cell differentiation. Differentiation of ES cells *in vitro* faces major challenges: the difficulty to establish complicated culture conditions that, in mice, usually involve a step of embryoid body formation, the combinatorial use of growth factors, and the heterogeneity of induced differentiation in terms of cell states and types. Often, a uniform population of cells differentiating along a particular cell lineage is difficult to obtain. Instead, the cell population is a mixture of undifferentiated ES cells and of cells differentiating along various lineages. As in mice, medaka ES cells undergo spontaneous differentiation into many different types of cells, including fibroblasts, neural cells, muscles, and pigment cells (Hong and Schartl,

Table III
Strain and Species Difference in ES Cell Derivation[a]

Species/strains	ES cell initiation[b]	ES cell serial culture[c]	Chimera formation[d]	Notes
O. latipes				
HNI	+++	+++	+++	
HB12A	+++	+++	+++	
HB32C	+++	+++	+++	
HB11A	+++	+++	ND	
HB32D	+++	+++	ND	
SOK	+++	+++	ND	
Carbio	+++	++	++	
Da	++	+	+	Differentiation at Day 7
Sakura	+	−	ND	Differentiation at Day 7
Kaga	+	−	ND	Differentiation at Day 3
Yokote	+	−	ND	Differentiation at Day 3
HB11C	−	−	ND	No attachment
O. celebensis	+	−	ND	Little attachment, vacuolization
O. minutillus	+++	+++	+++	
O. curvinotus	+++	+++	ND	
O. mekongenesis	+++	+++	ND	

[a]Cell initiation, serial culture, and chimera formation are considered as impossible (−), difficult (+), possible (++) and easy (+++). ND, not determined.
[b]Cell attachment and growth during first 3 days in primary culture.
[c]Cell attachment and growth during 4–12 days in subculture.
[d]Cells after culture for 3–12 days were tested for pigment chimera formation.

1996; Hong *et al.*, 1996). Although this demonstrates pluripotency *in vitro*, such a spontaneous process is hardly useful for biochemical and transcriptome/proteome analyses of the mechanisms underlying pluripotency and differentiation. Therefore, a system of directed differentiation of medaka ES cells was established. At first, the melanocyte-specific microphthalmia-associated transcription factor (mitf) was used as a master regulator to induce differentiation of medaka ES cells into pigment cells. After simple transient transfection with a mitf-expressing plasmid, medaka ES cells are able to give rise to functional melanocytes. This now provides a possibility to test functionally other genes of interest for their unknown or predicted differentiation potential and developmental biological function (Bejar *et al.*, 2003).

The crucial step in establishing ES technology is transmission of the retransplanted ES cells through the germline. Although several medaka ES lines have proved to be fully competent to contribute to all somatic cell lineages and organs of the adult fish, germline transmission has not been obtained so far. Germline contribution relies on the use of proper donor/recipient combinations and optimized protocols of chimera production. Most recently, we have succeeded in establishing a normal medakafish spermatogonial stem cell line capable of sperm production *in vitro* (Hong *et al.*, 2004b). It will be intriguing to determine whether *in vitro* spermatogenesis from this cell line can be programmed to provide a number of functional sperm sufficient for germline transmission by artificial insemination.

For the production of knockout fish, homologous recombination (HR) of endogenous sequences with altered transgenic DNA in the HR vector must occur. HR activity has been shown in cells of zebrafish embryos (Hagmann *et al.*, 1998). As such, HR events will be extremely rare, the altered ES cells with the desired genotype have to be enriched by selection procedures comparable to what is used in the mouse. Vector cassettes and the appropriate selection schemes for HR with medaka ES cells have been developed already (Chen *et al.*, 2003).

III. Current Status of Medaka Genomics

A. Expressed Sequence Tag (EST) Mapping, Sequencing, and Genetic Mapping

Large-scale medaka expressed sequence tag (EST) analysis and gene mapping are essential for positional cloning of the genes responsible for mutants and the genomewide comparison of linkage relationships among vertebrate species. More than 120,000 medaka EST sequences are registered in the public database. Sequence comparisons of orthologous loci showed that single-base-pair polymorphisms between two inbred strains from Southern and Northern Japanese populations, AA2 and HNI, respectively, were about 1% in coding regions, which makes it easy to find polymorphic linkage markers (Naruse *et al.*, 2000; Wada *et al.*, 1995). Currently, more than 1600 ESTs are mapped onto 24 linkage groups that correspond to the diploid chromosome number of medaka (http://mbase. bioweb.ne.jp/~dclust/medaka top.html). The total map length of all linkage groups (LGs) is about 1400 cM in male meiosis. If the total genome size of medaka is taken as 800 Mb, the estimated physical length of each LG would range from 19 to 59 Mb. Comparisons of marker distribution for anonymous DNA markers and EST markers suggest that distributions of genes are not uniform on each LG. For example, the gene density of LG2 is estimated to be 4.3 times lesser than that of LG 22.

This linkage map facilitates the positional cloning of genes responsible for mutant phenotypes. Fukamachi *et al.* (2001) identified the product encoded by the *b* locus. It encodes the membrane-associated transporter protein (MATP), which mediates melanin synthesis. It is the first successful positional cloning in medaka. Recent studies revealed that mutations in this gene also cause pigmentation disorders in other organisms: the mouse underwhite mutant; the human oculocutaneous albinism Type 4, OCA4; and the horse cream coat color (Mariat *et al.*, 2003; Newton *et al.*, 2001). The medaka ectodysplasin-A receptor (EDAR) gene was found to be responsible for the *rs-3* phenotype, which is a recessive mutation leading to an almost complete lack of scales (Kondo *et al.*, 2001). This is the first evidence for a genetic pathway essential for the formation of both fish scales and mammalian hair and that is also required for normal tooth development.

In zebrafish, seven Hox gene clusters were found, almost twice as many as in humans and mice. Sequence analysis revealed that of the seven Hox clusters, one is the single fish orthologue of one of the mammalian Hox clusters and the remaining six appeared to be the result of a duplication of the other three that are found in mammals. Mapping of a number of genes that flank these three Hox clusters in mammals that are duplicated in zebrafish revealed that some of these genes are also present in two copies and map to the same linkage groups as the corresponding Hox cluster duplicates in zebrafish. The genes that are not duplicated in zebrafish were mapped to one of the two linkage groups with paralogous Hox clusters in zebrafish (Amores et al., 1998; Postlethwait et al., 2000). Thus, not only the three Hox clusters but also the whole surrounding chromosome segments appeared to be duplicated. Extrapolating from the situation of the three duplicated Hox chromosome segments, it was suggested that major parts of the zebrafish genome are present in duplicate, possibly because of whole genome duplication in an ancestor of zebrafish. That one Hox cluster is not present in two versions was explained by a later loss of the second copy in the lineage leading to the zebrafish. But many questions such as when the whole genome duplication happened in fish lineage, how many fish duplicated genes remained, and how much two paralogous chromosomes are conserved among fish species were left to be answered.

A recent genomewide comparison of orthologous genes among medaka, zebrafish, and humans strongly indicates that the genome amplification was not partial, but involved the whole genome, and occurred before the last common ancestor of euteleosts (Mitani et al., in press; Naruse et al., in press). Many genes in fish are present as a single copy, and therefore most of the duplicated versions of genes must have degenerated since the initial duplication event (Taylor et al., 2003). However, the redundant genes produced by genome duplication might have evolved new functions that were necessary for fish diversity. The comparison of map position of medaka genes based on conserved syntenies with pufferfish and zebrafish with functional analysis is a powerful tool for understanding how duplicated fish genes have evolved or degenerated (Inoue et al., 2003; Okubo et al., 2003).

B. Bacterial Artificial Chromosome (BAC) Libraries and End Sequencing

The development of bacterial artificial chromosome (BAC) and P1-derived artificial chromosome (PAC) cloning systems has revolutionized genome analysis in vertebrates. BAC/PAC vectors permit the stable maintenance of large DNA fragments, exceeding 200 kb in size, in an E. coli host strain. At the same time, the BAC/PAC molecule can easily be separated from the E. coli genomic DNA, using a standard plasmid isolation protocol employing the alkaline lysis method. Both these features make BACs/PACs superior to yeast artificial chromosomes (YACs), which provide a larger cloning capacity (exceeding 1000 kb), but which are unstable, prone to rearrangements, and difficult to separate from the endogenous yeast chromosomal DNA.

High-quality medaka BAC libraries have been constructed from two inbred strains, Hd-rR and HNI. The Hd-rR library consists of ~92,000 BAC clones with average insert size of 210 kb, covering 24 times the medaka genome (Matsuda *et al.*, 2001). The HNI library consists of ~96,000 BAC clones with an average insert size of 160 kb, covering 20 times the medaka genome (Kondo *et al.*, 2002). These BAC clones were spotted on nylon membranes for colony hybridization screening. The third medaka BAC library was constructed from the Cab strain (C. Amemiya and J. Wittbrodt, unpublished).

A rapid screening system of the Hd-rR BAC library was established, using the 4D-PCR screening method with significant modification (Asakawa *et al.*, 1997). In brief, 30,720 BAC clones are arrayed in eighty 384-well microtiter plates and every four plates are pooled as a superpool, which contains 1536 clones. The first screening is against these 20 superpools, and the second screening is against the particular superpool identified as positive by the first screening. Because individual clones of each superpool are four-dimensionally addressed, a particular clone is readily identified by performing 28 PCR reactions (1D: 4 superpools, 2D: 4 superpools, 3D: 12 superpools, and 4D: 8 superpools). Using the 4D-PCR screening, a desired BAC clone can be obtained within 1 day.

BAC-end sequences have been determined for ~21,000 clones, including ~12,000 Hd-rR BAC clones, ~8000 HNI BAC clones, and ~1000 Cab BAC clones, and a database has been established. The BAC-end sequences are being used as markers for linkage and RH mapping to construct an integrated map.

The Hd-rR and HNI BAC libraries have been used for positional cloning of several genes, including the sex-determining gene *dmrt1bY/DMY* (Fukamachi *et al.*, 2001; Kondo *et al.*, 2001; Matsuda *et al.*, 2002; Nanda *et al.*, 2002) and for the genomic structural analysis of complex genes such as MHC (Matsuo *et al.*, 2002). In addition, the Hd-rR/HNI and Cab BAC libraries are being used to construct genomewide BAC contigs (Zadeh Khorasani *et al.*, 2004, see later). The Hd-rR BAC library is being used for sequencing the medaka chromosome LG22.

C. BAC Mapping

At present, three different medaka BAC libraries are available, generated from the three different inbred medaka strains Hd-rR, Cab, and HNI. (See Section III.B for details.) Because of the different strain backgrounds, the libraries can be used in a complementary way. HNI is a Northern strain, whereas Cab and Hd-rR both are Southern strains. Because a considerable genetic distance has been observed between Northern and Southern strains (Naruse *et al.*, 2000), a comparison of clones from different strains could be rewarding for the analysis of gene families or genomic regions that evolve rapidly, that, for example, are implicated in immune defense or speciation. The Cab strain background is at present used for generating mutations by using ethylnitrosourea (ENU). Cab clones will therefore be used preferentially when sequence analysis in candidate gene regions is necessary to identify a gene causing the mutant phenotype. Also, complementation of ENU-induced defects will preferentially be carried out by using transgenes based on the Cab genotype. Hd-rR has been selected as the reference genotype to determine the complete genomic sequence of the medaka, using both the whole genome shotgun (WGS) approach and BAC sequencing.

Different strategies can be employed for generating BAC maps. The fingerprinting approach was initially applied in *C. elegans*, using cosmids. Fingerprinting involves the isolation of DNA from thousands of genomic clones, followed by digestion with a restriction enzyme and size separation of the fragments. If two clones overlap, they will share restriction fragments of the same size that allow the overlap to be identified. The more clones the data set of fingerprints contains, the greater the likelihood that matching fingerprints will be identified, generating clone contigs. Conceptually, this approach will result in a restriction map for a complete genome, underlayed with genomic clones. The advantage of the strategy is that a set of clones that minimally overlaps can easily be extracted from the data set for subsequent characterization and sequencing. However, there are limitations to this approach. First, fingerprinting does not provide information on the location of the contigs in the genome. Second, contig assembly is difficult if the source material for libraries is from heterozygous individuals or from different genotypes.

Marker-based approaches for map generation can be implemented either by using PCR or hybridization. The advantage of marker-based strategies is the possibility to use probes with known locations in the genome, for example, with positions determined previously by genetic mapping. As a result, contigs will be anchored to the genome map early on. The data set for the current medaka BAC map has been generated by hybridization, using probes from medaka genes and BAC ends (Zadeh Khorasani *et al.*, 2004). At present, 2721 markers are on the map, representing 2510 different loci. Both synteny with the fugu genome and experimental data were exploited to arrange probes into map segments. In the most recent map build from January 8, 2004, 2721 probes were arranged into 902 map segments. The number of probes per map segment ranges from 1 to 27. The number of BAC clones on the map is 41,882 (65% of BACs used, assuming no empty plate positions). Between adjacent markers, 6795 BACs are shared, and 462 of map segments (51%) contain at least one genetically mapped marker from the medaka database, M-Base (http://mbase.bioweb.ne.jp/). Eleven map segments, and therefore a very small proportion of the data set, contain markers that genetically map to different medaka chromosomes, highlighting the overall excellent quality of the map. Contig data are available on a collaborative basis from H. Himmelbauer (himmelbauer@molgen.mpg.de).

D. Radiation Hybrid Mapping

Radiation hybrid (RH) mapping has made important contributions to genomic research, in particular for humans, mice, and zebrafish, facilitating mapping of a number of genes (Geisler *et al.*, 1999; Hukriede *et al.*, 2001). This system is now exploited in medaka genomics. OLF-136 fibroblast cells originating from the HB32 medaka (Southern Japan population) were transfected with a G418 resistance-conferring gene, irradiated, and fused with B78 mouse melanoma cells to produce hybrid lines carrying different sets of medaka chromosome fragments under selection with G418. A total of 270 lines were expanded to isolate genomic DNAs, and 140 clones were further characterized.

Framework RH maps are being constructed first by examining retention of markers in individual hybrids, taking advantage of the collection of 1138 genetically mapped markers developed for PCR typing (Naruse *et al.*, 2000). Among 837 markers tested, 638 have been successfully used for evaluation of marker retention in individual RH lines and for generating the framework RH map, using a couple of different computational programs.

Once the framework RH map is established, any unique sequences of a genome can be quickly mapped with the resolution of genomic span covered by a couple of BAC clones. Analysis by using this RH panel is supported by robotic handling of reagents by using Beckmann Biomek 2000 that can handle 6400 PCR reactions a day. RH mapping will be useful in various aspects of genome projects and facilitate identification of genes responsible for mutants in medaka.

E. Direct Visualisation of Medaka Linkage Groups by Fluorescence *In Situ* Hybridization (FISH) Analysis

Chromosome analysis of medaka started long before the emergence of mammalian cytogenetics (Goodrich, 1927). Despite this early beginning and the widely acknowledged use of medaka as a model for various genetic and biological experiments (Wittbrodt *et al.*, 2002; Yamamoto, 1975), progress in generating a physical map of the medaka genome has been made only recently. At present, a well-defined medaka recombination map comprising approximately 1400 markers, including several hundred ESTs distributed over 24 distinct linkage groups that correspond to the diploid 48 chromosomes, is available (Naruse *et al.*, 2000; http://mbase.bioweb.ne.jp/). In addition several hundred BAC clones have been isolated from gridded libraries, which are targeted to contain sequence-tagged sites (STSs). Ordering of these STSs within the linkage groups is underway in many laboratories, using genetic linkage or RH mapping (see previously). Because large-insert genomic clones are very suitable for fluorescence *in situ* hybridization (FISH), an additional approach is to map these well-characterized BACs to medaka metaphase chromosomes. This experimental strategy should allow individual linkage groups to be connected to specific chromosomes. Moreover, FISH, together with the cytogenetic map, is a useful tool in situations in which assignment of BAC contigs to linkage groups is difficult.

Figure 1 illustrates examples of FISH mapping on medaka chromosomes, using large-insert clones such as BACs and cosmids. By using a limited number of clones it was possible to assign clones from four linkage groups to four different chromosomes, which differ in size and shape. The resolution of chromosomal mapping by FISH is surprisingly high. By using six different BACs from LG1 (sex chromosomes) and two-color FISH, it was possible not only to confirm the orientation of

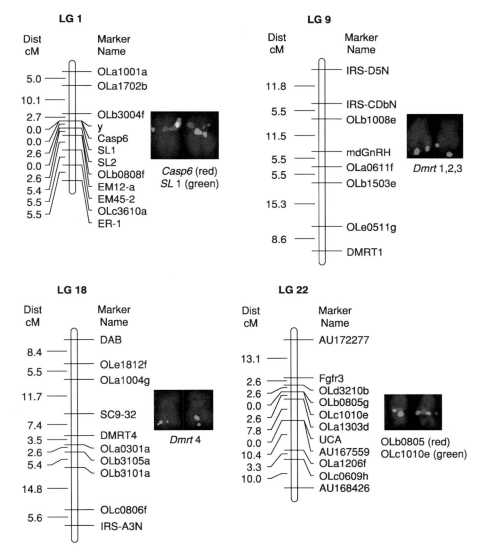

Fig. 1 Preliminary mapping of genomic clones to four different linkage groups in medaka. (See Color Insert.)

individual BACs as established through genetic mapping, but even to recognize the small Y-specific segment spanning nearly 260 kb on the sex chromosomes (see Section II.D.2). In addition, FISH mapping was useful to demonstrate the remarkable duplication of the *dmrt1* locus, and specifically to show *trans* as well as *cis* chromosomal locations of coduplicated genes on the same segment that might not be easily inferred from molecular experiments. With respect to the current efforts to obtain a complete sequence of LG22, two BACs OLb0805g and OLc1010e are mapped simultaneously on a small acrocentric LG22-equivalent chromosome. Consistent with the genetic map, the location of OLb0805 is found to be closer to the centromere than that of OLc1010e. Physical mapping with available BAC clones from LG22 is in progress to orient unmapped clones as well as to assign them to a specific region of the chromosome.

These results exemplify the efforts to assemble large-insert DNA clones of medaka chromosomes in order to prepare a cytogenetic map that should correspond to the genetic map, which is scaled in proportion to the relative frequency of recombination. A standard constitutive banding pattern specific to each chromosome common to all cold-blooded vertebrates is not feasible. Consequently, the definitive identification of all individual chromosomes within a mitotic chromosome spread can be problematic. Therefore, assigning genetic loci to chromosomal positions should involve the FLpter (ratio of the distance of the FISH signal to the telomere of the p-arm divided by the length of the whole chromosome) measurement. Furthermore, simultaneous FISH mapping with two independent BACs located at the both ends of each LGs will accelerate the establishment of a preliminary medaka cytogenetic map.

F. The Linkage Group 22 Sequencing Project

On the basis of available resources, which include a large number of ESTs, a linkage map of EST markers, BAC libraries from inbred strains, BAC-end sequences, BAC contig maps, and RH panels, the Medaka Genome Initiative has set an ambitious goal to determine the entire DNA sequence of the medaka genome through the conventional clone-by-clone (CBC) strategy. A feasibility study has begun that focuses on a particular medaka chromosome, linkage group 22 (LG22). The immediate goal of the LG22 project is to generate a large sequence scaffold covering the entire LG22 of ~20 Mb with a few gaps. It will provide important information about the characteristics of the medaka genome, including GC contents, features of repeats, gene density, promoters, and average intron sizes. The LG22 project will also facilitate comparative genomics and the discovery of genes responsible for medaka mutants.

Screening of the Hd-rR BAC library by hybridization of high-density colony filters with 17 marker probes identified 148 BAC clones, from which 10 contigs were made as anchor points (Himmelbauer *et al.*, unpublished; Mitani *et al.*, unpublished). For further screening, a chromosome walking was done by finding adjacent BAC clones with 4D-PCR screening and/or *in silico* search of the

BAC-end sequence database (Asakawa *et al.*, 1997). As of March 31, 2004, the extended BAC contigs cover ~75% of the LG22 and the accumulated high-quality sequence is ~10 Mb covering 50% of the LG22.

IV. Other Progress

Because of space limitations, this chapter can only give a comprehensive overview of the current progress in medaka genetics and genomics. Many other fields in which medaka is used as a model system could not be discussed. However, we would like to point to a few subjects relevant to the topic of this chapter that could not be included here:

- Production of see-through medaka strain (Wakamatsu *et al.*, 2001b)
- GFP expression under the control of a germline specific promoter in transgenic lines (Tanaka *et al.*, 2001)
- Nuclear transplantation for production of clonal medaka (Wakamatsu *et al.*, 2001a)

V. Internet Web Sites Relevant to Medaka Resources and Databases

- BAC library at RZPD: http://www.rzpd.de
- BAC libraries (Hd-rR and HNI) at Keio University: shimizu@dmb.med.keio.ac.jp, asa@dmb.med.keio.ac.jp
- BAC library (Hd-rR) at Nagoya University: hori@biol1.bio.nagoya-u.ac.jp
- BAC contig information: Himmelbauer@molgen.mpg.de
- Medaka Expression Pattern Database (MEPD): http://www.embl-heidelberg.de/mepd/,http://www.embl-heidelberg.de/wp/gsd/
- EST database at M-base: http://mbase.bioweb.ne.jp/~dclust/medakatop.html
- Information about LG22: http://medaka.dsp.jst.go.jp/MGI/
- General information in medakafish homepage: http://biol1.bio.nagoyau.ac.jp:8000/

Acknowledgments

We are indebted to Robert Geisler, Mario Chevrette, and Mark Ekker for their collaboration in radiation hybrid mapping of medaka. We thank Monika Niklaus-Ruiz for expert help in preparing the manuscript.

References

Alpen, E. L. (1998). "Radiation Biophysics" 2nd ed., Academic Press, San Diego.

Aizawa, K., Mitani, H., Kogure, N., Shimada, A., Hirose, Y., Sasado, T., Morinaga, C., Yasuoka, A., Yoda, H., Watanabe, T., Iwanami, N., Kunimatsu, S., Osakada, M., Suwa, H., Niwa, K., Deguchi, T., Henrich, T., Todo, T., Shima, A., Kondoh, H., and Furutani-Seiki, M. (2004). Identification of radiation-sensitive mutants in the medaka, *Oryzias latipes. Mech. Dev.* **121,** 895–902.

Amores, A., Force, A., Yan, Y. L., Joly, L., Amemiya, C., Fritz, A., Ho, R. K., Langeland, J., Prince, V., Wang, Y. L., Westerfield, M., Ekker, M., and Postlethwait, J. H. (1998). Zebrafish hox clusters and vertebrate genome evolution. *Science* **282,** 1711–1714.

Asakawa, S., Abe, I., Kudoh, Y., Kishi, N., Wang, Y., Kubota, R., Kudoh, J., Kawasaki, K., Minoshima, S., and Shimizu, N. (1997). Human BAC library: Construction and rapid screening. *Gene* **191,** 69–79.

Bejar, J., Hong, Y., and Alvarez, M. C. (2002). An ES-like cell line from the marine fish *Sparus aurata*: Characterization and chimaera production. *Transgenic Res.* **11,** 279–289.

Bejar, J., Hong, Y., and Schartl, M. (2003). Mitf expression is sufficient to direct differentiation of medaka blastula derived stem cells to melanocytes. *Development* **130,** 6545–6553.

Bradley, A., Evans, M., Kaufman, M. H., and Robertson, E. (1984). Formation of germ-line chimaeras from embryo-derived teratocarcinoma cell lines. *Nature* **309,** 255–256.

Brunner, B., Hornung, U., Shan, Z., Nanda, I., Kondo, M., Zend-Ajusch, E., Haaf, T., Ropers, H. H., Shima, A., Schmid, M., Kalscheuer, V. M., and Schartl, M. (2001). Genomic organization and expression of the doublesex-related gene cluster in vertebrates and detection of putative regulatory regions for dmrt1. *Genomics* **77,** 8–17.

Chen, S. L., Ye, H. Q., Sha, Z. X., and Hong, Y. H. (2003). Derivation of a pluripotent embryonic cell line from red sea bream blastulae. *J. Fish Biol.* **63,** 795–805.

Driever, W., Solnica-Krezel, L., Schier, A. F., Neuhauss, S. C., Malicki, J., Stemple, D. L., Stainier, D. Y., Zwartkruis, F., Abdelilah, S., Rangini, Z., Belak, J., and Boggs, C. (1996). A genetic screen for mutations affecting embryogenesis in zebrafish. *Development* **123,** 37–46.

Elmasri, H., Winkler, C., Liedtke, D., Sasado, T., Morinaga, C., Suwa, H., Niwa, K., Henrich, T., Hirose, Y., Yasuoka, A., Yoda, H., Watanabe, T., Deguchi, T., Iwanami, N., Kunimatsu, S., Osakada, M., Loosli, F., Quiring, R., Carl, M., Grabher, C., Winkler, S., Del Bene, F., Wittbrodt, J., Abe, K., Takahama, Y., Takahashi, K., Katada, T., Nishina, H., Kondoh, H., and Furutani-Seiki, M. (2004). Mutations affecting somite formation in the medaka (*Oryzias latipes*). *Mech. Dev.* **121,** 659–672.

Flajnik, M. F., and Kasahara, M. (2001). Comparative genomics of the MHC: Glimpses into the evolution of the adaptive immune system. *Immunity* **15,** 351–362.

Fukamachi, S., Shimada, A., and Shima, A. (2001). Mutations in the gene encoding B, a novel transporter protein, reduce melanin content in medaka. *Nat. Genet.* **28,** 381–385.

Furutani-Seiki, M., Sasado, T., Morinaga, C., Suwa, H., Niwa, K., Yoda, H., Deguchi, T., Hirose, Y., Yasuoka, A., Henrich, T., Watanabe, T., Iwanami, N., Kitagawa, D., Saito, K., Asaka, S., Osakada, M., Kunimatsu, S., Elmasri, H., Winkler, C., Ramialison, M., Loosli, F., Quiring, R., Carl, M., Grabher, C., Winkler, S., Del Bene, F., Shinomiya, A., Kota, Y., Yamanaka, T., Okamoto, Y., Takahashi, K., Todo, T., Abe K., Takahama, Y., Tanaka, M., Mitani, H., Katada, T., Nishina, H., Nakajima, N., Wittbrodt, J., and Kondoh, H. (2004). A systematic genomewide screen for mutations affecting organogenesis in Medaka, *Oryzias latipes. Mech. Dev.* **121,** 647–658.

Geisler, R., Rauch, G. J., Baier, H., van Bebber, F., Brobeta, L., Dekens, M. P., Finger, K., Fricke, C., Gates, M. A., Geiger, H., Geiger-Rudolph, S., Gilmour, D., Glaser, S., Gnugge, L., Habeck, H., Hingst, K., Holley, S., Keenan, J., Kirn, A., Knaut, H., Lashkari, D., Maderspacher, F., Martyn, U., Neuhauss, S., Haffter, P., *et al.* (1999). A radiation hybrid map of the zebrafish genome. *Nat. Genet.* **23,** 86–89.

Goodrich, H. B. (1927). A study of the development of Mendelian characters in *Oryzias latipes. J. Exp. Zool.* **49,** 261–287.

Goriely, A., McVean, G. A., Rojmyr, M., Ingemarsson, B., and Wilkie, A. O. (2003). Evidence for selective advantage of pathogenic FGFR2 mutations in the male germ line. *Science* **301**, 643–646.

Grabher, C., Henrich, T., Sasado, T., Arenz, A., Wittbrodt, J., and Furutani-Seiki, M. (2003). Transposon-mediated enhancer trapping in medaka. *Gene*. **322**, 57–66.

Haffter, P., Granato, M., Brand, M., Mullins, M. C., Hammerschmidt, M., Kane, D. A., Odenthal, J., van Eeden, F. J., Jiang, Y. J., Heisenberg, C. P., Kelsh, R. N., Furutani-Seiki, M., Vogelsang, E., Beuchle, D., Schach, U., Fabian, C., and Nusslein-Volhard, C. (1996). The identification of genes with unique and essential functions in the development of the zebrafish, *Danio rerio*. *Development* **123**, 1–36.

Hagmann, M., Bruggmann, R., Xue, L., Georgiev, O., Schaffner, W., Rungger, D., Spaniol, P., and Gerster, T. (1998). Homologous recombination and DNA-end joining reactions in zygotes and early embryos of zebrafish (*Danio rerio*) and *Drosophila melanogaster*. *Biol. Chem.* **379**, 673–681.

Hawkins, W. E., Walker, W. W., Fournie, J. W., Manning, C. S., and Krol, R. M. (2003). Use of the Japanese medaka (*Oryzias latipes*) and guppy (*Poecilia reticulata*) in carcinogenesis testing under national toxicology program protocols. *Toxicol. Pathol.* **31**(Suppl.), 88–91.

Henrich, T., Ramialison, M., Quiring, R., Wittbrodt, B., Furutani-Seiki, M., Wittbrodt, J., and Kondoh, H. (2003). MEPD: A medaka gene expression pattern database. *Nucleic Acids Res.* **31**, 72–74.

Henrich, T., Ramialison, M., Segerdell, E., Westerfield, M., Furutani-Seiki, M., Wittbrodt, J., and Kondoh, H. (2004). GSD: A genetic screen database. *Mech. Dev.* **121**, 959–964.

Henrich, T., and Wittbrodt, J. (2000). An *in situ* hybridization screen for the rapid isolation of differentially expressed genes. *Dev. Genes Evol.* **210**, 28–33.

Hong, Y., Liu, T., Zhao, H., Xu, H., Wang, W., Liu, R., Chen, T., Deng, J., and Gui, J. (2004b). Establishment of a normal medakafish spermatogonial germ cell line capable of sperm production *in vitro*. *Proc. Natl. Acad. Sci. USA.* **101**, 8011–8016.

Hong, Y., and Schartl, M. (1996). Establishment and growth responses of early medakafish (*Oryzias latipes*) embryonic cells in feeder layer-free cultures. *Mol. Mar. Biol. Biotechnol.* **5**, 93–104.

Hong, Y., Winkler, C., Liu, T., Cai, G., and Schartl, M. (2004a). Activation of the mouse Oct4 promoter in medaka embryonic stem cells and its use for ablation of spontaneous differentiation. *Mech. Dev.* **121**, 933–944.

Hong, Y., Winkler, C., and Schartl, M. (1996). Pluripotency and differentiation of embryonic stem cell lines from the medakafish (*Oryzias latines*). *Mech. Dev.* **60**, 33–44.

Hong, Y., Winkler, C., and Schartl, M. (1998a). Production of medakafish chimeras from a stable embryonic stem cell line. *Proc. Natl. Acad. Sci. USA* **95**, 3679–3684.

Hong, Y., Winkler, C., and Schartl, M. (1998b). Efficiency of cell culture derivation from blastula embryos and of chimera formation in the medaka (*Oryzias latipes*) depends on donor genotype and passage number. *Dev. Genes Evol.* **208**, 595–602.

Hukriede, N., Fisher, D., Epstein, J., Joly, L., Tellis, P., Zhou, Y., Barbazuk, B., Cox, K., Fenton-Noriega, L., Hersey, C., Miles, J., Sheng, X., Song, A., Waterman, R., Johnson, S. L., Dawid, I. B., Chevrette, M., Zon, L. I., McPherson, J., and Ekker, M. (2001). The LN54 radiation hybrid map of zebrafish expressed sequences. *Genome Res.* **11**, 2127–2132.

Inoue, K., Naruse, K., Yamagami, S., Mitani, H., Suzuki, N., and Takei, Y. (2003). Four functionally distinct C-type natriuretic peptides found in fish reveal evolutionary history of the natriuretic peptide system. *Proc. Natl. Acad. Sci. USA* **100**, 10079–10084.

Ishikawa, Y. (2000). Medakafish as a model system for vertebrate developmental genetics. *Bioessays* **22**, 487–495.

Ivics, Z., Hackett, P. B., Plasterk, R. H., and Izsvak, Z. (1997). Molecular reconstruction of Sleeping Beauty, a Tc1-like transposon from fish, and its transposition in human cells. *Cell* **91**, 501–510.

Iwamatsu, T. (1994). Stages of normal development in the medaka *Oryzias latipes*. *Zoo. Sci.* **11**, 825–839.

Iwanami, N., Takahama, Y., Kunimatsu, S., Li, J., Takei, R., Ishikura, Y., Suwa, H., Niwa, K., Sasado, T., Morinaga, C., Yasuoka, A., Deguchi, T., Hirose, Y., Yoda, H., Henrich, T., Ohara, O.,

Kondoh, H., and Furutani-Seiki, M. (2004). Mutations affecting thymus organogenesis in medaka, *Oryzias latipes. Mech. Dev.* **121,** 779–790.

Kawakami, K., Shima, A., and Kawakami, N. (2000). Identification of a functional transposase of the Tol2 element, an Ac-like element from the Japanese medaka fish, and its transposition in the zebrafish germ lineage. *Proc. Natl. Acad. Sci. USA* **97,** 11403–11408.

Kawase, E., Suemori, H., Takahashi, N., Okazaki, K., Hashimoto, K., and Nakatsuji, N. (1994). Strain difference in establishment of mouse embryonic stem (ES) cell lines. *Int. J. Dev. Biol.* **38,** 385–390.

Kitagawa, D., Watanabe, T., Saito, K., Asaka, S., Sasado, T., Morinaga, C., Suwa, H., Niwa, K., Yasuoka, A., Deguchi, T., Yoda, H., Hirose, Y., Henrich, T., Iwanami, N., Kunimatsu, S., Osakada, M., Winkler, C., Elmasri, H., Wittbrodt, J., Loosli, F., Quiring, R., Carl, M., Grabher, C., Winkler, S., Del Bene, F., Katada, T., Nishina, H., Kondoh, H., Furutani-Seiki, M. (2004). Genetic dissection of the formation of the forebrain in medaka, *Oryzias latipes. Mech. Dev.* **121,** 673–686.

Koga, A., and Hori, H. (1999). Homogeneity in the structure of the medaka fish transposable element Tol2. *Genet Res.* **73,** 7–14.

Koga, A., Hori, H., and Sakaizumi, M. (2002a). Gene transfer and cloning of flanking chromosomal regions using the medaka fish Tol2 transposable element. *Mar. Biotechnol.* **4,** 6–11.

Koga, A., Iida, A., Kamiya, M., Hayashi, R., Hori, H., Ishikawa, Y., and Tachibana, A. (2003). The medaka fish Tol2 transposable element can undergo excision in human and mouse cells. *J. Hum. Genet.* **48,** 231–235.

Koga, A., Sakaizumi, M., and Hori, H. (2002b). Transposable elements in medaka fish. *Zool. Sci.* **19,** 1–6.

Koga, A., Shimada, A., Shima, A., Sakaizumi, M., Tachida, H., and Hori, H. (2000). Evidence for recent invasion of the medaka fish genome by the Tol2 transposable element. *Genetics* **155,** 273–281.

Kondo, M., Froschauer, A., Kitano, A., Nanda, I., Hornung, U., Volff, J.-N., Asakawa, S., Mitani, H., Naruse, K., Tanaka, M., Schmid, M., Shimizu, N., Schartl, M., and Shima, A. (2002). Molecular cloning and characterization of DMRT genes from the medaka *Oryzias latipes* and the platyfish *Xiphophorus maculatus. Gene.* **295,** 213–222.

Kondo, M., Nanda, I., Hornung, U., Asakawa, S., Shimizu, N., Mitani, H., Schmid, M., Shima, A., and Schartl, M. (2003). Absence of the candidate male sex-determining gene dmrt1b(Y) of medaka from other fish species. *Curr. Biol.* **13,** 416–420.

Kondo, S., Kuwahara, Y., Kondo, M., Naruse, K., Mitani, H., Wakamatsu, Y., Ozato, K., Asakawa, S., Shimizu, N., and Shima, A. (2001). The medaka rs-3 locus required for scale development encodes ectodysplasin-A receptor. *Curr. Biol.* **11,** 1202–1206.

Kuroda, N., Naruse, K., Shima, A., Nonaka, M., and Sasaki, M. (2000). Molecular cloning and linkage analysis of complement C3 and C4 genes of the Japanese medaka fish. *Immunogenetics* **51,** 117–128.

Kuwahara, Y., Shimada, A., Mitani, H., and Shima, A. (2002). A critical stage in spermatogenesis for radiation-induced cell death in the medaka fish, *Oryzias latipes. Radiat. Res.* **157,** 386–392.

Kuwahara, Y., Shimada, A., Mitani, H., and Shima, A. (2003). Gamma-ray exposure accelerates spermatogenesis of medaka fish, *Oryzias latipes. Mol. Reprod. Dev.* **65,** 204–211.

Loosli, F., Del Bene, F., Quiring, R., Rembold, M., Martinez-Morales, J. R., Carl, M., Grabher, C., Iquel, C., Krone, A., Wittbrodt, B., Winkler, S., Sasado, T., Morinaga, C., Suwa, H., Niwa, K., Henrich, T., Deguchi, T., Hirose, Y., Iwanami, N., Kunimatsu, S., Osakada, M., Watanabe, T., Yasuoka, A., Yoda, H., Kondoh, H., Furutani-Seiki, M., Wittbrodt, J. (2004). Mutations affecting retina development in medaka. *Mech. Dev.* **121,** 703–714.

Loosli, F., Koster, R. W., Carl, M., Kuhnlein, R., Henrich, T., Mucke, M., Krone, A., and Wittbrodt, J. (2000). A genetic screen for mutations affecting embryonic development in medaka fish (*Oryzias latipes*). *Mech. Dev.* **97,** 133–139.

Loosli, F., Staub, W., Finger-Baier, K. C., Ober, E. A., Verkade, H., Wittbrodt, J., and Baier, H. (2003). Loss of eyes in zebrafish caused by mutation of chokh/rx3. *EMBO Rep.* **4,** 894–899.

Mariat, D., Taourit, S., and Guerin, G. (2003). A mutation in the MATP gene causes the cream coat colour in the horse. *Genet. Sel Evol.* **35,** 119–133.

Matsuda, M., Kawato, N., Asakawa, S., Shimizu, N., Nagahama, Y., Hamaguchi, S., Sakaizumi, M., and Hori, H. (2001). Construction of a BAC library derived from the inbred Hd-rR strain of the teleost fish, *Oryzias latipes. Genes Genet. Syst.* **76,** 61–63.

Matsuda, M., Sato, T., Toyazaki, Y., Nagahama, Y., Hamaguchi, S., and Sakaizumi, M. (2003). *Oryzias curvinotus has DMY*, a gene that is required for male development in the medaka, *O. latipes. Zool. Sci.* **20,** 159–161.

Matsuda, M., Nagahama, Y., Shinomiya, A., Sato, T., Matsuda, C., Kobayashi, T., Morrey, C. E., Shibata, N., Asakawa, S., Shimizu, N., Hori, H., Hamaguchi, S., and Sakaizumi, M. (2002). DMY is a Y-specific DM-domain gene required for male development in the medaka fish. *Nature* **417,** 559–563.

Matsuo, M. Y., Asakawa, S., Shimizu, N., Kimura, H., and Nonaka, M. (2002). Nucleotide sequence of the MHC class I genomic region of a teleost, the medaka (*Oryzias latipes*). *Immunogenetics* **53,** 930–940.

Mitani, H., Naruse, K., Tanaka, M., Shima, A. (in press) Medaka genome mapping for functional genomics. "Fish Development and Genetics: The Zebrafish and Medaka Models" (Z. Gong, and V. Korzh,). World Scientific Publishing Singapore.

Morinaga, C., Tomonaga, T., Sasado, T., Suwa, H., Niwa, K., Yasuoka, A., Henrich, T., Watanabe, T., Deguchi, Y., Yoda, H., Hirose, Y., Iwanami, N., Kunimatsu, S., Shinomiya, A., Tanaka, M., Kondoh, H., Furutani-Seiki, M. (2004). Mutations affecting gonadal formation in Medaka, *Oryzias latipes. Mech. Dev.* **121,** 829–840.

Nanda, I., Hornung, U., Kondo, M., Schmid, M., and Schartl, M. (2003). Common spontaneous sex-reversed XX males of the medaka, *Oryzias latipes. Genetics* **163,** 245–251.

Nanda, I., Kondo, M., Hornung, U., Asakawa, S., Winkler, C., Shimizu, A., Shan, Z., Haaf, T., Shimizu, N., Shima, A., Schmid, M., and Schartl, M. (2002). A duplicated copy of DMRT1 in the sex determining region of the Y chromosome of the medaka, *Oryzias latipes. Proc. Natl. Acad. Sci. USA* **99,** 11778–11783.

Naruse, K., Fukamachi, S., Mitani, H., Kondo, M., Matsuoka, T., Kondo, S., Hanamura, N., Morita, Y., Hasegawa, K., Nishigaki, R., Shimada, A., Wada, H., Kusakabe, T., Suzuki, N., Kinoshita, M., Kanamori, A., Terado, T., Kimura, H., Nonaka, M., and Shima, A. (2000). A detailed linkage map of medaka, *Oryzias latipes*: comparative genomics and genome evolution. *Genetics* **154,** 1773–1784.

Naruse, K., Tanaka, K., Mita, K., Shima, A., Postlethwait, J., and Mitani, H. (2004). A medaka gene map: The trace of ancestral vertebrate proto-chromosomes revealed by comparative gene mapping. *Genome Res.* **14,** 820–828.

Newton, J. M., Cohen-Barak, O., Hagiwara, N., Gardner, J. M., Davisson, M. T., King, R. A., and Brilliant, M. H. (2001). Mutations in the human orthologue of the mouse underwhite gene (uw) underlie a new form of oculocutaneous albinism, OCA4. *Am. J. Hum. Genet.* **69,** 981–988.

Okubo, K., Ishii, S., Ishida, J., Mitani, H., Naruse, K., Kondo, M., Shima, A., Tanaka, M., Asakawa, S., Shimizu, N., and Aida, K. (2003). A novel third gonadotropin-releasing hormone receptor in the medaka, *Oryzias latipes*: Evolutionary and functional implications. *Gene.* **314,** 121–131.

Postlethwait, J. H., Woods, I. G., Ngo-Hazelett, P., Yan, Y., Kelly, P. D., Chu, F., Huang, H., Hill-Force, A., and Talbot, W. S. (2000). Zebrarish comparative genomics and the origins of vertebrate chromosomes. *Genome Res.* **10,** 1890–1902.

Quiring, R., Wittbrodt, B., Henrich, T., Ramialison, M., Burgtorf, C., Lehrach, H., Wittbrodt, J. (2004). Large-scale expression screening by automated whole-mount in situ hybridization—a technical note. *Mech. Dev.* **121,** 971–976.

Samonte, I. E., Sato, A., Mayer, W. E., Shintani, S., and Klein, J. (2002). Linkage relationships of genes coding for alpha2-macroglobulin, C3 and C4 in the zebrafish: implications for the evolution of the complement and Mhc systems. *Scand. J. Immunol.* **56,** 344–352.

Sasado, T., Morinaga, C., Niwa, K., Shinomiya, A., Yasuoka, A., Suwa, H., Hirose, Y., Yoda, H., Henrich, T., Deguchi, T., Iwanami, N., Watanabe, T., Kunimatsu., Osakada, M., Okamoto, Y.,

Kota, Y., Yamanaka, T., Tanaka, M., Kondoh, H., Furutani-Seiki, M. (2004). Mutations affecting early distribution of primordial germ cells in medaka (*Oryzias latipes*) embryo. *Mech. Dev.* **121**, 817–828.

Shima, A., and Shimada, A. (1991). Development of a possible nonmammalian test system for radiation-induced germ-cell mutagenesis using a fish, the Japanese medaka (*Oryzias latipes*). *Proc. Natl. Acad. Sci. USA* **88**, 2545–2549.

Shima, A., and Shimada, A. (1994). The Japanese medaka, *Oryzias latipes*, as a new model organism for studying environmental germ-cell mutagenesis. *Environ Hlth. Perspect.* **102**(Suppl. 12), 33–53.

Shima, A., and Shimada, A. (2001). The medaka as a model for studying germ-cell mutagenesis and genomic instability. *Mar. Biotech.* **3**, 141–144.

Solnica-Krezel, L., Schier, A. F., and Driever, W. (1994). Efficient recovery of ENU-induced mutations from the zebrafish germline. *Genetics* **136**, 1401–1420.

Song, M., and Gutzeit, H. O. (2003). Primary culture of medaka (*Oryzias latipes*) testis: A test system for the analysis of cell proliferation and differentiation. *Cell Tissue Res.* **313**, 107–115.

Sprague, J., Clements, D., Conlin, T., Edwards, P., Frazer, K., Schaper, K., Segerdell, E., Song, P., Sprunger, B., and Westerfield, M. (2003). The Zebrafish Information Network (ZFIN): The zebrafish model organism database. *Nucleic Acids Res.* **31**, 241–243.

Tanaka, M., Kinoshita, M., Kobayashi, D., and Nagahama, Y. (2001). Establishment of medaka (*Oryzias latipes*) transgenic lines with the expression of green fluorescent protein fluorescence exclusively in germ cells: A useful model to monitor germ cells in a live vertebrate. *Proc. Natl. Acad. Sci. USA* **98**, 2544–2549.

Taylor, J. S., Braasch, I., Frickey, T., Meyer, A., and Van de Peer, Y. (2003). Genome duplication, a trait shared by 22000 species of ray-finned fish. *Genome Res.* **13**, 382–390.

Terado, T., Okamura, K., Ohta, Y., Shin, D. H., Smith, S. L., Hashimoto, K., Takemoto, T., Nonaka, M., Kimura, H., Flajnik, M. F., and Nonaka, M. (2003). Molecular cloning of C4 gene and identification of the class III complement region in the shark MHC. *J. Immunol.* **171**, 2461–2466.

The MHC sequencing Consortiuml(1999). Complete sequence and gene map of a human major histocompatibility complex. *Nature* **401**, 921–923.

Volff, J. N., Hornung, U., and Schartl, M. (2001). Fish retroposons related to the Penelope element of *Drosophila virili's* define a new group of retrotransposable elements. *Mol. Genet. Genom.* **265**, 711–720.

Wada, H., Naruse, K., Shimada, A., and Shima, A. (1995). A genetic linkage map of a fish, the Japanese medaka *Oryzias latipes*. *Mol. Mar. Biol. Biotech.* **4**, 269–274.

Wakamatsu, Y., Ju, B., Pristyaznhyuk, I., Niwa, K., Ladygina, T., Kinoshita, M., Araki, K., and Ozato, K. (2001a). Fertile and diploid nuclear transplants derived from embryonic cells of a small laboratory fish, medaka (*Oryzias latipes*) *Proc. Natl. Acad. Sci. USA* **98**, 1071–1076.

Wakamatsu, Y., Ozato, K., and Sasado, T. (1994). Establishment of a pluripotent cell line derived from a medaka (*Oryzias latipes*) blastula embryo. *Mol. Mar. Biol. Biotechnol.* **3**, 185–191.

Wakamatsu, Y., Pristyazhnyuk, S., Kinoshita, M., Tanaka, M., and Ozato, K. (2001b). The see-through medaka: A fish model that is transparent throughout life. *Proc. Natl. Acad. Sci. USA* **98**, 10046–10050.

Watanabe, T., Asaka, S., Kitagawa, D., Saito, K., Kurashige, R., Sasado, T., Morinaga, C., Suwa, H., Niwa, K., Henrich, T., Hirose, Y., Yasuoka, A., Yoda, H., Deguchi, T., Iwanami, N., Kunimatsu, S., Osakada, M., Loosli, F., Quiring, R., Carl, M., Grabher, C., Winkler, S., Del Bene, F., Wittbrodt, J., Abe, K., Takahama, Y., Takahashi, K., Katada, T., Nishina, H., Kondoh, H., and Furutani-Seiki, M., (2004). Mutations affecting liver development and function in medaka, *Oryzias latipes*, screened by multiple criteria. *Mech. Dev.* **121**, 791–802.

Wittbrodt, J., Shima, A., and Schartl, M. (2002). Medaka—a model organism from the Far East. *Nat. Rev. Genet.* **3**, 53–64.

Yamamoto, T. (1975). "Medaka (Killfish): Biology and Strains" Keigaku Publishing, Tokyo.

Yasuoka, A., Hirose, Y., Yoda, H., Morinaga, C., Aihara, Y., Suwa, H., Niwa, K., Sasado, T., Deguchi, T., Henrich, T., Iwanami, N., Kunimatsu, S., Abe, K., Kondoh, H., and Furutani-Seiki,

M. (2004). Mutations affecting the formation of posterior lateral line system in medaka, *Oryzias latipes*. *Mech. Dev.* **121,** 729–738.

Yoda, H., Hirose, Y., Yasuoka, A., Sasado, T., Morinaga, C., Henrich, T., Deguchi, T., Iwanami, N., Watanabe, T., Osakada, M., Kunimatsu, S., Wittbrodt, J., Suwa, H., Niwa, K., Okamoto, Y., Yamanaka, T., Kondoh, H., and Furutani-Seiki, M. (2004). Mutations affecting retinotectal pathfinding in medaka, *Oryzias latipes. Mech. Dev.* **121,** 715–728.

Zadeh Khorasani, M., Hennig, S., Imre, G., Asakawa, S., Palczewski, S., Berger, A., Hori, H., Naruse, K., Mitani, H., Shima, A., Lehrach, H., Wittbrodt, J., Kondoh, H., Shimizu, N., and Himmelbauer, H. (2004). A first generation physical map of the medaka genome in BACs essential for positional cloning and clone-by-clone based genomic sequencing. *Mech. Dev.* **121,** 903–914.

CHAPTER 11

Transgenesis and Gene Trap Methods in Zebrafish by Using the *Tol2* Transposable Element

Koichi Kawakami

Division of Molecular and Developmental Biology
National Institute of Genetics
Mishima
Shizuoka 411–8540, Japan

I. Introduction

Zebrafish (*Danio rerio*) has been used as a model animal to study vertebrate development by genetic approaches (Streisinger *et al.*, 1981). Large-scale screens for mutants by using a chemical mutagen have been performed, and hundreds of mutations affecting various processes of development have been isolated (Driever *et al.*, 1996; Haffter *et al.*, 1996). Cloning of these point mutations, however, has been laborious because it requires time-consuming positional cloning approaches. On the other hand, an insertional mutagenesis method that uses a pseudotyped retrovirus, which is composed of a Moloney murine leukemia virus vector and the envelope glycoprotein (G-protein) of the vesicular stomatitis virus, has been developed (Gaiano *et al.*, 1996a,b; Lin *et al.*, 1994). This method enabled a large-scale screen for insertional mutants to be performed and the mutated genes to be cloned rapidly (Amsterdam *et al.*, 1999; Golling *et al.*, 2002). Although these approaches have identified a number of genes important for development, genetic methodologies available in zebrafish are still limited. For instance, an enhancer trap or a gene trap method, which is powerful to study the function of developmental genes in *Drosophila* and mouse, has not been developed.

The *Tol2* transposable element, which is found in the genome of a small freshwater teleost, the Japanese medaka fish (*Oryzias latipes*), belongs to the hAT family of transposons that includes *hobo* of *Drosophila*, *Ac* of maize, and *Tam3* of snapdragon (Koga *et al.*, 1996). The zebrafish genome does not contain this element. We have identified an autonomous member of the *Tol2* element that encodes a gene for a fully functional transposase capable of catalyzing transposition in the zebrafish germ lineage and also in mouse ES cells (Kawakami and Noda, 2004; Kawakami and Shima, 1999; Kawakami *et al.*, 1998, 2000). To date, *Tol2* is the only natural DNA transposable element in vertebrates for which an autonomous member has been identified.

The *Tol2* element encodes a gene for an active transposase, which is composed of four exons and has the capacity to produce a protein of 649 amino acids. A transcript of 2156 nucleotides is synthesized in zebrafish embryos injected with the *Tol2* element, and the cDNA has been cloned (Kawakami and Shima, 1999). In previous studies, we coinjected a plasmid DNA containing a nonautonomous *Tol2* element and the transposase mRNA synthesized *in vitro* by using the cDNA as a template into fertilized eggs, and demonstrated that the nonautonomous element can transpose from the plasmid to the genome during embryonic development and the transposon insertions can be transmitted to the next generation through the germ lineage (Kawakami *et al.*, 2000). The transgenic frequency with this system had been, however, too low to generate a number of transposon insertions in a small laboratory.

Recently, we successfully improved the transgenic frequency by using the *Tol2* transposon system (Kawakami *et al.*, 2004). The frequency of obtaining founder fish was increased: about 50% of fish injected with a plasmid DNA containing a transposon vector and the transposase mRNA can transmit the transposon insertion to the progeny. Also, the number of transposon insertions transmitted by single founder fish was increased: 1 to more than 25, and on average 5.6 insertions per founder fish can be transmitted. This breakthrough dramatically increases the usefulness of the *Tol2* transposon system in zebrafish. First, construction of transgenic fish expressing a reporter gene such as the green fluorescent protein (GFP) or a gene of interest in a specific tissue or organ becomes much easier.

Second, because this transposon system allows us to create hundreds of transposon insertions rather efficiently, a gene trap method has been developed for the first time in zebrafish and fish expressing GFP in temporally and spatially restricted patterns have been obtained efficiently. I describe here how these methods should be performed.

II. Transgenesis by Using the *Tol2* Transposable Element in Zebrafish

A. Microinjection and Excision of *Tol2* in the Injected Embryo

1. Methods

a. Plasmids Containing Transposon Vectors

Plasmid DNA containing (*Tol2-tyr*)ΔRV (Kawakami *et al.*, 1998), T2KXIG, or T2KSAG (Kawakami *et al.*, 2004; Fig. 1) are used for microinjection. Plasmid DNA is prepared by using the QIAfilter Plasmid Maxi Kit (QIAGEN), purified once by phenol/chloroform extraction, precipitated with ethanol, and suspended in H_2O.

b. Synthesis of Transposase mRNA In Vitro

The transposase cDNA (Kawakami and Shima, 1999; Kawakami *et al.*, 2000; Fig. 1) was cloned into pCS2+ (Rupp *et al.*, 1994; Turner and Weintraub, 1994),

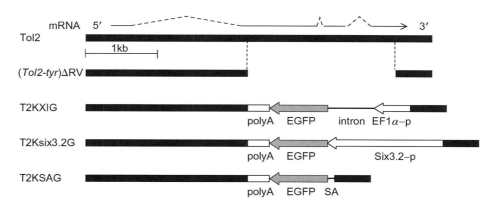

Fig. 1 Transposon constructs described in the chapter. Tol2, the full-length *Tol2* element. The thin line and dotted line above the *Tol2* element indicate mRNA encoding the transposase; (*Tol2-tyr*)ΔRV, a nonautonomous *Tol2* vector lacking the region indicated by the dotted lines; T2KXIG contains the green fluorescent protein (GFP) expression cassette composed of the *Xenopus* EF1α enhancer/ promoter, the rabbit β-globin intron, the EGFP gene, and the SV40 polyA signal; T2Ksix3.2G contains the *six3.2* promoter, the EGFP gene, and the SV40 poly A signal; T2KSAG, the promoter and the splice donor are removed from T2KXIG.SA, the splice acceptor.

resulting in pCS-TP (Kawakami *et al.*, 2004). pCS-TP is linearized by digestion with *Not*I, and mRNA is synthesized *in vitro* by using mMESSAGE mMACHINE SP6 Kit (Ambion, Inc.). The synthesized mRNA is purified by using 'quick spin columns for radiolabeled RNA purification' (Roche), precipitated, and suspended in H_2O.

c. Microinjection

Circular DNA of a plasmid containing a transposon vector and the transposase mRNA are mixed in a final concentration of 25 ng/μl each in 0.2 M KCl. Approximately 1 nl of the DNA/RNA mixture is injected into a fertilized egg. The injected embryos are incubated at 28 °C.

d. Polymerase Chain Reaction (PCR) Analysis of the Injected Embryo

About 8–10 h after the microinjection, embryos are transferred one by one to 0.2-ml strip tubes (eight tubes per strip) and lysed in 50 μl of 10 mM Tris-HCl pH 8.0, 10 mM EDTA, 200 μg/ml proteinase K at 50 °C for 2hrs to overnight. Proteinase K is inactivated at 95 °C for 5 min. The polymerase chain reaction (PCR) is performed as described (Kawakami and Shima, 1999; Kawakami *et al.*, 1998) with some modifications. One microliter of the DNA sample (about 100–150 ng) is used for PCR (35 cycles of 94 °C, 30 sec; 55 °C, 30 sec, and 72 °C, 30 sec). The following primers are used to detect excision of the transposon vector:

- BS1: 5′-AAC AAA AGC TGG AGC TCC ACC G-3′
- TYR1: 5′-AAG GCT CTT GGA TAC GAG TAC GCC-3′
- PCR products are analyzed on 2% agarose gel electrophoresis.

2. Results and Discussion

The transposase is synthesized from the injected mRNA, functions in *trans*, and catalyzes excision of the transposon vector from the plasmid DNA. The double-strand break created on the plasmid DNA is then repaired and religated. When all these processes are carried out properly in the injected embryo, PCR products of approximately 250 bp are amplified. The outline of this experiment is shown in Fig. 2.

This transient embryonic excision assay (TEEA) is important to test *cis* and *trans* activities of the transposon system. If the transposase mRNA is degraded for any reason, if there are any mistakes in construction of the transposon vectors, or if researchers fail to inject eggs properly, TEEA will not work. Therefore, it is strongly recommended that a researcher perform TEEA in every microinjection experiment. A researcher can inject more than 100 fertilized eggs with the plasmid and mRNA per day. When the injection has been done, four to eight embryos should be picked up for TEEA. On the same day, the researcher will know whether the excision occurred properly in the injected embryos. In case TEEA did not work, the injected embryos can be discarded to avoid wasting time and efforts to raise them.

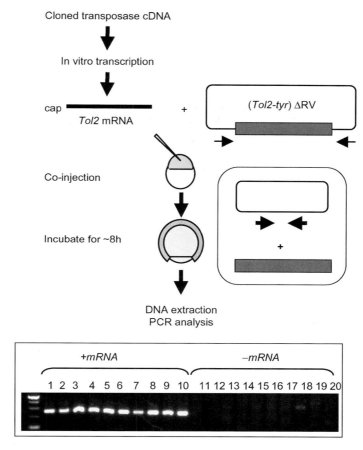

Fig. 2 Outline of the transient embryonic excision assay. The *Tol2* mRNA is synthesized *in vitro* by using the cloned transposase cDNA as a template. Arrows below (*Tol2-tyr*)ΔRV indicate directions and positions of primers used for PCR. The lanes of the electrophoresis gel represent DNA samples from individual embryos injected with or without the transposase mRNA. The bands shown on the gel are detected only in embryos coinjected with the plasmid DNA and the transposase mRNA.

B. Germline Transmission of *Tol2*

1. Methods

a. Plasmids

Plasmid DNA containing T2KXIG (Kawakami *et al.*, 2004; Fig. 1) is used for microinjection. T2KXIG contains the *Xenopus* EF1α enhancer/promoter cassette (Johnson and Krieg, 1994), the rabbit β-globin intron, the EGFP gene, and the SV40 polyA signal. The GFP gene is placed in the reverse orientation relative to the transposase gene to minimize possible interactions with the endogenous promoter activity located near the 5′ end of the *Tol2* element (Kawakami and Shima, 1999).

b. Transposase mRNA and Microinjection

Preparation of the transposase mRNA and microinjection are carried as described previously.

c. Microscopic Analysis of F1 Embryos

The injected fish are crossed with noninjected wild-type fish to obtain F1 embryos. The embryos are placed on a glass depression plate and GFP expression is examined by using MZ FL III and MZ 16 FA (Leica) fluorescent dissecting microscopes. GFP-positive embryos are picked up and raised.

d. PCR Analysis of Pooled Embryos

Fifty Day 1-embryos are collected in a 1.5 ml microtube and lysed in 500 μl of DNA extraction buffer (10 mM Tris-HCl, pH 8.2, 10 mM EDTA, 200 mM NaCl, 0.5% SDS, 200 μg/ml proteinase K) at 50 °C for 3 h to overnight. Genomic DNA is extracted with phenol/chloroform, precipitated with ethanol, and suspended in 50 μl of 10 mM Tris-HCl (pH 8.0), 1 mM EDTA. One microliter of the DNA sample is used for PCR (35 cycles of 94 °C, 20 sec; 56 °C, 20 sec; 72 °C, 20 sec). The following primers are used to detect transposon vectors carrying the GFP gene:

- EGFP/f1: 5′-CTC CTG GGC AAC GTG CTG GTT-3′
- EGFP/r1: 5′-GTG GTG CAG ATG AAC TTC AG-3′

e. Detection of the Transposon Insertions by Southern Blot Hybridization

Caudal fins of the F1 fish are clipped and lysed in 200 μl of DNA extraction buffer at 50 °C for 3 h to overnight. Genomic DNA is extracted with phenol/ chloroform, precipitated with ethanol, and resuspended in 50 μl of 10 mM Tris-HCl (pH 8.0), 1 mM EDTA. Approximately 20–30 μg DNA in total will be obtained. To perform Southern blot hybridization, 5 μg of the genomic DNA is digested with *Bgl*II, which cuts T2KXIG and T2Ksix3.2G once; separated on a 1% agarose gel; and transferred to Hybond-N+ (Amersham). The ~800 bp *Bam*HI-*Cla*I fragment from the T2KXIG plasmid containing the GFP gene is labeled with ^{32}P and used for hybridization. Images are analyzed by using BAS2500 (Fuji Photo Film). Because the backbone of the T2KXIG is pBluescript (Stratagene), linearized pBluescript DNA is used to detect bands containing the plasmid sequence.

2. Results and Discussion

Germline transmission of the transposon vector can be achieved by coinjecting plasmid DNA containing a transposon vector and the transposase mRNA. The experiment was carried out with T2KXIG (Kawakami *et al.*, 2004). Ten fish coinjected with the T2KXIG plasmid and the transposase mRNA were raised to adulthood and mated with noninjected fish. F1 embryos obtained from these crosses were analyzed for GFP expression. GFP-expressing embryos were identified in the progeny from five injected fish (50%; XIG-1~5). The GFP expression was rather ubiquitous as has been observed in transgenic fish expressing GFP under the control of the *Xenopus* EF1α enhancer/promoter, which were generated by microinjection of a plasmid DNA or infection of a pseudotyped retrovirus (Amsterdam *et al.*, 1995; Linney *et al.*, 1999).

The F1 embryos, which did not express GFP, were pooled and analyzed by PCR for the presence of the T2KXIG sequence. All of the GFP-negative embryos were also PCR-negative, indicating that transgenic fish carrying the T2KXIG insertion always express GFP.

F1 fish were analyzed by Southern blot hybridization and the number of transposon insertions transmitted to the F1 generation were counted. In one extreme case, 100% ($n = 259$) of F1 embryos from single founder fish (XIG-1) expressed GFP. Southern blot analysis of 14 F1 fish revealed that the founder fish transmitted more than 15 different insertions. In the other case (XIG-3), 27% of F1 fish expressed GFP and the founder fish transmitted nine insertions. The outline of this experiment is shown in Fig. 3.

The same membrane filters were used for Southern blot hybridization, using the plasmid probe. The probe did not hybridize to most of the bands, indicating that the transposon vector alone was integrated through transposition. Three bands in the XIG-1 F1 fish and one band in the XIG-3 F1 fish, however, hybridized to the plasmid probe, indicating that these bands probably represent integration of the entire plasmid DNA, which occurred concomitantly with transposition (XIG1~5). Excluding these four, 28 insertions were transmitted by the five founder fish. Thus, the average number of transposon insertions transmitted by single founder fish to the progeny is 5.6.

The founder frequency observed in our method, 50%, is higher than that observed in any other transgenesis method developed to express a transgene in fish, that is, injection of naked plasmid DNA (5%, Stuart *et al.*, 1990; 5–9%, Amsterdam *et al.*, 1995), the *Tc3* transposon system (7.5%, Raz *et al.*, 1998), a pseudotyped retrovirus expressing GFP (10%, Linney *et al.*, 1999), the *I-SceI* meganuclease system (30.5%, Thermes *et al.*, 2002), and the *Sleeping Beauty* transposon system (5–31%, Davidson *et al.*, 2003).

With a conventional method to generate transgenic zebrafish, a researcher needs to inject plasmid DNA to hundreds of fertilized eggs, raise the injected embryos to adulthood, and mate them to achieve more than 100 successful pair-crosses. With our transposon-mediated transgenesis method, 10 injected adult fish should be enough to obtain founder fish that can transmit insertions to F1 fish.

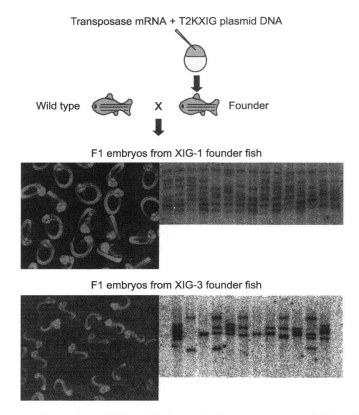

Fig. 3 Transgenesis by using T2KXIG. Fish injected with the transposase mRNA and the plasmid DNA containing the transposon vector was mated with noninjected wild-type fish. The GFP expression in F1 embryos from the XIG-1 and the XIG-3 founder fish and Southern blot hybridization analysis of the F1 fish by using the GFP probe are shown. (See Color Insert.)

C. Regulated Gene Expression in Transgenic Zebrafish

1. Methods

a. Plasmids

Plasmid DNA containing T2Ksix3.2G (Kawakami *et al.*, 2004; Fig. 1) is used for microinjection. T2Ksix3.2G contains the promoter region of the *six3.2* gene, the EGFP gene, and the SV40 polyA signal. To obtain the genomic DNA containing the *six3.2* promoter, a lambda phage genomic library (Stratagene) is screened by plaque hybridization, using the *six3.2* cDNA probe (Kobayashi *et al.*, 2001). Plasmids are constructed containing DNA fragments of various lengths of the promoter region and the GFP gene. The promoter activity is tested by microinjection of those constructs into fertilized eggs and by examining transient

GFP expression in injected embryos. A 1.6-kb *Hind*-III fragment thus identified is cloned into a transposon vector with the EGFP gene and the SV40 polyA signal.

b. Microscopic Analysis of F1 embryos by Southern Blot Hybridization

The T2Ksix3.2G insertions in F1 fish are analyzed by Southern blot hybridization as described previously.

2. Results

An important application of transgenesis in zebrafish is to construct transgenic fish expressing GFP in a specific tissue or organ (Higashijima *et al.*, 1997; Long *et al.*, 1997). To test whether the *Tol2* transposon system can be applied to this purpose, T2Ksix3.2G was constructed (Kawakami *et al.*, 2004). Seven fish coinjected with the pT2Ksix3.2G plasmid DNA and the transposase mRNA were raised to adulthood and mated with noninjected fish. GFP-expressing embryos were identified in the progeny from two of them (29%).

Although the founder frequency was a little lower than that observed in the experiment using T2KXIG, it is practically important that founder fish can be obtained by testing such a small number of injected fish. F1 fish from one of the founder fish were analyzed by Southern blot hybridization, using the GFP probe. Five different insertions were transmitted by the founder, and transgenic F1 fish with a single insertion could be identified (Fig. 4A). The F1 fish with the single insertion was crossed with wild-type fish. F2 embryos expressed GFP in the forebrain and eyes at 24 hpf. The expression pattern was similar to that of the endogenous *six3.2* mRNA as revealed by whole-mount *in situ* hybridization (Fig. 4B). This result indicates that the regulated gene expression can be recapitulated by transgenesis using the *Tol2* transposon system. Different levels of GFP expression were, however, observed in fish with insertions at different loci, suggesting that the locus where the transposon integrates might affect its expression.

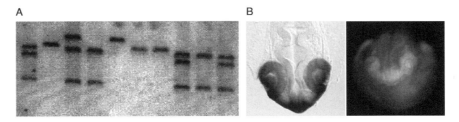

Fig. 4   Transgenesis by using T2Ksix3.2G. (A) Southern blot hybridization analysis by using the GFP probe of F1 fish from the founder fish injected with the transposase mRNA and the plasmid DNA containing T2Ksix3.2G. (B) Expression of the endogenous *six3.2* mRNA revealed by whole mount *in situ* hybridization (left) and GFP expression in the F2 embryo carrying a single T2Ksix3.2G insertion (right). (See Color Insert.)

III. A Gene Trap Approach that Uses the *Tol2* Transposon System in Zebrafish

A. A Transposon–Based Gene Trap Construct

1. Methods

a. Construction of a Gene Trap Vector

T2KXIG was digested with *Apa*I and self-ligated, resulting in T2KSAG (Kawakami *et al.*, 2004). T2KSAG contains a splice acceptor, a promoterless EGFP gene, and the SV40 polyA signal (Fig. 1). There is no stop codon between the splice acceptor and the ATG codon of the GFP gene in the same frame as GFP. Therefore, the ATG codon can serve either as an initiation codon or as an internal methionine. Although gene trap events are expected to occur less frequently than enhancer trap events, we constructed the gene trap vector because the trapped gene would be easily identified by 5' rapid amplification of cDNA ends (RACE).

b. Microinjection

T2KSAG was injected with the transposase mRNA to fertilized eggs as described previously.

2. Results and Discussion

Figure 5 shows embryos injected with the T2KXIG plasmid and the transposase mRNA, or the T2KSAG plasmid and the transposase mRNA. Whereas the former expressed GFP rather ubiquitously 24 h after the injection, the latter hardly expressed GFP (Fig. 5A). This indicates that T2KSAG lacks promoter activity. In some embryos injected with the T2KSAG plasmid and the transposase mRNA, some somatic cells, however, expressed GFP rather strongly (Fig. 5B). This suggests that gene trap events might have occurred in these somatic cells. Thus, T2KSAG is thought to function as we designed and is used for further studies.

B. Spatially and Temporally Regulated Green Fluorescent Protein (GFP) Expression in Transgenic Fish

1. Methods

a. Microinjection of the T2KSAG Gene Trap Construct

Micoinjection of the T2KSAG plasmid and the transposase mRNA is carried out as described previously. The injected embryos are raised to adulthood.

b. Analysis of GFP Expression in F1 Embryos

Injected fish are crossed with noninjected wild-type fish or with injected fish of the opposite sex to obtain F1 embryos. Embryos are placed on a glass depression plate, and GFP expression is examined by using an MZ 16 FA (Leica) fluorescent dissecting microscope. GFP-positive embryos are picked up and raised.

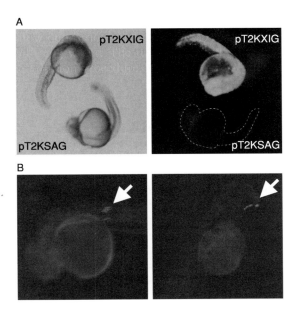

Fig. 5   Transient GFP expression by using T2KXIG and T2KSAG. (A) GFP expression in embryos injected with the T2KXIG plasmid DNA and the transposase mRNA or the T2KSAG plasmid DNA and the transposase mRNA. (B) GFP expression in some muscle cells at Day 1 (left) or some neurons at Day 2 (right) in embryos injected with the T2KSAG plasmid DNA and the transposase mRNA. (See Color Insert.)

2. Results

To determine whether T2KSAG can indeed capture endogenous transcripts, a pilot experiment for gene trapping was performed (Kawakami *et al.*, 2004). Circular DNA of the T2KSAG plasmid and the transposase mRNA were injected into fertilized eggs. A total of 156 injected fish were raised to adulthood (107 males and 49 females) and used for mating. Because more males were obtained than females, 62 fish (60 males and 2 females) were crossed with noninjected wild-type fish of the opposite sex and 94 fish (47 males and 47 females) were crossed as pairs. A variety of GFP expression patterns were observed in F1 embryos: some were weak and some were strong, or some were ubiquitous, and some temporally and spatially restricted patterns. This indicates that T2KSAG was inserted in various loci in the genome and GFP was expressed under the control of endogenous promoters. When no F1 progeny expressed GFP, at least 50 embryos from the cross were pooled and analyzed by PCR, using the primers EGFP/f1 and EGFP/r1 for the presence of the T2KSAG sequence. Although in the cross using the injected fish as a pair it was not determined whether the founder fish was a male or a female (or both), it was estimated that about 80 fish of the 156 injected fish were founder fish (51%) that could transmit the T2KSAG insertion to the progeny.

Thirty-six unique GFP expression patterns at Day 1 of development in F1 embryos were identified (Fig. 6A). Putative regions where GFP expression is observed include the forebrain, midbrain, hindbrain, midbrain–hindbrain boundary, heart, notochord, and neural tube. The frequency of obtaining such unique expression patterns is very high: one unique expression pattern can be obtained per four or five injected fish (23%; 36 patterns of 156 injected fish). Suppose that 5.6 insertions per founder fish were transmitted by single founder fish as described previously, it can be estimated that 8% of the T2KSAG insertions lead to such unique GFP expression patterns $[36/(80 \times 5.6) = 0.08]$.

The same founder fish are used for mating repeatedly to collect F1 embryos with the same GFP expression patterns. Germ cells of the founder fish are highly mosaic, and the frequencies to obtain the same pattern in the F1 embryos are between 0.6 and 45% (Kawakami et al., 2004). F1 fish are raised and mated to obtain F2 fish. The expression patterns are transmitted in a Mendelian fashion.

Not only zygotically expressed genes but also maternally expressed genes can be identified. GFP expression in the fertilized eggs was found, suggesting that the putative trapped gene is expressed during oogenesis (Fig. 6B).

C. Characterization of Gene Trap Insertions by Southern Blot, Inverse PCR, 5′ Rapid Amplification of cDNA Ends, (RACE) and RT-PCR

1. Methods

a. Southern Blot Hybridization

Genomic DNA prepared from tail fins of F1 fish with the GFP-expressing gene trap fish are analyzed by Southern blot hybridization, using the GFP probe as described previously.

b. Cloning of the Junction Fragments by Inverse PCR

One microgram of the genomic DNA is digested with MboI, HaeIII, or AluI in a 20-μl reaction solution. The digested DNA is diluted to 500 μl and self-ligated, using T4 DNA ligase (Takara, Japan). By adding 50 μl of 3 M sodium acetate and 1 ml of 100% ethanol, the ligated DNA is precipitated and used for PCR. To amplify the 5′ junctions, the first PCR is carried out by using the following primers:

- Tol2-5′/f1: 5′-AGT ACT TTT TAC TCC TTA CA-3′
- Tol2-5′/r1: 5′-GAT TTT TAA TTG TAC TCA AG-3′

Then the second PCR is carried out by using the nested primers:

- Tol2-5′/f2: 5′-TAC AGT CAA AAA GTA CT-3′
- Tol2-5′/r2: 5′-AAG TAA AGT AAA AAT CC-3′

To amplify the 3′ junctions, the first PCR is carried out by using the following primers:

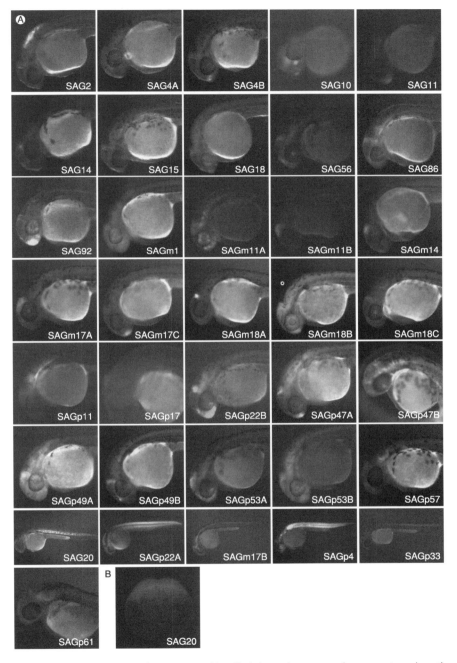

Fig. 6 Unique GFP expression patterns identified in embryos carrying gene trap insertions. (A) Unique GFP expression patterns in embryos carrying T2KSAG insertions at 30–36 hpf. The patterns are named SAGXX. (B) Maternal GFP expression in an SAG20 embryo at the two-cell stage. (See Color Insert.)

- Tol2-3′/f1: 5′-TTT ACT CAA GTA AGA TTC TAG-3′
- Tol2-3′/r1: 5′-CTC CAT TAA AAT TGT ACT TGA-3′

Then the second PCR is carried out by using the nested primers:

- Tol2-3′/f2: 5′-ACT TGT ACT TTC ACT TGA GTA-3′
- Tol2-3′/r2: 5′-GCA AGA AAG AAA ACT AGA GA-3′

The inverse PCR products are gel-extracted by using the QIAquick Gel Extraction Kit (QIAGEN) and sequenced with the BigDye Terminator v3.1 Cycle Sequencing Kit (Applied Biosystems), using the primers used for inverse PCR.

c. 5′ RACE Analysis of Trapped Transcripts

Fifty GFP-expressing embryos at Day 1 of development are homogenized and total RNA is prepared by using the TRIZOL reagent (Invitrogen). 5′ RACE is carried out by using 5 μg of total RNA and the 5′ RACE system (Invitrogen). The following primer is used for the first strand cDNA synthesis.

- EGFP/r2 primer: 5′-CTT GCC GTA GGT GGC ATC GCC CTC-3′

Amplification of the dC-tailed cDNA is carried out by using the Abridged Anchor Primer (Invitrogen) and the following primer:

- EGFP/r3 primer: 5′-GCT GAA CTT GTG GCC GTT TAC-3′

Finally, the nested amplification is carried out using the Abridged Universal Amplification Primer (Invitrogen) and the the following primer:

- EGFP/r4 primer: 5′-GAT GGG CAC CAC CCC GGT GA-3′

d. RT–PCR Analysis of Trapped Genes: Analysis of the Insertion in the hoxc Cluster as an Example

The trapped gene in the SAGp22A gene trap line was analyzed by RT-PCR. Oligo dT-primed cDNA was synthesized from 5 μg of total RNA from wild-type embryos. RT-PCR was carried out by using the cDNA as a template and a forward primer in the 5′ RACE sequence and reverse primers in members of the hoxc cluster, hoxc6a, hoxc5a, hoxc4a, and hoxc3a.

- SAGp22A/5′ RACE/f1: 5′-AAC AAG ACA CAA GGC AAG CAAC-3′
- hoxc3a/r1: 5′-GTC ACC AGT TTT CAG TTT TTC TG-3′
- hoxc4a/r1: 5′-CTT GCT GGC GAG TGT AAG CAG T-3′
- hoxc5a/r1: 5′-AAC TTT GAG TCC TTC TTC CAC T-3′
- hoxc6a/r1: 5′-GGT ATC TGG AGT AAA TCT GGC G-3′

Total RNA were prepared from wild-type, SAGp22A heterozygous, and SAGp22A homozygous embryos to perform semi-quantitative RT-PCR analysis (30 cycles of 94 °C, 30 sec; 55 °C, 30 sec; 72 °C, 30 sec), using the forward primer in exon1 of hoxc3a,

- hoxc3a/f1: 5′-AAC AAG ACA CAA GGC AAG CAAC-3′

and the reverse primer in exon4 of *hoxc3a*.

- hoxc3a/r2: 5′-TCA TCC AAG GGT ACT TCA TGG T-3′

The control RT-PCR (27 cycles of 94 °C, 30 sec; 55 °C, 30 sec; 72 °C, 30 sec) is performed by using the primers in the zebrafish EF1α gene.

- EF1α /exon3/f1: 5′-ACA TTG CTC TCT GGA AAT TCG AG-3′
- EF1α /exon6/r1: 5′-TGA CCT CAG TGG TTA CAT TGG C-3′

2. Results

Transposon insertions responsible for expression patterns are first analyzed by Southern blot hybridization. For instance, two distinguishable patterns, SAG4A (GFP expression in the heart) and SAG4B (GFP expression in the forebrain), were identified in F1 embryos from crosses using single founder fish (Fig. 6A). Southern blot analysis revealed that four different insertions were transmitted to these F1 fish (Fig. 7A). Further, 12 F1 fish with the SAG4A expression pattern were analyzed. All the F1 fish had one band in common and some of them did not have the other band, indicating that the former band is responsible for GFP expression in the heart (Fig. 7B).

Fish with the 36 GFP expression patterns were also analyzed by Southern blot hybridization. The number of insertions transmitted with the GFP expression pattern varied from 1 to more than 25. When fish with a single T2KSAG insertion are not identified in the F1 generation, the F1 fish are crossed with wild-type fish and fish with the single insertion are identified in the F2 or F3 generation.

From fish lines with single T2KSAG insertions, the DNA fragment containing the junction between genomic DNA and T2KSAG insertion can be amplified by inverse PCR. The gel-extracted inverse PCR products are used for cycle sequencing.

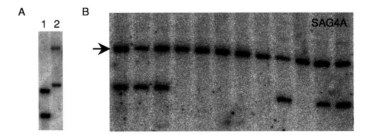

Fig. 7 T2KSAG insertions capture endogeneous transcripts. (A) Southern blot hybridization analysis of the SAG4A (1) and SAG4B (2) F1 fish. (B) Southern blot hybridization analysis of F1 fish with the SAG4A GFP expression pattern. An arrow indicates the insertion responsible for the unique expression.

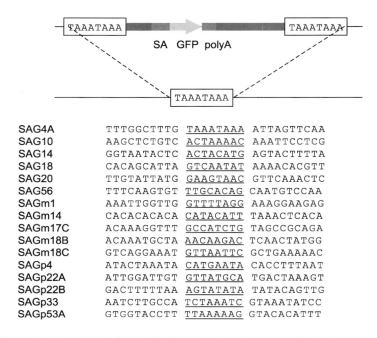

SAG4A	TTTGGCTTTG	TAAATAAA	ATTAGTTCAA
SAG10	AAGCTCTGTC	ACTAAAAC	AAATTCCTCG
SAG14	GGTAATACTC	ACTACATG	AGTACTTTTA
SAG18	CACAGCATTA	GTCAATAT	AAAACACGTT
SAG20	TTGTATTATG	GAAGTAAC	GTTCAAACTC
SAG56	TTTCAAGTGT	TTGCACAG	CAATGTCCAA
SAGm1	AAATTGGTTG	GTTTTAGG	AAAGGAAGAG
SAGm14	CACACACACA	CATACATT	TAAACTCACA
SAGm17C	ACAAAGGTTT	GCCATCTG	TAGCCGCAGA
SAGm18B	ACAAATGCTA	AACAAGAC	TCAACTATGG
SAGm18C	GTCAGGAAAT	GTTAATTC	GCTGAAAAAC
SAGp4	ATACTAAATA	CATGAATA	CACCTTTAAT
SAGp22A	ATTGGATTGT	GTTATGCA	TGACTAAAGT
SAGp22B	GACTTTTTAA	AGTATATA	TATACAGTTG
SAGp33	AATCTTGCCA	TCTAAATC	GTAAATATCC
SAGp53A	GTGGTACCTT	TTAAAAAG	GTACACATTT

Fig. 8 Inverse PCR analysis of the T2KSAG insertions. Sequences of the integration sites are shown. Junctions of both ends of insertions are cloned by inverse PCR. The 8-bp sequences underlined are duplicated on integration of T2KSAG.

In all cases, an 8-bp sequence at the integration site is duplicated at the both ends of the insertion. No obvious specificity has been observed in the 8-bp sequences at the DNA sequence level (Fig. 8). These genomic DNA sequences are analyzed by the BLAST search. Genes are not identified in those sequences, except for SAGm18C. In insertional mutagenesis in zebrafish by using the pseudotyped retrovirus, genes responsible for mutant phenotypes have been identified in ~50% of the cases immediately by sequencing the products of the first inverse PCR (Golling *et al.*, 2002). This difference in the frequencies of finding genes might be explained by the fact that four-base cutters (*Alu*I, *Hae*III, or *Mbo*I) are used in our present protocol for inverse PCR, whereas six-base cutters are usually used in the inverse PCR.

Fusion transcripts of endogenous sequences and the GFP sequence are identified by 5′ RACE (Fig. 9). In these cases, the endogenous transcripts jumped into the T2KSAG sequence precisely at the splice acceptor. Thus, the T2KSAG gene trap construct can capture endogenous transcripts in zebrafish. However, in about half the cases in which 5′ RACE was performed in the same condition, such fusion products were not amplified. This is not surprising because there might be several possibilities on why 5′ RACE did not work: (1) the amount of fusion transcript was too small to be detected by 5′ RACE, (2) the 5′ end of the

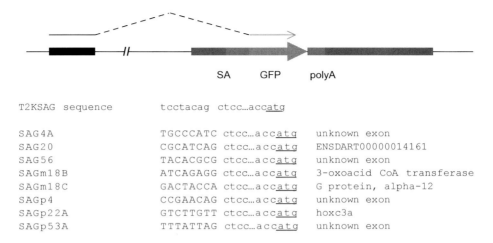

Fig. 9 5′ RACE analysis of genes trapped by T2KSAG insertions. Fusion transcripts identified by 5′ RACE. The T2KSAG sequence is shown in lower case. The ATG codon for the GFP protein is underlined. The endogenous exon sequence fused to the T2KSAG sequence is shown in upper case. ENSDART00000014161 is a predicted gene in the ensembl database.

fusion transcript was too far to be reached by 5′ RACE, or (3) the fusion transcript could not be reverse-transcribed because of its structure, e.g., GC-rich.

The sequences identified by 5′ RACE are analyzed by the BLAST search. Of the eight 5′ RACE sequences shown in Fig. 9, two are known genes. In SAGm18B, the T2KSAG insertion is located in an internal intron of a gene for succinyl CoA:3-oxoacid CoA-transferase (SCOT). The reading frame of the SCOT gene is maintained through the GFP gene, suggesting that a SCOT-GFP fusion protein is synthesized in the SAGm18B fish. In SAGm18C, the T2KSAG insertion is located within a gene for the guanine nucleotide-binding protein alpha-12 subunit and traps the first noncoding exon of the gene.

The GFP gene on the T2KSAG construct is designed to be expressed when it is inserted either (1) upstream of the initiation codon of a gene and downstream of either a promoter or exon(s) encoding the 5′ untranslated region in the proper orientation, or (2) downstream of the initiation codon of a gene, either in an exon or an internal intron, in the proper orientation and reading frame. In the four cases in which the trapped gene was identified (Fig. 9), three were the former (SAG20, SAGm18C, and SAGp22A) and one was the latter (SAGm18B). Thus, T2KSAG can function as a gene trap construct in zebrafish.

When the BLAST search does not pick up any gene in the database, 3′ RACE using primers in the 5′ RACE sequence and/or RT-PCR using primers in the 5′ RACE sequence and presumable downstream exon sequences, which are predicted based on the genomic sequence information in the database such as the ensembl database (http://www.ensembl.org/Danio_rerio/), should be carried out to identify the trapped gene. In SAGp22A, the insertion was mapped on a BAC clone (BX005254) containing the *hoxc* cluster (Amores *et al.*, 1998) by inverse PCR, between the *hoxc4a* and *hoxc5a* genes. However, the 5′ RACE product identified from the SAGp22A fish line was an unknown "orphan" exon, because it did not match with any transcribed sequence in the database. To test whether any of the downstream *hoxc* genes contains the orphan exon, RT-PCR analysis was performed. The amplified product was detected only when RT-PCR was performed using the forward primer and the reverse primer in the *hoxc3a* gene, indicating that it was the unidentified first exon of the *hoxc3a* gene. Thus, the SAGp22A insertion was found to capture the first noncoding exon of *hoxc3a*. Such a complicated transcript in the *hox* cluster has not been reported (Fig. 10A).

It is important to test whether the T2KSAG insertion can interfere with synthesis of a normally spliced transcript when it is inserted in an intron. Fish homozygous for the SAGp22A insertion are viable and fertile. They were crossed to each other or crossed to wild-type fish, and homozygous and heterozygous embryos were obtained. Total RNA prepared from the wild-type, heterozygous and

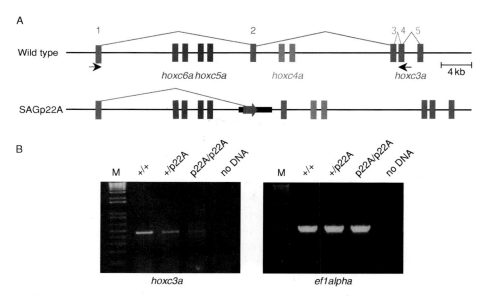

Fig. 10 5′ RACE analysis of genes trapped by T2KSAG insertions. (A) The structure of the *hoxc* cluster in wild-type and SAGp22A fish. Exons of the *hoxc3a* gene are numbered from the 5′ end. Arrows indicate the positions and directions of the primers used in (B). (B) RT-PCR analysis of the *hoxc3a* transcript in wild-type, SAGp22A heterozygous, and homozygous fish. The *hoxc3a* transcript is reduced to less than 25% in the SAGp22A homozygous embryo (left). RT-PCR using primers in the EF1alpha transcript was carried out as a positive control.

homozygous embryos were analyzed by RT-PCR, using the hoxc3a/f1 and hox-c3a/r2 primers. The *hoxc3a* mRNA was reduced in the heterozygous embryos and more greatly in the homozygous embryos (Fig. 10B). It was estimated that the amount of the *hoxc3a* transcript in the homozygous embryos was decreased to less than 25% of that of the transcript in wild-type embryos (Kawakami *et al.*, 2004). Thus, the T2KSAG insertion can markedly interrupt synthesis of the normally spliced transcript. These levels of decrease might cause hypomorphic mutations when T2KSAG is inserted in an intron of some essential genes. The T2KSAG insertion in the SAGp22A fish, however, does not abolish the wild-type transcript completely. This finding might account for the observation that we have not identified any zygotic lethal mutations by analyzing the gene trap fish lines up to now.

IV. Summary and Perspectives

A. Transgenesis by Using the *Tol2* Transposon System

The highly efficient transgenesis method that uses the *Tol2* transposon system, was described here. The frequency of obtaining founder fish reached to more than 50%, which is higher than that observed in any other transgenesis methods developed to express a transgene in fish. Transgenic zebrafish expressing GFP in specific tissues and organs have been useful to study vertebrate development (Higashijima *et al.*, 1997; Long *et al.*, 1997). Such studies will be speeded up by our method. The transgenic frequency achieved by using our method has been reproducibly high in ongoing transgenic studies in our lab and also in our collaborator's labs (personal communications).

Transgenesis by using the *Tol2* transposon system can have the following advantages. First, transgenic fish carrying a single-copy insertion can easily be isolated, whereas transgenic fish constructed by the plasmid DNA injection sometimes carry concatemers at a single locus (Stuart *et al.*, 1988). Second, the transposon insertion is clean and does not cause any gross rearrangement at the integration locus, which is sometimes associated with insertions created by the plasmid DNA injection (Cretekos and Grunwald, 1999). Third, expression of a transgene inserted by transposition might persist after the passage through generations. For now, GFP expression in our transgenic fish lines can be observed consistently up to the F4 generation.

B. The Gene Trap Approach That Uses the *Tol2* Transposon System

The gene trap approach that uses the *Tol2* transposon system will be useful to identify novel developmental genes and novel structures of known developmental genes. Although we have analyzed embryos homozygous for the 36 gene trap insertions, no obvious mutant phenotype has been detected. Our next goal will be to apply the gene trap approach to insertional mutagenesis. This can be achieved by modifying the gene trap construct, that is, by constructing gene trap constructs containing a stronger splice acceptor or a stronger poly A signal, because such vectors might be more mutagenic. Alternatively, a complete loss of the function of the trapped gene can be achieved by creating deletions on mobilization of the integrated existing transposon. At present, construction of transgenic fish with specific GFP expression is carried out as follows. First, a gene expressed in a specific tissue or organ is identified and the genomic DNA surrounding the gene is cloned. Second, plasmids containing various lengths of DNA of the promoter region and the GFP gene are constructed and the promoter activity is tested by a transient assay. Finally, transgenic fish are generated by a method with low efficiencies; that is, microinjection of the plasmid DNA. Thus, making one transgenic fish line with specific GFP expression will be a project of more than 1 year. I propose that the gene trap approach will be an alternative. Because one expression pattern can be isolated in every four or five injected fish (23%; 36 patterns out of 156 injected fish), a small lab can collect tens or hundreds of different expression patterns within 1 year, which might include the desired expression pattern. Further, in the future, a collaborative work by several laboratories will produce thousands of gene trap lines, which should be useful resources. A number of genes that might play important roles in vertebrate development will also be identified by analyzing those fish lines.

The *Tol2* transposon system should be applied to develop the Gal4-UAS system. It has been shown that Gal4 can activate expression of a gene placed downstream of UAS in transgenic zebrafish (Scheer and Campos-Ortega, 1999). By placing the Gal4 gene on a gene trap vector, we will be able to obtain a number of fish lines expressing Gal4 in various tissues and organs. Such Gal4 lines should allow a gene of interest to be expressed at the desired time and place.

This chapter described transposon-mediated methodologies in zebrafish. These methods should facilitate studies on the function of genes involved in vertebrate development and organogenesis and provide a basis for further development of useful genetic methodologies in zebrafish.

Acknowledgments

I thank my collaborators N. Kawakami, A. Shima, M. Mishina, H. Hori, A. Koga, H. Takeda, M. Kobayashi, and N. Matsuda. I thank the fish room technicians A. Ito, W. Nakashima, I. Watanabe, R. Sugisaki, and R. Watanabe. This work was supported by grants from NIH/NIGMS GM069382-01; The Naito Foundation; The Sumitomo Foundation; and the Ministry of Education, Culture, Sports, Science and Technology of Japan.

References

Amores, A., Force, A., Yan, Y. L., Joly, L., Amemiya, C., Fritz, A., Ho, R. K., Langeland, J., Prince, V., Wang, Y. L., Westerfield, M., and Postlethwait, J. H. (1998). Zebrafish hox clusters and vertebrate genome evolution. *Science* **282,** 1711–1714.

Amsterdam, A., Burgess, S., Golling, G., Chen, W., Sun, Z., Townsend, K., Farrington, S., Haldi, M., and Hopkins, N. (1999). A large-scale insertional mutagenesis screen in zebrafish. *Genes Dev.* **13**, 2713–2724.

Amsterdam, A., Lin, S., and Hopkins, N. (1995). The *Aequorea victoria* green fluorescent protein can be used as a reporter in live zebrafish embryos. *Dev. Biol.* **171**, 123–129.

Cretekos, C. J., and Grunwald, D. J. (1999). Alyron, an insertional mutation affecting early neural crest development in zebrafish. *Dev. Biol.* **210**, 322–338.

Davidson, A. E., Balciunas, D., Mohn, D., Shaffer, J., Hermanson, S., Sivasubbu, S., Cliff, M. P., Hackett, P. B., and Ekker, S. C. (2003). Efficient gene delivery and gene expression in zebrafish using the Sleeping Beauty transposon. *Dev. Biol.* **263**, 191–202.

Driever, W., Solnica-Krezel, L., Schier, A. F., Neuhauss, S. C. F., Malicki, J., Stemple, D. L., Stainier, D. Y. R., Zwartkruis, F., Abdelilah, S., Rangini, Z., Belak, J., and Boggs, C. (1996). A genetic screen for mutations affecting embryogenesis in zebrafish. *Development* **123**, 37–46.

Gaiano, N., Allende, M., Amsterdam, A., Kawakami, K., and Hopkins, N. (1996a). Highly efficient germ-line transmission of proviral insertions in zebrafish. *Proc. Natl. Acad. Sci. USA* **93**, 7777–7782.

Gaiano, N., Amsterdam, A., Kawakami, K., Allende, M., Becker, T., and Hopkins, N. (1996b). Insertional mutagenesis and rapid cloning of essential genes in zebrafish. *Nature* **383**, 829–832.

Golling, G., Amsterdam, A., Sun, Z., Antonelli, M., Maldonado, E., Chen, W., Burgess, S., Haldi, M., Artzt, K., Farrington, S., Lin, S. Y., Nissen, R. M., and Hopkins, N. (2002). Insertional mutagenesis in zebrafish rapidly identifies genes essential for early vertebrate development. *Nat. Genet.* **31**, 135–140.

Haffter, P., Granato, M., Brand, M., Mullins, M. C., Hammerschmidt, M., Kane, D. A., Odenthal, J., van Eeden, F. J. M., Jiang, Y.-J., Heisenberg, C.-P., Kelsh, R. N., Furutani-Seiki, M., Vogelsang, E., Beuchle, D., Schach, U., Fabian, C., and Nüsslein-Volhard, C. (1996). The identification of genes with unique and essential functions in the development of the zebrafish, *Danio rerio*. *Development* **123**, 1–36.

Higashijima, S., Okamoto, H., Ueno, N., Hotta, Y., and Eguchi, G. (1997). High-frequency generation of transgenic zebrafish which reliably express GFP in whole muscles or the whole body by using promoters of zebrafish origin. *Dev. Biol.* **192**, 289–299.

Johnson, A. D., and Krieg, P. A. (1994). pXeX, a vector for efficient expression of cloned sequences in *Xenopus* embryos. *Gene* **147**, 223–226.

Kawakami, K., Koga, A., Hori, H., and Shima, A. (1998). Excision of the *Tol2* transposable element of the medaka fish, *Oryzias latipes*, in zebrafish, *Danio rerio*. *Gene* **225**, 17–22.

Kawakami, K., and Noda, T. (2004). Transposition of the Tol2 element, an Ac-like element from the Japanese medaka fish *Oryzias latipes*, in mouse embryonic stem cells. *Genetics* **166**, 895–899.

Kawakami, K., and Shima, A. (1999). Identification of the *Tol2* transposase of the medaka fish *Oryzias latipes* that catalyzes excision of a nonautonomous *Tol2* element in zebrafish *Danio rerio*. *Gene* **240**, 239–244.

Kawakami, K., Shima, A., and Kawakami, N. (2000). Identification of a functional transposase of the *Tol2* element, an *Ac*-like element from the Japanese medaka fish, and its transposition in the zebrafish germ lineage. *Proc. Natl. Acad. Sci. USA* **97**, 11403–11408.

Kawakami, K., Takeda, H., Kawakami, N., Kobayashi, M., Matsuda, N., and Mishina, M. (2004). A transposon-mediated gene trap approach identifies developmentally regulated genes in zebrafish. *Dev. Cell.* **17**, 133–144.

Kobayashi, M., Nishikawa, K., Suzuki, T., and Yamamoto, M. (2001). The homeobox protein six3 interacts with the Groucho corepressor and acts as a transcriptional repressor in eye and forebrain formation. *Dev. Biol.* **232**, 315–326.

Koga, A., Suzuki, M., Inagaki, H., Bessho, Y., and Hori, H. (1996). Transposable element in fish. *Nature* **383**, 30.

Lin, S., Gaiano, N., Culp, P., Burns, J. C., Friedmann, T., Yee, J.-K., and Hopkins, N. (1994). Integration and germ-line transmission of a pseudotyped retroviral vector in zebrafish. *Science* **265,** 666–669.

Linney, E., Hardison, N. L., Lonze, B. E., Lyons, S., and DiNapoli, L. (1999). Transgene expression in zebrafish: A comparison of retroviral-vector and DNA-injection approaches. *Dev. Biol.* **213,** 207–216.

Long, Q., Meng, A., Wang, H., Jessen, J. R., Farrell, M. J., and Lin, S. (1997). *GATA-1* expression pattern can be recapitulated in living transgenic zebrafish using GFP reporter gene. *Development* **124,** 4105–4111.

Raz, E., van Luenen, H. G. A. M., Schaerringer, B., Plasterk, R. H. A., and Driever, W. (1998). Transposition of the nematode *Caenorhabditis elegans Tc3* element in the zebrafish *Danio rerio*. *Curr. Biol.* **8,** 82–88.

Rupp, R. A. W., Snider, L., and Weintraub, H. (1994). *Xenopus* embryos regulate the nuclear localization of XMyoD. *Genes Dev.* **8,** 1311–1323.

Scheer, N., and Campos-Ortega, J. A. (1999). Use of the Gal4-UAS technique for targeted gene expression in the zebrafish. *Mech. Dev.* **80,** 153–158.

Streisinger, G., Walker, C., Dower, N., Knauber, D., and Singer, F. (1981). Production of clones of homozygous diploid zebra fish (*Branchydanio rerio*). *Nature* **291,** 293–296.

Stuart, G. W., McMurray, J. V., and Westerfield, M. (1988). Replication, integration and stable germ-line transmission of foreign sequences injected into early zebrafish embryos. *Development* **103,** 403–412.

Stuart, G. W., Vielkind, J. R., McMurray, J. V., and Westerfield, M. (1990). Stable lines of transgenic zebrafish exhibit reproducible patterns of transgene expression. *Development* **109,** 577–584.

Thermes, V., Grabher, C., Ristoratore, F., Bourrat, F., Choulika, A., Wittbrodt, J., and Joly, J. S. (2002). I-SceI meganuclease mediates highly efficient transgenesis in fish. *Mech. Dev.* **118,** 91–98.

Turner, D. L., and Weintraub, H. (1994). Expression of achaete-scute homolog 3 in *Xenopus* embryos converts ectodermal cells to a neural fate. *Genes Dev.* **8,** 1434–1447.

PART II

The Zebrafish Genome and Mapping Technologies

CHAPTER 12

The Zebrafish Genome Project: Sequence Analysis and Annotation

Kerstin Jekosch

Wellcome Trust Sanger Institute
Cambridge CB10 1SA, United Kingdom

I. Introduction

Rapid advances in zebrafish genetics have led to an increasing need for a genome sequence to facilitate interpretation of data. To promote the use of the zebrafish (*Danio rerio*) as a model system for vertebrate biology, the Wellcome Trust Sanger Institute (www.sanger.ac.uk) started a project to sequence the zebrafish genome in Spring 2001. The project is targeted for completion by the end of 2005.

To make the product as useful as possible to the zebrafish community, the Sanger Institute has also committed to identify all zebrafish genes. An annotated zebrafish genome sequence is immensely informative for both forward and reverse genetics. It also provides the basis for extensive comparative genomics and hence the improvement of the annotation of already existing genomes from other model organisms, and is also a valuable tool for phylogenetic and evolutionary research.

Two approaches were chosen to obtain the genome sequence: a whole genome shotgun (WGS) assembly (Mullikin and Ning, 2003), with subsequent automated annotation in EnsEMBL (Table I, Clamp *et al.*, 2003; Hubbard *et al.*, 2002), and the classical clone mapping and clone-by-clone sequencing with subsequent manual annotation in Otter (Searle *et al.*, 2004), displayed by the Vega Web browser (Table I).

The WGS sequencing approach allows for rapid generation of sequence that can be immediately useful for experimental scientists. Assembly of the WGS reads provides a global view of the genomic landscape and a substrate for initial curation and identification of genome features.

The advantage of speed gained by the WGS approach is offset, however, by the quality of the sequence. Initial assemblies are beset with problems such as global and local misjoins, and over- or underrepresentation of certain areas (as discussed in Mouse Genome Sequencing Consortium, 2002). In addition, although the superimposed automated annotation is improving quite impressively, it will probably never reach the quality of manual curation (Ashurst and Collins, 2003).

Although in general the hierarchical mapping and clone-by-clone sequencing strategy (International Human Genome Sequencing Consortium, 2001; Waterston *et al.*, 2002) produces results more slowly than WGS sequencing, the emphasis is placed on the quality of the final product. Sequence emerging from the clone-by-clone project qualifies for the gold standard of genome sequence quality (95% of the euchromatin covered to an accuracy greater than 99.99%, Grafham and Willey, 2003).

There are additional advantages in pursuing such a strategy: the genome is made available in units of 100 to 200 kb clones in bacterial vectors which can be used in lab research; the generation of a clone fingerprint (FPC) map (Soderlund *et al.*, 1997) facilitates positional cloning; problems can be isolated and resolved more easily; and regions of interest can be focussed on by request. The superimposition of manual annotation undertaken to the level of gold standard (as defined by the HAWK committee for human annotation www.sanger.ac.uk/HGP/Havana/hawk.shtml) adds further value to the sequence. Feedback on the annotation from community experts contributes to the quality of the final product.

All these efforts are undertaken in collaboration and synchronization with the central zebrafish database ZFIN (www.zfin.org, Sprague *et al.*, 2003). The following chapters describe the current status of automated and manual annotation of the genome and different ways of mining these data. All relevant Web sites hosted by the Sanger Institute are listed in Table I.

Table I

Zebrafish Related Web Pages Provided by the Sanger Institute

Description	URL
Sanger homepage	www.sanger.ac.uk
Zebrafish homepage	www.sanger.ac.uk/Projects/D_rerio
User guide	www.sanger.ac.uk/Projects/D_rerio/faqs.shtml
Clone mapping project	www.sanger.ac.uk/Projects/D_rerio/mapping.shtml
Pre-EnsEMBL browser	pre.ensembl.org/Danio_rerio
EnsEMBL browser	www.ensembl.org/Danio_rerio
Vega browser	vega.sanger.ac.uk/Danio_rerio
WGS assembly FTP	ftp.ensembl.org/pub/assembly/zebrafish
WGS marker mapping	www.sanger.ac.uk/cgi-bin/Projects/D_rerio/mapsearch
WGS trace server	trace.ensembl.org

Note: The zebrafish homepage is the central page that provides links to all resources and data generated at the Sanger.

II. Automated Annotation of Whole Genome Shotgun (WGS) Assemblies

A. Generation of WGS Assemblies

WGS sequence data were generated from random DNA fragments varying in size from 2 to 10 kb and from the end sequences of fosmid and BAC clone libraries. The DNA was derived from Tuebingen strain fish, provided by R. Geissler and C. Nuesslein-Volhard. The reads were assembled using the Phusion assembler (Mullikin and Ning, 2003) when the sequence coverage of the genome, estimated to be 1.5–1.7 Gb in size, was greater than threefold. In the most recent assembly (Zv3), the WGS supercontigs have been mapped to the linkage groups. Detailed information about every assembly is published on our Web site; for the Zv3 assembly, see www.sanger.ac.uk/Projects/D_rerio/Zv3_assembly_ information.shtml. The assembly sequence and the tiling path file containing information about the placements of assembly supercontigs on the chromosomes can be downloaded from the FTP pages (Table I).

B. The EnsEMBL Pipeline

To identify features and genes in a genome sequence, numerous analyses have to be performed. EnsEMBL (Clamp *et al.*, 2003; Hubbard *et al.*, 2002) was chosen for this task, as it comprises a well-established and widely used system for storing and retrieving genome-scale data, a Web site for genome display, and an automated annotation method.

The first step in building an EnsEMBL database is the computational identification of a variety of features in the genome sequence. These include repeats, markers, tRNA genes, CpG islands, transcription start sites, and *ab initio* gene predictions by Genscan (Burge and Karlin, 1997). The Genscan exons are BLASTed against proteins in Swall (a nonredundant database combined from SwissProt, SPTrEMBL, and TrEMBLnew), all vertebrate mRNAs from EMBL/ GenBank and zebrafish EST clusters from collaborators to identify similarity matches. The restriction to Genscan exons was put into place for reasons of speed and manageability. It is justified by the fact that although Genscan has a high failure rate at correctly predicting complete gene structures, it has a very high success rate in predicting exons. In addition to the similarity searches, zebrafish ESTs are mapped to the best place in the whole of the genomic sequence.

Management of this large and diverse set of analyses is facilitated by the EnsEMBL analysis pipeline (Potter *et al.*, 2004). This software controls the distribution of work across a large number of computers, taking into account that some steps are dependent on the results of others. The pipeline is modular and other data can be added as necessary; for example, a mapping of the ZFIN expression data set to the genome assembly will be available soon. The results of these analyses are released in a pre-EnsEMBL database, where the features and sequence can be viewed and downloaded (Table I).

C. The EnsEMBL Gene Build

The aim of the EnsEMBL gene build is the automated generation of annotated gene structures on the genome sequence (described in detail in Curwen *et al.*, 2004). The method was developed by observing the way annotators use alignments of protein and cDNA sequences to create gene structures and condensing this process into a set of rules.

First, regions within the genome sequence that are likely to contain genes are located. This is done by generating approximate protein sequence alignments to the genome. To ensure that the known genes are found, all zebrafish protein sequences from SPTrEMBL (SwissProt and translated EMBL) are mapped by using Pmatch (R. Durbin, unpublished), a tool that quickly identifies exact matches between protein and genome. The regions containing known genes are supplemented by those with Genscan predictions with similarity to Swall proteins (see Section II-B).

This procedure serves to associate targeted genomic regions with a specific protein sequence. To identify the precise gene structures within these regions, Genewise (Birney *et al.*, 2004) is used. This program predicts genes by the highly accurate alignment of a protein to the genome sequence, explicitly accounting for splice sites, introns, and frameshifts. For reasons of efficiency, Genewise is not given the raw genomic sequence, but a shortened Miniseq, composed of regions around possible exons as identified by BLASTing the protein sequence to its associated genomic region (Fig. 1).

In a further step, Genscan-predicted exons from the pipeline are used to produce an additional set of gene predictions in regions still lacking genes. Here,

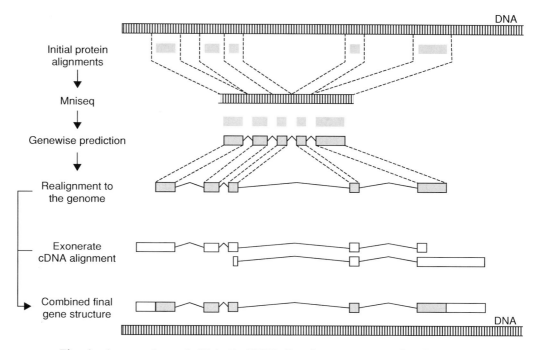

Fig. 1 Automated gene build in EnsEMBL. Protein sequences are aligned to the genome and Genewise is used to predict a gene structure on a shortened Miniseq containing the genomic regions around the probable exons. Untranslated regions (UTRs) are added by aligning cDNAs to the genome, using Exonerate. All predicted gene structures are merged in a final step to produce the resulting gene annotation.

two neighboring exons are required to be supported by BLAST matches to adjacent parts of the same protein sequence. These exon pairs are then recursively linked into transcripts. The number of genes contributed by this step is dependent on the available data sets. As more protein sequences from a certain species are published, the amount of sequence not covered by homology-based gene predictions decreases. In the human genome, for example, the Genscan-based predictions make no significant contribution to the gene set and are therefore no longer included. For a species such as zebrafish, however, for which gene sequence information is rather limited, these predictions are still responsible for about a third of the EnsEMBL genes.

To add untranslated regions (UTRs) and alternative splice variants to the existing set of gene structures, cDNA evidence is used. For this step, zebrafish cDNAs are downloaded from EMBL/Genbank and Exonerate (G. Slater, unpublished) is applied to accurately align these cDNAs to the genome. In the final gene build step, all predictions are gathered and merged into genes with multiple unique transcripts, where coding regions and untranslated regions can be distinguished

(Fig. 1). All predicted genes are checked for sensible intron lengths, translations, and similarity to the parent protein.

The EnsEMBL gene build now offers the ability to identify processed pseudogenes. Predicted genes that display the classic characteristics of pseudogenes (namely the presence of a poly-A sequence downstream of a single exon gene, the absence of an ATG-start codon, frameshifts, and the supporting evidence being aligned in a spliced fashion elsewhere in the genome) are tagged as pseudogenes. This process is still being tested for zebrafish, but will be applied in the future.

The protein sequences of the resulting genes are analyzed by a range of tools as described in Potter *et al.* (2004), compared to the zebrafish entries in SwissProt and RefSeq and named after the best matching entry with sufficient similarity. If such a match cannot be found, the gene is tagged as *novel*. Genes listed in the ZFIN database are tagged as *known* and crosslinked to ZFIN to enable the user to obtain all relevant zebrafish information instantly.

In addition to the described gene build, another based on all available zebrafish expressed sequence tags (ESTs) downloaded from EMBL/Genbank is performed. As ESTs are usually generated by high-throughput projects, the resulting sequence is often of lower quality or even contaminated with genomic sequence (for example). Therefore, the EST gene build is kept separate from the protein/cDNA gene build. The procedure resembles that described previously for cDNAs, with an additional step using the ClusterMerge algorithm (Eyras *et al.*, 2004). Genomewise (Birney *et al.*, 2004) is superimposed to find translations within the genes.

Whenever a new assembly is released and a new gene build performed, the resulting genes, transcripts, and exons are compared to the ones found by the gene build on the previous assembly and stable identifiers are applied where possible.

After completing the above analyses, the resulting genes are compared to each other and also to the genes of other organisms in EnsEMBL. The first step leads to the definition of gene families within the zebrafish gene set and the second to the identification of likely orthologs. In this context, an ortholog pair is defined as two proteins that reciprocally show each other as the best match in the appropriate genome (Ureta-Vidal *et al.*, 2003).

All the generated data are integrated and displayed in the EnsEMBL Web browser (Stalker *et al.*, 2004). A brief introduction to the use of the Web browser is given in Section IV.

III. Clone-by-Clone Analysis and Annotation

A. Mapping and Sequencing

The strategy to generate finished genomic sequence is based on the construction of a physical map of bacterial clone inserts and subsequent identification of a minimal overlapping set from which clones are selected for sequencing. The physical map for the zebrafish genome has been built from four clone libraries, using restriction digest fingerprinting and alignment to mapped markers as described for the human and mouse genomes (International Human Genome Sequencing Consortium, 2001; Waterston *et al.*, 2002). Further details and information about the used libraries are available from our Web site (Table I).

B. Automated Feature Collection

The resulting finished clone sequence is annotated manually, after performing a series of computational analyses similar to those performed by the automated EnsEMBL pipeline (see Section II-B). The main difference is that the BLAST searches are not restricted to Genscan predictions, but widened to the whole sequence for each clone.

Annotators use ACeDB (www.acedb.org) to visualize the results of the analyses and to create and edit gene structures (Fig. 2). The resulting structures are written

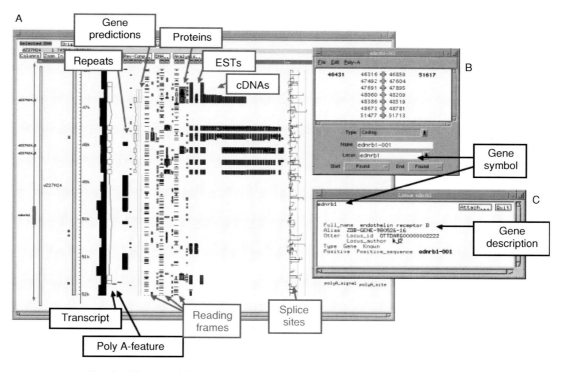

Fig. 2 The use of ACeDB, Spandit, and Blixem in manual annotation. The ACeDB display (A) shows the DNA in vertical orientation next to columns of different features and homology matches. All feature columns are collapsed by default but can be extended to display a single column for each hit (as shown for cDNAs). Protein hits can be shown grouped into the three different reading frames. All the data are used to deduce a correct transcript structure and the translation start and stop. The Spandit tool (B) is used to transfer gene symbol and coordinates that were confirmed through the use of Blixem and Dotter (Fig. 3) to the ACeDB database. The locus window (C) is used to enter gene description.

back to an Otter database. Otter is an extension of the EnsEMBL database scheme that allows storing of the extra textual information produced by manual annotation (Searle *et al.*, 2004).

C. Manual Annotation of Gene Structures

Manual annotation can be described as the process of inspecting the alignments of proteins, cDNAs, and ESTs to the genome sequence (performed automatically as explained previously) and inferring consensus gene structures from them. Typical activities include checking for correct exon boundaries and continuity of coordinates (e.g., missing or repeated exons) and consideration of alternative splice variants for a gene, whether coding or noncoding. Several tools can be used from within ACeDB to assist in finding the correct gene structures. For example, Blixem and Dotter (Sonnhammer and Durbin, 1994, 1995) are used to check sequence alignments, and Spandit (J. Gilbert, unpublished) is used to translate gene feature coordinates supplied by the annotator into ACeDB gene objects. Figures 2 and 3 show examples of the application of these tools.

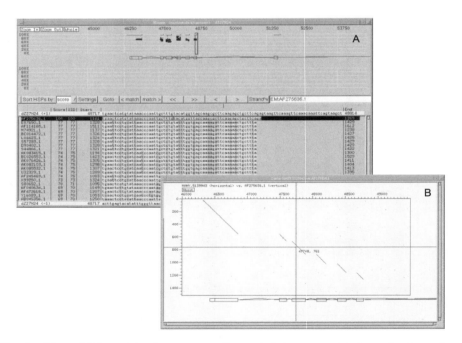

Fig. 3 Blixem (A) is used to view sequence alignments between a certain type of evidence (here cDNA) and the genomic sequence. The upper half of the tool displays the alignment of the supporting evidence in genomic context, and the lower part shows the local alignment as selected by the open box in the upper half. The annotators can obtain further information about a certain match by selecting a sequence and invoking either a dot-plot comparison of selected regions in Dotter (B) or downloading the appropriate EMBL/Genbank file.

In addition to the transcript structure, CDS coordinates (for coding transcripts) and poly-A features deduced from poly-A-containing EST/cDNA matches are annotated. The supporting evidence, which has been used to define gene structures, is stored. All transcripts that overlap by at least one exon are grouped into a gene.

During the annotation process, gene features are added only when completely supported by the above types of evidence. This criterion ensures that the resulting gene structures are of the highest quality. It can also, however, result in the annotation of partial structures when incomplete evidence is available at the time of annotation. Such structures can be corrected as additional cDNA and protein data become available.

D. Gene Classification and Functional Annotation

To provide information about biological function, each annotated gene is given a type and a name. Five different gene types are currently distinguished. A gene is classified as *known* if the best and nearly identical match is a zebrafish protein/cDNA listed by ZFIN. The gene then gets named after and linked to the corresponding ZFIN gene (Fig. 2C). If a protein-coding gene is based on a similar but not identical zebrafish match or on a match from a different organism, it is tagged as *novel CDS* and called *novel protein similar to*, followed by the full description of the match, including the organism it was found in and the appropriate gene symbol in brackets. This information is taken from LocusLink (www.ncbi.nlm.nih.gov/LocusLink/).

A non-protein-coding gene is tagged and named *novel transcript* whereas a pseudogene is tagged and named *novel x pseudogene*, where *x* is the description of the best match. If a non-protein-coding gene is based on weak evidence, for instance, a single EST with just two exons, it is tagged as *putative*. As just *known* genes can get a real gene symbol, it was agreed with the zebrafish nomenclature committee that all other genes get the symbol Sl:clone name.number, SI being the abbreviation for Sanger Institute (zfin.org/zf_info/nomen.html). ZFIN is actively taking part in the manual annotation by checking these as yet unknown genes and assigning functional annotation and approved gene symbols to them, based on the sequence matches and the available literature. For this, they are using the Otter system remotely (Searle *et al.*, 2004).

E. Making it Public

After finding all genes, naming them, and adding poly-A features, an EMBL file containing the clone sequence and all feature coordinates and descriptions is submitted to EMBL/Genbank. Also, the clone is flagged as *to be published* and integrated into the next release of the Vega annotation browser (Table I). Vega is a database built on the same underlying schema as EnsEMBL databases and hosts all the information gathered from the clone-by-clone sequencing and the manual annotation process. Because of the frequent release scheme of Vega, corrections/additions reported by community experts can be quickly applied and made visible.

IV. The Sanger Zebrafish Web Services

A. The Sanger Zebrafish Homepage

The Sanger zebrafish home page as depicted in Fig. 4 and listed in Table I serves as the main gateway to all our zebrafish-related services and data collections. It offers two main links: one to the WGS assembly and automated annotation part of the project and the other to the clone mapping, sequencing, and manual annotation part. These links take you to pages that provide brief overviews of the respective project as well as all further relevant links. All links are also available from the left panel on each side. A "Frequently Asked Questions" page and a helpdesk (zfish-help@sanger.ac.uk) offer a hand in finding the right bit of information.

B. WGS Assemblies and EnsEMBL

The main entry points for data retrieval regarding the WGS assembly are the FTP page for bulk download and the EnsEMBL Web browser for information about genome features, sequence similarity searches (BLAST or SSAHA), and the download of certain parts of the sequence (Table I). Stalker and Cox (2003) as well as the EnsEMBL online documentation describe browsing EnsEMBL including

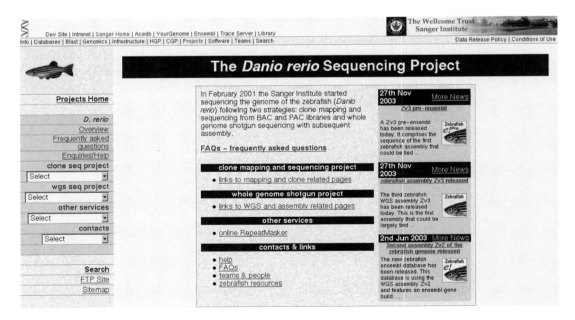

Fig. 4   The Sanger Institute zebrafish homepage at www.sanger.ac.uk/Projects/D_rerio.

data retrieval and using the EnsEMBL API in detail, and therefore the following should be understood as just a very brief introduction.

Assuming you have a certain cDNA sequence and want to find the corresponding gene in the zebrafish assembly, you have several options, starting with www.ensembl.org/Danio_rerio. First, you can use BLAST or SSAHA to align the cDNA sequence to the assembly. BLAST is recommended for nonzebrafish sequences, whereas SSAHA should be the tool of choice for zebrafish–zebrafish searches for reasons of accuracy and speed. Your second option is to use the general text search to find a zebrafish gene with an appropriate name. Yet another way to get to a result is to identify a gene in a different EnsEMBL organism and then go to the zebrafish homologue if listed under "Homology Matches" on the GeneView page. Or you might have found a gene of interest at ZFIN and can then use the EnsEMBL link provided.

If you cannot find an appropriate gene, this might be because the corresponding part of the genome sequence is absent from the current assembly. This will improve over time. Another reason might be that neither the cDNA/protein sequence from the gene you are looking for nor any homologous sequence from another species with high enough similarity was available at the time the gene build was performed. In this case, it might appear in subsequent releases and in the meantime, you can use the other displayed features in ContigView (BLAST matches, Genscan predictions) to get an idea of what the gene might look like.

If any of these searches are successful, you will be taken to the aforementioned ContigView and GeneView pages (Figs. 5 and 6). The ContigView page gives a *scrollable* and *zoomable* overview of the genomic region your match was found in, thereby displaying annotated genes and features. All view options are configurable by the user. You can also add data through a DAS server (Dowell *et al.*, 2001). This feature is used, for example, to show Vega genes that map to the region you are looking at.

The GeneView page lists the gene description, the structure of the transcripts and the genomic neighbourhood, other members of the same protein family within the same organism ("Protein Matches"), orthologuous genes in other EnsEMBL organisms ("Homology Matches"), the protein feature annotation, and the corresponding sequences. Corresponding entries in other databases, for example, ZFIN, can be found under "Similarity Matches." The page also provides links to the "Protein," "Transcript," and "Exon" pages, where you can find additional information, for example, the protein properties, supporting features or the sequence of the whole transcripts, translations, single exons, or exon–intron boundaries.

To download data, you can use the ExportView pages, which can be reached from all zebrafish EnsEMBL Web pages through "Export Data." You will be offered the option to export features and sequence in a range of different formats (www.ensembl.org/Danio_rerio/exportview). To download pregenerated data sets, choose "Download" from any zebrafish EnsEMBL Web page.

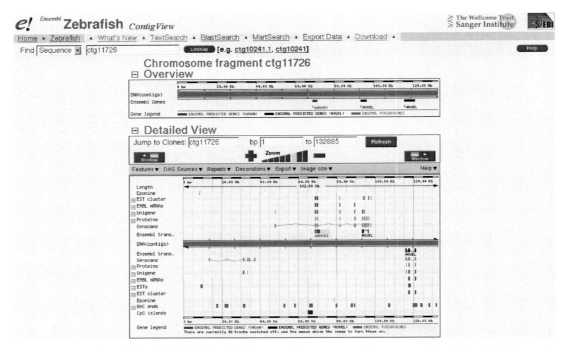

Fig. 5   The EnsEMBL ContigView page. The upper part shows an overview of the selected genomic region with known and novel genes (see Section II.C). The lower part shows a selection of features (here collapsed; can be extended to show each match within a feature group in a single row) aligned to the sequence. The user can customize the page by selecting the features, DAS, decorations, etc., to be displayed.

For more advanced types of data retrieval, the EnsMart tool is extremely useful (www.ensembl.org/Multi/martview; Kasprzyk *et al.*, 2004). EnsMart provides an easy and intuitive way of retrieving data according to a series of user-supplied criteria. You can, for instance, download the DNA sequences 750 bp upstream of the 5′ UTRs of genes with coding SNPs, or the protein sequences of all genes with a certain Interpro domain or any other data set filtered on combinations of your choice. EnsMart also provides different outputs formats, including plain text, HTML, and MS Excel.

C. Finding Markers in the WGS Assembly

To provide user-friendly searches for markers mapped to the WGS assembly, the mapsearch service (www.sanger.ac.uk/cgi-bin/Projects/D_rerio/mapsearch) was developed. Mapsearch contains the positions of STS markers, ESTs listed by ZFIN with map information, and BAC end sequences on the latest version of the WGS assembly. You can either look for a specified type of marker on a given

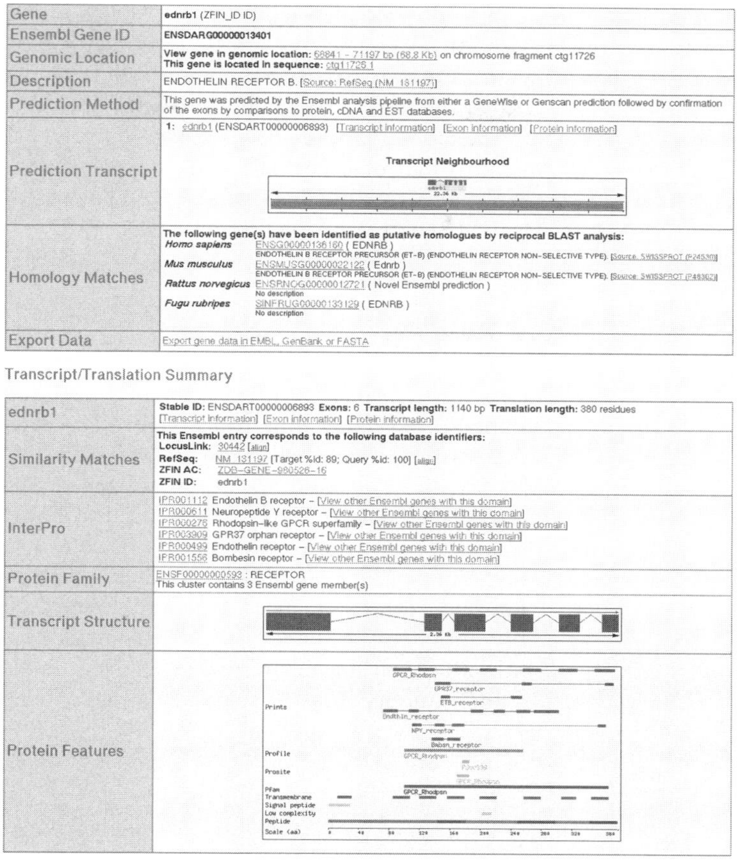

Fig. 6 The EnsEMBL GeneView page. This page lists all available annotations for a certain gene. It provides links to related pages such as ContigView and further information about transcripts and translation. Orthologous genes from other species are listed as well as other protein family members, links to representations of this gene in other databases, and the protein annotation are displayed.

assembly contig or for contigs that contain a specific marker. The results page allows you to access the available marker information and also to check the corresponding assembly contigs in EnsEMBL.

D. Clone Data Retrieval

There are multiple ways to access data from the clone mapping and sequencing project, again all starting from the Sanger zebrafish homepage (Fig. 4). The mapping project (Table I) provides links to overviews, information about the used libraries, contacts, and the FPC database itself. The finished and unfinished clone sequences can be downloaded from our FTP page (or from EMBL/Genbank) and searched with BLAST. The biological clones can be ordered from the providers listed on www.sanger.ac.uk/Projects/D_rerio/library_details.shtml.

Vega (Table I) offers a Web interface for browsing finished and annotated clones from the clone mapping and sequencing project. The interface is designed to be identical to that of EnsEMBL, although at the time of writing it lacked the EnsMart tool. You can use a DAS server to display genes from other sources (such as EnsEMBL) as described previously.

V. Future Releases

At present, the zebrafish genome sequence including annotation is available from either the WGS or the clone-by-clone project. Integration of data from both sources is performed by mapping the WGS assembly to the FPC map. In this "tying" process (M. Caccamo, Z. Ning, K. Jekosch, and T. Hubbard, unpublished), the placement of the BAC ends is used to link the assembly supercontigs to the FPC contigs and the clones within. If such a relationship can be found, the assembly supercontig is named after the corresponding FPC contig, and the available finished clone sequences are used to replace the appropriate part of the assembly supercontig. This process will improve the presented genome sequence continuously with every release until eventually all WGS sequence can be replaced by high-quality finished sequence data.

Acknowledgments

I thank Mario Caccamo, the EnsEMBL team (Sanger and EBI), the Sanger Web team, and the Sanger Finished Analysis team for all their help and for providing the code that made the work presented here possible. I also thank all users and collaborators, especially ZFIN, for feedback and discussions. The zebrafish genome project is funded by the Wellcome Trust. The manual annotation is partly funded by a supplementary grant to ZFIN (NIH P41 HG002659).

References

Ashurst, J. L., and Collins, J. E. (2003). Gene annotation: Prediction and testing. *Annu. Rev. Genom. Hum. Genet.* **4,** 69–88.

Birney, E., Clamp, M., and Durbin, R. (2004). Genewise and genomewise. *Genome Res.* **14,** 988–995.

Burge, C., and Karlin, S. (1997). Prediction of complete gene structures in human genomic DNA. *J. Mol. Biol.* **268,** 78–94.

Clamp, M., Andrews, D., Barker, D., Bevan, P., Cameron, G., Chen, Y., Clark, L., Cox, T., Cuff, J., Curwen, V., Down, T., Durbin, R., Eyras, E., Gilbert, J., Hammond, M., Hubbard, T., Kasprzyk, A., Keefe, D., Lehvaslaiho, H., Iyer, V., Melsopp, C., Mongin, E., Pettett, R., Potter, S., Rust, A., Schmidt, E., Searle, S., Slater, G., Smith, J., Spooner, W., Stabenau, A., Stalker, J., Stupka, E., Ureta-Vidal, A., Vastrik, I., and Birney, E. (2003). EnsEMBL 2002: Accommodating comparative genomics. *Nucleic Acids Res.* **31,** 38–42.

Curwen, V., Eyras, E., Andrews, T. D., Clarke, L., Mongin, E., Searle, S., and Clamp, M. (2004). The EnsEMBL automatic gene annotation system. *Genome Res.* **14,** 942–950.

Dowell, R. D., Jokerst, R. M., Day, A., Eddy, S. R., and Stein, L. (2001). The distributed annotation system. *BMC Bioinform.* **2,** 7–13.

Eyras, E., Caccamo, M., Curwen, V., and Clamp, M. (2004). ESTgenes: Alternative splicing from ESTs in EnsEMBL. *Genome Res.* **14,** 976–987.

Grafham, D., and Willey, D. (2003). Sequence finishing. *In* "Nature Encyclopedia of the Human Genome" (D. N. Cooper, ed.), Vol. 3, pp. 225–228. Macmillan Publishers, Nature Publishing Group, London.

Hubbard, T., Barker, D., Birney, E., Cameron, G., Chen, Y., Clark, L., Cox, T., Cuff, J., Curwen, V., Down, T., Durbin, R., Eyras, E., Gilbert, J., Hammond, M., Huminiecki, L., Kasprzyk, A., Lehvaslaiho, H., Lijnzaad, P., Melsopp, C., Mongin, E., Pettett, R., Pocock, M., Potter, S., Rust, A., Schmidt, E., Searle, S., Slater, G., Smith, J., Spooner, W., Stabenau, A., Stalker, J., Stupka, E., Ureta-Vidal, A., Vastrik, I., and Clamp, M. (2002). The EnsEMBL genome database project. *Nucleic Acids Res.* **30,** 38–41.

International Human Genome Sequencing Consortium. (2001). Initial sequencing and analysis of the human genome. *Nature* **409,** 860–921.

Kasprzyk, A., Keefe, D., Smedley, B., London, D., Spooner, W., Melsopp, C., Hammond, M., Rooca-Serra, P., Cox, T., and Birney, E. (2004). EnsMart—a generic system for fast and flexible access to biological data. *Genome Res.* **14,** 160–169.

Mouse Genome Sequencing Consortium. (2002). Initial sequencing and comparative analysis of the mouse genome. *Nature* **5,** 520–562.

Mullikin, J. C., and Ning, Z. (2003). The phusion assembler. *Genome Res.* **13,** 81–90.

Potter, S. C., Clarke, L., Curwen, V., Gilbert, J. G. R., Keenan, S., Mongin, E., Searle, S. M. J., Stabenau, A., Storey, R., and Clamp, M. (2004). The EnsEMBL analysis pipeline. *Genome Res.* **14,** 934–941.

Searle, S. M. J., Gilbert, J., Iyer, V., and Clamp, M. (2004). The Otter annotation system. *Genome Res.* **14,** 963–970.

Searle, S. M. J., Stabenau, A., Storey, R., and Clamp, M. (submitted). The EnsEMBL analysis pipeline. *Genome Res.*

Soderlund, C., Longden, I., and Mott, R. (1997). FPC: A system for building contigs from restriction fingerprinted clones. *Comput. Appl. Biosci.* **13,** 523–535.

Sonnhammer, E. L., and Durbin, R. (1994). A workbench for large-scale sequence homology analysis. *Comput. Appl. Biosci.* **10,** 301–307.

Sonnhammer, E. L., and Durbin, R. (1995). A dot-matrix program with dynamic threshold control suited for genomic DNA and protein sequence analysis. *Gene* **167,** 1–10.

Sprague, J., Clements, D., Conlin, T., Edwards, P., Frazer, K., Schaper, K., Segerdell, E., Song, P., Sprunger, B., and Westerfield, M. (2003). The Zebrafish Information Network (ZFIN): The zebrafish model organism database. *Nucleic Acids Res.* **31,** 241–243.

Stalker, J. W., and Cox, A. V. (2003). EnsEMBL: An open-source software tool for large-scale genome analysis. *In* "Introduction to Bioinformatics: A Theoretical and Practical Approach" (I. Krawetz and D. D. Woble, eds.), pp. 413–429. Humana Press, Totowa, NJ.

Stalker, J., Gibbins, B., Meidl, P., Smith, J., Spooner, W., Hotz, H.-R., and Cox, A. V. (2004). The EnsEMBL website: Mechanics of a genome browser. *Genome Res.* **14,** 951–955.

Ureta-Vidal, A., Ettwiller, L., and Birney, E. (2003). Comparative genomics: Genome-wide analysis in metazoan eukaryotes. *Nat. Rev. Genet.* **4,** 251–262.

CHAPTER 13

Molecular Cytogenetic Methodologies and a Bacterial Artificial Chromosome (BAC) Probe Panel Resource for Genomic Analyses in Zebrafish

Charles Lee[*,†] and Amanda Smith[*]

[*]Department of Pathology
Brigham and Women's Hospital
Boston, Massachusetts 02115

[†]Harvard Medical School
Boston, Massachusetts 02115

I. Introduction

The zebrafish genome contains 1.7×10^9 bp of DNA, approximately half the genome size of most mammals, organized into 25 pairs of chromosomes ($2n = 50$) (Hinegardner and Rosen, 1972). Classical cytogenetic studies of zebrafish have shown that individual chromosomes are difficult to unequivocally identify on the basis of chromosome size, morphology (i.e., position of the centromere), and banding pattern (Amores and Postlethwait, 1999). It is therefore not surprising that since 1968 at least 12 different zebrafish karyotypes have been published (Sola and Gornung, 2001). Because cytogenetics offers a unique perspective in genome analysis, by directly visualizing targeted chromosomal loci and large genomic alterations, it is important to advance zebrafish cytogenetics and develop the resources that will provide valuable complementary methods for genetic analyses in this important model organism.

The field of molecular cytogenetics emerged in the 1980s around a technique referred to as fluorescence *in situ* hybridization (FISH). By using FISH methodologies, a specific DNA sequence or collection of DNA fragments can be selectively labeled with a hapten molecule or fluorescent dye and hybridized to denatured chromosomes or interphase cells. DNA hybridization kinetics permit these labeled probes to anneal to their complementary sequences on metaphase chromosomes or in interphase cells. In this manner, one can directly visualize whether the sequence of interest is present in the genome being interrogated. If so, the relative chromosomal position of the sequence can also usually be ascertained.

In an effort to establish molecular cytogentic tools for zebrafish genomic analyses, we have developed a first-generation zebrafish BAC probe panel by using genetically-positioned and chromosomally-mapped bacterial artificial chromosome (BAC) clones. The following methods were used to chromosomally map the BAC probes in this panel. The same methods can also be used in experiments to assess for genomic imbalances, ploidy, and genomic instability in zebrafish.

II. Methods

A. Zebrafish Chromosome Preparations

1. Metaphase Chromosomes from Embryos

   1. Take ~100 embryos at 7 h postfertilization.
   2. Add colchicine (Acros Organics) to a final concentration of 1 mg/ml. Incubate at 28.5 °C for 16 h.
   3. Dechorionate embryos by pronase treatment (Westerfield, 2000).
   4. Rinse embryos three times with 0.48 mg/ml Instant Ocean (Aquarium Systems) fish water.
   5. Transfer embryos to a clean microfuge tube on ice. Remove all but 100 μl of fish water and homogenize (approximately 10 strokes) by using a pellet pestle (Kontes Glass Company).
   6. Add 1 ml of ice-cold 0.9 × PBS (ICN Biomedicals, Inc.), 10% fetal calf serum (Hyclone).
   7. Using a 5-ml syringe, filter sequentially through 105-μm and 40-μm nylon filters (Small Parts, Inc.) into a clean 15-ml centrifuge tube.

8. Add 5 ml of ice-cold $0.9 \times$ PBS, 10% fetal calf serum.

9. Centrifuge at 250 rcf for 10 min at 4 °C.

10. Discard the supernatant and gently resuspend the cells in 10 ml of 1.1% sodium citrate, 4 mg/ml colchicine.

11. Incubate at 25 °C for 25 min.

12. Add 1 ml of ice-cold fixative (3:1 methanol:glacial acetic acid).

13. Centrifuge at 450 rcf for 10 min at 4 °C.

14. Remove supernatant and add 10 ml of ice-cold fixative.

15. Centrifuge cells at 450 rcf for 10 min at 4 °C.

16. Repeat two more times with fresh ice-cold fixative.

17. Store cell pellet at -20 °C until required.

2. Metaphase Chromosomes from Established Adherent Cell Cultures

Several established zebrafish cell lines are now available from the American Type Culture Collection (http://www.atcc.org), including ZF4, SJD.1, AB.9, and ZFL. Metaphase chromosomes can be prepared from these and other adherent cell lines as follows:

1. Grow an established zebrafish cell line to ∼60% confluency.

2. Add colcemid (Irvine Scientific) to a final concentration of 0.1 mg/ml and incubate at 28.5 °C and 5% CO_2 for 17 h.

3. Check cells under an inverted microscope to determine whether more than 50% mitotic cells can be observed. (Mitotic cells will appear as smaller, round, and shiny cells.) If less than 50% mitotic cells are observed, continue incubation and check mitotic index every hour.

4. Hit the side of the flask to shake loose the mitotic cells into the media.

5. Transfer the cell-containing media to a clean 50-ml centrifuge tube.

6. Centrifuge cells at 250 rcf for 10 min at room temperature.

7. Remove the supernatant and gently resuspend cell pellet in 10 ml of 1.1% sodium citrate.

8. Incubate at room temperature for 10 min.

9. Add 1 ml of fixative (3:1 methanol:glacial acetic acid) and mix by inverting tube 10 times.

10. Spin at 450 rcf for 10 min at room temperature.

11. Remove supernatant and replace with 10 ml of fixative. Spin at 450 rcf for 10 min.

12. Repeat Step 11 two more times.

13. Store cell pellet at -20 °C until ready to use.

B. Preparation of Bacterial Artificial Chromosome (BAC) Probe DNA

1. Streak BAC clones from glycerol stocks onto LB-agar (EM Science) plates with the appropriate antibiotic.

2. Pick a single colony and inoculate a starter culture, 5 ml of TB media (American Bioanalytical) with the appropriate antibiotic. Incubate in a shaker incubator for 4 h at 37 °C and 270 rpm.

3. Inoculate 250 ml of TB media, with the appropriate antibiotic, using the starter culture. Incubate for 16 h at 37 °C and 270 rpm.

4. Isolate DNA by the plasmid purification Midi Kit (Qiagen No. 12643)

   a. Chill the P1 (50 mM Tris-HCl, pH 8, 10 mM EDTA) and P3 (3 M potassium acetate, pH 5.5) buffers at 4 °C.

   b. Centrifuge cells at 2500 rcf at 4 °C for 15 min in a clean, sterile 250-ml centrifuge bottle (Nalgene). Pour off the supernatant.

   c. Incubate cells at room temperature for 5 min.

   d. Resuspend cells in 25 ml of ice-cold P1 buffer with 100 mg/ml RNase A (Roche). Keep cells on ice.

   e. Add 25 ml of P2 buffer (2 mM NaOH, 1% SDS) at room temperature by slowly pouring the solution down the sides of the bottle without agitation. Mix gently by rotating the bottle 360° repeatedly and lay the bottle on its side for 2 min. Gently repeat rotation to ensure complete cell lysis. (Remember that BACs and other large clones are susceptible to DNA shearing.)

   f. Add 25 ml of ice-cold P3 buffer down the sides of the bottle, rotating the bottle slowly as the buffer is added.

   g. Gently stir the cells with the 25 ml serological pipette used to dispense buffer P3.

   h. Keep cells on ice for 20 min.

   i. Rotate again as in Step e and centrifuge at 12,000 rcf for 30 min. Filter the supernatant through a paper filter (Whatman) into a clean centrifuge bottle on ice.

   j. Equilibrate each Qiagen Midi-Tip with 4 ml of QBT buffer (750 mM NaCl, 50 mM MOPS, pH 7.0, 15% isopropanol, 0.15% Triton X-100).

   k. Apply supernatant to equilibrated Qiagen Midi-Tip column. Allow to drain completely.

   l. Wash tip twice with 10 ml of QC buffer (1 M NaCl, 50 mM MOPS, pH 7.0, 15% isopropanol).

   m. Elute BAC DNA from each tip into a clean 15-ml centrifuge tube with two washes of prewarmed (65 °C) QF buffer (1.25 mM NaCl, 50 mM Tris-HCl, pH 8.5, 15% isopropanol).

5. Precipitate DNA by adding 6 ml of isopropanol. Mix thoroughly but gently to minimize shearing of DNA.

6. Centrifuge at 3000 rcf for 30 min at 4 °C. Pour off supernatant.

7. Rinse pellet with ice-cold 70% ethanol. Spin at 3000 rcf for 15 min at 4 °C.

8. Air-dry pellet for 10 min at room temperature.

9. Resuspend each pellet in 150 μl of sterile ddH$_2$O. Incubate at 55 °C for 15 min.

10. Rehydrate DNA at 4 °C for 16 h and then transfer to a clean, sterile 1.5-ml microfuge tube.

C. Fluorescence *In Situ* Hybridization

1. Preparation of C$_0$t1 DNA

1. Take freshly anesthetized zebrafish and place in mortar. (The use of albino zebrafish for this purpose will result in DNA with less contaminating pigment.) Thoroughly freeze the zebrafish in liquid N$_2$.

2. Grind the animal to a fine pulp, intermittently adding liquid N$_2$ to carefully maintain the tissue as a slurry. Transfer the cells into a sterile 50-ml centrifuge tube and swirl to allow the remaining liquid N$_2$ to evaporate.

3. Add 20 ml of Proteinase K lysis solution [50 mM Tris-Cl, pH 8, 100 mM EDTA, 100 mM NaCl, 1% SDS, 100 μg/ml Proteinase K (Roche)] to the tube. Shake to resuspend pellet. Gently rock the solution in a 55 °C incubator for 24 h.

4. Extract with 10 ml of pH 8.0-buffered phenol. Gently mix the solution by rocking the tube 20 times.

5. Add 10 ml chloroform/isoamyl alcohol (24:1), and mix thoroughly as in Step 4.

6. Spin at 1500 rcf for 5 min at room temperature to separate the phases.

7. Transfer the upper aqueous phase into a new, sterile 50-ml centrifuge tube. Repeat Steps 4–6.

8. Transfer the aqueous phase to a new, sterile 50-ml centrifuge tube. Add 0.1 × 3 M sodium acetate, mix thoroughly, and then add 1 × isopropanol. Gently mix the solution until the genomic DNA starts to aggregate.

9. Spin down the DNA pellet at 3000 rcf for 10 min at 4 °C. Carefully pour off the supernatant.

10. Wash the DNA pellet with 20 ml of ice-cold 70% ethanol and spin down again as in Step 9.

11. Use a sterile rod to spool the DNA precipitate and transfer to a sterile microfuge tube.

12. Rehydrate the DNA pellet with 2 ml of sterile ddH$_2$O and incubate at 55 °C for 16 h.

13. Store at $-20\,°C$ until ready to proceed further.

14. Obtain DNA concentration by standard spectrophotometric measurements.

15. Shear genomic DNA in a sonicator (Sonics, Inc.) to an average fragment size of 400 bp.

16. Calculate the time required for the C_0t1 fraction of the given DNA sample to reanneal (time in minutes = 5.92/DNA concentration in mg/ml).

17. Denature the DNA at $100\,°C$ for 15 mins.

18. Place the DNA in a $65\,°C$ water bath for 4 mins. Add NaCl to a final concentration of 0.3 M. Allow the DNA to reanneal for the time calculated in Step 16.

19. Add $1 \times$ ice-cold $2 \times$ S1 nuclease buffer (Fisher Scientific). Add 1 unit of S1 nuclease (Fisher Scientific) for every microgram of genomic DNA.

20. Incubate at $37\,°C$ for 30 mins.

21. Add 10 ml phenol/chloroform/isoamyl alcohol (25:24:1), mix thoroughly, and spin at 1500 rcf for 5 min at room temperature to separate phases.

22. Transfer upper aqueous phase to a new, sterile 15-ml centrifuge tube. Add $0.1 \times 2.5\,M$ ammonium acetate, mix thoroughly, and then add $2.5 \times 100\%$ ethanol. Gently mix solution to precipitate DNA and then spin down the DNA pellet at 3000 rcf for 10 min at $4\,°C$.

23. Rehydrate pellet in an appropriate volume of sterile ddH_2O and store at $-20\,°C$ until ready to use.

2. Probe Synthesis

1. Mix the following in a microfuge tube:
   - $1\,\mu g$ of probe DNA.
   - $5\,\mu l$ of $10 \times$ nick translation buffer (500 mM Tris-HCl, pH 7.2, 100 mM $MgSO_4$, 1 mM dithiothreitol).
   - $5\,\mu l$ of 0.1 mM dTTP.
   - $10\,\mu l$ of dNTP mixture (0.1 mM dATP, dGTP, dCTP each).
   - $10\,\mu l$ of nick translation enzyme (5 units DNA polymerase I, 0.1 units DNase I; 50 mM Tris-HCl, pH 7.2, 10 mM $MgSO_4$, 0.1 mM dithiothreitol, 0.5 mg/ml nuclease-free bovine serum albumin).
   - $2.5\,\mu l$ of 0.2 mM cy3-11-dUTP (Amersham) or fluorescein-12-dUTP (Molecular Probes).

2. Add sterile ddH_2O to a final volume of $50\,\mu l$.

3. Mix and incubate for 12 h at $15\,°C$.

4. Take a 5-μl aliquot and run on a 1% agarose gel to check the probe size. The ideal probe should comprise a collection of fragments of less than 500 bp.

5. Purify the labeled DNA by a Sephadex-G50 column (Zymo).
6. Add $10\,\mu g$ of zebrafish C_0t1 DNA.
7. Spin probe + zebrafish C_0t1 DNA mixture in a Speed-Vac on medium heat until dry.
8. Resuspend pellet in $30\,\mu l$ of hybridization buffer (50% formamide; $2 \times$ SSC; 10% dextran sulfate).

3. Probe Hybridization

1. Apply $10\,\mu l$ of the probe mixture to the dehydrated slides.
2. Cover with a $22 \times 22\,mm^2$ coverslip and seal edges with rubber cement.
3. Codenature on a hot plate at $70\,°C$ for 3 min.
4. Incubate in a humidified chamber in the dark for 16 h at $37\,°C$.

Table I

Near–Telomeric and Near–Centromeric Bacterial Artificial Chromosome (BAC) Clones for Each Zebrafish Chromosome

LG	Short-arm telomere	Long-arm telomere	Centromere
1	zC093G23	zC141F18	zC022O06
2	zK014G06	zC009D09	zC039P05
3	zK007C07	zK005H01	zC115J06
4	zK030C13		zC132J14
5	zC087E10	zC150K20	zK007B18
6	zC051O03	zC060H08	zK011J04
7	zK009M06	zC128L16	zK014N10
8	zC069A12	zC027L17	zC103G04
9	zC115B08	zC012N08	zK001A09
10	zC128P08	zC022E09	zK030A08
11	zC108O08	zC115I06	zC042L14
12	zC121C04	zC086E02	zK022H21
13	zK006L12	zC065J24	zK011L06
14	zC117N19	zC125N22	zC117E17
15	zC055C01	zC059M05	zC125H09
16	zC127P21	zC121P03	zC132M17
17	zK013L17	zK014B13	zK006P15
18	zC095I06	zK014D24	zK005J13
19	zC039E15	zC036I10	zC132A16
20	zC118G14	zC134L13	zK015B08
21	zC122A16	zK014M09	zC065O02
22	zK002J07	zC009D01	zC132L16
23	zC041B11	zC059K08	zC051C19
24	zK022E19	zC118G02	zK001A04
25	zC096F02	zC087L10	zC059G12

4. Posthybridization Washes and Counterstain

    1. Gently peel away rubber cement and remove coverslip.

    2. Perform two 5-min washes in 50% formamide, 2 × SSC at 45 °C in coplin jars.

    3. Perform two 5-min washes in 2 × SSC at 45 °C in coplin jars.

    4. Wash slides in 4 × SSC, 0.05% Tween-20 (Sigma) for 8 min at 37 °C.

    5. Apply 30 μl of antifade mounting medium with DAPI (Vector Labs) and cover with a 24 × 60 mm^2 coverslip.

    6. Incubate in the dark at room temperature for 5 min.

    7. Gently press out excess mounting medium and seal the edges of the coverslip with nail polish.

    8. Analyze by fluorescence microscopy with appropriate filter sets.

III. First–Generation Zebrafish BAC Probe Panel

The development and application of molecular cytogenetic probes should begin to bridge the longstanding gap between the DNA sequenced-physical maps and the more well established genetic maps. It is anticipated that well-characterized, chromosomally mapped BACs will serve as cytogenetic anchors for the zebrafish genome sequencing efforts. Toward these goals, we have developed a first-generation zebrafish chromosome-specific BAC probe panel (Table I), consisting of FISH-verified BAC clones near the centromere and telomeres of each zebrafish chromosome. BAC clones were selected from the CHORI 211 or Danio Key BAC

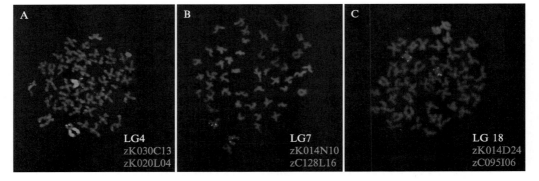

Fig. 1 Representative two-color FISH images. Two-color FISH images for (**A**) the near-telomeric short-arm probe (red) and long-arm heterochromatin probe (green) of linkage group 4, (**B**) the near-centromeric (green) and near-telomeric long-arm probe (red) of linkage group 7, (**C**) the near-telomeric short-arm probe (red) and near-telomeric long-arm probe (green) for linkage group 18. (See Color Insert.)

Table II
Flanking z Markers for Each BAC Clone in the Probe Panel

LG	BAC	Anchored marker—up	cM position	Anchored marker—down	cM position
1	zC093G23	z17325	18.8	z11369	20.1
	zC022O06	z1368	66.1	z22347	77.7
	zC141F18	z12049	85.0	z53427	80.2
2	zK014G06	z22747	24.8	NA	NA
	zC039P05	z4733	39.4	z4300	40.6
	zC009D09	z13579	68.7	z67174	75.6
3	zK007C07	z7486	100.6	z25778	107.5
	zC115J06	z3725	78.2	z20058	94.8
	zK005H01	z419	15.1	z872	22.3
4	zK030C13	z54409	0.0	z1389	5.9
	zC132J14	z10164	50.8	z26519	55.4
5	zC087E10	z6803	69.6	z26127	79.5
	zK007B18	z13641	57.5	z26603	59.2
	zC150K20	z6132	10.7	z68291	6.1
6	zC051O03	z11310	32.4	z51328	36.2
	zK011J04	z25827	43.1	z12094	49.0
	zC060H08	z4950	59.6	z7666	72.6
7	zK009M06	7z9249	81.3	z7875	92.5
	zK014N10	z8216	53.4	z54956	56.4
	zC128L16	NA	NA	z191	1.9
8	zC069A12	z14886	70.8	z51584	88.2
	zC103G04	z6764	54.1	z11946	54.1
	zC027L17	z9420	19.8	z1637	4.9
9	zC115B08	z22173	5.9	z15447	16.7
	zK001A09	z25375	44.0	z55183	50.0
	zC012N08	z22141	89.0	z4577	91.3
10	zC128P08	z14825	14.5	z7316	16.9
	zK030A08	z13685	48.7	z54048	44.7
	zC022E09	z6992	77.3	NA	NA
11	zC108O08	z12083	25.3	z13411	22.6
	zC042L14	z3412	39.3	z67494	41.1
	zC115I06	z6272	65.9	z8816	70.6
12	zC121C04	NA	NA	z7576	1.3
	zK022H21	z4847	43.3	z11518	45.7
	zC086E02	z4499	62.0	z13675	72.6
13	zK006L12	z8317	4.7	z11918	7.4
	zK011L06	z5608	45.3	z10963	45.3
	zC065J24	z10963	45.3	17223	51.5
14	zC117N19	z1536	23.3	z14423	27.0
	zC117E17	z23266	53.2	z9057	54.4
	zC125N22	z3984	86.7	z46595	89.0
15	zC055C01	z10289	7.3	z6722	15.4
	zC125H09	z49653	47.1	z4396	49.3
	zC059M05	z26441	73.2	z20632	85.2
16	zC127P21	NA	NA	NA	NA
	zC132M17	z9685	43.4	z5022	46.9
	zC121P03	z10256	63.7	z4678	59.0

(continues)

Table II (*continued*)

LG	BAC	Anchored marker—up	cM position	Anchored marker—down	cM position
17	zK013L17	z21144	81.1	z10387	82.3
	zK006P15	z62083	42.8	z62970	44.1
	zK014B13	z21151	0.0	z55648	3.6
18	zC095I06	z11854	61.9	z10691	63.4
	zK005J13	z3853	48.9	z8343	52.4
	zK014D24	z25142	19.8	z13329	26.2
19	zC039E15	z7450	22.9	z9468	29.6
	zC132A16	z5352	48.8	z1234	59.4
	zC036I10	z13292	71.8	z26695	76.6
20	zC118G14	z9962	17.6	z10901	21.2
	zK015B08	z22926	66.1	z49730	67.5
	zC134L13	z46013	101.3	z8554	107.2
21	zC122A16	z4074	120.5	z45254	128.5
	zC065O02	z42943b	49.3	z42626	73.0
	zK014M09	z3476	3.7	z20701	17.3
22	zK002J07	z3286	59.2	z4682	68.6
	zC132L16	z11679	29.1	z6805	30.3
	zC009D01	z9516	6.8	z11379	10.3
23	zC041B11	z13781	16.6	z20895	30.5
	zC051C19	z342	31.6	z15422	32.2
	zC059K08	z26329	52.0	z3532	53.2
24	zK022E19	z241	1.2	z10458	2.5
	zK001A04	z9852	37.0	z7967	42.9
	zC118G02	z6438	59.4	z9380	68.7
25	zC096F02	z43617	49.6	z15401	66.8
	zC059G12	z25717	43.0	z3490	40.3
	zC087L10	z6924	9.8	z22653	6.3

Note: NA: not available.

libraries, initially on the basis of their inferred proximity to the ends and centromere of each chromosome, using linkage map data. Each BAC clone was end-sequenced for clone identity verification and subsequently mapped by FISH to metaphase chromosomes from wild-type zebrafish embryos. For each FISH experiment, two different BAC clones for a given linkage group were simultaneously hybridized to metaphase chromosome spreads, providing visual confirmation of chromosomal synteny of the hybridized BAC clones (e.g., Fig. 1). A BAC clone mapping near the primary constriction of each chromosome was designated as the near-centromeric clone for that chromosome. The centromere divides each chromosome into two visually distinct chromosome arms, allowing each near-telomeric probe to be assigned to either the short arm or long arm of the respective chromosome. In addition, most of the BAC clones in this probe panel were anchored to flanking z-markers (Table II).

The BAC probe panel is complete with the exception of linkage group 4, where a suitable near-telomeric long arm probe has not yet been identified. Linkage group 4 corresponds to a zebrafish chromosome previously reported to contain substantial amounts of constitutive heterochromatin along most of the long arm (Sola and Gornung, 2001). Constitutive heterochromatin is commonly composed of highly repetitive DNAs (John, 1988) and some of the repetitive DNA here appears to be that of 5S rDNA (Phillips and Reed, 2000; Gornung et al., 2000). Five different BAC clones have already been chosen from this chromosome region, all showing a hybridization pattern (e.g., probe zK020L04 in Fig. 1A) similar to that observed with the 5S rDNA (Gornung et al., 2000; Phillips and Reed, 2000). Rather than the confined, locus-specific signal observed with the other BAC clones of this probe panel, each linkage group 4 long-arm probe has so far consistently hybridized along much of the length of this chromosome arm. Intriguingly, there appears to be approximately 30 cM of markers corresponding to this chromosome arm (Fig. 2). It is not known whether some unique sequences

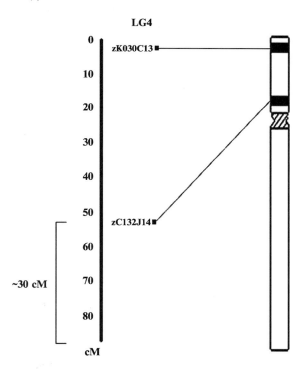

Fig. 2 Genetic map and chromosomal mapping of bacterial artificial chromosomes (BACs) on linkage group 4. A near-telomeric short-arm BAC probe (zK030C13) and a near-centromeric BAC probe (zC13J14) have been identified for the chromosome corresponding to linkage group 4. Approximately 30 cM of markers correspond to the long arm of this chromosome (indicated to the left).

are embedded in this chromosome region, which could be subsequently used as locus-specific FISH probes.

In humans, most centromeres contain repetitive sequences that are chromosome specific. These DNA sequences are ideal because they provide prominent hybridization signals for straightforward chromosomal identification in metaphase chromosome preparations and accurate chromosomal enumeration in interphase nuclei (Choo, 1997; Lee *et al.*, 1997a). Unfortunately, zebrafish centromeric DNAs characterized to date are similar to those of mouse (e.g., Wong and Rattner, 1988) and other non-primate vertebrate species (e.g., Lee *et al.*, 1997b), in that they do not appear to have chromosome-specific repetitive DNA sequences (Sola and Gornung, 2001). Therefore, chromosomal enumeration probes are alternatively chosen from loci near the primary constriction of each chromosome. The near-centromeric BAC clones chosen for this first generation probe panel have already been successfully used to determine ploidy in interphase cells of *retsina* mutants (Paw *et al.*, 2003) and *mps1*^{zp1} mutants (Poss *et al.*, 2004).

Zebrafish near-telomeric DNA probes can also be used for enumeration studies but are likely more useful in identifying and confirming interchromosomal

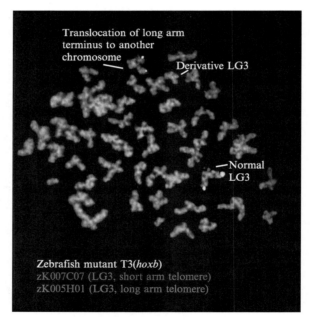

Fig. 3   FISH confirmation of a chromosome translocation in the zebrafish mutant, T3(*hoxb*). A two-colored FISH experiment shows syntenic hybridization of a near-telomeric short-arm probe and a near-telomeric long-arm probe on one normal LG3 chromosome. The nonsyntenic hybridization of these same probes to two different chromosomes in this zebrafish mutant is consistent with a chromosomal translocation involving one LG3 chromosome and another nonhomologous chromosome. (See Color Insert.)

translocations and terminal deletions (i.e., deletions that extend all the way to one of the chromosome ends). For example, the zebrafish mutant *T3(hoxb)* is thought to have a translocation involving the LG3 chromosome long arm (Fritz *et al.*, 1996). Two color FISH assays with near-telomeric probes for the LG3 short arm and long arm have confirmed the chromosomal rearrangement in this mutant (Fig. 3).

Both near-telomeric and near-centromeric BAC clones can be used to evaluate genomic instability in cancer and cell cycle mutants. At present, a commonly used indirect method for assessing genomic instability involves the telomeric recessive mutation *golden* (Driever *et al.*, 1996). Zebrafish mutants suspected of having genomic instability are crossed to *golden* heterozygotes. If the mutants promote genomic instability, some of the heterozygous progeny will develop a chromosomal aberration that deletes the one normal *golden* locus in the heterozygotes, leading to a lack of eye and body pigmentation. Unfortunately, this method for assessing genomic instability assesses only one chromosomal locus and therefore conceivably misses genomic instability in certain mutants and underestimates the extent of genomic instability in other mutants. In addition, this method for assessing genomic instability is incompatible with early embryonic lethals. Such limitations are overcome by multicolor FISH experiments that use several BAC probes simultaneously, providing rapid and accurate assessment of genomic instability in zebrafish (e.g., Fig. 4).

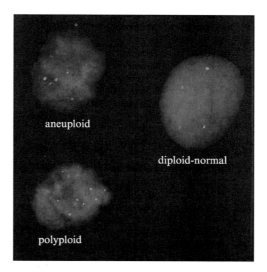

Fig. 4 Use of BAC probes to assess genomic instability. Normal hybridization patterns would show two copies of each BAC probe in a given nucleus (diploid-normal). A polyploid cell would be expected to have the same number of signals for each BAC probe, provided there are three or more signals for each probe in a given nucleus. The polyploid cell shown has four green and four red signals, consistent with a tetraploid cellular content. Aneuploid cells would be expected to have a different number of signals for each BAC probe in a given nucleus. The aneuploid cell shown here has two copies of the green-labeled BAC probe and four copes of the red-labeled probe. (See Color Insert.)

The molecular cytogenetic methods described in this chapter should provide complementary assays for genetic analysis of zebrafish mutants. In addition, the development of a BAC probe panel resource provides an initial cytogenetic framework for the accurate assembly of the zebrafish genome sequence that is independent of existing genetic and radiation hybrid maps. Clearly, accurate assembly of the zebrafish genome will accelerate individual positional cloning projects directed at understanding fundamental and highly conserved molecular mechanisms involved in vertebrate development and cell maintenance (Postlethwait and Talbot, 1997).

References

Amores, A., and Postlethwait, J. H. (1999). Banded chromosomes and the zebrafish karyotype. *Methods Cell Biol.* **60,** 323–338.

Choo, K. H. A. (1997). "The Centromere." Oxford, Oxford University Press.

Driever, W., Solnica-Krezel, L., Schier, A. F., Neuhauss, S. C., Malicki, J., Stemple, D. L., Stainier, D. Y., Zwartkruis, F., Abdelilah, S., Rangini, Z., Belak, J., and Boggs, C. (1996). A genetic screen for mutations affecting embryogenesis in zebrafish. *Development* **123,** 37–46.

Fritz, A., Rozowski, M., *et al.* (1996). Identification of selected gamma-ray induced deficiencies in zebrafish using multiplex polymerase chain reaction. *Genetics* **144**(4), 1735–1745.

Gornung, E., De Innocentiis, S., *et al.* (2000). Zebrafish 5S rRNA genes map to the long arms of chromosome 3. *Chromosome Res.* **8**(4), 362.

John, B. (1988). The biology of heterochromatin. *In* "Heterochromatin: Molecular and Structural aspects" (R. S. Verma, ed.), pp. 1–30. Cambridge, Cambridge University Press.

Hinegardner, R., and Rosen, D. E. (1972). Cellular DNA content and the evolution of the teleostean fishes. *Am. Nat.* **166,** 621–644.

Lee, C., Wevrick, R., Fisher, R. B., Ferguson-Smith, M. A., and Lin, C. C. (1997a). Human centromeric DNAs. *Hum. Genet.* **100,** 291–304.

Lee, C., Court, D. R., *et al.* (1997b). Higher-order organization of subrepeats and the evolution of cervid satellite I DNA. *J. Mol. Evol.* **44**(3), 327–335.

Paw, B. H., Davidson, A. J., *et al.* (2003). Cell-specific mitotic defect and dyserythropoiesis associated with erythroid band 3 deficiency. *Nat. Genet.* **34**(1), 59–64.

Phillips, R. B., and Reed, K. M. (2000). Localization of repetitive DNAs to zebrafish (*Danio rerio*) chromosomes by fluorescence in situ hybridization (FISH). *Chromosome Res.* **8**(1), 27–35.

Poss, K. D., Nechiporuk, A., Stringer, K. F., Lee, C., and Keating, M. T. (2004). Germ cell aneuploidy in zebrafish with mutations in the mitotic checkpoint gene mps1. *Genes Dev.* **18,** 1527–1532.

Postlethwait, J. H., and Talbot, W. S. (1997). Zebrafish genomics: from mutants to genes. *Trends Genet.* **13**(5), 183–190.

Sola, L., and Gornung, E. (2001). Classical and molecular cytogenetics of the zebrafish, *Danio rerio* (Cyprinidae, Cypriniformes): an overview. *Genetica* **111**(1–3), 397–412.

Westerfield, M. (2000). "The Zebrafish Book. A Guide for the Laboratory Use of Zebrafish (*Danio rerio*)," 4th ed. University of Oregon Press, Eugene, OR.

Wong, A. K., and Rattner, J. B. (1988). Sequence organization and cytological localization of the minor satellite of mouse. *Nucleic Acids Res.* **16**(24), 11645–11661.

CHAPTER 14

Automated Analysis of Conserved Syntenies for the Zebrafish Genome

John Postlethwait,* Victor Ruotti,[†] Michael J. Carvan,[‡] and Peter J. Tonellato[†]

*Institute of Neuroscience
University of Oregon
Eugene, Oregon 97403

[†]Human & Molecular Genetics Center
Medical College of Wisconsin
Milwaukee, Wisconsin 53226

[‡]Great Lakes WATER Institute
University of Wisconsin–Milwaukee
Milwaukee, Wisconsin 5320

I. Introduction

To understand the origin of a large variety of biological processes and to optimize the utility of model organisms, researchers seek to maximize connections among species. Biological and functionally useful connections among genomes are especially important because genomes provide a record of change over time that documents, underlies, and directs those elements that are conserved among species and those factors that account for each species' unique qualities. A key question for connecting zebrafish to the biology of humans and other mammals is: To what degree are gene structure and gene orders conserved between zebrafish and mammalian genomes? In this chapter, we first discuss general concepts regarding conservation of syntenies and then present several examples connecting zebrafish biology with human biology. Finally, we discuss the progress and problems in automated analysis of conserved syntenies between zebrafish and mammalian genomes.

II. The Conservation of Zebrafish and Mammalian Genomes

Investigation of the conservation of genomes occurs at several levels.

A. Orthologs

Investigating genome conservation begins by identifying orthologs. Orthologs are pairs of genetic elements, one in each of two different species that are descended from a single genetic element in the last common ancestor of the two species. Orthology involves only evolutionary history and not gene function. Thus, phylogenetic analysis is key to the identification of orthologs. Because orthology depends on where a gene comes from and not what the gene does, the term *functional ortholog* can be misleading. Orthologies of very ancient or rapidly evolving genetic elements or gene duplication events can sometimes present difficulties for analysis because of ambiguity in remaining phylogenetic signal (i.e., compare Robinson-Rechavi *et al.*, 2001 and Van de Peer *et al.*, 2002).

B. Syntenies

Two loci are syntenic ("syn," same; "tene," thread) if they occupy the same chromosome, the same DNA thread. Mammalian geneticists originally used the term for somatic genetic investigations (Kucherlapati and Ruddle, 1975). In a meiotic mapping cross, two genes can segregate independently according to Mendel's second law, but the genes can still be syntenic if they are located distantly on the same chromosome. According to these concepts, a gene in one species cannot be syntenic with a gene in another species, because two genes in two different species clearly cannot be on the same chromosome. Nevertheless, one often hears people speak inappropriately of a gene in zebrafish being syntenic with a human gene.

C. Shared Syntenies

If two genes occupy a single chromosome in one species and the orthologs of those two genes inhabit a single chromosome in a second species, these pairs of orthologs show a shared synteny. For example, *hoxb1a* and *dlx4a* on linkage group (LG) 3 of zebrafish and *HOXB1* and *DLX4* on chromosome 17 (Hsa17) in humans represent a shared synteny.

D. Conserved Syntenies

A subset of shared syntenies will be conserved syntenies, cases in which the shared synteny exists because in the last common ancestor of the two species had the ancestors of these two pairs of orthologous genes resided on a single chromosome. An alternative explanation for a shared synteny is that the last common ancestor of the two compared species had the orthologs of the two genes under discussion on two different chromosomes; then, independently in the two lineages under discussion, chromosome translocations by chance brought the two genes together onto a single chromosome. Because zebrafish and humans each have a large number of chromosomes, it is unlikely that shared syntenies will occur very often by chance translocations. Instead, shared syntenies will usually be conserved syntenies. If three or more pairs of orthologs share syntenies, it becomes highly unlikely that the shared synteny is because of chromosome rearrangements, but increasingly more evident that it is because of conservation of syntenies.

E. Conserved Chromosome Segments

Conserved syntenies involve whole chromosomes; therefore, two pairs of orthologs could exhibit a conserved synteny even if they are close together on a chromosome in one species and located far from each other but still on the same chromosome in the other species. A finer level of genome comparison is the conserved chromosome segment, cases in which all genes in a portion of a chromosome in one species have orthologs in a single chromosome segment in another species. An ortholog or two might be missing from the conserved chromosome segment in one or the other species, and the order of orthologs might be different, but if no genes from outside the segment are present, this is a conserved chromosome segment.

F. Conserved Gene Orders

The most rigorous level of long-range genome conservation is the conservation of gene orders within conserved chromosome segments. In a segment with conserved gene orders, three or more orthologs are aligned in the same order and are transcribed in the same direction. Chromosome segments with conserved gene order in two species have been inherited without rearrangement from the last common ancestor of the two species.

Fundamentally, these sequence, chromosome, and evolutionary-based connections provide insight into the genomic relationships, function, and diversity among organisms. The first question is: How frequently does the zebrafish genome conserve orthologies, syntenies, chromosome segments, and gene orders with other vertebrates, especially humans and other mammals? Here we report progress in the automation of the analysis of conserved syntenies between zebrafish and mammals.

III. Using Genome Conservation in Zebrafish Research

Identifying genome conservation at various levels helps facilitate zebrafish research in several ways.

A. Orthologs

Because orthologous genes are descended from a common ancestral gene, many of their functions are often conserved among species. Thus, a function demonstrated for a zebrafish gene provides a hypothesis for the function of that gene's ortholog in other species. For instance, consider *ferroportin 1*, a gene whose function was shown by mutation in zebrafish to involve transport of iron from the yolk sac to the embryo (Donovan *et al.*, 2000). This finding made the human ortholog a candidate for a previously unidentified iron transporter, and the revelation from the zebrafish mutant led to the discovery that the human gene is mutated in hemochromatosis disease patients (Wallace *et al.*, 2002).

B. Conserved Syntenies

The genomewide organization of conserved syntenies reveals the history of genome change since the divergence of the two compared species. Global conserved syntenies reveal the evolutionary pattern of chromosome rearrangements, the duplication of genes and chromosome segments, and the origin of polyploidization events. The organization of conserved syntenies was crucial in the discovery that a genome duplication event occurred in the zebrafish lineage (Amores *et al.*, 1998; Barbazuk *et al.*, 2000; Postlethwait *et al.*, 1998; Woods, 2000), and that it was likely shared by all Euteleost fish (Amores *et al.*, 2004; Meyer and Schartl, 1999; Naruse *et al.*, 2004; Taylor *et al.*, 2001, 2003; Van de Peer *et al.*, 2003; Vandepoele *et al.*, 2004; Vogel, 1998; Wittbrodt *et al.*, 1998).

Comparing conserved syntenies in two duplicated zebrafish chromosomes reveals chromosome rearrangements that occurred before and after the genome duplication event. For example, LG3 and LG12 are duplicates, containing duplicated regions orthologous to most of human chromosome (Hsa) 17 and portions of Hsa16, 19, and 22. In addition, LG12 has a large portion orthologous to the long arm of Hsa10 (Hsa10q), but the duplicate copy of Hsa10q is on LG13. Thus, the Hsa10q ortholog was either attached to the parent chromosome of LG3 and LG12 and subsequently translocated away from LG3 to LG13, or was originally attached to another chromosome and translocated to LG12 (Postlethwait *et al.*, 1998, 2000; Woods, 2000). Interestingly, medaka has orthologs of most zebrafish chromosomes, including LG3 and LG12, and the medaka LG12 ortholog also contains a copy of Hsa10q material (Naruse *et al.*, 2004). This shows that the chromosome rearrangement occurred after the genome duplication but before the divergence of zebrafish and medaka lineages.

Conserved syntenies can help arbitrate ambiguities in ortholog assignments arising from phylogenetic analysis of genes with ambiguous or confusing phylogenetic signal. For example, phylogenetic analysis did not fully clarify the origin of the *EVX* family gene *eve1*, but conserved synteny analysis showed that it originated at the time of the original vertebrate genome expansion and was later lost in the mammalian lineage, but was maintained in the zebrafish lineage (Joly *et al.*, 1993; Postlethwait *et al.*, 1998). This analysis showed that *eve1* has no human ortholog.

C. Conserved Chromosome Segments with Conserved Gene Orders

Identifying segments with conserved gene orders can facilitate the cloning of mutants and the identification of candidates for genetic regulatory elements. If in a positional cloning project, a marker closely linked to a phenotypic locus resides in a region with substantial conserved synteny with mammals or in a segment with conserved gene order, then the mammalian genes can become candidates for the gene disrupted by the mutation (i.e., Burgess *et al.*, 2002; Iovine and Johnson, 2002; Jensen and Westerfield, 2001; Katoh, 2003; Yoder and Litman, 2000).

Within a conserved chromosome segment with conserved gene orders, the regions between adjacent genes should have been uninterrupted by chromosome rearrangement. Because regulatory elements frequently occupy these intergenic regions as well as introns, researchers can look confidently at the flanks of genes embedded within conserved gene orders for conserved nongenic sequences (Aparicio *et al.*, 1995; Dermitzakis *et al.*, 2003, 2004; Goode *et al.*, 2003; Lettice *et al.*, 2003; Nolte *et al.*, 2003). These conserved nongenic sequences will be candidates for genetic regulatory elements. If, however, the human orthologs of an adjacent trio of zebrafish genes are not also adjacent, then the intergenic region of the middle zebrafish gene might not represent the complete orthologous region compared to the human, and the identification of conserved, potential regulatory regions becomes more problematic.

IV. Genomic Connections Between Zebrafish and Tetrapods Facilitate Disease Research

Both basic and model organism research in zebrafish can benefit from the information provided by conserved genomic evidence. Zebrafish are more genetically accessible than other vertebrate model organisms and can therefore be used to dissect molecular pathways and test the functions of numerous candidate genes related to specific phenotypes. Knowledge of conserved syntenic relationships between zebrafish and regions containing quantitative trait loci (QTLs) identified in other organisms can allow the systematic evaluation of the function of each candidate gene by overexpression or morpholino knockdown. The following examples describe the power of analyzing homologous genes in zebrafish and other organisms to determine gene functions and develop hypotheses as to their functions in other vertebrates. We also provide an example of a homologous gene in which gene function is not conserved, illustrating the limitations of using sequence comparison alone to assign functional homology.

A number of zebrafish genes with known function provide hypotheses of similar function of homologous genes from other organisms. Two examples are *deafness autosomal dominant 5* (*dfna5*) and *heart of glass* (*heg*). Defects in the *DFNA5* gene can result in the production of a truncated DFNA5 protein and lead to a nonsyndromic autosomal dominant form of hearing loss in humans (De Leenheer *et al.*, 2002; Van Laer *et al.*, 1998). The *DFNA5* protein shares no obvious homology to other known proteins and, because there had been no adequate animal model, its normal function was unknown prior to its functional analysis in zebrafish. The zebrafish *dfna5* cDNA was cloned by BLAST (Altschul *et al.*, 1997) searches, using the human protein against EST databases (Busch-Nentwich *et al.*, 2004). *In situ* hybridization experiments showed that zebrafish *dfna5* is ubiquitously expressed prior to 20 h postfertilization (hpf) and is expressed in the ear (predominantly in the projections of the developing semicircular canals) between 48 and 72 hpf. Morpholino antisense oligonucleotides targeted to recapitulate the loss of exon 8 as seen in human *DFNA5* mutations associated with hearing loss lead to disruption of the epithelial monolayer and epithelial basement membrane in the developing semicircular canals (Busch-Nentwich *et al.*, 2004). The anatomy of the vestibular inner ear, and many aspects of the development and morphogenesis of semicircular canals, is highly conserved among vertebrates (Riley and Phillips, 2003). It is highly likely that further investigation into the function of *dfna5* in zebrafish will lead to a better understanding of its role in human hearing loss.

The *heg* mutation appeared in a large-scale mutagenic screen as a recessive embryonic lethal mutation with cardiac growth abnormalities leading to an enlarged heart with no apparent extracardiac defects (Mably *et al.*, 2003). The *heg* gene was identified by positional cloning and shown to encode a protein of 977 amino acids with a predicted C-terminal membrane-spanning domain, a glycosylated extracellular domain containing two EGF repeats, and an N terminus containing a putative signal peptide with cleavage site (Mably *et al.*, 2003). Three alternative splice variants were isolated by RT-PCR; the two smaller transcripts result in peptides that lack the membrane-spanning domain, resulting in soluble proteins. The smallest transcript also lacks the C-terminal peptide domains. The most abundant transcript encodes the full-length transmembrane form of *heg* and was shown to be critical for normal growth patterning of the heart by specifically targeted morpholinos. Expression of *heg* is primarily restricted to the endocardium and presumably signals from the endocardium to the myocardium, regulating global patterning of the heart without affecting cell number (Mably *et al.*, 2003).

A number of genes from other organisms demonstrate homology with zebrafish and provide insight into the likely function of the zebrafish gene, including *nuclear factor erythroid 2* (*nrf2*) a member of the Cap 'n' Collar family of transcription factors with a basic region leucine zipper structure, and *kelch-like ECH-associated protein 1* (*keap1*), a cytoskeleton-associated protein (Itoh *et al.*, 1999). It has been shown in knockout mice that regulation of oxidative stress response genes through electrophile response elements (EPREs, also known as antioxidant response elements) is dependent on both *Nrf2* (Itoh *et al.*, 1997) and *Keap1*, which sequesters Nrf2 to the cytoplasm under normal redox conditions (Itoh *et al.*, 1999; Kwak *et al.*, 2002). Dissociation of Nrf2 from Keap1 is apparently regulated by protein kinase C phosphorylation of specific residues on Nrf2 and by direct interactions between electrophiles and the sulfhydryl groups on Keap1 (Dinkova-Kostova *et al.*, 2002; Huang *et al.*, 2002). The transcriptional activity of Nrf2 also appears to be influenced by a number of other coactivators that presumably bind to EPRE-specific transcription factors and enhance transcription of the oxidative stress responsive genes by enabling chromatin remodeling (Zhu and Fahl, 2001). The balance between Nrf2 and Keap1 is critical to maintain normal gene expression levels. Overexpression of Nrf2 will activate gene expression through EPRE sequences in the absence of inducer, and overexpression of Keap1 will suppress normal oxidative stress-responsive gene induction in murine cell cultures (Itoh *et al.*, 1999).

Kobayashi *et al.* (2002) provided elegant molecular and genetic support for the function of the Nrf2–Keap1 system in zebrafish. They verified that several genes known to be regulated by EPRE sequences in humans and rodents are induced by *t*-butylhydroquinone (tBHQ). tBHQ has previously been shown to activate EPRE sequences in cultured zebrafish cells (Carvan *et al.*, 2000, 2001). Morpholino studies demonstrate that gene induction through EPRE sequences in zebrafish larvae is dependent on both nrf2 and keap1 and that the zebrafish Nrf2 and Keap1 proteins interact with each other in a yeast two-hybrid assay (Kobayashi *et al.*, 2002). The functional protein domains of the zebrafish Nrf2 and Keap1 are well conserved, showing 45–70% identity in the deduced amino acid sequence to the mouse counterparts, and mutation of specific residues in the ETGE (Glu-Thr-Gly-Glu) motif of murine Nrf2 or the double glycine repeat (DGR) motif of murine Keap1 abolished their interaction and disrupted normal EPRE-mediated gene regulation in cultured mouse cells. Zebrafish Nrf1 and Keap1 (and their respective ETGE or DGR mutants) are also able to disrupt normal gene regulation *in vivo*. Overexpression of Nrf2 in zebrafish larvae induces gene expression and addition of overexpressed Keap1 returns expression to normal. Zebrafish Nrf1 with a mutation in the ETGE motif do not induce gene expression when overexpressed *in vivo*, and overexpressed Keap1 with a mutation in the DGR motif do not suppress gene expression induced by excessive Nrf2 *in vivo* (Kobayashi *et al.*, 2002). The molecular dissection of the Nrf2–Keap1 interactions by using a combination of knockout mice, cultured murine cells, yeast two-hybrid, and zebrafish demonstrates the accessibility of zebrafish to genetic manipulation and the evolutionary conservation of this transcriptional regulation system.

In a number of instances, gene homology does not equate to identical function. One excellent example is the aromatic hydrocarbon receptor (*Ahr*), which codes for a ligand-activated transcription factor and is duplicated in the zebrafish genome as *ahr1* and *ahr2*. The most potent ligand for Ahr in virtually all species examined thus far is 2,3,7,8-tetrachlorodibenzo-*p*-dioxin (TCDD). The ligand-activated receptor forms a heterodimeric complex with the Ahr nuclear translocator protein that binds aromatic hydrocarbon DNA response elements (AHREs) upstream of numerous genes. The two zebrafish *ahr* cDNAs show high similarity (>60%) in the deduced amino acid sequence to each other and to the human *AHR* (Andreasen *et al.*, 2002; Tanguay *et al.*, 1999). Sequence comparisons show that Ahr1 is more closely related to the mammalian AHRs, and zebrafish Ahr2 is more closely related to other fish Ahr2s (Andreasen *et al.*, 2002). This might be an example like *eve1* cited previously, in which an ancient duplicate is lost in the mammalian lineage. Interestingly, there are striking functional differences between the two zebrafish Ahr proteins. Ahr1 is expressed primarily in the liver, whereas Ahr2 is expressed in nearly all tissues. Abnet *et al.* (1999) revealed that overexpressed zebrafish Ahr2 is able to drive ligand-dependent expression of a luciferase reporter gene regulated by trout AHREs in monkey COS-7 cells. However, the zebrafish Ahr1 is unable to bind ligand (Andreasen *et al.*, 2002) and thus does not directly participate in ligand-activated cell signaling. Expression of *ahr1* and *ahr2* mRNA is detected by RT-PCR at 24 hpf, and both messages are detected throughout development (Abnet *et al.*, 1999; Tanguay *et al.*, 1999). It is likely that the multiple roles of the human AHR are partitioned between the fish Ahr1 and Ahr2 (Hahn, 2001) as predicted by subfunction partitioning (Force *et al.*, 1999). The more ancient functions of Ahr likely involve developmental regulation, whereas the adaptive responses to TCDD were derived more recently (Hahn, 2002). Zebrafish Ahr2 binds dioxin and presumably participates in gene induction through interaction with AHREs in much the same manner as murine and human AHR. The function of Ahr1 in zebrafish is currently unknown, but its lack of TCDD binding suggests that it might provide the ancestral functions of the AHR family without those functions that evolved more recently.

Regions with conserved syntenies among rats, mice, and humans help focus the identification of numerous candidate genes and develop useful laboratory models for human disease. The following are a few examples of candidate genes for human disease, generally identified by linkage analysis, being studied in rodent models to elucidate the mechanisms of disease progression. *Disrupted-in-Schizophrenia* (*Disc1*) is a candidate gene for schizophrenia. Identification of the gene in rodents facilitates the investigation of *Disc1* function and creation of mouse models of *DISC1* disruption. Cross-species analysis reveals conservation of the leucine zipper and coiled-coil domains in *DISC1* orthologs (Ma *et al.*, 2002). The DISC1 protein is found in the brain, heart, liver, and kidney (Ozeki *et al.*, 2003), and yeast two-hybrid analysis reveals association of DISC1 with a number of proteins of the cytoskeleton and centrosome (Morris *et al.*, 2003; Ozeki *et al.*, 2003). Truncation of DISC1 in a region predicted to be deleted in humans with DISC1 translocations results in the loss of interactions with NUDEL and MIPT3 and a reduction in neurite outgrowth when transfected into PC12 cells (Miyoshi *et al.*, 2003; Morris *et al.*, 2003; Ozeki *et al.*, 2003). Use of whole animal models carrying either disruptions in DISC1 or mutations similar to those associated with schizophrenia in humans will lead to a much more complete understanding of its role in normal biology and disease progression.

Coumarin derivatives such as warfarin target blood coagulation by inhibiting the vitamin K epoxide reductase multiprotein complex (VKOR; Rost *et al.*, 2004). A number of heritable blood disorders in humans have been linked to alteration of the VKOR complex, including combined deficiency of vitamin-K-dependent clotting factors type 2, resistance to coumarin-type anticoagulant drugs, and Familial multiple coagulation factor deficiency (Fregin *et al.*, 2002; Rost *et al.*, 2004). The human disorders map to a region of human chromosome 16, with homologous regions in rats and mice that are linked to warfarin resistance. The vitamin K epoxide reductase complex subunit 1 (VKORC1) gene was cloned by using linkage information from both humans and rodents and was shown to code for a small transmembrane protein of the endoplasmic reticulum. VKORC1 contains missense mutations in both human disorders and in a warfarin-resistant rat strain. Overexpression of wild-type VKORC1, but not VKORC1 carrying the VKCFD2 mutation, leads to a marked increase in VKOR activity, which is sensitive to warfarin inhibition (Rost *et al.*, 2004). The long history of warfarin use as an anticoagulant in both humans and rodents has resulted in a substantial biochemical knowledge base from which to rapidly understand the role of VKORC1, the mechanistic basis for the heritable disorders associated with its alteration, and the components of the multiprotein VKOR complex.

The examples discussed in this section show the power of conserved synteny in identifying candidate disease genes within known disease loci how analyzing their function in laboratory animals further focuses the gene list, and how going back to human populations looking for correlations between gene sequence changes can contribute to understanding human disease.

V. Bioinformatic Approaches to Automating Zebrafish Connections and Comparative Maps

The automated identification of conserved genomic regions is the basis for creating comprehensive comparative maps among zebrafish, other model organisms, and humans. Most of the results presented previously were derived from analysis of individual candidate genes, a time consuming and tedious task. Our goal is to rapidly and comprehensively identify all homologies and conserved syntenic regions between multiple organisms. We automate the process of identifying homologues and conserved syntenic regions by adapting the fundamental definitions presented in Section I to algorithmic analysis of appropriate data sets. As in all analyses, the automation, algorithmic analysis, and results are only as valuable as the data provided. Consequently, great care must be taken to minimize simultaneously type I and type II errors (false-positive and false-negative data, respectively). In addition, the algorithms must accurately model biological phenomena and take account of the peculiarities that exist between organisms.

Automation has been of value to a wide spectrum of gene, genome, and genome comparative analysis. Early studies were based on pure sequence alignment and included simple DNA base-pair comparison between sequences. Results of these alignments were originally used to align separate sequence-calling runs from single bacterial artificial chromosome (BAC) sequence reads useful for quality control and full-length BAC sequence construction. More sophisticated methods created platforms for full-genome builds, single nucleotide polymorphism (SNP) discovery, organism–organism comparisons, homology detection, and comparative map construction. Once accomplished with fidelity and robustness, the automation, coupled with visualization and data mining techniques, creates a Web-based environment that a broad community of investigators can use with ease. In addition, the automated system can be updated as new data become available.

A. Automated Map Development and Comparisons

The comparative map development process presented here was used to create comparison maps of human–mouse and rat (Kwitek-Black and Jacob, 2001) and to create a refined comparative disease gene hunting environment based on improved data and analysis (Twigger *et al.*, 2004). Details of the methodology, database, and environment are presented in those two papers. Consequently, only a brief review is presented here along with modifications and preliminary results associated with zebrafish analysis.

The first step in the analysis is to develop correct alignment criteria for testing pairs of DNA sequences for potential homology. To do so, 1000 UniGene clusters from humans, mice, and rats and 45 unigenes selected from a collection of novel zebrafish–human homologous genes were tested by using the gapped BLAST program (Altschul *et al.*, 1997). Common orthologs were selected between each species (100 for mouse–human, rat–human, and rat–mouse, 27 for zebrafish–human).

Potential confounding issues were addressed by including 10 putative paralogous genes, each corresponding to 1 of the 100 orthologous genes. Another 890 randomly chosen UniGenes not found in the selected genes were also tested. BLAST parameters that minimized false-positive and maximized true-positive predictions as well as provided consistency between previous results (Makalowski and Boguski, 1998; HomoloGene) were determined to be 85% over a 100-bp (ungapped) stretch.

Construction of the comparative maps of zebrafish and rats, mice, and humans was accomplished by aligning all ESTs sequence identity by using the optimal parameters. A compression and scoring algorithm (for complete description see Kwitek-Black and Jacob, 2001) was then applied to predict homologous pairs between each organism. The algorithm demonstrated a 91% accuracy for predicting the known orthologs (based on the 1027 gene test set). All one-to-one homologous UniGenes were then used to construct the comparative maps. The most recent release of the NCBI UniGene builds (Table I) was incorporated into the analysis. The approach has been used to create a publicly available resource located at the Rat Genome Database (www.rgd.mcw.edu). A total of 3247 homologies across humans, mice, and rats are currently integrated into the comparative map. A similar resource for zebrafish investigators is the goal of this investigation.

One advantage of creating a computer-based comparative map is the ability to virtually map previously unmapped genes and ESTs. This is accomplished by identifying conserved segments between two organisms. If two UniGenes lie within an uninterrupted conserved segment in one species, additional one-to-one homologous UniGenes between those flanking markers are virtually mapped, based on the map position of the homolog in the other species. If a UniGene defines a potential evolutionary breakpoint, additional one-to-one homologous UniGenes can be predicted upstream and/or downstream of that marker. In this case, homologous UniGenes directly upstream or downstream (depending on which end of the conserved segment is considered) of the UniGene flanking the breakpoint are identified and prioritized for wet-lab mapping to either confirm a segment defined by a single anchor or extend and better define the evolutionary breakpoint. Predictions were made for all three species' backbones as described previously for rat.

Table I

NCBI UniGene Builds Used in Automated Map Construction, Including Number of Sequences and UniGene Clusters in Each Organism

NCBI UniGene	Build	Number of UniGenes	Number of sequences
Human	148	95,928	3,115,711
Mouse	104	83,530	2,268,927
Rat	100	59,882	308,877
Zebrafish	29	14,893	159,261

B. Adjustments for Zebrafish

Both the relational clustering algorithm and the scoring methods described previously are applicable to the zebrafish genome. However, modifications to the analysis and to the algorithm must be developed further to compensate for the history of the zebrafish genome, in particular, those issues reviewed in Section II. Of particular importance is the major genome duplication events resulting from evolutionary readjustments of the genome. Gene duplication events add unique complexity to the automatic identification and resolution of the so-called one-to-one and many-to-many candidate homologues assigned by the analysis. Organisms without large duplicative events are less likely to provide complex candidate multialigned gene identifications. In the analysis of zebrafish against humans, mice, and rats, this complexity has hindered the unambiguous assignment of homologies and complicated the alignment of syntenic regions. In addition, data that support the comparative analysis described previously do not necessarily accurately reflect the duplicative events. Future efforts will provide the refinements required to create an automated algorithm that more accurately identifies and assigns these homologies associated with duplication.

The scoring algorithm that identifies the unique one-to-one homologies was created to obtain the best alignment from many-to-many relationships when rat, human, and mouse comparative maps were constructed. Taking into account the duplication events that took place in zebrafish, a modification of the scoring algorithm will allow us to identify duplicated genes in the zebrafish genome based on the clusters created by this algorithm. These duplicated genes could then be compared with the list of known duplicated genes. In addition, we will better identify syntenic regions in the zebrafish genome that are highly conserved in humans, mice, and rats.

VI. A Conserved Syntenic Map for Zebrafish: Preliminary Results

Preliminary analysis of the entire UniGene data sets with mapped information (Table II) was conducted between zebrafish and humans to test and further develop the automated algorithm. In total, 785 homologies were detected between zebrafish and humans, about 35% of which were mapped on one or the other organism. Figure 1 shows a view of a region of zebrafish chromosome 7 identified by the automated process showing conserved regions to rat chr 1 (Rno1) and human chr 11 (Hsa11). The region contains the pyruvate carboxylase gene. This gene maps in Hsa11 and occupies a region of conserved synteny on LG7 along with several other well-characterized genes (*fth1, slc3a2, men1, fgf3,* and *cycd1*) in zebrafish (Yoder and Litman, 2000). In addition to confirming this region between zebrafish and humans, the automated system has identified a conserved segment associated with rat with the same genes.

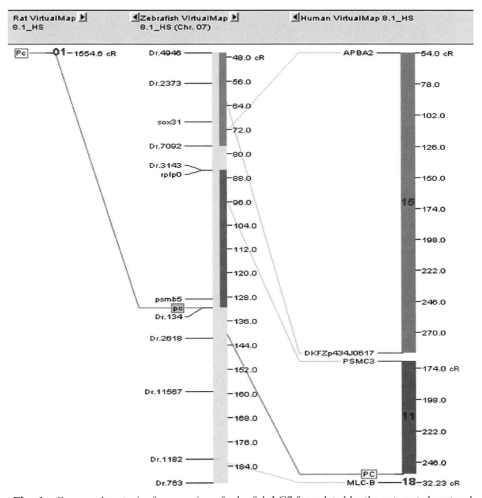

Fig. 1 Conserved syntenies for a portion of zebrafish LG7 formulated by the automated protocol.

VII. Conclusion

Comparative mapping is a powerful approach for systematic creation of auto-matically detected orthologous genes and construction of comparative maps from conserved syntenic regions identified by using common mapped information

Table II
Maps and Number of Mapped Markers Used in Automated Construction

Species	Maps	Number of markers
Rat	MCW RH V.3.2	14,172
Mouse	MIT Genetic R9	13,532
Human	GeneMap99	45,756
Zebrafish	HS Genetic	3,732

across multiple organisms. Automation of the analysis frees investigators to focus on the potential biological and functional interpretation. However, special problems arising from genome duplication require further refinement of the current algorithms before genomewide results become fully available.

Acknowledgments

We thank NIH grants R01RR10715 and P01HD22486 for support.

References

Abnet, C. C., Tanguay, R. L., Heideman, W., and Peterson, R. E. (1999). Transactivation activity of human, zebrafish, and rainbow trout aryl hydrocarbon receptors expressed in COS-7 cells: Greater insight into species differences in toxic potency of polychlorinated dibenzo-*p*-dioxin, dibenzofuran, and biphenyl congeners. *Toxicol. Appl. Pharmacol.* **159,** 41–51.

Altschul, S. F., Madden, T. L., Schaffer, A. A., Zhang, J., Zhang, Z., Miller, W., and Lipman, D. J. (1997). Gapped BLAST and PSI-BLAST: A new generation of protein database search programs. *Nucleic Acids Res.* **25,** 3389–3402.

Amores, A., Force, A., Yan, Y.-L., Joly, L., Amemiya, C., Fritz, A., Ho, R. K., Langeland, J., Prince, V., Wang, Y.-L., Westerfield, M., Ekker, M., and Postlethwait, J. H. (1998). Zebrafish *hox* clusters and vertebrate genome evolution. *Science* **282,** 1711–1714.

Amores, A., Suzuki, T., Yan, Y. L., Pomeroy, J., Singer, A., Amemiya, C., and Postlethwait, J. H. (2004). Developmental roles of pufferfish Hox clusters and genome evolution in ray-fin fish. *Genome Res.* **14,** 1–10.

Andreasen, E. A., Hahn, M. E., Heideman, W., Peterson, R. E., and Tanguay, R. L. (2002). The zebrafish (*Danio rerio*) aryl hydrocarbon receptor type 1 is a novel vertebrate receptor. *Mol. Pharmacol.* **62,** 234–249.

Aparicio, S., Morrison, A., Gould, A., Gilthorpe, J., Chaudhuri, C., Rigby, P., Krumlauf, R., and Brenner, S. (1995). Detecting conserved regulatory elements with the model genome of the Japanese puffer fish, *Fugu rubripes. Proc. Natl. Acad. Sci. USA* **92,** 1684–1688.

Barbazuk, W. B., Korf, I., Kadavi, C., Heyen, J., Tate, S., Wun, E., Bedell, J. A., McPherson, J. D., and Johnson, S. L. (2000). The syntenic relationship of the zebrafish and human genomes. *Genome Res.* **10,** 1351–1358.

Burgess, S., Reim, G., Chen, W., Hopkins, N., and Brand, M. (2002). The zebrafish *spiel-ohne-grenzen* (*spg*) gene encodes the POU domain protein Pou2 related to mammalian *Oct4* and is essential for formation of the midbrain and hindbrain, and for pre-gastrula morphogenesis. *Development* **129,** 905–916.

Busch-Nentwich, E., Sollner, C., Roehl, H., and Nicolson, T. (2004). The deafness gene dfna5 is crucial for ugdh expression and HA production in the developing ear in zebrafish. *Development* **131,** 943–951.

Carvan, M. J., 3rd, Dalton, T. P., Stuart, G. W., and Nebert, D. W. (2000). Transgenic zebrafish as sentinels for aquatic pollution. *Ann. N. Y. Acad. Sci.* **919,** 133–147.

Carvan, M. J., 3rd, Sonntag, D. M., Cmar, C. B., Cook, R. S., Curran, M. A., and Miller, G. L. (2001). Oxidative stress in zebrafish cells: potential utility of transgenic zebrafish as a deployable sentinel for site hazard ranking. *Sci. Total Environ.* **274,** 183–196.

De Leenheer, E. M., van Zuijlen, D. A., Van Laer, L., Van Camp, G., Huygen, P. L., Huizing, E. H., and Cremers, C. W. (2002). Clinical features of DFNA5. *Adv. Otorhinolaryngol.* **61,** 53–59.

Dermitzakis, E. T., Kirkness, E., Schwarz, S., Birney, E., Reymond, A., and Antonarakis, S. E. (2004). Comparison of human chromosome 21 conserved nongenic sequences (CNGs) with the mouse and dog genomes shows that their selective constraint is independent of their genic environment. *Genome Res.* **14,** 852–859.

Dermitzakis, E. T., Reymond, A., Scamuffa, N., Ucla, C., Kirkness, E., Rossier, C., and Antonarakis, S. E. (2003). Evolutionary discrimination of mammalian conserved non-genic sequences (CNGs). *Science* **302,** 1033–1035.

Dinkova-Kostova, A. T., Holtzclaw, W. D., Cole, R. N., Itoh, K., Wakabayashi, N., Katoh, Y., Yamamoto, M., and Talalay, P. (2002). Direct evidence that sulfhydryl groups of Keap1 are the sensors regulating induction of phase 2 enzymes that protect against carcinogens and oxidants. *Proc. Natl. Acad. Sci. USA* **99,** 11908–11913.

Donovan, A., Brownlie, A., Zhou, Y., Shepard, J., Pratt, S. J., Moynihan, J., Paw, B. H., Drejer, A., Barut, B., Zapata, A., Law, T. C., Brugnara, C., Lux, S. E., Pinkus, G. S., Pinkus, J. L., Kingsley, P. D., Palis, J., Fleming, M. D., Andrews, N. C., and Zon, L. I. (2000). Positional cloning of zebrafish ferroportin 1 identifies a conserved vertebrate iron exporter. *Nature* **403,** 776–781.

Force, A., Lynch, M., Pickett, F. B., Amores, A., Yan, Y.-L., and Postlethwait, J. (1999). Preservation of duplicate genes by complementary, degenerative mutations. *Genetics* **151,** 1531–1545.

Fregin, A., Rost, S., Wolz, W., Krebsova, A., Muller, C. R., and Oldenburg, J. (2002). Homozygosity mapping of a second gene locus for hereditary combined deficiency of vitamin K-dependent clotting factors to the centromeric region of chromosome 16. *Blood* **100,** 3229–3232.

Goode, D. K., Snell, P. K., and Elgar, G. K. (2003). Comparative analysis of vertebrate Shh genes identifies novel conserved non-coding sequence. *Mamm. Genome* **14,** 192–201.

Hahn, M. E. (2001). Dioxin toxicology and the aryl hydrocarbon receptor: Insights from fish and other non-traditional models. *Marine Biotech.* **3,** S224–S238.

Hahn, M. E. (2002). Aryl hydrocarbon receptors: diversity and evolution. *Chem. Biol. Interact.* **141,** 131–160.

Huang, H. C., Nguyen, T., and Pickett, C. B. (2002). Phosphorylation of Nrf2 at Ser-40 by protein kinase C regulates antioxidant response element-mediated transcription. *J. Biol. Chem.* **277,** 42769–42774.

Iovine, M. K., and Johnson, S. L. (2002). A genetic, deletion, physical, and human homology map of the long fin region on zebrafish linkage group 2. *Genomics* **79,** 756–759.

Itoh, K., Chiba, T., Takahashi, S., Ishii, T., Igarashi, K., Katoh, Y., Oyake, T., Hayashi, N., Satoh, K., Hatayama, I., Yamamoto, M., and Nabeshima, Y. (1997). An Nrf2/small Maf heterodimer mediates the induction of phase II detoxifying enzyme genes through antioxidant response elements. *Biochem. Biophys. Res. Commun.* **236,** 313–322.

Itoh, K., Wakabayashi, N., Katoh, Y., Ishii, T., Igarashi, K., Engel, J. D., and Yamamoto, M. (1999). Keap1 represses nuclear activation of antioxidant responsive elements by Nrf2 through binding to the amino-terminal Neh2 domain. *Genes Dev.* **13,** 76–86.

Jensen, A. M., and Westerfield, M. (2001). Zebrafish *mosaic eyes* gene is required for tight junction formation in the retinal pigmented epithelium. *Dev. Biol.* **235,** 181(#46).

Joly, J.-S., Joly, C., Schulte-Merker, S., Boulekbache, H., and Condamine, H. (1993). The ventral and posterior expression of the zebrafish homeobox gene eve1 is perturbed in dorsalized and mutant embryos. *Development* **119,** 1261–1275.

Katoh, M. (2003). Evolutionary conservation of CCND1-ORAOV1-FGF19-FGF4 locus from zebrafish to human. *Int. J. Mol. Med.* **12,** 45–50.

Kobayashi, M., Itoh, K., Suzuki, T., Osanai, H., Nishikawa, K., Katoh, Y., Takagi, Y., and Yamamoto, M. (2002). Identification of the interactive interface and phylogenic conservation of the Nrf2-Keap1 system. *Genes Cells* **7,** 807–820.

Kucherlapati, R. S., and Ruddle, F. H. (1975). Mammalian somatic hybrids and human gene mapping. *Ann. Intern. Med.* **83,** 553–560.

Kwak, M. K., Itoh, K., Yamamoto, M., and Kensler, T. W. (2002). Enhanced expression of the transcription factor Nrf2 by cancer chemopreventive agents: Role of antioxidant response element-like sequences in the nrf2 promoter. *Mol. Cell. Biol.* **22,** 2883–2892.

Kwitek-Black, A. E., and Jacob, H. J. (2001). The use of designer rats in the genetic dissection of hypertension. *Curr. Hypertens. Rep.* **3,** 12–18.

Lettice, L. A., Heaney, S. J., Purdie, L. A., Li, L., de Beer, P., Oostra, B. A., Goode, D., Elgar, G., Hill, R. E., and de Graaff, E. (2003). A long-range Shh enhancer regulates expression in the developing limb and fin and is associated with preaxial polydactyly. *Hum. Mol. Genet.* **12,** 1725–1735.

Ma, L., Liu, Y., Ky, B., Shughrue, P. J., Austin, C. P., and Morris, J. A. (2002). Cloning and characterization of Disc1, the mouse ortholog of DISC1 (Disrupted-in-Schizophrenia 1). *Genomics* **80,** 662–672.

Mably, J. D., Mohideen, M. A., Burns, C. G., Chen, J. N., and Fishman, M. C. (2003). heart of glass regulates the concentric growth of the heart in zebrafish. *Curr. Biol.* **13,** 2138–2147.

Makalowski, W., and Boguski, M. S. (1998). Evolutionary parameters of the transcribed mammalian genome: An analysis of 2820 orthologous rodent and human sequences. *Proc. Natl. Acad. Sci. USA* **95,** 9407–9412.

Meyer, A., and Schartl, M. (1999). Gene and genome duplications in vertebrates: The one-to-four (-to-eight in fish) rule and the evolution of novel gene functions. *Curr. Opin. Cell Biol.* **11,** 699–704.

Miyoshi, K., Honda, A., Baba, K., Taniguchi, M., Oono, K., Fujita, T., Kuroda, S., Katayama, T., and Tohyama, M. (2003). Disrupted-In-Schizophrenia 1, a candidate gene for schizophrenia, participates in neurite outgrowth. *Mol. Psychiat.* **8,** 685–694.

Morris, J. A., Kandpal, G., Ma, L., and Austin, C. P. (2003). DISC1 (Disrupted-In-Schizophrenia 1) is a centrosome-associated protein that interacts with MAP1A, MIPT3, ATF4/5 and NUDEL: Regulation and loss of interaction with mutation. *Hum. Mol. Genet.* **12,** 1591–1608.

Naruse, K., Tanaka, M., Mita, K., Shima, A., Postlethwait, J., and Mitani, H. (2004). A medaka gene map: The trace of ancestral vertebrate proto-chromosomes revealed by comparative gene mapping. *Genome Res.* **14,** 820–828.

Nolte, C., Amores, A., Nagy Kovacs, E., Postlethwait, J., and Featherstone, M. (2003). The role of retinoic acid response element in establishing the anterior neural expression border of Hoxd4 transgenes. *Mech. Dev.* **120,** 325–335.

Ozeki, Y., Tomoda, T., Kleiderlein, J., Kamiya, A., Bord, L., Fujii, K., Okawa, M., Yamada, N., Hatten, M. E., Snyder, S. H., Ross, C. A., and Sawa, A. (2003). Disrupted-in-Schizophrenia-1 (DISC-1): Mutant truncation prevents binding to NudE-like (NUDEL) and inhibits neurite outgrowth. *Proc. Natl. Acad. Sci. USA* **100,** 289–294.

Postlethwait, J., Yan, Y., Gates, M., Horne, S., Amores, A., Brownlie, A., Donovan, A., Egan, E., Force, A., Gong, Z., Goutel, C., Fritz, A., Kelsh, R., Knapik, E., Liao, E., Paw, B., Ransom, D., Singer, A., Thomson, M., Abduljabbar, T., Yelick, P., Beier, D., Joly, J., Larhammar, D., Rosa, F., Westerfield, M., Zon, L., Johnston, S., and Talbot, W. (1998). Vertebrate genome evolution and the zebrafish gene map. *Nat. Genet.* **18,** 345–349.

Postlethwait, J. H., Woods, I. G., Ngo-Hazelett, P., Yan, Y.-L., Kelly, P. D., Chu, F., Huang, H., Hill-Force, A., and Talbot, W. S. (2000). Zebrafish comparative genomics and the origins of vertebrate chromosomes. *Genome Res.* **10,** 1890–1902.

Riley, B. B., and Phillips, B. T. (2003). Ringing in the new ear: Resolution of cell interactions in otic development. *Dev. Biol.* **261,** 289–312.

Robinson-Rechavi, M., Marchand, O., Escriva, H., and Laudet, V. (2001). An ancestral whole-genome duplication may not have been responsible for the abundance of duplicated fish genes. *Curr. Biol.* **11,** R458–R459.

Rost, S., Fregin, A., Ivaskevicius, V., Conzelmann, E., Hortnagel, K., Pelz, H. J., Lappegard, K., Seifried, E., Scharrer, I., Tuddenham, E. G., Muller, C. R., Strom, T. M., and Oldenburg, J. (2004). Mutations in VKORC1 cause warfarin resistance and multiple coagulation factor deficiency type 2. *Nature* **427,** 537–541.

Tanguay, R. L., Abnet, C. C., Heideman, W., and Peterson, R. E. (1999). Cloning and characterization of the zebrafish (*Danio rerio*) aryl hydrocarbon receptor. *Biochim. Biophys. Acta* **1444,** 35–48.

Taylor, J., Braasch, I., Frickey, T., Meyer, A., and Van De Peer, Y. (2003). Genome duplication, a trait shared by 22,000 species of ray-finned fish. *Genome Res.* **13,** 382–390.

Taylor, J. S., Van de Peer, Y., and Meyer, A. (2001). Revisiting recent challenges to the ancient fish-specific genome duplication hypothesis. *Curr. Biol.* **11,** R1005–R1008.

Twigger, S. N., Nie, J., Ruotti, V., Yu, J., Chen, D., Li, D., Mathis, J., Narayanasamy, V., Gopinath, G. R., Pasko, D., Shimoyama, M., De La Cruz, N., Bromberg, S., Kwitek, A. E., Jacob, H. J., and Tonellato, P. J. (2004). Integrative genomics: In silico coupling of rat physiology and complex traits with mouse and human data. *Genome Res.* **14,** 651–660.

Van de Peer, Y., Frickey, T., Taylor, J., and Meyer, A. (2002). Dealing with saturation at the amino acid level: A case study based on anciently duplicated zebrafish genes. *Gene* **295,** 205–211.

Van de Peer, Y., Taylor, J. S., and Meyer, A. (2003). Are all fishes ancient polyploids? *J. Struct. Funct. Genom.* **3,** 65–73.

Van Laer, L., Huizing, E. H., Verstreken, M., van Zuijlen, D., Wauters, J. G., Bossuyt, P. J., Van de Heyning, P., McGuirt, W. T., Smith, R. J., Willems, P. J., Legan, P. K., Richardson, G. P., and Van Camp, G. (1998). Nonsyndromic hearing impairment is associated with a mutation in DFNA5. *Nat. Genet.* **20,** 194–197.

Vandepoele, K., De Vos, W., Taylor, J. S., Meyer, A., and Van de Peer, Y. (2004). Major events in the genome evolution of vertebrates: Paranome age and size differ considerably between ray-finned fishes and land vertebrates. *Proc. Natl. Acad. Sci. USA* **101,** 1638–1643.

Vogel, G. (1998). Doubled genes may explain fish diversity. *Science* **281,** 1119, 1121.

Wallace, D. F., Pedersen, P., Dixon, J. L., Stephenson, P., Searle, J. W., Powell, L. W., and Subramaniam, V. N. (2002). Novel mutation in ferroportin1 is associated with autosomal dominant hemochromatosis. *Blood* **100,** 692–694.

Wittbrodt, J., Meyer, A., and Schartl, M. (1998). More genes in fish? *BioEssays* **20,** 511–515.

Woods, I. G., Kelly, P. D., Chu, F., Ngo-Hazelett, P., Yan, Y.-L., Huang, H., Postlethwait, J. H., and Talbot, W. S. (2000). A comparative map of the zebrafish genome. *Genome Res.* **10,** 1903–1914.

Yoder, J. A., and Litman, G. W. (2000). The zebrafish fth1, slc3a2, men1, pc, fgf3 and cycd1 genes define two regions of conserved synteny between linkage group 7 and human chromosome 11q13. *Gene* **261,** 235–242.

Zhu, M., and Fahl, W. E. (2001). Functional characterization of transcription regulators that interact with the electrophile response element. *Biochem. Biophys. Res. Commun.* **289,** 212–219.

CHAPTER 15

Update of the Expressed Sequence Tag (EST) and Radiation Hybrid Panel Projects

Yi Zhou

Division of Hematology/Oncology, Children's Hospital Boston
Dana-Farber Cancer Institute and Harvard Medical School
Boston, Massachusetts 02115

Zebrafish has rapidly become an important genetic model organism for studying early vertebrate development and human genetic diseases. To take advantage of this powerful forward genetic system, genomic resources are necessary for positional and candidate gene cloning projects. Characterizing the molecular nature of zebrafish mutants provides important biological information on vertebrate gene function *in vivo* in higher vertebrates.

In 1996, the zebrafish research community appealed to the National Institutes of Health (NIH) to establish a transinstitution funding source to generate necessary genomic information and build the genomic tools for zebrafish development and genetic studies. Seven centers were awarded under this effort. These centers were established to build a high-density microsatellite genetic map, a comprehensive meiotic genetic map, an expressed sequence tag (EST) information database, two comprehensive radiation hybrid (RH) maps, a panel of deletion mutants, a

METHODS IN CELL BIOLOGY, VOL. 77

zebrafish stock center, and the zebrafish information network. The large-scale EST sequencing project was performed at the Washington University Genome Center, and an EST sequence information database was established in Steve Johnson's laboratory. To further use EST sequence information, all cDNA clones were sequenced from both the 5' and 3' ends to facilitate RH mapping ESTs. In 1998, three centers—Children's Hospital Boston and the University of Ottawa; Washington University and the National Institute of Diabetes and Digestive and Kidney Diseases (NIDDK); and the Max-Planck Institute at Tuebingen—were funded to construct RH maps of the zebrafish genome, using ESTs as markers. Since the beginning, I have been directing the RH mapping project in Leonard Zon's laboratory at the Children's Hospital Boston.

I. Expressed Sequence Tag (EST) Projects

ESTs are sequence information obtained by sequencing individual cDNA clones. This single-pass sequence information of transcripts serves as an efficient means to discover gene information in an organism (Adams *et al.*, 1993). To maximize the ratio of gene discovery, cDNA clones were collected from a variety of cDNA libraries (Table I). Because ESTs are primarily sequences of expressed gene transcripts, they are very useful in identifying zebrafish orthologs of genes initially described in other species, such as mice and humans, and as markers on genomic maps (Barbazuk *et al.*, 2000; Geisler *et al.*, 1999; Hukriede *et al.*, 1999, 2001; Postlethwait *et al.*, 2000; Woods *et al.*, 2000). Their genomic locations provide useful information in studying gene orthology, function, and evolution. ESTs with map locations also serve as markers and potential candidates for cloning genetic mutants. Furthermore, EST sequence information is also used to

Table I
Libraries Used in the Expressed Sequence Tag (EST) Project and ESTs Generated by 2001

Library	ESTs (5' + 3')
Late somitogenesis and liver	26,419
Adult 5' enriched	19,040
Fin growth and regeneration	6,520
C32 caudal fin	9,006
Adult brain	3,953
Olfactory epithelium	3,846
Adult kidney	11,035
Shield embryonic	2,755
15–19 hr embryonic	1,529
Adult retina	10,596
Miscellaneous libraries	627
Total	95,326

refine algorithm development for predicting genes and their genome organizations and will also be used to guide the final assembly of the zebrafish genome (Hudson *et al.*, 1995, 2001; Olivier *et al.*, 2001). As such, sequencing and positioning ESTs provide essential genomic tools used by the entire zebrafish field.

To date, more than 200,000 EST sequences have been generated at the Washington University Genome Sequencing Center (WUGSC) and deposited into GenBank at the National Center for Biotechnology Information (NCBI). In the Washington University Zebrafish EST Database (http://zfish.wustl.edu/), 95,326 zebrafish ESTs are available for searches and have been derived from a variety of cDNA libraries (Table I).

ESTs can also be used to identify zebrafish genes. ESTs from the WUGSC have been grouped into 25,601 WashU Zebrafish EST assemblies (WZs), with each EST assembly containing one or more EST sequences. Preliminary BLAST comparisons of the EST assemblies to known zebrafish genes identified roughly 61% of the latter (S. Johnson, unpublished), which suggests that the total EST cluster collection covers about 61% of all potential zebrafish genes. Thus, one can estimate that the zebrafish genome has about 28,500 genes, a number similar to that in the human genome.

In addition to the EST project organized by the Trans-NIH Zebrafish Genome Project Initiatives, large numbers of ESTs from different organ-specific cDNA libraries have also been sequenced. Gene expression profiles in specific tissues have been used to shed light on organ development. For example, the laboratories of Mark Fishman and C. C. Liew are studying the regulation of embryonic heart development, to which end they have sequenced clones from a cDNA library derived from 3-day-old embryonic hearts. A total of 5102 ESTs have been generated from this library (Ton *et al.*, 2000). These ESTs form 4049 unique EST assemblies, and about 102 of these EST clusters have been RH mapped on the T51 RH panel (Geisler *et al.*, 1999).

Zebrafish kidney marrow is the site of adult hematopoiesis, thereby performing a function similar to that of bone marrow in mammals. Leonard Zon's laboratory has constructed a cDNA library from the adult kidney. This library was made into a lambda phage vector by using the Lambda Express Vector system from Stratagene, Inc. Later, phage clones were *in vivo* excised into the pBK-CMV phagemid and arrayed in 384-well plates. From this kidney library, 11,053 ESTs were sequenced by WUGSC and an additional 27,872 ESTs were generated by Zhu Chen and Huaidong Song's laboratory at the Shanghai National Genome Center. The latter set of ESTs formed 7742 unique EST clusters, 44% of these being uncharacterized ESTs.

The group of Jinrong Peng at the Institute of Molecular and Cell Biology and the National University of Singapore studies vertebrate organogenesis, using zebrafish as a model system. cDNA libraries were generated from whole embryos and adult fish. These libraries were normalized to increase the ratio of new gene discovery in this EST project and to facilitate the fabrication of nonredundant cDNA microarray chips for expression profile studies. The libraries were then sequenced to give 26,927 EST sequences, representing 15,590 unique EST assemblies (Lo *et al.*, 2003).

By using sequence information obtained through the WUGSC EST project, Zhirong Bao (WUGSC) and Rick Waterman in Steve Johnson's laboratory were able to compile libraries of zebrafish repetitive sequences. These sequences are useful in masking repetitive sequence regions when analyzing the genome. The repeat mask program using this information and the repeat sequences submitted to RepBase is available online at http://www.sanger.ac.uk/Projects/D_rerio/fishmask.shtml. This resource is especially useful when designing primers for PCR amplification of genomic sequences and for designing unique oligonucleotides for genotyping and hybridization experiments used in chromosome walking projects.

As of March 5, 2004, 10,630 zebrafish genes and 450,652 ESTs have been deposited into GenBank, including 5905 full-length cDNA sequences. From these sequences, 17,925 unigenes have been compiled and are available in the Unigene database at the NCBI.

In the summer of 2002, Compugen released an oligonucleotide library for zebrafish gene expression profile studies, and MWG Biotech AG (MWG) began selling its oligonucleotide microarray sets in the spring of 2003. The array oligonucleotides were designed by using publicly available zebrafish gene and EST sequences. Compugen and MWG independently performed their own bioinformatic analyses of the sequences and used proprietary algorithms and computational strategies for selecting gene-specific oligonucleotides. The Compugen set has 16,339 oligonucleotides and covers 15,806 genes based on Compugen's transcriptome prediction program LEADS. The MWG set has 14,240 oligonucleotides and covers a minimum of 14,067 genes according to MWG's bioinformatic analysis. Compugen oligonucleotides are 65-mers and were synthesized and validated by Sigma-Genosys. The oligonucleotides were arranged according to gene ontology and their predicted biological functions. The MWG oligonucleotides are 50-mers. These oligonucleotides were purified by using HPSFR® technology and quality controlled by MALDI-TOF mass spectrometry. MWG demonstrated that the 50-mers offer the same, if not better, specificity and signal strength than 70-mers in microarray analysis (http://www.mwg-biotech.com/docs/discovery/an_arrays2_014.pdf).

NIH, NCBI, and Affymetrix designed and produced the first zebrafish Affy-chip array in the summer of 2003, a year after the gene and EST sequence release used by Compugen and MWG for oligonucleotide microarray design. At that time, the entire zebrafish community was given a chance to submit sequences of its favorite genes to NCBI prior to the selection of gene clusters for designing the Affymetrix chip. Notably, this gene chip set includes EST sequence information from normalized zebrafish libraries from Singapore and an adult kidney cDNA library. The chip contains more than 14,900 gene-specific oligonucleotide sets for detecting gene expression profiles, with each gene set containing 16 pairs of oligonucleotides (wild-type sequence vs. single base-pair mismatch) of 25 nucleotides.

At the Workshop on Genomic and Genetic Tools for the Zebrafish in 2002, the zebrafish community designated the full-length cDNA project as the top priority for NIH funding and this project started in August 2002. These full-length sequences will (1) assist assembly and annotation of the genome sequences as part of the Zebrafish Genome Project supported by the Sanger Centre of the Wellcome Trust and by the NIH, (2) enable functional analysis of these full-length sequenced genes by using genetics and overexpression, and (3) provide useful sequence information for either studying a particular gene or for developing tools for studying multiple genes. Given the availability of rapid gene knockdown by using morpholino antisense oligonucleotides, sequence information surrounding AUG regions of genes is of particular interest. Thus, the NIH funded a zebrafish gene collection (ZGC) program as part of the ongoing Mammalian Gene Collection (MGC) project, aimed to sequence approximately 10,000 full-length cDNAs from different tissue-cDNA libraries. In this effort, the ZGC project collected different zebrafish tissues and organs and constructed full-length enriched cDNA libraries from these tissues. Preliminary 5′-end sequence analysis of clones from these libraries was performed and full-length clones identified from this initial analysis were selected for sequencing fully. The sequence information was directly deposited into GenBank and can be searched at http://zgc.nci.nih.gov/. In addition, these full-length clones are available through the IMAGE consortium network. In the United States, clones can be purchased from the American Type Culture Collection (ATCC) at http://www.atcc.org/ (ZGC clone numbers available through ZGC at its Web site), or from Open Biosystems at http://www.openbiosystems.com/query.php. Outside the United States, distributors include the U.K. Human Genome Mapping Project Resource Center at http://www.hgmp.mrc.ac.uk/geneservice/index.shtml and the Resource Center of the German Human Genome Project at http://www.rzpd.de/. As of March 1, 2004, ZGC had 3473 full-length clones and 3007 nonredundant genes.

II. Radiation Hybrid (RH) Panel Projects

RH mapping panels are constructed by fusing lethally irradiated cells containing the genome (e.g., zebrafish) of interest to rodent hybridoma cells. During fusion, a given host cell will incorporate a random subset of zebrafish genomic DNA fragments from its irradiated zebrafish cell partner. Therefore, each host cell contains but a fraction of the genome of interest. Because fusion is random, the zebrafish genomic DNA fragments in one cell hybrid might be unique or might overlap with the DNA fragments present in other hybrid cells. Fused cells having the highest proportion of donor genome fragments are cultured in large quantities, and genomic DNA from each hybrid clone is isolated to establish the RH mapping panel. RH mapping works under the following assumptions: (1) the probability of inducing an irradiation break between two points (markers) on a given chromosome is linearly dependent on the physical distance between the two points and (2) two markers residing on the same DNA fragment have a greater chance of being taken into the same host cell than two markers distributed on different DNA fragments. Thus, two markers close together have a higher chance of remaining on the same DNA fragment after a given dose of irradiation and would have a more similar RH distribution pattern within a mapping panel than two points further apart. By using statistical algorithms, computer programs have been developed to predict the physical relationship between two markers (sequences) on a genome according to the distribution patterns of the markers on an RH mapping panel. The two most popular academic RH mapping software packages are SAMapper 1.0 developed by David Cox's group at the Stanford Human Genome Center (Boehnke *et al.*, 1991) and RHMAPPER developed by Eric Lander's group at the Whitehead Genome Institute (Stein *et al.*, 1995). Recently, Richa Agarwala and colleagues at the NCBI have developed a new RH mapping package (RHmap; Agarwala *et al.*, 2000), and this program is available on the NCBI site (ftp:// ftp.ncbi.nlm.nih.gov/pub/agarwala/rhmapping/rh_tsp_map.tar.gz).

To facilitate PCR analyses, RH panels normally consist of 96 DNA samples, including one positive and one negative control. At present, two RH mapping panels are being used for constructing high-density radiation maps of the zebrafish genome. The Goodfellow T51 RH panel (Kwok *et al.*, 1998) was made in Peter Goodfellow's lab at Oxford and was commercially available from Research Genetics, Inc. (now Invitrogen, Inc.); unfortunately stocks are currently exhausted. The Ekker LN54 RH panel (Ekker *et al.*, 1999; Hukriede *et al.*, 1999) was constructed in Marc Ekker's laboratory in Ottawa, Canada, and was distributed free of charge. Both panels used the same zebrafish fin fibroblast AB9 cell line (Paw and Zon, 1999) established by Barry Paw in Leonard Zon's lab for the genome contribution. Pascal Haffter and Robert Geisler completed the initial characterization of the Goodfellow T51 RH panel and assembled the initial T51 panel RH map (Geisler *et al.*, 1999). The laboratories of Igor Dawid, Marc Ekker, Steve Johnson, Mike McPherson, and Len Zon built the initial RH map for the Ekker LN54 panel (Hukreide *et al.*, 1999). Table II compares these two RH mapping panels.

Table II
Comparison Between two Zebrafish Radiation Hybrid (RH) Panels

	Goodfellow panel (T51)	Ekker panel (LN54)
Number of hybrids in mapping panels	94	93
Average retention rate	18.4%	22%
Average size of zebrafish fragments	6.1 Mb	14.8 Mb
Relationship of X-ray breakage to distance	1 cR = 61 kb	1 cR = 162 kb
Average resolution of comprehensive maps	350 kb	500 kb
Mapping success for random markers	87%	90%
Publication date of the zebrafish genome maps	9/1999	8/1999
Distribution of panels	Unknown	50 labs

A. Establishing a High-Throughput RH Mapping Facility

To establish an efficient RH mapping facility in the Zon laboratory at Children's Hospital, we visited major RH mapping centers around the world. This included visits to Pascal Haffter's group (in particular, Robert Geisler) in Tuebingen, Germany; to the Johnson/McPherson group at the Genome Sequencing Center in St. Louis, MO; and to David Cox's group at Stanford Human Genome Center in Palo Alto, CA. After examining our needs and the RH mapping experience of the Zon lab, we decided to introduce a 384-well RH mapping system based on a more flexible robotic system. This RH mapping system provided us with the speed and quality required for the RH mapping project.

Our high-throughput RH mapping facility includes two Genesis RSP150 robots from TECAN, one Robbins 96 Hydra 96-channel liquid handling robot (Robbins, Inc.), seven dual 384-well GeneAmp 9700 PCR machines (Applied Biosystems, Inc.), sixteen A3.1 agarose gel apparatuses (OWL Separations), a gel imaging system (Ultralum, Inc.), and a gel-scoring software program that we developed in our laboratory (Fig. 1). With this system, a four-person team can RH map 100 ESTs per week. The flow of our high-throughput RH mapping system is described in Fig. 2.

B. Characterizing the Ekker LN54 RH Panel and the Goodfellow T51 RH Panel

A quality EST RH map is a useful resource for candidate and positional cloning projects. In addition, it also aids the annotation and assembly of the zebrafish genome. As such, RH panel resolution and its capacity to order markers correctly are critical factors when researchers choose and use a panel. To evaluate these parameters for the two available panels, we compared the Goodfellow T51 RH panel and the Ekker LN54 RH panel by using our high-throughput mapping system. Over the last 4 years, we have conducted seven positional cloning projects in our laboratory and have finished six of them. The loci are distributed randomly throughout the zebrafish genome, and most of the genetic and physical distances are known for the markers used in chromosomal walks within these regions. The markers from these positional cloning projects allowed us to evaluate the resolution and arrangement of known markers on both RH panels. The Goodfellow T51 panel correctly predicted the markers for four of the chromosomal walks, including those for *weissherbst, moonshine, chablis,* and *sauternes* (Yi Zhou *et al.,* unpublished). The Ekker LN54 panel was not as predictive of the order of these markers as the Goodfellow T51 RH panel. However, LN54 was more accurate in ordering the markers in the *retsina* position cloning project. Because these two panels are complementary, we believe it is advantageous to have both panels available, and we have mapped approximately 600 EST markers to both. With data obtained from our group and from the Johnson, Dawid, McPherson, and Ekker laboratories, we have finished a framework map with 703 markers and placed more than 1000 EST markers on the Ekker LN54 panel (Hukreide *et al.,*

Four sets of RH mapping templates are arrayed in a 384-well PCR plate on a Robbins Hydra 96 Robot, dried at 55°C, and then stored at –20°C.

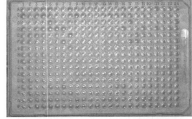

In 90 min, PCR reactions of 24 markers can be assembled in duplicates in 12 384-well PCR plates on the chilled deck of *GENESIS RSP 150* robot (TECAN).

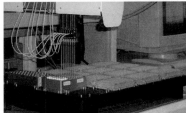

In 2.5 h, PCR reactions of twelve 384-well plates can be completed using six GenAmp 9700 PCR machines (Applied BiosystemsInc.) (Total time = 4 h).

In 1 h, PCR reactions of each 384-well PCR plates (two-marker worth PCR reactions) can be analyzed on a 2% agarose gel in 1XTBE (total time 5 h).

Gel images are acquired and RH scores generated by using a Windows-based image processing software developed in the lab.

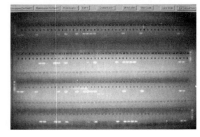

Fig. 1   A high-throughput system for radiation hybrid panel mapping. (See Color Insert.)

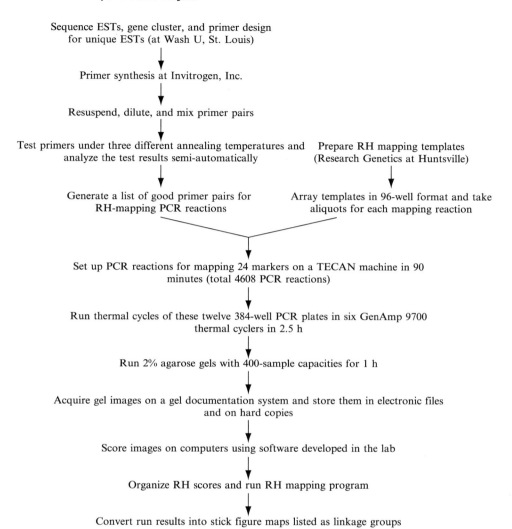

Sequence ESTs, gene cluster, and primer design
for unique ESTs (at Wash U, St. Louis)

Primer synthesis at Invitrogen, Inc.

Resuspend, dilute, and mix primer pairs

Test primers under three different annealing temperatures and   Prepare RH mapping templates
analyze the test results semi-automatically   (Research Genetics at Huntsville)

Generate a list of good primer pairs for   Array templates in 96-well format and take
RH-mapping PCR reactions   aliquots for each mapping reaction

Set up PCR reactions for mapping 24 markers on a TECAN machine in 90
minutes (total 4608 PCR reactions)

Run thermal cycles of these twelve 384-well PCR plates in six GenAmp 9700
thermal cyclers in 2.5 h

Run 2% agarose gels with 400-sample capacities for 1 h

Acquire gel images on a gel documentation system and store them in electronic files
and on hard copies

Score images on computers using software developed in the lab

Organize RH scores and run RH mapping program

Convert run results into stick figure maps listed as linkage groups

Fig. 2 High-throughput expressed sequence tag (EST) radiation hybrid (RH) mapping.

1999). From our analysis, in collaboration with Neil Hukreide in Igor Dawid's group, more than 95% of the same markers have been mapped on both panels to the same linkage group and to similar locations. Typing the same markers on both RH mapping panels will help assemble both panel maps into an integrated RH map of the zebrafish genome. Given the limited funds available, we have chosen the higher-resolution Goodfellow T51 panel as the primary panel for mapping genes and ESTs.

C. Designing and Testing RH Mapping Primers

Dan Fisher and Frank Li in Steve Johnson's group kindly designed primers by using the Primer 3 software package (Rozen and Skaletsky, 2000) for most of EST markers on the RH maps. For RH mapping at the Boston center, the primer facility of Lifetech, Inc. (now Invitrogen, Inc.) synthesized these EST primers and provided them in a 96-well format. The primers were resuspended to a concentration of 200 μM as stocks and subsequently diluted to 20 μM for use in PCR reactions. The best annealing temperature for a particular primer is determined through a three-temperature primer test on positive-control DNA (from AB9 cells or AB adult fish) and negative-control DNA samples (hamster genomic DNA and mouse genomic DNA) of both zebrafish RH panels. The overall primer-test success rate of primers on our three-temperature primer test was approximately 80%, and the average success rate of mapping PCR reactions on the Goodfellow panel was 75–80%.

The primer testing process has also been automated. One plate of 96-primer pairs can be tested in one 384-well PCR plate on arrays of four testing controls (buffer, hamster genomic DNA, mouse genomic DNA, and AB cell line DNA or adult AB fish DNA) at a given annealing temperature (50, 55, or 60 °C). PCR products from these tests are separated on 2% agarose gels, and the gel images are stored both as electronic files and photographic images. Electronic images are scored by using the Windows-based scoring software developed by Dr. Anhua Song (Children's Hospital Boston), and the results of the primer test are electronically loaded into the primer database. Primers that passed the test are queued for RH mapping.

D. Mapping Markers on the T51 RH Panel

The T51 RH panel template consists of 94 hybrid DNA samples, one positive control (DNA from an AB cell line or from whole adult AB fish), and one negative control (hamster W3G cell line DNA). The DNA samples are distributed in 96 two-milliliter screw cap vials at a concentration of 25 ng/μl and are stored at −20 °C. Before use, these samples are thawed at 4 °C and each is centrifuged briefly to collect the DNA solution at the bottom from the cap and wall in the vial before they are diluted to a concentration of 12.5 ng/μl and arranged into a 96-well deep-well plate, using a TECAN Genesis liquid handling robot. Dilution of DNA and the reordering of samples on a 96-well plate are necessary to allow accurate distribution of DNA samples robotically to PCR reaction plates and to maintain the numeric order of the samples for subsequent agarose gel electrophoresis and mapping analysis. The format for template deposition in the 96-well plates is shown in Fig. 3. Two microliters of DNA at 12.5 ng/μl for each of the 94 zebrafish RH samples and the two controls are transferred to wells in the four quadrants of 384-well PCR plates, using a Hydra 96 multipipettor (Robbins Scientific). The DNA aliquots in the 384-well plates are then dried in a 55 °C incubator, which are then stored in clean plastic bags at −20 °C. Storing the

	1	2	3	4	5	6	7	8	9	10	11	12
A	1	3	5	7	9	11	13	15	17	19	21	23
B	2	4	6	8	10	12	14	16	18	20	22	24
C	25	27	29	31	33	35	37	39	41	43	45	47
D	26	28	30	32	34	36	38	40	42	44	46	48
E	49	51	53	55	57	59	61	63	65	67	69	71
F	50	52	54	56	58	60	62	64	66	68	70	72
G	73	75	77	79	81	83	85	87	89	91	93	95
H	74	76	78	80	82	84	86	88	90	92	94	96

Fig. 3 The arrangement of cell hybrid DNAs in a 96 format. This arrangement was designed to accommodate the spacing between the tips of a multichannel pipettor and between the wells of a 400-lane agarose gel used in our mapping system. By loading A rows first into every other well on the gel and then loading B rows into the wells between A row samples, and so on, the numeric order of the cell hybrids on the panel is restored for generating RH scores in the determined order.

reaction-ready PCR template in PCR plates is very reliable and avoids cross-contamination of samples that can occur when aqueous DNA samples are stored frozen.

Two TECAN Genesis RSP 150 liquid handling robots are used to set up the PCR reactions in 10.0-μl reaction mixes containing 1× PCR buffer; 200 μM each of dATP, dCTP, dGTP, and dTTP; 200 nM each of forward and reverse primers; and 0.20 unit of Taq polymerase. 1× PCR buffer contains 10 mM Tris-HCl (pH 8.3), 50 mM KCl, 1.5 mM MgCl$_2$, 0.01% (w/v) gelatin, 0.01% NP-40, and 0.01% Tween 20. The reaction mix for one duplicate mapping of a primer pair is reconstituted to a volume of 1100 μl and then 10 μl of this mixture is distributed into each well in a 96-well quadrant in a 384-well plate. (Each well contains 25 ng of dry RH mapping template). To automate this process, the PCR buffer mix is made at 1.38× concentration (containing all four dNTPs) and then aliquoted into reagent tubes in 96 format racks for access by the machines. PCR buffer aliquots are stored at −20 °C, thawed, and then maintained at 4 °C before use. For each RH mapping duplicate, 800 μl of 1.38× buffer is mixed with 286 μl of H$_2$O, 11 μl of 20 μM forward and reverse primers, and 22 units of Taq DNA polymerase. After carefully mixing all reaction components (except DNA templates), 10 μl of this mixture is distributed to each well of the 384-well plates containing the dried DNA template. Finished 384-well plates are briefly centrifuged at 4 °C, using a Eppendorf 5810 R centrifuge (A-4-62 rotor) at 3000 rpm to ensure contact between the PCR reaction components and the DNA templates on the bottom of each well. PCR reactions for mapping 24 markers in duplicate can be set up in 1.5 h, using the TECAN system. All our PCR reagents are maintained at 4 °C throughout the pipetting process. PCR reactions are performed by using ABI GeneAmp 9700 thermocyclers with dual 384-well tops. The cycling conditions are as follows:

- Initial denaturing at 94 °C for 2 mins.
- Thirty-five cycles of denaturing at 94 °C for 30 secs, annealing at 50, 55, or 60 °C (dependent on the primer test results) for 30 secs and extension at 72 °C for 1 min.
- Final extension at 72 °C for 10 min.

Finished reactions are stored at 10 °C on PCR machines and subsequently at 4 °C until assay by gel electrophoresis. Three microliters of 5× gel-loading dye containing 15% glycerol in 1× TE buffer (pH 8.0) is added to each PCR reaction, and the reactions are loaded onto 2% agarose gels (in 0.5× TBE), each having a 400-lane capacity (eight rows of 50 lanes). With a 12-channel pipettor, we can load all samples from a 384-well plate plus molecular weight standards on a single gel. For each set, two lanes on one side of each row are loaded with 200 ng of pBR322-MspI size standards (New England Biolabs). Samples in agarose gels are electrophoresed in 0.5× TBE buffer for 1 h at 300 V. Gel images are captured both as electronic files and on thermal printer paper.

Electronic images are processed semiautomatically, using our in-house image scoring software. Even very weak bands are scored as positive if they can be distinguished from the background. Given that the Ekker and Goodfellow Panel retention rates are 20% and 18%, respectively, a marker must have a minimum of 10 positive scores to be considered viable for producing an accurate map position. In the event of any disagreement between duplicate images (i.e., a hybrid scored as positive in one instance and negative in another), the images are rescored. If more than three inconsistencies cannot be resolved by rescoring the images, the PCR reaction is repeated.

As of last year, the T51 panel DNA samples are no longer commercially available to individual research laboratories, although both the Children's Hospital, Boston, and Tuebingen groups have secured limited amounts of the panel DNA stocks to support fingerprinted BAC end RH mapping. Both groups are providing RH mapping services to individual researchers based on requests at http://zfrhmaps.tch.harvard.edu/ZonRHmapper/mapService.htm and http://wwwmap.tuebingen.mpg.de/.

E. Mapping Markers on the LN54 RH Panel

The Ekker LN54 panel is distributed by Marc Ekker's laboratory (mekker@lri.ca) at the University of Ottawa. This panel was made by fusing irradiated zebrafish AB9 cells with mouse B78 cells. It has 93 cell hybrid lines, one positive control (genomic DNA of AB9 cells or AB adult fish), one mix control containing one tenth of positive-control DNA and nine tenths of negative-control DNA, and one negative control (genomic DNA of mouse B78 cells). The procedure for mapping on the LN54 RH panel is very similar to that described for T51. The LN54 panel DNAs are provided at concentrations of 100 ng/μl. These samples are diluted 4× to 25 ng/μl and are arranged in 96-well deep-well plates as shown in

Figure 3. Two microliters of each DNA sample of the panel are aliquoted into quadrants of a 384-well plate, thus depositing 50 ng of template per well. PCR reactions are set up and PCR products analyzed by using the same system designed for the T51 RH panel.

To date, 17,861 markers have been typed on the T51 RH mapping panel, including 2492 single strand length polymorphism (SSLP) markers, 7020 ESTs, 349 genes, 2812 sequence tagged site (STS) markers, and 5188 bacterial artificial chromosome (BAC) ends. Of these markers, 14,989 (83.9%) have been placed on the current map by using the RHmap software and 12,785 (71.6%) by using SAMapper 1.0. On the LN54 panel, 4192 markers, including 3049 ESTs, 394 genes, 749 SSLPs, and 391 STSs, have been placed by using RHMAPPER.

It is important for individual researchers to understand that the radiation map is calculated based on a statistical analysis of possible positional relationships for any given points on the genome. For this reason, the programs present the highest possible linear relationships of markers on the genome and not absolute physical relationships. Thus, different statistical analyses of RH result do not produce identical maps. It is helpful to provide researchers with a diverse view of maker orders on the genome before the map matures. Between Robert Geisler's group and our group, we currently are presenting two different analyses of RH mapping results. On the Tuebingen site, the map was calculated by using the RHmap software. This software package takes advantage of the algorithm developed to solve the famous traveling salesman problem (TSP) and is capable of building a dense RH map. Although RHmap is able to place more markers on the map (82.5% vs. 72.6% for SAMapper), it does so at a lower resolution, with more markers being grouped together at the same genomic location. We continue to use SAMapper 1.0 to calculate the T51 RH map based on the same data set that was used for RHmap calculation. The SAMapper 1.0 software package was used to initially characterize the T51 panel and to build the first T51 panel RH map in 1999 (Geisler et al., 1999). The higher resolution of SAMapper 1.0 makes it useful for positional cloning of mutant genes. Recently, through discussions between Robert Geisler, ZFIN, and our group, the T51 panel map calculated by the RHmap has been chosen for releasing to the ZFIN and is available through the ZFIN site (http://zfin.org).

F. RH Mapping Services

In addition to this large-scale EST mapping effort, we also provide RH mapping services to individual researchers in the field. Researchers can submit their mapping requests in three ways. The first entails providing a sequence from which we design, synthesize, and test the primers. Successful primers are put on the Goodfellow panel and RH scores are generated from PCR reactions. The second method is to submit primer sequences that are designed by individual researchers for use in their own research projects or solely for RH mapping. We either obtain an aliquot of the investigator's primers or synthesize them ourselves. Lastly, if individual researchers have access to the panel and can perform their own RH mapping PCR reactions, we offer to run the mapping program using their data and inform them of their results.

In instances in which we design the primers, the Oligo 6 program is used to select the best primers to specifically amplify a marker region from a pool of zebrafish and host cell genomic DNA fragments. Because most markers requested by individual researchers for RH mapping are genes and ESTs (not genomic sequences), we first analyze the submitted sequence by sequence similarity comparison to the whole zebrafish genome shotgun sequence, using the BLAST server. This step helps tremendously in several ways. First, it confirms the submitted sequence information with the Sanger Centre's genomic sequence effort and allows us to pick reliable regions for primer design (regions designated as those having a perfect match between the submitted sequence and the Sanger sequence). Second, it provides us with intron–exon boundary locations. Because the PCR product size is limited, it is critical that the primers do not flank a large intron region such that the PCR product is too large to amplify efficiently. Finally, the homology search also identifies regions of repetitive sequences such that we can purposely avoid those repeat regions in the primer design processes. We also encourage researchers to use an online repeat masker program at the Sanger Center Web site (http://www.sanger.ac.uk/Projects/D_rerio/fishmask.shtml) to filter out repeats in marker sequences. For genes and ESTs, we normally use 3′-UTR regions to design primer pairs for mapping, because they are typically unique to the cDNA species and have fewer introns. As such, primers derived from these regions are more likely to generate unique PCR products and to reduce PCR failures caused by introns in the flanked region. In those instances in which no suitable noncoding regions for primer design are available, we design one primer from the 3′-coding region and the other primer from the 3′-UTR region. One can also design both forward and reverse primers in two adjacent exons if the intron is small. Because the intron size of the orthologous region of the host genome would most likely be different from that of the zebrafish, nonspecific amplification of the host genome would be apparent by PCR products of different sizes or by the failure of amplification from the host genome because of large introns. Of course, given the availability of the zebrafish genomic sequence, one can tailor primers to a specific genomic region. In this case, extra attention is needed both to ensure that the sequence being mapped is part of the expected gene and to avoid genomic repeats in the region. These strategies can reduce the chance of amplifying conserved orthologs from either the host cell (hamster or mouse) or family members (homologs) in zebrafish. In summary, BLAST analysis of submitted sequences and careful primer design allows us to increase the success rate and shorten the assay time for customer RH mapping.

Figure 4 shows an example of a BLAST search of the 3′-UTR region of unigene Dr.17339 (fx75 g04.y1) against the zebrafish whole-genome shotgun sequences available on the Zon Lab BLAST Server. Because the genome sequence assembly is not mature and contains errors, it is better to perform the sequence similarity search against the random shotgun sequences (available on the Web site of the

Children's Hospital Boston Zebrafish Genome Project Initiative at http://zfrhmaps.tch.harvard.edu/zfblast/blastforpublic.htm). The shotgun sequences were derived from single-sequence reads of a continuous genomic region, whereas preliminary assemblies may contain inaccurate junctions between different sequence reads. Because the shotgun sequence database provides four to five times genome coverage, it is likely that multiple sequences will match with high similarity to your query sequence (Fig. 4A). However, if one of your sequence regions should match more than 10 different shotgun sequences, primers should not be designed to this region. Figure 4B shows the alignments between the unigene sequence (query on top) and a few shotgun sequences (subject on bottom). These sequences possess the much higher BLAST scores and smaller e-values than other sequences in the genome. These sequences should be selected for designing primers to the genomic region of the unigene. When discrepancies occur in small regions between the two sequences, one cannot determine whether the shotgun sequence or the query sequence is correct. Thus, only the regions with perfect sequence matches should be used for selecting primers. In this particular case (Fig. 4B), one would use the 1-171 region for designing the forward primer and the 208–314 region for designing the reverse primer. When one cannot find a sufficiently long, perfect sequence match between the query sequence and the shotgun candidates, the matched shotgun sequence can be used to extend the query sequence to aid in primer design. However, this matched sequence needs to be compared to other shotgun sequences to ensure that there are no other potential matches. It is most important to avoid single-nucleotide mismatched regions for primer design. For example, the query sequence positions at 62 and 305 (Fig. 4B and 4C) should be avoided because these single-nucleotide differences could be because of sequencing errors or single-nucleotide polymorphisms among fish strains. Primers designed against perfectly matched regions are much more likely to lead to successful PCR amplification.

To provide individual researchers with a fast RH mapping tool, Anhua Song (of our RH mapping team) developed an Instant RH mapping program, using the same algorithm as SAMapper 1.0. This mapping program is available online at http://zfrhmaps.tch.harvard.edu/zonrhmapper/instantmapping.htm. If researchers choose to carry out their own RH mapping PCR reactions, they can input their RH mapping scores into this online program to calculate the relative distance between the input markers and the existing markers on the map. The Web page for using the Instant Mapping program is shown in Fig. 5. For this program, users need to provide their own mapping data and input the PCR results of the cell hybrids as shown in Fig. 5A. Because the scores of positive and negative controls are not included in the calculations, only 94 digits should be put into the result box, with 1 being a positive result (specific PCR product), 0 being negative (no specific PCR product), and 2 being questionable and/or no duplicates. It is useful to provide your e-mail information to help us track the usage of this service. After a user clicks the submit button, the results are quickly returned in the format shown in Fig. 5B. The first part of the result is a summary of the submitted request, including the name of a marker and its RH score on the T51 panel. The

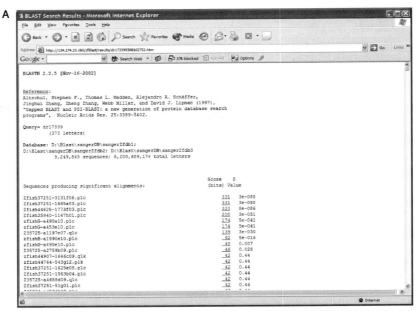

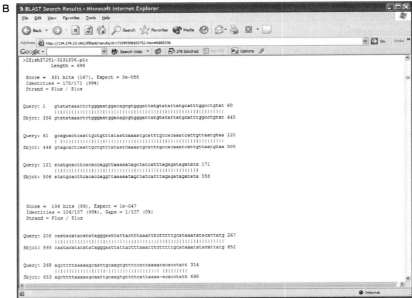

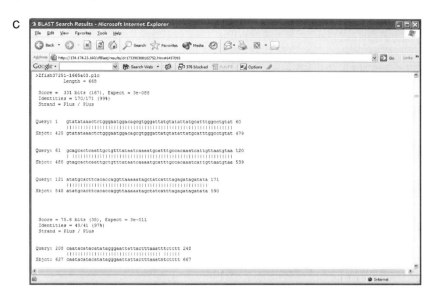

Fig. 4 Examples of BLAST seach results used for primer design. (A) The BLAST search result of unigene Dr. 17339 against the Sanger Center zebrafish whole genome shotgun sequences. (B) Nucleotide sequence alignment between Dr. 17339 and a shotgun sequence, Zfish37251-3131f06.plc. (C) Nucleotide sequence alignment between Dr. 17339 and another shotgun sequence, Zfish37251-1665a03.plc.

second half of the results is a list of closely linked markers, with distance being measured in centiRays (cR). Based on statistical analysis, one centiRay on the T51 panel is roughly 61 kb. The LOD score measures the significance of the linkage between two markers, with larger scores indicating more reliable linkage between the markers.

We have developed an interactive Web site for releasing our mapping data and RH mapping services to individual researchers worldwide (http://zfrhmaps.tch. harvard.edu/ZonRHmapper/Default.htm). This Web site is updated periodically with newly mapped markers. To date, we have had inquires from Europe, Australia, New Zealand, Singapore, Japan, China, Canada, and the United States. In the past year, researchers have visited our Web site more than 6500 times.

The Sanger Genome Centre is funded by the Wellcome Trust to sequence the zebrafish genome and anticipates releasing a draft zebrafish genome sequence in 2005. After examining the human genome and other sequencing projects, the Sanger Centre decided to use a hybrid approach to sequence the zebrafish genome. The first approach is to generate a 3–5× genome coverage, using a whole-genome random shotgun sequencing with subsequent assembly. The second relies on construction of a BAC contig map using the fingerprinting method of BACs with large inserts and end sequenced, followed by the shotgun sequencing of individual BAC clones that cover the contigs. After sequencing, clusters of computers will be used to calculate all available genetic and genomic information, including the

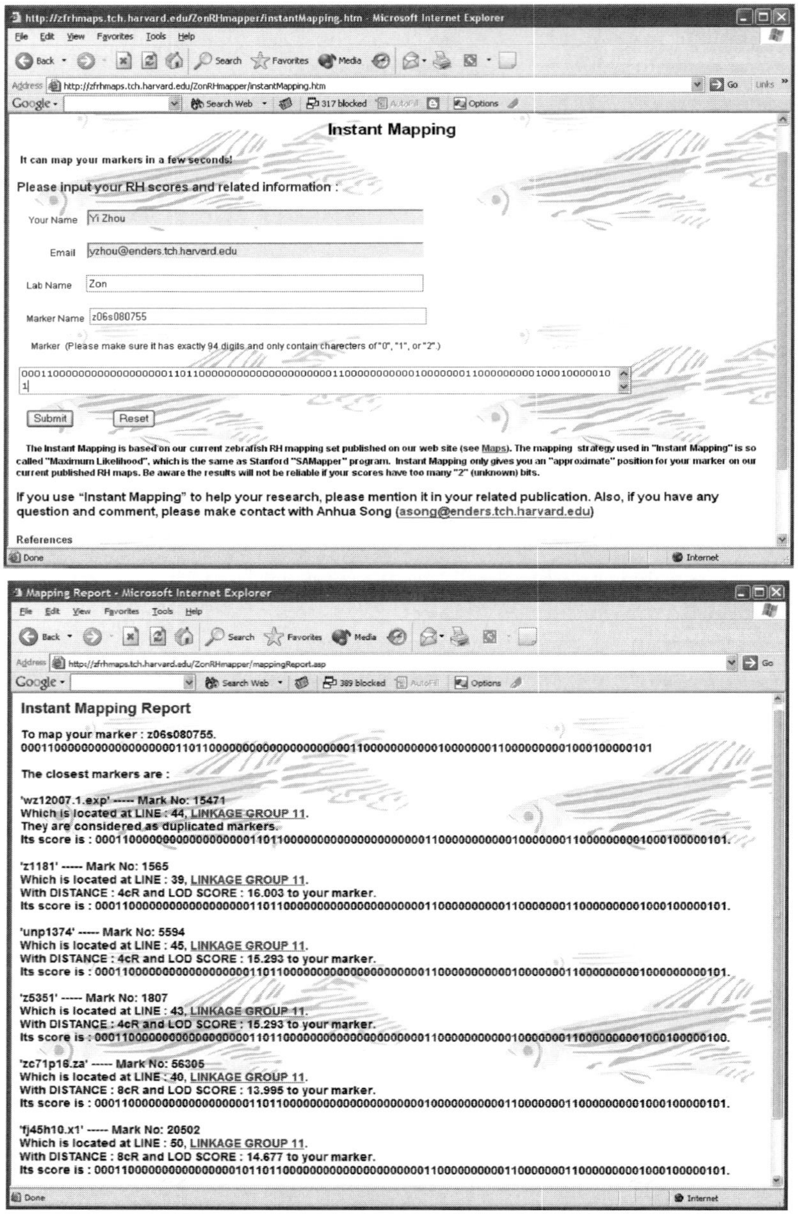

Fig. 5   Instant Mapping program. (A) The online tool for calculating positions on the T51 panel RH map. (B) The result page of the Instant Mapping software.

dense RH map that we and others have generated. The assembly of the zebrafish genome sequence will be of tremendous help when selecting candidate genes from genomic intervals known to encompass a particular mutant. In the future, chromosomal walking could be done electronically rather than through the tedious steps of a BAC/PAC walk.

To this end, the Children's Hospital Boston center and the Max-Plank Institute are mapping thousands of BAC ends that represent individual fingerprinted contigs. At present, there are about 3775 fingerprinted contigs assembled by using a stringent automated assembly program, and we have already typed 5188 BAC ends, including ones within the fingerprinted contigs on the T51 RH panel. A total of 3972 (76.6%) have been positioned on the map updated on January 2, 2004, at the Tuebingen site at http://wwwmap.tuebingen.mpg.de/and 3474 (67%) at the Children's Hospital Boston site at http://zfrhmaps.tch.harvard.edu/ZonRHmapper/Default.htm.

III. Future Directions

At the beginning of the Trans-NIH Zebrafish Genome Project Initiative, three RH mapping centers focused on building a preliminary genomic map, using two different RH mapping panels. Subsequently, both the Children's Hospital Boston and the Tuebingen group focused their effort on the T51 panel in order to establish a high-density map of the zebrafish genome by using ESTs. As a result of the zebrafish genome sequencing project, we have refocused our RH mapping efforts to assigning fingerprinted BAC ends to the T51 RH panel.

Although the current RH maps of the zebrafish genome are not perfect, both the Boston and the Tuebingen groups are working to improve the map quality by using the following measures. First, new anchor markers have been chosen to reduce the number of gaps on the map and to increase the mapping ratio on the T51 panel. Second, questionable anchor markers and markers that increase instability in mapping are being retyped and/or replaced with better markers to improve map quality. As such, the current map is built by using 624 accurately typed anchor markers, whereas previously we used 902 anchor markers (some positioned inaccurately) to construct the T51 RH map.

Another strategy for generating a higher-quality RH map of the zebrafish genome is to integrate the two (the T51 and the LN54) RH maps. The only way to achieve this goal is to increase the density of shared markers that have been mapped to the two panels. At the very least, we need to ensure that all backbone markers of the two RH maps are mapped on both panels with similar genome locations. At present, less than 50% of backbone microsatellite markers are shared. In addition, increasing the number of overlapping markers between the two RH maps will not only ease the process of integrating the two maps but also aid in the quality control of the entire RH mapping effort. This better map will in turn produce a higher-quality genome sequence assembly, providing more accurate information of genomic regions of interest to researchers.

The RH software that has been used in the RH mapping projects is freeware and easily accessible. However, there is limited documentation on how to use this software and the software is often not well maintained or updated. In addition, no statistical method is perfect; each algorithm has intrinsic shortcomings. Of all the RH mapping software packages, the RHmap software package (NCBI) is the most recent. This software is very powerful and has been used to recalculate both the human and mouse RH mapping data. It will also be used to build the ultimate RH map of the zebrafish genome. The reason the RH map program has not yet produced a higher-resolution map of the T51 RH panel is that the data on this panel have not yet been optimized and are insufficient to provide all the necessary information to use fully the power of the program. The zebrafish RH mapping centers are now making a special effort to provide mapping of the well-characterized anchor markers required by RHmap to generate higher-quality maps.

Although our mapping priorities have changed, the Children's Hospital Boston RH mapping center will continue to map those additional ESTs and genes requested by individual researchers in the community. At the same time, we will type as many fingerprinted BAC contigs to the T51 panel as funds permit, both before and after the whole sequence assembly of the zebrafish genome. RH mapping centers will maintain collaborations with the Sanger Centre, thus helping eliminate gaps in genome assembly to improve its quality.

Acknowledgments

The NIH funded the majority of zebrafish EST and RH mapping projects. This article could not be written without critical comments on the manuscript from Jeffrey Guyon. I thank Nelson Hsia for his help in improving this manuscript.

References

Adams, M. D., Kerlavage, A. R., Fields, C., and Venter, J. C. (1993). 3,400 new expressed sequence tags identify diversity of transcripts in human brain. *Nat. Genet.* **4**(3), 256–267.

Agarwala, R., Applegate, D. L., Maglott, D., Schuler, G. D., and Schaffer, A. A. (2000). A fast and scalable radiation hybrid map construction and integration strategy. *Genome Res.* **10**, 350–364.

Barbazuk, W. B., Korf, I., Kadavi, C., Heyen, J., Tate, S., Wun, E., Bedell, J. A., McPherson, J. D., and Johnson, S. L. (2000). The syntenic relationship of the zebrafish and human genomes. *Genome Res.* **10**(9), 1351–1358.

Boehnke, M., Lange, K., and Cox, D. (1991). Statistical methods for multipoint radiation hybrid mapping. *Am. J. Hum. Genet.* **49**, 1174–1188.

Ekker, M., Ye, F., Joly, L., Tellis, P., and Chevrette, M. (1999). Zebrafish/mouse somatic cell hybrids for the characterization of the zebrafish genome. *Methods Cell Biol.* **60**, 303–321.

Geisler, R., Rauch, G. J., Baier, H., van Bebber, F., Bross, L., Dekens, M. P., Finger, K., Fricke, C., Gates, M. A., Geiger, H., Geiger-Rudolph, S., Gilmour, D., Glaser, S., Gnugge, L., Habeck, H., Hingst, K., Holley, S., Keenan, J., Kirn, A., Knaut, H., Lashkari, D., Maderspacher, F., Martyn, U., Neuhauss, S., Neumann, C., Nicolson, I, Pelegri, T., Ray, R., Rick, J. M., Roehl, H., Roeser, T., Schauerte, H. E., Schier, A. F., Schonberger, U., Schonthaler, H. B., Schulte-Merker, S., Seydler, C.,

Talbot, W. S., Weiler, C., Nusslein-Volhard, C., Haffter, P., *et al.* (1999). A radiation hybrid map of the zebrafish genome. *Nat. Genet.* **23**(1), 86–89.

Hudson, T. J., Stein, L. D., Gerety, S. S., Ma, J., Castle, A. B., Silva, J., Slonim, D. K., Baptista, R., Kruglyak, L., Xu, S. H., Hu, X., Colbert, A. M. E., Rosenberg, C., Reeve-Daley, M. P., Rozen, S., Hui, L., Wu, X., Westergaard, C., Wilson, K. M., Bae, J. S., Maita, S., Ganiatsas, S., Evans, C. A., DeAngelis, M. M., and Ingalls, K. A. (1995). An STS-based map of the human genome. *Science* **270**(5244), 1945–1954.

Hudson, T. J., Church, D. M., Greenaway, S., Nguyen, H., Cook, A., Steen, R. G., Van Etten, W. J., Castle, A. B., Strivens, M. A., Trickett, P., Heuston, C., Davison, C., Southwell, A., Hardisty, R., Varela-Carver, A., Haynes, A. R., Rodriguez-Tome, P., Doi, H., Ko, M. S.,, Pontius, J., Schriml, L., Wagner, L., Maglott, D., Brown, S. D., Lander, E. S., Schuler, G., and Denny, P. (2001). A radiation hybrid map of mouse genes. *Nat. Genet.* **29**(2), 201–205.

Hukriede, N., Fisher, D., Epstein, J., Joly, L., Tellis, P., Zhou, Y., Barbazuk, B., Cox, K., Fenton-Noriega, L., Hersey, C., Miles, J., Sheng, X., Song, A., Waterman, R., Johnson, S. L., Dawid, I. B., Chevrette, M., Zon, L. I., McPherson, J., and Ekker, M. (2001). The LN54 radiation hybrid map of zebrafish expressed sequences. *Genome Res.* **11**(12), 2127–2132.

Hukriede, N. A., Joly, L., Tsang, M., Miles, J., Tellis, P., Epstein, J. A., Barbazuk, W. B., Li, F. N., Paw, B., Postlethwait, J. H., Hudson, T. J., Zon, L. I., McPherson, J. D., Chevrette, M., Dawid, I. B., Johnson, S. L., and Ekker, M. (1999). Radiation hybrid mapping of the zebrafish genome. *Proc. Natl. Acad. Sci. USA* **96**(17), 9745–9750.

Kwok, C., Korn, R. M., Davis, M. E., Burt, D. W., Critcher, R., McCarthy, L., Paw, B. H., Zon, L. I., Goodfellow, P. N., and Schmitt, K. (1998). Characterization of whole genome radiation hybrid mapping resources for non-mammalian vertebrates. *Nucleic Acids Res.* **26**(15), 3562–3566.

Lo, J., Lee, S., Xu, M., Liu, F., Ruan, H., Eun, A., He, Y., Ma, W., Wang, W., Wen, Z., and Peng, J. (2003). 15,000 unique zebrafish EST clusters and their future use in microarray for profiling gene expression patterns during embryogenesis. *Genome Res.* **13**(3), 455–466.

Olivier, M., Aggarwal, A., Allen, J., Almendras, A. A., Bajorek, E. S., Beasley, E. M., Brady, S. D., Bushard, J. M., Bustos, V. I., Chu, A., Chung, T. R., De Witte, A., Denys, M. E., Dominguez, R., Fang, N. Y., Foster, B. D., Freudenberg, R. W., Hadley, D., Hamilton, L. R., Jeffrey, T. J., Kelly, L., Lazzeroni, L., Levy, M. R., Lewis, S. C., Liu, X., Lopez, F. J., Louie, B., Marquis, J. P., Martinez, R. A., Matsuura, M. K., Misherghi, N. S., Norton, J. A., Olshen, A., Perkins, S. M., Perou, A. J., Piercy, C., Piercy, M., Qin, F., Reif, T., Sheppard, K., Shokoohi, V., Smick, G. A., Sun, W. L., Stewart, E. A., Fernando, J., Tejeda, Tran, N. M., Trejo, T., Vo, N. T., Yan, S. C., Zierten, D. L., Zhao, S., Sachidanandam, R., Trask, B. J., Myers, R. M., and Cox, D. R. (2001). A high-resolution radiation hybrid map of the human genome draft sequence. *Science* **291**(5507), 1298–1302.

Paw, B. H., and Zon, L. I. (1999). Primary fibroblast cell culture. *Methods Cell Biol. 1999* **59,** 39–43.

Postlethwait, J. H., Woods, I. G., Ngo-Hazelett, P., Yan, Y. L., Kelly, P. D., Chu, F., Huang, H., Hill-Force, A., and Talbot, W. S. (2000). Zebrafish comparative genomics and the origins of vertebrate chromosomes. *Genome Res.* **10**(12), 1890–1902.

Rozen, S., and Skaletsky, H. J. (2000). Primer3 on the WWW for general users and for biologist programmers. *In* "Bioinformatics Methods and Protocols: Methods in Molecular Biology" (S. Krawetz and S. Misener, eds.), pp. 365–386. Humana Press, Totowa, NJ. Source code available at http://frodo.wi.mit.edu/primer3/primer3_code.html.

Stein, S., Kruglyak, L., Slonim, D., and Lander, E. (1995). RHMAPPER, unpublished software, Whitehead Institute/MIT Center for Genome Research. Available at http://www.genome.wi.mit.edu/ftp/pub/software/rhmapper/. Also available via anonymous ftp to ftp.genome.wi.mit.edu, directory/pub/software/rhmapper.

Ton, C., Hwang, D. M., Dempsey, A. A., Tang, H. C., Yoon, J., Lim, M., Mably, J. D., Fishman, M. C., and Liew, C. C. (2000). Identification, characterization, and mapping of expressed sequence tags from an embryonic zebrafish heart cDNA library. *Genome Res.* **10**(12), 1915–1927.

Woods, I. G., Kelly, P. D., Chu, F., Ngo-Hazelett, P., Yan, Y. L., Huang, H., Postlethwait, J. H., and Talbot, W. S. (2000). A comparative map of the zebrafish genome. *Genome Res.* **10**(12), 1903–1914.

CHAPTER 16

Bacterial Artificial Chromosome (BAC) Clones and the Current Clone Map of the Zebrafish Genome

Romke Koch,[*] **Gerd-Jörg Rauch,**[†] **Sean Humphray,**[‡]
Robert Geisler,[†] **and Ronald Plasterk**[*]

[*]Hubrecht Laboratory
Uppsalalaan 8
3584 CT Utrecht, The Netherlands

[‡]Wellcome Trust Sanger Institute
Wellcome Trust Genome Campus
Hinxton, Cambridge CB10 1SA, United Kingdom

[†]Max-Planck-Institut für Entwicklungsbiologie
D-72074 Tübingen, Germany

I. Introduction

Once the zebrafish genome sequence is complete, there will be only one physical map, and that will be the genome sequence. All clones mapped to the genome sequence will be visible as annotations, and any new clone or sequence from an individual researcher can in principle be mapped by comparing a short stretch of sequence to that of the entire genome. The publication of the genome sequence (expected in 2005) might not result in that blissful situation immediately, because there will probably still be areas that are not completely finished or not fully covered by known clones. In any case, the genome sequence is not yet complete, and therefore it is useful for zebrafish investigators to appreciate the current physical map for what it is: a collection of large plasmid clones [mostly bacterial artificial chromosome (BAC) clones] and of short sequenced clones, which are often linked into contigs (contiguous clone sets). This thousand-island genome can be accessed by hybridization to BAC clone filters, DNA sequence comparison, inspection of overlapping BAC clones in the physical map database on the Internet, and local assembly of shotgun sequence tags into a crude local genome sequence. There are two commonly used entries into the physical map: knowledge of a stretch of DNA sequence and DNA hybridization. (There is a third entry, discussed later, and is a DNA restriction pattern of a large clone, usually a BAC. Such a pattern will usually not be taken as an entry point, because one needs comparison to specific size markers under well-described electrophoresis conditions, and thus it is unlikely that an individual investigator will use this entry point in a single case. We will discuss here the approach under the assumption that a set of potentially overlapping BAC clones, a contig, is available and needs to be verified.) We illustrate both the common entry strategies, using an example of our own recent work, the *dicer1* gene of zebrafish.

The genome project that is undertaken for the zebrafish combines three branches. The first is the maintenance and refinement of the genetic map and markers. The second is the generation of a fully contiguated physical map of the chromosomes by using BAC clones. The third is the sequencing of the whole genome by whole-genome shotgun sequencing (WGS) and sequencing of a minimal set of overlapping clones from the physical map. This chapter gives some insight into how the physical map of the zebrafish genome is created. It describes how the current builds are made and how these data can be used in zebrafish genetics, for example, in positional cloning projects.

II. Physical Map of Fingerprinted Clones

The generation of a fingerprinted physical clone map is a collaborative effort of three European laboratories (R. Geisler at the Max-Planck-Institut in Tübingen, S. J. Humphray at the Wellcome Trust Sanger Institute in Hinxton, and R. H. A. Plasterk at the Hubrecht Laboratory in Utrecht). The project was started in 2000 and has led to a preliminary clone map of roughly 3500 contigs, which cover the complete zebrafish genome. More information on the progress of creation of the physical clone map can be found at the Zebrafish Fingerprinting Project Web site (http://www.sanger.ac.uk/Projects/D_rerio/mapping.shtml).

Several genomic DNA clone libraries have been described for zebrafish, including plasmids, BACs, and yeast artificial chromosomes (YACs) (Amemiya *et al.*, 1999; Barth *et al.*, 1997; Zhong *et al.*, 1998). BAC libraries are the first choice for generating physical maps because they provide large-insert clones. Their large insert sizes make them very suitable to cover large regions of the genome and cover whole genes. Furthermore, they can be grown by using fairly simple culturing techniques. This is in contrast to YACs, which are more difficult to culture.

The physical map is mainly built on BAC and Pl-derived artificial chromosome (PAC) clones. BACs and PACs contain relatively large insert sizes ranging from 75 to 250 kb. The clones used to create the physical map can contribute to available techniques for positional cloning and rescue experiments. Furthermore, the map is used as a framework for the sequencing project of zebrafish.

A. Genomic Clone Libraries

With the generation of a physical map in mind, two large-insert BAC libraries and their pilot libraries were developed (Table I). One was generated by the laboratory of Pieter de Jong at BACPAC Resources in Oakland and the other was generated by Keygene in Wageningen in collaboration with our group.

The main goal was to use libraries with large inserts, each providing an $\sim$10$\times$ coverage of the zebrafish genome. The DNA for both libraries was obtained from testis of the Tübingen zebrafish strain and provided by R. Geisler. The inserts of the CHORI-211 library were cloned by using *Eco*RI and have an average insert size of $\sim$165 kb. The mean insert size of the Keygene library is slightly higher (175 kb) and inserts were cloned by using *Hin*dIII. The Keygene library also contains ligations with extra-large inserts: sizes can go up to 230 kb. Use of different restriction enzymes should reduce representation biases in the library due to base composition. Also, data from the BUSM-1 PAC library created by C. Amemiya were used. This library is based on genomic DNA isolated from blood in the AB strain. The library has an average insert size of 115 kb and a 7$\times$ redundancy.

The genomic coverage of the libraries fingerprinted exceeds 20$\times$ redundancy. The libraries are publicly available and can be screened by using filters and PCR pools.

B. Restriction Digest Fingerprinting

The overlaps of the clones were established by using a fingerprinting technique described by Marra (1997). The method relies on the determination of the overlaps between randomly selected clones by analyzing shared restriction fragments. Figure 1 shows a schematic overview of the principle of restriction digest fingerprinting.

Table I
Libraries Used to Construct the Physical Map of the Zebrafish Genome

Library name	CHORI-211 Bac library	Daniokey Bac library	BUSM1 Pac library	RPCI-71 Bac library	Daniokey Pilot Bac library
Prefix	zC	zK	dZ	bZ	zKp
Zebrafish strain	Tübingen	Tübingen	AB	Tübingen	Tübingen
Vector	pTARBAC2.1	pIndigoBAC-536	pCYPAC6	pTARBAC2	pIndigoBAC-536
Source	Male testis DNA	Male testis DNA	Blood cells	Mixed-gender DNA	Male testis DNA
No. of clones	105,907	104,064	104,064	33,408	11,808
Estimated insert size	165 kb	175 kb	115 kb	85 kb	130 kb
Redundancy	10×	10×	7×	1.7×	1×
Originators	P. de Jong, R. Geisler	R. Plasterk, Keygene N.V.	C. Amemiya	P. de Jong, R. Geisler	R. Plasterk, Keygene N.V.
Distributor	BACPAC Resources http://bacpac.chori.org/	RZPD, Germany http://www.rzpd.de	RZPD, Germany http://www.rzpd.de	BACPAC Resources http://bacpac.chori.org/	—

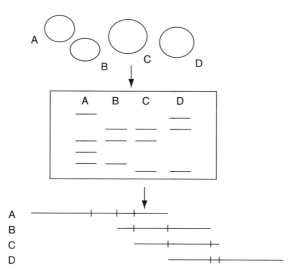

Fig. 1 Schematic overview of the principle of restriction digest fingerprinting. Clones are finger-printed by using a restriction enzyme. Overlapping clones share a number of the restriction fragments, which can be used to determine the overlap between clones.

All clones were cultured in 96-well format and BAC DNA was isolated by using a modified alkaline lysis protocol. BAC fingerprinting was done by restriction digesting of each clone with *Hin*dIII. The fragments were separated by agarose gel electrophoresis and scanned by a Fluorimager after staining with a fluorescent dye. Subsequently, gels were imported into Image, software developed at the Sanger Institute. This software is used to assign the lanes on the gels, recognize bands, and determine band sizes (band calling).

C. Contiguation of Fingerprinted Clones

After band calling, fragments were normalized to the marker lanes and exported to FPC (Finger Printed Contigs). The FPC software (Soderlund *et al.*, 2000) is used to calculate the overlaps between the clones and the relative order, based on the overlap of restriction fragments. FPC uses an algorithm to cluster clones in contigs based on their probability of coincidence score and provides a detailed visualization of the clone overlaps. First, the clones are placed in contigs, using stringent criteria to prevent incorrect joins. Subsequently, the stringency is lowered and clones are manually added and contigs are merged.

At present, more than 200,000 clones have been added to the fingerprinting database and manual curation is performed to link the contigs together. However, several thousand separate contigs are expected to remain. Efforts are therefore underway to anchor the majority of contigs on the T51 radiation hybrid map of the zebrafish genome by radiation hybrid mapping of selected BAC ends (in a collaboration of R. Geisler and Y. Zhou, Children's Hospital Boston). This procedure will yield a coherent physical map of the zebrafish genome by the end of 2004. Concurrently, a minimal tiling path is determined from the clones, which acts as a template for clone-by-clone sequencing of the genome.

III. Screening Methods and Utilization of the Current Zebrafish Clone Map

We describe a few strategies that combine the available repositories and can help the zebrafish researcher solve biological questions. Then we give an example of how we were able to identify the zebrafish *dicer1* gene making use of the physical map and BAC libraries. This method can be altered depending on what information is available for your gene.

A. Screening Genomic Libraries

To find out on which BAC clones the region of interest is located, one can use two strategies: hybridization to spotted BAC filters or screening pools that are generated from the libraries through PCR. Hybridization is mostly achieved by labeling a specific probe for the region of interest and incubation with the filters to allow hybridization to the positive BAC clones. Positive clones can then be recognized on the autoradiogram. Colonies on the filters are spotted in duplicate, using a specific spotting pattern to allow recognition of positive clones from the background. Because the genomic clone filters can contain around 25,000 clones per filter, screening a complete library with a probe can be done rapidly. A disadvantage is the cost of production of filters.

Another method to screen genomic libraries is PCR analysis of pooled DNA samples in the library. These DNA pools are made in a hierarchical fashion so that by performing two to three sets of PCRs on the pools, the coordinates of the positive clones can be found (Barillot *et al.*, 1991). When, for example, a positive DNA pool is found in the first dimension, only that pool is screened for the second, and, if necessary, third dimension. This allows one to rapidly zoom in on the clones of interest. Both hybridization filters and DNA pools are available for the two BAC libraries from its distributors.

B. Screening Sequence Databases

When the sequence is known for the region of interest, one can make use of the data that have been released from the sequencing projects. Sequences from whole-genome shotgun as well as from the clone sequencing project can be searched.

These databases can be queried by using nucleotide alignment search tools such as BLAST and SSAHA. It is also possible to search with translated sequences, such as tblastn, allowing one to pick up homologs and/or orthologs. Databases of finished and unfinished sequences of BAC and PAC clones can be searched by BLAST (http://www.sanger.ac.uk/cgi-bin/blast/submitblast/d_rerio). The Zv3 assembly contains the sequence data available from the sequencing of shotgun clones (WGS) and BAC ends. The supercontigs assembled in this build are tied to the contigs from the FPC map, using the BAC ends. The WGS trace files can be searched by SSAHA (http://trace.ensembl.org/perl/ssahaview?server= danio_rerio).

When a matching sequence is found, one can determine whether it is linked to a clone that is present in a contig of the FPC fingerprint database. This will allow one to find neighboring clones or to find a BAC clone that spans a region that was sequenced in the WGS only.

C. Use of Genetic Marker Data

When other data, such as cDNA sequences or genetic marker data, are available, one can also choose to use the annotated databases that are created for the clone sequencing and WGS project.

A useful interface to access the data of the Zv3 assembly is the Ensembl Web page (http://www.ensembl.org/Danio_rerio/). A contig view display provides a schematic overview of the sequenced contigs, which incorporates gene predictions, ESTs, and other features. The Vega database contains the manually annotated data obtained from the clone-based sequencing of BAC and PAC clones after being selected for sequencing from the FPC contigs. The Vega database can be searched by using the Vega Web interface at http://vega.sanger.ac.uk/Danio_rerio/. Depending on where a match is found and the sequence coverage in that region, one can decide how to continue to examine the nucleotide sequences in the region or try to find other BAC clones in that region.

D. Fingerprinted Contigs (FPC) Search

The FPC database, which allows one to browse the physical map, can be useful if one has located a BAC clone by using one of the methods described previously. Web interface has been created to access the FPC database of fingerprinted clones (http://www.sanger.ac.uk/Projects/D_rerio/WebFPC/zebra /large.shtml). This interface can be used to search the FPC database by clone name or by marker (Fig. 2). The contig display shows information on the number of (sequenced) clones in a contig, the presence of markers, and the number of overlapping BAC clones in a contig. When a contig is selected, a new window is presented with a graphical representation of all clones in that contig. This gives the opportunity to select a tiling path of clones that might also contain the region of interest (Fig. 3). Furthermore, it is indicated which clones in the contig are selected for sequencing and what the status is of those clones. Contigs identified as of special interest to the zebrafish community can be prioritized for sequencing (See the Web site for details.)

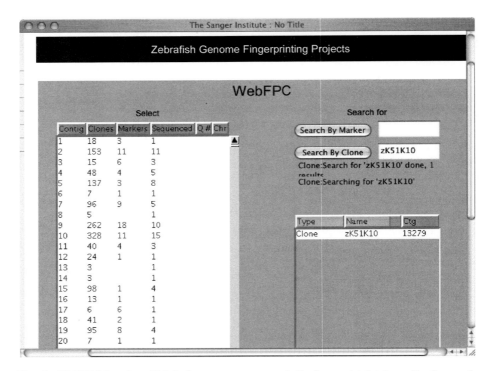

Fig. 2 WebFPC interface. This is the start page to search the fingerprint database. Contigs can be searched for markers and clones. The left panel shows information on the contig number, the number of clones in a contig, and marker and sequence data.

IV. Example: Discovery of the Zebrafish *dicer1* Gene

To illustrate how one can use the methods discussed previously, we describe how one of our labs has mapped the *dicer1* gene by using data from the clone map (Wienholds, 2003). We started by querying the zebrafish EST database with the mRNA sequence of the human Dicer1 homologue, Helicase-MOI. This resulted in three EST clones expressed from the 3′ part of *dicer1*. We continued with clone fc39d11.×1 and could determine an insert size of 2600 bp. To obtain the sequence of the complete cDNA, clone deletion plasmids were created and sequencing was performed on the inserts. Using three primers, we were able to close the gaps in the cDNA sequence and join several smaller contigs into one contig of 2594 bp. The EST clone contained homology to exon 23 to 28 of the human homologue of *dicer1*, Helicase-MOI. Blast analysis resulted in homology with Dicer1 homologues in humans, *C. elegans*, and *Drosophila*.

We used reverse transcriptase PCR to extend the cDNA sequence of *dicer1* in the 5′ direction. For this we needed to design primers in the 5′ region. This was

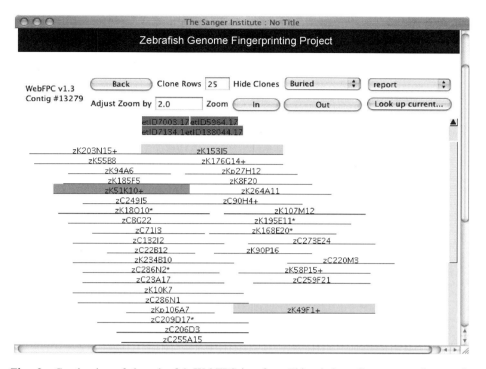

Fig. 3 Contig view of the zebrafish WebFPC interface. This window allows one to browse the fingerprinted clones. Finished clones and clones submitted for sequencing are color coded. This contig contains the clones positive for dicer. The sequence of zK51K10 provided the basis for finding the exons of the gene.

accomplished by making a local trace database of sequences from the WGS project, using software developed in the laboratory (Berezikov *et al.*, 2002). Genotrace identifies the genomic organization for a cDNA, using raw data from genome sequencing projects in progress (trace archives). Local genomic contigs are generated, allowing, for example, the design of PCR primers in intronic sequences to amplify coding regions of a gene. We were able to amplify and sequence products ranging from 1000 to 1500 bp. Combining these data resulted in a contig of 1473 bp, extending the cDNA sequence to 4000 bp, resulting in homology to exon 17 to 28 of the human homologue of *dicer1*.

To complete the coding sequence of *dicer1* in zebrafish, we also needed to identify homology to the first 17 exons of the human homologue. The size of human Helicase-MOI exceeds 43 kb of genomic sequence. At this point we needed to make use of large-insert clones to cover such a region. For that reason, we screened filters of the Daniokey BAC library to find BAC clones containing the zebrafish homologue of Dicer. As a probe we used the complete insert of the EST clone fc39d11.×1. Hybridization of this radioactively labeled probe resulted in 17 positive clones.

Next we used the FPC interface to examine whether the clones were fingerprinted and whether information could be found on how the clones overlapped. It appeared that one contig was found to contain eight positive BAC clones for *dicer1* (Fig. 3). Because at the time no sequence information was available for the BACs, we submitted one of the *dicer1*-positive clones for sequencing at the Sanger Institute. When the BAC was sequenced, we were able to use the amino acid sequence of the human Helicase-MOI in a pairwise tblastn blast search against the BAC sequence data to identify exon 3 to 16. Exon 1 and 2 could not be identified in the BAC sequence because they are both located in the 5′ UTR. By using various other methods we were able to predict the genomic structure of the *dicer1* gene. The predicted cDNA encodes exon 3 to 28, with a length of 6524 nt and a predicted ORF of 1855 amino acids. The domains are highly homologous to known Dicer1 homologues.

The current physical map being generated from fingerprinted clones can be a valuable resource for zebrafish researchers. In combination with other available resources and tools, ordered clones will function as a good starting point to examine the genomic organization of genes in zebrafish.

References

Amemiya, C. T., Zhong, T. P., Silverman, G. A., Fishman, M. C., and Zon, L. I. (1999). Zebrafish YAC, BAC, and PAC genomic libraries. *Methods Cell Biol.* **60**, 235–258.

Barth, A. L., Dugas, J. C., and Ngai, J. (1997). Noncoordinate expression of odorant receptor genes tightly linked in the zebrafish genome. *Neuron* **19**, 359–369.

Barillot, E., Lacroix, B., and Cohen, D. (1991). Theoretical analysis of library screening using a *N*-dimensional pooling strategy. *Nucleic Acids Res.* **19**, 6241–7247.

Berezikov, E., Plasterk, R. H., and Cuppen, E. (2002). GENOTRACE: cDNA-based local GENOme assembly from TRACE archives. *Bioinformatics* **18**(10), 1396–1397.

Marra, M. A., Kucaba, T. A., Dietrich, N. L., Green, E. D., Brownstein, B., Wilson, R. K., McDonald, K. M., Hillier, L. W., McPherson, J. D., and Waterston, R. H. (1997). High throughput fingerprint analysis of large-insert clones. *Genome Res.* **7**(11), 1072–1084.

Soderlund, C., Humphray, S., Dunham, A., and French, L. (2000). Contigs built with fingerprints, markers and FPC V4.7. *Genome Res.* **10**, 1772–1787.

Wienholds, E., Koudijs, M. J., van Eeden, F. J., Cuppen, E., and Plasterk, R. H. (2003). The microRNA-producing enzyme Dicer1 is essential for zebrafish development. *Nat. Genet.* **35**(3), 217–218.

Zhong, T. P., Kaphingst, K., Akella, U., Haldi, M., Lander, E. S., and Fishman, M. C. (1998). Zebrafish genomic library in yeast artificial chromosomes. *Genomics* **48**, 136–138.

CHAPTER 17

The Zon Laboratory Guide to Positional Cloning in Zebrafish

Nathan Bahary,⋆ Alan Davidson,⋆,† David Ransom,⋆ Jennifer Shepard,⋆ Howard Stern,⋆ Nikolaus Trede,⋆ Yi Zhou,⋆ Bruce Barut,† and Leonard I. Zon⋆,†

⋆Division of Hematology/Oncology
Children's Hospital Boston
Boston, Massachusetts, 02115

†Howard Hughes Medical Institute
Harvard Medical School
Boston, Massachusetts, 02115

I. Introduction

The process of positional cloning involves unique issues for all organisms. Success is usually based on experience. Because the Leonard Zon laboratory at the Children's Hospital of Boston has used positional cloning to isolate more than 10 zebrafish genes, the laboratory team has accumulated significant experience in the field. We have used this experience to write a guide that will be helpful for the community of researchers who study and work with zebrafish.

II. Mapping Strains

Many of the problematic issues concerning positional cloning in zebrafish (*Danio rerio*) arise from the genetic polymorphisms in the individual strains of zebrafish. In the Zon Laboratory, we have typically used the AB or TU (Tübingen) strains for mutagenesis screens. A mutant should be maintained on the laboratory strain of zebrafish on which it was created. These and all other widely available zebrafish strains are not entirely inbred. Genetic polymorphisms may be present in a given family of fish. One cannot assume that a family of zebrafish is as isogenic as inbred strains of mice. Therefore, it is important to track polymorphisms in a mutant family by obtaining grandparent and parent DNA, usually from tail clips. This tracking will enable the examination of polymorphisms in progeny that subsequently become important in the genetic mapping of a mutant gene.

WIK and SJD, two zebrafish strains, are polymorphic with respect to the AB and TU strains and can therefore be used for genetic mapping of mutants on AB and TU backgrounds. Strains used in mutagenesis and mapping can be interchanged; that is, mutations can be created on WIK or SJD, and mapping can be done with AB or TU. For our mutations on the AB background, a heterozygote carrier is mated to WIK, and a mapping family is generated.

III. Families and Genetic Markers

Once map crosses have been created and heterozygote mapping pairs have been identified from these tanks, it is important to tail clip and store the DNA from these mapping pairs and from the parents of these fish [e.g., the AB mutant heterozygote and WIK wild type that created the map cross (the grandparents) as well as the AB/WIK heterozygotes (the parents)]. Parent DNA and grandparent DNA are helpful for analyzing polymorphisms in subsequent mapping. Once flanking markers have been found in the initial mapping stage, it is important to test the parents and the grandparents to determine which families are polymorphic and to segregate the marker in an easily interpretable manner. In other words, it is best to collect embryos with suitable allele systems for high-resolution mapping. Once polymorphism has been determined, the mapping heterozygotes that are polymorphic can be selectively used to create mapping panels that will also be polymorphic for the flanking markers.

IV. Crosses for Line Maintenance and Mapping

For the purposes of this discussion, we should note that mutagenesis is performed with AB, and WIK is the polymorphic strain used for mapping. In addition, the reader should assume that the mutation of interest is embryonic lethal, and lines must be maintained as heterozygotes.

Definitions

Incross: Sibling cross

Outcross: AB(mutant)/AB heterozygote × AB wild type

Mapcross: AB(mutant)/AB heterozygote × WIK wild type

Backcross: AB(mutant)/WIK heterozygote × WIK wild type

For long-term line maintenance, we keep the mutation on the same strain in which mutagenesis was originally performed (in this case AB) so that we do not jeopardize future mapping efforts. One could perform either an AB(mutant)/AB heterozygote incross or an outcross. Outcrosses are generally preferable because they help dilute out other recessive mutations acquired during the ENU mutagenesis that are not linked to the phenotype of interest. Using mapcrosses or backcrosses for line maintenance is problematic because recombination can result in a loss of distant polymorphisms that could be critical for future low-resolution mapping.

For mapping, polymorphic hybrid strains must be created. To do this we perform a mapcross. The offspring are raised, half of which will be heterozygotes.

Heterozygotes are identified by multiple random incrosses, and identified heterozygote pairs are set aside for ongoing incross embryo collection. If we cannot collect enough embryos in that generation and the close flanking markers are not available, we turn to the newest generation of the pure original strain [AB(mutant)/AB] and perform another mapcross to repeat the process. We do not raise embryos from an AB(mutant)/WIK incross because recombination in this generation renders the next generation of embryos useless for mapping. If, however, close markers flanking the mutation are available, a backcross can be performed. If the markers are agarose scorable, the potential exists to identify large numbers of heterozygotes by tail clipping instead of mating.

Using a high-throughput PCR format, we have identified up to 50 heterozygote pairs in a week. It is critical to tail clip the parents used in the backcross as well as tail clip the grandparents used for the mapcross so the allele system can be accurately followed in the next generation. To facilitate this process, it is useful for a laboratory to have isolated several (about 10) tail-clipped wild-type fish (in this example, WIK is used). Before the backcross is performed, the tail-clipped WIK wild types are screened with close flanking markers along with tail-clip DNA from the AB(mutant)/WIK heterozygote to establish which wild type has the best allele system for following the mutation. Those wild types are then used for the backcross.

When identifying heterozygotes by tail-clip DNA, it is important to remember that there will be a defined error rate due to recombination. The magnitude of the error depends on the distance between the marker and the mutation. The recombination rate in males is about tenfold lower than in females; therefore, if flanking markers are still somewhat far from the mutation, one should consider doing a backcross in which the AB(mutant)/WIK is male and the WIK wild type is female.

A. Choosing Grandparents and Parents for Better High-Resolution Mapping

Usually after low-resolution mapping, our laboratory segregates individual WIK fish and then genotypes them. The goal is to define the allele in the system that represents the best advantage for our high-resolution mapping purposes.

B. Microsatellite Markers—Agarose Scorable

We have recently undertaken a large-scale approach to evaluate microsatellites (available from the Fishman laboratory) for their ability to be scorable on an agarose gel (Massachusetts General Hospital, 2001). (http://zebrafish.mgh.harvard.edu/mapping/ssr_map_index.html).

Through this analysis, we realized that SJD fishes were mostly isogenic; however, the strain has some regions with polymorphisms. SJD allows for easier mapping than WIK, but the strain is very difficult to use and cannot be propagated in our laboratory. SJD males can be obtained and used to fertilize eggs *in vitro* from heterozygote females for creating mapping families. The WIK strain works well, but it is very polymorphic between individual fish (not inbred). We use WIK in most of our mapping. We are in the process of developing microsatellites for use on a capillary system, such as the ABI 3730 system. These microsatellites should be very useful for high-throughput mapping using different strains.

V. Preparation of the DNA

If the embryos have not already hatched, our laboratory dechorionates them before freezing and prepping. We place the embryos individually into wells of a 96-well plate. We then remove excess buffer, store them dry or in methanol and make sure they are kept at $-20\,^{\circ}$C.

When working with embryos, plates should be kept on ice unless otherwise noted. To prepare the embryos, remove all methanol from the wells. All of the following incubation steps can be carried out in a PCR machine:

1. Add 50 μl of lysis buffer (composition follows) to each well and incubate at 98 $^{\circ}$C for 10 min to lyse cells. Quench on ice or 4 $^{\circ}$C in the PCR machine.
2. Add 5 μl of Proteinase K (10 mg/mL stock) to remove proteins.
3. Incubate at 55 $^{\circ}$C for at least 2 h. We recommend at least one mixing during the PK incubation to increase consistency of the DNA preparations. This incubation can also be left to run overnight. The longer the incubation time, the cleaner the DNA tends to be.
4. Incubate at 98 $^{\circ}$C for 10 min to destroy Proteinase K. Quench on ice (or 4 $^{\circ}$C sink in PCR machine).
5. Spin down lysed embryo debris at 4000 rpm for 10 min.
6. Draw off supernatant into a clean 96-well plate.
7. Dilute as necessary.

Embryo Lysis Buffer
Solution: 1× PCR buffer made to 0.3% Tween 20 (10% stock) and 0.3% NP40 (10% stock)
For 10 ml of buffer, use the following:

- 10 ml PCR buffer (see following composition)
- 300 μl NP40, 10% stock
- 300 μl Tween 20, 10% stock

PCR Buffer
Solution: 10 mM Tris-HCl, pH 8.3, and 50 mM KCl
For 50 ml of buffer, use the following:

- 500 μl 1 M Tris, pH 8.3 (autoclaved)
- 2.5 ml 1 M KCl
- 47 ml sterile ddH$_2$O

VI. Mapping Genes

A. Low-Resolution Mapping

Two preferred methods exist for low-resolution mapping of a gene to a particular chromosome. The first method makes use of scanning microsatellite markers throughout the genome (Knapik *et al.*, 1998; Shimoda *et al.*, 1999). The second method, called half-tetrad analysis (Kane *et al.*, 1999), makes use of early pressure-treated embryos to evaluate the mutated chromosome.

1. Half-Tetrad Analysis

Although the zebrafish is a diploid organism, haploids can live for several days. Maternally homozygous diploid fish can be produced by applying early pressure (EP) to inhibit the second meiotic division after fertilization with UV-inactivated sperm. Analogous to the creation of the maternal diploids, diploid androgenotes can be obtained by UV inactivation of the egg, subsequent fertilization by normal sperm, and application of EP to inhibit the second meiotic division. The ability to create gynogenetic diploids allows the rapid assignment of a gene to a particular chromosome while obtaining information about its distance from the centromere. To genetically localize a mutation, female offspring of an AB mutant × AB (wild type) outcross are squeezed and fertilized with sperm from a genetically unrelated male (also wild type). The F1 heterozygous females are identified by random matings between the F1 offspring. The F1 heterozygous females are subsequently squeezed, and gynogenetically diploid embryos are derived. The distance between the mutation and the centromere can be approximately calculated using the following equation:

Distance [cM] = 50[1 − (2 × Mutant Number/Total Number of Embryos)].

Because the second meiotic division was inhibited in creating gynogenetic diploids, the region of the chromosome between the centromere and the mutation could not recombine in mutant embryos. Hence, markers proximal to the mutant (so-called centromeric markers), when polymorphic between the background and wild-type strain, are homozygous for the background strain allele in mutants and have the wild-type allele in unaffected embryos. Because centromeric markers have been defined for all 25 zebrafish chromosomes, chromosomal localization and distance from the centromere can rapidly be assigned (Kane *et al.*, 1999).

In our laboratory's early work with zebrafish, we used half-tetrad analysis. As we developed robotics in the laboratory, it became easier for us to develop the scanning method described in the following paragraphs. Recently, with the advent of many polymorphic markers that are agarose-scorable, we favor the scanning method.

2. Low-Resolution Scanning

The first step in mapping a recessive mutation to a chromosome is the generation of mapping hybrids (AB(mutant)/WIK). To make this mapping cross, a heterozygous AB carrier (AB(mutant)/AB) is mated to a wild-type WIK fish, and the resulting F1 generation is raised. Practically, we generate two to four mapcross families with different WIK founders to ensure an informative allele system is obtained (because the WIK strain is not completely inbred). To identify heterozygous F1 individuals, these AB/WIK hybrids are mated to each other (incrossed), and their clutches scored for the mutant phenotype. Once a pair of heterozygous hybrids are found, they are mated, and their wild-type and mutant progeny are collected.

In addition, we tail clip the mapping heterozygotes to obtain DNA for analysis of the alleles carried. For the initial low-resolution mapping, we use 40 mutants and 40 wild types. Two mutant pools of 20 and two wild-type pools of 20 are made from these stocks. To make the pools, we take 8 μl of the individual stock DNA from each of the 20 individuals (mutant embryos for the mutant pool or wild-type embryos for the wild-type pool) to give 160 μl, and we increase the volume to 1.2 ml with water. Last, we use the agarose-scorable microsatellite markers to scan the genome for linkage to the phenotype in bulk segregant analysis.

Bulk segregant analysis uses 239 agarose-scorable microsatellites that are typed on a set of DNA samples from wild-type and mutant embryos. Each set contains two pools of twenty embryos from wild types and mutants. PCR products are run on 3% agarose gels at 200 V for 2 h to separate bands. Thus far, most polymorphisms encountered have been subtle; hence, running the gels longer than necessary is always better. Linkage is assumed when a band present in the wild-type pools is absent in the mutant pools. The mutant band(s) may also have a size shift when compared with the wild types. This observation may also indicate linkage. It is best that individuals for this stage as well as for the intermediate mapping stage (see next section) come from the same family. Occasionally, our laboratory is not able to map a mutation in one family, and we have to test another family. It is critical that the 80 embryos come from the same family. Introducing individuals with a different set of alleles might lead to false positives. One might assume linkage to a particular chromosome based on the pattern obtained with the additional family. Very frequently, we find linkage to about three chromosomes, but only one of these linkages is real. To determine which microsatellite is truly linked, each positive marker must be tested on individuals. By testing individuals, chromosomal linkage is confirmed, and the distance between markers can be evaluated.

B. Intermediate-Resolution Mapping

The purpose of intermediate-resolution mapping is to position the gene between flanking markers that are scorable on an agarose gel. This technique allows us to do high-resolution mapping with 1500 embryos with relative ease.

Mapping with this number of embryos is not always possible, but it is a goal worth striving for.

We collect 8 wild types and 88 mutants for intermediate-resolution mapping. To clone a mutant gene, flanking markers should ideally be less than 10 cM apart. We scan microsatellites on the chromosome by ordering roughly six microsatellites on the chromosome arm. If these microsatellites are not polymorphic, we test another six markers until the mutation is linked to microsatellites on the chromosome arm. Based on this recombination mapping strategy, it should be possible to define the flanking microsatellite markers. Markers that are far away from the mutation should yield more recombinants than markers that are close to the mutation. Markers that are on opposite sides of the mutation should give different sets of recombinants, and markers that are on the same side of the mutation should share recombinants. We can narrow down the region by studying more microsatellites until we have markers that are 10 cM (or less) apart. When we are able to define microsatellite markers that are polymorphic on our agarose gel and that are close enough to use as flanking markers, we set up mapping crosses with zebrafish that have this allele system. These new stocks of fish and validated flanking markers are then used for the high-resolution mapping.

C. Fish Husbandry

The number of tanks needed to map a mutant varies based on zebrafish sex ratio in the tanks and on the ease of scoring the phenotype. We will typically generate eight mapcross tanks or backcross tanks for a genetic mapping. Ultimately, we will sacrifice most of these fish, but the goal is to have at least five pairs of fish with an advantageous allele system so that genetic mapping can be conducted very quickly.

D. Tempo

While working out the flanking polymorphisms and genotyping, it is important to continue collecting mutant embryos. We find that between 1500 and 2000 embryos are required for positional cloning. Assuming an interval of 600 kb/cM and a meiotic recombination frequency of one per embryo (this is true for haploid individuals; diploids have an average of 1.3 to 1.5 meioses/individual—one from the mother and 0.3 to 0.5 from the father) will give a resolution of close to 30 kb per meiosis event. This resolution allows the positioning of the mutant gene on a BAC or PAC clone.

The characterization of markers between the flanking polymorphic markers is important. The number of recombinants obtained with each of the flanking markers [which should be placed on the radiation hybrid (RH) map if this information is not available], from intermediate-resolution mapping is considered a guide for estimating the position of the mutated gene. The RH map contains a multitude of expressed sequence tags (ESTs) and BAC ends, which can be used as markers on the walk toward the mutated gene. Typically, our laboratory will pick three to four ESTs in proximity of the estimated gene location and first check if the 3'-UTRs of these ESTs are polymorphic in our mapcross. Frequently, the primers used to map the ESTs are indicated on the Washington University in St. Louis Web site created for EST projects (Washington University in St. Louis, 2004a). If this is not the case, single-stranded conformation polymorphism (SSCP) primers are obtained from a noncoding DNA sequence in the region. In our experience, one out of four ESTs is polymorphic in a given mapcross. If a polymorphism with the chosen ESTs is not found, more ESTs can be tested for polymorphism, or EST primers can be used to isolate PACs. Sequencing the ends of PACs will increase the likelihood of identifying polymorphic markers. Recently, we have found that BLAST analysis of the Sanger Institute genome sequence (The Sanger Institute, 2004a) with an EST will give more extensive sequence with introns and UTRs. These sequences can be searched for polymorphisms. Introns have a higher rate of polymorphism than 3'-UTRs. In addition, CA repeats or contigs can be used to search for polymorphic markers. Once polymorphic markers are identified, the panel of recombinants identified at each of the flanking markers is tested.

Example

Problem: A total of 30 recombinants were identified with the left flanking marker and 35 with the right flanking marker. If 1500 mutant embryos have been collected as part of a recombinant panel, the distance is estimated to be 2 cM from the left flanking marker and 2.3 cM from the right flanking marker (too far to initiate a walk). Two ESTs are identified in the estimated region of the mutated gene. Both are polymorphic. Test the panels of recombinants from either side with these markers as shown in Fig. 1.

Solution: The mutated gene is situated between EST1 and 2. The estimated distance from the closer EST1 is 0.26 cM. This distance is sufficiently small to initiate a chromosome walk using PAC or BAC clone. To solve this problem, one can also use the Sanger Institute genome sequence. A comparison is made of contigs that contain ESTs from the RH panels. If concordance is found, this improves the confidence of the genetic interval.

		EST1		EST2		
Flanking Marker Left			|		Flanking marker Right	
			|			
Recom-binants	30	4	|	0	0	
	0	0	|	16	35	Recom-binants
			|			
			GENE			

Fig. 1 Mapping of mutant genes.

E. High-Resolution Mapping

We have traditionally collected between 1500 and 2000 mutant embryos in an effort to clone these positionally. In high-resolution mapping, these mutant embryos, arrayed in a 96-well format, are tested with the flanking markers found in the low and intermediate stages of mapping. It is critical that every recombination event is scored in this step, assuming a two-allele system. If a three- or four-allele system is used for mapping, some recombination events will be missed. Therefore, it is recommended that the families used for collecting the embryos be chosen according to the most useful allele system as well as the correct polymorphic mapping strain. Furthermore, it is advantageous to limit the number of families used to collect the mutant embryos. Although collecting embryos from only two or three pairs of fish may lengthen the time needed to reach the target number of embryos, this step will simplify further steps in positional cloning. We recently moved our high-resolution mapping to the ABI 3730.

F. Three-Allele Systems Versus Four-Allele Systems

Typing mapcross parents is essential to evade the "allele traps" encountered by this type of bulk segregant analysis. Once flanking markers have been found, mapcross parents should be tested for the proper allele segregation. Three-allele systems are quite common in the AB/WIK crosses used in our laboratory. We often see that the flanking markers do not segregate similarly between different families. This becomes a problem with the high-resolution scan that includes individuals from all mapcrosses. One of the wild-type alleles sometimes has the tendency to migrate the same as the mutant alleles in the agarose gel, so when a high-resolution scan is performed, recombinants are missed because they look like a mutant embryo. A simple way to avoid these bad-allele systems is to type all the parents and grandparents before collecting all 1500 embryos. While the low and intermediate stages of mapping are being performed, embryos can be collected. One should remove those crosses that have bad allele migration from the mapping collection. Four-allele systems can be just as confusing if not fully investigated before the high-resolution phase. Generally, four alleles can be tracked with ease. Problems, however, can arise when an AB allele segregates with the mutant WIK allele. Heterozygotes would be counted as homozygous mutants and not recombinants. The same situation is seen with three-allele systems (Fig. 2).

G. Collecting Mutant Embryos

With the conclusion of high-resolution mapping, a number of recombinants have been identified. These recombinants are rearrayed on a new plate to create the recombinant panel. During this process, it is advisable to continue collecting mutant embryos from the mapping strain beyond the initial collection number of 1500 to 2000 because more embryos may be required later in the process. The recombinant panel is now used in positional cloning, with the number of recombination events lessening as the mutation is neared.

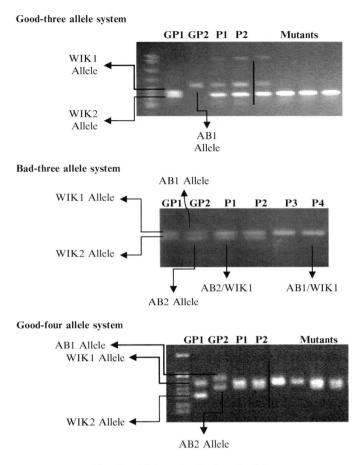

Fig. 2 Allele system in the zebrafish.

H. Chromosomal Walking

After collecting 1500 to 2000 mutant embryos, you can begin chromosomal walking (Fig. 3). Start with flanking the microsatellite markers that are agarose scorable and linked to your gene. The walk starts from an internal marker between two flanking polymorphic microsatellites. Based on the internal marker sequence, an overgo marker (a double-stranded 40-mers) can be designed. Typically, our laboratory uses a 3′-UTR of an EST or a marker from a P1-derived artificial chromosome (PAC) or a bacterial artificial chromosome (BAC) clone end screened with one of the microsatellites or ESTs. Using this internal marker, orientation of the walk relative to the gene is established by studying meiotic recombination in F2 embryos. By taking the marker that is closest to the gene, a

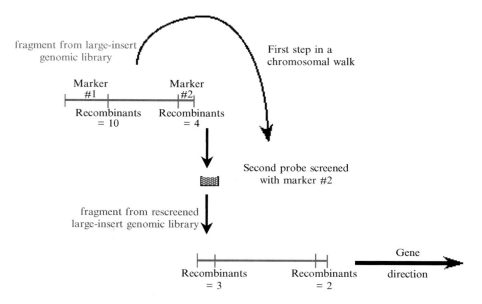

Fig. 3 Chromosomal walking. (See Color Insert.)

large-insert genomic library (PAC or BAC) is screened and the new genomic DNA clones are sequenced from both ends. These end sequences are used to generate new overgo probes. These overgoes are used to screen the large-insert genomic libraries, and, therefore, a "walk" has been established. When starting a chromosomal walk, keep in mind that the narrower the genetic interval, the faster the positional cloning will go.

I. Screening the PAC or BAC Libraries

There are various methods for screening the PAC and BAC libraries. We use a hybridization strategy, using a prediction program to make overgoes (Washington University at St. Louis, 2004b). The overgoes are individually hybridized onto filters. The availability of "double-positives" (positives with two independent overgoes designed against the same sequence fragment) is an advantage for isolating true positives. We typically isolate between 5 to 15 clones per hybridization.

The first PAC library for zebrafish was originally isolated from AB zebrafish red cells (Amemiya and Zon, 1999). The PAC library has an insert size of roughly 100 kb that encompasses 5× coverage or 250 384-well plates. The BAC library is made from a single AB fish and has clones that are only 82 kb on average. Both libraries are assembled in a similar format. Both libraries are in-house and readily available to us for our everyday work. These BAC and PAC libraries are also available at the Deutsches Ressourcenzentrum für Genomforschung (RZPD), translated as the Resources Center of the Human Genome Project in Berlin, Germany.

J. New BAC Libraries

A new BAC library was prepared by Chris Amemiya's laboratory. The library was created by isolation of DNA from blood pooled from 10 male and 10 female Tubingen fish, for a total of 20 fish. The BAC vector backbone, pBACe3.6, has the advantage of being far smaller than the PAC vector. This is advantageous when shotgun sequencing a single large-insert genomic clone (more clones will represent your insert and not vector). The average insert size is approximately 150 kb, with a large number of 200 kb clones represented in the library. Researchers P. de Jong and R. Plasternack have also constructed two additional libraries, CHORI 211 and *Danio* Key Zebrafish BAC. These two libraries are the backbone of the Sanger Institute's *Danio rerio* Sequencing Project and are available at the Children's Hospital Oakland Research Institute (BACPAC Resources Center, 2004) (http://bacpac.chori.org) and the Deutsches Ressourcenzentrum für Genomforschung (RZPD, 2004).

K. Care of the BAC and PAC Libraries

Condensation and cross-contamination are sources of problems for maintenance of the glycerol stocks of the PAC and BAC libraries. To reduce cross-contamination and maintain viability of the cultures, the 384-well plates are handled carefully to limit defrosting. Plates are removed from −80 °C storage and are allowed to warm at room temperature for approximately 5 min. A sterile pipette tip is used to remove a chip of the frozen culture, and subsequent streaking is performed on an agar plate or inoculation of a broth culture. The PAC clones are kanamycin resistant (25 ug/ml), and the BAC clones are chloramphenicol resistant (34 ug/ml). The inner sides of the lids are wiped with ethanol if necessary, and the plates are replaced in storage.

VII. Overgo Strategy for Rapidly Doing Chromosomal Walks and Positional Cloning

Overgoes are extremely useful for designing hybridizations. We have found that the strength of the hybridization signal is equal to that of cDNAs and certainly much better than that of individual oligonucleotides. Based on this excellent hybridization, our approach to positional cloning is to isolate ESTs that are close to a region and isolate clones using the hybridization strategy of overgoes.

Overgoes are generated to two separate sequences on the BAC, and double-positive BAC clones are thus picked for further analysis. We use these clones to evaluate the fingerprinted BAC contigs on the Sanger Institute Web site, and then we see if the walk is extended. If it is not, individual clones are sequenced from both ends, and overgoes are designed to all the clones. In a mass hybridization of all the overgoes, we repeat the process, often with four to six overgoes at a time. Before overgoes are designed, we use Repeat Masker and BLAST sequences to see if there is a repetitive sequence represented by a newly designed overgo. In addition, we may test the overgoes on mock filters from the BAC library to see if there is a low copy number repeat. Checking this number helps prevent blackening of filters and allows the walk to move forward.

Hybridization of multiple overgoes produces many clones, perhaps 40 clones at a time. These are individually picked and gridded out on a filter. We then hybridize each of the overgoes separately to define which of the ends are represented on which clones. This hybridization strategy is extremely efficient because it builds the contig. The strategy is also useful for verifying contigs from the Sanger Institute. Once we determine the direction of walk based on polymorphic markers, we repeat the process again and then assemble the contigs.

VIII. Protocol for Overgo Probing of High-Density Filters

The overgo should be two 24-mers with an 8-bp overlap in the middle; the final product size is 40 bp. Note that there exists online software called Overgo Maker to help you design the overgo that can be found on the Washington University Genome Sequencing Center Web site (Washington University in St. Louis, 2004c). http://genome.wustl.edu/gsc/overgo/overgo.html.

A. Overgo Labeling

The following list is the procedure used for overgo labeling.

1. Heat the stock solution (10 pmol/μl) of mixed oligos at 80 °C for 5 min, then at 37 °C for 10 min, and store on ice.
2. Use the labeling reaction (10 μl): oligo + H_2O = 5.5 μl. Note that doubling the amount of probe (2× this latter reaction, which makes 20 μl) results in more consistent high-quality hybridizations. Incubate at room temperature for a minimum of 1 h.

Oligonucleotides	10 pmol each (1 μl in this case)
BSA (2 mg/ml)	0.5 μl
OLB (-A, -C, -N6)	2.0 μl
^{32}P-dATP[a]	0.5 μl
^{32}P-dCTP[a]	0.5 μl
H_2O	to 10 μl (4.5 μl in this case)
Klenow fragment	1 μl (2 U/μl)

[a]Concentration 3000 Ci/m*mol*.

3. Remove unincorporated nucleotides using Sephadex G50 columns. Denature probes before adding to blots.

B. Preparation of OLB (-A, -C, -N6) Solutions

OLB (-A, -C, -N6) = A:B:C (1:2.5:1.5)
Solution O

- 1.25 M Tris-HCL, pH8
- 125 mM $MgCl_2$

Solution A

- 1 ml solution O
- 18 μl 2-mercaptoethanol
- 5 μl 0.1 M dTTP
- 5 μl 0.1 M dGTP

Solution B

- 2M HEPES-NaOH, pH 6.6

Solution C

- 3 mM Tris-HCl, pH 7.4
- 0.2 mM EDTA
- Aliquot and store at $-20\,°C$.

C. Hybridization Procedure

It is important to note that the Zon Laboratory team performs hybridizations in a hybridization oven at 58 °C. This temperature works well for overgoes with a GC content between 40 and 60%. Using overgoes that are AT-rich may require lowering the hybridization temperature.

1. A warmed 20-ml hybridized solution is added to a 30 cm × 4 cm hybridized bottle. Filters are first wet down with warmed 2 × SSC, are sandwiched between mesh, and then are inserted into the bottle. With the cap on, the bottle is rotated to allow the filter to unroll slowly and prevent the trapping of trap air bubbles. A maximum of eight filters are placed in the same bottle.

Filters are prehybridized for 4 h the first time they are used and 1 to 2 h thereafter. Sheared salmon testes DNA is not used.

2. After prehybridization of the filters, the labeled oligos are denatured at 90 °C for 10 min. They are then added to each bottle. Probes are typically allowed to hybridize overnight; however, a hybridization of 2 days gives somewhat stronger signals. This longer procedure time may be useful when working with older filters.

Hybridization Solution

Composition	For 1000 ml[a]
1% BSA (Fraction V, Sigma)	10 g
mM EDTA	2 ml of 0.5 M EDTA (pH 8.0)
7% SDS (use 99.9% pure SDS)	70 g
0.5 M sodium phosphate[b]	500 ml of 1 M sodium phosphate[b]

[a]To make 1000 ml of solution, use autoclaved deionized H_2O.

[b]An amount of 1 M sodium phosphate must be made as follows (this is really 1 M with respect to sodium but follows the referenced nomenclature): use 134 g of $Na_2HPO_4 \cdot 7H_2O$, and add 4 ml of 85% H_3PO_4; make to 1000 ml.

Using the information in the table entitled Composition, follow these steps:

1. Heat 100 ml H_2O for 15 sec in microwave on high. Add 10 g BSA, and then stir to dissolve.
2. To 500 ml 1 M sodium phosphate, add 200 ml of H_2O, 2 ml of 0.5 M EDTA, and 70 g of SDS. Stir until the SDS is dissolved (approximately 1 h). Keep in mind that the 1 M solution must be made as follows: use 134 g of $Na_2HPO_4 \cdot 7H_2O$, and add 4 ml of 85% H_3PO_4; make to 1000 ml.
3. Add the dissolved BSA, and make volume to 1000 ml.
4. Filter the hybridization solution, and store at 37 °C, preventing the SDS from precipitating.

D. Washing

1. Remove the hybridized solution, and fill the bottle to the 2/3 line with room temperature 2× SSC, 0.1% SDS. Return the bottle to the oven, and rotate the bottle for about 30 min.
2. Transfer the filters to a larger tub on a rotary platform, and wash them as follows:

2 L 1.5× SSC, 0.1% SDS	58 °C	30 min
2 L 0.5× SSC, 0.1% SDS	58 °C	30 min

E. Autoradiography

1. Seal the filters in plastic, and expose them using XAR5 film at −70 °C.
2. Expose the filters overnight, which is usually all the time that is required. This time period can be increased as the filters age (typically a 3-day exposure).

IX. General Flow of Information from the Radiation Hybrid Panel Maps, the Sanger Institute Sequencing Project, and Fingerprinting the BACs

The Sanger Institute has released version 4 of the zebrafish genome assembly, called Zv3. This version includes the whole genome shotgun sequence and is interfaced with sequence from the BAC libraries. The BAC sequence is generated from end sequencing and also from fingerprinted BAC shotgun sequencing projects that are ongoing at the Sanger Institute. Entire BAC contigs are being sequenced and interfaced with the whole genome shotgun sequence. These large contigs provide an incredibly useful point to start looking at your interval. Our laboratory traditionally examines markers from the radiation hybrid panel maps that are in the critical genetic interval, and we BLAST those to obtain the Sanger contigs. Very often, we will find that the same contig contains two individual ESTs on the RH maps in the region. This process establishes that the contig does in fact represent a sequence between the flanking markers.

The Zon Laboratory has also developed two independent programs, available on the laboratory Web site, that evaluate the contigs. The first program involves BLASTing a human gene query to the assembly contigs (Zon Laboratory, 2004a). This procedure will allow any human gene in an interval to be BLASTed in order to find a contig that represents that human gene. Using BLAST in this manner is very useful for examining conserved synteny relationships in zebrafish and human genomes because other genes in the interval may be present. In addition, for our second program, we are using a reverse BLAST in which we have examined the Sanger Institute V4 assemblies BLASTed to known human proteins (Zon Laboratory, 2004b). The same can be done between zebrafish and Japanese pufferfish (*Fugu rubripes*) genomes as well as between human and *Fugu* genomes. Reverse BLASTing allows an investigator to see if other genes are present on the contig and can be used again to make a synteny story. In addition, the Ensemble database at the Sanger Institute has gene annotation in an individual contig, which is a tremendously helpful resource (Sanger Institute, 2004b). We find that in our laboratory, we need to use both Ensemble and our own databases because even though the Ensemble database provides genes, the Web site is difficult to use for directly finding human proteins. Our Web site compares sequences by a BLAST statistical number and allows this type of analysis to occur very rapidly by providing information that is preBLASTed for the entire assembly against the human protein database. From this information, we have determined that a contig exists within our critical interval.

Our laboratory studies contigs that are very close to our gene based on meiotic recombination. We use the assembly sequence, find CA repeats of greater than 12 nucleotides, and use flanking primers in single-stranded length polymorphism (SSLP) analysis to determine whether linkage is available. This process is much better than looking for SSCPs that are present within the interval (although in important areas, we do use SSCPs). By finding CA repeats on a contig in the region, we can quickly study the critical interval and then genotype more embryos that will narrow this interval. Candidate genes on contigs can be tested by rescue assays or by morphilino analysis.

A. PCR Screening of the Pooled PAC and BAC Libraries

In addition to screening the PAC and BAC libraries by hybridizing to filters, we have found it useful to screen a pooled library by PCR (Amemiya and Zon, 1999). It is often helpful to do a PCR reaction using two separate sets of 20-mers oligos (to screen sequences). This process confirms true positives and reduces the number of false positives. There are 270 plates (384-well) in the PAC library. All of the wells from one plate are combined to make 270 plate pools. The 33 superpools are created by pooling either eight (superpools 1–27) or nine (superpools 28–33) plate pools.

The first step is to PCR screen the 33 superpools. It is good to include a positive control (such as zebrafish genomic DNA) and a negative control (either random plasmid or water). The positive control should have strong bands that preferably amplify with more than one primer pair. Once a superpool positive has been found, the next step is to screen the plate pools and row/column DNA. There are six 384-well plates that are divided into 33 sections corresponding to the 33 superpools. Each section contains the eight or nine plate pools that correspond to the superpool and row/column DNA. There are 16 rows of DNA pools (for rows A–P) and 24 columns of DNA pools. For example, a well may contain all the A rows from the eight or nine plates that are in the superpool. The 48 or 49 wells in the section that correspond to the superpool positive should be screened by PCR. This section should be diluted before use (~1:30 in TE). This PCR should yield three positive wells: one that corresponds to a plate pool (giving you the plate number), one that corresponds to a row pool (giving you the row letter), and one that corresponds to the column number (giving you the column number). An example is shown in Fig. 4.

	1	2	3	4	5	6
A	plate 1	pp1 row A	pp1 row I	column 1	column 9	column 17
B	plate 2	pp1 row B	pp1 row J	column 2	column 10	column 18
C	plate 3	pp1 row C	pp1 row K	column 3	column 11	column 19
D	plate 4	pp1 row D	pp1 row L	column 4	column 12	column 20
E	plate 5	pp1 row E	pp1 row M	column 5	column 13	column 21
F	plate 6	pp1 row F	pp1 row N	column 6	column 14	column 22
G	plate 7	pp1 row G	pp1 row O	column 7	column 15	column 23
H	plate 8	pp1 row H	pp1 row P	column 8	column 16	column 24

This shows the upper left section of a 384 well plate containing the plate pools and row/column pools corresponding to superpool 1.

Fig. 4 Plate orientation.

These "positive" clones can then be retrieved from the PAC library and streaked out on LB/Kan plates. One should confirm that this is an overlapping clone by direct PCR of this clone or by sequencing with a sequence-specific primer.

B. Walking and Establishing Contigs by Sequencing PACs

A situation might occur in which taking the next step in a walk becomes impossible because of a lack of SSCP or SSLP markers. Assume that you isolated PAC 1, sequenced both ends, found polymorphic SSCP markers on both ends, and oriented the walk by identifying recombinants on both ends (Fig. 5a). According to the data you have acquired, you would make the correct decision to walk to the left (T7 end) because this end has fewer recombinants and is therefore closer to the gene than the Sp6 end is.

1. You now isolate the PACs with the primers from the T7 end. You obtain PACs 2, 3, 4, 5, and 6 (Fig. 5b).

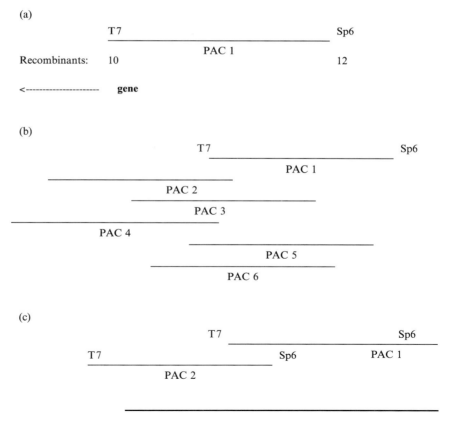

Fig. 5 Chromosomal walking using genomic libraries.

2. You then sequence the ends of all five PACs (although in reality, you may have obtained fewer than five PACs, or you did not acquire reliable sequence from all ends). You may now find that none of the primers have yielded reliable SSCP polymorphisms (although this is quite rare if sequence is available from all 10 ends).

3. To address this poor yield, you can take the forward (or reverse) primers from the T7 and Sp6 ends of PAC 2 (or 3, 4, 5, or 6), 5'-end radiolabel them, and then use them as priming oligos to sequence PAC 1. Only one reaction will work (e.g., the Sp6 end of PAC 2 sequenced off PAC 1). You should not walk from the Sp6 end of PAC 2, however, because it points in the wrong direction (Fig. 5c). From your results, you gather that the T7 end points toward he gene of interest.

4. From this point, you can proceed in one of two ways:

   a. You can sequence the remaining PACs (3, 4, 5, and 6) with the T7 end of PAC 2. Positive results here indicate that the sequenced PAC extends farther toward the gene than PAC 2 (in our example, this would be true for PAC 4). Negative results indicate that PAC 2 extends farther toward the gene than the sequenced (negative) PACs (in our example, these are PACs 3, 5, and 6). You now use PAC 4 for further walking. To determine the end of PAC 4 that is closer to the gene, use both ends and sequence off PAC 2 (or any other PAC). The end that yields a negative result is the one that is closer to the gene. You can now continue the process as described in the following possible Step 4b.

   b. You can use the end that is closer to the gene to isolate another set of PACs (see step 1 described previously). Sequence the ends and attempt to generate polymorphic SSCP markers. If this is not possible, go to step 2 described previously and continue the process from that step. Depending on your estimated distance from the gene (see recombinants on PAC 1), you can take several steps before attempting to re-estimate the distance to the gene by testing recombinants on polymorphic SSCP markers from the PAC ends.

When referring to step 4a, note that instead of sequencing PAC 1 using PAC 2 primers as described, you can attempt the PCR using the primers derived from the ends of PAC 2 and DNA from PAC 1 as a template. This works in many cases, but it requires stringent controls (positive and negative controls) because PCR is not as specific as sequencing.

In reference to step 4b, note that if you do a number of sequencing reactions simultaneously, you can feasibly only run the "G" (or A or T or C) lane because knowing the exact sequence is not as important as ascertaining that the reaction worked.

Because the third version of the genome has been released, our laboratory will typically BLAST V4 with BAC or PAC end sequences. This often yields more sequences for finding polymorphisms. In addition, we examine the fingerprints of

BAC and PAC contigs to order the region. Note that we still require SSCPs at some point to orient our walk.

X. Synteny Between Human, Zebrafish, and *Fugu* Genomes

The Department of Energy (DOE) Joint Genome Institute Japanese pufferfish (*Fugu rubripes)* genome Web site has been updated in 2004 and is a wonderful resource for those involved in positional cloning and comparative genomics (DOE Joint Genome Institute, 2004). The *Fugu* genome has been shotgun sequenced at a 4× the coverage level. This allows assembly of 10- to100-kb scaffolds of genomic sequence. Because of the relatively small size of the *Fugu* genome, synteny is extremely helpful for positional cloning in zebrafish. We make use of this DOE site by first finding a zebrafish EST near a mutant gene. We then look at the human genome site set up by the University of California at Santa Cruz (University of California at Santa Cruz, 2004). And attempt to establish if there is zebrafish and human synteny. We look at all the adjacent genes in the human and see if there are zebrafish orthologs. We then use TBLASTN analysis of the *Fugu* genome using the predicted peptide sequence of either human or zebrafish genes. Synteny can usually be found on a scaffold. Each scaffold is nicely annotated on the DOE site, and the sequence can be downloaded and used for BLAST analysis.

Alternatively, one can use gene prediction programs, such as BLAST P and genescan. These programs allow investigators to probe the syntenic relationships. When one clicks on the BLAST P or genescan program, individual predicted exons are revealed at the bottom of the page. Clicking on the exons gives the predicted homology between *Fugu* and human genomes, and this result can be cross-referenced to the zebrafish. Clicking on the upper bar gives the predicted peptide of the *Fugu* sequence. Using the predicted peptide of the *Fugu* sequence, we use our TBLASTN server and establish if the Sanger Institute sequence has the zebrafish ortholog of the *Fugu* gene. Putative orthologous genes are mapped using the zebrafish radiation hybrid panels to confirm if syntenic relationships occur. We have found that for most chromosomal walks, synteny does exist. These sites are also extremely useful for isolating zebrafish orthologs of human or *Fugu* genes. The reader can browse the Zon laboratory's comparative genomics data (e.g, for human, zebrafish, and *Fugu* genomes] for more information (Zon Laboratory, 2004c).

The Sanger Institute, along with scientists Robert Geisler and Ron Pasternk, has recently undertaken a large-scale fingerprinting project for the CHORI 211 and *Danio* Key Zebrafish BAC libraries using the Hind III enzyme. A physical assembly of the zebrafish genome has been created using fingerprinting software called FPC. This software is available on the Sanger Institute Web site (Sanger Institute, 2004a), and tiles of fingerprinted genomic clones can be visualized. The sequence of a BAC can be entered, and a contig may be available. As more BACs are fingerprinted, this physical map will eventually fill in and eliminate the need for chromosomal walking in many instances. The Sanger Institute has recently annotated these tiles with BACs that are to be end-sequenced. It will be useful for investigators to use this end-sequenceing to establish polymorphic markers in the region. Using the radiation hybrid panels, these BAC contigs are being mapped on the zebrafish genome by the Zon Laboratory and the Robert Geisler Laboratory, providing an excellent resource for positional cloning (Zon Lab website).

A number of sites have been created to help with positional cloning. These sites are listed on the ZFIN database at The Zebrafish Model Organism Database Web site (ZFIN, 2004). We have provided ZFIN with a short description of each of the Web sites.

In a positional cloning project, we find a close polymorphic marker after the high-resolution map, and we proceed to evaluate the fingerprint and the end sequences available. We then derive markers from those end sequences and determine where on the chromosomal walk the gene may lie. This process will be used increasingly in the future as chromosomal walking will become something of the past.

Sean Humphray from the Sanger Institute is willing to extend contigs for individuals. An interested individual can simply contact Humphray and give him information regarding a chromosomal walk. Humphray can lower the stringency of the FPC software and thereby extend walks over many more contigs. The fingerprinting software is very specific for matches of Hind III fragments. By relaxing the stringency (where not all Hind III fragments must match), one can extend contigs, but one can also encounterless certainty that the contigs are truly forming a physical linear assembly. Nevertheless, these BACs can be tested by polymorphisms or RH panel mapping of markers within these regions.

XI. Proving a Candidate Gene Is Responsible for the Mutant Phenotype

What is termed the *rescue* of the mutant phenotype is the gold standard for confirming that a candidate gene is responsible for the mutant phenotype. One can demonstrate that the wild-type gene can rescue the genetic defect (assuming that the defect has compromised the function of the gene rather than generated a gain-of-function mutation). To do this, both wild-type and mutant cDNAs should be subcloned into a vector that is suitable for the synthesis of capped mRNA (i.e., an RNA polymerase site at the 5'-end of the cDNA, a stable 3'-untranslated region containing a polyadenylation signal, and unique restriction enzyme sites for linearization of the plasmid). Commonly used vectors include pCS2+, pSP64T, and pXT7. Both wild-type and mutant cDNA plasmids should be used to generate an *in vitro* translated protein (i.e., 35S-met labeled), which can be resolved on acrylamide minigels and detected by autoradiography. This assay provides an assessment of protein size, stability, and translational efficacy of the wild-type and mutant constructs.

"Run-off" synthetic mRNA transcripts can be generated from the linearized template using commercially available kits, purified, and then resuspended in either sterile distilled water or in a 1× Danieau's solution (typically at a stock concentration of 1–3 μg/μl). Diluted mRNA (typically 5–500 pg/μl, depending on the potency and toxicity of the encoded protein) is then injected into one to four cell-stage embryos produced by parents carrying the mutation. Amelioration of the phenotype (either by morphology or biochemical assay) is then assessed at the appropriate developmental stage. Generally, the later the onset of the mutant phenotype, the more difficult it is to *rescue*, owing to the degradation of the injected mRNA and protein during development. Thus, it is unreasonable to expect every mutant embryo to be rescued. In addition, many proteins, particularly transcriptional regulators, can have severe effects on early embryonic patterning events, thereby making it impossible to rescue later developmental pathways. Injection of mRNA synthesized from the mutant cDNA plasmid permits an assessment of the severity of the mutation [i.e., *no rescue* (or morphological effect on development) would suggest a complete loss of function, whereas *partial rescue* may indicate a hypomorphic mutation].

XII. Morpholinos

To evaluate candidate genes, one morpholino is used against the ATG region to prevent translation, and another morpholino is designed against a splice site. Typically, we use the splice donor because this seems to create aberrant splicing. An RT-PCR reaction can be extremely useful in showing that there is no normal splice form.

A. Allele-Specific Oligonucleotide Hybridization

Oligonucleotide hybridization illustrates that the genetics for a positional cloning project are correct. Polymorphism between mapping strains for markers recovered from a chromosomal walk, such as ends of BAC, PAC, and YAC clones, can be converted into an allele-specific oligonucleotide (ASO) hybridization assay for a meiotic map. The ASO assay (Farr *et al.*, 1988; Wood *et al.*, 1985) is capable of detecting single nucleotide polymorphic differences between mapping strains. The basis for the assay is PCR amplification from genomic DNA for the corresponding BAC, PAC, or YAC end, dotting the PCR product on nylon membrane, differential hybridization with 5'-labeled 19-mers containing the single nucleotide difference present in the mapping strains, and autoradiography of the washed nylon filter. All the hybridizations of the 5'-labeled ASO primers are performed in tetramethylammonium (TMA) buffer (Farr *et al.*, 1988) at 45°C, and they are washed at 55° to 56°C in TMA for any oligo length of 19-mers, irrespective of GC-content.

To use this assay, you need to know the sequences for the corresponding BAC, PAC, or YAC end from your two mapping strains to design ASO primers. You can use the PCR primers to amplify the fragment from your homozygous mutants and homozygous wild-type embryos (these have been ascertained by other microsatellite markers). The PCR fragments are then directly sequenced or subcloned to determine their single nucleotide differences, which are used to design your ASO 19-mers. The PCR fragments are immobilized onto nylon membrane in duplicate. For one set of membrane, you would hybridize with an ASO primer for one strain, and then you would hybridize the duplicate membrane with the second ASO in a separate container. After hybridization at 42 °C for minimum of 4 h, wash the filters in 2 × SSC + 0.1% SDS at room temperature for 20 min, and then wash them in a TMA wash buffer at 55° to 56 °C for specificity. Then you can expose the filters for autoradiography.

Once the ASO has been shown to work in the mapping cross, genotyping of large numbers of embryos can happen in a high-throughput manner. Because the dot blot manifold can accommodate 96 samples, you can array 96 samples of PCR products from individual embryos/filters. Thus, using this assay, you could genotype approximately 1000 embryos with 10 to 11 filters in only one experiment.

Note the supplies needed for the ASO assay:

• ASO: You will need to design an oligonucleotide of 19 nucleotides in length. This oligonucleotide has a single nucleotide difference from your mapping strains located at the center of the 19-mers to maximize their Tm differences.

• Dot/slot blot manifold (Schleicher and Schuell, YEAR; Biorad, YEAR): You will need this manifold that facilitates application of PCR products onto a nylon membrane.

• Tetramethylammonium (TMA) chloride solution: This solution is the key component to the ASO assay because it equalizes the Tm differences based on GC content of the 19-mers. As a result, the Tm stability is a function of length that exactly matches to the target DNA. A premade 5-M stock solution is available from Sigma (Cat. #T3411).

• $[\gamma\text{-}^{32}P]ATP$ of 3000–6000 Ci/mmol specific activity:

Acknowledgments

The original text for the section entitled Protocol for Overgo Probing of High-Density Filters was courtesy of John D. McPherson at Washington University in St. Louis, Missouri.

References

Amemiya, C. T., Alegria-Hartman, M. J., Aslanidis, C., Chen, C., Nikolic, J., Gingrich, J. C., and de Jong, P. J. (1992). A two-dimensional YAC pooling strategy for library screening via STS and Alu-PCR methods. *Nucleic Acids Res.* **25;20**(10), 2559–2563.

Amemiya, C. T., and Zon, L. I. (1999). Generation of a zebrafish P1 artificial chromosome library. *Genomics* **1;58**(2), 211–213.

BACPAC Resources Center (2004). BACPAC home page. http://bacpac.chori.org DOE Joint Genome Institute (2004). Japanese pufferfish *Fugu rubripes*. http://genome.jgi-psf.org/fugu6/fugu6. home.html

Farr, C. J., Saiki, R. K., Erlich, H. A., McCormick, F., and Marshall, C. J. (1988). Analysis of RAS gene mutations in acute myeloid leukemia by polymerase chain reaction and oligonucleotide probes. *Proc. Natl. Acad. Sci. USA* **85**, 1629–1633.

Johnson, S. L., Africa, S., Horne, S., and Postlethwait, J. H. (1995). Half-tetrad analysis in zebrafish: Mapping the *ros* mutation and the centromere of linkage group I. *Genetics* **139**(4), 1727–1735.

Kane, Zon, and Detrich (1999). *The zebrafish: Genetics and genomics.* Vol. 60. Academic Press, San Diego.

Knapik, E. W., Goodman, A., Ekker, M., Chevrette, M., Delgado, J., Neuhauss, S., Shimoda, N., Driever, W., Fishman, M. C., and Jacob, H. J. (1998). A microsatellite genetic linkage map for zebrafish (*Danio rerio*). *Nat. Genet.* **18**(4), 338–343.

Massachusetts General Hospital (2001). Zebrafish microsatellite map server. http://zebrafish.mgh. harvard.edu/mapping/ssr_map_index.html.

RZPD (2004). Deutsches Ressourcenzentrum für Genomforschung Gmbh home page. http:// www.rzpd.de/

Shimoda, N., Knapik, E. W., Ziniti, J., Sim, C., Yamada, E., Kaplan, S., Jackson, D., de Sauvage, F., Jacob, H., and Fishman, M. C. (1999). Zebrafish genetic map with 2000 microsatellite markers. *Genomics* **15;58**(3), 219–232.

The Sanger Institute (2004a). Sequencing projects BLAST search services. http://www.sanger.ac.uk/ DataSearch/

The Sanger Institute (2004b). Pre! Project Ensemble. http://pre.ensembl.org

University of California at Santa Cruz (2004). UCSC Genome Bioinformatics. http://genome.ucsc.edu/ index.html

Washington University in St. Louis (2004a). GSC EST projects. http://genome.wustl.edu/est/

Washington University in St. Louis (2004b). Tools, protocols, and technical information. http:// genome.wustl.edu/tools/

Washington University in St. Louis (2004c). Genome sequencing center. http://genome.wustl.edu

Wood, W. I., Gitschier, J., Lasky, L. A., and Lawn, R. M. (1985). Base composition-independent hybridization in tetramethylammonium chloride: A method for oligonucleotide screening of highly complex gene libraries. *Proc. Natl. Acad. Sci. USA* **82**, 1585–1588.

ZFIN (2004). The zebrafish model organism database. http://zfin.org/zf_info/dbase/db.html

Zon Laboratory (2004a). Human BLAST against zebrafish assembly 3. http://134.174.23.160/ humanblastzv3

Zon Laboratory (2004b). Zebrafish assembly 3 BLAST human gene. http://134.174.23.160/ zv3blasthuman/

Zon Laboratory (2004c). Zon lab comparative genomics. http://zfblasta.tch.harvard.edu/ CompGenomics/

PART III

Transgenesis

CHAPTER 18

Lessons from Transgenic Zebrafish Expressing the Green Fluorescent Protein (GFP) in the Myeloid Lineage

Karl Hsu,★,† **A. Thomas Look,**★ **and John P. Kanki**★

★Department of Pediatric Oncology
Dana–Farber Cancer Institute
Harvard Medical School
Boston, Massachusetts 02115

†Division of Hematology/Oncology
Beth Israel Deaconess Medical Center
Harvard Medical School
Boston, Massachusetts 02115

METHODS IN CELL BIOLOGY, VOL. 77

I. Introduction

This chapter provides a useful reference for those interested in myeloid development in zebrafish. Previous studies of zebrafish hematopoiesis have revealed that it is similar to mammals and other higher vertebrates regarding representative blood cell types, including the erythroid, thrombocytic, myeloid, and lymphoid lineages (reviewed in Berman *et al.*, 2003). The optically clear zebrafish embryo is particularly amenable to *in vivo* analyses and the use of green fluorescent protein (GFP) expression for monitoring hematopoietic development in transgenic zebrafish. The relative stability of the GFP protein allows it to be used as a cell-tracking marker in order to experimentally analyze hematopoietic cell differentiation and movements, correlated with developmental changes in gene expression patterns. These studies can contribute in a major way to our understanding of the genetic regulatory mechanisms underlying the hierarchical progression of hematopoietic development. The *pu.1* gene is expressed early in myelopoiesis, and this chapter focuses on the use of stable transgenic zebrafish lines, expressing GFP under the control of the *pu.1* promoter, to examine myeloid cell differentiation and development.

II. Myeloid Cells in Zebrafish

Zebrafish and mammals share many of the morphological and cytological features of myelopoietic cell types, including granulocytes and monocyte/macrophages (Bennett *et al.*, 2001). One of the granulocytic lineages in the zebrafish resembles mammalian neutrophils, and maturation of the zebrafish neutrophil precursors reiterates numerous elements of this developmental process in humans. As in mammalian myelopoiesis, large granulocytic promyelocytes in the zebrafish become neutrophils, with a segmented nucleus and granular cytoplasm, although the nucleus comprises two or three lobes rather than the four or five lobes typically seen in mammals. Zebrafish neutrophils exhibit myeloperoxidase and acid phosphatase activity, but unlike human neutrophils, are not positively stained in periodic acid-Schiff (PAS) reactions.

Another zebrafish granulocytic lineage possesses highly granular cytoplasm resembling that of basophils and eosinophils in humans, but lacks the segmented nuclei characteristic of these cell types. In contrast to human eosinophils, the granules of zebrafish cells are peroxidase negative and have a positive PAS reaction. The ultrastructural features of these granules resemble those of human basophils and mast cells. Similar cells have been identified in other teleosts, and because they exhibit attributes of both eosinophils and basophils they have been termed basophil/eosinophils (Bennett *et al.*, 2001). This cell type might reflect part of the evolutionary divergence of mammals from teleosts and might represent a common precursor that evolved into two distinct cell lineages in mammals.

Monocyte/macrophages are also found in zebrafish originating from the anterior lateral plate mesoderm (ALPM) during embryogenesis (Herbomel, 1999). As in other species, zebrafish macrophages are highly motile with large, dynamic lamellipodia, filopodia, and complex pseudopodia. These embryonic macrophages also exhibit phagocytosing properties much like their mammalian counterparts (Takahashi, 1996).

Myelopoietic gene expression in zebrafish exhibits many of the patterns discerned for mammalian myelopoietic orthologs (Bennett *et al.*, 2001). As in humans, the cloned zebrafish *myeloperoxidase* (*mpo*) and *leucocyte-specific plastin* (*l-plastin*) genes are respectively expressed in granulocytes and monocyte/macrophages, as demonstrated by mRNA *in situ* hybridization assays. By 20 h postfertilization (hpf) expression of *mpo* was detected in cells of the posterior intermediate cell mass (ICM) and both *mpo* and *l-plastin* were detected in cells developing in the ALPM and migrating over the yolk cell. Between 2 and 4 days postfertilization (dpf), these cells had entered the circulation and were distributed throughout the embryo. Other orthologs of known mammalian genes active during myelopoiesis, such as *cebpα*, *lysozyme-c*, *L-plastin*, *coronin*, and *phox47* are also expressed in zebrafish myeloid cells (J. Rhodes and T. Liu, personal communication); (Bennett *et al.*, 2001; Herbomel *et al.*, 1999). As in other teleosts, the kidney serves as the principal hematopoietic organ in adult zebrafish and staining of sections through this organ demonstrates the presence of *mpo*-expressing cells.

III. Expression of *pu.1* in Myeloid Development

A. Expression in Vertebrates

The PU.1 protein belongs to the ets family of transcription factors and plays an early role in myelopoiesis that is essential for the development of both myeloid (granulocytes and monocytes/macrophages) and lymphoid cells (Akashi *et al.*, 2000; Hromas *et al.*, 1993; Klemsz *et al.*, 1990). In transgenic mouse studies, embryos in one *PU.1*-deficient mouse line died in late gestation and exhibited an array of functional deficiencies in macrophages, granulocytes, and progenitors of B and T lymphocytes (Scott *et al.*, 1994). Another *PU.1*-deficient mouse line produced live pups that had impaired myeloid development, lacked B-lymphocytes, and died soon after birth from septicemia (McKercher *et al.*, 1996). Analysis of the promoter and enhancer sequences of genes expressed by hematopoietic cells have defined *PU.1*-dependent regulatory elements, including components of the B-cell receptor and an array of adhesion molecules, growth factor receptors, and lysozymal enzymes expressed by myeloid cells (Tenen *et al.*, 1997).

In developing vertebrates, common myeloid progenitors give rise to separate precursors for cells of the granulocyte/macrophage and erythrocyte/megakaryocyte lineages. Evidence suggests that *PU.1* has a major role in determining myeloid cell fate. Upregulation of *PU.1* during hematopoiesis leads to stem cell commitment toward the myeloid lineage, whereas downregulation results in erythroid commitment (Voso *et al.*, 1994; reviewed in Cantor *et al.*, 2002). Recent studies in *Xenopus* have shown that *PU.1* blocks erythroid differentiation by directly antagonizing GATA-1 activity (Rekhtman *et al.*, 1999; Zhang *et al.*, 2000). The *PU.1* promoter is regulated by PU.1 itself in an autoregulatory loop, and C/EBPα can induce *PU.1* gene expression, suggesting a direct role for both proteins in the regulation of the *PU.1* gene (Wang *et al.*, 1999). *PU.1* might also play an important role in the development of myeloid leukemias. Overexpression of *PU.1* in transgenic mice, driven by the spleen focus-forming virus (SFFV) long terminal repeat (LTR) results in erythroleukemia (Moreau-Gachelin *et al.*, 1996).

B. Expression in Zebrafish

Consistent with observations in mammals, zebrafish *pu.1* expression is detectable by whole-mount mRNA *in situ* hybridization in hematopoietic cells between 12 and 30 hpf, but not in older embryos (Bennett *et al.*, 2001; Lieschke *et al.*, 2002). Expression of *pu.1* is observed by 12 hpf in the anterior lateral plate mesoderm of the developing embryo, followed by transient expression in the ICM. The posterior expression is significantly decreased by 20 hpf, whereas anterior expression persists and is observed in cells spreading anteriorly over the yolk, until 28–30 hpf. As in mammals, *pu.1*-expressing cells appear later in development than those expressing *scl*, *lmo2*, and *gata2* and before more mature markers of myeloid development such as *mpo* and *l-plastin*, suggesting that this transcription factor regulates the differentiation of hematopoietic stem cells along the myeloid pathway. Thus, development of a transgenic zebrafish expressing *EGFP* under control of the zebrafish *pu.1* promoter would be a useful tool to study the regulation of myeloid cell differentiation *in vivo*.

IV. Analysis of the Zebrafish *pu.1* Locus

A. Genomic Structure

Genomic bacterial artificial chromosome (BAC) clones harboring the zebrafish *pu.1* locus have been identified by hybridization, using a 1034-bp *pu.1* cDNA fragment (Hsu *et al.*, in press; Ward *et al.*, 2003). Comparison of genomic and cDNA sequences showed that the *pu.1* gene is distributed over six exons encompassing approximately 15 kb (Fig. 1B); this pattern is similar to findings with other *pu.1* orthologs, except for the presence of an extra exon (exon 2; Ward *et al.*, 2003). However, RT-PCR analysis of zebrafish *pu.1* mRNA indicates an alternatively spliced product that skips exon 2 (Ward *et al.*, 2003).

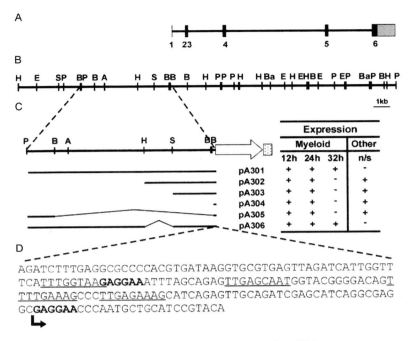

Fig. 1 Analysis of the zebrafish *pu.1* locus and the promoter region. (A) Intron–exon structure of the *pu.1* gene. Solid bars represent the translated exonic sequence, and the shaded bar shows the 3′ UTR. The 5′ UTR is too small to be visualized at this scale. (B) Linear depiction of the *spi1* genomic locus. Letters indicate restriction enzyme sites (A = *Apa*I, B = *Bgl*II, Ba = *Bam*HI, E = *Eco*RI, H = *Hin*DIII, P = *Pst*I, S = *Sph*I). (C) Activity of *spi1* promoter fragments in transient expression assays. Embryos injected with the constructs indicated at the left were examined at 12, 24, and 32 hpf for patterns of enhanced green fluorescent protein (EGFP) expression that largely recapitulate endogenous *pu.1* expression pattern. Results were scored as present (+) or absent (−). Nonspecific (n/s) expression at other sites (typically the skeletal muscle and the eye) was similarly scored. All constructs were injected at least twice, with 20–40% of injected embryos showing EGFP fluorescence in each case. Promoter-only injection controls yielded no fluorescence. (D) Sequence analysis of the core *pu.1* promoter. Bases correspond to the *pu.1* promoter region cloned upstream of *EGFP* in pA304, which produced early myeloid expression. Consensus *pu.1* and *c/ebpα* sites are shown in boldface and underlined, respectively; the transcription start is indicated by an arrow. From Ward *et al.* (2003). The zebrafish Spi1 promoter drives myeloid specific expression in stable transgenic fish. *Blood* **102**, 3238–3240, with permission.

B. Zebrafish *pu.1* Promoter

Various fragments of the genomic sequence upstream of the *pu.1* start codon have been cloned in front of an EGFP reporter gene to drive its expression in myeloid cells (Hsu *et al.*, 2004; Ward *et al.*, 2003). Promoter analysis by Ward and coworkers suggests that a 181-bp core promoter is capable of providing early myeloid-specific expression, whereas other regions extending up to 5.3 kb upstream are required for later expression in circulating cells and for suppressing nonspecific expression (Ward *et al.*, 2003). The murine *PU.1* promoter possesses similar elements with segments of the murine *PU.1* promoter as small as 334 bp conferring myeloid-specific gene expression in transient transfection assays *in vitro* (Chen *et al.*, 1995, 1996). However, these small genomic fragments, as well as longer ones extending up to 2.1 kb, could not drive reporter gene expression in stable lines of transgenic mice (D. Tenen, personal communication). Recently, investigators have generated transgenic mice in which a 91-kb murine *PU.1* genomic DNA fragment directed reporter gene expression in a pattern similar to that observed for the endogenous *PU.1* gene (Li *et al.*, 2001).

Analysis of the zebrafish 181-bp core region revealed the presence of putative Pu.1-binding sites, one beginning at base pair 93 and another immediately after the transcription start site, with the latter arrangement conserved in both the murine and human promoters (Chen *et al.*, 1995, 1996; Ward *et al.*, 2003). Sites for C/Ebpα binding were also identified, as had been postulated for the murine promoter, although no Pu.1 or octamer-binding sites were observed (Wang *et al.*, 1999).

V. Germline Expression of Enhanced Green Fluorescent Protein (EGFP) Under Control of the Zebrafish *pu.1* Promoter

In two independent laboratories, stable transgenic fish lines were established by injecting one-cell-stage embryos with either linearized 9- or 5.3-kb enhanced GFP (EGFP) fragments and grown to adulthood (Hsu *et al.*, 2004; Ward *et al.*, 2003). With the 9-kb fragment, 90 adults were analyzed and seven transgenic founders were identified that produced offspring expressing the *EGFP* transgene within 24 hpf. With the 5.3-kb fragment, 210 adults were screened, resulting in one transgenic founder. The transmission rates for the 9.0-kb fragment varied from 4 to 50% among the progeny of the transgenic founders, indicating germ cell mosaicism in genomic transgene integration, consistent with previous transgenic zebrafish reports (Gong and Hew, 1995; Stuart *et al.*, 1988). All the transgenic founders produced embryos that expressed *EGFP* at levels detectable with a dissecting microscope equipped with epifluorescence. However, individual founder lines displayed different levels of *EGFP* expression, and some lines exhibited either ectopic expression in the brain and eyes or muscle expression. Fortunately, expression of EGFP in these ectopic tissues did not interfere with the analysis of EGFP-expressing hematopoietic cells. Presumably, this variation is due to the differences in transgene integration sites within the genome. The rate of transgene transmission in the F2 incross of F1 siblings, each harboring the transgene, suggested a dominant Mendelian inheritance ratio (approximately 75%), consistent with the integration of the transgene into a single chromosome locus.

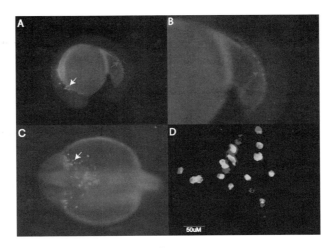

Fig. 2 Fluorescent images of *TG(pu.1:EGFP)*^{df5} embryos at 22 hpf. (A) Lateral view of GFP+ cells. (B) Magnified lateral view of the posterior tail in (A). (C) Dorsal view of the anterior head region. White arrow, *pu.1*-expressing myeloid cells migrating from the anterolateral mesoderm over the yolk; red arrow, *pu.1* cells in the ICM. (D) Confocal image of two-color coexpression assays on cells migrating over the anterior yolk at 22 hpf using a *pu.1* RNA probe (red) and an antibody to EGFP (green). Yellow indicates coexpression. From Hsu, K. *et al.* (2004). The pu.2 promoter drives myeloid gene expression in zebrafish. *Blood* **104**, 1291–1297, with permission. (See Color Insert.)

A. Analysis of *EGFP* Expression in *pu.1*-Transgenic Embryos

EGFP expression in the transgenic fish lines was detected at low levels as early as the six-somite stage as bilateral stripes within the anterior-lateral plate mesoderm (ALPM), when the endogenous *pu.1* transcript is first detectable by *in situ* hybridization (Hsu *et al.*, 2004; Lieschke *et al.*, 2002). These cells then converge toward the midline and by 22 hpf have begun to migrate away from this region, spreading over the yolk cell and into the head (J. Rhodes, personal communication). *pu.1* is also transiently expressed in the ICM at this time, and together the EGFP expression pattern in hematopoietic cells is indistinguishable from the distribution of endogenous *pu.1* mRNA (Fig. 2A–C). Confocal coexpression studies with an anti-GFP antibody and an antisense zebrafish *pu.1* mRNA probe revealed that more than 95% of the cells expressed both EGFP and *pu.1* mRNA at 22 hpf (yellow cells, Fig. 2D). This observation indicates that EGFP expression reliably represents this hematopoietic population of *pu.1*-expressing cells *in vivo*. Furthermore, the ability to perform mRNA coexpression studies on GFP-expressing cells allows the analysis of their changes in gene expression throughout development.

At early timepoints, when EGFP-positive cells are within bilateral clusters of the ALPM along the lateral sides of the developing head, they coexpress orthologs of mammalian stem cell genes, such as *scl*, *gata-2*, and *lmo2*. All EGFP-positive cells move ventromedially, clustering at the midline, and lose the expression of these genes (J. Rhodes, personal communication). However, these genes continue to be expressed in other EGFP-negative cells within these clusters that remain within the developing brain. Cytological examination of EGFP-expressing cells isolated from transgenic embryos at this stage by fluorescence-activated cell sorting (FACS) revealed a homogenous population of intermediate-sized cells with round, slightly irregular nuclei consisting of fine, dispersed chromatin and featureless cytoplasm, lacking distinct granularity. These cells morphologically resembled early immature myeloblasts seen in mammals.

pu.1 mRNA levels were undetectable by 30 hpf, but EGFP-positive cells were observed much later, indicating the stability of the EGFP protein. This feature of the transgenic line allowed the GFP expression to serve as a cell-specific lineage marker demonstrating the continued survival of these *pu.1*-expressing cells as they differentiated into more mature myeloid cell types. As the GFP-expressing cells dispersed away from the midline, they became irregular in shape and exhibited pseudopodial extensions (Herbomel *et al.*, 1999). As these cells actively migrated out over the yolk, these GFP-expressing cells decreased their expression of *pu.1* mRNA and increased coexpression with more mature myeloid cell markers such as *mpo, l-plastin* and other mature myeloid genes such as *coronin, lysozyme-c,* and *phox 47.* These observations suggest that at this stage, the GFP-expressing cells undergo the transition from myeloid progenitor cells to mature myeloid cells (J. Rhodes, personal communication). Consistent with these findings, the GFP-expressing cells isolated by FACS during these later stages represent a heterogenous population of hematopoietic cells consisting of intermediate-sized cells, as well as larger cells with condensed nuclei, increased cytoplasmic granularity, and some exhibiting nuclear indentation. These cells represented a range of maturing myeloid cell morphologies through the metamyelocyte cell stage.

Recent studies showed that the use of morpholinos (Nasevicius and Ekker, 2000) designed to specifically block protein expression of the zebrafish *pu.1* gene resulted in a substantial decrease, or complete loss, of myeloid gene expression, demonstrating its requirement for myeloid differentiation. Because the *pu.1* morpholino did not directly inhibit GFP expression, the subsequent fate of these cells could be analyzed. Interestingly, the inhibition of endogenous *pu.1* protein expression in EGFP-transgenic embryos led to a loss of cell motility in GFP-expressing cells and the expression of erythroid genes, indicating their developmental switch from myeloid to erythroid cell lineages. Coexpression analysis of GFP with *gata-1* and *α-globin* in these embryos confirmed that EGFP-expressing cells underwent this transition and demonstrated their bipotential capacity. These experiments identified myeloerythroid progenitor cells (MPCs) in the zebrafish that are likely to represent the functional equivalent of common myeloid progenitor cells in mammals (J. Rhodes, personal communication). These *in vivo* results are consistent with studies from mammals and *Xenopus*, indicating that *pu.1* plays a major role determining myeloid vs. erythroid cell fate in progenitor cells. Furthermore, transplantation assays placing EGFP-positive cells isolated by FACS into the

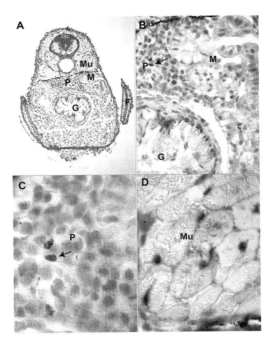

Fig. 3 Immunohistochemistry of *TG(pu.1:EGFP)^{df5}* 20-day-old transgenic larvae. (A) Transverse section at the level of the pectoral fin (F), showing the pronephros (P), mesenephros (M), muscle (Mu), and gut (G). Enlarged views of the (B) mesonephros and pronephros junction, (C) pronephros, and (D) muscle. (A) 100×, (B) 200×, (C, D) 1000×. From Hsu, K. *et al.* (2004). The pu.1 promoter drives myeloid gene expression in zebrafish. *Blood* **104**, 1291–1297, with permission.

posterior erythroid-forming compartment of wild-type host embryos resulted in their expression of *gata-1*, indicating the non-cell–autonomous control of MPC lineage determination (Rhodes, personal communication). These studies exemplify the utility of EGFP-transgenic zebrafish in the analysis of embryonic hematopoietic development.

B. Analysis of *EGFP* Expression in Larvae and Adult Transgenic Zebrafish

The major site of definitive hematopoiesis in the zebrafish beginning at 7–10 dpf and continuing through adulthood lies within the interstitium of the kidney, or the kidney marrow (Zapata, 1979). In the free-swimming larvae of fish and amphibians, the pronephros forms first, consisting of bilateral tubules and glomeruli (reviewed in Drummond, 2003). Embryonic *pu.1* expression was not detected by mRNA *in situ* hybridization assays after 30 hpf (Bennett *et al.*, 2001; Lieschke *et al.*, 2002). However, immunohistochemical examination of tissue sections from 20-day-old transgenic larvae revealed EGFP expression. EGFP was expressed by cells in the kidney and ectopically in other tissues, such as CNS and muscle (Fig. 3A). The expression in nonhematopoietic tissues of the larvae was consistent

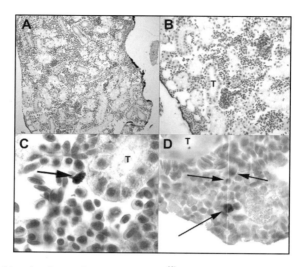

Fig. 4 Immunohistochemistry of *TG(pu.1:EGFP)*^{dfs} adult kidney marrow. (A–B–C) Transverse sections showing EGFP$^+$ cells in the kidney marrow of transgenic adults at progressively increased magnifications and (D) a transverse section of the kidney of transgenic fish analyzed with zebrafish *pu.1* antisense RNA. T, renal tubule. (A) 100×, (B) 200×, (C, D) 1000×. From Hsu, K. *et al.* (2004). The pu.1 promoter drives myeloid gene expression in zebrafish. *Blood* **104**, 1291–1297, with permission.

with ectopic embryonic expression of EGFP in this particular transgenic line. EGFP-positive hematopoietic cells were found in the interstitial kidney marrow cells located within the developing glomeruli of the pronephros (Fig. 3B, C), a region also shown to contain *gata1*-positive cells (Long *et al.*, 1997).

When tissue sections of adult (six-month-old) zebrafish were also examined by immunohistochemistry (Fig. 4A), *EGFP* expression was restricted to a small fraction of cells in the kidney marrow. At this stage, development of the zebrafish kidney is complete, and hematopoietic cells are situated between the renal tubules (Fig. 4B). A small number of hematopoietic cells were found to be EGFP-positive (Fig. 4C), similar to the number of *pu.1*-positive cells observed by RNA *in situ* hybridization on adjacent sections from the same fish (Fig. 4D).

Hematopoietic cells from the kidney, spleen, and blood of adult transgenic fish were isolated by FACS and analyzed. In these fish, cells expressing *EGFP* accounted for $1.8 \pm 0.3\%$ ($n = 8$) of all hematopoietic cells in the kidney and enriched in the myeloid and "lymphoid" cell light scatter gates (Fig. 5A,B). In the spleen, *EGFP*-expressing cells accounted for $1.1 \pm 0.4\%$ ($n = 6$) of hematopoietic cells, with the majority of these cells falling within the myeloid scatter fraction (Fig. 5C, D). In the blood, only a very small fraction (0.0086%) of cells expressed *EGFP*, predominantly within the myeloid cell population (Fig. 5E, F). Cell sorting was used to analyze the morphology of the EGFP-positive kidney marrow cells in the 'lymphoid' (Fig. 6A) and myeloid (Fig. 6B) compartments. By analysis of

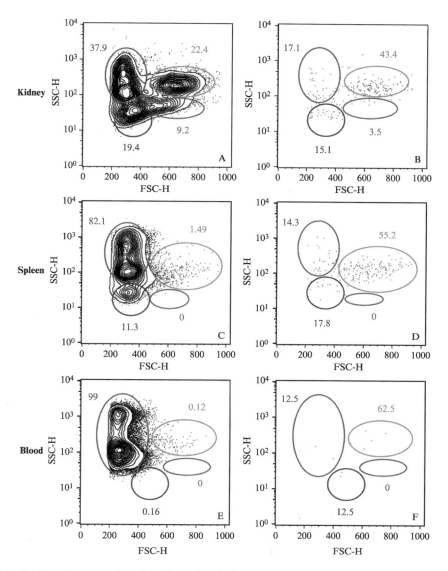

Fig. 5 Fluorescent activated cell sorting (FACS) analysis of hematopoietic cells from *TG(pu.1:EGFP)^{df5}* adult fish. Cells from kidney (A, B), spleen (C, D), and blood (E, F) of adult transgenic zebrafish were isolated and analyzed by FACS by light-scatter gating (A, C, E, total cellular subfractions; B, D, F, same subfractions of EGFP+ cells). Gated populations are as follows: erythrocytes (red), lymphocytes (blue), granulocytes and monocytes (green), and blood cell precursors (purple). Cell size is represented by forward scatter (FSC; abscissa), and granularity by side scatter (SSC; ordinate). The mean percentage of cells is indicated for each gated subpopulation. From Hsu, K. *et al.* (2004). The pu.1 promoter drives myeloid gene expression in zebrafish. *Blood* **104**, 1291–1297, with permission.

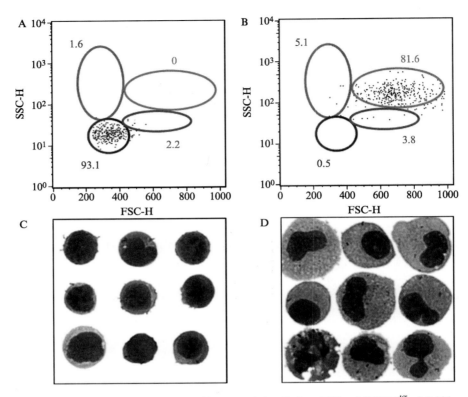

Fig. 6   FACS analysis and morphology of hematopoietic cells from *TG(pu.1:EGFP)*^*df5* adult kidney. GFP-sorted cells were further separated into "lymphoid" (A) and myeloid (B) compartments, and the morphology of the "lymphoid" (C) and myeloid (D) sorted cells was evaluated by cytospin methods. Gated populations as described previously. Populations of cells within each gate are described as mean percentages of total cells. From Hsu, K. *et al.* (2004). The pu.1 promoter drives myeloid gene expression in zebrafish. *Blood* **104,** 1291–1297, with permission.

May-Grunwald staining reactions, cells in the "lymphoid" compartment had the appearance of lymphoid or undifferentiated early progenitor cells (Fig. 6C), whereas those in the myeloid compartment were predominantly monocytic or bi-lobed neutrophils (Fig. 6D).

Cells in the myeloid compartment expressed both *pu.1* and *mpo*, as demonstrated by RT-PCR at the 10- (Fig. 7A, Lanes 4 and 5) and 100-cell level. The observation that EGFP-positive cells in the "lymphoid" compartment expressed *pu.1* (Fig. 7B, Lane 4) but lacked expression of zebrafish *mpo* or the lymphoid markers *lck, rag2,* and *Ig light chain* (Langenau *et al.,* 2004) at the 10- (Fig. 7B, Lanes 1–3) and 100-cell level (data not shown) suggests that these cells represent early hematopoietic or immature lymphoid cells. GFP-positive cells in the "lymphoid" compartment were further analyzed by examining the expression of the

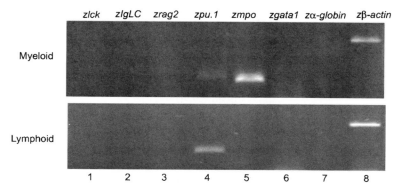

Fig. 7 RT-PCR analysis of hematopoietic cells from the *TG(pu.1:EGFP)^{df5}* adult kidney. GFP-sorted cells from the myeloid and "lymphoid" light-scatter compartments analyzed for lineage-specific expression of the zebrafish *lck*, *IgLC*, *rag2*, *pu.1*, *mpo*, *gata1*, *α-hemaglobin*, and *β-actin* genes. From Hsu, K. *et al.* (2004). The pu.1 promoter drives myeloid gene expression in zebrafish. *Blood* **104,** 1291–1297, with permission.

stem cell marker *scl*, which was not detected at either the 10- or 100-cell level. Thus, aside from *pu.1* expression, the GFP-positive cells with light scatter properties similar to those of lymphoid cells lacked expression of any of the markers of differentiated hematopoietic cells that were tested, suggesting that they represent immature lymphoid cells or, in the absence of functional or cell surface marker analysis, early immature (*scl*−) hematopoietic progenitor cells. These results might be consistent with a putative role of *pu.1* in early mammalian stem cells and their development into common myeloid and lymphoid progenitor cells (Akashi *et al.*, 2000).

VI. Conclusions

Zebrafish is a powerful vertebrate system for both its forward genetic potential and for the *in vivo* analysis of organogenesis, such as hematopoiesis. Stable transgenic zebrafish lines specifically expressing *EGFP* in subsets of hematopoietic cells can play an important role in elucidating the mechanisms that regulate vertebrate hematopoiesis. Stability of the EGFP protein allows its expression to be used as a cell lineage marker and the combined analysis of both transgenic EGFP expression and gene expression permits examination of the developmental progression of hematopoietic differentiation. Because embryonic zebrafish is amenable to experimental manipulations, such as transplantation and targeted gene knockdown assays, questions of hematopoietic stem and progenitor cell regulation and lineage determination can be addressed *in vivo*. In addition, the ability to sort these living cells by fluorescence allows their morphological and genetic analysis and transplantation studies by using purified population of cells isolated at different stages of development.

We are only beginning to understand the molecular mechanisms regulating the differentiation of pluripotential cells into the wide variety of blood cells types during hematopoiesis. Combined with the capacity of the zebrafish model to accommodate forward genetic mutagenesis screens, the *pu.1* transgenic lines described here can be very useful in identifying genes that influence myeloid-erythroid and lymphoid/ progenitor cell development as well as the differentiation of other myeloid cell types, including granulocyte and monocyte/macrophage lineages. Furthermore, transgenic zebrafish lines have already proved valuable for dissecting molecular pathways involved in leukemogenesis (Langenau *et al.*, 2003). Thus, the promoter fragments described here that can drive *pu.1* expression in the myeloid cells of adult zebrafish should provide useful tools for regulating the expression of oncogenes during myelopoiesis, in order to develop zebrafish models of acute myeloid leukemia.

Acknowledgments

This work was supported by CA93152 (ATL), CA96785 (KH), and the Dana-Farber/Harvard Cancer Center support grant CA006516, all from the National Institutes of Health, and the Leukemia and Lymphoma Society SCORE support grant.

References

Akashi, K., Traver, D., Miyamoto, T., and Weissman, I. L. (2000). A clonogenic common myeloid progenitor that gives rise to all myeloid lineages. *Nature* **404,** 193–197.

Bennett, C. M., Kanki, J. P., Rhodes, J., Liu, T. X., Paw, B. H., Kieran, M. W., Langenau, D. M., Delahaye-Brown, A., Zon, L. I., Fleming, M. D., and Look, A. T. (2001). Myelopoiesis in the zebrafish, *Danio rerio. Blood* **98,** 643–651.

Berman, J., Hsu, K., and Look, A. T. (2003). Zebrafish as a model organism for blood diseases. *Br. J. Haematol.* **123,** 568–576.

Cantor, A. B., and Orkin, S. H. (2002). Transcriptional regulation of erythropoiesis: An affair involving multiple partners. *Oncogene* **21,** 3368–3376.

Chen, H., Ray-Gallet, D., Zhang, P., Hetherington, C. J., Gonzalez, D. A., Zhang, D. E., Moreau-Gachelin, F., and Tenen, D. G. (1995). PU.1 (Spi-1) autoregulates its expression in myeloid cells. *Oncogene* **11,** 1549–1560.

Chen, H., Zhang, P., Radomska, H. S., Hetherington, C. J., Zhang, D. E., and Tenen, D. G. (1996). Octamer binding factors and their coactivator can activate the murine PU.1 (spi-1) promoter. *J. Biol. Chem.* **271,** 15743–15752.

Drummond, I. (2003). Making a zebrafish kidney: A tale of two tubes. *Trends Cell Biol.* **13,** 357–365.

Gong, Z., and Hew, C. L. (1995). Transgenic fish in aquaculture and developmental biology. *Curr. Top. Dev. Biol.* **30,** 177–214.

Herbomel, P., Thisse, B., and Thisse, C. (1999). Ontogeny and behaviour of early macrophages in the zebrafish embryo. *Development* **126,** 3735–3745.

Hromas, R., Orazi, A., Neiman, R. S., Maki, R., Van Beveran, C., Moore, J., and Klemsz, M. (1993). Hematopoietic lineage- and stage-restricted expression of the ETS oncogene family member PU.1. *Blood* **82,** 2998–3004.

Hsu, K., Traver, D., Kutok, J. L., Hagen, A., Liu, T. X., Paw, B., Rhodes, J., Berman, J., Zon, L., Kanki, J. P., and Look, A. T. (2004). The pu.1 promoter drives myeloid gene expression in zebrafish. *Blood* **104,** 1291–1297.

Klemsz, M. J., McKercher, S. R., Celada, A., Van Beveren, C., and Maki, R. A. (1990). The macrophage and B cell-specific transcription factor PU.1 is related to the ets oncogene. *Cell* **61**, 113–124.

Langenau, D. M., Ferrando, A. A., Traver, D., Kutok, J. L., Hezel, J.-P. D., Kanki, J. P., Zon, L. I., Look, A. T., and Trede, N. S. (2004). *In vivo* tracking of T cell development, ablation, and engraftment in transgenic zebrafish. *Proc. Natl. Acad. Sci. USA* **101**, 7369–7374.

Langenau, D. M., Traver, D., Ferrando, A. A., Kutok, J. L., Aster, J. C., Kanki, J. P., Lin, S., Prochownik, E., Trede, N. S., Zon, L. I., and Look, A. T. (2003). Myc-induced T cell leukemia in transgenic zebrafish. *Science* **299**, 887–890.

Li, Y., Okuno, Y., Zhang, P., Radomska, H. S., Chen, H., Iwasaki, H., Akashi, K., Klemsz, M. J., McKercher, S. R., Maki, R. A., and Tenen, D. G. (2001). Regulation of the PU.1 gene by distal elements. *Blood* **98**, 2958–2965.

Lieschke, G. J., Oates, A. C., Paw, B. H., Thompson, M. A., Hall, N. E., Ward, A. C., Ho, R. K., Zon, L. I., and Layton, J. E. (2002). Zebrafish SPI-1 (PU.1) marks a site of myeloid development independent of primitive erythropoiesis: Implications for axial patterning. *Dev. Biol.* **246**, 274–295.

Long, Q., Meng, A., Wang, H., Jessen, J. R., Farrell, M. J., and Lin, S. (1997). GATA-1 expression pattern can be recapitulated in living transgenic zebrafish using GFP reporter gene. *Development* **124**, 4105–4111.

McKercher, S. R., Torbett, B. E., Anderson, K. L., Henkel, G. W., Vestal, D. J., Baribault, H., Klemsz, M., Feeney, A. J., Wu, G. E., Paige, C. J., and Maki, R. A. (1996). Targeted disruption of the PU.1 gene results in multiple hematopoietic abnormalities. *EMBO J.* **15**, 5647–5658.

Moreau-Gachelin, F., Wendling, F., Molina, T., Denis, N., Titeux, M., Grimber, G., Briand, P., Vainchenker, W., and Tavitian, A. (1996). Spi-1/PU.1 transgenic mice develop multistep erythroleukemias. *Mol. Cell. Biol.* **16**, 2453–2463.

Nasevicius, A., and Ekker, S. C. (2000). Effective targeted gene 'knockdown' in zebrafish. *Nat. Genet.* **26**, 216–220.

Rekhtman, N., Radparvar, F., Evans, T., and Skoultchi, A. I. (1999). Direct interaction of hematopoietic transcription factors PU.1 and GATA-1: Functional antagonism in erythroid cells. *Genes Dev.* **13**, 1398–1411.

Scott, E. W., Simon, M. C., Anastasi, J., and Singh, H. (1994). Requirement of transcription factor PU.1 in the development of multiple hematopoietic lineages. *Science* **265**, 1573–1577.

Stuart, G. W., McMurray, J. V., and Westerfield, M. (1988). Replication, integration and stable germ-line transmission of foreign sequences injected into early zebrafish embryos. *Development* **103**, 403–412.

Takahaski, K., Naito, M., and Takeya, M. (1996). Development and heterogenity of macrophages and their related cells through their differentiation pathways. *Pathology International* **46**, 473–485.

Tenen, D. G., Hromas, R., Licht, J. D., and Zhang, D. E. (1997). Transcription factors, normal myeloid development, and leukemia. *Blood* **90**, 489–519.

Voso, M. T., Burn, T. C., Wulf, G., Lim, B., Leone, G., and Tenen, D. G. (1994). Inhibition of hematopoiesis by competitive binding of transcription factor PU.1. *Proc. Natl. Acad. Sci. USA* **91**, 7932–7936.

Wang, X., Scott, E., Sawyers, C. L., and Friedman, A. D. (1999). C/EBPalpha bypasses granulocyte colony-stimulating factor signals to rapidly induce PU.1 gene expression, stimulate granulocytic differentiation, and limit proliferation in 32D cl3 myeloblasts. *Blood* **94**, 560–571.

Ward, A. C., McPhee, D. O., Condron, M. M., Varma, S., Cody, S. H., Onnebo, S. M., Paw, B. H., Zon, L. I., and Lieschke, G. J. (2003). The zebrafish spi1 promoter drives myeloid-specific expression in stable transgenic fish. *Blood* **102**, 3238–3240.

Zapata, A. (1979). Ultrastructural study of the teleost fish kidney. *Dev. Comp. Immunol.* **3**, 55–65.

Zhang, P., Zhang, X., Iwama, A., Yu, C., Smith, K. A., Mueller, B. U., Narravula, S., Torbett, B. E., Orkin, S. H., and Tenen, D. G. (2000). PU.1 inhibits GATA-1 function and erythroid differentiation by blocking GATA-1 DNA binding. *Blood* **96**, 2641–2648.

CHAPTER 19

Sleeping Beauty Transposon for Efficient Gene Delivery

Spencer Hermanson, Ann E. Davidson, Sridhar Sivasubbu, Darius Balciunas, and Stephen C. Ekker

The Arnold and Mabel Beckman Center for Transposon Research
Department of Genetics, Cell Biology and Development
University of Minnesota Medical School
Minneapolis, Minnesota 55455

I. Introduction

The core use of the *Sleeping Beauty* (SB) transposon (Ivics *et al.*, 1997) in zebrafish has been previously published (Davidson *et al.*, 2003). The advantages of using SB over DNA injection methods include a higher rate of transgenesis, single-copy integrations with reproducible and long-term Mendelian expression of the transgene cassette, an increase in the total number of chromosomes modified within a particular founder animal strain, and an enzymatically precise engineering of the resulting host chromosome. SB is a potent and reliable gene transfer and expression method for zebrafish.

SB is used as a two-component system in zebrafish embryos: a transposon DNA vector and transposase-encoding mRNA. The transposon consists of DNA cargo of choice flanked by palindromic sequences of inverted and direct repeats (IR/DRs). The transposase is delivered as synthetic mRNA, and the embryo's potent early translational machinery processes the mRNA into the active transposase enzyme. The transposase enzyme then binds the terminal IR/DR sequences and moves the transposon with its cargo by a cut-and-paste mechanism. The resulting modified chromosome can be readily characterized by standard molecular biology techniques.

II. Transgenesis Constructs

A. Transposon

Two versions of IR/DR sequences have been tested in zebrafish: the original pT and a modified pT2 (Davidson *et al.*, 2003). Even though both sequences transpose in zebrafish, we currently use pT2 sequences for all of our transposon vectors because of increased transposition rates as compared with those of pT. We have developed modular pT2-based transgenesis vectors for use in zebrafish (Davidson *et al.*, 2003). We recommend cloning the cargo of interest between IR/DR(L) and the GFP expression cassette in these vectors. One drawback of published vectors was that they were not suitable for translational fusions with GFP because of the presence of an in-frame stop codon between ATG of GFP and polylinker. We now have a version of pT2/s1EF1α-GFP suitable for translational fusions readily available (Balciunas and Ekker, unpublished). Transposon vectors are injected along with transposase mRNA (see later) to achieve germline integration. We typically use a nominal injection volume of ~3 nl for the delivery of this nucleic acid cocktail.

Studies conducted with the original SB vectors show an inverse relationship of size to activity (Geurts *et al.*, 2003). This work demonstrated that a 5.5-kb transposon functions at 50% of the activity of a 2-kb transposon in tissue culture cells. In contrast, larger transposons (7–10 kb) no longer follow a linear relationship of size to activity, but instead continue to be active at 30% of the rate of a 2-kb transposon. This rate of transposition for large transposons still conveys an advantage over DNA alone (Davidson *et al.*, 2003). Our strategy has been to build the smallest practical transposon vectors that meet our needs.

In comparison with several methods of preparing transposon DNA for injection, we obtained the best injection results by using Qiaprep spin minipreps (Qiagen, Cat. No. 27106). There are two key factors in isolating high-quality plasmid DNA for injections by using this protocol. First, 5-ml overnight cultures are prepared to achieve high concentration yields (approximately 500 ng/μl).

Second, the additional PB buffer wash of the spin column is included in the protocol to exclude residual nucleases before the ethanol wash. These two important steps assist in increasing embryo survival and expression frequency, both determinants of successful injections.

B. Transposase

The original SB10 transposase has proven to be effective at transposition in zebrafish embryos, particularly when used with the pT2 transposon vectors. SB10-encoding synthetic mRNA is generated from the SBRNAX plasmid (Dupuy *et al.*, 2002) and prepared by using an *in vitro* transcription kit (Ambion, Cat. No. 1344). Using the supplied reagents, set up a double reaction per tube and proceed according to the kit instructions. Once the reaction is complete, the protocol offers two alternative methods for recovering the mRNA. We chose the phenol/chloroform method for purification. The mRNA is resuspended in nuclease-free water, followed by careful gel quantification of the transposase mRNA. After the concentration has been determined, aliquots of the stock are made to preserve the stability of the mRNA. The transposase mRNA is used at a concentration of 100 ng/μl in the final DNA/mRNA injection cocktail; this is half the dose of transposase mRNA required for LD50.

The transposon DNA (8.3 ng/μl) and transposase mRNA (100 ng/μl) are mixed together on ice in an eppendorf tube just prior to injection. We suspend this mix of DNA and mRNA in nuclease-free water (Ambion, Cat. No. 9930). This ensures that the transposase mRNA does not get degraded prior to injection. As a quality check, it is always encouraged to run the leftover mixture after injection on a gel and verify that the mRNA is still intact.

III. Microinjection of the Zebrafish Embryo

A. Microinjection Station

The microinjection apparatus we use is shown in Fig. 1 (Hyatt and Ekker, 1999). There have been no significant changes made to the core system, which has proven to be durable and highly consistent with minimal maintenance. Any quality dissection microscope will fulfill the basic microinjection needs, and other volume regulator options are also available.

B. The Needle

Needles used for microinjection are made by pulling glass capillaries (World Precision Instruments, Cat. No. 1B100F-4) in a Sutter P87 instrument. The Sutter instrument is programmable and capable of producing a wide range of needle shapes, and the user-defined programs work well from machine to machine. The major variables include the capillary glass selected and the filament style used by a particular instrument.

Microinjection Apparatus

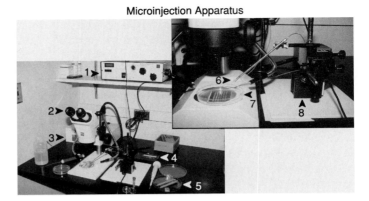

Fig. 1 Microinjection apparatus. Salient features (and manufacturer) are (1) nitrogen tank driven microinjector volume controller (PLI90, Medical Systems Corp.)—nitrogen tank and regulator are not shown; (2) dissecting microscope with heat-shielded light source (Zeiss Stemi 2000 and Fostec AceI light bundle with heat filter); (3) pipettetips for backfilling injection needle (Eppendorf Microloader); (4) fine forceps for needle calibration (Sigma #5, 110 mm); (5) microinjection needle tray—open for photography; (6) loaded microinjection needle in holder; (7) embryos in agarose injection tray; (8) micromanipulator, three axis (Narishige M-152) on metal plate and stand. (From Hyatt, T. M., and Ekker, S. C. (1999). Vectors and techniques for ectopic gene expression in zebrafish. *Methods Cell Biol.* **59**, 117–126, with permission.)

The taper of the needles is important for a successful injection. A balance needs to be struck between strength and flexibility. If the taper of the needle is too steep it will leave large holes in the embryo and chorion, causing embryo death and decreasing efficiency. In contrast, a long, thin needle with a shallow taper will be too flexible to penetrate the chorion of the embryo. Properly formed needles will easily puncture through the chorion into the embryo and exit without picking up the embryo or leaving large holes in the chorion. We suggest making a range of needles varying in shape and testing several to find a needle shape that meets the needs of your injection style.

The needle is backloaded with 2–3 μl of injection solution by using elongated pipette tips (Eppendorf, Cat. No. 5242 956.003). After the injection solution is drawn into the pipette, the pipette is inserted into the back of the needle and the solution is dispensed at the front of the needle. Avoid making air bubbles and especially gaps when backloading the needle because they can affect the accuracy of the injection. If an air bubble forms, hold the needle with the point down and gently flick the needle to allow the bubbles to float to the surface. Backloading the needle is quick and uses less injection solution than do alternative loading methods such as capillary action.

We calibrate each needle prior to injection using the pico injector to regulate drop size. The pico injector is set for a defined time (such as a 1 sec) pulse, and then the end of the needle is clipped with a jewelers forceps (Sigma-Aldrich, Cat. No.

F6521-1EA) to form an opening. It is best to break off small pieces of needle at a time, approximately equal to the width of the end of the forceps. Test the drop size after each break until the desired calibration is achieved. After opening the end of the needle, transfer the drop solution into a microcapillary tube (Drummond, Cat. No. 1-000-0010) to quantify the drop volume. The microcapillary tube holds approximately 30 nl in 1 mm. Continue breaking the needle until the transferred drop reaches 1 mm. The pico injector controls are set for a 100-msec pulse to attain a nominal 3-nl drop size. Make a table of the time settings for a range of distances up the capillary for quick reference when calibrating. Calibrating each needle by this method provides consistency in volume delivered from experiment to experiment.

C. Injecting the Embryo

Embryos are collected within 15 min of spawning and are placed onto cooled agarose loading trays as described (Westerfield, 1995). The agarose loading tray provides a soft and moist surface for the embryos (Fig. 2A). The plates are prechilled to 4 °C to slow the initial embryonic cleavages. A truncated Pasteur pipette, in combination with a pipette pump, is used to draw up embryos and quickly dispense them into the agarose wells. The embryos are loaded with their chorions intact, and all injections are done through the chorion to facilitate efficiency.

Accurately piercing embryos with any speed can be very challenging. We use two general approaches. The first uses the micromanipulator to control all movements of the needle while the tray of embryos remains relatively still. In the second

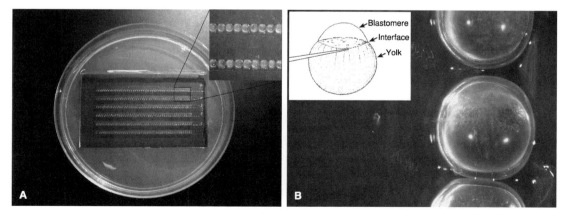

Fig. 2 Microinjection process for the generation of transgenic fish, using the *Sleeping Beauty* transposon. (A) Agarose microinjection tray is loaded with one-cell-stage zebrafish embryos. (B) Transposon DNA/transposase RNA injection solutions are injected at the yolk–blastomere interface in one-cell zebrafish embryos. In this example, a tracer dye is included to visualize the solution transfer. (See Color Insert.)

approach, the tray of embryos is held at an angle and embryos are pushed into a stationary needle. Both methods have proven to be quick and successful at delivering the injection by a proficient scientist.

Injection of naked DNA solution into zebrafish embryos results in mosaic inheritance of the DNA (as viewed by GFP expression in Fig. 3A and B; Finley *et al.*, 2001). In contrast, injected RNA will readily distribute in early zebrafish embryos (Ekker *et al.*, 1995; Hyatt and Ekker, 1999). Injecting embryos at the one-cell stage increases the percent of injected embryos that take up and express both DNA and RNA (see Fig. 3C). We have also found that injection into the

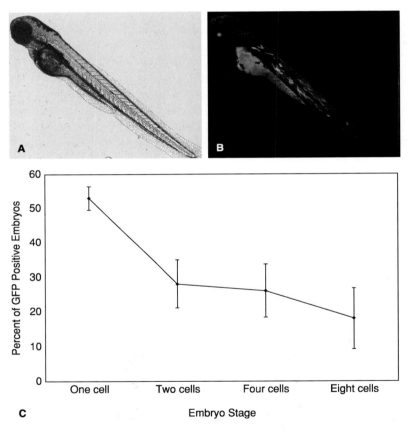

Fig. 3 Maximal delivery of DNA by microinjection is obtained in one-cell zebrafish embryos. DNA encoding a ubiquitous green fluorescent protein (GFP) expression cassette (EF1α-GFP) was microinjected into the yolk–blastomere interface into embryos at the indicated stages and assayed for expression at 48 h, using fluorescein to this cyanate (GFP) fluorescence. An example embryo from a one cell injection is shown. (A) Bright-field image. (B) GFP image. (C) Penetrance summary example (average plus standard error) demonstrating the relative inheritance of the injected DNA as a function of embryo stage.

yolk–blastomere interface produces the most consistently even distribution of DNA in zebrafish embryos. (See Fig. 2B for description of injection site.) Increasing both the percent of fish inheriting the injection solution and the distribution of the solution in the embryos are key points in producing transgenic fish at efficient rates by increasing the odds that a transposition event will occur in a germline cell.

IV. Raising Injected Embryos

Successfully rearing injected embryos is a critical step in the transgenesis process. In general, embryos that have been manipulated have a reduced survival rate over their unperturbed siblings. Right after injection, embryos are washed off the agarose trays into disposable plastic petri dishes using fish water (Westerfield, 1995). Embryos are kept at low densities, about 100–150 embryos per 10-cm dish. At the end of Day 0, living embryos are transferred to a new dish with fresh fish water to separate healthy embryos from dead embryos and the toxic film that they produce. It is recommended to continue this step each day until the embryos hatch, usually 2–3 days at 30 °C. We have found that a certain percentage of transposase RNA-injected embryos that survive to Day 3 are grossly malformed. These malformed embryos are removed after hatching.

Depending on the properties of the injected construct, embryos can be scored for fluorescence (see Section VI) at 1–4 days. In our experience, scoring at 3 days is most efficient, because embryos are fairly large and no longer in chorions, making it easier to observe mosaic fluorescent reporter expression (Fig. 3B). The drawback is that significant autofluorescence in the GFP channel appears after Day 1 and progressively increases as the animals mature. Only embryos scored positive for expression are raised to ensure delivery of the injection solution. In contrast to a published report in medaka fish (Grabher *et al.*, 2003), we do not find a correlation between the abundance of GFP expression and the probability of germline transmission in a given F_0 embryo.

At Day 5, the embryos are transferred to tanks for rearing using paramecia. Mating cultures of *Paramecium multimicronucleatum* (CE-13-1558), wheat seed (CE-13-2425), and protozoan pellets (CE-13-2360) were obtained from Carolina Biological. Essentially, any large holding container will work such as 2-L polycarbonate beakers with lids from Nalgene.

We make two cultures of paramecia for each day of the week. Each culture media is prepared by boiling 1000 ml of water treated for fish use in an Erlenmeyer flask. A mortar and pestle is used to homogenize one protozoan pellet. This and 20 wheat seeds are added to freshly boiled water and then covered. This media is left overnight to cool to room temperature, otherwise it will kill or stunt the growth of the paramecia. After a day of cooling, the culture is poured into a 2-L beaker; any container with a cover can also be used. Then the paramecium culture is added to the media, covered, and allowed to grow for 1 week.

We use established cultures in a feeding rotation. A pair of cultures is designated for use on each day of the week to ensure that cultures are given enough time to grow before they are harvested. The day before the paramecia are harvested, 1000 ml of culture media is prepared for each culture to be used the following day. To remove wheat seed and other debris, the paramecium culture is poured through a brine shrimp net and the flow-through that contains the paramecia is collected. Flow-through (300 ml) is set aside to be used later to inoculate the culture for the following week. A dissecting microscope can be used to observe the health and the density of the paramecia. If the paramecia appear lethargic, ill, or are contaminated with parasites, another culture should be used to start the culture for the following week. If the culture is not dense enough, more than 300 ml of culture might be needed when starting the culture for the following week.

The paramecia hatchery should then be thoroughly cleaned with water and a brush. It is important not to use any bleach or other cleaner because these can be toxic to the paramecia. If parasite contamination is discovered, the polycarbonate beakers can be sterilized in an autoclave prior to their next use. The clean hatchery is then filled with the media prepared on the previous day and the 300 ml of culture set aside earlier. Be sure to cover the hatchery once it is filled.

After starting the cultures for the next week there should be approximately 1000 ml of each culture left to feed the larvae. Using this rearing method, the amount of paramecia given corresponds to the number of larvae in the tank, with each receiving 1–2 ml of paramecium culture. They receive paramecia once a day for 1 week, after which they are fed brine shrimp in the morning and paramecia in the afternoon for an additional week. Once the larvae start eating the brine shrimp their bellies will appear full and pink. All larvae should be feeding on brine shrimp by the third week of feeding. At this time they no longer need to be given paramecia and need only be fed brine shrimp twice daily with fresh water changes as needed.

V. Identifying Transgenic Founders

Injected embryos are fragile and do not have high survival rates. With proper care, 30–40% of positive-scored injected embryos survive to sexual maturity and are fertile. Adult F_0 fish are outcrossed at ~10–12 weeks of age. We often use a commercially available allele of the *brass* locus at this step to reduce fish husbandry errors. The resulting F_1 embryos are screened for reporter expression. Because of germline mosaicism, there are often only a handful of F_1 embryos containing a transposon insertion in any given clutch (Davidson *et al.*, 2003).

One popular alternative screening process by using PCR-based methods is, in our experience, prone to identifying multicopy genetic loci, chromosomes that are enriched for epigentically silenced expression transgene cassettes. Screening for reporter expression at the outset, in contrast, selects for expressing transgenic chromosomes with a greater likelihood of multigeneration Mendelian inheritance expression patterns.

VI. Visualizing Fluorescent Reporters

To visualize expression of fluorescent reporters, we use a Zeiss Axioscope 2 compound microscope with Zeiss 5X Fluar and Zeiss 10X Plan-apochromat objectives (Fig. 4). These objectives have increased sensitivity when viewing fluorescence, high numerical apertures, and extended working distances that make them ideal for fluorescent imaging of zebrafish embryos. We use the 5× Fluar objective for scoring, because its working distance is large enough to view embryos directly in a standard 10-cm petri dish (Fig. 4B). This capability greatly facilitates speed when attempting high-throughput screening techniques compared with individually loading embryos onto slides or trays for screening. Typically, we score at 1 and/or 3–4 dpf. We swirl the embryos to the middle of the dish and scan through them by moving the petri dish under the objective. One-day-old embryos are scored inside the chorion. Tricaine (Sigma, Cat. No. A5040) is administered to later-stage embryos to stop them from swimming away from the excitation light. A dose of 300 μl of 4 mg/ml tricaine solution is added to one 10-cm petri dish containing approximately 30 ml of water. This dose is effective at immobilizing embryos without causing harm, and the embryos quickly recover when placed in fresh fish water. A Zeiss Axiocam digital camera is used

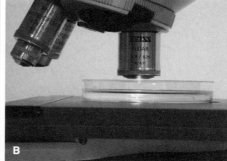

Fig. 4 Upright compound fluorescence imaging station. This instrument is based on a Zeiss Axioskop II (Carl Zeiss, Inc.) microscope with a mercury light source and fish-specific filters (Finley *et al.*, 2001) for fluorescent imaging applications. (A) Salient features: (1) ergonomic head for image visualization—compound microscopes commonly invert the image field; (2) low light capability digital camera for fluorescence documentation; (3) high-quality monitor for real-time image preview during photo documentation. (B) A higher-magnification image of the imaging station setup for the active screening process. The Zeiss 5 × FLUAR objective has sufficient working distance to allow direct screening of zebrafish embryos housed in standard petri plates (see text for details). Note also that for convenience, we remove objectives in adjacent slots for maximal working space.

in conjunction with Zeiss Axiovision software to document the expression of fluorescent reporters in living zebrafish.

VII. Molecular Characterization of the Transposon Integration Site

A. Genomic DNA Isolation

Genomic DNA is typically isolated from pools of 20–50 outcross embryos harvested 5–10 days after fertilization. Finclips can also be a source of tissue for this procedure, but the use of embryos reduces the number of class B animals required for this research. Embryos are homogenized in 0.75 ml lysis buffer (0.1 M Tris, pH 9, 0.1 M NaCl, 0.05 M EDTA, 0.2 M sucrose, and 0.5% SDS) as described (Davidson *et al.*, 2003). The dounce homogenizer consists of a 2.0-ml tissue grind tube (Kontes Glass Company, Cat. No. 885303–0002) and a loose tissue grind pestel (Kontes Glass Company, Cat. No. 885301–0002). The tissue is ground, the lysed tissue transferred to a 50-ml centrifuge tube, and lysis buffer added up to 20 ml. Proteinase K is added to a final concentration of 0.01 μg/μl and incubated at 65 °C for 30 min. Next, 1.5 ml of 8.0 M potassium acetate solution is added and immediately incubated on ice for 30 min, spun at 12,000 rpm for 10 min, and supernatant transferred. Isopropanol (0.6×) is added, incubated at room temperature for 10 min, and spun at 15,000 rpm for 10 min. The pellet is washed with 100% ethanol and resuspended in 200 μl TE. This is incubated overnight at 4 °C for optimal yields.

B. Southern Blots

Genomic DNA (5 μg) is digested overnight to completion with 40 units of restriction enzyme. The digested genomic DNA is separated by standard gel electrophoresis on a 1% gel in 1 × TAE buffer. The gel is usually run at 60–70 V for approximately 5 h and then photographed under UV light. The gel is cut to the desired size and soaked in alkali solution (1.5 M NaCl, 0.5 M NaOH) for 45 min and followed by 1.5 h in neutralizing solution (1.5 M NaCl, 1 M Tris-HCl, pH 8.0) on an orbital shaker. In the meantime, a Hybond N+ membrane (Amersham Biosciences, Cat. No. RPN303B) is cut to the desired size and allowed to soak in distilled water for 15 min, followed by a 30-min soak in the blotting buffer (10× SSC). The transfer of the DNA from the gel to the membrane is done by upward capillary transfer method (Sambrook and Russel, 2001). The blotting is usually done overnight, after which the membrane is lifted off the blotting apparatus by using a blunt forceps, flipped over, and labeled. The membrane is then UV cross-linked by using a UV crosslinker (Fisher Biotech, Cat. No. FB-UVXL-1000). Prior to hybridization, the membrane is soaked in 0.1× SSC, 1% SDS, for 1 h at 65 °C and the membrane is not allowed to get dry after this step. The membrane is gently rolled and transferred to a hybridization tube. Prehybridization buffer (20 ml, 4× SSCP, 1× Denhardt's, 1% SDS, and ~100 μg/ml salmon sperm DNA) is added and the tube incubated in a hybridization oven for 2 h at 65 °C.

Probe labeling is done by using the Prime-A-Gene labeling system (Promega, Cat. No. U1100) following the manufacturer's protocol. The probe consists of an insert-specific DNA, such as a PCR-generated GFP-specific fragment (700 bp). After labeling, remove the unincorporated labeled nucleotides from the reaction, using Probe Quant G-50 microcolumns (Amersham Biosciences, Cat. No. 27-5335-01) following the manufacturer's protocol. Denature the probe by heating for 5 min at 100 °C followed by quick chill on ice for 5 min.

The prehybridization buffer is poured off quickly and 12.5 ml of fresh hybridization buffer (1% SDS, 10% dextran sulfate, 1× Denhardt's, 3.8× SSCP) is added along with the denatured radiolabeled probe and denatured salmon sperm DNA (1.2 mg) into the hybridization tube. The tube is sealed and replaced in the hybridization oven. The membrane and the probe are incubated for 12–24 h at 65 °C.

Before beginning the washes to remove unbound probe, the wash buffers are prewarmed at 65 °C. The membrane is removed from the hybridization tube and excess hybridization buffer drained off. The membrane is placed in a tray and washed twice with 1× SSCP, 0.1% SDS, for 30 min each. It is checked with a survey meter as to how hot the membrane is and the amount of SSCP decreased in the washes as required. We usually do our final wash with 0.25× SSCP, 0.1% SDS, for 15 min at 65 °C.

When the washes are done, the membrane is placed on a Saran Wrap sheet. Edges are covered and sealed by Saran Wrap to prevent contamination of the film holder. The membrane is exposed to an X-ray film or phosphoimager screen for 12 h at −70 °C to obtain an autoradiographic image.

C. Inverse PCR

Of the multiple methods for the recovery of genomic sequences, we regularly use inverse PCR (iPCR) because of its ability to recover sequences that flank both sides of the transposon. The inverse PCR reaction consists of three steps (Fig. 5). The first step is digestion of genomic DNA by restriction endonucleases, the second is circularization of digested DNA, and the third is amplification of flanking sequences by PCR by outward facing primers. Each step is discussed in detail.

The choice of restriction endonuclease depends on the transposon used. For pT2/s1EF1α-GFP, we suggest *Ase*I (Fig. 5, left) because it does not cut inside the transposon and its recognition sequence is frequent in zebrafish genomic DNA. The frequency in genomic DNA is important because rare cutters will on average produce very large fragments that will fail to amplify. We consider 4 kb to be the practical upper limit of this inverse PCR protocol. One can use *Nsi*I instead of *Ase*I, but *Nsi*I sites are somewhat less frequent in the zebrafish genome. An

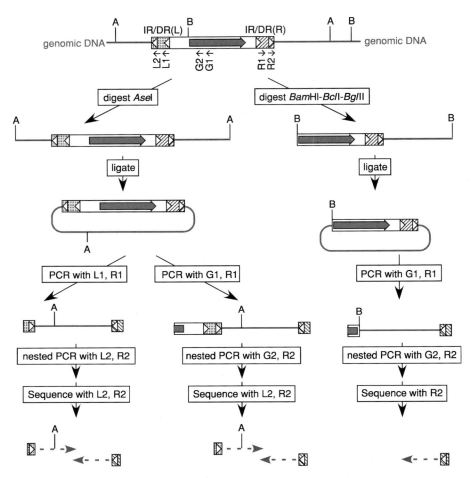

Fig. 5 Diagram of nested inverse PCR. Top, schematic representation of pT2/S1EF1α-GFP transposon integrated into genomic DNA (grey line). The GFP expression cassette is depicted as an open box with a solid arrow. It is flanked by IR/DR sequences shown as filled boxes with open triangles. IR/DR(L) is shown dotted, and IR/DR(R) is shown dashed. Actual (inside transposon) and putative (genomic) positions of restriction endonuclease sites are shown on top. A, *Ase*I; B, *Bam*HI, *Bcl*I, or *Bgl*II. The relative positions of PCR primers L1, L2, G1, G2, R1, and R2 are shown by small arrows. On the bottom of the figure, sequence reading into the genomic DNA is shown as a dashed arrow.

alternative is to use enzymes that cut at a defined place in the transposon in combination with enzymes that yield compatible cohesive ends. One such combination is *Bam*HI–*Bgl*II–*Bcl*I (Fig. 5, right). *Bam*HI cuts just upstream of GFP in pT2/s1EF1α-GFP, and neither *Bgl*II nor *Bcl*I have sites inside the transposon. An alternative combination is *Xba*I–*Nhe*I–*Avr*II–*Spe*I. Both these combinations remove part of the transposon upstream of GFP and allow sequencing only from IR/DR(R). This strategy can be especially useful if a large promoter and/or cargo is cloned into the transposon upstream of GFP. We digest 1 μg of genomic DNA in a 30 μl final reaction volume, using appropriate manufacturer's buffer with 10 units of (each) enzyme for 4 h at 37 °C. Ten microliters of the reaction is used to confirm digestion by gel electrophoresis. The remaining 20 μl is frozen, thawed, and ligated in a final volume of 110 μl with 15 μl 10× ligation buffer and 1 μl high-concentration ligase (Roche, Cat. No. 799 009). The ligation reaction is incubated overnight at 4 °C and then for 1 h at room temperature. As a template, 1 μl and 3 μl of each ligation reaction is used in a 25 μl PCR reaction, using the Roche Expand High Fidelity PCR System (Cat. No. 1 732 641) for amplification. If the selected restriction enzymes do not cut inside the transposon, outward-facing IR/DR(L)- and IR/DR(R)-specific primers are used for amplification (Fig. 5, left lower path). The IR/DR(L)-specific primer can be replaced with a reverse GFP primer (Fig. 5, center lower path). This strategy adds several hundred base pairs to the amplified fragment, but also greatly increases the specificity of the PCR reactions. For DNA digested with either enzyme combination, reverse GFP and outward IR/DR(R) primers should be used (Fig. 5, right lower path). One microliter of the first PCR reaction is used as a template for a second (nested) PCR reaction with primers positioned just outside the first set. A PCR program with an extension time of 6 min or more is used to ensure amplification, even if the chosen enzyme(s) did not cut close to the transposon. The second (nested) PCR reaction is run out on a gel, and bands are excised and sequenced by using appropriate IR/DR primer(s).

Acknowledgments

We thank M. Pat Cliff and Jenny Shaffer for paramecia culturing protocols. Thanks also to Dr. Perry Hackett and members of his research group, with whom these protocols have been developed over the years. This work was supported by grants to S.C.E from the Arnold and Mabel Beckman Foundation and the National Institutes of Health (DA14546).

References

Davidson, A. E., Balciunas, D., Mohn, D., Shaffer, J., Hermanson, S., Sivasubbu, S., Cliff, M. P., Hackett, P. B., and Ekker, S. C. (2003). Efficient gene delivery and gene expression in zebrafish using the Sleeping Beauty transposon. *Dev. Biol.* **263,** 191–202.

Dupuy, A. J., Clark, K., Carlson, C. M., Fritz, S., Davidson, A. E., Markley, K. M., Finley, K., Fletcher, C. F., Ekker, S. C., Hackett, P. B., Horn, S., and Largaespada, D. A. (2002). Mammalian germ-line transgenesis by transposition. *Proc. Natl. Acad. Sci.* **99,** 4495–4499.

Ekker, S. C., Ungar, A. R., Greenstein, P., von Kessler, D. P., Porter, J. A., Moon, R. T., and Beachy, P. A. (1995). Patterning activities of vertebrate hedgehog proteins in the developing eye and brain. *Curr. Biol.* **5,** 944–955.

Finley, K. R., Davidson, A. E., and Ekker, S. C. (2001). Three-color imaging using fluorescent proteins in living zebrafish embryos. *Biotechniques* **31,** 66–70, 72.

Geurts, A. M., Yang, Y., Clark, K. J., Liu, G., Cui, Z., Dupuy, A. J., Bell, J. B., Largaespada, D. A., and Hackett, P. B. (2003). Gene transfer into genomes of human cells by the sleeping beauty transposon system. *Mol. Ther.* **8,** 108–117.

Grabher, C., Henrich, T., Sasado, T., Arenz, A., Wittbrodt, J., and Furutani-Seiki, M. (2003). Transposon-mediated enhancer trapping in medaka. *Gene.* **322,** 57–66.

Hyatt, T. M., and Ekker, S. C. (1999). Vectors and techniques for ectopic gene expression in zebrafish. *Methods Cell Biol.* **59,** 117–126.

Ivics, Z., Hackett, P. B., Plasterk, R. H., and Izsvak, Z. (1997). Molecular reconstruction of Sleeping Beauty, a Tc1-like transposon from fish, and its transposition in human cells. *Cell* **91,** 501–510.

Sambrook, J., and Russel, W. D. (2001). "Molecular Cloning: A Laboratory Manual." Cold Spring Harbor Laboratory Press, Cold Spring Harbor, NY.

Westerfield, M. (1995). "A Guide for the Laboratory Use of Zebrafish (*Danio rerio*)" University of Oregon Press, Eugene OR.

CHAPTER 20

Transgene Manipulation in Zebrafish by Using Recombinases

Jie Dong and Gary W. Stuart

Department of Life Sciences
Indiana State University
Terre Haute, Indiana 47809

I. Introduction

A. Transgenesis and Recombination

The use of conventional genetic engineering technology in the production of transgenic animals has provided an important avenue for (1) understanding gene function and regulation during differentiation and development, (2) production and analysis of human disease models, (3) overproduction of useful gene products, and (4) production of new, improved strains of organisms having one or more desired characteristics. However, with the notable exception of mouse, current transgenic model systems generally rely on the random integration of transgenes into the recipient genome (Chan, 1999; Niemann and Kues, 2003). Consequently, expression of a given transgene can vary widely among transgenic lines. This variation frequently depends on the site of integration (position effect) or the copy number of the transgene (Chan, 1999; Dobie et al., 1997; Garrick et al., 1998). In most cases, the level of expression of the transgene is not closely related to the number of copies, but is subject to the random effects of elements at the site of integration (Chan, 1999; Niemann and Kues, 2003). In addition, our current understanding of RNA interference (RNAi) mechanisms indicates that portions of the RNAi pathway participate in DNA modification, such as DNA methylation and chromatin remodeling (Hall et al., 2003; Volpe et al., 2002; Zilberman et al., 2003). Hence, RNAi produced from randomly rearranged multicopy inserts expressing an antisense RNA could provide a pathway for recognizing and suppressing transgenes. To avoid these pitfalls, exogenous recombination systems, such as the *Sleeping Beauty* (*SB*) transposon system or the bacteriophage P1 Cre-loxP system, can be used to exercise greater control over transgene copy number and position (Branda and Dymecki, 2004; Izsvak and Ivics, 2004).

B. Nonspecific Recombination: Tc1 and *Sleeping Beauty (SB)*

Transposons are repetitive elements capable of moving from one chromosomal location to another. The fundamental components of transposable elements are (1) a gene encoding the transposase that is necessary for transposition and (2) flanking sequences required for recognition by the transposase. The Tc1/mariner transposable elements are members of a large superfamily of transposons. Homologs of Tc1 and those of the related mariner transposon are widespread in animals, including vertebrates (Plasterk et al., 1999). Tc1/mariner elements are about 1300–2400 bp long and consist of a single gene encoding a transposase enzyme that is flanked by terminal inverted repeats. These transposable elements transpose by a cut-and-paste mechanism through a DNA intermediate, using an element-encoded transposase. They have short, inverted terminal repeats and duplicate a TA target site on insertion. Beyond this simple dinucleotide specificity, insertion site preference is, from a practical standpoint, essentially random (Izsvak and Ivics, 2004; Vigdal et al., 2002).

Transposons have several advantages for the production of transgenic animals: a relatively wide host range, the ability to produce single-copy integration of transgenes, and the ability to facilitate the stable maintenance of faithful transgene expression throughout multiple generations of transgenic cells and organisms (Davidson et al., 2003; Dupuy et al., 2002; Fischer et al., 2001; Izsvak et al., 2000; Plasterk et al., 1999). The extremely broad range of these elements suggested that they could be used as generalized DNA vectors, a suggestion eventually shown to be correct (Izsvak et al., 2000). For instance, the reconstructed fish transposon *SB*, a member of the Tc1/mariner superfamily of transposable elements, is found to mediate efficient and precise cut-and-paste transposition in cells of a variety of species, including mouse embryonic stem cells (Fischer et al., 2001; Luo et al., 1998) and human cells (Geurts et al., 2003; Ivics et al., 1997). *SB* is the first active member of the Tc1 family of transposons discovered in vertebrates. The *SB* transposon system has potential for use in insertional mutagenesis because of its germline transposition (Carlson *et al.*, 2003; Ivics *et al.*, 2004). Furthermore, the *SB* transposon has been shown to efficiently insert transgenes into vertebrate chromosomes *in vivo* with long-term, possibly lifelong, transgene expression (Belur *et al.*, 2003; He *et al.*, 2004; Yant *et al.*, 2000, 2002). Because of its stable and efficient integration, *SB* is at present being engineered for use as a nonviral vector for gene therapy (Dupuy *et al.*, 2001, 2002; Izsvak and Ivics, 2004; Kaminski *et al.*, 2002; Mikkelsen *et al.*, 2003; Richardson *et al.*, 2002).

C. Site–Specific Recombination: Cre and PhiC31

Unlike most transposon systems, the Cre-loxP system can mediate site-specific gene insertion or replacement of transgenes. Cre recombinase is a 38-kDa protein isolated from the bacteriophage P1. It catalyzes the site-specific recombination of DNA by recognizing specific sites known as lox P sites (Abremski *et al.*, 1983, 1984). The loxP site is a 34-bp DNA sequence that possesses two 13-bp inverted repeats (Cre-recognition sites) separated by an 8-bp asymmetrical spacer. The spacer region determines the directional nature of the loxP site. Cre recombinase can mediate either excision or inversion between two target lox sites on the same DNA molecule, depending on their relative arrangement. Recombination between two loxP sites present in opposite orientation leads to inversion, whereas recombination between two loxP sites present in direct orientation results in deletion (Hoess and Abremski, 1984). Functional variants of the loxP site can be made by changing only the sequence of the spacer; only lox sites with matching spacer sequence can recombine efficiently (Hoess *et al.*, 1986). Lox sites containing different spacer sequences should not recombine with each other and are referred to as heterospecific lox sites (Siegel *et al.*, 2001). This feature can be used to generate preferential recombination events (Bethke and Sauer, 1997; Feng *et al.*, 1999; Siegel *et al.*, 2001).

Site-specific recombination mediated by Cre-loxP has allowed researchers to not only control site-specific integration and copy number of transgenes but also replace or delete precisely any sequence within a target gene (Babinet *et al.*, 1997; Gu *et al.*, 1994; Lasko *et al.*, 1996). Moreover, the site-specific recombination mediated by Cre-loxP has been used to manipulate gene expression in a specific cell type or at a specific stage of development (Babinet *et al.*, 1997; Metzger and Feil, 1999; Schwenk *et al.*, 1995). The Cre-loxP system has also been used to induce a variety of chromosome rearrangements in mouse ES cells. Cre-mediated chromosomal rearrangements in mice have resulted in large deletions, inversions, duplications, and translocations (Liu *et al.*, 2002, 2003; Nishijima *et al.*, 2003; Yu and Bradley, 2001). These chromosomal rearrangements have been used to model certain human genetic diseases and to enable a fine genetic dissection of their causes (Branda and Dymecki, 2004; Rabbitts *et al.*, 2001; Yu and Bradley, 2001). However, although the Cre-loxP system is a powerful tool for targeting the integration of foreign DNA, it is not easily used to specifically mobilize a transgene to other chromosomal locations.

Although Cre recombinase efficiently performs excision in mammalian cells, the Cre-mediated integration frequency is low because of the excisive back reaction (Sauer and Henderson, 1990). A promising alternative for Cre-loxP system has recently arisen through establishment of the *Streptomyces* phage derived φC31 SSR system (Belteki *et al.*, 2003). Unlike Cre recombinase, the φC31 integrase performs only the integration reaction and requires accessory factors for the reverse reaction (Thorpe and Smith, 1998). φC31 integrase mediates recombination between attachment sites on the phage and bacterial genomes, known as attP (39 bp) and attB (34 bp), respectively (Groth *et al.*, 2000). The product sites, attL and attR, are not substrates for recombination by the integrase, and therefore the reaction is unidirectional (Belteki *et al.*, 2003). The recombination between an attP and an attB site can vary based on the location and orientation of these two sites. Two sites located on the same molecule in the same orientation will lead to a simple deletion, whereas two sites in the opposite orientation will lead to a stable inversion. When two sites are located on different linear molecules, recombination will result in a reciprocal exchange, or if one molecule is circular, a stable insertion. It has been noted that an attB-bearing plasmid inserts into a genomic attP site more readily than an attP-containing plasmid inserts into a genomic attB site (Belteki *et al.*, 2003; Thyagarajan *et al.*, 2001). Also, φC31 integrase appears to be able to mediate efficient integration of attB-bearing plasmids into genomic locations using endogenous pseudo attP sites (Thyagarajan *et al.*, 2001). A similar tendency has also been observed with Cre (Bethke and Sauer, 1997; Lee and Saito, 1998). Hence, some caution is required when exposing complex genomes to integrases and recombinases. Overall, the φC31 integrase and target sites appear to have great potential as an SSR system for manipulating genomes and can prove advantageous as a replacement for Cre-loxP, depending on the application.

D. The Zebrafish Model

Transgenic fish are an interesting and valuable alternative to the use of transgenic mice as models for genome manipulation. Unlike mice, large numbers of fish can be maintained in a small space at an affordable price. Federal regulations concerning the care and maintenance of fish are minimal and easily satisfied. Although they are not mammals, fish are complex vertebrates with the same general body plan, most of the same organs, and most of the same cell types as humans. Large regions of human and zebrafish chromosomes show conservation of gene order. Mutations in zebrafish genes have provided models for several human genetic disorders (North and Zon, 2003; Penberthy *et al.*, 2002; Stern and Zon, 2003; Zhong *et al.*, 2000). The zebrafish has become a popular developmental-genetic system because of its distinct features, which include short generation time, *in vitro* development, high fecundity, optically transparent embryos, and efficient mutagenesis and screening methods. Although a workable system for targeted homologous recombination in zebrafish is currently absent, such a system is likely to be available soon (Fan *et al.*, 2004). In addition, the zebrafish genome sequence is near completion. A complete genome sequence will likely facilitate the systematic analysis of all or most of the genetic functions regulating vertebrate development and differentiation.

II. Recombinase–Mediated Transgene Exchange and Mobilization

To improve the efficiency, flexibility, and reproducibility with which genetically engineered organisms are produced, we are attempting to exploit two well-characterized single-copy recombination systems: the *SB* transposon system and the Cre-loxP system from the bacteriophage P1. The potential specificity and efficiency of a combined system will make it ideal for applications (e.g., gene therapy) in which specific, well-controlled genome modifications are required. Moreover, because these systems require no exogenous factors other than their respective recombinases, it should be possible for transgenes in any organism to be easily introduced and mobilized to alternative chromosome locations by using *SB* transposase and specifically deleted or replaced using Cre recombinase. The following specific subgoals were adopted for the development of this combined system for use in transgenic zebrafish: (1) evaluate the activity and compatibility of variant lox sites, (2) demonstrate Cre-mediated transgene deletion and/or replacement, and (3) demonstrate *SB*-mediated transgene deletion and/or insertion (i.e., mobilization).

A. Variant Lox Site Activity and Compatibility

To test the potential utility of modified lox sites for chromosomal engineering in fish, the activity and compatibility of functional lox variants have been examined in *E. coli*. We have incorporated a cluster of five different lox sites (loxP, loxA, loxB, loxC, and loxY) into a single plasmid construct. In some cases, these variant sites correspond to similar sites investigated by other laboratories (Albert *et al.*, 1995; Kolb, 2001; Langer *et al.*, 2002; Lee and Saito, 1998). Ten additional plasmid constructs were also made, each containing one additional copy (either in the same or in the opposite orientation) of each of these lox sites (Fig. 1). We then individually transformed these 10 plasmid constructs into Cre-expressing *E. coli* and verified exclusive recombination between the two homologous lox sites present in each construct by specific restriction enzyme analysis. Both deletions and inversions were detected for most lox sites except for loxY (Table I). In almost all cases, deletions and inversions involved homotypic sites exclusively. The occurrence of deletion or inversion between two wild-type loxP sites was 100%. A comparison of variant lox sites indicated that the occurrence of deletion or inversion between two loxA sites was much higher than that observed with the other variant lox sites.

loxP: ATGTATGC (spacer)
loxA: ATGTGTAC (spacer)
loxB: GGATACTT (spacer)
loxC: TTATATAT (spacer)
loxY: GAAAGGTA (spacer)

Two loxA sites in the same orientation

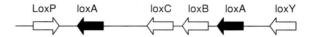

Two loxA sites in the opposite orientation

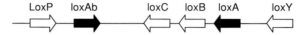

Fig. 1   Two of ten plasmid constructs used for testing the functional lox sites.

Table I
Summary of Recombination Percentage ($n = 32$)

	LoxP		LoxA		LoxB		LoxC		LoxY	
	Del	Inv	Del	Inv	Del	Inv	Del	Inv	Del	Inv
LoxP	100	100	0	0	0	0	0	0	0	0
LoxA			86.9	34.7	0	0	3.1	0	0	0
LoxB					25	20	0	0	0	0
LoxC							27.7	0	0	0
LoxY									0	0

Note: Del, Deletion; Inv, Inversion.

Consequently, we have designed our initial zebrafish transformation vectors to contain primarily loxP and loxA sites in order to facilitate specific gene deletion and/or exchange. In the future, several compatible variant lox sites might be used simultaneously to build artificial chromosomal loci (Fig. 2).

B. Recombinase Target Gene Design and Construction

1. pFRMwg

The vector FRMwg was created by Patrick Gibbs (Gibbs and Schmale, 2000) and was used unaltered in our experiments as a positive control for the production of transgenic fish and as a source construct for a fish expression cassette driven by the carp β-actin promoter. It contains the carp β-actin promoter and *gfp* sequence encoding green fluorescent protein (GFP; Fig. 3A).

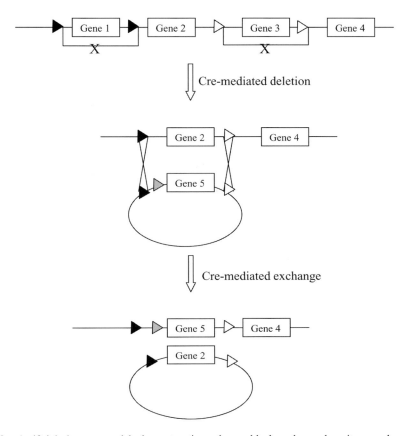

Fig. 2 Artificial chromosomal loci construction scheme; black and grey lox sites can be used to target additional genes to the locus by using a reiterative replacement strategy.

2. pBa-loxP$_2$-gfp

The vector Ba-loxP$_2$-gfp was created in our laboratory. It contains the carp β-actin promoter and the marker *gfp* gene flanked by two identical loxP sites (Fig. 3B). This marker gene was created to test the ability of introduced Cre recombinase to precisely delete a specific length of sequence and eliminate the activity of the target gene in transgenic fish.

3. pBa-loxP$_2$-rfp-gfp

With just a single marker gene, a mosaic pattern of expression would be expected following a Cre-mediated deletion event that occurred in a subset of cells during development. Patches of cells lacking GFP expression would perhaps be difficult to observe in a background of cells providing a high level of GFP

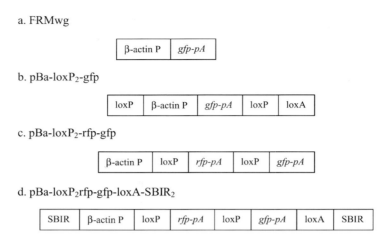

Fig. 3 Structure of recombinase target gene constructs.

expression. We therefore designed an additional construct to allow expression of a second marker gene following deletion of the first marker gene. This construct consists of the carp β-actin promoter, the marker *rfp* gene, and SV40 poly(A) signal flanked by two identical loxP sites (Fig. 3C). Also, a GFP coding sequence is included just downstream of the red fluorescent protein (RFP) reporter gene; it can only be transcribed when the coding sequence for RFP is deleted by Cre-catalyzed recombination. This construct will allow us to easily detect Cre-catalyzed deletion based on the appearance of GFP expression.

4. pBa–loxP$_2$rfp–gfp–loxA–SBIR$_2$

This vector contains two loxP sites flanking the *rfp* gene and SV40 poly(A) signal. The *gfp* coding sequence is placed immediately downstream of the floxed *rfp* reporter gene (Fig. 3D). SBIRs flank the entire expression cassette and can therefore be used to test the ability of introduced *SB*-recombinase to catalyze *de novo* integration of the coinjected target gene. Subsequent *SB*-mediated deletions could also potentially result in the mobilization of the expression cassette to randomly chosen alternative sites in the genome. This construct could also provide a target substrate for Cre-mediated deletion of the *rfp* gene, allowing *gfp* gene expression to signal the presence of active Cre. In addition, the presence of the loxA site should eventually allow Cre-mediated specific gene replacement. This construct represents our prototype vector using a combination of Cre-loxP and SB recombination systems.

C. Production of Transgenic Zebrafish with Recombinase Target Genes

Transformation vectors containing each of the recombinase target genes described previously were microinjected into zebrafish embryos by using already established procedures (Stuart *et al.*, 1988). Briefly, several hundred embryos were injected with a supercoiled plasmid construct (e.g., pBa-loxP$_2$-gfp) within a few hours. Injection of 5 pl of double-stranded DNA at 50 μg/ml allows the survival of about 25% of the injected embryos (Stuart *et al.*, 1988).

Stable lines of transgenic fish are established when the injected transgenes are inherited by the progeny of injected survivors. Hence, the injected fish were raised in our facility until they became sexually mature (about 3 months) and then bred to uninjected fish (outcrossed) to produce the F1 (first filial) generation. Fish fry were raised in fish system water on a mixture of dehydrated algae, paramecia, and finely ground baby fish food until they were old enough to eat freshly hatched brine shrimp. After about 1 month, they were old enough to be moved from their isolated rearing buckets into a facility tank and fed on our normal fish diet (fish flakes and brine shrimp). Adult fish were bred in specially designed breeding tanks or isolated buckets in which a 2-mm mesh barrier protects the eggs that fall to the bottom of the vessel from being eaten by the adult fish. Fertilized eggs were collected, rinsed, and held in fry rearing buckets until the fry hatched (2–3 days).

After 24 h of development, embryos microinjected with plasmid vector were examined for green or red fluorescence under a fluorescence microscope by using the appropriate filter set. The bandpass GFP filter set was used to detect GFP expression (excitation 450–490 nm and emission 500–550 nm). The rhodamine filter set was used to detect RFP expression (excitation 546 $\pm$ 10 nm and emission 570 $\pm$ 10 nm; Finley *et al.*, 2001). Anaesthetized adult fish were scored visually with a handheld, long-wave UV source.

Embryos microinjected with either linear fragment or supercoiled plasmid gave similar results and almost always produced a mosaic pattern of GFP (or RFP) expression (for comparison, see Gibbs and Schmale, 2000). After 24 h of development, most embryos had several to many fluorescent cells, most of which were muscle cells. From then on, fluorescent marking was generally observed to be stable and intense. Approximately 10–23% of microinjected embryos survived to maturity. The frequency with which the mosaic founder fish were found to be capable of producing GFP-expressing progeny ranged from 3 to 10% (Table II). Overall, GFP expression was found to be robust and inherited in a Mendelian fashion through multiple generations (Fig. 4).

D. Cre-Mediated Deletion and Replacement

To demonstrate the utility of the Cre-loxP system, fish embryos with stably integrated transgene targets (pBa-loxP$_2$-gfp) were microinjected with mRNA encoding the Cre recombinase. Cre mRNA was synthesized from SpeI-linearized Cre template DNA by using T7 RNA polymerase. Homozygous GFP fish were outcrossed to wild-type fish to produce heterozygous fish embryos for injection. Approximately 50 μg/ml of Cre recombinase mRNA in a 5-pl volume was injected into one-cell- or two-cell-stage embryos at the blastoderm–yolk interface. GFP expression was generally expected to appear in a mosaic pattern because of the chance failure of Cre-mediated deletion in some cells. Recombination

Table II
Summary of the Generation of Transgenic Zebrafish Lines

Injected construct	Survival rate (%)	Transgenesis frequency (%)	F_0 germline mosaicism[a] (%)
pFRMwg (7 Kb)	12 (47/393)	3 (1/36)	3 (4/119)
Linearized FRMwg (7 Kb)	10 (29/287)	5 (1/20)	7 (6/82)
pBa-gfp-loxP$_2$ (8 Kb)	18 (88/487)	9 (6/65)	9 (10/106)
pBa-rfp-loxP$_2$-gfp (8 Kb)	23 (66/286)	9 (4/44)	10 (10/98)

[a]Only one founder fish was tested for each construct.

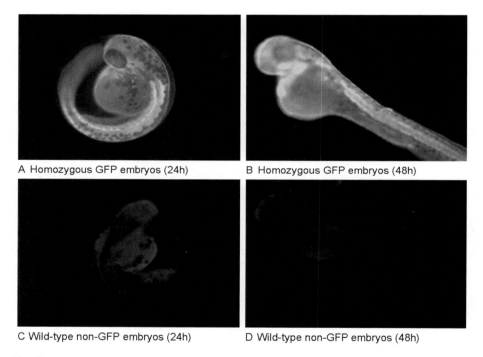

A Homozygous GFP embryos (24h)   B Homozygous GFP embryos (48h)

C Wild-type non-GFP embryos (24h)   D Wild-type non-GFP embryos (48h)

Fig. 4 Embryos with germline integration of ubiquitous green fluorescent protein (GFP) expression at 24 hpf and 48 hpf.

efficiency was measured as the approximate fraction of the embryo in which GFP-fluorescence appeared to be absent (Fig. 5). Recombination efficiencies were estimated at 75–100%, depending on the concentration of Cre mRNA. The precise sequence structure predicted for a Cre-mediated deletion was subsequently verified by PCR and DNA sequencing.

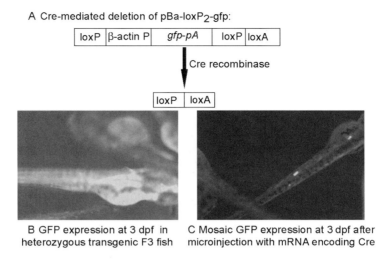

A Cre-mediated deletion of pBa-loxP$_2$-gfp:

| loxP | β-actin P | *gfp-pA* | loxP | loxA |

Cre recombinase

| loxP | loxA |

B GFP expression at 3 dpf in   C Mosaic GFP expression at 3 dpf after
heterozygous transgenic F3 fish   microinjection with mRNA encoding Cre

Fig. 5 Cre-mediated deletion of *gfp* gene in heterozygous transgenic F3 fish.

An alternative approach is to use a transient expression assay. In this assay, the recombination of target genes was examined by merely coinjecting the pBa-loxP$_2$-rfp-gfp target vector with Cre mRNA or Cre-expressing plasmid. To identify the optimum amount of injected Cre mRNA, different concentrations of Cre mRNA solutions (e.g. 100, 75, or 50 ng/μ) were microinjected into one-cell embryos. The appearance of green fluorescence was used as an indicator of *rfp* gene deletion. After 3 days of development, the fish injected with Cre mRNA showed a significantly higher number of GFP-positive fish as compared to fish injected with Cre-recombinase plasmid DNA. The recombination efficiency after coinjection of Cre mRNA at 50 ng/μl was 75% as compared with 30% for coinjection with Cre-recombinase plasmid DNA at 50 ng/μl (Table III). Because we have employed an RFP-GFP switch gene reporter system, the appearance of green fluorescence and simultaneous disappearance of red fluorescence can both be observed by fluorescence microscopy (Fig. 6). The majority of embryos were completely converted from red to green fluorescence when injected with high concentration of Cre mRNA. However, a small number remained positive for both GFP and RFP. The precise sequence structure predicted for a Cre-mediated deletion event was subsequently verified by PCR and DNA sequencing. The switch gene reporter will soon be tested in recently produced stable germline transformants (Fig. 7).

E. *SB*-Mediated Mobilization

Efficient detection of *SB*-mediated mobilization (i.e., reinsertion of a deleted gene into an alternative locus within the genome) requires a sophisticated design. In this case, we plan to produce a transheterozygote transgenic fish in which an

Table III
Summary of Efficiency of Cre–Mediated Deletion

Concentration of mRNA or DNA	Recombination efficiency (%)
Cre mRNA (100 ng/μl)	100 (20/20)
Cre mRNA (75 ng/μl)	80 (16/20)
Cre mRNA (50 ng/μl)	75 (12/16)
Plasmid DNA encoding Cre (50 ng/μl)	30 (6/20)

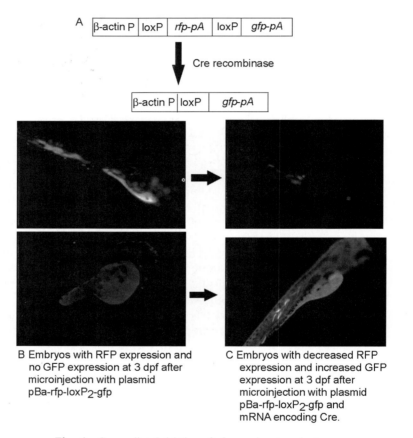

B Embryos with RFP expression and no GFP expression at 3 dpf after microinjection with plasmid pBa-rfp-loxP2-gfp

C Embryos with decreased RFP expression and increased GFP expression at 3 dpf after microinjection with plasmid pBa-rfp-loxP2-gfp and mRNA encoding Cre.

Fig. 6   Cre-mediated deletion of *rfp* gene in a transient assay.

SBIR-flanked *rfp* gene (construct 'D'; Fig. 3) is expressed from one chromosome and an SBIR-*gfp* gene (construct D after *rfp* deletion; Fig. 3) is expressed from the same locus, but on the other homologous chromosome. In control experiments, when this fish is outcrossed in the absence of a recombinase, the offspring must

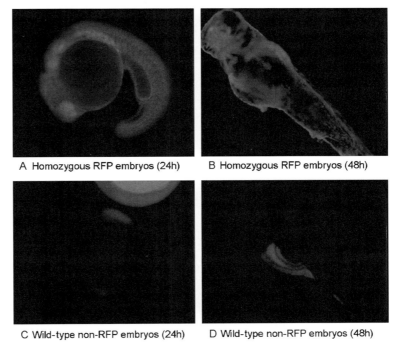

A Homozygous RFP embryos (24h)   B Homozygous RFP embryos (48h)

C Wild-type non-RFP embryos (24h)   D Wild-type non-RFP embryos (48h)

Fig. 7 Embryos with germline integration of ubiquitous RFP expression at 24 and 48 hpf.

inherit either the *rfp* gene (red) or the *gfp* gene (green), but never both. However, if either gene becomes mobilized to another chromosome following the injection of *SB* mRNA, some outcrossed offspring would be able to inherit both genes simultaneously about 50% of the time, producing offspring with both red and green fluorescence as an easy visual marker of mobilization. This procedure could also be adopted for use as a simple screen for generating a large number of potential insertion mutations.

III. Summary and Conclusion

Although much remains to be done, our results to date suggest that efficient and precise genome engineering in zebrafish will be possible in the future by using Cre recombinase and SB transposase in combination with their respective target sites. In this study, we provide the first evidence that Cre recombinase can mediate effective site-specific deletion of transgenes in zebrafish. We found that the efficiency of target site utilization could approach 100%, independent of whether the target site was provided transiently by injection or stably within an integrated transgene. Microinjection of Cre mRNA appeared to be slightly more effective for this purpose than microinjection of Cre-expressing plasmid DNA. Our work has not yet progressed to the point where *SB*-mediated mobilization of our transgene constructs would be observed. However, a recent report has demonstrated that *SB* can enhance transgenesis rates sixfold over conventional methods by efficiently mediating multiple single-copy insertion of transgenes into the zebrafish genome (Davidson *et al.*, 2003). Therefore, it seems likely that a combined system should eventually allow both *SB*-mediated transgene mobilization and Cre-mediated transgene modification.

Our goal is to validate methods for the precise reengineering of the zebrafish genome by using a combination of Cre-loxP and *SB* transposon systems. These methods can be used to delete, replace, or mobilize large pieces of DNA or to modify the genome only when and where required by the investigator. For example, it should be possible to deliver particular RNAi genes to well-expressed chromosomal loci and then exchange them easily with alternative RNAi genes for the specific suppression of alternative targets. As a nonviral vector for gene therapy, the transposon component allows for the possibility of highly efficient integration, whereas the Cre-loxP component can target the integration and/or exchange of foreign DNA into specific sites within the genome. The specificity and efficiency of this system also make it ideal for applications in which precise genome modifications are required (e.g., stock improvement). Future work should establish whether alternative recombination systems (e.g., ϕC31 integrase) can improve the utility of this system. After the fish system is fully established, it would be interesting to explore its application to genome engineering in other organisms.

References

Abremski, K., and Hoess, R. (1984). Bacteriophage P1 site-specific recombination: Purification and properties of the Cre recombinase protein. *J. Biol. Chem.* **259,** 1509–1514.

Abremski, K., Hoess, R., and Sternberg, N. (1983). Studies on the properties of P1 site-specific recombination: Evidence for topologically unlinked products following recombination. *Cell* **32,** 1301–1311.

Albert, X., Dale, E. C., Lee, E., and Ow, X. W. (1995). Site-specific integration of DNA into wild-type and mutant lox sites placed in the plant genome. *Plant J.* **7,** 649–659.

Babinet, C. (1997). Transgenic strategies for the study of mouse development: An overview. *In* "Transgenic Animals: Generation and Use" (L. M. Houdebine, ed.), pp. 371–386. Overseas Publishers Association, Amsterdam.

Belteki, G., Gertsenstein, M., Ow, D. W., and Nagy, A. (2003). Site-specific cassette exchange and germline transmission with mouse ES cells expressing phiC31 integrase. *Nat. Biotechnol.* **21,** 321–324.

Belur, L. R., Frandsen, J. L., Dupuy, A. J., Ingbar, D. H., Largaespada, D. A., Hackett, P. B., and Scott Melvor, R. (2003). Gene insertion and long-term expression in lung mediated by the Sleeping Beauty transposon system. *Mol. Ther.* **8,** 501–507.

Bethke, B., and Sauer, B. (1997). Segmental genomic replacement by Cre-mediated recombination: Genotoxic stress activation of the p53 promoter in single-copy transformants. *Nucleic Acids Res.* **25,** 2828–2834.

Branda, C. S., and Dymecki, S. M. (2004). Talking about a revolution: The impact of site-specific recombinase on genetic analyses in mice. *Dev. Cell* **6,** 7–28.

Carlson, C. M., Dupuy, A. J., Fritz, S., Roberg-Perez, K. J., Fletcher, C. F., and Largaespada, D. A. (2003). Transposon mutagenesis of the mouse germline. *Genetics* **165,** 243–256.

Chan, W. S. A. (1999). Transgenic animals: Current and alternative strategies. *Cloning* **1,** 25–46.

Davidson, A. E., Balciunas, D., Mohn, D., Shaffer, J., Hermanson, S., Sivasubbu, S., Cliff, M. P., Hackett, P. B., and Ekker, S. C. (2003). Efficient gene delivery and gene expression in zebrafish using the Sleeping Beauty transposon. *Dev. Biol.* **263,** 191–202.

Dobie, K., Mehtali, M., McClenaghan, M., and Lathe, R. (1997). Variegated gene expression in mice. *Trends Genet.* **13,** 127–130.

Dupuy, A. J., Clark, K., Carlson, C. M., Fritz, S., Davidson, A. E., Markley, K. M., Finley, K., Fletcher, C. F., Ekker, S. C., Hackett, P. B., Horn, S., and Largaespada, D. A. (2002). Mammalian germ-line transgenesis by transposition. *Proc. Natl. Acad. Sci. USA* **99,** 4495–4499.

Dupuy, A. J., Fritz, S., and Largaespada, D. A. (2001). Transposition and gene disruption in the male germline of the mouse. *Genesis* **30,** 82–88.

Fan L., Alestrom A., Alestrom P., Collodi P. (2004). Production of zebrafish germ-line chimeras from cultured cells. *Methods in Molecular Medicine* **254,** 289–300.

Fan, L., and Collodi, P. (2002). Progress towards cell-mediated gene transfer in zebrafish. *Brief. Func. Genom. Proteom.* **1,** 131–138.

Feng, Y. Q., Seibler, J., Alami, R., Eisen, A., Westerman, K. A., Leboulch, P., Fiering, S., and Bouhassira, E. E. (1999). Site-specific chromosomal integration in mammalian cells: Highly efficient CRE recombinase-mediated cassette exchange. *J. Mol. Biol.* **292,** 779–785.

Finley, K. R., Davidson, A. E., and Ekker, S. C. (2001). Three-color imaging using fluorescent proteins in living zebrafish embryos. *Biotechniques* **31,** 66–70, 72.

Fischer, S. E. J., Wienholds, E., and Plasterk, R. H. (2001). Regulated transposition of a fish transposon in the mouse germ line. *Proc. Natl. Acad. Sci. USA* **98,** 6759–6764.

Garrick, D., Fiering, S., Martin, D. I., and Whitelaw, E. (1998). Repeat-induced gene silencing in mammals. *Nat. Genet.* **18,** 56–59.

Geurts, A. M., Yang, Y., Clark, K. J., Liu, G., Cui, Z., Dupuy, A. J., Bell, J. B., Largaespada, D. A., and Hackett, P. B. (2003). Gene transfer into genomes of human cells by the sleeping beauty transposon system. *Mol. Ther.* **8,** 108–117.

Gibbs, P. D. L., and Schmale, M. C. (2000). GFP as a genetic marker scorable throughout the life cycle of transgenic zebrafish. *Mar. Biotechnol.* **2,** 107–125.

Groth, A. C., Olivares, E. C., Thyagarajan, B., and Calos, M. P. (2000). A phage integrase directs efficient site-specific integration in human cells. *Proc. Natl. Acad. Sci. USA* **97,** 5995–6000.

Gu, H., Marth, J. D., Orban, P. C., Mossmann, H., and Rajewsky, K. (1994). Deletion of a DNA polymerase beta gene segment in T cell using cell type specific gene targeting. *Science* **265,** 103–106.

Hall, I. M., Noma, K., and Grewal, S. I. (2003). RNA interference machinery regulates chromosome dynamics during mitosis and meiosis in fission yeast. *Proc. Natl. Acad. Sci. USA* **100,** 193–198.

He, C. K., Shi, D., Wu, W. J., Ding, Y. F., Feng, D. M., Lu, B., Chen, H. M., Yao, J. H., Shen, Q., Lu, D. R., and Xue, J. C. (2004). Insulin expression in livers of diabetic mice mediated by hydrodynamics-based administration. *World J. Gastroenterol.* **10,** 567–572.

Hoess, R. H., and Abremski, K. (1984). Interaction of the bacteriophage P1 recombinase Cre with the recombining site loxP. *Proc. Natl. Acad. Sci. USA* **81,** 1026–1029.

Hoess, R. H., Wierzbicki, A., and Abremski, K. (1986). The role of the loxP spacer region in P1 site-specific recombination. *Nucleic Acids Res.* **14,** 2287–2300.

Ivics, Z., Hackett, P. B., Plasterk, R. H., and Izsvak, Z. (1997). Molecular reconstruction of sleeping beauty, a Tc1-like transposon from fish, and its transposition in human cells. *Cell* **91,** 501–510.

Ivics, Z., Kaufamn, C. D., Zayed, H., Miskey, C., Walisko, O., and Izsvak, Z. (2004). The Sleeping Beauty transposable element: Evolution, regulation and genetic applications. *Curr. Issues Mol. Biol.* **6,** 43–55.

Izsvak, Z., and Ivics, Z. (2004). Sleeping Beauty transposition: Biology and application for molecular therapy. *Mol. Ther.* **9,** 147–156.

Izsvak, Z., Ivics, Z., and Plasterk, R. H. (2000). Sleeping Beauty, a wide host-range transposon vector for genetic transformation in vertebrates. *J. Mol. Biol.* **8,** 93–102.

Kaminski, J. M., Huber, M. R., Summers, J. B., and Ward, M. B. (2002). Design of a nonviral vector for site-selective, efficient integration into the human genome. *FASEB J.* **16,** 1242–1247.

Kolb, A. F. (2001). Selection-market-free modification of the murine beta-casein gene using a lox2272 [correction of lox2272] site. *Anal. Biochem.* **290,** 260–271.

Langer, S. J., Ghafoori, A. P., Byrd, M., and Leinwand, L. (2002). A genetic screen identifies novel non-compatible loxP sites. *Nucleic Acids Res.* **30,** 3067–3077.

Lasko, M., Pichel, J. G., Gorman, J. R., Sauer, B., Okamoto, Y., Lee, E., Alt, F. W., and Westphal, H. (1996). Efficient *in vivo* manipulation of mouse genomic sequences at the zygote stage. *Proc. Natl. Acad. Sci. USA* **93,** 5860–5865.

Lee, G., and Saito, I. (1998). Role of nucleotide sequences of loxP spacer region in Cre-mediated recombination. *Gene* **216,** 55–65.

Liu, P., Jenkins, N. A., and Copeland, N. G. (2002). Efficient Cre-loxP-induced mitotic recombination in mouse embryonic stem cells. *Nat. Genet.* **30,** 66–72.

Liu, P., Jenkins, N. A., and Copeland, N. G. (2003). A highly efficient recombineering-based method for generating conditional knockout mutations. *Genome Res.* **13,** 476–484.

Luo, G., Ivics, Z., Izsvak, Z., and Bradley, A. (1998). Chromosomal transposition of a Tc1/mariner-like element in mouse embryonic stem cells. *Proc. Natl. Acad. Sci. USA* **95,** 10769–10773.

Metzger, D., and Feil, R. (1999). Engineering the mouse genome by site-specific recombination. *Curr. Opin. Biotechnol.* **10,** 470–476.

Mikkelsen, J. G., Yant, S. R., Meuse, L., Huang, Z., Xu, H., and Kay, M. A. (2003). Helper-independent Sleeping Beauty transposon-transposase vectors for efficient nonviral gene delivery and persistent gene expression *in vivo. Mol. Ther.* **8,** 654–665.

Niemann, H., and Kues, W. A. (2003). Application of transgenesis in livestock for agriculture and biomedicine. *Anim. Reprod. Sci.* **79,** 291–317.

Nishijima, I., Mills, A., Qi, Y., Mills, M., and Bradley, A. (2003). Two new balancer chromosomes on mouse chromosome 4 to facilitate functional annotation of human chromosome 1p. *Genesis* **36,** 142–148.

North, T. E., and Zon, L. I. (2003). Modeling human hematopoietic and cardiovascular diseases in zebrafish. *Dev. Dyn.* **228,** 568–583.

Penberthy, W. T., Shafizadeh, E., and Lin, S. (2002). The zebrafish as a model for human disease. *Front. Biosci.* **7,** d1439–d1453.

Plasterk, R. H., Izsvak, Z., and Ivics, Z. (1999). Resident aliens: Tc1/mariner superfamily of transposable elements. *Trends Genet.* **15,** 326–332.

Rabbitts, T. H., Appert, A., Chung, G., Collins, E. C., Drynan, L., Forster, A., Lobato, M. N., McCormack, M. P., Pannell, R., Spandidos, A., Stocks, M. R., Tanaka, T., and Tse, E. (2001). Mouse models of human chromosomal translocations and approaches to cancer therapy. *Blood Cells Mol. Dis.* **27,** 249–259.

Richardson, P. D., Augustin, L. B., Kren, B. T., and Steer, C. I. (2002). Gene repair and transposon-mediated gene therapy. *Stem Cells* **20,** 105–118.

Sauer, B., and Henderson, N. (1990). Targeted insertion of exogenous DNA into the eukaryotic genome by the Cre recombinase. *New Biol.* **2,** 441–449.

Schwenk, F., Baron, U., and Rajewsky, K. (1995). A cre-transgenic mouse strain for the ubiquitous deletion of loxP-flanked gene segments including deletion in germ line. *Nucleic Acids Res.* **23,** 5080–5081.

Siegel, R. W., Jain, R., and Bradbury, A. (2001). Using an *in vivo* phagemid system to identify non-compatible loxP sequences. *FEBS Lett.* **505,** 467–473.

Stern, H. M., and Zon, L. I. (2003). Cancer genetics and drug discovery in the zebrafish. *Nat. Rev. Cancer* **3,** 533–539.

Stuart, G. W., McMurray, J. V., and Westerfield, M. (1988). Replication, integration and stable germ-line transmission of foreign sequences injected into early zebrafish embryos. *Development* **103**, 403–412.

Thorpe, H. M., and Smith, M. C. M. (1998). *In Vitro* site-specific integration of bacteriophage DNA catalyzed by a recombinase of the resolvase/invertase family. *Proc. Natl. Acad. Sci. USA* **95**, 5505–5510.

Thyagarajan, B., Olivares, E. C., Hollis, R. P., Ginsburg, D. S., and Calos, M. P. (2001). Site-specific genomic integration in mammalian cells mediated by phage phiC31 integrase. *Mol. Cell Biol.* **21**, 3926–3934.

Vigdal, T. J., Kaufman, C. D., Izsvak, Z., Voytas, D. F., and Ivics, Z. (2002). Common physical properties of DNA affecting target site selection of Sleeping Beauty and other Tc1/mariner transposable elements. *J. Mol. Biol.* **323**, 441–452.

Volpe, T. A., Kidner, C., Hall, I. M., Teng, G., Grewal, S. I., and Martienssen, R. A. (2002). Regulation of heterochromatic silencing and histone H3 lysine-9-methylation by RNAi. *Science* **297**, 1833–1837.

Yant, S. R., Ehrhardt, A., Mikkelsen, J. G., Meuse, I., Pham, T., and Kay, M. A. (2002). Transposition from a gutless adeno-transposon vector stablizes transgene expression *in vivo*. *Nat. Biotechnol.* **20**, 999–1005.

Yant, S. R., Meuse, L., Chiu, W., Ivics, Z., Izsvak, Z., and Kay, M. A. (2000). Somatic integration and long-term transgene expression in normal and haemophillic mice using a DNA transposon system. *Nat. Genet.* **25**, 35–41.

Yu, Y. J., and Bradley, A. (2001). Engineering chromosomal rearrangements in mice. *Nat. Genet.* **2**, 780–790.

Zilberman, D., Cao, X., and Jacobsen, S. E. (2003). ARGONAUTE4 control of locus-specific siRNA accumulation and DNA and histone methylation. *Science* **299**, 716–719.

Zhong, T. P., Rosenberg, M., Mohideen, M. A. P. K., Weinstein, B., and Fishman, M. C. (2000). *gridlock*, an HLH gene required for assembly of the aorta in zebrafish. *Science* **287**, 1820.

CHAPTER 21

Highly Efficient Zebrafish Transgenesis Mediated by the Meganuclease I-SceI

Clemens Grabher,★ Jean-Stephane Joly,[†] and Joachim Wittbrodt★

★Developmental Biology Program
European Molecular Biology Laboratory (EMBL)
69117-Heidelberg, Germany

[†]INRA, Institute de Neurobiologie A. Fessard
CNRS
91198 Gif-Sur-Yvette, France

I. Introduction

A. The Purpose of Transgenesis

Few technical achievements in biological sciences have opened up such possibilities as transgenesis technologies have. The ability to change selectively the genetic composition of multicellular organisms and thereby permanently alter the activity of particular proteins has important bearing on all areas of biological research. Transgenic animals have been instrumental in providing new insights into mechanisms of development and developmental gene regulation, the action of oncogenes, and intricate cell interactions within the immune and nervous systems. Moreover, transgenic technology offers exciting possibilities for generating precise animal models for human genetic diseases.

Fish are excellent candidates for the production of transgenics for two important reasons. First, fish represent the largest and most diverse group of vertebrates and provide an advantageous system for *in vivo* studies of developmental processes to gain knowledge of gene regulation and the action of gene products in vertebrates. Second, conventional selective breeding of fish for improved growth or other characteristics is a very slow process. By contrast, transgenic fish technology has the potential to improve genetic traits such as increased growth potential, disease resistance, improved feed conversion efficiency, or other desirable genetic traits for aquaculture in one generation.

The establishment of methods for successful transgenesis is one of the basic criteria for an organism to be referred to as model organism. To clarify the terms used in the field of transgenic research, we propose to distinguish the superior term *transgenesis* from the term *germline integration*. The term *germline integration* includes (1) introduction of exogenous DNA into a host organism and (2) stable integration of the foreign DNA into the host genome. The term *transgenesis* should only be used when describing the successful achievement of all the following: (1) introduction of exogenous genes into a host organism, (2) transmission of these genes to the next generation (germline integration), and (3) appropriate expression of this transgene in the host organism. The separate use of these terms will also facilitate to compare results of different transgenesis approaches in various species.

B. Methods of Gene Delivery: Overview

Transgenesis was first applied to fish in the mid-1980s (Zhu *et al.*, 1985). Since then, transgenic fish have been widely used in both basic and applied research. Several techniques that yielded significant success in the generation of stable transgenic zebrafish have been developed in recent years. Pseudotyped retrovirus infection has been used to generate single-copy insertions of transgenes (Gaiano *et al.*, 1996; Lin *et al.*, 1994a; Linney *et al.*, 1999). Despite some drawbacks in the past, this method is highly efficient with present protocols (see Amsterdam and Hopkins, Chapter 1). However, construction, packaging, titering, and infection are laborious processes that require considerable expertise. Small laboratories with the intention to generate only few transgenic fish might thus want to use simpler approaches.

Since 1990, electroporation has been applied for transgenesis in fish and some success has been reported (Inoue *et al.*, 1990; Ono *et al.*, 1997). However, in recent years, electroporation of fertilized fish eggs has been mostly used for transient expression of exogenous genes rather than to generate transgenic animals (Sussman, 2001; Tawk *et al.*, 2002). To facilitate transgenesis, electroporating sperm before fertilization represents an interesting variation to the electroporation technique (Muller *et al.*, 1992; Sin *et al.*, 2000). However, integration of the foreign DNA occurs infrequently and expression of the exogenous genes is poor. The usefulness of sperm-mediated gene transfer as a routine protocol for mass transgenesis in fish will depend on the improvement of integration and expression of the foreign gene.

Particle bombardment, originally developed for plant transgenesis, has been adapted to fertilized zebrafish eggs in which transgenes have been successfully delivered and expressed in the targeted embryos (Zelenin *et al.*, 1991). Only more recently, transmission of transgenic green fluorescence protein (GFP) to the germline of medaka embryos has been achieved, resulting in true transgenic F1 offspring (Yamauchi *et al.*, 2000). However, available data do not allow definitive comparison to other techniques.

In medaka, ES cells (Mes1, medaka embryonic stem cells) have been established (Hong *et al.*, 1996) and were found to contribute to organs of all three germ layers in chimeras (Hong *et al.*, 1998). However, generation of stable transgenic fish has not been successful because of the failure of ES cells to contribute to the germline. Although cell cultures exhibiting characteristics of ES cells have been described in zebrafish, only short-term cell cultures, which must be maintained in the presence of cells from the rainbow trout, have produced germline chimeras (Ma *et al.*, 2001). It remains to be determined whether these cells will contribute to the germline after long-term culture, which is required for genetic manipulations involving homologous recombination and selection.

As an alternative to embryonic stem cells, cultured somatic cells offer the possibility of producing cloned animals with targeted genetic manipulations (Lai *et al.*, 2002; McCreath *et al.*, 2000). Fish nuclei of blastula cells from different species have been transplanted into enucleated eggs to study the nucleocytoplasmic interaction (Zhu and Sun, 2000). Wakamatsu *et al.* (2001) demonstrated that diploid fertile medaka could be produced by nuclear transfer by using blastula cells as donors. These findings show that nuclei prepared from fresh blastula cells can be reprogrammed in fish to support embryonic and adult development. In 2002, the first cloned zebrafish using long-term cultured cells that was amenable to genetic manipulation was established (Lee *et al.*, 2002). Although, the current success rate of ~2% does not represent an improvement to transgenesis, the potential availability of cell cultures that can be used for homologous recombination could pave the way for gene targeting in lower vertebrates.

In 1980, Gordon *et al.* (1980) demonstrated that exogenous DNA could be introduced into the mouse genome simply by physical injection of DNA solution into the zygote. At present, microinjection is still the most widely used method of germline transgenesis in several vertebrate species, including fish.

C. Methods of Gene Delivery: Microinjection

For medaka and zebrafish, a finely drawn glass needle loaded with DNA solution is used for injection. Under a standard dissecting microscope, with the aid of a micromanipulator, fertilized eggs are penetrated with the needle. The injection needle is guided through the chorion into the cytoplasm of the cell of an embryo or the yolk of the egg at the one-cell stage. Once the tip of the needle has entered the cytoplasm (yolk), approximately 1–2 nl of DNA solution containing 10^5–10^7 DNA molecules is injected.

The first transgene to be delivered into fish (medaka) embryos was the δ-crystalline gene of chicken (Ozato *et al.*, 1986). Transient expression of the transgene occurred in a mosaic manner, but no germline transmission was observed. It was only in 1988 that transgenesis by microinjection was successfully performed, including transgene expression and transmission to the next generation in a teleost genetic model system (zebrafish, Stuart *et al.*, 1988). At present, microinjection provides the fastest and simplest means for germline transgenesis and transient expression studies in fish (Chou *et al.*, 2001; Lin, 2000).

However, major drawbacks, e.g., mosaic transgene expression in G0, low insertion frequency, and mosaic germline distribution, have not yet been overcome (Fig. 1). Moreover, transgenesis frequencies on microinjection are still very variable, depending on the vector used and on the skills of the injector. Average stable transgenesis frequencies range from 1 to 10% (Collas and Alestrom, 1998; Culp *et al.*, 1991; Lin *et al.*, 1994b; Stuart *et al.*, 1988, 1990; Tanaka and Kinoshita, 2001), only exceptionally reaching more than 20% (Higashijima *et al.*, 1997). Similarly, efficiencies of transient expression of a transgene in the G0 generation vary between 10 and 50% (Chou *et al.*, 2001; Higashijima *et al.*, 1997), but are invariably mosaic. Generally, comparison between different reports is difficult because of the differences of promoters and/or vector design.

1. General Fate of Injected DNA

To develop strategies that overcome the drawbacks, one has to consider the fate of injected DNA inside a cell [reviewed in more detail in Hackett (1993) and Iyengar *et al.* (1996)]. When plasmid DNA is injected into fish embryos, three

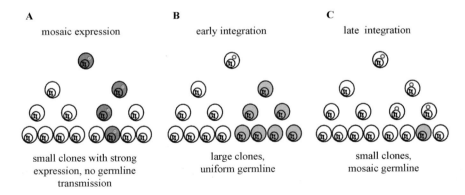

A	B	C
mosaic expression	early integration	late integration

small clones with strong expression, no germline transmission

large clones, uniform germline

small clones, mosaic germline

Fig. 1 Fate of injected DNA. Injected DNA can meet three different fates. (A) DNA stays episomal (probably in concatemers) and is expressed in small bright clones, because of the high number of DNA molecules and their uneven segregation. (B) DNA integrates early in development (one to two-cell stage). Depending on the copy number of inserted transgenes, their expression level might vary but the germline will be uniform, resulting in a large proportion of transgenic F1 progeny. (C) DNA integrates at later stages. Mosaicism of both transgene expression and germline depends on the time point of insertion.

alternative fates can occur: (1) persistence and replication in an extrachromosomal state within the cell and its descendants for several cell divisions; (2) integration into the chromosomal DNA of the cell; (3) loss of plasmid DNA.

Commonly, the first two fates lead to embryos that are mosaics with respect to the presence of plasmid DNA. The uneven distribution and replication of the episomal DNA among daughter cells result in mosaic expression. Nearly all (90–99%) fish that have integrated transgenes will also be mosaic for its integration and/or level of expression, because most insertion events occur after the one-cell stage (Hackett, 1993). Both transient (G0) expression of the transgene and germline transmission critically depend on the time point of integration. The earlier an integration event occurs, the more the descendents of this transgenic cell will contain and express the transgene. Consequently, such a transgenic animal will exert broad transgene expression in G0, whereas transgene integration at a later time point will affect fewer cells and thus result in mosaic expression in G0. Similarly, if integration occurs at a later developmental stage, only few primordial germ cells might have integrated the injected DNA, leading to a mosaic germline. Thus, the proportion of transgenic F1 progeny depends on the degree of mosaicism.

Only genomic integration at the one-cell stage will lead to a fully transgenic germline in which 50% of the F1 offspring inherit the transgene (Jowett, 1999).

2. Strategies to Improve Transgenesis by Microinjection

To avoid position effects mediated by DNA methylation or changes of chromatin state that affect transgene expression, it is desirable to provide the transgene with regulatory sequences (enhancers, etc). Moreover, expression domains (genes) are believed to remain insulated from neighboring sequences by boundary regions. A feature commonly linked to such boundary elements is the ability to protect from influences that affect gene expression, namely position effects (Kellum and Schedl, 1991; Noma et al., 2001; Stief et al., 1989). Inverted terminal repeats (ITRs) from adeno-associated virus (AAV) have been used to improve transient transgene expression and insertion in mammalian cell culture, frog, and fish (Chou et al., 2001; Fu et al., 1998; Hsiao et al., 2001; Philip et al., 1994). In addition, Noma et al. (2001) identified inverted repeats acting as barriers for heterochromatin spreading in fission yeast. A key to transgenesis lies in the efficient uptake of foreign DNA by the cell nucleus. Recent studies have shown that a limiting step in fish transgenesis resides in slow nuclear import of DNA (Collas and Alestrom, 1997, 1998), which might favor late and mosaic transgene integration into the germline (Culp et al., 1991; Stuart et al., 1988). Improvements in nuclear uptake of DNA have resulted from the use of protein–DNA complexes. Noncovalent attachment of DNA to karyophilic proteins including NLS peptides (CGGPKKKRKVG-NH$_2$) has been shown to enhance nuclear import and expression of DNA in cultured mammalian cells and zebrafish (Collas and Alestrom,

1997; Fritz *et al.*, 1996; Kaneda *et al.*, 1989). However, in medaka and zebrafish, reports on the use of noncovalent NLS peptide applications are somewhat conflicting. Whereas Collas and coworkers reported enhanced integration and expression, other authors found no evidence for an enhancing effect of NLS peptides (Higashijima *et al.*, 1997). Another promising technology that was initially applied to *Dictyostelium* (Kuspa and Loomis, 1992) and *Xenopus* (Kroll and Amaya, 1996; Kroll and Gerhart, 1994) involves the use of restriction endonucleases. A similar approach has been adapted for zebrafish (Jesuthasan and Subburaju, 2002), in which sperm nuclei has been injected into unfertilized zebrafish eggs. These eggs have performed normal cleavage and further developed to fertile adults. By preincubating the sperm nuclei with linearized DNA, transgenic fish with widespread transgene expression have been obtained. However, this technique requires more expertise than simple microinjections do.

Two novel approaches have been reported that use enzymes that bind to DNA directly and mediate entry into the nucleus. One technique involves the coinjection of a meganuclease (I-SceI), which is discussed in detail in Section II (Thermes *et al.*, 2002). Second, the application of transposons has recently been shown to enhance transgenesis in fish (medaka and zebrafish) significantly (Davidson *et al.*, 2003; Fadool *et al.*, 1998; Grabher *et al.*, 2003; Kawakami *et al.*, 2000; Raz *et al.*, 1998). The use of the *Sleeping Beauty* transposon system (Ivics *et al.*, 1997) is discussed in Chapter 19.

II. Transgenesis by Meganucleases

A. Meganucleases

Several endonucleases (meganucleases) encoded by introns and inteins have been shown to promote homing (lateral transfer) of their respective genetic elements into intron- or inteinless homologous allelic sites. By introducing site-specific double-strand breaks (DSBs) in intronless alleles, these nucleases create recombinogenic ends that engage in gene conversion, resulting in duplication of the intron.

In the 1970s, the genetic marker ω in *S. cerevisiae* was found to transfer to strains lacking the marker when crossed to ω^+ strains (Coen *et al.*, 1970). This marker corresponded to a 1-kb group I intron of the large rRNA gene of the mitochondrial genome. An open reading frame in this intron has been further shown to encode a site-specific endonuclease capable of recognizing and cleaving the intronless allele, thereby initiating the homing event (Colleaux *et al.*, 1988; Jacquier and Dujon, 1985; Macreadie *et al.*, 1985). This protein, now called I-SceI, was the first of more than 250 homing endonucleases since identified (Belfort and Roberts, 1997). It recognizes an 18-bp sequence with little tolerance to degeneration, thereby representing one of the most specific meganucleases. Meganucleases are divided into four major groups characterized by the sequence motifs

LAGLIDADG, GIY-YIG, H-N-H, and His-Cys box (Belfort and Roberts, 1997; Chevalier and Stoddard, 2001).

B. The I-SceI Meganuclease

The protein I-SceI is a member of the largest class of homing enzymes, characterized by the presence of either one or two conserved amino acid residue sequence motifs (LAGLIDADG). Most of these proteins, like I-SceI, carry the motif in duplicate and are endonucleases. I-SceI has been purified as a monomeric globular protein of 235 amino acids (Monteilhet *et al.*, 1990). Its endonuclease activity requires Mg^{2+} or Mn^{2+} to asymmetrically cleave DNA within its recognition sequence (TAGGGATAACAGGGTAAT) and leaves a 4-bp overhang with a 3′-hydroxyl terminus (Monteilhet *et al.*, 1990). The enzyme displays a low turnover, which is likely to result from slow release of the reaction product because of its strong affinity to the larger half-site (Fig. 2; Perrin *et al.*, 1993).

I-SceI is one of the most specific meganucleases because of its long recognition site and little tolerance of degeneracy within this sequence (Colleaux *et al.*, 1988). To add to the biochemical analyses of I-SceI cleavage reactions, the crystal structure of I-SceI has recently been resolved (Moure *et al.*, 2003). The DNA recognition sequence is bound by an I-SceI monomer in the crystal structure (Fig. 3). The sequence specificity is due to numerous direct phosphate and base-specific contacts in addition to water-mediated interactions.

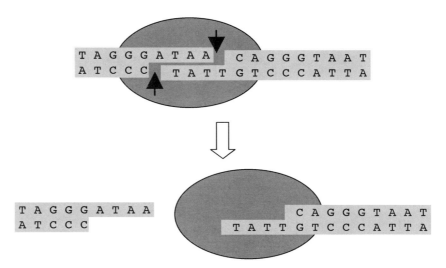

Fig. 2 Mechanism of cleavage of the I-SceI meganuclease. The meganuclease acts in a monomeric form; it recognizes and cleaves an 18-bp recognition sequence in an asymmetrical fashion. It exhibits a low turnover because of its strong association to the larger half-site.

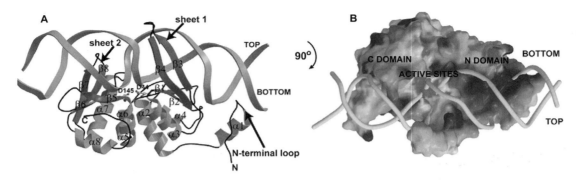

Fig. 3 Structure of the I-SceI protein–DNA complex. (A) The secondary structure elements of the protein interacting with DNA (cyan ribbon) have been labeled as sheet 1 and sheet 2 (major groove contacts) and the N-terminal loop (minor groove contacts). α-Helices are depicted in green and β-strands in magenta. The catalytic aspartate residues are represented as ball-and-stick models. (B) GRASP potential surface area of the protein. The DNA is shown as a cyan ribbon. Positive charges are more abundant in the N domain. This figure was obtained by rotating the model shown in (A) by 90° about a horizontal axis, from Moure, C. M. *et al.* (2003). The crystal structure of the gene targeting homing endonuclease I-SceI reveals the origins of its target site specificity. Reprinted from *J. Mol. Biol.* **334,** 685–695, with permission. (See Color Insert.)

The 18-bp recognition site of I-SceI is expected to be found only once in 7×10^{10} bp of random sequence. Consequently, such a site has not been found in any vertebrate genome to date. While restriction endonuclease mediated integration (REMI) carries the intrinsic risk to fractionate the host genome, extremely rare cutting meganucleases such as I-SceI can be employed for transgenesis, because they only act on sites engineered into the donor construct. Preliminary experiments had shown that cotransfection of plasmids bearing meganuclease recognition sites with expression vectors encoding the corresponding meganuclease efficiently led to stably transfected cell lines with single-copy integrations (Thermes *et al.*, 2002). Based on this, a simple, fast, and efficient technique was established that allows the generation of stable transgenic medaka lines by coinjection of the I-SceI protein with reporter vectors that are flanked at both ends by the corresponding recognition sites. Coinjection of DNA with I-SceI results in enhanced promoter-dependent transgene expression in G0 whereby both the number of expressing embryos and number of expressing cells per embryo are increased. Furthermore, injected embryos show reduced mosaicism. Moreover, transgenesis frequencies are improved two- to three fold compared with control injections. In the meantime, the meganuclease approach has been used successfully in several fish species (medaka, stickleback, zebrafish), amphibia (axolotl, *Xenopus*), and ascidians (*Ciona*). To optimize the meganuclease injection protocol, we have sent out questionnaires to laboratories that use the meganuclease technique. Based on their input and our experience, we present an optimized protocol for meganuclease coinjections. In addition to the basic protocol, crucial steps and specific hints are discussed and highlighted in the checklist (Table I).

Table I
Meganuclease Transgenesis Checklist

- Aliquot meganuclease on arrival, store at $-80\,°C$ and thaw on ice shortly before injection.
- Carefully adjust DNA concentration (start with 10–30 ng/μl).
- Prepare fresh injection solution shortly before injection.
- Keep injection solution on ice while injecting.
- Use only one-cell-stage embryos.
- Inject into the cytoplasm of the cell.
- Assure that the injection volume does not exceed 10% of the cell volume.

C. Meganuclease Transgenesis Protocol (Zebrafish)

1. Materials

a. Equipment

1. Flaming/Brown micropipette puller (e.g., P-87), Sutter Instrument Company.
2. Borosilicate glass capillaries [e.g., GC100F-10 (1 mm OD $\times$ 0.58 mm ID, Clark Electromedical Instruments)].
3. Micromanipulator [e.g., Leica (manual), Eppendorf InjectMan NI2 (automatic)].
4. Microinjector (e.g., Microinjector 5242, Eppendorf).
5. Stereomicroscope (e.g., Stemi 2000, Zeiss) and/or fluorescence stereomicroscope (e.g., MZFLIII, Leica) with appropriate filtersets (GFP, DSR, CFP, UV) equipped with a digital camera (e.g., DC500, Leica).
6. Cold light source (e.g., KL 1500 electronic, Schott).
7. Microloader tips (e.g., Microloader, Eppendorf)
8. Pasteur pipettes.
9. Petri dishes (100 mm $\times$ 20 mm).
10. Casting molds.
11. Forceps.

b. Reagents and Buffers

1. Agarose.
2. ddH$_2$O.
3. Plasmid DNA preparation kit for highly purified DNA (e.g., QiaFilter Plasmid Maxi Kit, Qiagen).
4. DNA vector including two I-SceI restriction sites flanking a multiple cloning site and appropriate reporters (GFP, lacZ, etc. e.g., I-SceI backbone vector; Thermes et al., 2002).
5. I-SceI meganuclease (aliquot 2 μl each on arrival and store at $-80\,°C$) and I-SceI buffer (e.g., Roche, New England Biolabs).

6. 0.3× Danieau's solution (30×, 1.74 M NaCl, 21 mM KCl, 12 mM MgSO$_4$, 18 mM Ca(NO$_3$)$_2$, 150 mM HEPES, 1% Pen/Strep, pH 7.6).

2. Methods

a. Preparation of Plasmid DNA

The vector to be injected should be designed such that the expression cassette of interest (e.g., including promoter, transgene, and polyadenylation signal) is flanked by two I-SceI recognition sites. Because of the asymmetrical cleavage and the strong association of the enzyme with the larger half-site after cleavage, orientation of the I-SceI sites is inverted (facing the larger half-site to the expression cassette). Thus, the meganuclease stays associated on the side of the insert rather than the vector backbone. The plasmid DNA should be prepared and purified by a high-purity plasmid preparation kit (see Section II-C-1-b). DNA concentration and purity can be checked by spectrometry. The ratio of A$_{260}$/A$_{280}$ should be between 1.8 and 2.0.

b. Preparation of Microinjection Plates

Several types of microinjection plates can be used (e.g., Culp *et al.*, 1991). The type used in our laboratory is a petri dish (100 mm × 20 mm). Agarose (1.5%) is prepared with tap water. Warm agarose solution is poured into a petri dish and a plastic injection mold put on the top of the agarose solution (swimming) and left at room temperature until the agarose solidifies. A schematic drawing of the injection mold is shown in Fig. 4. Prior to injection, the mold is removed from the solid agarose that is overlaid with ddH$_2$O and stored in a refrigerator. On the day of injection, the plate is equilibrated to room temperature for 1 h and the water is exchanged with 0.3× Danieau's solution.

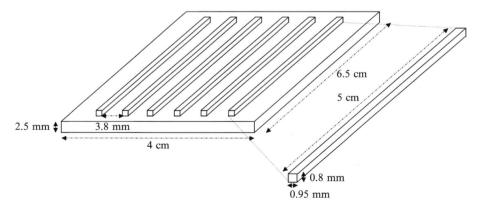

Fig. 4 Schematic representation of an injection mold. Structure and dimensions of a type of injection mold suitable for zebrafish embryos are indicated. The mold should be made of thermoresistant plastics.

c. Preparation of Embryos

Matings are set up as described (Westerfield, 1995). Embryos are collected approximately 20 min after allowing a female and a male zebrafish to mate (or as soon as eggs are laid and fertilized). Single embryos are transferred and aligned into the trenches of the injection plate by using a Pasteur pipette (approximately 20 embryos/trench). To aim for transient assays or for the generation of stable transgenic lines, the embryos *must* be at the one-cell stage for consistent results (Fig. 5).

d. I-SceI Microinjection

Because of the low stability of the meganuclease, aliquots of enzyme solution should be prepared (e.g., 2 μl) on arrival and stored at $-80\,°$C. The microinjection solution should be prepared shortly before injection and kept on ice.

Injection solution	Final concentration	Volume (μl)
DNA (1 μg/μl)	10–30 ng/μl	0.3–0.9
I-SceI buffer (10×)	0.5× (1:20)	1.5
I-SceI enzyme (5 U/μl)	0.3 U/μl (1:20)	2
ddH$_2$O		26.2–25.6

Preincubation of the injection solution did not improve results significantly in our hands. However, one user gave the feedback that short preincubation (15 min at 37 °C) enhanced transient transgene expression. Microinjection needles are prepared as described (Meng *et al.*, 1999), backfilled with injection solution (4 μl), mounted to a micromanipulator, and connected to a microinjector. The orientation of the embryos can be adjusted by using forceps as shown in Fig. 5. If the needle is closed, the tip of the needle has to be broken with forceps. To inject, the chorion and the membrane of the cell are penetrated with the open tip of the needle. The injection volume should not exceed 10% of the total cell volume. Larger volumes will result in increased mortality rates of injected embryos. Although, to obtain expression on injection of RNA and DNA it is sufficient to inject into the yolk, for consistent results by using the meganuclease approach it is mandatory to inject *directly* into the *cytoplasm* of the cell. According to information we gathered from the meganuclease poll, injection of the DNA–enzyme mix into the yolk did not result in significant improvement of transient expression of transgenes or transgenesis rates (Fig. 5). It is important to carefully adjust the DNA concentration of the injection solution. We have experienced a dramatic increase in mortality of injected embryos if the DNA concentration exceeds 30 ng/μl (Table I). The concentration window of 10–30 ng/μl is a good starting point, but the DNA concentration resulting in best transient expression and highest transgenesis rates might depend on the type of DNA (promoter, regulatory elements, transgene, etc.) and should be optimized empirically in case of unsatisfactory results. Embryos should be raised at the appropriate temperature after injection. Leaving the injected embryos aligned in the injection plates will facilitate

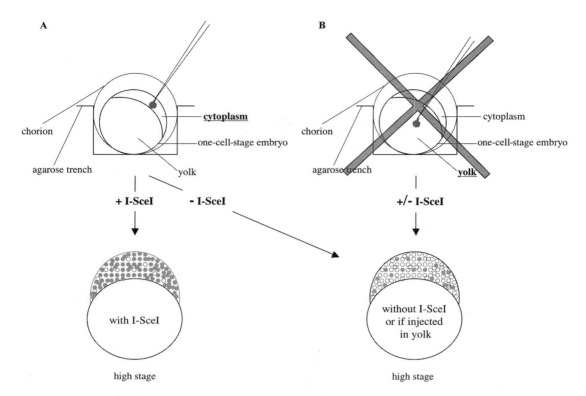

Fig. 5 Schematic representation of zebrafish microinjection. Microinjection for meganuclease-mediated transgenesis should be performed as depicted. One-cell-stage zebrafish embryos should be oriented as indicated. The injection volume must not exceed 10% of the cell volume. (A) Injection is performed directly into the cytoplasm of the cell. On coinjection of DNA with I-SceI, this procedure significantly enhances transient transgene expression and transgenesis frequency. Injection without I-SceI results in highly mosaic transient transgene expression and low transgenesis frequency even if injected into the cytoplasm. (B) Injection into the yolk of a one-cell-stage zebrafish abolishes the enhancing effect of I-SceI. Transgene expression and transgenesis frequency are similar to those in conventional microinjection. Therefore, injection into the yolk only should be avoided.

monitoring of results. To avoid bacterial contamination, add penicillin/streptomycin to the embryo-rearing medium and change the buffer daily. Select transgene-expressing embryos as putative founders 1–3 days post fertilization (depending on the transgene).

3. Results

By using conventional DNA microinjection, *in vivo* analysis of gene expression at late stages of embryogenesis or in adult stages has been difficult because of the relatively low number of transgene expressing cells. Moreover, transgenes are

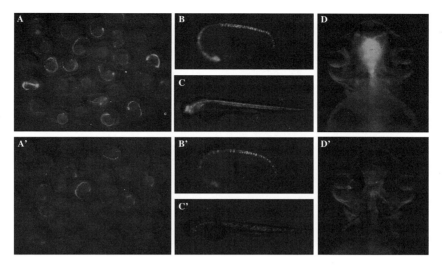

Fig. 6 G0 expression of *shh*-GFP in zebrafish on microinjection with and without I-SceI (also see Table II). Circular vector containing an expression cassette [green fluorescent protein (GFP) driven by the *zfshh* promoter] flanked by I-SceI recognition sites (15 ng/μl) was injected into the cytoplasm of one-cell-stage zebrafish embryos with (A–D) or without (A′–D′) I-SceI meganuclease (0.3 U/μl). (A, A′) Overview of GFP-expressing embryos at 24 npf. Total numbers of GFP-expressing embryos increased on coinjection of meganuclease (A) compared to injection of DNA alone (A′). (B, B′) Representative samples of whole embryos exhibiting uniform promoter-dependent GFP expression at 24 hpf. Coinjection (B) of meganuclease yielded more transgene-expressing cells within the primary expression domain (notochord) than injection without I-SceI (B′), resulting in less mosaic GFP expression. (C, C′) The same representative samples as in B and B′ are shown at 48 hpf. The meganuclease coinjected embryos retained uniform GFP expression (C), but expression of GFP was greatly reduced in embryos injected without I-SceI (C′). (D,D′) A close-up of the head region of embryos at 48 hpf showing primary and secondary SHH expression domains. Overall GFP expression levels in meganuclease-injected embryos were still strong, and secondary domains (retinal ganglion cells and amacrine cells) also showed uniform expression of GFP (D). GFP expression levels in conventionally injected embryos were poor and highly mosaic in primary and secondary domains of SHH expression (D′). (See Color Insert.)

often transcribed from extrachromosomal plasmid DNA. Because of the uneven segregation and frequent loss of episomal DNA, expression is highly mosaic and transient, which further hampers the detection of secondary expression domains during development. Following the optimized meganuclease protocol improves the results in transient assays also (Fig. 6). First, it improves the number of fish expressing a transgene in a promoter-dependent manner. Second, the number of expressing cells within the correct domains is augmented, resulting in less mosaic and more stable expression. In turn, it facilitates detection of secondary expression domains in transient assays. Evaluation of the meganuclease protocol in injected zebrafish embryos is illustrated in Fig. 6 and Table II. Expression was compared on injection of a zebrafish *shh*-promoter GFP vector with or without meganuclease (Loosli and Wittbrodt, unpublished; modified from Neumann and

Table II
Transient Transgenesis in Zebrafish Mediated by I–SceI

	− I-SceI (%)	+ I-SceI (%)
Total GFP-expressing embryos (24 hpf)—*poor p. founders*[a]	34	58
Uniform promoter-dependent GFP-expressing embryos (24 hpf)—*good p. founders*[b]	15	48
Uniform promoter-dependent GFP-expressing embryos (48 hpf)—*excellent p. founder*[c]	2	27

Note: Circular vector containing an expression cassette (GFP driven by the *zfshh* promoter) flanked by I-SceI recognition sites was injected into the cytoplasm of one-cell-stage zebrafish embryos with or without I-SceI meganuclease. (Also see Fig. 6.)

[a]These fish include all embryos exhibiting any kind of GFP expression, ranging from highly mosaic (single cells) to uniform promoter-dependent expression. The probability that all these fish will transmit a functional transgene to the F1 generation is very low.

[b]These fish exhibit uniform promoter-dependent expression in primary expression domains (early in embryogenesis). The probability that all these fish will transmit a functional transgene to the F1 generation is moderate.

[c]These fish exhibit uniform promoter-dependent expression in all expected expression domains. The probability that all of these fish will transmit a functional transgene to the F1 generation is very high.

Nuesslein-Volhard, 2000). Injection of plasmid DNA without meganuclease resulted in 34% of injected embryos showing GFP expression at 24 hpf. More than half exhibited a highly mosaic pattern, whereas only 15% showed an almost uniform pattern within the expected primary expression domain. One day later, at 48 hpf, only 2% were still expressing GFP uniformly in the primary expression domain but were only mosaic in secondary domains. In contrast, on coinjection of the reporter vector with meganuclease, 58% of injected fish expressed GFP at 24 hpf and 48% showed uniform expression in the primary domain. At 48 hpf, uniform GFP expression in the primary and later expression domains was detectable in 27% of injected fish, representing excellent candidates for putative founders of a stable transgenic fish line. In addition to elevated numbers of expressing embryos, the increased number of expressing cells per expression domain is clearly visible in the meganuclease coinjected embryos.

Based on the detailed evaluation of stable transgenesis in medaka and results obtained with the ascidian *Ciona savignyi* (Fig. 7), improvement of stable transgenesis in zebrafish was expected (Deschet *et al.*, 2003; Thermes *et al.*, 2002). Although, statistical evaluation of the meganuclease protocol for the generation of stable transgenic zebrafish lines has not been performed so far, shared experience from a number of zebrafish laboratories that use the meganuclease protocol; from laboratories working with axolotl, medaka, stickleback, or *Xenopus*; and our own experience with zebrafish transgenesis strongly supports this. Individual findings

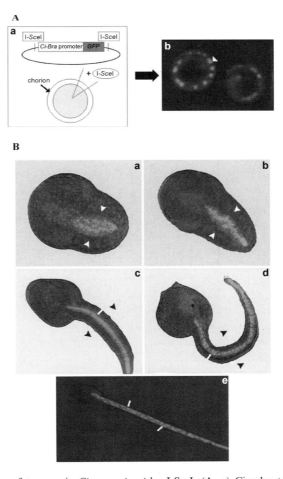

Fig. 7 Generation of transgenic *Ciona savignyi* by I-SceI. (A, a) Circular transgene containing flanking I-SceI recognition sites was coinjected with meganuclease. (b) G0 late tailbud stage embryos. GFP expression in the notochord is evident. Arrowhead points to a GFP-positive notochord cell. (B) GFP-expressing embryos and larva derived from a transgenic founder. (a, b) Early tailbud stage embryos. Notochord cells are converging toward the midline (white arrowheads). (c) Mid-tailbud stage embryo. Mediolateral intercalation is complete and the notochord now consists of a single row of GFP-positive cells (white arrow). (d) Late tailbud stage embryo. Note the discontinous GFP signal due to formation of vacuoles in the notochord cells (white arrow). (e) GFP-positive swimming larva. Note the position of GFP-labeled cell nuclei (white arrows). From Deschet, K. *et al.* (2003). Generation of Ci-Brachyury-GFP stable transgenic lines in the ascidran *Ciona savignyi*. *Genesis* **35,** 248–259. Reprinted by permission of Wiley-Liss, Inc., a subsidiary of John Wiley & Sons, Inc. (See Color Insert.)

differed considerably, ranging from no enhancement to 2- to 15-fold enhanced transgenesis rates. Optimization of the protocol will help improve the procedure and increase the enhancement factor for all investigators. A functional insertion event occurring early in embryogenesis improves both expression in G0 as well as transgenesis rates. As an indirect measure for the time point of integration, the germline transmission frequency (number of transgene expressing offspring of an identified founder) can be used. A transmission frequency of 50% strongly suggests an ideal integration event at the one-cell stage. However, similar (or higher) ratios can also occur in case of several independent insertions during later developmental stages. This results in a mosaic germline by which the transgenes are inherited independently to different F1 offspring. However, expression in G0 will still be highly mosaic in case of several independent insertions at later stages. Thus, uniform G0 expression in combination with an elevated germline transmission frequency is a good indicator for an early insertion event. Meganuclease transgenesis in medaka and *Ciona savignyi* has resulted in germline transmission rates of 25–50%, that together with enhanced G0 expression show that I-SceI can facilitate early functional integration of the transgene (one- to two-cell stage).

For the reliable use of a transgenic line, its expression pattern and levels must be stable for many generations. Several parameters determine the expression pattern and level of a given transgene. These include the position effect of the specific genomic locus of insertion, the number of loci of independent insertions, and the number of copies in a given tandem array. Given the high transgenesis frequency when using the meganuclease protocol, the probability to generate only transgenic lines that are strongly affected by position effects is relatively low. Multiple insertions at independent loci can hamper the use of a transgenic line as this requires several generations of outcrossing to separate them into single insertion lines. Moreover, the existence of multiple independent insertions can only be ruled out consistently on Mendelian segregation in the F2 generation. It is thus preferable to generate transgenic fish that harbor a transgene insertion at a single locus. All transgenic fish and ascidians generated thus far have shown Mendelian segregation of transgene expression in F2, indicating that meganuclease-mediated transgenesis preferentially yields transgene integration at a single genomic locus. Another important difference from conventional plasmid injections is the copy number of transgenes inserted into the host genome. Systematic studies performed in medaka and ascidians have demonstrated transgene insertions as single inserts mediated by I-SceI or as head-to-tail tandem arrays of low copy number (1–10). Thus, the size of tandem arrays is considerably lesser than that observed in standard plasmid injections (up to 2000 copies). This is important because fewer copies might result in lower expression levels of a transgene. However, in transient assays, I-SceI coinjection results in a more uniform (higher number of expressing cells) expression. Furthermore, stable transgenic fish did not show detectable reduction of transgene expression levels when compared to conventionally generated transgenics. Long transgene concatemeres have been reported to favor variegated expression and gene silencing in successive generations (Garrick *et al.*, 1998; Kelly *et al.*, 1997). In contrast, stable transgene expression (intensity and absence of variegation) has been observed through six successive generations in several independent transgenic fish lines generated with the meganuclease protocol.

Besides the apparent enhancing effects of I-SceI on transgenesis in fish and other organisms, its role in mechanistic terms is not very well understood. I-SceI creates DSBs; this feature has been applied to study homologous recombination in mammalian systems and *Xenopus* (Johnson and Jasin, 2001; Segal and Carroll, 1995). Thus, meganuclease might create a DSB in the host genome and the injected plasmid DNA, thereby enhancing the integration by a repair mechanism. However, this hypothesis is without support because a single endogenous I-SceI recognition site in a host genome would favor integration at this very site with high frequency. This is in contrast to the observation of independent insertion sites in several medaka and *Ciona* transgenic lines. The existence of multiple endogenous sites is highly unlikely to occur because of its length (18-bp), which is expected only once in 7×10^{10} bp of random sequence. Thus, there is no evidence to date that I-SceI introduces DSBs into a vertebrate genome by specific cleavage. A potential nuclear targeting activity mediating enhanced transgenesis frequencies has been ruled out by recognition site mutation and deletion experiments in medaka. The occurrence of insertions as short tandem repeats can be attributed to an activity of I-SceI counteracting endogenous ligase and replicase activity that are thought to be responsible for the strong concatemerization of conventionally injected DNA. The transgene would remain as short fragments, exposing more recombinogenic ends to favor integration. Probably by association with one of its cleavage products, I-SceI exerts a low turnover. This could accomplish inhibition of endogenous ligases or replicases by both, cleavage of generated concatemeres, and/or protection of cleaved recombinogenic ends from degradation or ligation. I-SceI-induced DSBs allow recombination in the mammalian system at high frequency (Choulika *et al.*, 1995) and the natural homing process is thought to be finished by the host DSB repair system. As association of homing endonucleases with DSB repair is evolutionarily ancient it is possible that direct interactions of meganucleases with components of the host DSB repair system have evolved. Thus, enhanced integration frequency might not only be initiated by DSBs, but also be activated by direct interaction of I-SceI with the double-strand break repair machinery.

4. Future Prospects

The I-SceI system has been used as a tool in mammalian cells (e.g., Choulika *et al.*, 1995) and *Drosophila* (e.g., Rong and Golic, 2000). The meganuclease can also be used in fish for comparative studies of *cis*-acting regulatory elements and homologous recombination (HR).

Investigating the activity of regulatory elements *in vivo* is an attractive opportunity. However, direct comparison of regulatory elements is often hampered by effects on transgene expression mediated by the specific locus of insertion and/or distribution of episomal plasmid DNA. Taking advantage of the rare occurrence of I-SceI recognition sites in vertebrate genomes, a meganuclease site engineered into the host genome might serve as a unique docking station for transgene integration. Once a locus allowing transgene expression that is unaffected by the host genomic environment is found, it can be used for any kind of transgenesis and, in particular, to perform comparative studies on regulatory elements under controlled conditions. Meganuclease transgenesis as described in this chapter could be further improved, including insulators upstream and downstream of the DNA of interest to protect the transgene from influences of heterochromatin and epigenetic control of surrounding genomic sequences.

Application of homologous recombination is highly desirable in fish too. It has been made possible in mouse by the ES cell technology, which is currently under development for fish. In *Drosophila*, which lacks the ES cell technology, it is difficult to introduce a linear DNA molecule into germ cells. Recently, a method to generate such a linear fragment *in vivo* has been reported, accompanied by the demonstration of gene targeting. For targeting, the FLP recombinase and I-SceI meganuclease expression are induced to generate DSBs that stimulate HR. Additional studies have also shown that in principle gene targeting in *Drosophila* could also be achieved by using I-SceI alone, although at lower efficiencies (Gong and Golic, 2003). These low efficiencies have been coupled with a highly efficient repair of I-SceI-mediated DSBs. Results obtained on transgenesis frequency applying the meganuclease in fish suggest that I-SceI actively participates in an integration event. Not only the linearization step itself, producing DSBs promotes integration into the genome as injection of *in vitro* linearized DNA fragments results in lower transgenesis frequencies. This is indicative of an additional function performed by the meganuclease, as discussed earlier. I-SceI meganuclease thus provides potential to be used for gene targeting also in fish.

Acknowledgments

We thank M. Carl and members of our laboratory for comments on the manuscript and F. Loosli for sharing plasmids prior to publication. We thank all participants of the meganuclease transgenesis poll for sharing their experience to help optimize the approach. We highly appreciate the ideas of A. Choulika to establish the meganuclease protocol for fish transgenesis.

References

Belfort, M., and Roberts, R. J. (1997). Homing endonucleases: Keeping the house in order. *Nucleic Acids Res.* **25,** 3379–3388.

Chevalier, B. S., and Stoddard, B. L. (2001). Homing endonucleases: Structural and functional insight into the catalysts of intron/intein mobility. *Nucleic Acids Res.* **29,** 3757–3774.

Chou, C. Y., Horng, L. S., and Tsai, H. J. (2001). Uniform GFP-expression in transgenic medaka (*Oryzias latipes*) at the F0 generation. *Transgenic Res.* **10,** 303–315.

Choulika, A., Perrin, A., Dujon, B., and Nicolas, J. F. (1995). Induction of homologous recombination in mammalian chromosomes by using the I-SceI system of *Saccharomyces cerevisiae*. *Mol. Cell Biol.* **15,** 1968–1973.

Coen, D., Deutsch, J., Netter, P., Petrochilo, E., and Slonimski, P. P. (1970). Mitochondrial genetics. I. Methodology and phenomenology. *Symp. Soc. Exp. Biol.* **23,** 449–496.

Collas, P., and Alestrom, P. (1997). Nuclear localization signals: A driving force for nuclear transport of plasmid DNA in zebrafish. *Biochem. Cell Biol.* **75**, 633–640.

Collas, P., and Alestrom, P. (1998). Nuclear localization signals enhance germline transmission of a transgene in zebrafish. *Transgenic Research* **7**, 303–309.

Colleaux, L., D'Auriol, L., Galibert, F., and Dujon, B. (1988). Recognition and cleavage site of the intron-encoded omega transposase. *Proc. Natl. Acad. Sci. USA* **85**, 6022–6026.

Culp, P., Nusslein-Volhard, C., and Hopkins, N. (1991). High-frequency germline transmission of plasmid. DNA sequences injected into fertilized zebrafish eggs. *Proc. Natl. Acad. Sci. USA* **88**, 7953–7957.

Davidson, A. E., Balciunas, D., Mohn, D., Shaffer, J., Hermanson, S., Sivasubbu, S., Cliff, M. P., Hackett, P. B., and Ekker, S. C. (2003). Efficient gene delivery and gene expression in zebrafish using the Sleeping Beauty transposon. *Dev. Biol.* **263**, 191–202.

Deschet, K., Nakatani, Y., and Smith, W. C. (2003). Generation of Ci-Brachyury-GFP stable transgenic lines in the ascidian *Ciona savignyi*. *Genesis* **35**, 248–259.

Fadool, J. M., Hartl, D. L., and Dowling, J. E. (1998). Transposition of the mariner element from *Drosophila* mauritiana in zebrafish. *Proc. Natl. Acad. Sci. USA* **95**, 5182–5186.

Fritz, J. D., Herweijer, H., Zhang, G., and Wolff, J. A. (1996). Gene transfer into mammalian cells using histone-condensed plasmid DNA. *Hum. Gene Ther.* **7**, 1395–1404.

Fu, Y., Wang, Y., and Evans, S. M. (1998). Viral sequences enable efficient and tissue-specific expression of transgenes in *Xenopus*. *Nat. Biotechnol.* **16**, 253–257.

Gaiano, N., Allende, M., Amsterdam, A., Kawakami, K., and Hopkins, N. (1996). Highly efficient germline transmission of proviral insertions in zebrafish. *Proc. Natl. Acad. Sci. USA* **93**, 7777–7782.

Garrick, D., Fiering, S., Martin, D. I. K., and Whitelaw, E. (1998). Repeat-induced gene silencing in mammals. *Nat. Genet.* **18**, 56–59.

Gong W. J., Golic K. G. (2003). Ends-out, or replacement, gene targeting in *Drosophila*. *Proc. Natl. Acad. Sci. USA* **100**(5), 2556–2561.

Gordon, J. W., Scangos, G. A., Plotkin, D. J., Barbosa, J. A., and Ruddle, F. H. (1980). Genetic transformation of mouse embryos by microinjection of purified DNA. *Proc. Natl. Acad. Sci. USA* **77**, 7380–7384.

Grabher, C., Henrich, T., Sasado, T., Arenz, A., Furutani-Seiki, M., and Wittbrodt, J. (2003). Transposon-mediated enhancer trapping in medaka. *Gene* **322**, 57–66.

Hackett, P. B. (1993). The molecular biology of transgenic fish. *In* "Biochemistry and Molecular Biology of Fishes" (H. A. Mommsen, ed.), pp. 207–240. Elsevier Science, Amsterdam.

Higashijima, S., Okamoto, H., Ueno, N., Hotta, Y., and Eguchi, G. (1997). High-frequency generation of transgenic zebrafish which reliably express GFP in whole muscles or the whole body by using promoters of zebrafish origin. *Dev. Biol.* **192**, 289–299.

Hong, Y., Winkler, C., and Schartl, M. (1996). Pluripotency and differentiation of embryonic stem cell lines from the medakafish (*Oryzias latipes*). *Mech. Dev.* **60**, 33–44.

Hong, Y., Winkler, C., and Schartl, M. (1998). Production of medakafish chimeras from a stable embryonic stem cell line. *Proc. Natl. Acad. Sci. USA* **95**, 3679–3684.

Hsiao, C. D., Hsieh, F. J., and Tsai, H. J. (2001). Enhanced expression and stable transmission of transgenes flanked by inverted terminal repeats from adeno-associated virus in zebrafish. *Dev. Dynam.* **220**, 323–336.

Inoue, K., Yamashita, S., Hata, J., Kabeno, S., Asada, S., Nagahisa, E., and Fujita, T. (1990). Electroporation as a new technique for producing transgenic fish. *Cell Differ Dev.* **29**, 123–128.

Ivics, Z., Hackett, P. B., Plasterk, R. H., and Izsvak, Z. (1997). Molecular reconstruction of Sleeping Beauty, a Tc1-like transposon from fish, and its transposition in human cells. *Cell* **91**, 501–510.

Iyengar, A., Muller, F., and Maclean, N. (1996). Regulation and expression of transgenes in fish—a review. *Transgenic Res.* **5**, 147–166.

Jacquier, A., and Dujon, B. (1985). An intron-encoded protein is active in a gene conversion process that spreads an intron into a mitochondrial gene. *Cell* **41**, 383–394.

Jesuthasan, S., and Subburaju, S. (2002). Gene transfer into zebrafish by sperm nuclear transplantation. *Dev. Biol.* **242**, 88–95.

Johnson, R. D., and Jasin, M. (2001). Double-strand-break-induced homologous recombination in mammalian cells. *Biochem Soc. Trans.* **29**, 196–201.

Jowett, T. (1999). Transgenic zebrafish. *Meth. Mol. Biol.* **97**, 461–486.

Kaneda, Y., Iwai, K., and Uchida, T. (1989). Increased expression of DNA cointroduced with nuclear protein in adult rat liver. *Science* **243**, 375–378.

Kawakami, K., Shima, A., and Kawakami, N. (2000). Identification of a functional transposase of the Tol2 element, an Ac-like element from the Japanese medaka fish, and its transposition in the zebrafish germ lineage. *Proc. Natl. Acad. Sci. USA* **97**, 11403–11408.

Kellum, R., and Schedl, P. (1991). A position-effect assay for boundaries of higher order chromosomal domains. *Cell* **64**, 941–950.

Kelly, W. G., Xu, S., Montgomery, M. K., and Fire, A. (1997). Distinct requirements for somatic and germline expression of a generally expressed *Caernorhabditis elegans* gene. *Genetics* **146**, 227–238.

Kroll, K. L., and Amaya, E. (1996). Transgenic *Xenopus* embryos from sperm nuclear transplantations reveal FGF signaling requirements during gastrulation. *Development* **122**, 3173–3183.

Kroll, K. L., and Gerhart, J. C. (1994). Transgenic *X. laevis* embryos from eggs transplanted with nuclei of transfected cultured cells. *Science* **266,** 650–653.

Kuspa, A., and Loomis, W. F. (1992). Tagging developmental genes in *Dictyostelium* by restriction enzyme-mediated integration of plasmid DNA. *Proc. Natl. Acad. Sci. USA* **89**, 8803–8807.

Lai, L., Kolber-Simonds, D., Park, K. W., Cheong, H. T., Greenstein, J. L., Im, G. S., Samuel, M., Bonk, A., Rieke, A., Day, B. N., Murphy, C. N., Carter, D. B., Hawley, R. J., and Prather, R. S. (2002). Production of alpha-1,3-galactosyltransferase knockout pigs by nuclear transfer cloning. *Science* **295**, 1089–1092.

Lee, K.-Y., Huang, H., Ju, B., Yang, Z., and Lin, S. (2002). Cloned zebrafish by nuclear transfer from long-term-cultured cells. *Nat. Biotechnol.* **20**, 795–799.

Lin, S. (2000). Transgenic zebrafish. *Methods Mol. Biol.* **136**, 375–383.

Lin, S., Gaiano, N., Culp, P., Burns, J. C., Friedmann, T., Yee, J. K., and Hopkins, N. (1994a). Integration and germline transmission of a pseudotyped retroviral vector in zebrafish. *Science* **265**, 666–669.

Lin, S., Yang, S., and Hopkins, N. (1994b). lacZ Expression in germline transgenic zebrafish can be detected in living embryos. *Dev. Biol.* **161**, 77–83.

Linney, E., Hardison, N. L., Lonze, B. E., Lyons, S., and DiNapoli, L. (1999). Transgene expression in zebrafish: A comparison of retroviral-vector and DNA-injection approaches. *Dev. Biol.* **213**, 201–216.

Ma, C. G., Fan, L. C., Ganassin, R., Bols, N., and Collodi, P. (2001). Production of zebrafish germline chimeras from embryo cell cultures. *Proc. Natl. Acad. Sci. USA* **98**, 2461–2466.

Macreadie, I. G., Scott, R. M., Zinn, A. R., and Butow, R. A. (1985). Transposition of an intron in yeast mitochondria requires a protein encoded by that intron. *Cell* **41**, 395–402.

McCreath, K. J., Howcroft, J., Campbell, K. H., Colman, A., Schnieke, A. E., and Kind, A. J. (2000). Production of gene-targeted sheep by nuclear transfer from cultured somatic cells. *Nature* **405**, 1066–1069.

Meng, A., Jessen, J. R., and Lin, S. (1999). Transgenesis. *Methods Cell Biol.* **60,** 133–148.

Monteilhet, C., Perrin, A., Thierry, A., Colleaux, L., and Dujon, B. (1990). Purification and characterization of the *in vitro* activity of I-Sce I, a novel and highly specific endonuclease encoded by a group I intron. *Nucleic Acids Res.* **18**, 1407–1413.

Moure, C. M., Gimble, F. S., and Quiocho, F. A. (2003). The crystal structure of the gene targeting homing endonuclease I-SceI reveals the origins of its target site specificity. *J. Mol. Biol.* **334**, 685–695.

Muller, F., Ivics, Z., Erdelyi, F., Papp, T., Varadi, L., Horvath, L., and Maclean, N. (1992). Introducing foreign genes into fish eggs with electroporated sperm as a carrier. *Mol. Marine Biol. Biotechnol.* **1**, 276–281.

Neumann, C. J., and Nuesslein-Volhard, C. (2000). Patterning of the zebrafish retina by a wave of sonic hedgehog activity. *Science* **289**, 2137–2139.

Noma, K., Allis, C. D., and Grewal, S. I. (2001). Transitions in distinct histone H3 methylation patterns at the heterochromatin domain boundaries. *Science* **293**, 1150–1155.

Ono, H., Hirose, E., Miyazaki, K., Yamamoto, H., and Matsumoto, J. (1997). Transgenic medaka fish bearing the mouse tyrosinase gene: Expression and transmission of the transgene following electroporation of the orange-colored variant. *Pigment Cell Res.* **10**, 168–175.

Ozato, K., Kondoh, H., Inohara, H., Iwamatsu, T., Wakamatsu, Y., and Okada, T. S. (1986). Production of transgenic fish: Introduction and expression of chicken delta-crystallin gene in medaka embryos. *Cell Differ.* **19**, 237–244.

Perrin, A., Buckle, M., and Dujon, B. (1993). Asymmetrical recognition and activity of the I-SceI endonuclease on its site and on intron-exon junctions. *EMBO J.* **12**, 2939–2947.

Philip, R., Brunette, E., Kilinski, L., Murugesh, D., McNally, M. A., Ucar, K., Rosenblatt, J., Okarma, T. B., and Lebkowski, J. S. (1994). Efficient and sustained gene expression in primary T lymphocytes and primary and cultured tumor cells mediated by adeno-associated virus plasmid DNA complexed to cationic liposomes. *Mol. Cell. Biol.* **14**, 2411–2418.

Raz, E., van Luenen, H. G., Schaerringer, B., Plasterk, R. H. A., and Driever, W. (1998). Transposition of the nematode *Caenorhabditis elegans* Tc3 element in the zebrafish *Danio rerio*. *Curr. Biol.* **8**, 82–88.

Rong, Y. S., and Golic, K. G. (2000). Gene targeting by homologous recombination in *Drosophila*. *Science* **288**, 2016–2018.

Segal, D. J., and Carroll, D. (1995). Endonuclease-induced, targeted homologous extrachromosomal recombination in *Xenopus* oocytes. *Proc. Natl. Acad. Sci. USA* **92**, 806–810.

Sin, F. Y., Walker, S. P., Symonds, J. E., Mukherjee, U. K., Khoo, J. G., and Sin, I. L. (2000). Electroporation of salmon sperm for gene transfer: Efficiency, reliability, and fate of transgere. *Mol. Reprod. Dev.* **56**, 285–288.

Stief, A., Winter, D. M., Stratling, W. H., and Sippel, A. E. (1989). A nuclear DNA attachment element mediates elevated and position-independent gene activity. *Nature* **341**, 343–345.

Stuart, G. W., McMurray, J. V., and Westerfield, M. (1988). Replication, integration and stable germline transmission of foreign sequences injected into early zebrafish embryos. *Development* **103**, 403–412.

Stuart, G. W., Vielkind, J. R., McMurray, J. V., and Westerfield, M. (1990). Stable lines of transgenic zebrafish exhibit reproducible patterns of transgene expression. *Development* **109**, 577–584.

Sussman, R. (2001). Direct DNA delivery into zebrafish embryos employing tissue culture techniques. *Genesis* **31**, 1–5.

Tanaka, M., and Kinoshita, M. (2001). Recent progress in the generation of transgenic medaka (*Oryzias latipes*). *Zool. Sci.* **18**, 615–622.

Tawk, M., Tuil, D., Torrente, Y., Vriz, S., and Paulin, D. (2002). High-efficiency gene transfer into adult fish: A new tool to study fin regeneration. *Genesis* **32**, 27–31.

Thermes, V., Grabher, C., Ristoratore, F., Bourrat, F., Choulika, A., Wittbrodt, J., and Joly, J.-S. (2002). J-SceI meganuclease mediates highly efficient transgenesis in fish. *Mech. Dev.* **118**, 91–98.

Wakamatsu, Y., Ju, B. S., Pristyaznhyuk, I., Niwa, K., Ladygina, T., Kinoshita, M., Araki, K., and Ozato, K. (2001). Fertile and diploid nuclear transplants derived from embryonic cells of a small laboratory fish medaka (*Oryzias latipes*). *Proc. Natl. Acad. Sci. USA* **98**, 1071–1076.

Westerfield, M. (1995). "The Zebrafish Book." University of Oregon Press, Eugene, OR.

Yamauchi, M., Kinoshita, M., Sasanuma, M., Tsuji, S., Terada, M. M. M., and Ishikawa, Y. (2000). Introduction of a foreign gene into medakafish using the particle gun method. *J. Exp. Zool.* **287**, 285–293.

Zelenin, A. V., Alimov, A. A., Barmintzev, V. A., Beniumov, A. O., Zelenina, I. A., Krasnov, A. M., and Kolesnikov, V. A. (1991). The delivery of foreign genes into fertilized fish eggs using high-velocity microprojectiles. *FEBS Lett.* **287**, 118–120.

Zhu, Z., Li, G., He, L., and Chen, S. (1985). Novel gene transfer into fertilized eggs of goldfish (*Carassius auratus* L. 1758). *Z. Angew. Ichtyol.* **1**, 31–34.

Zhu, Z. Y., and Sun, Y. H. (2000). Embryonic and genetic manipulation in fish. *Cell Res.* **10**, 17–27.

CHAPTER 22

Cloning Zebrafish by Nuclear Transfer

Bensheng Ju, Haigen Huang, Ki-Young Lee, and Shuo Lin

Department of Molecular, Cellular, and Developmental Biology
University of California Los Angeles
Los Angeles, California 90095

I. Introduction

The zebrafish, *Danio rerio*, possesses some unique features such as small body size, short reproductive cycle, large egg cluster size, and, most importantly, transparent and *in vitro* embryogenesis. These features have made zebrafish an excellent vertebrate model for extensive studies in various fields of biology such as genetics, development, behavior, and even human diseases (Dooley and Zon, 2000; Zon, 1999). Further enhancing zebrafish's status as an important vertebrate

METHODS IN CELL BIOLOGY, VOL. 77

model is the relative ease with which transgenesis, cell labeling, and transplantation techniques can be applied to this species. Despite these favorable characteristics, the zebrafish lacks a practical method to disrupt a specific gene in order to study its *in vivo* gene function, such as the knockout approach routinely applied to mouse through homologous recombination in embryonic stem (ES) cells. Although tremendous efforts have been made (Ma *et al.*, 2001), so far no ES cells equivalent to those of mouse are available in zebrafish. As the scheduled genome project is to be completed soon, readily available genome sequences and gene structures call for a targeted genetic manipulation in zebrafish to understand gene function. In mammals lacking ES cells, targeted gene disruption in fibroblast cells coupled with animal cloning by nuclear transfer has been successfully demonstrated (Lai *et al.*, 2002; McCreath *et al.*, 2000). To test the feasibility of a similar approach in zebrafish, we developed the nuclear transfer technology in zebrafish, using long-term cultured embryonic fibroblast cells (Lee *et al.*, 2002).

Cloning of zebrafish by nuclear transfer is a complex procedure involving cell culture, egg selection, and micromanipulation of the eggs and cells, each step affecting the overall efficiency of cloning. Here we provide a detailed description of the nuclear transfer procedure developed in our laboratory.

II. Recipes for Cell Culture and Nuclear Transfer

Hank's solution was used throughout the experiment as the buffer for nuclear transfer. The formula can be found in Westerfield (1995). Hank's work solution can also be purchased from Gibco or Cellgro.

Holtfreter's solution was used to dechorionate recipient eggs. It contains 3.5 g NaCl, 0.2 g NaHCO$_3$, 0.12 g CaCl$_2 \cdot$ 2H$_2$O, 0.05 g KCl in 1 l of distilled water. Adjust pH to 6.5–7.1.

Zerafish embryo extracts were used for cell culture. About 100–150 twenty-four-hour-old embryos were homogenized in 1 ml PBS (phosphate buffer saline, Cellgro, Mediatech), followed by filtration (0.45 μm, Millipore). Embryo extracts were aliquoted and stored at $-80\,^{\circ}$C.

III. Cell Culture

A. Medium for Cell Culture

The DMEM (Dulbecco's modification of Eagles's medium) was used as the basic medium for zebrafish cell culture. This medium contains L-glutamine, 4.5 g/l glucose, but lacks sodium pyruvate (Gibco BRL, Rockville, MD, and Cellgro, Mediatech). The medium was supplemented with 15% fetal bovine serum (FBS, Gibco BRL, Rockville, MD), 1% trout serum (SeaGrow, East Coast Biologics), bovine insulin (10 μg/ml, Sigma), and 0.5% (v/v) zebrafish embryo extracts. For primary cell culture, bovine basic fibroblast growth factor (bbFGF, 20–50 ng/ml,

Sigma) was added to the regular medium described previously to inhibit melanocyte formation during the first 2 weeks. The antibiotics penicillin (100 units/ml) and streptomycin (100 μg/ml) were also included in the medium until the establishment of long-term cultured cells.

B. Primary and Long-Term Cell Culture

To make primary cells, 20–30 five to fifteen-somite-stage embryos were dechorionated by protease treatment (Sigma, Cat. No. P5147) for 10 min at 28 °C. The protease was prepared in distilled water (30 mg/ml) and then diluted in Holtfreter's solution to the working concentration of 10 mg/ml. Embryos were washed with Holtfreter's solution, homogenized, and dissociated in trypsin/EDTA (0.25% trypsin/1 mM EDTA in PBS) at 37 °C for 5–10 min. The cells were washed with PBS and centrifuged at 1000 rpm for 5 min and then suspended in DMEM medium until nuclear transfer. Fresh primary cells could be used for nuclear transfer up to 3 h.

To establish long-term cultured cells, 200–300 eggs were collected from multiple pairs of zebrafish and kept at room temperature (approximately 23 °C) until the following day. Embryos were used for cell culture when they were at the 10- to 15-somite stage. Healthy-looking embryos with intact chorions were selected under a dissection microscope and dechorionated in protease for about 10 min. Dechorionated embryos were washed six to eight times with sterile Holtfreter's solution, followed by another six times with sterile PBS in a tissue culture hood and then disinfected with 0.04% bleach (Aldrich, 4% sodium hypochlorite) for exactly 3 min. Treated embryos were washed four times with PBS to remove residual bleach. Primary cells obtained from these embryos were cultured in a DMEM-based medium at 28–29 °C with 5% CO_2. Cells were not disturbed for the initial 48 h, after which one third of the medium was changed. bbFGF (Sigma, 20–50 μg/ml) was included in the medium for the first 2 weeks to inhibit pigment cell formation. After 8 weeks and about 13 subcultures, the cells were considered as long-term cultured cells. These cells can be infected with pseudotyped retroviruses or transfected with exogenous DNA constructs.

For nuclear transfer, long-term cultured cells were subjected to serum starvation by culturing them in DMEM medium supplemented with 0.5% FBS for 4 days and were then dissociated with trypsin/EDTA (0.25% trypsin/1 mM EDTA in PBS), washed once with PBS, and centrifuged at 1000 rpm for 5 min. Cells were suspended in DMEM containing 0.5% FBS and kept on ice until nuclear transfer.

IV. Nuclear Transfer

A. Micromanipulation Equipment

The Narishige micromanipulation system (NT-188NE) mounted on an Axiovert 200 microscope (Carl Zeiss) was used for nuclear transfer. Microinjection capillaries for making transfer and holding needles were purchased from Harvard Apparatus Ltd. (Cat. No. GC100-10). We used a model P-97 micropipette puller from Sutter Instrument Co. to make injection needles. Pulled glass capillaries were broken by briefly and abruptly touching two capillaries at the ends to make injection needles with inner diameters of 10–12 μm. Holding needles were made by flaming the capillaries on an alcohol lamp. The narrow end was cut with a small Tungsten Carbide Pencil (Fisher) to make an opening ~260 μm in inner diameter and then fire polished to smoothen the surface. The relative size of the holding and injection needle is shown in Fig. 1.

B. Preparation of Recipient Eggs

Zebrafish were kept on a 14-h light/10-h dark cycle. A pair of male and female fish was placed in a mating cage, separated by a central divider. Usually 10–12 pairs were prepared in the afternoon the day before the nuclear transfer. The following morning, the divider in one mating cage was removed to allow the male to chase the female. Mating activity was closely watched in a way to allow the male to touch the female, but spawning was prevented by separating the fish immediately with a fish net. After the male fish touched the female three to four times, the female was immediately removed from the cage and anesthetized for approximately 1 min in 0.1% tricaine solution (Sigma). The fish was wiped dry with Kimwipes and gently squeezed from the urogenital opening to obtain unfertilized eggs. Good-quality matured eggs are slightly granular and yellowish in color, whereas immature eggs appear whitish or withered. Good-quality eggs were directly placed in Holtfreter's solution and dechorionated with protease. After a brief washing with Holtfreter's solution four times, the eggs were immediately transferred into precooled (4 °C) Hank's solution supplemented with 1.5% BSA

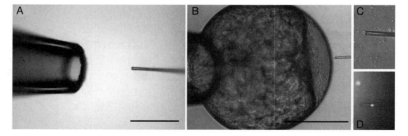

Fig. 1 Enucleation and nuclear transfer in zebrafish. (A) Holding and nuclear transfer needles in relative sizes. (B) The nuclear transfer needle approaches the second polar body to remove the egg's pronucleus. (C) Picking up a cell by the nuclear transfer needle. (D) The donor cell nucleus in the transfer needle is fluorescent when observed under the fluorescent microscope because donor cells were infected with retrovirus containing GFP with a nuclear localization signal. Bar = 300 μm (A, B), 100 μm (C, D). Adapted from Lee, K. Y. *et al.* (2002) Cloned zebrafish by nuclear transfer from long-term cultured cells. *Nat. Biotechnol.* **20**, 795–799, with permission.

(w/v, fraction V, heat shocked, Roche). These eggs were used as recipients for nuclear transfer up to 1 h. If the eggs obtained were of poor quality, another pair of fish was used. Usually, we obtained three or four batches of good-quality eggs from 10–12 pairs of fish.

C. Enucleation and Nuclear Transfer

A critical step for nuclear transfer is to completely remove the maternal pronucleus from the recipient egg. Unlike in mammals, the maternal pronucleus of a living fish egg is not visible under the microscope. However, the location of the maternal pronucleus can be revealed by staining the unfertilized eggs with Hoechst 33342 (Sigma), which is a membrane-permeable fluorescent DNA dye intercalating in A-T regions of DNA. Dechorionated eggs were fixed for 1 h at room temperature in 4% (w/v) paraformaldehyde, stained for 10 min in Hoechst 33342 (1 mg/ml), and washed 10 times in PBS. The stained eggs could be visualized under UV light by using a fluorescent microscope. As shown in Fig. 2A, Hoechst 33342 staining revealed two bright spots, which were the second polar body in upper

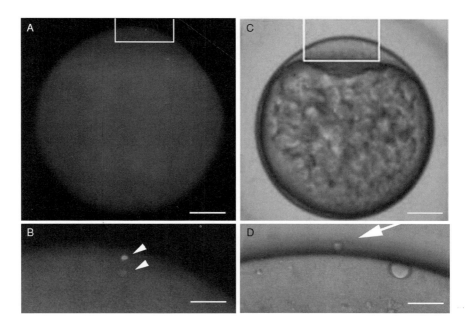

Fig. 2 Locating the recipient egg's pronucleus. (A) Hoechest 33342 staining of an unfertilized egg. (B) Inset from (A) showing both maternal pronucleus and the polar body (arrow heads). The egg nucleus is located just underneath the egg surface against the polar body. (C) Bright-field view of an unfertilized egg. (D) Inset from (C) showing the second polar body (arrow). Bar = 150 μm (A, C), 50 μm (B, D). Adapted from Lee, K. Y. *et al.* (2002). Cloned zebrafish by nuclear transfer from long-term cultured cells. *Nat. Biotechnol.* **20**, 795–799, with permission.

location and maternal pronucleus in lower location. Live zebrafish eggs are activated and begin to release their second polar bodies on contacting with water. The polar body is visible as a small transparent ball under a phase-contrast microscope (∼8 μm diameter; Fig. 2B), which provides a reference point for locating the maternal pronucleus as Hoechst 33342 staining shows that the pronucleus is just beneath the polar body against the egg membrane.

During nuclear transfer, both the recipient eggs and donor cells were placed into an inverted cover of a Falcon Tissue culture dish (Cat. No. 353004). The eggs were kept in a large drop of precooled Hank's solution supplemented with 1.5% BSA and the donor cells were kept nearby in a drop of precooled DMEM medium supplemented with 0.5% FBS (v/v). A recipient egg was held at the tip of the holding needle and appropriately positioned to allow the animal pole to face the transfer needle. Using the polar body as a reference, the pronucleus was removed by sucking out a very small amount of cytoplasm just below the polar body. To avoid compromising the egg's developmental potential, the nucleus was removed in as small a volume as possible. Donor cells of appropriate size (Fig. 1C), which are round in shape, were picked up by the transfer needle and slightly ruptured by repeated aspiration inside the needle. Donor cells must be gently operated and slightly broken to avoid damaging the nuclei. The slightly ruptured donor cells were gently transplanted into the cytoplasm of the enucleated egg at the exact location of enucleation. An experienced researcher can operate six to eight eggs from each batch and three to four batches in a typical morning, which would add up to an average of 20 eggs performed each day.

D. Embryo Maintenance

Transplanted eggs were transferred from Hank's solution to small containers, such as 60-mm or 100-mm Falcon tissue culture dishes, containing Holtfreter's solution and maintained in a 28–29 °C incubator. Because the freshly transplanted eggs are extremely delicate, avoid unnecessary movements for the first 24 h. The developing embryos did not need to be fed for the first 3 days, after which they were transferred to a mouse cage filled with fish water, fed with paramecia for 10 days, and then switched to both paramecia and live brine shrimps. After another 10–15 days, they were transferred to regular fish tanks until they reached sexual maturity.

V. Summary of Nuclear Transfer

We normally perform nuclear transfers in the morning and count each day's operation as one experiment. When dissociated embryonic cells were used as donors, the embryos were homozygous for a transgene expressing the green fluorescent protein (GFP), so GFP expression served as a donor marker to help us determine the origin of the developing embryos. Overall, approximately 80% of experiments never yielded any developing embryos, most likely because of poor egg quality. For those experiments that produced developing nuclear transplants, the embryos exhibited various degrees of abnormity, but normal individuals were also obtained. In a series of eight successful nuclear transfer experiments involving 67 transplanted eggs, 20 (30%) embryos reached the blastula stage, 12 (18%) embryos hatched, and 11 (16%) of them survived to adulthood. All the hatched embryos expressed GFP.

For nuclear transfer involving long-term cultured cells, embryos were disaggregated and cultured initially for 8 weeks and then a concentrated stock of pseudotyped retroviral vector containing GFP reporter gene driven by the *Xenopus* elongation 1 alpha (XeX) promoter (Linney *et al.*, 1999) was used to infect these cells. GFP-positive cells (Fig. 1D) were then used as donors about 4 weeks later. As experienced in our initial study using primary cells as donors, more than 80% of experiments failed to produce developing embryos. From 10 experiments that produced embryos that went through cell cleavages, 34 (36%) embryos reached the blastula stage, 15 (16%) embryos in six experiments hatched, and 11 (12%) of these embryos reached adulthood. All the hatched embryos we obtained expressed GFP, again confirming that donor cells contributed to their development.

Using both dissociated embryonic cells and long-term cultured cells as donors, successful nuclear transplants represented approximately 2% of total embryos operated. Nine adult fish from long-term cultured cells were mated with wild-type fish; 50% of the offspring expressed GFP, suggesting that the GFP donor marker gene was transmitted to the subsequent generation in a Mendelian fashion. We also performed Southern blot analyses on eight cloned fish. Each fish had a junction fragment that was different from the others, indicating that they were derived from different donor cells.

VI. Potential Applications of Zebrafish Cloning

We have established the procedure for cloning the zebrafish by nuclear transfer by using long-term cultured cells. The cells can be cultured for up to 26 weeks, frozen, and thawed, and their capacity for producing viable nuclear transplants remains. This long window period provides ample opportunity for various genetic manipulations to the cells, such as proviral infection and DNA transfection. The availability of such a cloning technology can have many potential applications.

A. Produce Transgenic Fish Through Cloning

Although transgenic fish can be produced by direct injection of DNA into fertilized eggs, it requires screening a large number of founder fish to identify germline transmission. By nuclear transplantation, donor cells containing stable transgene integrations can be selected in cell culture prior to fish cloning, and we have achieved this by obtaining heterozygous transgenic fish at the first generation.

B. Develop Techniques for Generating Zebrafish from Cultured Cells Carrying Gene Traps

Gene trapping is a method of random insertional mutagenesis that uses a fragment of DNA coding for a reporter or selectable marker gene as a mutagen (Friedrich and Soriano, 1993). Gene trappings can be carried out in cultured zebrafish cells, and the trapped genes can be isolated and structurally analyzed. Cells that harbor interesting trapped genes can then be used to obtain fish clones. By studying the cloned fish, functions of the genes of interest can be revealed.

C. Study Effects of Cloning on Animal Development

Cloned zebrafish can be excellent models for studying effects of cloning on animal development. Developmental abnormalities in cloned zebrafish can be easily found because of *in vitro* and transparent embryogenesis. Short generation time and easy access to a large number of progenies mean that effects of cloning can be monitored thoroughly in multiple generations in a relatively short time.

D. Develop Techniques for Targeted Mutagenesis in Zebrafish

This involves designing targeting constructs, selecting cells carrying homologous recombination events, and cloning zebrafish by using these cells. There are two challenges for developing such a technology in zebrafish. First, we need to find out whether DNA homologous recombination can be achieved in our cultured cells, and second, if homologous recombination is achievable, whether cells carrying a homologous recombination event still have the competency to generate normal cloned zebrafish. If successful, zebrafish will have all the genetic tools available to the mouse system, and we can fully realize its potential as a excellent vertebrate model to study gene function and human diseases.

Acknowledgments

We thank members of our laboratory for technical assistance and discussion. This work was supported by a grant from the National Institutes of Health (R01 RR13227) to S. L.

References

Dooley, K., and Zon, L. I. (2000). Zebrafish: A model system for the study of human disease. *Curr. Opin. Genet. Dev.* **10,** 252–256.

Friedrich, G., and Soriano, P. (1993). Insertional mutagenesis by retroviruses and promoter traps in embryonic stem cells. *Methods Enzymol.* **225,** 681–701.

Lai, L., Kolber-Simonds, D., Park, K. W., Cheong, H. T., Greenstein, J. L., Im, G. S., Samuel, M., Bonk, A., Rieke, A., Day, B. N., Murphy, C. N., Carter, D. B., Hawley, R. J., and Prather, R. S. (2002). Production of alpha-1,3-galactosyltransferase knockout pigs by nuclear transfer cloning. *Science* **295,** 1089–1092.

Lee, K. Y., Huang, H., Ju, B., Yang, Z., and Lin, S. (2002). Cloned zebrafish by nuclear transfer from long-term cultured cells. *Nat. Biotechnol.* **20,** 795–799.

Linney, E., Hardison, N. L., Lonze, B. E., Lyons, S., and DiNapoli, L. (1999). Transgene expression in zebrafish: A comparison of retroviral-vector and DNA-injection approaches. *Dev. Biol.* **213,** 207–216.

Ma, C., Fan, L., Ganassin, R., Bols, N., and Collodi, P. (2001). Production of zebrafish germ-line chimeras from embryo cell cultures. *Proc. Natl. Acad. Sci. USA* **98,** 2461–2466.

McCreath, K. J., Howcroft, J., Campbell, K. H., Colman, A., Schnieke, A. E., and Kind, A. J. (2000). Production of gene-targeted sheep by nuclear transfer from cultured somatic cells. *Nature* **405,** 1066–1069.

Westerfield, M. (1995). "The Zebrafish Book. A Guide for the Laboratory Use of Zebrafish (*Danio rerio*)" University of Oregon Press, Eugene, OR.

Zon, L. I. (1999). Zebrafish: A new model for human disease. *Genome Res.* **9,** 99–100.

PART IV

Informatics and Comparative Genomics

CHAPTER 23

Data Mining the Zebrafish Genome

Lynn M. Schriml* and Judy Sprague[†]

*National Center for Biotechnology Information (NCBI)
National Institutes of Health
Bethesda, Maryland 20894

[†]Zebrafish Information Network (ZFIN)
University of Oregon
Eugene, Oregon 97403

I. Introduction

A wide variety of Web-based resources provide access to the wealth of data available for zebrafish researchers. The suite of tools continues to grow and change to meet the needs of the genomics communities. With a click of a mouse, users can navigate between disparate types of data, exploring connections, for example, between publications, sequence, diseases, and homology, thereby opening the door to exciting discoveries.

In this chapter, users find a guide to mining zebrafish data at the National Center for Biotechnology Information (NCBI; http://www.ncbi.nlm.nih.gov, Section II) and the Zebrafish Information Network (ZFIN; http://zfin.org, Section III), with suggestions of ways to search for publications, gene information, homology, sequence, map, structure, or expression data and methods for navigating among the different types of available data and resources. Following the NCBI and ZFIN sections in this chapter, we have included a set of sample questions as a tutorial (Section IV). In addition, Table I includes the URLs for the resources discussed here.

This chapter is intended as an overview and therefore includes only highlights of selected resources and tools. A comprehensive guide to NCBI tools can be found in the online version of the "NCBI Handbook." In addition, a site search of NCBI can be done either at NCBI's home page by selecting "NCBI Web Site" from the search bar at the top of the page or by submitting an Entrez Global Query (Fig. 1., Section II.A.3) and viewing the "Site Search" results. A site search of ZFIN is also available through the Zebrafish search machine at http://zfin.org/zf_info/SEARCH_SITE/searchcrit.html.

II. National Center for Biotechnology Information (NCBI) Tools, Resources, and Data Sets

NCBI provides a queryable interface that enables navigation between interconnected data types and between information for multiple genomes. NCBI brings together the power of large-scale computational analyses and an integrated system of data retrieval with detailed information on maps, sequence, expression, structure, genomes, genes, diseases, phenotypes, publications, protein domains, and structures. NCBI provides connections between these disparate types of data by computation (e.g., HomoloGene, UniGene, Related Sequences, and Domains) and curation (Wheeler *et al.*, 2004).

The connectivity of NCBI resources helps users find the information of interest, whether the search begins by looking at genes (Section II.B), sequences (Section II.C), publications (Section III.D), or map data (Section II.E).

To become acquainted with the set of tools and databases and types of data available at NCBI for the zebrafish community, users can begin by looking at the Zebrafish Genome Resources page (Section II.A.1), the Taxonomy Database page (Section II.A.2) for *Danio rerio*, or submit a search against all Entrez databases through Entrez's Global Query (Section II.A.3).

In addition to the online resources, NCBI also provides unrestricted access to NCBI's software and genome data from an FTP site (www.ncbi.nlm.nih.gov/Ftp/index.html). For example, daily updates for all NCBI Reference Sequences are available as a single file or as separate files for each genome, including the

Table I

Web Resources for Zebrafish Genome Data Mining

Web resources	URLs
National Center for Biotechnology Information (NCBI)	www.ncbi.nlm.nih.gov
NCBI Handbook	www.ncbi.nlm.nih.gov/entrez/query.fcgi?db=Books
Entrez's GQuery	www.ncbi.nlm.nih.gov/gquery/gquery.fcgi
Entrez search fields and qualifiers	www.ncbi.nlm.nih.gov/entrez/query/static/help/Summary Matrices.html#Search Fields and Qualifiers
PubMed	www.ncbi.nlm.nih.gov/entrez/query.fcgi?db=PubMed
Zebrafish Genome Resources	www.ncbi.nlm.nih.gov/genome/guide/zebrafish
UniGene	www.ncbi.nlm.nih.gov/entrez/query.fcgi?db=unigene
ProtEST	www.ncbi.nlm.nih.gov/UniGene/protest.shtml
HomoloGene	www.ncbi.nlm.nih.gov/entrez/query.fcgi?db=homologene
dbSNP	www.ncbi.nlm.nih.gov/SNP/
UniSTS	www.ncbi.nlm.nih.gov/entrez/query.fcgi?db=unists
e-PCR	www.ncbi.nlm.nih.gov/genome/sts/epcr.cgi
CDD	www.ncbi.nlm.nih.gov/entrez/query.fcgi?db=cdd
GEO	www.ncbi.nlm.nih.gov/geo/
MapViewer	www.ncbi.nlm.nih.gov/mapview/
BLAST	www.ncbi.nlm.nih.gov/BLAST/
Zebrafish genome BLAST	www.ncbi.nlm.nih.gov/genome/seq/DrBlast.html
Trace MegaBLAST	www.ncbi.nlm.nih.gov/blast/tracemb.shtml
BLink	www.ncbi.nlm.nih.gov/sutils/blink.cgi?pid=23943785
CDD	www.ncbi.nlm.nih.gov/Structure/cdd/cdd.shtml
CDART	www.ncbi.nlm.nih.gov/Structure/lexington/lexington. cgi?cmd=rps
Clone Registry	www.ncbi.nlm.nih.gov/genome/clone/
Trans-NIH Model Organism Initiative	www.nih.gov/science/models/index.html
Zebrafish Information Network (ZFIN)	www.zfin.org
Zebrafish Nomenclature Guidelines	zfin.org/zf_info/nomen.html
ZFIN Expression Search	zfin.org/cgi-bin/webdriver?MIval=aa-xpatselect.apg
HGNC Gene Grouping/Family Nomenclature	www.gene.ucl.ac.uk/nomenclature/genefamily.shtml
Gene Ontology Consortium	www.geneontology.org/
	www.godatabase.org/cgi-bin/go.cgi
GOA at EBI	www.ebi.ac.uk/GOA/
VEGA annotation browser	vega.sanger.ac.uk/Danio_rerio/
WashU	zfish.wustl.edu/
Zebrafish Gene Collection (ZGC)	zgc.nci.nih.gov/
FishBase	www.fishbase.org/search.cfm
PROW	www.ncbi.lm.nih.gov/prow

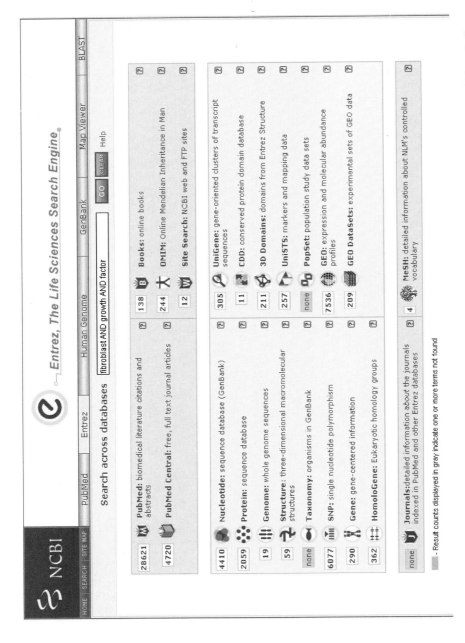

Fig. 1 The Entrez Cross-Database search page. Representative result using the Entrez cross-database global query (GQuery) allows simultaneous searches of all Entrez databases as demonstrated by the query: fibroblast AND growth AND factor. For each database, the number of hits is located in the box to the left of the database icons.

zebrafish mRNA and protein reference sequences (ftp://ftp.ncbi.nih.gov/refseq/ D_rerio/mRNA_Prot). Questions regarding NCBI resources can be directed to NCBI's Help Desk (info@ncbi.nlm.nih.gov).

A. Getting Started: Where to begin Your Search of NCBI Resources

1. Zebrafish Genome Resources Page at NCBI

The Zebrafish Genome Resources page (Fig. 2) was created to provide a gateway to Web resources for the zebrafish community and includes links to a number of NCBI sites as well as links to external resources that might be of interest to members of the zebrafish community. These resources include ZFIN, Sanger's *Danio rerio* Sequencing Project, expressed sequence tag (EST) data at Washington University's Zebrafish Genome Resources Project (WashU), and cDNA sequences produced from the Zebrafish Gene Collection (ZGC) Project. On the Zebrafish Genome Resources Web page, searches of other NCBI resources can be initiated through the query bar at the top of the page or by following the provided links to the resource home pages. For example, users can begin to search map data by either following the provided link to NCBI's Map Viewer home page or by choosing any of the LG links in the "Jump to the Genome" figure to go to the zebrafish Map Viewer directly and view all markers mapped for that linkage group.

2. Taxonomy Database

Entrez's Taxonomy page for *Danio rerio* provides another entry point to the set of available data for zebrafish. From the Taxonomy home page, users can choose the *Danio rerio* (zebrafish) link to view the *Danio rerio* Taxonomy page. This page includes a table reporting the current number of zebrafish entries in each of the Entrez databases. The number of entries is linked to the individual databases. In addition, the *Danio rerio* taxonomy page includes the list of centers submitting Trace records. For each type of Trace record (Clone end, EST, Finishing, Shotgun, WGS, and ALL), the user can view the number of records and follow links for each type to the Trace Archive. The *Danio rerio* Taxonomy page also includes a list of LinkOuts (see Section II.A.4).

3. Entrez and GQuery

All Entrez databases can be searched simultaneously with a single query via the Entrez Global Query (GQuery) system. This tool allows complex queries of one or more terms that include boolean operators (AND, OR, NOT) and searches of specified fields by including a qualifier such as [organism], [ORGN], [taxonomy_id], [keyword], or [KYWD] following the term. A link to the complete list of the available search fields, their descriptions, the databases at which the search fields can be used, and the qualifier is included in Table I. As shown in Fig. 1, the

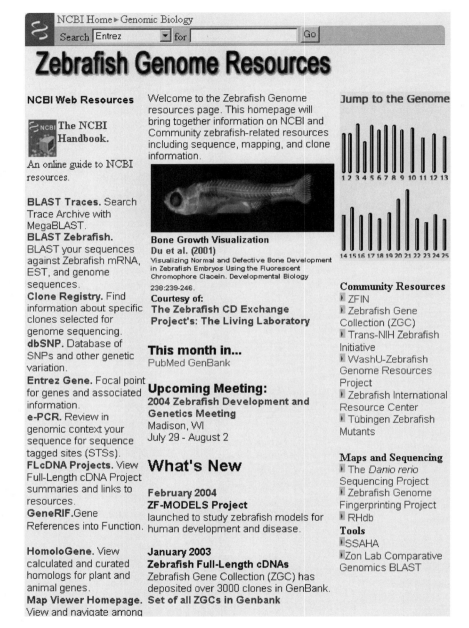

Fig. 2 The Zebrafish Genome Resources page at the National Center for Biotechnology Information (NCBI). Find links to NCBI and community resources, search among NCBI resources, and view PubMed IDs and GenBank accessions new this month.

GQuery result page displays the number of records returned from the search of each Entrez database. By clicking on the number of records or the database name, users can link to the individual database and view the detailed results for that database. If the box containing the number of hits is gray, it means that an exact match to the query was not found.

Note that Entrez queries that include one or more spaces or a colon in the text, for example, Danio rerio, LG 8, wu:fa01a01, or zgc:55283, should be enclosed within double quotes. Search terms that include a space and are not within quotes will be searched as two independent words. Double quotes are necessary because Entrez queries follow specific rules: terms are split by spaces, terms included within double quotes are searched as a phrase, and an unquoted colon is treated as a range operator. By using the range operator, users can search for a range of accession by using the following format: BC066000:BC066100[accession] or search for all zebrafish mRNAs longer than 800 bp but shorter than 2 kb with the query "Danio rerio" [organism] AND biomol_mrna[properties] AND 800:2000[sequence length].

Although not all Entrez databases currently contain zebrafish-specific data sets, these resources are still valuable for data mining of information related to zebrafish orthologs. For example, the Gene Expression Omnibus (GEO) database is a public respository for expression and hybridization data generated by high-throughput microarray experiments. At present, GEO does not contain any *Danio rerio* data sets; however, queries for data sets in a number of other organisms are an additional source of functional information.

4. LinkOut

To provide further utility to the data in Entrez, LinkOut provides connections between data elements in Entrez databases and non-Entrez databases. Users can view the list of LinkOut connections by choosing "LinkOut" in the Links pull-down menu on any Entrez page, including PubMed IDs, Entrez Gene IDs, and nucleotide and protein accessions. For example, LinkOut links to ZFIN are maintained on GenBank nucleotide and protein records. Submitting the query by using the LinkOut provider filter for ZFIN, loprovzfin[filter], to Entrez's GQuery identifies the set of nucleotide records common to both ZFIN and Entrez Gene. The Taxonomy database provides additional LinkOuts for *Danio rerio* to FishBase, ZFIN, and the Trans-NIH Model Organism Initiative.

B. Searching for Gene Data at NCBI

1. Entrez Gene and the Reference Sequence Project

Entrez Gene was developed to expand the scope of NCBI's gene-oriented LocusLink database to include all genomes from the Reference Sequence (RefSeq) collection. The functionality of LocusLink is retained in Entrez Gene. GeneIDs, as with LocusIDs, are stable identifiers that can be tracked over time.

NCBI's RefSeq project (Pruitt *et al.*, 2003) provides a nonredundant reference collection of DNA, transcript (mRNA), and protein sequences. RefSeq accessions are included on the Entrez Gene pages and can also be retrieved with the GQuery "Danio rerio" [organism] AND biomol_mrna[properties] AND srcdb_refseq. The vast majority of zebrafish RefSeqs are reported as "predicted" records. A complete list of the RefSeq accession formats can be found at http://www.ncbi.nlm.nih.gov/RefSeq/key.html#accession. Predicted RefSeq records are created by a series of computational steps designed to choose full-length representative mRNAs for a GeneID. As additional curation is done by the RefSeq staff, and additional evidence of function is identified and added to the RefSeq records, the status is upgraded to provisional.

Data exchanges between NCBI and ZFIN, as part of the ongoing collaboration, improve the representation of data at both sites. In addition, NCBI identifies new accessions that can define novel genes or add value to existing genes defined only by partial transcripts.

2. Zebrafish Gene Collection Collaboration

In addition to NCBI collaborations with ZFIN and other model organism databases, NCBI plays an active role in collaborations with large-scale cDNA projects, including the Mammalian Gene Collection (MGC), which includes the Zebrafish Gene Collection. As part of the MGC collaboration, NCBI performs protein and nucleotide comparisons, determines coding potential, annotates protein domains (Conserved Domain Database (CDD)) , and provides initial product names and periodic reanalysis and updates for each cDNA clone.

NCBI and ZFIN staff identify and create novel Entrez Gene zebrafish records and insert ZGC accessions to existing Entrez Gene records based on sequence and homology analyses for the Entrez Gene and RefSeq projects. The novel records created are given a symbol that begins with "zgc." To search for all Entrez Gene records created from this collaboration, submit the GQuery search "Danio rerio"[organism] AND MGC[keyword] or the search "Danio rerio"[organism] AND zgc*. The current set of ZGC accessions in GenBank can be identified in GQuery by the search "Danio rerio"[organism] AND biomol_mrna [properties] AND MGC [keyword] or by a clicking on the "Set of all ZGCs in GenBank" link provided on NCBI's Zebrafish Genome Resources Guide page.

3. GeneRIF: Gene References into Function

GeneRIF submissions are another valuable resource for mining functional information and connections provided at NCBI. The GeneRIF function allows connections to be made between PubMed citations and functional information related to a publication for records in Entrez Gene.

Members of the scientific community are able to submit a GeneRIF for a GeneID by follow the "Submit GeneRIF" link found on each Entrez Gene report page. The majority of GeneRIF entries are submitted by staff at the National Library of Medicine (NLM). For example, GeneID 30501 (otx2, orthodenticle homolog 2) contains the GeneRIF "In vertebrates, the Otx2 promoter acquires multiple, spatiotemporally specific cis-regulators in order to precisely control highly coordinated processes in head development," which is associated with PMID 14645121).

Entrez Gene NEWENTRY records are another source of mining connections between publications and functional information. For each organism in Entrez Gene, one identifier (symbol: NEWENTRY) has been created to allow GeneRIFs to be submitted for genes that have yet to be assigned an identifier by NCBI. For zebrafish, this record is Entrez GeneID: 192346. Users can browse through the PubMed IDs and GeneRIFs in the NEWENTRY record or submit a query to mine this set of GeneRIFs. For example, submitting the query NEWENTRY [symbol] AND notochord to Entrez's GQuery returns the zebrafish Entrez Gene NEWENTRY record that contains the GeneRIF "requirement for laminin beta1 and laminin gamma1 in the formation of a specific vertebrate organ and show that laminin or the laminin-dependent basement membrane is essential for the differentiation of chordamesoderm to notochord (SLY, GUP)." This GeneRIF is associated with PMID:12070089 entitled "Zebrafish mutants identify an essential role for laminins in notochord formation."

C. Searching for Highly Related Sequences at NCBI

1. Precomputed Sequence Comparisons

Precomputed sequence comparisons provided at NCBI enable users to quickly find highly related nucleotides, proteins, protein domains, and single-nucleotide polymorphisms (SNPs). Without having to submit a single BLAST job, users can easily identify highly similar nucleotide (by nucleotide neighbors, HomoloGene, UniGene, or UniSTS) or protein sequences (by protein neighbors, BLink, or UniGene's ProtEST). These analyses also provide computationally derived annotated CDD protein domains on RefSeq (mRNAs and proteins) and MGC GenBank records and placement of independently identified SNPs.

a. UniGene and HomoloGene Provide Precomputed Nucleotide Comparisons

The redundancy of cDNA sequences in GenBank continues to increase as high-throughput cDNA projects produce large volumes of data in the continuing search for novel genes. UniGene was developed to reduce this redundancy by providing gene-oriented sets of cDNA sequences through an automated comparison of EST and mRNA sequences. For example, in the February 2004 build (Build #66), UniGene reduced 364,067 zebrafish sequences (including 9122 mRNAs) to 17,890 nonredundant sets with 4432 sets containing at least one mRNA.

As cDNA and genome sequences have become available for more organisms, cross-species sequence analysis has become increasingly useful for identifying putative orthologs. To provide the means for these discoveries, HomoloGene was developed to produce precomputed comparisons between multiple eukaryotic organisms. The build procedure for HomoloGene uses DNA sequence to identify closely related orthologs and then looks for the more distant relationships (orthologs or paralogs) by protein comparisons (blastp). As a result, at this time, zebrafish sequences are not included in the initial build procedure and are not connected by Entrez GeneID but are included when zebrafish UniGene sets are found to share significant similarity to the initial HomoloGene set. Submitting the search Danio OR rerio, to find the HomoloGene sets containing zebrafish UniGene sets. HomoloGene sets include links to the Entrez Gene records for human, mouse, *Drosophila melanogaster, Anopheles, C. elegans*, the related UniGene sets, including *D. rerio*, and the HomoloGene set statistics.

b. BLAST Link (BLink) Reports Highly Related Proteins

Another option for finding precomputed protein relationships is to use the BLink viewer of proteins neighbored by BLAST. BLink provides a graphical alignment of up to 200 protein BLAST hits for each protein sequence in Entrez protein. Results can be ordered by either the BLAST score or by taxonomy grouping with additional options for display available from the header bar, including All Hits, Best Hits, Common Tree, Taxonomy Report, 3D Structures, CDD Search, and GI List.

Exploring data connected to related proteins identified in BLink, including publications and nomenclature, is a valuable way to take advantage of the wealth of data already stored for other organisms. For example, following the BLink link from the Entrez protein (AAH55600) record for the ZGC cDNA BC055600, the highest-scoring human BLAST hit, at 72%ID, is human cortactin (Entrez GeneID: 2017).

c. Protein Domains: Inferring Function from Conserved Domains

The collection of protein domains in the CDD is used to computationally identify and annotate probable protein domains on proteins produced from the RefSeq and MGC projects. The CDD domains are also viewable and searchable in Entrez. For example, users can follow the CDD link (pfam04004: Leo1-like protein) from the ZGC clone BC066443 record to view an alignment of proteins also containing pfam04004, related publications via the "References" link, or follow the "Proteins" link to CDART (Conserved Domain Architecture Retrieval Tool) to view a summary of 21 sequences known to contain pfam04004.

2. Finding Nucleotide and Protein Sequences via BLAST

The volume of sequence data in GenBank continues to expand at a rapid rate. To meet the need for ever more similarity searches both within and across organisms, NCBI provides a suite of BLAST tools. The BLAST tools (Fig. 3) include, for example, searches against nucleotide or protein databases, translational

BLAST

| Entrez | BLAST | OMIM | Taxonomy | Structure |

NEW **15 November 2003** The BLAST databases in FASTA format will move from .Z to .gz compression. Read more...

Nucleotide

- Discontiguous megablast
- Megablast
- Nucleotide-nucleotide BLAST (blastn)
- Search for short, nearly exact matches
- Search trace archives with megablast or discontiguous megablast

Protein

- Protein-protein BLAST (blastp)
- PHI- and PSI-BLAST
- Search for short, nearly exact matches
- Search the conserved domain database (rpsblast)
- Search by domain architecture (cdart)

Translated

- Translated query vs. protein database (blastx)
- Protein query vs. translated database (tblastn)
- Translated query vs. translated database (tblastx)

Genomes

- Human, mouse, rat
- Fugu rubripes, zebrafish
- Insects, nematodes, plants, yeasts, malaria
- Microbial genomes, other eukaryotic genomes

Special

- Align two sequences (bl2seq)
- Screen for vector contamination (VecScreen)
- Immunoglobin BLAST (IgBlast)

Meta

- Retrieve results by RID
- Get this page with javascript-free links

Fig. 3 The NCBI BLAST home page. View NCBI's suite of nucleotide, protein, and translational BLAST tools, including the zebrafish genome BLAST page.

searches (protein vs. nucleotide), comparing two sequences by BLAST 2 Sequences (Align two sequences: bl2seq), and genome-specific BLAST pages.

From the Zebrafish Genome BLAST page (Fig. 4), users can submit searches against zebrafish mRNAs, ESTs, RefSeq mRNAs and Proteins, HTGS, WGS or EST Traces. This page can be accessed through links on the Zebrafish Genome Resource page or the BLAST home page.

The "Options for advanced blasting" section on the BLAST query page allows searches to be performed against two organisms with the AND option or to search for hits in either organism with the OR option. For example, to find either Japanese medaka or fugu sequences similar to a zebrafish sequence of interest, select the organisms to BLAST against by typing Oryzias latipes in the Options "Limit by entrez query" box, select the OR option, and then select Takifugu rubripes [ORGN] from the second Options box.

The tblastx Translational BLAST (nucleotide to protein translation) is an additional way to find highly related non-zebrafish sequences in which both the query and the database are translated. Submitting a nucleotide sequence search against the Trace Archives by discontinuous MegaBLAST can also yield further results.

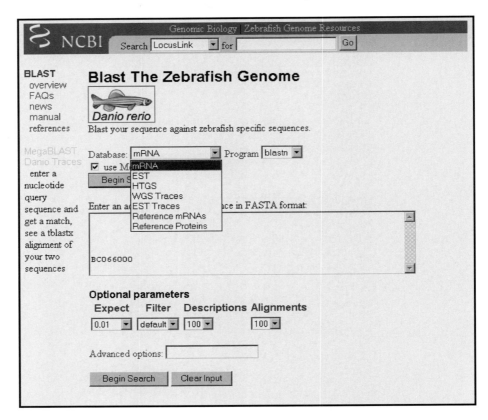

Fig. 4 The NCBI zebrafish genome BLAST page. BLAST zebrafish mRNAs, ESTs, RefSeq mRNAs, RefSeq proteins, HTGS, WGS Traces, or EST Traces.

D. Mining Publication Data via PubMed and Books

PubMed enables retrieval of information contained in publications as well as publications that share a common topic. Searches can be limited to a specific author, journal, and/or publication year. For example, users can start at any paper of interest and view other related publications by clicking on the "Related Articles" link.

Precomputed connections to books on NCBI's Bookshelf are provided for terms in PubMed titles and abstracts. These links can be mined to clarify a term or to find related studies described in other PubMed IDs. Users can choose the "Books" link for PubMed ID:14757435 and click on the highlighted word "prothrombin" in the abstract to see a list of related items found in the book "Cancer Medicine". One of the items found in "Cancer Medicine," Section 40 "(Complications of Cancer and its treatment)" links to additional details about the levels of prothrombin following L-asparaginase chemotherapy (PubMed ID: 2939229).

E. Mining Map Data via Map Viewer

Map Viewer was developed to enable users to search for information by map position. The Map Viewer home page (Fig. 5) allows users to view the diverse set of organisms represented in NCBI, to navigate to organism-specific BLAST pages, or click on an organism name to initiate a search on the genome view page. Users can choose a linkage group or chromosome or submit a text query (e.g., symbol, alternative symbol, names, parts of names, or accessions) in the "Search for" box in the blue header bar to look for a marker or gene of interest. Users can identify a region defined by two markers or sequences located on the same linkage group or

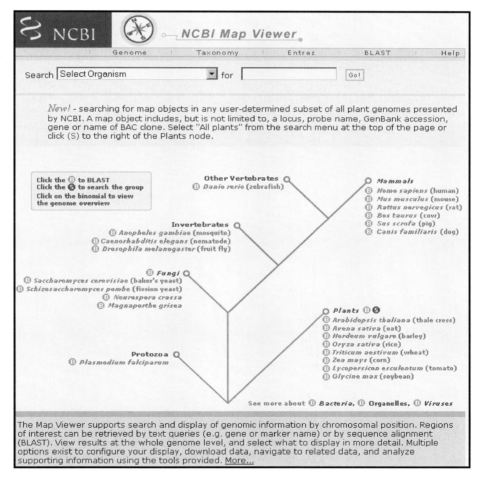

Fig. 5 The NCBI Map Viewer home page. View the tree of organisms available in Map Viewer, and link to organisms BLAST and Map Viewer pages.

chromosome by using the OR term (marker1 OR marker2). Additional details on the types of information that can be searched for each genome are available from the Map Viewer Help Documentation by clicking on the "Help" link at the top of each page.

Search results are displayed on the genome view page, with hits indicated by red tick marks on the ideogram for each linkage group ideogram. Details for each result are listed at the bottom of the genome view page under the "Search results for query" heading. Following the links for the "Map Element" or "Maps," users go to the Map View page to see the marker placement, adjacent markers, connections between maps, and the provided links. For zebrafish, links are provided to ZFIN, Entrez Gene, UniGene, or UniSTS.

Users can add or remove maps and change the display options by opening the "Maps & Options" window, from which users can choose to show or not show connections between maps, choose the verbose mode to see or hide available marker details, change compress map to view (on) or hide (off) marker labels, and adjust page length.

The *Danio rerio* Map Viewer includes the T51 and LN54 radiation hybrid maps and the MOP, MGH, GAT, HS, ZMAP, and SNP-HS genetic maps. These maps are provided in collaboration with ZFIN and members of the zebrafish research community. The SNP-HS map is produced at NCBI from SNPs in dbSNP mapped to the Heat Shock (HS) panel (Stickney *et al.*, 2002).

III. Zebrafish Information Network (ZFIN)

ZFIN, the zebrafish model organism database, provides the central location for the curation and integration of zebrafish genetic, genomic, and phenotypic data and for their subsequent integration with other model organism databases (Fig. 6) (Sprague *et al.*, 2003). ZFIN maintains both experimental data and data about the research community itself. Data are updated daily by professional scientific curators who extract relevant data from literature. Uploads of large data sets from zebrafish labs; data exchanges with other organizations such as NCBI, the Sanger Institute, and SWISS-PROT, and direct user submissions provide additional data to ZFIN. Data are attributed to their original source (Table II).

ZFIN facilitates integrated studies of zebrafish functional genomics by providing query interfaces for mutant, gene, marker, clone, mapping, and expression data. Integration of data within ZFIN allows easy navigation between related data as evidenced by the results from a gene search, which guide the user to relevant expression, clone, and mutant data, as well as to gene-specific data.

ZFIN is accessible to the public at http://zfin.org. Data found in ZFIN can be downloaded at http://zfin.org/zf_info/downloads.html. Questions and comments should be directed to zfinadmn@zfin.org.

ZFIN
The Zebrafish Information Network

General Information
Positions with ZFIN
About ZFIN
Citing ZFIN
Helpful Hints
User Support
Site News
Glossary
Download Data

Genomics
Zebrafish Genome Resources
Trans NIH
Zebrafish Initiative
Other Genomes

Information and News
Anatomical Atlases
Meetings / Jobs /
News
The Zebrafish Book
The Zebrafish Science Monitor
Zebrafish Newsgroup
Zebrafish for K-12

Nomenclature
Laboratory
Allele Designations
Nomenclature Conventions
Obtaining Approval for Gene Names

Zebrafish Resource Center
Info, Strains, Probes
Pathology Services
Disease Manual

Search This Site

Mutants / Transgenics	Search for mutations / transgenic lines by gene name, map location or phenotype.
Wild-Type Stocks	Zebrafish wild-type lines.
Genes / Markers / Clones	Search for genes, markers and clones by name, accession number, LG, vector type or sequence type.
Gene Expression	Search for gene expression patterns by gene name, developmental stage, anatomical structure, developmental or physiological process.
Genetic Maps	Generate graphical views of genetic, radiation hybrid or consolidated maps.
Mapping Panels	Summary listing of zebrafish mapping panels.
Accession #	Search ZFIN by data accession number.
Publications	Search for zebrafish research publications by author, title or citation.
People	Search for zebrafish researchers by name or address.
Laboratories	Search for laboratories by name, address or research interests.
Companies	Search for companies supplying zebrafish reagents.

Login: [   ] Password: [   ] [Log in]
(Login required only to update personal records)

Fig. 6 The Zebrafish Information Network (ZFIN) home page. The zebrafish model organism database, ZFIN, is the zebrafish community online resource for laboratory, genetics, genomics, and developmental information.

Table II
ZFIN Database Content Statistics (as of 29 January 2004)

Category	Number
Publications	5,043
Genes	12,634
Genetic markers	21,730
Alleles	2,360
Curated zebrafish/human orthologs	1,100
Curated zebrafish/mouse orthologs	872
ZFIN markers with GenBank sequence accessions	25,350
Genes with Swiss-Prot protein sequences	1,675
Genes with expression patterns	1,360
Markers with links to Sanger FPC or VEGA	6,810

Note: ZFIN database contents are updated daily.

A. Mutants

Efficient methods for generating, isolating, and characterizing mutants make zebrafish a powerful organism for studies of gene function. Curation of literature, collaborations with investigators performing large-scale mutagenesis screens, and personal communications with individual investigators allow ZFIN to acquire large amounts of data on fish lines. These data can be accessed by using the ZFIN mutant search form, which allows the retrieval of mutant data based on name, selected structures and defects, LG, mutagen, and mutagen type. For example, queries can be formulated to find all deficiency mutants on LG13, all alleles for the cyclops mutant, or all mutants affecting a particular anatomical structure. Mutant data include name, abbreviated name, images, discoverer, current availability, parental lineage, segregation, and phenotype. Mutagenesis protocols, mutagen type, linkage group, and known linkages are also provided. Links to corresponding genes supply mapping information, related clones, and sequence data.

B. Genes/Markers/Clones

To facilitate genomic research, ZFIN works closely with the Sanger Institute and NCBI to maintain extensive links between these sites, thus providing users with a wide array of genetic data from genome location to mutant phenotypes and expression patterns associated with a particular gene. A search for a gene, marker, or clone at ZFIN begins by specifying a name, an accession number, LG, or sequence type. A search specifying a gene name will return links to the ZFIN gene page as well as links to associated bacterial artificial chromosomes (BACs), P1-derived artificial chromosomes (PACs), and ESTs giving access to expression data and Sanger sequence data. The resulting gene data page displays the approved nomenclature name and symbol as well as previous names and nomenclature history. Importantly, relationships between genes and molecular segments (physical pieces of DNA such as BACs, PACs, and ESTs) are described. These relationships help identify links between genes and mutants to molecular segments, genetic maps, and ultimately the genome. Gene Ontology (GO) annotations describe gene products. The use of these evolving controlled-vocabulary GO terms describing molecular function, biological process, and cellular component facilitate the comparison of zebrafish gene and gene products with those of other species. A code describing the supporting evidence is provided for each term to allow researchers to assign a level of confidence to the annotation.

Mapping details from the six zebrafish mapping panels and from literature citations are provided. A graphical map viewer provides a means for linking mutants with genes and other markers, thus facilitating positional cloning.

Links to InterPro, PROSITE, and Pfam databases provided access to protein family and domain information. Similarly, links to RefSeq, GenBank, WashU, UniSTS, and Sanger provide nucleotide sequence data. Protein sequence information is provided via links to RefSeq, GenPept, and SWISS-PROT. Links to UniGene and WashU provide sequence cluster data. Gene and segment pages at LocusLink, Sanger (VEGA and FPC), and Ensembl may also be accessed from ZFIN gene pages. Access to relevant mutant and expression data in ZFIN is provided by links on these pages.

To facilitate an understanding of relationships between gene and gene functions in zebrafish and other organisms, ZFIN curators capture data pertaining to orthologous human, mouse, *Drosophila*, and yeast genes. Correct nomenclature and links to orthologous gene records at OMIM, MGI, Flybase, SGD, and EntrezGene/LocusLink are provided. Evidence codes describing orthology assertions are given. Links to the supporting publication are included. ZFIN gene searches support the use of nomenclature approved symbols and names from other organisms. Clone data are described by library, cloning site, vector, digest, tissue, and strain.

C. Expression

Gene expression data offer powerful insights into understandings of biological processes and gene function. To aid in these understandings, support for gene expression in ZFIN continues to expand. Large data sets of high-quality annotated images from laboratories performing large-scale *in situ* hybridizations as well as data submitted by individual investigators are incorporated into ZFIN routinely. In addition, gene expression cited in literature is curated to allow searches of expression of specific genes in specific mutations. An expression search form allows complex queries based on name, anatomical structure, LG, developmental stage, assay type, and mutant background (Fig. 7). This form can be used to obtain an overview of expression of a particular gene, to learn what genes are expressed in a given structure at a particular time stage of development, or to

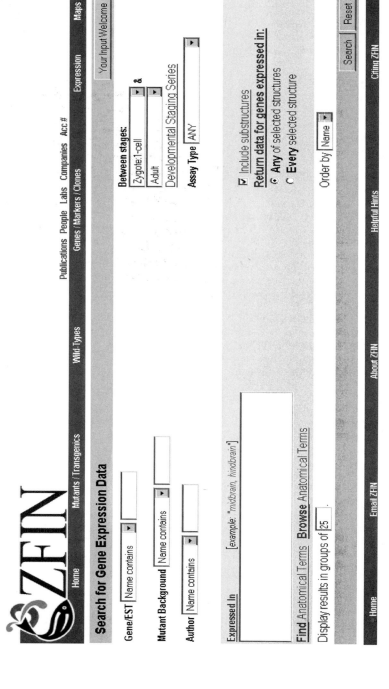

Fig. 7 ZFIN Gene Expression Search Form. Search for gene expression patterns by gene name, developmental stage, anatomical structure, or developmental or physiological process.

locate a probe for an anatomical structure. ZFIN displays images, anatomical dictionary keywords, and text summaries describing the expression patterns. Links provide easy navigation between expression patterns, genes, probes, and their sources.

D. Publications

An important feature of ZFIN is an up-to-date bibliography of zebrafish publications. Curators annotate ZFIN records with data cited in these publications. Publications can be searched by author, title, abstract contents, PubMed ID, year, and keywords. Publication abstracts can be viewed from search results or by following links on associated ZFIN data, person, and laboratory pages. Links on the publication abstract page provide easy access to ZFIN data pages, which contain data attributed to the specified publication.

E. Research Community

To foster communication within the zebrafish community, ZFIN maintains data about the zebrafish research community, including investigators, laboratories, and companies. Contact information for investigators and labs can easily be found by querying on name, address, telephone, fax, e-mail, and research interests. Contact information for companies supplying materials and reagents listed in ZFIN is also searchable.

Nomenclature conventions, anatomical atlases, developmental staging series, and an online version of "The Zebrafish Book" are available. ZFIN also maintains lists of relevant job opportunities and meetings.

IV. Tutorial

We include next a set of questions involving genome, gene, map, homology, and expression data. For each question, we have included one or two possible ways to find the answers.

A. Genome and Genes

Is my gene annotated on the current whole-genome assembly?

The Sanger Institute initiated the zebrafish whole-genome sequencing and annotation project in 2001. Project status updates are available at the Sanger Web site. Curated VEGA annotated gene reports are available through the Sanger Zebrafish Annotation Browser by searching for an EST, gene, mRNA, peptide, or sequence. Links to VEGA annotations are included, when available, on ZFIN gene and Entrez Gene pages.

The zebrafish genome sequence will become available in NCBI's Map Viewer when the zebrafish assembly sequences are accessioned in GenBank. At that time, the zebrafish genome sequence will be added to the set of organisms included in NCBI's genome pipeline.

What other genes are annotated near my gene or mutant?

The map viewers at ZFIN and NCBI can be used to identify candidate genes for a region of interest. Links to ZFIN gene pages provide access to information that can be used for a mini chromosome walk. ESTs and BAC clones associated with the gene are specified. The genes contained in the BAC, the EMBL description of the BAC, and other BACs that overlap the BAC are provided.

Links to NCBI's Map Viewer are provided from the drop-down "Links" menu on the Entrez Gene pages to allow users to view the placement of their gene on the zebrafish genome.

Alternatively, beginning at the ZFIN gene page or the Entrez Gene page and following the link to VEGA's Zebrafish Annotation Browser, users can navigate to VEGA's ContigView to examine genes curated in the same region.

How can I find the zebrafish homolog of a recently cloned mouse gene?

If searches of the precomputed comparisons in HomoloGene or BLink do not yield any likely candidates, another approach is to submit a search of the mouse gene symbol to Entrez's GQuery to determine whether a similarly named zebrafish gene exists.

Additional connections can be found through homology data in HomoloGene. For example, if the search begins at a mouse Entrez Gene page, follow the HomoloGene Link to view the precomputed cross-species comparisons. Although the zebrafish gene might not yet be identified in Entrez Gene, it might be present in the UniGene dataset as ESTs or uncharacterized cDNAs. Go to UniGene to see the ESTs and mRNAs in the UniGene set. If a ZFIN marker or Entrez Gene record contains these uncharacterized sequences, a link to ZFIN or Entrez Gene will be at the top of the UniGene page.

This problem can also be approached by submitting a search at ZFIN, in which gene searches include zebrafish names and symbols as well as names and symbols of known homologs. Enter the nomenclature approved name or symbol in the name field of the Genes/Markers/Clones search form. The corresponding zebrafish gene will be returned.

Are there any single nucleotide polymorphisms (SNPs) associated with my gene?

To answer this question, submit a text query to dbSNP. Alternatively, a query in NCBI's Map Viewer by the gene symbol will allow users to see whether a gene or an EST known to be associated with their gene is mapped to the SNP-HS map. Submitting a search using the Entrez qualifier gene_snp[filter] to query Entrez Gene or snp_gene[filter] to query Entrez SNP will provide the set of records that contain data in both databases.

How can I identify Bacterial Artificial Chromosome (BAC) and P1-derived Artificial Chromosome (PAC) clones that would aid me in mapping genomic clones to chromosomes?

ZFIN maintains data for BACs and PACs from the Sanger sequencing and annotation initiative. This data can be searched by linkage group. Genes contained within the BAC or PAC and available mapping information are provided.

Another approach to find all mapped PACs or BAC_ENDs in NCBI's Map Viewer is to submit the search based on the naming schema for zebrafish PACs and BACs, for PACs search by busm1* and for BAC_ENDs search by bz*.

NCBI's clone registry of genomic clones can be mined to view sequence, library, map, and distributor information. Precomputed links from the Accession on the Clone page and links provided in other resources enable users to mine the data stored in other NCBI databases such as GenBank, and via links provided, for example, UniSTS, UniGene, Map Viewer, and Entrez Gene.

B. Expression/Function

What genes would make good markers for a study of notochord development?

ZFIN annotates expression data using structures and developmental stages from the zebrafish anatomical ontology. The use of a standardized vocabulary for annotation provides a powerful search tool. Genes with expression in a particular structure at a particular stage of development can be identified by using the ZFIN expression form.

Another approach is to submit an Entrez GQuery such as notochord AND development AND "Danio rerio"[organism] to view related records in Entrez Gene.

I am using a gene as a marker. Its expression is altered in my phenotype. How can I find expression patterns for my marker in early wild-type development to determine whether the observed change is due to misexpression or a delay in development?

ZFIN provides access to expression data that have been annotated by using terms and developmental stages from the zebrafish anatomical ontology. Images, text summaries, and links to publications are provided.

What are other possible ways to infer function for my gene?

The Gene Ontology Consortium has developed a dynamic controlled vocabulary describing molecular function, cellular components, and biological processes. Zebrafish genes are being annotated with these terms. Annotations can be viewed on Entrez Gene and ZFIN gene pages. Gene Ontology (GO) links provide detailed GO information at the Gene Ontology Consortium AmiGO or at EMBL-EBI's GOA.

ZFIN fosters an understanding of gene function by linking genotype, phenotype, and gene expression to gene sequence and gene models.

How do I find cDNAs derived from a specific library?

One approach is to go to the Zebrafish Gene Collection home page under the heading "ZGC Full-length Clone Information" and select the "Table" or "Full Text" buttons to view the ZGC library list details (Library, Tissue, Vector, or Number of Clones). On the Library List page, choose the "Library" name to go to the "Library Info Page" or the "Clones" link to view the list of IMAGE ids, GenBank accessions, symbols, and GenBank definition that correspond to each sequenced clone.

Another approach is to begin in UniGene and navigate to the details for each EST by clicking on the EST accession number to view the Sequence Information page. By clicking on the linked Library ID, users can view the Library Description, the Sequence Submitters, the number of UniGene clusters containing members of the library, and the number of sequences in each cluster. The list of cDNAs for each cluster can be seen by then clicking on the UniGene cluster number.

What genes have been found to be expressed in the eye?

ZFIN provides access to expression data that have been annotated by using terms and developmental stages from the zebrafish anatomical ontology. Expression data can be searched by specifying a structure from the anatomical dictionary.

Expression studies included in the GEO database can be queried by gene name, organism, or tissue and provide an additional source of functional information for multiple organisms.

Also, connections between expression and gene data can be found by a global query by using NCBI's GQuery to find related Entrez Genes and GeneRIFs.

Acknowledgments

We recognize staff members of NCBI and ZFIN for their creation of these resources and continued support of the zebrafish community. We thank Greg Schuler, Monte Westerfield, and Donna Maglott for their critical comments, which greatly improved the quality of this manuscript. At NCBI, we recognize the efforts of Donna Maglott, Kim Pruitt, Deanna Church, Lukas Wagner, Joan Pontius, Greg Schuler, Wonhee Jang, Cliff Claussen, Richa Agarwala, Tatiana Tatusova, Ron Edgar, Steve Sherry, and the Maps, RefSeq, BLAST, and SNP groups. At ZFIN, we recognize Monte Westerfield, Dave Clements, Tom Conlin, Pat Edwards, David Fashena, Ken Frazer, Doug Howe, Prita Mani, Sridhar Ramachandran, Kevin Schaper, Erik Segerdell, Peiran Song, Brock Sprunger, Sierra Taylor, and Ceri Van Slyke. ZFIN is supported by NIH NHGRI P41HG002659.

References

Pruitt, K. D., Tatusova, T., and Maglott, D. R. (2003). NCBI Reference Sequence project: Update and current status. *Nucleic Acids Res.* **31,** 34–37.

Sprague, J., Clements, D., Conlin, T., Edwards, P., Frazer, K., Schaper, K., Segerdell, E., Song, P., Sprunger, B., and Westerfield, M. (2003). The Zebrafish Information Network (ZFIN): The zebrafish model organism database. *Nucleic Acids Res.* **31,** 241–243.

Stickney, H. L., Schmutz, J., Woods, I. G., Holtzer, C. C., Dickson, M. C., Kelly, P. D., Myers, R. M., and Talbot, W. S. (2002). Rapid mapping of zebrafish mutations with SNPs and ligonucleotide microarrays. *Genome Res.* **12,** 1929–1934.

Wheeler, D. L., Church, D. M., Edgar, R., Federhen, S., Helmberg, W., Madden, T. L., Pontius, J. U., Schuler, G. D., Schriml, L. M., Sequeira, E., Suzek, T. O., Tatusova, T. A., and Wagner, L. (2004). Database resources of the National Center for Biotechnology Information: Update. *Nucleic Acids Res.* **32,** D35–D40.

CHAPTER 24

The Zebrafish DVD Exchange Project: A Bioinformatics Initiative

Mark S. Cooper,★ Greg Sommers–Herivel,★ Cara T. Poage,★ Matthew B. McCarthy,★ Bryan D. Crawford,[†] and Carey Phillips[‡]

★Department of Biology and Center for Developmental Biology
University of Washington
Seattle, Washington 98195

[†]Department of Biology
University of Alberta
Edmonton
Alberta T6G 2E9, Canada

[‡]Department of Biology
Bowdoin College
Brunswick, Maine 04011

I. Introduction

The zebrafish community relies heavily on photomicrographs, time-lapse recordings, and PowerPoint presentations to convey concepts among its members.
Each year, an individual zebrafish laboratory is capable of generating gigabytes
(GB) of information in the form of digital micrographs or terabytes (TB) if the
laboratory is making 4D confocal time-lapse recordings of GFP-transgenic embryos. However, only a small amount of visual information is exchanged directly
between laboratories, primarily because of the cost and difficulty of distributing
visual information through published research articles. How can we manage to
distribute and share this ever-increasing mass of visual information?

One way of accomplishing this is to place visual data on inexpensive high-capacity
random-access media such as digital video discs or digital versatile discs (DVD). Our
laboratories are starting to coordinate a collective effort called the Zebrafish DVD
Exchange Project. We will be collecting visual information from the zebrafish
community, arranging this information in hyper text markup language (HTML)
as well as eXtensible markup language (XML) format, replicating the information
on DVDs, and then disseminating the DVDs at zebrafish meetings.

Large amounts of information can be easily archived on DVDs and subsequently searched and accessed by using Web-browser tools. Once Internet2 becomes
widely available, the material on the DVDs can be migrated to various servers for
fast and long-term access. In this regard, the Zebrafish DVD Exchange Project can
serve as enabling technology for future bioinformatics efforts.

In this chapter, we describe both the technical challenges and potential benefits
of generating a mechanism for mass transfer of visual information within the
zebrafish community by DVD exchange. In doing so, we discuss the ability to
develop a virtual Intranet to serve our immediate and future visual bioinformatics
needs.

II. Economic Aspects of Visual Bioinformatics

Worldwide, the equivalent of more than $500,000,000 has probably been spent
on zebrafish research during the past 20 years. The magnitude of this expenditure
reflects the perceived importance of basic and biomedically oriented zebrafish
research to the scientific community as well as society as a whole. As the pace of

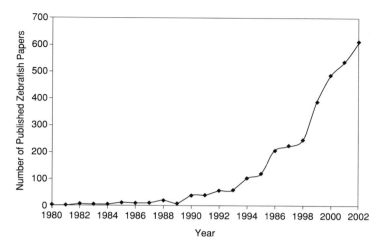

Fig. 1   Number of zebrafish publications from 1980 to 2002.

zebrafish research increases (Fig. 1), the zebrafish research community's need to exchange the ever-expanding mass of visual information that arises from this research will increase.

New data mining strategies are needed to breech the limitations of computational resources (Wegman, 2003). We also need to evaluate the technologies that currently exist for information exchange in order to determine alternatives that provide cost-effective and robust transmission of image data (Ouzounis and Valencia, 2003). To gauge the impact of visual information exchange within the zebrafish community, one must employ a mode of analysis that is commonly used by information technology industries: the economics of information transfer bandwidth.

Currently, no absolute quantifications of the output of images and visualizations are arising from the zebrafish community. However, some generalizations can be made to discuss the magnitude of expenditures necessary to generate and publish images produced by a given laboratory. For instance, how much money does it cost to produce *and* disseminate individual research images and other visualizations? A qualitative answer to this question lies in an assessment of research output—in this case, the number of images from a laboratory that are used to convey scientific findings from a research laboratory each year.

An average laboratory can spend approximately $100,000 in direct costs per year. The scientific results from that laboratory emerge as two to four research papers, with 4–10 figures per paper. Each figure might be 1–10 images (4–6 image panels more typically). Overall, the range of production is based on publishing roughly 10–200 images per year.

We have confirmed this rough estimate by analyzing 40 randomly selected papers from the zebrafish literature published in 2002. The range of visual images per paper was 2–106, with the mean number of images being 31 $\pm$ 24 ($N = 40$ papers). By analyzing the zebrafish literature, we have found that the majority of data presented in zebrafish research publications are in the form of images rather than graphs or tables. Dividing $100,000 per year by the mean number of images published per year (31), one obtains the approximate cost input into generating and disseminating a visual image by means of a published journal article. The final expenditure is roughly $3200 per image. This number represents the combined cost for *producing* the image data and *disseminating* the image data. By similar reasoning, the cost of publishing a QuickTime movie in a journal is similar to that for a figure.

Albeit imprecise, this empirical estimation offers an insight into the economics of visual data exchange within the zebrafish community. Technologies that decrease the cost of data exchange and increase information bandwidth (i.e., the effective rate of data transfer) will synergistically accelerate the total research capacity of our community.

III. Building Infrastructures to Access Images

The actual worth of a given image or visualization is almost impossible to gauge. A single image (e.g., the double helix) can profoundly transform the course of scientific inquiry.

There are very important images from older teleost embryological literature that need to be archived and disseminated. Consider the image shown in Fig. 2. The illustration shows the internal cellular structure of the blastula and gastrula of the bowfin (*Amia calva*), a bony ganoid fish whose ancestral lineage first arose in the early Mesozoic. This living fossil exhibits early cleavage patterns and gastrulation mechanics that are transitional between basal actinopterygian (ray-finned) fishes and advanced teleosts (the final bony fishes) (Dean, 1896), such as the zebrafish.

The illustration comes from a research paper published in 1906 in the *American Journal of Anatomy*. Very few libraries in the world have collections of these old journal volumes. Libraries that do have such older holdings are now financially pressed to save them. Moreover, the collections that are available are often difficult to access and search and therefore are often not used.

Through digital scanning, we can preserve these early papers in the form of PDF files and the illustrations as high-resolution JPEG images. These digital files can then be assembled on DVDs for dissemination to the zebrafish community. By doing so, every zebrafish lab can have access to an easily searchable library comprising more than 150 years of teleost embryology.

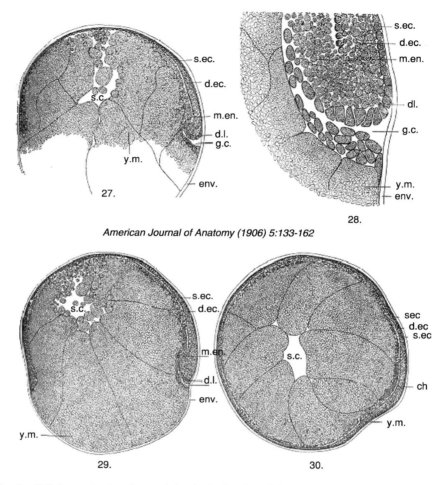

American Journal of Anatomy (1906) 5:133-162

Fig. 2 Cellular mechanics of gastrulation in the bowfin, a living fossil neopterygian fish. Plate from Eycleshymer and Wilson (1906). Gastrulation and embryo formation in *Amia calva. Am. J. Anat.* **5,** 133–162.

IV. Prior Experiences with Visual Data Dissemination (Mark Cooper)

The Cooper laboratory began disseminating images online in April 1995, using a Web site called FishScope. Within 6 months, more than 50,000 downloads were recorded from the site. Most of the movies on FishScope were QuickTime movies approximately 0.5–3.0 megabytes (MB) in size. FishScope cost approximately

$300 per year, mainly to rent the necessary server space on a university computer, which is connected to a fast Internet line.

In 1995, we noted from our server logs that certain laboratories returned repeatedly to FishScope to download specific movies. We surmised that the QuickTime movies archived on our Web site were probably being discussed within individual laboratories and that multiple people from those laboratories were individually exploring the site. We were gratified to see that rather than downloading each of the movies and storing these movies on their own computers, the users of FishScope were treating the Web site as a flexible database rather than as solely a single-use archive from which to download movies. Most Web users remember or bookmark the location of useful Web sites and view them as a resource that can be called on when needed.

What happens, however, when data transfer times from Internet archives and Web sites become too long? Clearly, the utility and functionalities of Web-based databases become limited. To circumvent these difficulties, we decided to shift our information dissemination efforts to CD (compact disc) technologies.

The first output of the Zebrafish DVD Exchange Project was a CD called *Zebrafish: The Living Laboratory*. Copies of this CD were made available to all participants of the 5th International Meeting on Zebrafish Development and Genetics at Madison, WI, in June 2002. Our motivation to create a CD-ROM for the zebrafish community was stimulated by the utility of another CD-ROM "GFP in Motion," produced for the cell biology community by Beat Lukin and Andrew Matus.

The *Zebrafish: The Living Laboratory* CD contains visualizations from more than 10 laboratories, methodology papers, as well as several freeware imaging programs (Figs. 3 and 4). By future standards, this first CD project (nearly 1.5GB) will look like a very small compilation of visual information.

The next project we are working on is a DVD compendium called "Zebrafish: 4D," which will focus on the use of GFP-transgenic technologies to visualize zebrafish development. *The Zebrafish: 4D* DVD compendium will be a single packet of six double-sided DVDs. With each DVD containing 8.5 GB of information, the entire DVD six-pack will contain approximately 50 GB of information.

Once the master copies of the DVD templates are generated, the manufacturing cost of producing an individual DVD six-pack should be about $20. Considering the massive amount of data transmission available, the DVD six-pack will be a powerful new means for organizing and disseminating visual information within the zebrafish community.

In terms of information transfer, it is possible to equate a DVD to a modern commercial container ship (hereafter the container ship and DVD will be referred to as *vehicles*). Place cargo (i.e., data) in a standardized container on a transport vehicle (in this case a standardized file on a DVD or CD) and then move the cargo-containing vehicles to a common destination. Unload the containers from the vehicles, and arrange the containers in a searchable fashion. Finally, replicate the compendium of data and then disseminate the final compendium as cheaply as possible. We estimate that 100 GB of visual information represents about 1% of the total visual

Fig. 3   Compact discs (CDs) and digital video discs (DVDs) serve as efficient vehicles for collecting visual data into a common visualization database. (See Color Insert.)

information generated by the zebrafish community in a given year. In Section V, we discuss potential applications of mass information exchange through the Zebrafish DVD Exchange Project.

V. Goals of the Zebrafish DVD Exchange Project

The Internet has provided a revolutionary transformation of information exchange. However, despite its great technical innovations, the Internet is currently unable to serve as a vehicle for the easy exchange of massive amounts of visual information.

In terms of informatics technology, the Zebrafish DVD Exchange Project represents the generation of a Virtual Intranet (Fig. 5): virtual in the sense that there are no actual communication lines between our community's computers and Intranet in the sense that our community's computers are still interlinked for data exchange in a secure fashion.

Fig. 4 "Zebrafish: The Living Laboratory." This CD visualization database was compiled from contributions from more than 10 laboratories. As a portable compendium, CDs and DVDs provide easily accessible repositories for large quantities of visualizations and other visual data. (See Color Insert.)

By increasing the information bandwidth through DVD exchange, the following uses (i.e. functionalities) become possible (Fig. 6):

1. Archiving literature in PDF.
2. Digital images and movies for PowerPoint presentations.
3. Instructional QuickTime tutorials on experimental procedures.
4. Conference proceedings/abstracts.
5. Portable visual database.
6. Raw 3D and 4D image data for computer-rendered visualizations.
7. Freeware image and visualization software.
8. Anatomical image atlas.
9. Class lectures and seminars.

Dissemination of ideas and knowledge within the zebrafish community hinges on information accessibility. We need to assess qualitatively the economics of visual bioinformatic exchange mechanisms within our research community. In Section VI, we discuss the costs and logistics of producing the first output of our new bioinformatics initiative: The "Zebrafish: The Living Laboratory" CD.

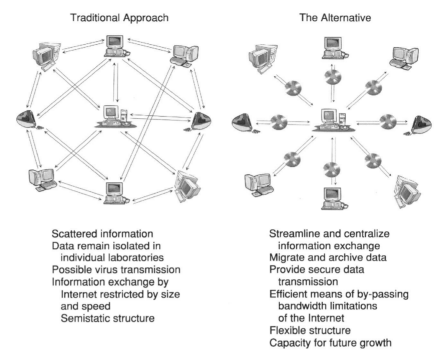

Traditional Approach   The Alternative

Scattered information
Data remain isolated in
  individual laboratories
Possible virus transmission
Information exchange by
  Internet restricted by size
  and speed
Semistatic structure

Streamline and centralize
  information exchange
Migrate and archive data
Provide secure data
  transmission
Efficient means of by-passing
  bandwidth limitations
  of the Internet
Flexible structure
Capacity for future growth

Fig. 5 The DVD Exchange Project represents a virtual Intranet for mass visual data exchange. (See Color Insert.)

VI. Production of the Compact Disc (CD) Set: "Zebrafish: The Living Laboratory" (Greg Sommers-Herivel)

Although we had early success with our Web site FishScope, a new methodology was required for sharing data as our data files grew in size. The Internet, with all its excellent functionalities, could not easily transmit the large amounts of data our laboratory and other laboratories were generating. The question arose: How do we continue to share the visual data we are acquiring?

CDs were the ideal answer, owing to their capacity, stability, and portability. Through the "Zebrafish: The Living Laboratory" CD set, we were able to compile nearly 1.5 GB of data from multiple labs.

Our research community, however, cannot rely on CDs as vehicles for mass information exchange. Our image data sets are growing larger than CDs can handle. An individual time-lapse series of a GFP-transgenic zebrafish embryo can be several hundred megabytes in size. Owing to this increase in information load, DVDs have become the new vehicle of choice for mass data transfer. Although DVDs may not yet be universally used in all laboratories, with the cost

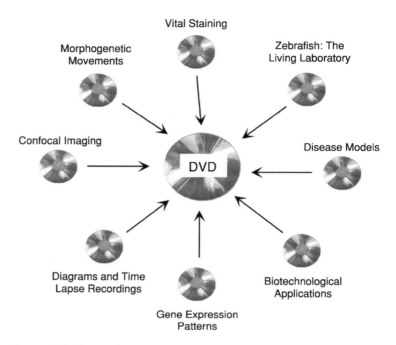

Fig. 6 Because of its increased storage capacity, approximately 10-GB DVDs will soon become a major archival medium for visual information. (See Color Insert.)

of DVD-ROM and DVD writable drives decreasing, it will not take long before most laboratories begin routinely using DVDs.

A. Collecting Data

Because the idea of the *Zebrafish DVD Exchange Project* is to gather and disseminate visualizations, a method for gathering the data was needed. The most important aspect of this was to maintain our user-friendly approach by simplifying the data submission process. Although we had a preferred format for data to be submitted, there were no specific requirements, except that a brief description be included with the visualization. After receiving submitted material, we changed the visualizations to the format that would best suit the needs of the project. Because the "Zebrafish: The Living Laboratory" CD was our initial venture, it was necessary to acquire images by contacting laboratories and requesting material to be sent on a CD. We hope that as the DVD Exchange Project is adopted and endorsed by members of our research community, submissions to the DVD Exchange Project will become more spontaneous. In an effort to make submission more conducive to this, we have set up an FTP server at which labs can send files through the Internet. We encourage researchers to submit full image data sets as well as final computer-rendered visualizations.

B. Presentation

After data are collected, they must be collated and organized into a format that is usable. For "Zebrafish: The Living Laboratory" CD project, this was the most time-consuming aspect. What made proper data presentation particularly important was that at the time it was necessary to truncate the file names to the earlier Microsoft OS convention of "eight.three" characters, which left the names of the files as somewhat meaningless to a user. Although we could have simply thrown all the files into segregated folders, this seemed awkward, requiring the user to wade through large quantities of files. In the end, we decided on a far more amicable interface modeled after the familiar landscape of Web site architecture. Using HTML, we created a friendly graphical interface, which not only allows users to peruse the CD contents in a manner that is familiar, but also creates a cross-platform format. This also made the CD reachable to a broader audience. Although the CD was produced for researchers in the zebrafish community, the CD design allows students and other interested parties to use the CD and not be overwhelmed by the contents. This was also the reasoning behind the artwork and exterior design. We wanted to make something that would not only appeal to the zebrafish community, but also attract people from outside the community. Incoming students would find the artwork and the overall design engaging, bringing them into the science.

For the DVD Exchange Project, our plan is to create multiple interface designs so that the user has many choices on how to navigate the data. Ideas from a comprehensive alphabetical index to an index based off of a laboratory researcher being considered, as is having a searchable database, so that users can go directly to the file they want. Our idea is to make future DVD compendiums as easy to use and as engaging as possible.

C. Archive

Another goal of the DVD Exchange Project is to create an archive of papers written by contributors and other scientists. "Zebrafish: The Living Laboratory" CD contains 11 PDF versions of published research papers. This is a tiny fraction of what we envision for the future. We hope to create a searchable database of current and classic papers, which will give laboratories a strong reference resource. Most importantly, we can put the images into high-resolution PDFs so that the images within can be examined in greater detail. This archive is not designed to replace journals or journal submission, but is designed to provide a rapid research context for the visualizations stored on the DVDs.

The publishers of the articles we used for the "Zebrafish: The Living Laboratory" CD granted us a license to put in PDFs of their articles. The publishers still hold the copyrights, and the permission we received pertains only for use in the CD Exchange Project. It was necessary to spell this out on the project as well, to ensure that everyone using the CDs was aware of this. For example, we obtained a picture of an oarfish that we found printed in a National Geographic issue. When we asked for permission to use the picture, we discovered that the copyrights for it were owned by Associated Press. We had to purchase a license from the Associated Press to use the image in our project. Associated Press still retains copyright of the image.

Having a centralized archive for data and papers saves the research community time and resources. For each laboratory to disperse its own data by a CD or DVD to every interested laboratory, it would take a significant amount of time and resources. Having a large selection of papers also can save time. By using the DVD Exchange Project, laboratories can send in their recent data and visualizations to one location and know that it will be used widely throughout the zebrafish community (Fig. 2).

D. Software

Another aspect of the DVD Exchange Project that needs to be highlighted is the value of distributing OpenSource software and other less well known data analysis tools. Programs such as NIH image and ImageJ were distributed within the project, giving laboratories access to analysis software they might not have been aware of. Because academic programmers generally write software for their own purposes and do not have massive marketing budgets, their projects often remain unknown to many who would find them useful. The DVD Exchange Project can be a valuable means for academic programmers to distribute their freeware to the zebrafish community.

E. Debugging

Probably the most intensive part of "Zebrafish: The Living Laboratory" CD project was quality control. The interface had to be tested and retested in an effort to make it compatible with Apple and Windows machines, as well as Netscape and Internet Explorer. We also worked with a colleague to test overseas application. We had to confirm that each link was functional prior to replication and that every movie and image worked as expected. With nearly 300 links and visualizations, it was an intensive process to ensure a working CD.

A major advantage of using CD/DVD media for data distribution is the ability to ensure a clean, virus-free exchange. We rigorously scanned the CD with several up-to-date virus softwares before sending it out to be replicated.

F. Duplication Costs

Besides costs for copyright permissions, duplication represents the only other hard cost investment in the project. The hard cost of production includes

CD replication.
CD silkscreen.
CD silkscreen preflight.

CD packages (materials and printing).

CD package preflight artwork.

Assembly.

The total cost for the CD exchange project, for 2500 packages containing 1.4 GB, was U.S. $8387, or U.S. $5591/GB. We estimate the DVD six-pack to be about $35,000 in hard costs, including all replication and copyright expenses, for 5000 packages containing approximately 51 GB. This equates to about $695/GB.

G. Sponsorship

We found a way to offset some of the production costs by approaching biomedical and imaging companies. Nearly every company we requested support from contributed generously to the project, offsetting the total project costs significantly.

H. Distribution

With most large-scale projects such as this, distribution is the most logistically difficult and often the most expensive part. Fortunately, we already have a low-cost distribution network available to us: international and regional zebrafish meetings. By shipping the CDs and DVDs to zebrafish meetings, these archives can be easily distributed among the attendees at the conference. This saves a tremendous amount of time and money and helps assure a wide distribution. So far, we have distributed the "Zebrafish: The Living Laboratory" CD at two international zebrafish meetings, as well as a number of zebrafish workshops and courses worldwide.

VII. Visual Bioinformatics and Databases

The Zebrafish Information Network (ZFIN) is an Internet database created by a team headed by Drs. Monte Westerfield and Judy Sprague (www.zfin.org). It is an excellent resource that provides genomic maps; search engines for zebrafish labs, researchers, and publications; anatomical atlases; a listing of zebrafish mutations; and the entire "The Zebrafish Book" online. Another useful feature of ZFIN is the list of meetings and current events in the community.

ZFIN's largest archive is that of gene expression and genetic mapping. It has an extensive library of maps and resources on this material. Other databases exist similar to ZFIN, many of which focus on macromolecular structure and genomic data (Golding, 2003; Golovin et al., 2004; www.biocam.ac.uk/compmolb.html; www.sacs.uscf.edu/resources/biolinks.html). The "Zebrafish: The Living Laboratory" CD is a perfect complement to ZFIN and similar sites as it has a variety of time-lapse and still images, which can be correlated to genetic information. In particular, morphological phenomenon can be linked to gene expression patterns, allowing for correlative analysis of morphology and genetics (Goesmann *et al.*, 2003; see also Section X).

The difficulty Web-based archives continue to face is the same—limited information bandwidth. With Internet2 around the corner, the bandwidth bottleneck will hopefully be reduced. Currently, however, sharing large data files is not practical for a Web-based database. Long data transfer times over the present Internet impede the transfer of high-resolution visualizations (e.g., 0.5–5 GB). During a long transfer, the possibility of the data file getting corrupted increases.

The Zebrafish DVD Exchange Project is able to transmit masses of large data files with high fidelity. This material can then be downloaded anywhere from the DVDs and entered into visual bioinformatics databases. The assembly of visual bioinformatics databases is now underway in many biological fields. These developing bioinformatics databases can be used by anyone in the scientific community, from researchers to students (Honts, 2003).

VIII. Internet2

Internet2 was started under the same philosophy as that for the present Internet: to support the exchange of data among the academic and research community. With the bandwidth of the present Internet being far too limiting for the rapid exchange of information demanded by present research, Internet2 has been developed to create an information bandwidth orders of magnitude higher than that for the traditional Internet. In a recent speed record, Caltech transferred 1.1 TB to CERN in 27 min (www.internet2.edu). This speed will not be necessary for all laboratories, for which a fraction of this information bandwidth would be sufficient for current needs.

Internet2 is a high-bandwidth vehicle designed with research needs in mind. Internet2 can transfer gigabytes of information in seconds, making it excellent for information exchange. The goals of Internet2 are to create not only a fast pathway where files can be shared but also an environment for advanced real-time multimedia interactions.

Concurrent with the creation of Internet2 is the next-generation Internet (NGI). NGI is similar in design to Internet2, but is designed for the government and corporate worlds. It is only a matter of time before the general public has access to the bandwidth of these new Internet technologies.

Until Internet2 becomes readily available to all, the DVD project can be used as the vehicle to transport information to different laboratories. In addition, the DVD project provides a fully portable and searchable library of data, visualizations, and papers, which Internet2 cannot do. In Section IX, we discuss the utility of using a new markup language, XML, as a means of archiving and interlinking zebrafish visual data files for storage on DVDs.

IX. Hyper Text Markup Language (HTML) and eXtensible Markup Language XML (Bryan Crawford)

Zebrafish databases need to connect visual images with text data. This can be accomplished by languages such as HTML and XML. HTML was designed to present information consistently on a variety of platforms. HTML consists of a set of tags that specify how information is to be presented. These tags are defined by a set of standards that have been agreed on by the World Wide Web Consortium (W3C; http://www.w3.org/) and that define how software browsers such as Internet Explorer (IE), Netscape, and Safari should behave. Unfortunately, not all Web designers, and/or the HTML editors they use, follow these standards. Nonstandard HTML has resulted in browser incompatibilities on many Web pages.

As a result, simple HTML and stand-alone browsers such as Netscape and IE are going extinct. (Microsoft has stopped all development of IE.) Web integration at the OS level is becoming standard, so no "browser" application will exist in the future. XML is likely to become the dominant format for data exchange, and its interpretation will be integral to the OS of all computer systems. This is already true of MacOS X, in which system and application preferences, as well as many other data files, are stored as XML.

XML is a standards-based system for encapsulating information along with the algorithms necessary for presenting, manipulating, or interpreting it into a platform-independent package. As such, it easily accommodates data and the Java, Perl, etc., applications necessary for working with the data, the metadata describing what portions of the file do what, and useful, human-readable, searchable text in one self-explanatory file. Thus, XML can be used to encapsulate HTML-based presentations, SQL-based databases, images, sound, VRML/QuickTime VR/Java3D virtual reality visualizations, physiology-data, or any other data, making it viewable and searchable on any platform. Although XML is not a panacea—algorithms for presenting and searching the visual data must still be devised and will still limit the ways in which users can interact with the data—it offers exciting possibilities for cross-platform data sharing.

XML is becoming increasingly popular with scientists interested in sharing data within their community. Some projects that use XML-derived data handling for sharing large datasets within the scientific community include the Grid eXtensible Data and eXtensible Data Format projects at NASA (http://people.nas.nasa.gov/~pv/gxd/), (http://xml.gsfc.nasa.gov/XDF/XDF_home.html), the Distributed Annotation Project (biodas.org), and the eXtensible Scientific Interchange Language Project (http://www.cacr.caltech.edu/SDA/xsil/index.html).

All these projects, as well as the Zebrafish DVD Exchange Project described in this chapter, share the common objective of facilitating the sharing of vast amounts of heterogeneously formatted data among scientists who will need to be able to search these archives in unanticipated ways. Because XML files can include the code necessary for viewing or otherwise working with the data, continuously changing software and hardware does not affect the utility of the archive. Should more sophisticated tools be developed, the raw data can be accessed independently of the algorithms included within the archive. Furthermore, because of XML's inherent tolerance for user-defined extensions, unanticipated and/or nonstandard extensions of the language will not hamper the use of XML-based archives as they have hampered the use of HTML-based data.

Future archives of scientific data we produce are likely to be generated by using XML, and extensions of it, as much as reasonably feasible. In Section X we discuss uses of virtual reality and interlinked visual data sets to portray connections between patterning and morphogenetic events.

X. Virtual Reality and Data Visualization Tools (Carey Phillips)

New technologies developed over the last few years enable us to collect an impressive amount of data in a relatively short amount of time. Laboratories worldwide are collecting data about when and where genes are being expressed much faster than our present abilities to analyze it. This phenomenon is, in part, responsible for the development of new fields in science, such as bioinformatics. Faculty working in informatics departments are designing new tools and methods for organizing, analyzing, and visualizing vast amounts of information (Quon *et al.*, 2003; Vernikos *et al.*, 2003; Xavier *et al.*, 2003).

One of the challenges is to create tools that allow users to interactively organize information accessed from online databases of mixed data types so that a variety of information types can be simultaneously visualized and modeled. We also need tools enabling us to alter conditional parameters on all aspects of the model, from developmental morphology to related patterns of gene expression. Our goal in this area is to create a tool that is readily accessible to researchers online, provide a useful way to store and interact with topic-related information of all media types, and become the framework for a series of integrated science information archives. One possible solution is to create an online dynamic 3D visualization tool, based in virtual reality, which functions as an interface to spatial and temporal information accessed through a variety of databases.

We have been experimenting with different ways to create such a tool that could accommodate at least most of the criteria listed previously. One such experiment involves creating animated 3D models of developing embryos that are exported as Virtual Reality Modeling Language (VRML) files (Fig. 7). The VRML models provide the potential for a number of interesting behaviors. Objects within the model can be mapped with a deformable matrix with resolutions up to $1000 \times 1000 \times 1000$. Temporal and spatial information is dynamically displayed within the matrix cells as the virtual embryo develops, essentially playing a movie of the temporal/spatial information on the surface of the model. Users can dynamically pick the information they wish to visualize from a list of available data. There are

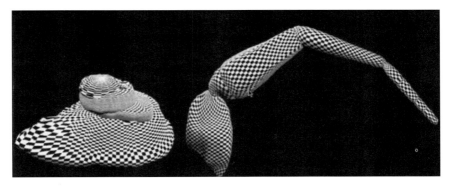

Fig. 7 Example of a novel application of virtual reality in embryological visualizations. Two stages of imaginal disc development in *Drosophila* showing the deformation of a grid on the developing limb. Each unit of the grid acts as an interactive window to a database(s) that can display a variety of types of information. Combinations or groups of grids can display patterns of gene expression, for example. The resolution of the grid can be as high as $1000 \times 1000 \times 1000$ for each of the five segments show here. The animated development of the insect leg can be stopped at any stage in the virtual world version and the user can "paint" patterns of gene expression on the limb. This information can be saved to a file, and all saved patterns of gene expression can then be viewed by selecting the ones you want to see. The user can also spin the limb around and view its development from any perspective.

a number of ways by which scientists can enter data into the virtual embryo database. In our first iteration, we created tools so that one can "paint" a pattern of gene expression on the developing embryo at each stage of development. This information is saved, or "published," to a SQL database along with the appropriate metadata providing information about the name of the gene, upstream and downstream controlling genes, the author, and where this information is published. Later iterations of this tool will include methods to input magnetic resonance (MR) or confocal data directly once the data is spatially rectified to fit a normalized embryonic coordinate system. In addition, one can treat each matrix cell, or selected groups of cells, as the spatial/temporal access point to microarray or massively parallel signature sequence (MPSS) data sets.

The following represents a scenario of how we envision such a tool might be used in the future. A scientist logs-in to an online virtual world, where he or she can interact with a 3D model of a developing organism. For example, our scientist selects a virtual world model on limb development in *Drosophila*. The virtual world displays the portion of the embryonic ectoderm straddling the parasegmental boundary that gives rise to the selected limb. A query is initiated and patterns of gene expression that determine the development of this imaginal disk at that spot are given. The scientist moves around the 3D object, viewing gene expression patterns and morphology from any perspective, adjusting transparencies, peeling off layers, or viewing multiple combinations of known gene expression patterns. The scientist "runs" development forward or backward, viewing development of the imaginal disk as it becomes a complete limb while displaying the overlapping RNA expression patterns of several selected genes occurring during specific periods of development. Our scientist "creates" mutants by knocking out genes or changing upstream gene expression patterns and visualizes the morphological consequences and subsequent downstream changes in the patterns of a number of genes of interest. Clicking on a portion of the developing disk or limb opens searchable databases of published reference materials related to that portion of the morphological structure around a designated time of development. These reference materials might include papers, microscopy images, lectures, URLs, animations, or movies.

Our scientist then accesses a portion of the developing virtual limb, opening a dependant virtual world, to interact with a much higher resolution model in which cytoskeletal rearrangements are projected onto a pattern of specified limb cells as a function of the mutants created in the parent world. Our scientist "meets" his or her students and a collaborator from France in the one of the virtual reality worlds to demonstrate the results of the latest experiment. Each person, represented by an avatar, can see the other guest avatars and communicate with each other in real time by talking into a microphone on his or her computer. The French collaborator shares new experimental data by entering a confocal movie into the database, where it automatically links to the appropriate morphology and developmental time, providing access to this information to all those online in the virtual world. This data is discussed, tested, and modeled by our scientist and students before being made accessible to the entire community. Our scientist returns to the laboratory for more hands-on work.

Systems more complex than the *Drosophila* model discussed previously will require additional modifications to study the internal developing structures. My laboratory is creating zebrafish development models that enable a user to "pull out" and view each developing organ system separately or in any combination. We intend to have these models and animations available for general use by the end of 2004.

XI. Summary

Scientists who study zebrafish currently have an acute need to increase the rate of visual data exchange within their international community. Although the Internet has provided a revolutionary transformation of information exchange, the Internet is at present unable to serve as a vehicle for the efficient exchange of massive amounts of visual information. Much like an overburdened public water system, the Internet has inherent limits to the services it can provide.

It is possible, however, for zebrafishologists to develop and use virtual intranets (such as the approach we outlined in this chapter) to adapt to the growing informatics need of our expanding research community. We need to assess qualitatively the economics of visual bioinformatics in our research community and evaluate the benefit:investment ratio of our collective information-sharing activities.

The development of the World Wide Web started in the early 1990s by particle physicists who needed to rapidly exchange visual information within their collaborations. However, because of current limitations in information bandwidth, the World Wide Web cannot be used to easily exchange gigabytes of visual information. The Zebrafish DVD Exchange Project is aimed at by-passing these limitations.

Scientists are curiosity-driven tool makers as well as curiosity-driven tool users. We have the capacity to assimilate new tools, as well as to develop new innovations, to serve our collective research needs. As a proactive research community, we need to create new data transfer methodologies (e.g., the Zebrafish DVD Exchange Project) to stay ahead of our bioinformatics needs.

Acknowledgments

We thank the following companies for their generous financial support of the Zebrafish DVD Exchange Project: Bio-Rad, Nikon, Inc., Chroma, Marine Biotech, Meridian Instruments, and Molecular Probes. This work was supported by NSF grants IBN-9808224 and IBN-0212258 (M. S. C) and CCLI 021017 (C. P), FIPSE grant PR116B00550 (C. P.), and NIH grant GM62283 (M. S. C).

References

Dean, B. (1896). The early development of Amia. *Q. J. Microsc. Sci.* **38**, 413–451.

Eycleshymer, A. C., and Wilson, J. M. (1906). The gastrulation and embryo formation in *Amia calva*. *Am. J. Anat.* **5**, 133–162.

Goesmann, A., Linke, B., Ruppa, O., Krause, L., Bartels, D., Dondrup, M., McHardya, A. C., Wilke, A., Pühler, A., and Meyer, F. (2003). Building a BRIDGE for the integration of heterogeneous data from functional genomics into a platform for systems biology. *J. Biotechnol.* **106**, 157–167.

Golding, G. B. (2003). DNA and the revolutions of molecular evolution, computational biology, and bioinformatics. *Genome* **46**(6), 930–935.

Golovin, A., Oldeld, T. I., Tate, J. G., Velankar, S., Barton, G. J., Boutselakis, H., Dimitropoulos, D., Fillon, J., Hussain, A., Ionides, J. M. C., John, M., Keller, P. A., Krissinel, E., McNeil, P., Naim, A., Newman, R., Pajon, A., Pineda, J., Rachedi, A., Copeland, J., Sitnov, A., Sobhany, S., Suarez-Uruena, A., Swaminathan, G. J., Tagari, M., Tromm, S., Vranken, W., and Henrick, K. (2004). E-MSD: An integrated data resource for bioinformatics. *Nucleic Acids Res.* **32**, D211–D216.

Honts, J. E. (2003). Evolving strategies for the incorporation of bioinformatics. *Cell Biol. Educ.* **2**, 233–247.

Ouzounis, C. A., and Valencia, A. (2003). Early bioinformatics: The birth of a discipline—a personal view. *Bioinformatics* **19**(17), 2176–2190.

Quon, G. T., Gordon, P., and Sensen, C. W. (2003). 4D bioinformatics: A new look at the ribosome as an example. *IUBMB Life* **55**(4/5), 279–283.

Vernikos, G. S., Gkogkas, C. G., Promponas, V. J., and Hamodrakas, S. J. (2003). Gene ViTo: Visualizing gene-product functional and structural features in genomic datasets. *BMC Bioinform.* **4**(1), 53.

Wegman, E. J. (2003). Visual data mining. *Stat. Med. B.* **22**(9), 1383–1397.

Xavier, J. B., White, D. C., and Almeida, J. S. (2003). Automated biofilm morphology quantification from confocal laser scanning microscopy imaging. *Water Sci. Technol.* **47**(5), 31–37.

CHAPTER 25

Comparative Genomics—An Application for Positional Cloning of the *weissherbst* Mutant

Anhua Song[*,†] and Yi Zhou[*,†,‡]

[*]Children's Hospital Boston
Boston, Massachusetts 02115

[†]Harvard Medical School
Boston, Massachusetts 02115

[‡]Dana-Farber Cancer Institute
Boston, Massachusetts 02115

I. Introduction

Comparative genomic resources are helpful to further establish zebrafish as an excellent model organism for studying vertebrate development and genetics and for modeling human biological processes and diseases. Genomic comparison between zebrafish and other vertebrate model organisms provides useful information that one can use to understand the evolution of each vertebrate species. At the same time, this comparison facilitates the use of genomic information obtained in other vertebrate organisms for zebrafish research projects and vice versa. It is therefore important to create and use comparative genomic tools and resources to accelerate research projects in zebrafish and other vertebrates.

II. Resources for Comparative Genomics Studies

The Children's Hospital Boston Zebrafish Genome Initiative, in the laboratory of Leonard I. Zon, has been constructing and maintaining a comparative genomics resource for the past 3 years (http://zfBlastA.tch.harvard.edu/CompGenomics/). This site provides comparative BLAST search results between (1) the zebrafish genome assembly (zV3) and all known human gene and mRNA sequences (6 August 2001 release), (2) the zebrafish genome assembly (zV2) and all known human gene and mRNA sequences, (3) the fugu genome assembly (V3.0) and all known human gene and mRNA sequences, (4) the zebrafish genome assembly (zV3 and zV2) and predicted fugu proteins, and (5) zebrafish whole-genome shotgun sequences and all known human gene and mRNA sequences. These BLAST search results are stored in a database and are quickly retrievable at http://zfBlastA.tch.harvard.edu/comp-Genomics/by searching key words in the definition lines of the GenBank entries and by knowing chromosomal locations of genes in the human genome. This site also provides sequence retrieve services. Users can click the "Retrieve Sequences in Databases" link and go to http://zfblasta./humanBlastZv3/Retrieve.htm. Sequence names of interest obtained through BLAST searches can be entered with respective databases chosen. Because the BLAST searches were performed on a specific date, the databases might not contain homology searches of sequences submitted post the BLAST searches. The amount of computing time needed to update these databases is demanding, but we are continually updating these databases as soon as new searches are completed. The goal of this effort is to provide individual researchers with convenient bioinformatic tools to identify orthologous genes in zebrafish, fugu, and humans by greatly reducing the need to perform individual BLAST searches.

In instances in which specific homology or orthology searches are necessary, a BLAST search server is provided at http://134.174.23.160/zfBlast/PublicBlast.htm. This BLAST server allows sequence similarity searches by using standard BLAST algorithms against 10 separate nucleotide databases containing zebrafish genomic and expressed sequence tag (EST) sequences, and three independent protein databases containing known human genes/mRNAs and fugu proteins. Five different BLAST search functions allow users to enter either protein or nucleotide sequences as queries for searching all databases on this server. Protein sequences are recommended in BLAST searches whenever possible because protein-coding sequences are more likely to be conserved among species and are derived from a small portion of genomic sequences. Thus, protein sequences permit more specific and faster searches of orthologs between different species. In BLAST searches that involve many different databases and require input sequences from the previous steps of searches, Direct DB BLAST simplifies the process. Sequence names instead of the actual sequences can be used to achieve desired BLAST searches, allowing users to escape the sequence retrieval steps before searching additional databases. This function reduces the need of opening multiple browser windows and also allows easier tracking of searches performed.

In addition, researchers can also use online services provided by the National Center for Biotechnology Information (http://www.ncbi.nlm.nih.gov/) and the Wellcome Trust Sanger Institute (the Sanger Center at http://www.sanger.ac.uk/ and Ensemble at http://www.ensembl.org/). Moreover, the human and mouse genome browser (http://genome.ucsc.edu) at the University of California at Santa Cruz (UCSC) and the fugu genome browser (Fugu v3.0) at the Department of Energy Joint Genome Institute (http://genome.jgi-psf.org/fugu6/fugu6.home.html) provide convenient user interfaces for researchers to browse through available genomic information necessary for comparative genomic studies.

III. An Example of Using Comparative Genomic Resources in a Positional Cloning Project

To demonstrate how one can use the available comparative genomic studies resources, the rest of the chapter presents an updated version of the bioinformatic method used in the positional cloning of the zebrafish hypochromic mutant *weissherbst* (*weh*) (Donovan *et al.*, 2000). The summary of this process is described in Fig. 1. When the *weh* gene was cloned in 1999, there were relatively few zebrafish genomic sequences in public databases. At present, the Sanger Center has sequenced most of the zebrafish genome, although these sequences will not be fully assembled until 2005. To demonstrate the use of current databases, it is necessary to describe a similar but not identical bioinformatic research process by using current genomic resources under the assumption that the *weh* gene and its family members are not in the databases and the genomic region of *weh* has not been fully sequenced and assembled.

Embryos with homozygous *weh* mutations develop microcytic hypochromic anemia during early embryogenesis. Mutant embryos that were examined at 27 and 33 h postfertilization (hpf) had very few mature red blood cells in circulation. This anemic phenotype was lethal, and embryos died sometime between 7 and 14 days postfertilization (dpf). To discover the molecular nature of the *weh* mutation, a positional cloning approach was used to identify the affected gene. This cloning process began with a half-tetrad linkage analysis and was followed by a fine genetic mapping. Two close flanking random amplified polymorphic DNA (RAPD) markers, 4K1300 and 6Q1300, were identified from RAPD analysis of the mutant families. Amplified fragment length polymorphism (AFLP) analysis was then used to scan approximately 10,000 polymorphic loci. At the end, AFLP marker I36, which is 0.13 cM from the *weh* locus on the centrameric side, was identified as the closest link to the mutation. The closest distal flanking marker, 4K1300 (no closer polymorphic marker was identified at this end), was 2.6 cM

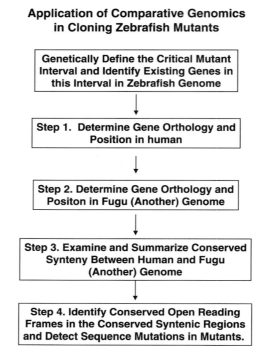

Fig. 1 An outline of a comparative genomic approach in positional cloning of the *weh* mutant.

from the locus. A subsequent chromosome walk was initiated by using I36 as the starting probe, leading to the identification of the *weh* locus on PAC clone 170G3. Single-strand conformation polymorphisms (SSCPs) were found on both ends of 170G3 PAC when tested on the *weh* walking panel, and genetic analysis by using these new markers confirmed that the two ends were on different sides of the *weh* locus. Thus, the *weh* gene was on this PAC clone.

The PAC 170G3 insert was isolated, radioactively labeled, and hybridized onto a set of high-density filters spotted with cDNA clones from a zebrafish adult kidney cDNA library. The kidney is the site of definitive hematopoiesis in adult zebrafish and contains genes functioning in red cell specification and differentiation pathways. Positive cDNA clones from the hybridization were isolated and sequenced. These cDNA clones clustered into three individual cDNA species, including a novel cDNA clone that, based on a sequence similarity search, was named wc1 (*weh* candidate 1). *In situ* hybridization and sequencing analysis of the wc1 gene was then performed in both wild-type and *weh* mutant embryos. Unfortunately, subsequent genetic and biological information obtained suggested that wc1 was not the *weh* gene. Based on homology, the two other cDNAs located on PAC 170G3 were identified as glutaminase C and stat1, and their known functions suggested that they were not candidates for the *weh* gene. Thus, the *weh* gene could not be identified on this PAC by a traditional cDNA selection approach. Because it was possible that other genes were positioned on PAC 170G3 but were undetected in our cDNA selection analysis, a bioinformatics analysis was performed to test this possibility by using the steps outlined next.

A. Step 1. Confirmation of wc1-glutaminase-stat1 Synteny in Humans

Synteny describes the distribution pattern of genes on a chromosome. This pattern of gene locations can be conserved such that genes positioned near each other on the genome in one species are likely to be found close to each other on a single chromosome in evolutionarily related species. The isolation of wc1, stat1, and glutaminase in the cDNA selection assay suggests that these three genes are syntenic on the zebrafish genome, a result that was confirmed by their locations on the T51 RH map. It is now important to know whether this zebrafish gene synteny is conserved in another vertebrate genome. Because the human genome assembly was available at the time, this allowed us to ask electronically whether the human orthologs of these three genes are syntenic on a human chromosome. More importantly, because the human genome had been assembled at a relatively high quality, it was better to compare gene synteny in the human equivalent region of the zebrafish *weh* critical interval. If the synteny is conserved, it would allow an analysis of the conserved synteny in another vertebrate and a prediction of gene orders in the region. The conserved synteny and gene order within the syntenic region between two vertebrate genomes will help predict other genes or open reading frames (ORFs) in the zebrafish region (zebrafish PAC 170G3).

1. Search for Human Orthologs

The first task was to identify the human orthologs of glutaminase, stat1, and wc1. Importantly, the scores and e-values of a BLAST search measure both the degree of sequence homology and the random chance of finding similar match in a given sized database—lower the chance, higher the significance of the match. In other words, the same exact match will have different BLAST scores and e-values if those searches are performed by using different-sized databases. Therefore, in addition to evaluating the scores and e-values of a BLAST search, it is very important to also examine the direct sequence alignment between the two sequences to identify which sequences are most homologous. The online BLAST server (http://134.174.23.160/zfBlast/PublicBlast.htm) was used to BLAST search zebrafish glutaminase 5′ sequences against the database of known human genes and mRNA sequences. Because it was a nucleotide query against a pool of protein sequences, we needed to choose the BLASTx function. The returned results indicated seven hits shared the same BLAST scores and e-values. After examining the actual protein sequence alignments (zebrafish sequence on top and human sequences on bottom of the alignments) between zebrafish glutaminase and its

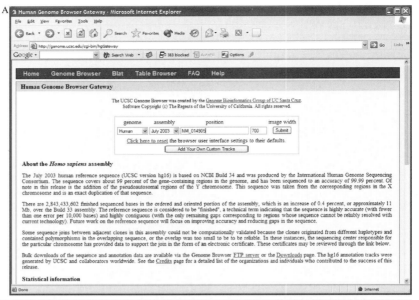

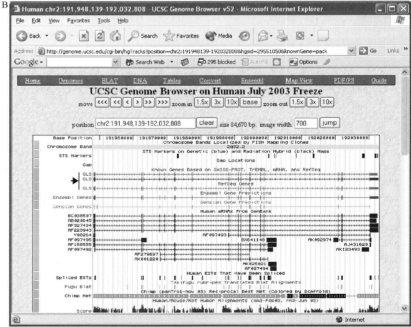

human BLAST hits, we believe that the top seven hits were against the same human gene because of very similar sequence matches of these top BLAST hits. Glutaminase C (GLS, NM_014905) was chosen as a representative sequence for later synteny analysis. The genome position information of these sequences on the UCSC genome browser also supported our claim, because all seven sequences were positioned to an almost identical location in the human genome. Similarly, human orthologs of stat1 and wc1 were also identified by using a similar process. Three hits encoded the human STAT1 protein and two hits represented the human hypothetical protein FLJ12519. Based on our sequence homology analysis, representative orthologs, STAT1 (NM_007315) and FLJ12519 (AK022581), respectively, from these hits were chosen to represent human orthologs of zebrafish stat1 and wc1.

2. Positioning Human Orthologs on the Genome

After identifying human orthologs for the three zebrafish cDNAs identified in the cDNA selection experiments, the UCSC human genome browser (http://genome.ucsc.edu) was used to examine whether a syntenic relationship between these zebrafish genes was conserved in humans. On the genome browser front page (Fig. 2A), the GenBank accession number of glutaminase C, NM_014905, was entered into the query window and the "Submit" button clicked. Within a few seconds, the human glutaminase C gene was localized on the human genome in the format shown in Fig. 2B. The base position of GLS was between chr2: 191,949,139–192,032,808. The same process was repeated to find the genomic locations of human STAT1 and FLJ12519, with base positions at chr2: 192,043,077–192,077,272 and chr2:190,508,701–190,542,807, respectively.

From the previous analysis, we first noticed that all three human orthologs are located on human chromosome 2 and their base pair locations on this chromosome are very similar. It is clear that these three human genes are located very close to each other on the same chromosome. To view their relative positions on the genome in one window, the browser can be zoomed outward 100× to reveal the entire FLJ12519-GLS-STAT1 region (Fig. 3). This analysis shows that synteny between the zebrafish wc1-glutaminase-stat1 and the human FLJ12519-GLS-STAT1 genes are maintained across these two species.

B. Step 2. Comparative Synteny Study on the Fugu Genome

Conserved synteny between two species does not necessarily mean that the order of genes on a chromosome is maintained. In this particular case, we now know that the *weh* gene is positioned near wc1, glutaminase, and stat1 in zebrafish,

---

Fig. 2 Genome location of human glutaminase C (NM_014905). (A) The University of California at Santa Cruz (UCSC) human genome browser front page. Queries such as GenBank accession numbers and gene name can be entered into the query box to identify its genome location. (B) The query result for glutaminase C (NM_014905).

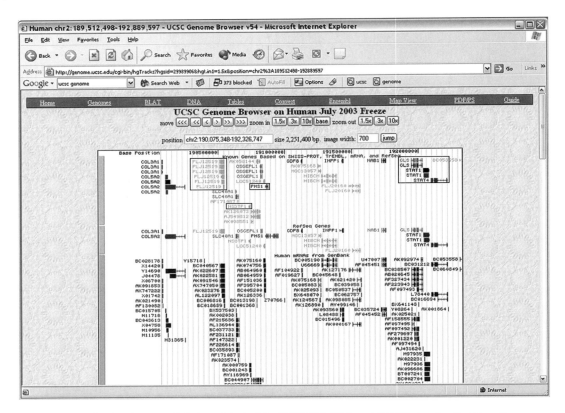

Fig. 3 An expanded view of the loci for FLJ12519, GLS, and STAT1 showing that these three genes reside very closely to each other on human chromosome 2.

based on both genetic and physical mapping data [on the same piece of genomic DNA and radiation hybrid (RH) mapping results]. Interestingly, the syntenic region in humans shows that FLJ12519 and the location of GLS and STAT1 are separated by a relatively large distance (approximately 1.5 Mb). Based on recombination frequencies and average insert size of the zebrafish PAC library (where 170G3 PAC was isolated), the same zebrafish orthologs are separated in the zebrafish genome by less than 100 kb. This discrepancy can be because of many inversions and rearrangements of gene loci found between the zebrafish and human genomes (Barbazuk *et al.*, 2000; Liu *et al.*, 2002; Postlethwait *et al.*, 2000). Thus, the *weh* gene could be inside or outside of the region gated by wc1 (at one end) and gls and stat1 (at the other end) as it was in the human genome. Many genes and ORFs (and/or hypothetical proteins) were distributed within this interval, and it would be difficult to test them individually as candidate *weh* genes. To shorten a list of candidate *weh* genes in this genomic interval, the genome sequence information of fugu fish was analyzed to better predict the location of candidate *weh* gene in relationship to the three identified genes.

The Japanese puffer fish (*Fugu rubripes*) has a very condensed genome with gene-rich sequences. Its genome is only one seventh the size of the human genome and approximately one third of the zebrafish genome. Because fugu fish shares commonalities of all the basic biological and anatomical features of vertebrates, its genome is thought to contain most of the genes necessary for vertebrate function and biology. As such, the difference in genome size is more likely the result of downsizing repeat-rich regions in the genome rather than due to loss of critical genes. The fugu fish therefore offers numerous advantages, including that it is easy to sequence and assemble its genome sequences. In addition, gene predictions are easy to make on this intron scarce genome.

To begin an analysis of the fugu genome for new candidate genes in the region, it was first necessary to test whether gene synteny for the *weh* region was conserved in the fugu genome. If conservation exists, synteny can then be used to identify closely associated candidates for the *weh* gene, and, in the case of novel ORFs, help predict zebrafish sequences of the novel genes. Regions encoding conserved predicted protein sequences can then be identified and used to design PCR primers for amplifying potential zebrafish orthologs. Again, this analysis is done under the assumption that we do not have enough genomic sequence information near the zebrafish genome region to search the zebrafish genome directly.

An example comparative genome sequence analysis has been completed using the UCSC genome browser (Fig. 4). On this browser page, annotation information of the genome sequences can be selectively displayed by choosing different options on the bottom of the browser. These options are names as tracks in the browser display. These tracks are grouped into six different categories: (1) mapping and sequencing tracks, (2) genes and gene prediction tracks, (3) mRNA and EST tracks, (4) expression and regulation tracks, (5) comparative genomics tracks, and (6) encode tracks. To help highlight the comparison between the human and fugu genome assemblies, only the human genome assembly base position information (one of the mapping and sequencing tracks), human known genes (one the genes and gene prediction tracks), and the fugu blat (one of the comparative genomics tracks) are selected for display. The resultant output (obtained by clicking the "Refresh" button), shown in Fig. 4, displays the comparative genomic analysis of the syntenic region between humans and fugu.

The current version of fugu genome was assembled mainly through a whole-genome shotgun approach. Although the fugu genome was relatively easy to assemble from random shotgun sequences, the assembly itself contained computation errors and could have redundantly represented certain genome sequence regions. It should not be surprising to have a gene positioned in different scaffolds. As predicted, many fugu scaffolds, including scaffold_1520 and scaffold_1729, showed sequence homology to this region. Subsequently, wc1, gls, and stat1 sequences were BLAST searched against the fugu genome assembly to identify that scaffold_1520 contained orthologs of GLS and STAT1 and a possible

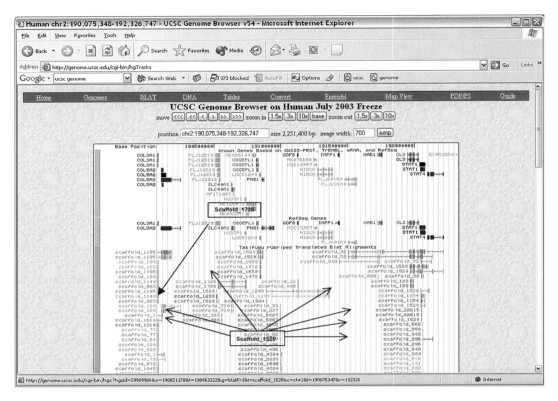

Fig. 4 Comparative genomic functions available in the UCSC human genome browser. (A) Option selection buttons for controlling information displayed in the browser window. (B) A browser window showing selected information for human genome nucleotide bases, human known genes, and comparative genomic matches between humans and the fugu fish. (See Color Insert.)

ortholog of FLJ12519 (wc1; Fig. 5), whereas scaffold_1729 contained sequences homologous to only the fugu ortholog of FLJ12519 (Fig. 6). FLJ12519 matched significantly to only two scaffolds of the fugu genome: scaffold_1520 and scaffold_1729. Scaffold_1520 contained the N-terminal 250 amino acids of FLJ12519, and scaffold_1729 covered the rest of C-terminal 250–830 amino acids. This suggests an inaccurate assembly of this region in the fugu genome. The GLS BLAST search result showed sequence matches to scaffold_375, scaffold_1520, scaffold_189, and scaffold_32. Scaffold_1520 and scaffold_32 had higher sequence homology to human GLS than the other two scaffolds did. The matches of scaffold_32 to human GLS seemed in the opposite direction in genomic sequence as matches between scaffold_1520 and human GLS. However, scaffold_32 did not contain matches to FLJ12519. Human STAT1 showed homology to many more scaffolds, including scaffold_1520 and scaffold_32 (with the highest BLAST scores

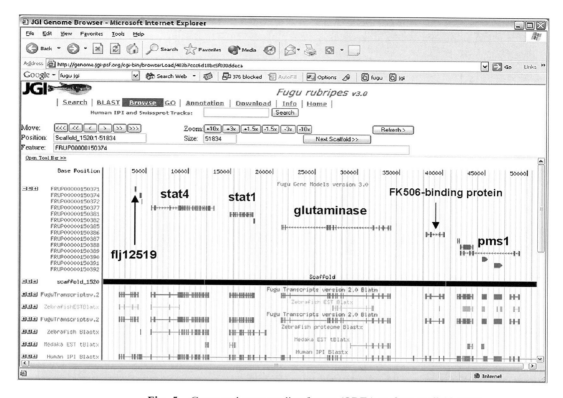

Fig. 5 Genes and open reading frames (ORFs) on fugu scaffold_1520.

and e-values) because of family members in the genome. Therefore, scaffold_1520 and scaffold_1729 have the best overall conservation of the human FLJ12519-GLS-STAT1 region and were chosen for further synteny analysis. Figure 7 summarizes the entire synteny analysis between the human and fugu intervals.

To confirm our findings by using the human genome browser at the UCSC, scaffold_1520 and scaffold_1729 were also examined on the fugu genome browser at the JGI (Figs. 5 and 6). Annotations on these two scaffolds again suggested that the fugu orthologs of wc1, GLS, and STAT1 did not exist on a single fugu genome contig (scaffold). As found by using the UCSC browser, the majority of fugu wc1 ortholog sequences were positioned on scaffold_1729 (Figs. 6 and 7), whereas full-length orthologs of GLS and STAT1 were located on scaffold_1520 (Figs. 5 and 7). Although sequence homology suggested that a smaller region of wc1 could be on scaffold_1520, the homology was less significant than that on scaffold_1729. The partial sequence of fugu wc1 on scaffold_1520 could be because of a sequence inaccuracy of fugu genome or could encode a homologous FLJ12519 protein family member.

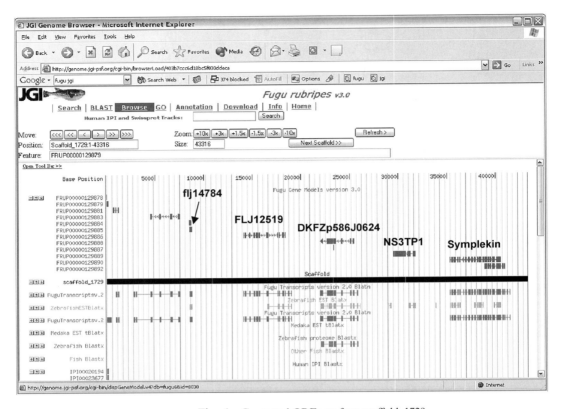

Fig. 6   Genes and ORFs on fugu scaffold_1729.

The above analyses suggests the partial syntenic conservation of the wc1-gls-stat1 region in zebrafish, fugu, and humans. The finding that these genes are separated by 1.5 Mb in humans but only 100 kb in zebrafish indicates that various chromosomal inversions and rearrangements occurred in the zebrafish with respect to human chromosome 2. Because synteny was better conserved between fugu and zebrafish because of closer in evolution, fugu scaffolds were used to investigate potential zebrafish gene arrangements in the *weh* gene critical interval.

C. Step 3. Examine and Summarize the Conserved Synteny Between Humans and Fugu

Because it was possible that scaffold_1520 contained all three ortholog genes (wc1, gls, and stat1) identified on the zebrafish PAC clone, this scaffold was first selected for bioinformatics analysis. The gene orders on scaffold_1520 (Fig. 5) indicated that stat4 and stat1 were closer to the flj12519 ortholog than were pms1 and gls, whereas FLJ12519 was closer to PMS1 and GLS than were STAT1 and STAT4 on human chromosome 2 (Figs. 5 and 7). This observation suggests that

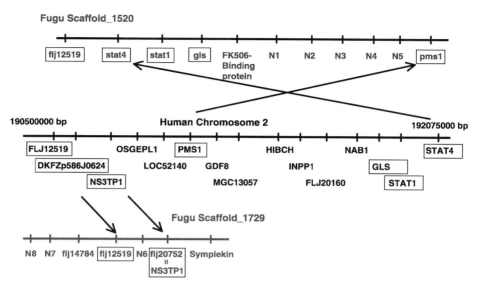

Fig. 7 Summary of the synteny comparison of the *weh* locus equivalent regions in humans and fugu. The synteny between the FLJ12519-GLS-STAT1 interval and fugu scaffolds does not appear to be intact, but parts of this synteny interval do. (See Color Insert.)

the fugu genome has a chromosome inversion within this region relative to humans. In fugu fish, no candidate ORFs were identified between the FLJ12519 ortholog and stat4 on scaffold_1520. Importantly, the other end of this scaffold ended at the pms1 gene and the human ortholog of this gene positioned four genes away from FLJ12519 on human chromosome 2. Given that the region between the stat4 and pms1 genes was relatively intact (although inverted) in fugu fish and that it did not encode other human orthologs in its middle, it was unlikely that any other ORFs found on this scaffold would serve as *weh* gene candidates.

As mentioned previously, it is likely that the true fugu ortholog of FLJ12519 was positioned on scaffold_1729 (Figs. 5 and 6). When analyzing scaffold_1729, there was one ORF, the ortholog of FLJ14784 positioned on the left side of FLJ12519. In humans, this hypothetical protein was positioned on chromosome 3, indicating that the synteny between the fugu orthologs of human FLJ12519 and FLJ14784 was not maintained in the human genome. On the other hand, two other ORFs on the right hand side of FLJ12519 were identified. One was orthologous to FLJ20752 (also known as NS3TP1) and the other was FRUP00000129888. Both these genes are syntenic to FLJ12519 on human chromosome 2 (Figs. 2 and 6). The FRUP00000129888 ORF had very high homology to DKFZp586J0624; a full-length human cDNA and the alignment between these two ORFs is shown in Fig. 8.

Fig. 8 The alignment of protein sequences between fugu FRUP00000129888 and human DKFZp586J0624 (AL136944), which is a complete cDNA sequence.

D. Step 4. Isolation of Zebrafish Candidate Genes and Identification of Mutations by Sequence Analysis

The closely associated synteny and high orthology suggested further analysis of this novel ORF. PCR primers were designed against the fugu FRUP00000129888 nucleotide sequence within a predicted coding exon and a proper-sized PCR fragment was specifically amplified from another zebrafish PAC clone (211O13) in the *weh* gene critical region. This fragment was radioactively labeled and hybridized to high-density membrane filters spotted with kidney cDNA clones. A full-length zebrafish cDNA clone was isolated and named *ferroportin 1* after its function was identified.

Subsequent sequence analysis of both the weh^{th238} and weh^{tp85c} mutant alleles identified point mutations of C to A at codon 361 and of G to T at codon 167, respectively. The mutation in the weh^{th238} allele generated a premature termination codon for *ferroportin 1*, whereas the weh^{tp85c} mutant allele resulted in a single amino acid change at position 167 from Leu to Phe. The identification of a premature stop codon in the weh^{th238} mutant allele argued strongly that the hypochromia phenotype of the weh^{th238} mutants was caused by a malfunctional *ferroportin 1* protein. That the phenotype could be rescued by overexpressing the full-length cDNA in an expression vector also supported that the mutation resided within the *ferroportin 1* gene. Functional analysis of *ferroportin 1* protein suggested a role in transporting iron across cell membranes. In addition, *in situ* hybridization analysis of *ferroportin 1* mRNA indicated that it was expressed in the yolk syncytial layer of developing embryos, a location at which the maternal source of iron is stored.

As described in the previous example, a vast amount of genomic information is available to the zebrafish community. To take advantage of these resources, researchers will need to be able to use comparative genomic analysis as a tool for identifying potential mutant genes in positional and candidate cloning projects of zebrafish genetic mutants.

Acknowledgments

Our research is funded by the National Institutes of Health. We thank Jeffery Guyon for his critical input and Nelson Hsia for his helpful comments during the preparation of this manuscript.

References

Barbazuk, W. B., Korf, I., Kadavi, C., Heyen, J., Tate, S., Wun, E., Bedell, J. A., McPherson, J. D., and Johnson, S. L. (2000). The syntenic relationship of the zebrafish and human genomes. *Genome Res.* **10**(9), 1351–1358.

Donovan, A., Brownlie, A., Zhou, Y., Shepard, J., Pratt, S. J., Moynihan, J., Paw, B. H., Drejer, A., Barut, B., Zapata, A., Law, T. C., Brugnara, C., Lux, S. E., Pinkus, G. S., Pinkus, J. L., Kingsley, P. D., Palis, J., Fleming, M. D., Andrews, N. C., and Zon, L. I. (2000). Positional cloning of zebrafish ferroportin1 identifies a conserved vertebrate iron exporter. *Nature* **403**(6771), 776–781.

Liu, T. X., Zhou, Y., Kanki, J. P., Deng, M., Rhodes, J., Yang, H. W., Sheng, X. M., Zon, L. I., and Look, A. T. (2002). Evolutionary conservation of zebrafish linkage group 14 with frequently deleted regions of human chromosome 5 in myeloid malignancies. *Proc. Natl. Acad. Sci. USA* **99**(9), 6136–6141.

Postlethwait, J. H., Woods, I. G., Ngo-Hazelett, P., Yan, Y. L., Kelly, P. D., Chu, F., Huang, H., Hill-Force, A., and Talbot, W. S. (2000). Zebrafish comparative genomics and the origins of vertebrate chromosomes. *Genome Res.* **10**(12), 1890–1902.

CHAPTER 26

Comparative Genomics in Erythropoietic Gene Discovery: Synergisms Between the Antarctic Icefishes and the Zebrafish

H. William Detrich, III and Donald A. Yergeau

Department of Biology
Northeastern University
Boston, Massachusetts 02115

METHODS IN CELL BIOLOGY, VOL. 77

I. Introduction

During erythropoiesis, hematopoietic stem cells generate erythroid-committed progenitors that subsequently differentiate into mature erythrocytes. This genetic program is regulated through a complex matrix of transcription factors, growth factors, and signaling pathways (Cantor and Orkin, 2002). Model vertebrate genetic systems, such as the mouse and the zebrafish (Orkin and Zon, 1997), have contributed greatly to our understanding of the erythropoietic program, but each has its drawbacks. Until the recent advent of random genetic screens (Rossant and McKerlie, 2001), the murine system was largely limited to reverse genetic knockout and knockin strategies (Pandolfi, 1998) applied to previously discovered genes. Large-scale mutagenesis of the zebrafish, by contrast, has provided a random approach to the discovery of new hematopoietic genes (Orkin and Zon, 1997; Parker *et al.*, 1999; Paw and Zon, 2000), but the generation of mutants, the recovery of mutated genes, and the functional analysis of these genes are laborious processes.

To complement these approaches while overcoming some of their limitations, we have developed a multimodel comparative genomics strategy, which first exploits the erythrocyte-null condition of the Antarctic icefishes to scan the vertebrate genome for new genes involved in erythropoiesis. These natural knockouts of the erythroid lineage are a unique resource for analyzing the genetic program of erythropoiesis by subtractive genomic strategies. However, because of their long generation times (many years to reproductive maturity), Antarctic fishes are not suitable subjects for functional analysis of the genes so discovered. The zebrafish (*Danio rerio*), our second model system, reproduces rapidly with high fecundity and is widely used for analysis of gene function during vertebrate development (Eisen, 1996; Grunwald, 1996). Thus, our overall strategy is to use Antarctic icefishes to isolate potential erythropoietic genes, to clone the zebrafish orthologs of these novel genes, and then to determine the functions of the genes in zebrafish embryos by reverse genetic technologies.

To identify erythropoietic genes, we applied a subtractive technology—cDNA-based representational difference analysis (RDA)—to the hematopoietic transcriptomes of two closely related Antarctic fishes, the red-blooded rockcod, *Notothenia coriiceps*, and the white-blooded erythrocyte-null icefish, *Chaenocephalus aceratus*. First developed to isolate genomic DNA differences (Lisitsyn *et al.*, 1993) and later modified for organ-, tissue-, or cell-specific cDNA populations (Hubank and Schatz, 1994), RDA has been used successfully to identify genes that are differentially expressed in various cancers (Braun *et al.*, 1995; Gress *et al.*, 1997). The strategy of RDA entails the selective enrichment of gene fragments unique to, or heavily overrepresented in, a tester representation by favorable hybridization kinetics and exponential PCR amplification, and the concomitant removal of gene fragments shared by both tester and driver cDNA representations. In this study, the tester representation was obtained from the head kidney of *N. coriiceps*, whereas the driver representation was from the head kidney of *C. aceratus*.

By using this subtractive approach, we isolated 45 gene-fragment contigs that are differentially expressed by the hematopoietic kidneys of the red- and white-blooded fishes (Section II). Of the 45 contigs, 28 were identifiable as previously known or probable blood-related genes and further analysis was not pursued (Section III). The remaining 17 RDA fragments were used to screen an *N. coriiceps* spleen cDNA library to obtain full-length cDNA clones of potentially novel erythropoietic genes, some of which are partially described in Section IV. Finally, we demonstrate the functional analysis of one newly discovered gene, *bloodthirsty*, by model hopping to the zebrafish system. These protocols, which are based on the novel hematopoietic phenotype of the Antarctic icefishes, provide an alternative approach to the discovery of candidate erythroid genes for functional studies in zebrafish and other higher vertebrates.

II. Representational Difference Analysis (RDA) Protocol

Because Antarctic icefishes do not express the entire erythroid program, application of cDNA RDA (Hubank and Schatz, 1994, 1999) to a major hematopoietic organ, the head kidney, from an icefish species and from a related red-blooded species should permit the recovery of the erythropoietic genes expressed by the latter. cDNA fragments unique to the tester representation or amplicon here derived from the red-blooded Antarctic fish species *N. coriiceps* are enriched selectively by favorable hybridization kinetics and exponential PCR amplification. Fragments common to the tester and driver representations (driver DNA is from the erythrocyteless icefish *C. aceratus*) are eliminated by enzymatic degradation. Because typical cells express approximately 10,000–15,000 genes, which constitute a small fraction of the total genomic DNA, the representations generated by cDNA RDA are designed to conserve sequence complexity as much as possible (Hubank and Schatz, 1999). The preservation of sequence complexity, the amplification of differences, and the subtraction of common cDNAs together produce high sensitivity and permit rare transcripts (<1 copy/cell) to be detected. Figure 1 outlines the strategy for cDNA RDA in which the tester representation is from *N. coriiceps* and the driver amplicon is from *C. aceratus*.

A. Materials

1. Poly(A)$^+$ RNA Isolation

- Oligo-dT$_{15}$ cellulose (Pharmacia Biotech)
- Diethylpyrocarbonate (DEPC, Sigma)
- Guanidium isothiocyanate (Sigma)
- Phenol (Fisher)
- Chloroform (Fisher)

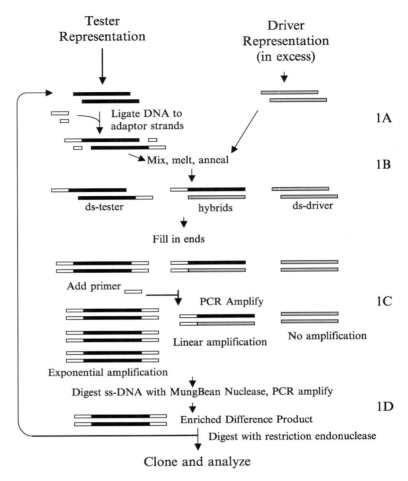

Tester
Representation

Driver
Representation
(in excess)

Ligate DNA to
adaptor strands   1A

Mix, melt, anneal   1B

ds-tester   hybrids   ds-driver

Fill in ends

Add primer

PCR Amplify   1C

Linear amplification   No amplification

Exponential amplification

Digest ss-DNA with MungBean Nuclease, PCR amplify

1D

Enriched Difference Product

Digest with restriction endonuclease

Clone and analyze

Fig. 1 Representational difference analysis. The tester population is designated by black boxes and the driver representation by gray boxes. In the present study, the tester representation was obtained from head kidney cDNA of the red-blooded rockcod *N. coriiceps*, whereas the driver representation was produced from head kidney cDNA of the white-blooded icefish *C. aceratus*. Three cycles were performed by using increasing ratios of driver to tester representations during annealing. Adapted from Lisitsyn *et al.* (1993). Cloning the differences between two complex genomes. *Science* **259**, 946–951, with permission.

2. cDNA Synthesis

- Oligo dT$_{15}$ primer (Pharmacia Biotech)
- Superscript II reverse transcriptase (Invitrogen)
- Dithiothreitol (DTT)

- RNasin (10 U/μl) (Promega)
- BSA: Bovine serum albumin, molecular biology grade (Sigma)
- β-NAD+ (4 mM; Sigma)
- RNase H (4 U/μl; Invitrogen)
- *E. coli* DNA ligase (10 U/μl; New England BioLabs)
- *E. coli* DNA polymerase I (10 U/μl; New England BioLabs)

3. RDA Reagents

- *Dpn*II (10 U/μl; New England BioLabs)
- 10× *Dpn*II buffer (New England BioLabs)
- Primers for representational difference analysis:

    R-Bgl-12: 5′ GATCTGCGGTGA 3′ (1 mg/ml in sterile water)
    R-Bgl-24: 5′ AGCACTCTCCAGCCTCTCACCGCA 3′ (2 mg/ml in sterile water)
    J-Bgl-12: 5′ GATCTGTTCATG 3′ (1 mg/ml in sterile water)
    J-Bgl-24: 5′ ACCGACGTCGACTATCCATGAACA 3′ (2 mg/ml in sterile water)
    N-Bgl-12: 5′ GATCTTCCCTCG 3′ (1 mg/ml in sterile water)
    N-Bgl-24: 5′ AGGCAACTGTGCTATCCGAGGGAA 3′ (2 mg/ml in sterile water)

- 5× PCR buffer: 330 mM Tris-HCl (pH 8.8), 20 mM MgCl$_2$, 80 mM $(NH_4)_2SO_4$, 165 μg/ml BSA
- 3 mM deoxynucleotide mix (dATP, dCTP, dGTP, dTTP; Promega)
- Klentaq DNA polymerase (5 U/μl; Clontech)
- T$_4$ DNA ligase (40 U/μl; New England BioLabs)
- 10× T4 DNA ligase buffer (New England BioLabs)
- 25:24:1 Phenol:chloroform:isoamyl alcohol (United States Biochemicals)
- EPPS: *N*-(2-Hydroxyethyl)piperazine-*N*′-3-propanesulfonic acid (Sigma)
- 3× EE buffer: 30 mM EPPS (pH 8.0), 3 mM EDTA
- Qiaex II Gel Extraction Kit (Qiagen)
- Agarose, molecular biology grade (Fisher)
- 50× TAE: 242 g Tris base, 57.1 ml glacial acetic acid, 100 ml 0.5 M EDTA (pH 8.0) per liter stock solution
- Yeast tRNA (Sigma type IV)
- MBN: Mung bean nuclease (10 U/μl; New England BioLabs)
- 10× MBN buffer (New England BioLabs)
- *Bam*HI (20 U/μl; New England Biolabs)

- 10× *Bam*HI buffer (New England Biolabs)
- pBluescript KS+ cloning vector (Stratagene)
- Competent JM109 *E. coli* cells
- Bio-Rad GenePulser electroporation apparatus or equivalent
- X-gal: 5-bromo-4-chloro-3-indoyl-β-D-galactoside (20 mg/ml in dimethylformamide; Sigma)
- IPTG: isopropyl thio-β-D-galactoside (25 mg/ml in water; Fisher)
- Ampicillin, 100 mg/ml (Fisher)
- LB broth: 10 g tryptone, 5 g yeast extract, 10 g NaCl per liter (Fisher)
- LB agar plates: LB broth plus 15 g/l of agar (Difco); 100 mm petri dishes
- Selective LB agar for blue/white screening: LB agar plates containing 100 μg/ml ampicillin plus 80 μg X-gal and 250 μg IPTG/plate

4. Blotting Materials and Reagents

- Magnagraph uncharged nylon membrane (Micron Separation, Inc.)
- [^{32}P]dCTP: 10 mCi/ml (ICN or equivalent)
- NEBlot Nick Translation Labeling Kit (New England Biolabs)
- Church–Gilbert solution: 7% sodium lauryl sulfate, 1% BSA in 0.5 M phosphate buffer (pH 7.4)

5. DNA Sequencing Reagents and Equipment

- T7 universal sequencing primer: 5′ TAATACGACTCACTATAGGG 3′
- Automated DNA sequencing apparatus with appropriate sequencing chemistry
- Wizard Miniprep Kit (Promega)
- Other molecular materials and methods (Sambrook and Russell, 2001)

B. Preparation of Total RNA, Poly(A)$^+$ RNA, and Double-Stranded cDNA

1. Preparation of Total RNA from Head Kidney and Other Tissues and Organs

Collect head kidneys from euthanized specimens of *N. coriiceps* and *C. aceratus* at Palmer Station, Antarctica. Either use organs immediately to prepare total RNA or freeze in liquid nitrogen and store at −80 °C. Isolate total RNA from fresh or frozen kidney by use of a modified acid guanidinium isothiocynate/phenol/chloroform extraction method (Puissant and Houdebine, 1990) and store in ethanol at −80 °C. (For analysis of the tissue specificity of expression of the RDA gene fragments, also prepare total RNA from blood cells, brain, gill, heart, liver, spleen, and trunk kidney.) Tissue samples are available in limited quantities from the Detrich laboratory.

2. Isolation of Poly(A)$^+$ RNA by Oligo–dT Cellulose Chromatography

Isolate poly(A)$^+$ RNA from head kidneys of the two fish species by affinity chromatography on oligo-dT cellulose (Kingston, 1993). After isolation, precipitate the poly(A)$^+$ RNA. Dissolve the poly(A)$^+$ RNA in DEPC-treated TE (pH 8.0) to give a final concentration of 1 μg/μl. Check the quality of the poly(A)$^+$ RNA by loading 1–5 μg on a 1.25% agarose gel containing 2.2 M formaldehyde as denaturant and 0.01% ethidium bromide as staining agent. High-quality poly(A)$^+$ RNA appears as a diffuse smear spanning 200 bp to 15 kb. From 1 mg of total RNA, one should obtain approximately 20–50 μg poly(A)$^+$ or 2–5% of the starting material.

3. Preparation of Double–Stranded cDNA

Using 10 μg of head kidney poly(A)$^+$ RNA from each fish species, prepare blunt-ended double-stranded cDNA by the method of Klickstein *et al.* (1995). Perform first-strand synthesis using an oligo-dT$_{15}$ primer (Pharmacia Biotech) and Superscript II reverse transcriptase. Following second-strand synthesis, resuspend the double-stranded cDNA to a final concentration of 0.5 μg/μl in sterile water.

C. Representational Difference Analysis

1. Generation of Representations, Driver, and Tester

Representational difference analysis is performed essentially as described by Hubank and Schatz (1994). To create representations of the head kidney cDNAs of the two fishes, digest the double-stranded cDNAs (2 μg in 10 μl 1× *Dpn*II buffer) from *N. coriiceps* (tester) and from *C. aceratus* (driver) with 5 U each of the enzyme *Dpn*II (a 4-bp cutter that recognizes GATC and creates a 4-bp 5' overhang) and isolate the resulting fragments by ethanol precipitation in the presence of carrier glycogen (−20 °C overnight) followed by centrifugation at 14,000 rpm (Brinkman microfuge) for 30–60 min at 4 °C. For each digest, resuspend the cDNA fragments in 10-μl aliquots of sterile TE. Add the two R oligonucleotides R-Bgl-12 (2 μg) and R-Bgl-24 (4 μg) to 12 μl of 10× T4 DNA ligase buffer and dilute to a final volume of 98 μl with sterile water. Combine each cDNA digest (10 μl) with 49 μl of R-oligonucleotide solution. Heat the solutions to 50 °C for 1 min in a PCR machine and then cool to 10 °C at 1°/min. Add 40 U (1 μl) of T$_4$ DNA ligase to each reaction and incubate overnight at 16 °C. Because the oligonucleotides are nonphosphorylated, the ligation will covalently couple only the 24-mer to the 5' overhang of the *Dpn*II site of the digested DNA; the 12-mer remains annealed to the 24-mer. Dilute each ligation reaction in TE (pH 8.0) to a final volume of 200 μl prior to PCR amplification.

For each fish species, prepare 30 PCR reactions as follows. To 145 μl of sterile water, add 4 μl of the ligation products, 40 μl of 5× PCR buffer, 8 μl of the 3 mM deoxynucleotide mix, and 2 μl of the R-Bgl-24 primer. To melt away the R-Bgl-12-mer, incubate the reactions at 72 °C for 3 min. Next, add 5 U Klentaq DNA polymerase (1 μl) to each tube and incubate at 72 °C for 5 min to fill in the ends. Amplify the cDNA substrates for 20 cycles, each consisting of denaturation at 95 °C for 1 min and primer annealing/substrate elongation at 72 °C for 3 min. After 20 cycles, program a 10-min extension reaction at 72 °C, and then cool the samples to 4 °C. Combine five individual PCR reactions and extract sequentially with phenol/chloroform/isoamyl alcohol and phenol/chloroform. Repeat in groups of five for the remaining 25 PCR reactions. Precipitate the representations by adding sodium acetate to 0.3 M [from a 10× stock (pH 5.2)] followed by three volumes of 100% ethanol. Centrifuge the tubes to pellet the DNA, resuspend the DNAs in TE, combine all samples, and adjust to a final concentration of 0.5 μg/μl.

To create the *C. aceratus* driver DNA population, digest 300 μg of its amplified cDNA representation with *Dpn*II (300 U) for 4 h, extract the DNA fragments with an equal volume of phenol/chloroform/isoamyl alcohol, and precipitate the DNA by adding sodium acetate to 0.3 M (pH 5.2) followed by three volumes of 100% ethanol. Centrifuge to pellet the DNA and resuspend the *C. aceratus* representation in TE to give a concentration of 0.5 μg/μl. To create the tester population (Fig. 1, Step 1A), digest 20 μg of the *N. coriiceps* cDNA representation with *Dpn*II, electrophorese the digest on a 1.2% agarose gel in 1× TAE to separate the amplicon from the R linkers, purify the fragments following the protocol for the Qiaex II gel extraction kit (Qiagen), and resuspend the DNA in 40 μl TE to a final concentration of 0.5 μg/μl. Take 4 μl of the purified tester DNA solution, add the J-Bgl-12 and J-Bgl-24 oligos (1 and 2 μg, respectively), and then anneal the oligos and ligate them to the tester DNA in a PCR machine under the same conditions used for the R-Bgl adapters. Dilute the tester population with TE to a final concentration of 10 ng/μl.

2. Subtractive Hybridization and Amplification: Difference Product 1

For the first subtractive hybridization, combine 40 μg of the digested driver DNA with 0.4 μg of J-ligated tester (100:1 driver:tester ratio), extract the pooled DNAs with phenol/chloroform/isoamyl alcohol, and collect the DNA by sodium acetate/ethanol precipitation and centrifugation. Resuspend the pellet containing both driver and tester DNAs in 4 μl of 3× EE buffer. Heat the driver/tester DNA solution to 37 °C for 5 min, then overlay the solution with mineral oil. Denature the DNA by incubation for 5 min at 98 °C in a PCR machine and cool to 67 °C. Next, add NaCl (1 μl of 5 M) to the solution (total volume now 5 μl) and incubate the samples at 67 °C for 20 h (Fig. 1, Step 1B) to allow single-stranded DNAs to anneal. [Three classes of hybrids will form, in order of abundance: (1) driver–driver hybrids that lack linkers and will therefore not be amplified because of the absence of primer binding sites; (2) driver–tester hybrids in which only the tester strand provides a primer binding site, which will result in linear amplification of these products; and (3) tester–tester hybrids (i.e., products reannealed from the unique cDNA strands of the tester that lack counterparts in the driver), which possess primer binding sites on both strands and will amplify exponentially.] Remove the mineral oil from the 5-μl sample, add 8 μl of 5 mg/ml yeast tRNA in TE followed by 25 μl of TE, mix vigorously by pipetting, and then dilute to a final volume of 400 μl by adding 362 μl TE.

Amplify the hybridized sample by PCR (Fig. 1, Step 1C), using the J-Bgl-24 primer set as follows. Prepare four reactions, each containing 20 μl of the hybridized sample, 40 μl of 5× PCR buffer, 8 μl of the deoxynucleotide stock, and 130 μl of sterile water. Heat the sample at 72 °C for 3 min to melt away the 12-mer, add 1 μl Klentaq (5U), heat again for 5 min at 72 °C to fill in the single-stranded ends, and then add 1 μl of J-Bgl-24 primer (2 μg). Perform PCR for 10 cycles, using the following cycling parameters: denaturation at 95 °C for 1 min and primer annealing/substrate elongation at 72 °C for 3 min. Combine the four reactions, extract the pooled DNA with phenol/chloroform/isoamyl alcohol, and precipitate the DNA by sodium acetate/ethanol precipitation and centrifugation. Resuspend the pellet in 40 μl of 0.2× TE. To remove single-stranded DNA (including linear amplified products and driver DNA), combine 20 μl of the PCR product, 20 U MBN (2 μl of 10 U/μl), 4 μl 10× MBN buffer, and 14 μl of sterile water; incubate at 30 °C for 35 min (Fig. 1, Step 1D). Stop the MBN reaction by adding 160 μl TE and heating the sample to 98 °C for 5 min. After stopping the MBN reaction, prepare four reactions, each containing 20 μl MBN-digested DNA, 2 μg J-Bgl-24 primer (1 μl of 2 mg/ml), 8 μl of the deoxynucleotide stock, and 170 μl of sterile water. Heat the four samples at 95 °C for 1 min, cool the samples to 80 °C, and add 1 μl Klentaq polymerase (5 U) to each. Perform PCR for 18 cycles, each consisting of denaturation at 95 °C for 1 min and primer annealing/substrate elongation at 70 °C for 3 min. Combine the PCR samples (800 μl total volume), extract with phenol/chloroform/isoamyl alcohol, and precipitate the DNA by adding sodium acetate to 0.3 M (80 μl of 3 M stock, pH 5.2) and 880 μl of isopropanol. Centrifuge to collect the DNA, and resuspend it to a concentration of ∼0.5 μg/μl in TE to yield the first difference product, termed DP1.

3. Difference Products 2 and 3

Difference products 2 and 3 (DP2 and DP3) are generated by minor modifications of the protocol for DP1. These include a change in adapters for DP2 (N-Bgl-12 and N-Bgl-24), the return to J linkers for DP3, and increases in the driver:tester ratios (800:1 for DP2 and 40,000:1 for DP3). Furthermore, PCR for DP3 is performed for 22 cycles. After phenol/chloroform/isoamyl alcohol extraction and sodium acetate/ethanol precipitation, the DP3 DNA is resuspended in TE to a final concentration of 0.5 μg/μl. The sizes of the DP3 RDA products should fall in the range of 200–500 bp.

D. Subcloning, Sequencing, and Preliminary Analysis of RDA Products

Digest the DP3 products with *Dpn*II, electrophorese the products on a 1.2% agarose gel in 1× TAE to separate the DP3 DNA from the J linkers, and purify the fragments following the protocol for the Qiaex II gel extraction kit (Qiagen).

Ligate the DP3 fragments into the *Bam*HI site of pBluescript KS(+) II plasmid, electroporate the plasmid pool into competent JM109 cells, select transformants on LB agar plates (100 ng/μl ampicillin, IPTG, and X-Gal), and identify potential recombinant plasmids by blue/white screening. Pick white colonies, grow them in 1× LB broth plus 100 μg/ml ampicillin, and isolate plasmid DNA by the boiling lysis protocol (Sambrook and Russell, 2001).

RDA subtractions often yield large numbers of DNA fragments corresponding to well-known, and thus dispensable, genes. For example, one should anticipate that many of the RDA products obtained from the *C. aceratus* driver/*N. coriiceps* tester subtraction would be α- and β-globin DNA fragments. Preliminary sequence analysis of the RDA population is likely to reveal other highly represented clones. To eliminate abundant clones from the RDA products, spot the clones individually (1 μl of clone DNA per 1 × 1 cm^2) onto a Magnagraph nylon membrane and allow the membrane to dry. Screen the membrane-bound clones by Southern hybridization (Southern, 1975) to cDNA probes for the dispensable genes (e.g., α- and β-globin cDNAs, others identified in first-pass sequencing) and eliminate positive RDA clones from further analysis.

Sequence the remaining RDA clones by use of an automated DNA sequencer and appropriate sequencing chemistry. To determine whether the RDA products correspond to known genes or, alternatively, might represent novel genes, submit the sequences for BLAST comparison to public nucleotide and protein databases, such as GenBank (www.ncbi.nlm.nih.gov), ExPASy (us.expasy.org; Appel *et al.*, 1994; Bairoch *et al.*, 2003), GENOMEnet (www.genome.ad.jp), and the Erythropoiesis Database (EpoDB, www.cbil.upenn.edu/EpoDB; Salas *et al.*, 1998; Stoeckert *et al.*, 1999).

E. Results

1. Proof of Principle: Recovery of Globin Contigs by cDNA RDA

A total of 316 RDA clones were obtained from the *C. aceratus/N. coriiceps* head kidney subtraction. Table I shows that clones corresponding to the α- and β-globin cDNAs (12 and 16.5%, respectively) were major components of the

Table I
Common RDA Clones

RDA clone	Protein product	Number	Percent total clones ($n = 316$)
1	β-Globin	52	16.5
48	α-Globin	38	12
2	NADH dehydrogenase subunit 4	33	10.5
5	NADH dehydrogenase subunit 5	8	2.5
65	Cytochrome c oxidase	6	1.9

RDA pool. Recovery of the globin contigs by the cDNA RDA strategy validates the utility of this method for isolating genes that are expressed differentially in the head kidney of the red-blooded species. Furthermore, their isolation is consistent with the disruption of the globin genes in the driver species, *C. aceratus* (Cocca *et al.*, 1995; Zhao *et al.*, 1998). Three mitochondrial genes, cytochrome c oxidase (1.9%), NADH dehydrogenase subunit 5 (2.5%), and NADH dehydrogenase subunit 4 (10.5%) were also abundantly represented.

2. Assembly and Bioinformatic Analysis of RDA Contigs

The sequences of the 316 clones were assembled to yield 45 distinct contigs by use of the Seqman (DNASTAR) alignment program. [We anticipate that comparable or higher contig yields could be obtained by repetition of the RDA protocol, with modifications (e.g., spiking the driver with unwanted DNAs, sampling alternative amplicons by use of other four-base-cutting endonucleases), using head kidney (this work) and spleen (as yet untested).] By BLAST analysis, we placed the 45 contigs into three categories based on similarity scores: (1) known genes, with an E score $< e^{-10}$; (2) uncertain genes, $e^{-10} < E \leq 0$; and (3) unknowns, $E > 0$ (Table II). Table III presents examples of the contigs (10, 22% of total) that fall in the category of known genes, and Table IV gives examples of the contigs (18, 40%) with uncertain orthologies. The remaining 38% of the contigs gave no matches to sequences in any of the public databases and were classified as unknowns. Some of the unknown contigs might correspond to the 3′-untranslated regions of cDNAs, which are often divergent between organisms and therefore would not be expected to recover matching genes. However, 10 of the 17 unknown contigs possessed open reading frames (ORFs) of >100 codons, which suggests strongly that at least some of them represent unknown, and potentially significant, erythropoietic genes.

III. Analysis of Unidentified RDA Contigs

The most interesting category of RDA contigs is clearly the group for which no protein-coding sequence matches could be found. Here we describe methods for characterizing the expression patterns of the unidentified RDA contigs by

Table II
Bioinformatic Parsing of RDA Contigs[a]

	Identified	Uncertain	Unidentified
n	10	18	17
Percent total contigs	22	40	38

[a]RDA products were assembled into 45 contigs, using the Seqman alignment program and then parsed into three categories based on similarity scores (E) to sequences in public databases: Identified, $E < e^{-10}$; uncertain, $e^{-10} \leq E \leq 0$; unidentified, $E > 0$.

Table III
Examples of Identified RDA Clones

Contig number	Protein product
1	β-Globin
8	Erythrocyte β-tubulin
12	α-Amylase
29	12-Lipoxygenase (platelet)
36	S-Adenylhomocysteinase
38	F-actin capping protein
48	α-Globin

Table IV
RDA Clones with Uncertain Orthologies

Contig number	Match?
3	H-Rev 107 protein homolog
4	HLA Class II histocompatibility antigen
7	Perioxisomal protein PMP47
13	DEZ receptor
20	Ying and Yang Factor-1
23	Moesin (Band 4.1 family)
27	Creb-binding protein
35	Proteosome component C13
45	MSP receptor (CD136 antigen)

Northern slot blot analysis and by *in situ* hybridization. The results are then used to guide the selection of genes for functional analysis.

A. Materials

1. Northern Slot Blot Analysis

- *Pst*I (20 U/μl; New England BioLabs)
- 10× NEB React buffers 2 and 3 (New England BioLabs)
- *Xba*I (20 U/μl; New England BioLabs)
- Low-melt agarose (Ameresco)
- Bio-Dot SF microfiltration apparatus (BioRad)
- 20× SSC: 175.3 g NaCl and 88.6 g sodium citrate per liter of water, pH 7.0
- 10% sodium dodecyl sulfate (SDS), molecular biology grade (Sigma)
- Spectra-Linker UV Crosslinker (Fisher)

- Formaldehyde (37%; Fisher)
- Formamide (Fisher)
- 50× Denhardt's solution: 1% w/v Ficoll 400, 1% w/v polyvinylpyrrolidone, 1% w/v bovine serum albumin (Sigma, Fraction V)
- 10× MOPS buffer: 0.2 M MOPS {3-[N-morpholino]propanesulfonic acid (Sigma)}, 20 mM sodium acetate, 10 mM EDTA in DEPC-treated water
- Hybridization buffer for Northern blots: 50% formamide (Sigma), 5×SSC, 5× Denhardt's solution, 0.2% SDS
- Quick Spin Column (Roche)
- PerfectHyb+ hybridization buffer (Sigma)
- Northern wash buffer: 0.2× SSC, 0.1% SDS
- Kodak X-OMAT X-ray film (or equivalent)

2. *In Situ* Hybridization

- Microscope slides, 25 × 75 × 1 mm
- *Sma*I (10 U/μl; New England Biolabs)
- 10× NEB React buffer 4 (New England BioLabs)
- 10× digoxigenin-UTP (DIG-UTP; Roche Biochemicals)
- T7 RNA polymerase (20 U/μl; Roche)
- T3 RNA polymerase (20 U/μl; Roche)
- 10× Transcription buffer (Roche)
- 0.5 M EDTA
- 4 M LiCl
- Microscope slides of blood smears and tissue prints
- *In situ* hybridization buffer (Hyb−): 50% formamide, 5× SSC, 0.1% Tween 20, 5 mg/ml yeast tRNA, 50 μg/ml heparin
- RNasin (Roche)
- Blocking reagent (Roche)
- Detection buffer: 100 mM Tris-HCl (pH 9.5), 150 mM NaCl, 50 mM $MgCl_2$; prepare on day of use
- Anti-DIG-alkaline phosphatase conjugate, Fab fragments (from sheep; Roche)
- BCIP: 5-bromo-4-chloro-3-indolyl-phosphate solution (Roche)
- NBT: 4-nitroblue tetrazolium chloride solution (Roche)
- Levamisole (Sigma)
- 1% Methylene green in water (w/v; Fisher)

B. Northern Slot Blots

To determine the tissue specificity of expression of the unknown RDA contigs, hybridize them to slot blots of total RNA from multiple tissues, both hematopoietic and nonhematopoietic.

1. Preparation of Radiolabeled RDA Probes

To create an RDA DNA probe, digest a clone with the restriction enzymes *Pst*I and *Xba*I, whose sites flank the *Bam*HI cloning site. Incubate the reaction (5 μg recombinant plasmid, 5 μl 10× appropriate restriction buffer, 10 U enzyme, sterile water to 50 μl) at 37 °C for 2 h. Electrophorese the digested sample on a 1% low-melt agarose gel containing 1× TAE, and excise the insert from the gel. Transfer the gel slice to a 1.5-ml microfuge tube, and add 200-μl sterile water. Melt the agarose and denature the DNA fragment at 100 °C for 5–10 min. Prepare the DNA fragment for labeling by combining 33 μl (~150 ng) of the low-melt preparation, 5 μl of [^{32}P]dCTP, and NEBlot labeling components according to the manufacturer's directions. Incubate the reaction at 37 °C for 1 h. Purify the labeled probe by centrifugation on a Quick Spin Column (Roche). Denature the labeled probe for 10 min at 100 °C, and place it on ice until used.

2. Northern Slot Blot Analysis

To prepare the slot blots, denature 5-μg aliquots of total RNA from tissues of *N. coriiceps* and *C. aceratus* (brain, head kidney, heart, liver, spleen, trunk kidney, gill, peripheral blood, etc.) in 50% formamide, 2 M formaldehyde, 1× MOPS at 65 °C for 15 min, and then place the samples immediately on ice. Wet a piece of Magnagraph nylon membrane in 10× SSC and apply it to the Bio-Dot SF microfiltration apparatus (BioRad). Apply a vacuum to the slot blot apparatus. Rinse the slots twice with 200 μl of 10× SSC (room temperature), and then remove the vacuum. Load the denatured RNA samples into the appropriate slots and reapply the vacuum to transfer the sample onto the membrane. Wash the wells twice with 200 μl of 10× SSC. Break the vacuum, remove the membrane and allow it to dry, and then covalently attach the RNA to the membrane by UV-cross-linking (Spectra-Linker, 45 s). Store blots between two pieces of Whatman 3MM paper in sealed bags in the dark.

Prehybridize blots in heat-sealable bags with 10–15 ml Northern blot hybridization buffer or PerfectHyb hybridization buffer at 65 °C for ≥2 h. Add the radioactive probe to the bag and incubate overnight at 65 °C with gentle shaking. Wash the blot twice at high stringency (0.2× SSC, 0.1% SDS, 65 °C) for 15 min. Expose the blot to Kodak X-OMAT X-ray film (or equivalent) with an amplifying screen at −80 °C. After an appropriate interval, develop the film by using an X-ray film processor. Membranes can be reused a maximum of three times after stripping the probe by washing in 0.1% SDS, 0.1× SSC at 100 °C for 10 min.

C. *In Situ* Hybridization to Tissue Prints and Smears

The cellular specificity of the RDA contigs is most easily determined by *in situ* hybridization of cRNA probes to tissue prints or blood smears prepared on microscope slides. Antisense cRNAs would be expected to bind to their cellular mRNAs, whereas sense cRNAs should serve as nonhybridizing controls. For RDA contigs with large ORFs, determination of sense and antisense strands is straightforward.

1. Preparation of Sense and Antisense Probes

Digest RDA clones with either *Xba*I or *Sma*I to prepare linear templates for labeling with DIG-UTP by use of T7 or T3 RNA polymerases as described previously (Chen *et al.*, 2002). Prepare *in vitro* transcription reactions by combining 1 μg of linearized plasmid DNA, 2 μl 10× transcription buffer, 2 μl 10× DIG nucleotide mix, 0.5 μl (20 U) RNasin, 1 μl desired RNA polymerase (T7 or T3; 40 U/μl), and sterile water to produce a final volume of 20 μl. Incubate the preparations at 37 °C for 1 h in an open-air incubator to minimize evaporation. Stop the reactions by adding 0.5M EDTA (0.8 μl). Precipitate the labeled cRNA by adding 2.5 μl LiCl (4 M) and 75 μl 100% ethanol; incubate at −70 °C overnight. Centrifuge the probes in a microfuge at maximum speed for 15 min at 4 °C. Resuspend the probes in 100 μl of DEPC-treated water (final concentration ∼100 ng/μl) and store in 10-μl aliquots at −70 °C until use.

2. *In Situ* Hybridization

Prepare tissue prints of head kidney or spleen from *C. aceratus* and from *N. coriiceps* by touching a portion of fresh tissue to multiple spots on a glass microscope slide; this deposits a monolayer of cells at each location. For blood smears, place a drop of blood at one end of a slide, take a second slide and touch its end perpendicularly to the long axis of the first at an oblique angle and then smoothly spread the drop along the first slide. [Because the blood of the icefish is quite dilute, concentrate the cells ∼10-fold by centrifugation in a clinical centrifuge and resuspension in a small volume of ice-cold Notothenioid Ringer's solution (260 mM NaCl, 5 mM KCl, 2.5 mM MgCl$_2$, 2.5 mM CaCl$_2$, 2 mM NaHCO$_3$, 2 mM NaH$_2$PO$_4$, 5 mM glucose) prior to deposition on the slide.] Immediately fix the cells by placing the slides in a Coplin jar containing 100% MeOH (5 min), and then store the slides at room temperature.

Place the slides, specimen face up and elevated on two sterile swabs, in a humidified Tupperware dish (moist paper towel in a corner of the dish), cover the tissue prints or blood smears with 250 μl of Hyb− solution, and incubate at 37 °C for 1 h. Remove the Hyb− buffer, taking care to drain the slide thoroughly. Add ∼200 ng DIG-labeled cRNA per ml of Hyb− buffer (now termed Hyb+), and apply 250-μl Hyb+ to each slide. Incubate overnight at 37 °C in the humidified chamber. Remove the Hyb+ probe and discard. Wash the slides (1) twice with 250 μl of 2× SSC at 37 °C (5 min per wash), (2) thrice with 250 μl of 60% formamide/0.2× SSC at 37 °C (5 min per wash), and (3) twice with 250 μl of

2× SSC at room temperature (5 min per wash), removing as much buffer as possible after each wash. Equilibrate the slides with 250 μl of 100 mM Tris-HCl (pH 7.5) plus 150 mM NaCl for 5 min at room temperature. Overlay the microscope slides with 250 μl of blocking reagent [prepared to saturation in 100 mM Tris-HCl (pH 7.5), 150 mM NaCl], and incubate at room temperature for 30 min. Meanwhile, dilute the alkaline-phosphatase-conjugated sheep anti-DIG Fab fragments 1:200 in blocking reagent. Remove the blocking reagent from the slides and replace with the diluted antibody solution; incubate for 2 h at room temperature. Remove the antibody solution, and wash the slides twice in 100 mM Tris-HCl (pH 7.5), 150 mM NaCl in Coplin jars (5 min/wash) and then once in detection buffer (10 min). Place the slides in the color reagent solution [detection buffer containing BCIP (0.18 mg/ml), NBT (0.34 mg/ml), and levamisole (0.24 mg/ml)] in a foil-wrapped Coplin jar. Incubate the slides in the color reagent for 4–18 h, and monitor color development (deposition of brownish black pigment) at regular intervals. Stop the color reaction by washing the slides in TE for 5 min. Rinse the slides with distilled water and counterstain the specimens with 1% methylene green for 10 min. Examine the slides under a high-quality brightfield microscope (e.g., Nikon Eclipse E800) and record micrographs either on color film or by use of a digital camera system (e.g., the SPOT32 from Diagnostic Instruments, Inc.).

D. Results

1. Proof of Methodologies: Differential Expression of the α-Globin mRNA by Hematopoietic Tissues of *C. aceratus* and *N. coriiceps*

To validate the methodologies for characterizing the unknown RDA contigs, we performed pilot studies with the α-globin gene (contig 48). Figure 2 shows that the α-globin message was abundant in the hematopoietic tissues (blood, head kidney, and spleen) of *N. coriiceps*; a positive signal was also found in the heart, most likely due to carry-over of blood in this organ. In contrast, α-globin mRNA was not detectable in the tissues of the icefish, as expected (Cocca *et al.*, 1995; Zhao *et al.*, 1998). Figure 3 demonstrates that normoblasts and erythrocytes in the peripheral blood of *N. coriiceps* were positive for the α-globin mRNA (brownish-black cytoplasmic signal) when hybridized to the antisense (experimental) probe (Fig. 3A), whereas the sense (control) probe gave no discernable signal (Fig. 3B). Neither the experimental (antisense) nor the control (sense) probes bound to the blood cells of the icefish (Fig. 3C and D, respectively). Thus, both the Northern slot blot and the *in situ* hybridization assays appear to be robust, specific, and suitable for analysis of the unknown RDA contigs.

2. Tissue Specificity of Expression of Unidentified RDA Contigs

Putative erythroid genes isolated by RDA applied to the head kidneys of *C. aceratus* and *N. coriiceps* might, nonetheless, be expressed in other tissues of either species. Therefore, we examined the tissue specificity of expression of

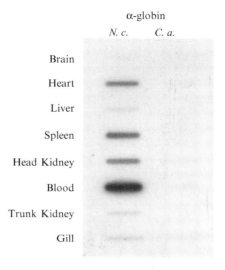

Fig. 2 Proof of principle for cDNA representational difference analysis (RDA): Northern slot blot analysis of α-globin gene expression by tissues of *C. aceratus* and of *N. coriiceps*. Total RNAs from the brain, heart, liver, spleen, head kidney, peripheral blood, trunk kidney, and gill tissues (5 μg per tissue) of *N. coriiceps* (*N.c.*) and *C. aceratus* (*C.a.*) were applied to a nylon membrane by using a BioRad Slot Blot vacuum manifold, and the membrane was incubated with a radiolabeled *N. coriiceps* α-globin cDNA probe.

each RDA contig by hybridization to Northern slot blots of total RNAs from multiple tissues of each species.

Figure 4 shows a Northern slot blot panel probed with five unknown RDA products. Four (23, 34, 197, 263) of the five products were expressed preferentially by the head kidney of *N. coriiceps*, with three (34, 197, 263) being essentially undetectable in *C. aceratus* kidney RNA. These are clear-cut examples of the difference products that our protocol was designed to recover. Furthermore, the four kidney-specific RDA fragments were also expressed in the spleen and peripheral blood of *N. coriiceps*, which further supports their potential involvement in erythropoiesis. Other RDA products (15, 23) show more complex expressions patterns. For example, RDA 15 RNA was present in nearly all tissues of both fishes, with the exception of peripheral blood, wherein the transcript was present only in *N. coriiceps*. Two explanations can account for the apparent global expression pattern of RDA 15: (1) the RDA probe detects related transcripts produced by a family of paralogous genes, one of which might be erythroid specific; or (2) this RDA product might not have been removed efficiently by the head kidney RNA subtraction.

Having established the Northern slot blot technology as an efficient screening tool, we proceeded to analyze representative RDA products corresponding to each of the 45 contigs. Based on this first-pass analysis, we selected a subset of

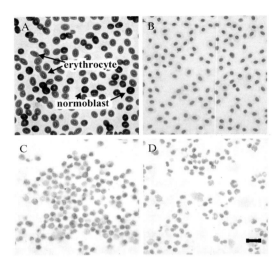

Fig. 3 *In situ* hybridization of α-globin cRNA to blood cells from *N. coriiceps* and *C. aceratus*. Blood smears from *N. coriiceps* (A, B) and *C. aceratus* (C, D) were hybridized either to antisense (A, C) or sense (B, D) DIG-labeled α-globin RDA transcripts. Following application of alkaline-phosphatase-conjugated anti-DIG Fab fragments, the bound RNAs were detected as a brownish-black reaction product resulting from enzymatic action on the color reagents NBT and BCIP. After stopping the reaction, cells were counterstained with methylene green. Bar 25 μm. (See Color Insert.)

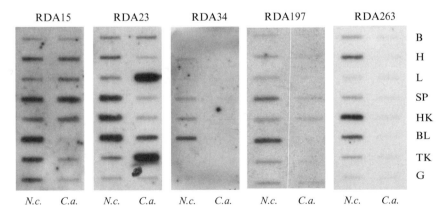

Fig. 4 Northern slot blot analysis of the expression of several RDA products. Total RNAs (5 μg) from either *N. coriiceps* (*N.c.*) or *C. aceratus* (*C.a.*) tissues were blotted to nylon membranes and hybridized to the RDA products indicated. Tissues: brain (B), heart, (H), liver (L), spleen (SP), head kidney (HK), peripheral blood (BL), trunk (excretory) kidney (TK), and gill (G).

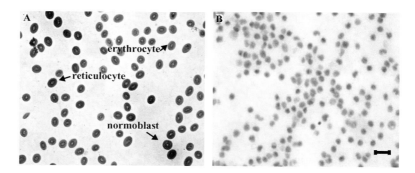

Fig. 5 *In situ* hybridization of RDA 197 antisense cRNA to blood cells. (A) *N. coriiceps.* (B) *C. aceratus.* Sense controls did not produce a signal. Bar 25 μm. (See Color Insert.)

the contigs for subsequent analysis by *in situ* hybridization. Next, we present one example that shows a typical result for the RDA subtraction products. We performed these studies on blood smears because the blood of fishes is an important site of erythropoiesis (Rowley *et al.*, 1988) and contains blast stages, reticulocytes, and mature erythrocytes. The blood of the Antarctic rockcod *N. coriiceps* is especially rich in immature red cell stages (H.W. Detrich, III, unpublished observations).

3. Cellular Specificity of Expression of One Unidentified RDA Contig

Figure 5 presents the hybridization of RDA 197 cRNA to blood smears of the red-blooded rockcod and the white-blooded icefish. The 197 difference product represents a gene whose expression was clearly specific to the red-blooded fish (Fig. 4), with its mRNA found primarily in the blood, spleen, and head kidney. There appears to be a gradient of staining intensity, which decreases with progressive maturity of the cells from the normoblast to reticulocyte to mature erythrocyte (Fig. 5A). Little, if any, staining was observed in the white cells of *C. aceratus* (Fig. 5B), and sense controls were negative for both fishes. These results suggest that the RDA 197 gene might function in terminal differentiation of blast stages to the mature erythrocyte.

IV. Isolation of Full-Length cDNAs Corresponding to RDA Products

The ultimate goals of our RDA subtraction are the isolation of novel erythroid genes from *N. coriiceps* (this section) and the functional analysis of their zebrafish orthologs (Section V). Thus, we screened an *N. coriiceps* spleen cDNA library by hybridization to the 17 RDA fragments whose genes could not be immediately identified by bioinformatic methods. Following sequence analysis of the cDNAs, we were able to focus on the subset of truly new genes, to scan the novel genes for protein motifs and domains, and to eliminate other putative unknowns that had been described previously.

A. Materials

- *N. coriiceps* spleen cDNA library in lambda gt10; inserts ligated into *Eco*RI site
- C_{600} *E. coli* bacteria
- 1 M $MgSO_4$
- NZY broth: 10 g N-Z-Amine A (Sigma), 5 g NaCl, 5 g yeast extract (Difco), 2 g $MgSO_4 \cdot 7H_2O$ per liter of water
- NZY agar plates: NZY broth + 15 g/l agar (Difco) in 100-mm plates
- NZY top agarose (0.7%): NZY broth + 7 g/l agarose (Amresco)
- Nylon membranes, 100 mm circular (Osmonics, Inc.)
- Denaturation solution (1.5 M NaCl, 0.5 M NaOH)
- Renaturation solution [1.5 M NaCl, 0.5 M Tris-HCl (pH 8.0)]
- Hybridization solution (6× SSC, 0.5% SDS, 50 μg/ml heparin, 0.1% sodium diphosphate)
- Wash buffer (3× SSC, 0.5% SDS, 1 mM EDTA)
- SM buffer: 5.8 g NaCl, 2 g $MgSO_4 \cdot 7H_2O$, 50 ml 1 M Tris-HCl (pH 7.5), 5 ml 2% (w/v) gelatin solution, water to 1 liter
- Lambda Mini Phage Kit (Qiagen)
- *Eco*RI (20 U/μl, New England BioLabs)
- 10× *Eco*RI restriction endonuclease buffer (New England BioLabs)
- Other standard materials (Sambrook and Russell, 2001)

B. Isolation of Full-Length cDNA Clones

1. Screening an *N. coriiceps* spleen cDNA library

Plate the bacteriophage library and transfer the plaques to nylon membranes as described by Quertermous (1996). To 10-ml LB broth, add a single colony of C_{600} bacteria from a freshly streaked LB plate and incubate for approximately 4 h at 37 °C with shaking (225 rpm). Collect the bacteria by centrifugation at 2000 rpm in a clinical centrifuge for 10 min, and resuspend the pellet in 10 mM $MgSO_4$ to a final OD_{600} of 0.5. Infect 200 μl of the C_{600} cell preparation with 2×10^4 recombinant bacteriophage. Incubate at 37 °C for 15 min without shaking. Add 3 ml of NZY top agarose (warmed to 50 °C) to the infected bacteria, mix, and pour the solution onto prewarmed 100-mm NZY agar plates. Allow the top agarose to solidify for 15 min at room temperature. Invert the plates and incubate them overnight at 37 °C.

After overnight growth, remove the plates from the incubator and cool them to 4 °C for a minimum of 4 h. Place a nylon membrane on each NZY plate and mark the position of the membrane asymmetrically, using India ink and a 22-gauge needle. After 2 min, remove the membrane with forceps. If necessary, prepare duplicate plaque lifts by placing a second membrane on each plate for 5–7 min. Allow the membranes to dry briefly. Float the membranes (plaque side up) in denaturation solution for 2 min. Remove the membranes with forceps and transfer to renaturation solution (plaque side up) for 5 min. Remove and place in $2 \times$ SSC for 30 s. Place the membranes on Whatman 3 MM paper to dry. Cross-link the plaque DNA to the membrane by UV-crosslinking (Spectra-Linker, 45 s).

Prehybridize the filters (no more than 15 filters per heat-sealable bag) in 50 ml of hybridization buffer at 60 °C for 1 h. Remove the prehybridization buffer, replace with fresh hybridization buffer, and add the radiolabeled probe (prepared as described in Section III-B-1). Incubate the filters and probe at 60 °C overnight with gentle agitation. Wash the filters twice with 250 ml $3\times$ SSC, 0.5% SDS, 1 mM EDTA at 60 °C, 15 min per wash. Monitor the radioactivity of the filters by a Geiger counter during the wash stages to ensure that plaque-bound probe is not entirely removed. Expose the filters to Kodak X-Ray film overnight. Position each plate over its autoradiograph and remove ('plug") positive plaques with a sterile pipet tip. Transfer the plug to a 1.5-ml microfuge tube containing 500 μl of SM buffer and 20 μl of chloroform. Invert tubes to mix the contents and store them at 4 °C overnight to allow the phage to elute from the agarose. To perform secondary and tertiary screens, dilute the eluted phage particles in SM buffer (1:100 or 1:1000, depending on the titer of the library) and apply to NZY agar plates as for the primary screen. To ensure that individual recombinant clones are obtained, screen each primary isolate through to the tertiary level.

2. Purification of Plaque DNA and DNA Sequence Analysis

Purify the DNA of recombinant phage clone(s) containing the cDNA of interest. Mix 100–200 μl of a tertiary isolate in SM with 200 μl of freshly grown C_{600} bacteria (OD_{600} 0.5) and incubate at 37 °C for 30 min. Add 10 ml of sterile LB broth and shake the solution overnight at 37 °C at 225 rpm. Remove the bacterial debris by centrifugation and isolate the phage DNA from the supernatant by use of a Qiagen Miniprep Phage kit. Sequence the cDNA inserts directly in the phage DNA by use of phage-specific primers and an automated DNA sequencer with appropriate sequencing chemistry.

To excise the insert cDNA, digest recombinant clones by combining 5 μl phage DNA with 40 U (2 μl) EcoRI, 2 μl 10$\times$ EcoRI restriction enzyme buffer, and 11 μl sterile water in a microfuge tube and incubating at 37 °C for 4 h. Electrophorese the samples on a 1% agarose gel in 1$\times$ TAE plus ethidium bromide and detect the insert bands by placing the gel on a UV transilluminator. For subsequent analysis, *Eco*RI inserts can be subcloned into pBluescript KS+ vector.

C. Typical Results

Using RDA probes from the unknown category, we were able to recover full-length cDNAs clones from the *N. coriiceps* spleen cDNA library. [The spleen library was chosen instead of a head kidney library because most of the RDA products are expressed by this tissue and because this library yields a larger proportion of full-length clones.] Some of these clones could be readily identified, whereas others were recognized only as putative genes in the human genome. Several examples are described next.

1. Clones Encoding Major Histocompatibility Complex (MHC) Antigens

Two cDNAs that corresponded to unknown RDA fragments, cDNAs 15 and 197, were found to encode major histocompatibility complex (MHC) molecules of classes I and II, respectively. Their RDA fragments revealed comparable patterns of expression on multitissue Northern slot blots (cf. RDA 15, Fig. 4) and also recognized mRNAs in maturing red cells (cf. RDA 197; Fig. 5). Although the erythrocytes of higher vertebrates express MHC class I molecules at very low levels (Gabbianelli *et al.*, 1990; Spack and Edidin, 1986), class I antigen production by the red cells of lower vertebrates, such as the trout (Sarder *et al.*, 2003), *Xenopus* (Flajnik *et al.*, 1984), and the chicken (Sgonc *et al.*, 1987), is readily detectable. Furthermore, Sarder *et al.* (2003) have shown that the MHC class I antigen is the major determinant of rejection of *in vivo* grafted trout erythrocytes. Hence, the recovery of the MHC class I cDNA by our RDA subtraction protocol is not surprising. By contrast, MHC class II expression by late-stage erythroid cells has not been observed previously. In higher vertebrates, MHC class II mRNA is transcribed by proerythroblasts and subsequently down-regulated (Falkenburg *et al.*, 1984; Gabbianelli *et al.*, 1990; Greaves *et al.*, 1985; Sieff *et al.*, 1982; Sparrow and Williams, 1986). Thus, the presence of MHC class II antigen mRNA in maturing erythrocytes of *N. coriiceps* (Fig. 5) raises the possibility of this antigen playing a role in regulating the differentiation of nonlymphoid hematopoietic lineages in some fish species.

2. Erythroid RhoGDI

cDNA 295 encoded a protein of 205 amino acids, which on BLAST analysis showed strong similarity to RhoGDI, the Rho GDP dissociation inhibitor (Sasaki and Takai, 1998). Three members of the Rho GDI family have been isolated. Rho GDIα is expressed ubiquitously, whereas Rho GDIβ (also known as Ly-GDI) is expressed exclusively in hematopoietic cells (Gorvel *et al.*, 1998; Groysman *et al.*, 2000; Scherle *et al.*, 1993) and Rho GDIγ is expressed in the brain and pancreas (Adra *et al.*, 1997). We propose that cDNA 295 encodes a Rho GDIβ ortholog, but further research will be required to confirm or reject this hypothesis. Rho GDIβ (Ly-GDI) might interact with proteins of the ezrin/radixin/ moesin (ERM) family, thereby initiating activation of Rho subfamily members to regulate actin filament organization and/or to regulate hematopoietic signaling pathways (Bretscher, 1999; Sasaki and Takai, 1998; Takahashi *et al.*, 1997). Recovery of the potential Ly-GDI cDNA and identification of ERM RDA products among the known RDA products (data not shown) suggest that our subtractive screen might lead to identification of novel signaling molecules and pathways.

3. A BTB/POZ Transcription Factor?

RDA/cDNA 213 was expressed ubiquitously in tissues of *N. coriiceps* but was absent from *C. aceratus* (data not shown). The cDNA encoded a predicted ORF of 220 amino acids that showed strong similarity to KIAA1317, a predicted protein annotated in the human genome. The 213 protein was found to contain the Broad-complex/Tram-track/Bric-a-Brac (BTB, also known as POZ) domain near its amino terminus. The BTB/POZ domain, first identified in *Drosophila* (Zollman *et al.*, 1994), is usually found in transcription factors together with Krüppel-like zinc finger domains. Although the 213 protein lacked identifiable Krüppel-like zinc fingers, it is possible that the protein is a bona fide transcription factor or, potentially, a regulator of other transcription factors.

4. True Unknowns

There are at least five genes for which we have no indication of functional identity (cDNAs 34, 222, 263, 269, and 276). The success of our RDA subtraction in identifying already known erythropoietic genes suggests that these unknowns might encode novel erythroid proteins that await discovery and analysis.

D. *N. coriiceps bloodthirsty*, a Novel Gene Related to RDA 23

The original *N. coriiceps* RDA 23 fragment was a large chimera (1 kb) composed of two gene fragments joined by a *Dpn*I site (all RDA products possess *Dpn*I overhangs). The first encoded an RNA-binding protein designated CIRP (cold-inducible ribosomal protein; also recovered as RDA 147; Nishiyama *et al.*, 1997) and the second encoded an ~100 amino acid fragment of a 170-residue protein domain termed B30.2 (Vernet *et al.*, 1993). We generated a PCR probe corresponding to the latter fragment for use in the isolation of B30.2-encoding cDNAs from the *N. coriiceps* spleen cDNA library.

The B30.2 domain is present in a wide variety of proteins whose functions range from transcription factors to signaling molecules (Henry *et al.*, 1997, 1998). This protein–protein interaction domain (Henry *et al.*, 1998; Seto *et al.*, 1999) is also the first to be found in proteins that occupy all cellular and extracellular compartments. Thus, it was not surprising that two cDNAs, the first containing the B30.2 probe sequence and the second a variant thereof, were recovered from the *N. coriiceps* cDNA library. [We describe the second clone here because our understanding of it is most complete.] The variant cDNA encoded a protein of the RING, B-box, coiled-coil family [RBCC, reviewed by Borden (1998); also known as the tripartite motif (TRIM) (Reymond *et al.*, 2001)] that terminated in a B30.2 domain. *In situ* hybridization of antisense cRNA to head kidney and spleen prints of *N. coriiceps* and *C. aceratus* demonstrated that this gene was preferentially expressed in proerythroblasts of the red-blooded fish. The paucity of expression by *C. aceratus* led us to name the gene *bloodthirsty* (gene *bty*, protein Bty), in recognition of the erythrocyte-null condition of the icefish family.

V. Model Hopping: Functional Analysis of the Zebrafish Ortholog of the Novel Antarctic Fish Gene *bloodthirsty*

The proof of our protocol lies in the ability to recover genes that play a demonstrable role in erythropoiesis. For reasons described previously (see introduction), we pursued the functional analysis of the *bloodthirsty* gene in the zebrafish model. First, we cloned the zebrafish ortholog of *bty*. Second, we used a reverse-genetic strategy employing antisense, morpholino-modified oligonucleotides (MOs; Nasevicius and Ekker, 2000) to determine the function of *bty* in developing zebrafish embryos. Complete details are presented elsewhere (Yergeau, Zhou, and Detrich, III, submitted).

A. Materials

See Section IV.

- Zebrafish head kidney oligo-dT-primed cDNA library in Lambda ZAP Express (Thompson *et al.*, 1998)
- *N. coriiceps bloodthirsty* cDNA
- Zebrafish embryos from mating of a wild-type strain
- Fluorescein isothiocyanate-(FITC-) tagged antisense MO Zebb302 (Gene Tools, LLC); 5′–CAGTGGATTACTGGAGGAGGACAT–3′
- FITC-tagged control MO 5′–CAGTGAATCACTGGAAGAAGACAT–3′, a 4-bp mismatch (base changes underlined) version of the experimental MO
- PLI-100 Picoinjector (Medical Systems Corporation) or equivalent
- Narishige Micromanipulator or equivalent
- Nikon SMZ-U dissecting microscope (or equivalent) equipped for epifluorescence
- Cooled CCD digital camera for dissecting microscope

B. Isolation of Zebrafish *bloodthirsty*

The isolation of the zebrafish ortholog of *N. coriiceps bloodthirsty* followed the procedures outlined in Sections IV-B-1 and IV-B-2. The *N. coriiceps bty* cDNA was used as the probe. The vector Lambda ZAP Express permits the recovery of the *bty* cDNA by *in vivo* excision, using helper phage to generate subclones in the plasmid pBK-CMV.

C. Functional Analysis of Zebrafish *bloodthirsty*

The function of *bty* during development was assessed by injecting embryos (one- to four-cell stage) with antisense MOs targeted to two contiguous sites at the 5′-end of the *bty* mRNA. Here we present the results for one antisense MO, Zebb302, and its 4-bp mismatch control MO, Zebb302b. Microinjection is performed as described by Westerfield (2000).

Figure 6 shows a 32-h wild-type zebrafish embryo (A) and age-matched embryos that were injected with the antisense MO or its mismatch control (B and C, respectively). Embryos injected with Zebb302 (10–15 ng) showed a slight delay in development without gross physical abnormalities when compared to embryos that received the control MOs or were not injected. Treatment of embryos with the antisense MO caused suppression of the production of hemoglobin and almost

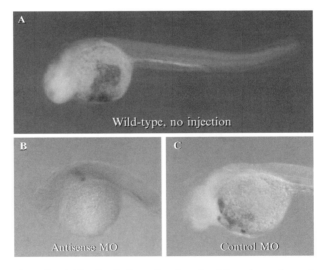

Fig. 6 Suppression of red blood cell formation in zebrafish embryos by antisense morpholino oligonucleotides targeted to the *bty* mRNA. (A–C) Hemoglobin detection by *o*-dianisidine. (A) Uninjected wild-type embryo. The circulation stains reddish brown when reacted with *o*-dianisidine, indicating the presence of hemoglobin-expressing red cells. (B) Antisense MO Zebb302. Note the nearly complete absence of red blood cells. (C) Control MO Zebb302b. Red blood cells were present at near wild-type levels. Embryos were age matched (32 hpf) and micrographed in 70% glycerol/PBS, using a Nikon dissecting microscope with digital imaging system. (See Color Insert.)

complete failure to produce erythrocytes in the circulation (Fig. 6B). There was no evidence of hemoglobin-positive erythrocytes pooling elsewhere in the experimental embryos (data not shown). By contrast, control-MO-injected embryos (Panel C) expressed near normal levels of hemoglobin and red cells were abundant in the circulation (compare Panels A and C). Thus, disruption of Bty translation compromises the differentiation of erythroid progenitors into primitive-lineage erythrocytes. Companion studies showed that Bty knockdown did not affect the synthesis of mRNA markers for early erythroid differentiation (*gata1*), myelopoiesis (*pu1*), and vasculogenesis (*flk1*), whereas production of the late-differentiation marker, α-globin, was markedly reduced (data not shown). Taken together, the most plausible interpretation of these results is that Bty is required, at a point yet to be determined, to progress through the late stages of erythroid differentiation.

VI. General Considerations

The methods described here exploit the unique erythrocyte-null phenotype of the Antarctic icefishes to discover new candidate erythroid genes for functional studies in the zebrafish and other higher vertebrates. They are robust and have led to the isolation of several unknown genes whose functions in erythropoiesis remain to be determined. Furthermore, the methods can be extended to other tissues of Antarctic fishes, such as the spleen, can be modified to eliminate recovery of common but unwanted products, and can be adapted to sample alternative amplicons.

In addition to its role in blood cell formation, the spleen carries out clearing functions in the erythroid system, including culling (destruction of erythrocytes undergoing senescence or damaged by pathological conditions), pitting (removal of inclusions from within erythrocytes with release of the cell back to the circulation), and polishing (removal of excess membrane and pocks or pits) (Shurin, 1995). Thus, the application of cDNA RDA to the spleens of *N. coriiceps* and *C. aceratus* could reveal new genes involved in not only erythrocyte formation but also erythrocyte senescence and clearance, processes that have received relatively little attention.

The results shown in Table I make clear that genes abundantly expressed by the tester species alone (e.g., the globins), as well as other products of little interest (e.g., the three mitochondrial clones), will be recovered in abundance in the RDA products. To eliminate these expected and irrelevant differences, which might obscure rarer but more interesting products, one can spike the *C. aceratus* driver with the unwanted cDNAs. Alternatively, one can employ iterative cDNA RDA (Hubank and Schatz, 1999), in which successive rounds of RDA are spiked with previously cloned differences and unwanted products so that rare products concealed by more readily amplified clones can be recovered. Use of these competitive strategies should enhance the rate of discovery of the desired new, unknown erythropoietic cDNAs.

The representations generated by cDNA RDA, in which a 4-bp cutting restriction endonuclease is used, should preserve most of the sequence complexity of the driver and tester transcript populations. Nevertheless, it is possible that some of the cDNAs of the tester that represent true differences might not be cut by the chosen restriction enzyme (*Dpn*II in this case) and thus will not generate amplifiable difference products. To recover these "missing" genes, one could perform cDNA RDA with other 4-bp cutting enzymes that cleave at different recognition sites to leave 4-bp overhangs. Two possibilities are *Tsp*509 I (recognizes AATT, leaves a 4-bp 5′ overhang) and *Nla*III (site CATG, leaves 4-bp 3′ overhang) (Hubank and Schatz, 1999).

Acknowledgments

This work was supported by NSF grants OPP-9815381 and OPP-0089451 (H.W.D.).

References

Adra, C. N., Manor, D., Ko, J. L., Zhu, S., Horiuchi, T., Van Aelst, L., Cerione, R. A., and Lim, B. (1997). RhoGDIgamma: a GDP-dissociation inhibitor for Rho proteins with preferential expression in brain and pancreas. *Proc. Natl. Acad. Sci. USA* **94**, 4279–4284.

Appel, R. D., Bairoch, A., and Hochstrasser, D. F. (1994). A new generation of information retrieval tools for biologists: The example of the ExPASy WWW server. *Trends Biochem. Sci.* **19**, 258–260.

Bairoch, A., Gasteiger, E., Gattiker, A., Hoogland, C., Lachaize, C., Mostaguir, K., Ivanyi, I., and Appel, R. D. (2003). The ExPASy proteome WWW server in 2003. pp. 1–6. Available online at us.expasy.org.

Borden, K. L. B. (1998). RING fingers and B-boxes: Zinc-binding protein-protein interaction domains. *Biochem. Cell Biol.* **76**, 351–358.

Braun, B. S., Frieden, R., Lessnick, S. L., May, W. A., and Denny, C. T. (1995). Identification of target genes for the Ewing's sarcoma EWS/FLI fusion protein by representational difference analysis. *Mol. Cell. Biol.* **15**, 4623–4630.

Bretscher, A. (1999). Regulation of cortical structure by the ezrin-radixin-moesin protein family. *Curr. Opin. Cell Biol.* **11**, 109–116.

Cantor, A. B., and Orkin, S. H. (2002). Transcriptional regulation of erythropoiesis: An affair involving multiple partners. *Oncogene* **21**, 3368–3376.

Chen, M. C., Zhou, Y., and Detrich, H. W., III (2002). Zebrafish mitotic kinesin-like protein 1 (Mklp1) functions in embryonic cytokinesis. *Physiol. Genom.* **8**, 51–66.

Cocca, E., Ratnayake-Lecamwasam, M., Parker, S. K., Camardella, L., Ciaramella, M., di Prisco, G., and Detrich, H. W., III (1995). Genomic remnants of α-globin genes in the hemoglobinless antarctic icefishes. *Proc. Natl. Acad. Sci. USA* **92**, 1817–1821.

Eisen, J. S. (1996). Zebrafish make a big splash. *Cell* **87**, 969–977.

Falkenburg, J. H., Jansen, J., van der Vaart-Duinkerken, N., Veenhof, W. F. J., Blotkamp, J., Goselink, H. M., Parlevliet, J., and van Rood, J. J. (1984). Polymorphic and monomorphic HLA-DR determinants on human hematopoietic progenitor cells. *Blood* **63**, 1125–1132.

Flajnik, M. F., Kaufman, J. F., Riegert, P., and Du Pasquier, L. (1984). Identification of class I major histocompatibility complex encoded molecules in the amphibian *Xenopus*. *Immunogenetics* **20**, 433–442.

Gabbianelli, M., Boccoli, G., Cianetti, L., Russo, G., Testa, U., and Peschle, C. (1990). HLA expression in hemopoietic development. Class I and II antigens are induced in the definitive erythroid lineage and differentially modulated by fetal liver cytokines. *J. Immunol.* **144**, 3354–3360.

Gorvel, J. P., Chang, T. C., Boretto, J., Azuma, T., and Chavrier, P. (1998). Differential properties of D4/LyGDI versus RhoGDI: Phosphorylation and rho GTPase selectivity. *FEBS Lett.* **422,** 269–273.

Greaves, M. F., Katz, F. E., Myers, C. D., Davies, L., and Sieff, C. (1985). Selective expression of cell surface antigens on human haemopoietic progenitor cells. *In* "Hematopoietic Stem Cell Physiology" (J. Palek, ed.), pp. 301–315. Alan R. Liss, New York.

Gress, T. M., Wallrapp, C., Frohme, M., Muller-Pillasch, F., Lacher, U., Friess, H., Buchler, M., Adler, G., and Hoheisel, J. D. (1997). Identification of genes with specific expression in pancreatic cancer by cDNA representational difference analysis. *Genes Chromosomes Cancer* **19,** 97–103.

Groysman, M., Russek, C. S., and Katzav, S. (2000). Vav, a GDP/GTP nucleotide exchange factor, interacts with GDIs, proteins that inhibit GDP/GTP dissociation. *FEBS Lett.* **467,** 75–80.

Grunwald, D. J. (1996). A fin-de-siècle achievement: Charting new waters in vertebrate biology. *Science* **274,** 1634–1635.

Henry, J., Ribouchon, M. T., Offer, C., and Pontarotti, P. (1997). B30.2-like domain proteins: A growing family. *Biochem. Biophys. Res. Commun.* **235,** 162–165.

Henry, J., Mather, I. H., McDermott, M. F., and Pontarotti, P. (1998). B30.2-like domain proteins: Update and new insights into a rapidly expanding family of proteins. *Mol. Biol. Evol.* **15,** 1696–1705.

Hubank, M., and Schatz, D. G. (1994). Identifying differences in mRNA expression by representational difference analysis of cDNA. *Nucleic Acids Res.* **22,** 5640–5648.

Hubank, M., and Schatz, D. G. (1999). cDNA representational difference analysis: A sensitive and flexible method for identification of differentially expressed genes. *Methods Enzymol.* **303,** 325–349.

Kingston, R. E. (1993). Preparation of poly(A)$^+$ RNA. *In* "Current Protocols in Molecular Biology" (F. M. Ausubel, R. Brent, R. E. Kingston, D. D. Moore, J. G. Seidman, J. A. Smith, and K. Struhl, eds.), pp. 4.5.1–4.5.3. Wiley Interscience, New York.

Klickstein, L. B., Neve, R. L., Golemis, E. A., and Gyuris, J. (1995). Conversion of mRNA into double-stranded cDNA. *In* "Current Protocols in Molecular Biology" (F. M. Ausubel, R. Brent, R. E. Kingston, D. D. Moore, J. G. Seidman, J. A. Smith, and K. Struhl, eds.), pp. 5.5.2–5.5.14. Wiley Interscience, New York.

Lisitsyn, N., Lisitsyn, N., and Wigler, M. (1993). Cloning the differences between two complex genomes. *Science* **259,** 946–951.

Nasevicius, A., and Ekker, S. C. (2000). Effective targeted gene 'knockdown' in zebrafish. *Nat. Genet.* **26,** 216–220.

Nishiyama, N., Itoh, K., Kaneko, Y., Kishishita, M., Yoshida, O., and Fujita, J. (1997). A glycine-rich RNA-binding protein mediating cold-inducible suppression of mammalian cell growth. *J. Cell Biol.* **137,** 899–908.

Orkin, S. H., and Zon, L. I. (1997). Genetics of erythropoiesis: Induced mutations in mice and zebrafish. *Annu. Rev. Genet.* **31,** 33–60.

Pandolfi, P. P. (1998). Knocking in and out genes and trans genes: The use of the engineered mouse to study normal and aberrant hemopoiesis. *Semin. Hematol.* **35,** 136–148.

Parker, L. H., Zon, L. I., and Stainier, D. Y. (1999). Vascular and blood gene expression. *In* "Methods in Cell Biology" (H. W. Detrich, III, M. Westerfield, and L. I. Zon, eds.), Vol. 59, pp. 313–336. Academic Press, San Diego.

Paw, B. H., and Zon, L. I. (2000). Zebrafish: A genetic approach in studying hematopoiesis. *Curr. Opin. Hematol.* **7,** 79–84.

Puissant, C., and Houdebine, L. M. (1990). An improvement of the single-step method of RNA isolation by acid guanidinium thiocyanate-phenol-chloroform extraction. *Biotechniques* **8,** 148–149.

Quertermous, T. (1996). Plating libraries and transfer to filter membranes. *In* "Current Protocols in Molecular Biology" (F. M. Ausubel, R. Brent, R. E. Kingston, D. D. Moore, J. G. Seidman, J. A. Smith, and K. Struhl, eds.), pp. 6.1.1–6.1.4. Wiley Interscience, New York.

Reymond, A., Meroni, G., Fantozzi, A., Merla, G., Cairo, S., Luzi, L., Riganelli, D., Zanaria, E., Messali, S., Cainarca, S., Guffanti, A., Minucci, S., Pelicci, P. G., and Ballabio, A. (2001). The tripartite motif family identifies cell compartments. *EMBO J.* **9,** 2140–2151.

Rossant, J., and McKerlie, C. (2001). Mouse-based phenogenomics for modeling human disease. *Trends Mol. Med.* **7**, 502–507.

Rowley, A. F., Hunt, T. C., Page, M., and Mainwaring, G. (1988). Fish. *In* "Vertebrate Blood Cells" (A. F. Rowley and N. A. Ratcliffe, eds.), pp. 19–127. Cambridge University Press, Cambridge.

Salas, F., Haas, J., Brunk, B., Stoeckert, C. J., Jr., and Overton, G. C. (1998). EpoDB: A database of genes expressed during vertebrate erythropoiesis. *Nucleic Acids Res.* **26**, 288–289.

Sambrook, J., and Russell, D. W. (2001). "Molecular Cloning: A Laboratory Manual," 3rd ed., Cold Spring Harbor Laboratory Press, Cold Spring Harbor, NY.

Sarder, M. R. I., Fischer, U., Dijkstra, J. M., Kiryu, I., Yoshiura, Y., Azuma, T., Köllner, B., and Ototake, M. (2003). The MHC class I linkage group is a major determinant of the *in vivo* rejection of allogeneic erythrocytes in rainbow trout (*Oncorhynchus mykiss*). *Immunogenetics* **55**, 315–324.

Sasaki, T., and Takai, Y. (1998). The Rho small G protein family-Rho GDI system as a temporal and spatial determinant for cytoskeletal control. *Biochem. Biophys. Res. Commun.* **245**, 641–645.

Scherle, P., Behrens, T., and Staudt, L. M. (1993). Ly-GDI, a GDP-dissociation inhibitor of the RhoA GTP-binding protein, is expressed preferentially in lymphocytes. *Proc. Natl. Acad. Sci. USA* **90**, 7568–7572.

Seto, M. H., Liu, H. L., Zajchowski, D. A., and Whitlow, M. (1999). Protein fold analysis of the B30.2-like domain. *Proteins* **35**, 235–249.

Sgonc, R., Hala, K., and Wick, G. (1987). Relationship between the expression of class I antigen and reactivity of chicken thymocytes. *Immunogenetics* **26**, 150–154.

Shurin, S. B. (1995). Disorders of the spleen. *In* "Blood: Principles and Practice of Hematology" (L. I. Handin, S. E. Lux, and T. P. Stossel, eds.), pp. 1359–1380. J. B. Lippincott, Philadelphia.

Sieff, C., Bicknell, D., Caine, G., Robinson, J., Lam, G., and Greaves, M. F. (1982). Changes in cell surface antigen expression during hemopoietic differentiation. *Blood* **60**, 703–713.

Southern, E. M. (1975). Detection of specific sequences among DNA fragments separated by gel electrophoresis. *J. Mol. Biol.* **98**, 503–517.

Spack, E., Jr., and Edidin, M. (1986). The class I MHC antigens of erythrocytes: A serologic and biochemical study. *J. Immunol.* **136**, 2943–2952.

Sparrow, R. L., and Williams, N. (1986). The pattern of HLA-DR and HLA-DQ antigen expression on clonable subpopulations of human myeloid progenitor cells. *Blood* **67**, 379–384.

Stoeckert, C. J., Jr., Salas, F., Brunk, B., and Overton, G. C. (1999). EpoDB: A prototype database for the analysis of genes expressed during vertebrate erythropoiesis. *Nucleic Acids Res.* **27**, 200–203.

Takahashi, K., Sasaki, T., Mammoto, A., Takaishi, K., Kameyama, T., Tsukita, S., and Takai, Y. (1997). Direct interaction of the Rho GDP dissociation inhibitor with ezrin/radixin/moesin initiates the activation of the Rho small G protein. *J. Biol. Chem.* **272**, 23371–23375.

Thompson, M. A., Ransom, D. G., Pratt, S. J., MacLennan, H., Kieran, M. W., Detrich, H. W., III, Vail, B., Huber, T. L., Paw, B. H., Brownlie, A., Oates, A. C., Fritz, A., Gates, M. A., Amores, A., Bahary, N., Talbot, W. S., Her, H., Beier, D. R., Postlewait, J. H., and Zon, L. I. (1998). The *cloche* and *spadetail* genes differentially affect hematopoieis and vasculogenesis. *Dev. Biol.* **197**, 248–269.

Vernet, C., Boretto, J., Mattei, M. G., Takahashi, M., Jack, L. J., Mather, I. H., Rouquier, S., and Pontarotti, P. (1993). Evolutionary study of multigenic families mapping close to the human MHC class I region. *J. Mol. Evol.* **37**, 600–612.

Westerfield, M. (2000). "The Zebrafish Book: A Guide for the Laboratory Use of Zebrafish (*Danio rerio*)" University of Oregon Press, Eugene, OR.

Zhao, Y., Ratnayake-Lecamwasam, M., Parker, S. K., Cocca, E., Camardella, L., di Prisco, G., and Detrich, H. W., III. (1998). The major adult α-globin gene of antarctic teleosts and its remnants in the hemoglobinless icefishes. Calibration of the mutational clock for nuclear genes. *J. Biol. Chem.* **273**, 14745–14752.

Zollman, S., Godt, D., Prive, G. G., Couderc, J. L., and Laski, F. A. (1994). The BTB domain, found primarily in zinc finger proteins, defines an evolutionarily conserved family that includes several developmentally regulated genes in *Drosophila*. *Proc. Natl. Acad. Sci. USA* **91**, 10717–10721.

CHAPTER 27

Spatial and Temporal Expression of the Zebrafish Genome by Large-Scale *In Situ* Hybridization Screening

Bernard Thisse, Vincent Heyer, Aline Lux, Violaine Alunni, Agnès Degrave, Iban Seiliez, Johanne Kirchner, Jean-Paul Parkhill, and Christine Thisse

Institut de Génétique et de Biologie Moléculaire et Cellulaire
UMR 7104 CNRS/INSERM/ULP
67404 Illkirch Cedex, France

I. Introduction and Goals

Whole-mount *in situ* hybridization is a method widely used to describe the expression patterns of developmentally regulated genes. Use of a highly sensitive *in situ* hybridization assay allows for reliable visualization of gene expression, including genes expressed at low levels. Here we describe a technique that employs *in vitro* synthesized RNA tagged with either digoxigenin (DIG) or fluorescein uridine-5′-triphosphate (UTP) to determine gene expression patterns in whole-mount embryos. Following hybridization, the transcript is visualized immunohistochemically, using an antidigoxygenin (or antifluorescein) antibody conjugated to alkaline phosphatase, the substrate of which is chromogenic.

The RNA *in situ* hybridization technique can be used to establish gene expression profiles, allowing for establishment of the tissue and cell specificity and for the time course of its expression during embryo differentiation. This technique serves as the gateway to determining which genes are affected by a given mutation. To explain further, if the analysis of a mutant reveals a gene showing disrupted expression as determined by *in situ* hybridization, the next step is to establish whether there is a link between this gene and the mutation. This is done by comparing the map position of the mutation and the gene of interest on the chromosomes. If they map to the same locus, it suggests that the gene is a good candidate for the gene altered in the mutation. Further studies, such as sequencing the gene in the mutant (for identification of molecular lesions) and rescuing the mutant phenotype by injecting the corresponding RNA, demonstrate the identity between the gene identified based on its expression pattern and the gene inactivated in the mutant (e.g., Donovan *et al.*, 2002; Kikuchi *et al.*, 2001; Schmid *et al.*, 2000).

The *in situ* hybridization technique described here is also important for defining synexpression groups. Synexpression analysis can reveal that a group of genes share temporal and spatial expression patterns, suggesting that they might be controlled by the same signaling pathways. On the basis of similarities in their expression patterns during zebrafish embryonic development, five genes that define a synexpression group have been identified: *fgf8, fgf3, sprouty2, sprouty4,* and *sef*. Further functional studies have shown that *sproutys* and *sef* are feedback-induced antagonists of the ras/raf/MEK/MAPK-mediated FGF signaling (Fürthaur *et al.*, 1997, 2001, 2002; Tsang *et al.*, 2002). Lastly, the *in situ* hybridization technique permits large-scale analysis of the spatial and temporal expression of the zebrafish genome, allowing for identification of a large collection of tissue- and cell-specific markers important for distinct developmental stages.

II. Preparation of Antisense Digoxigenin (DIG)-Labeled RNA Probes

A. Isolation and Preparation of the DNA Template

This method is used to prepare large amounts of antisense RNA probes and is divided into two steps: preparation of DNA and synthesis of antisense RNA probe.

For preparation of the DNA template, $5 \mu g$ of DNA is linearized in a 2-h digestion, using the appropriate restriction enzyme under appropriate conditions of salt and temperature. Care must be taken to ensure that the direction of insert of interest is known to allow for the correct production of the antisense and sense probes. Once the reaction is complete, a phenol/chloroform extraction is used to remove the enzyme. DNA is then purified from the aqueous phase by using Microcon YM-50 columns (Millipore, Cat. No. 42415). After a 5-min centrifugation at 10,000g (the column should be dry), 100 μl of sterile-filtered H_2O is added to the column and it is then centrifuged a second time for 5 min at 10,000g (again the column should be dry). The Microcon column is put in a new eppendorf, 20 μl of sterile water is added, it is briefly vortexed, and the Microcon device is put upside down and centrifuged for 1 min at 5000g. On a 1% agarose gel, 2 μl is tested to check whether linearization is complete.

Following successful production of template, the next step is the *in vitro* synthesis of the antisense RNA in a 2-h incubation at 37 °C with the following transcription mix:

Linearized DNA, 1 μg

Transcription buffer (Promega), 4 μl

NTP-DIG-RNA (Boehringer), 2 μl

RNase inhibitor (35 units/μl, Promega), 1 μl

T3 or T7 RNA polymerase (20 units/μl, Stratagene), 1 μl

Sterile water to make up the volume to 20 μl

Following the initial incubation, the DNA template is digested by adding 2 μl RNase-free DNase (Roche, Cat. No. 776785) for 15 min at 37 °C. The digestion reaction is stopped by adding 1 μl of 0.5 M EDTA, pH 8.0. Synthesized RNA is then precipitated by adding 2.5 μl of 4 M LiCl and 75 μl cold 100% ethanol, followed by incubation at −70 °C for 10 min and then centrifugation at 4 °C for 30 min at 10,000 g. Finally, the pellet is washed with 70% ethanol, dried, and resuspended in 20 μl sterile water. Alternatively, an RNA purification kit can be used. A Sigmaspin Post Reaction Purification column (Sigma, Cat. No. S5059) is placed in a microfuge tube and centrifuged for 15 sec at 750g. The base of the column is broken and the top removed and then recentrifuged for 2 min at 750g. The column is placed in a new tube, and the RNA sample added on top of the resin. The tube is centrifuged for 4 min at 750g and the column discarded. To the RNA sample, 1 μl of 0.5 M EDTA and 9 μl RNAlater (Sigma, Cat. No. R-0901) are added. The sample is stored at −20 °C. One tenth of the synthesized RNA is on a visualized 1% agarose gel to determine whether the procedure was successful.

B. PCR Generation of Template as an Alternative to Linearization and Purification of the DNA Template

The advantage of the PCR amplification method is that it is fast, can be used for large-scale *in situ* analysis, and is a viable method when no RNA polymerase promoters are available. For example, when the RNA polymerase promoter is determined by the oligonucleotide at the 3′ of the probe because of the poor incorporation of DIG-3′ UTP by the SP6 RNA polymerase. Therefore, only T3 or T7 RNA polymerases will be chosen. The PCR amplification method can also be used when no unique site is usable to linearize the DNA with a restriction enzyme at the 5′ of the cDNA.

For PCR amplification, 100 ng of purified DNA (or 1 μl of an overnight culture of the bacteria containing the plasmid carrying the cDNA), 0.5 μl Primer 1 (0.5 μg/ μl), 0.5 μl Primer 2 (0.5 μg/μl), 50 μl PCR master mix (Promega, Cat. No. M7505), and up to 100 μl sterile water in a 0.5-ml sterile tube are mixed. The mixture is denatured at 95 °C for 4 min, followed by 35 cycles of 95 °C for 30 sec, 55 °C for 30 sec, 72 °C for 3 min (at least 1 min/kb), and a final extension at 72 °C for 7 min. The product is then stored at −20 °C.

For PCR product purification, a microcon PCR device (Genomics Millipore, UFC7PCR50) is placed on the provided eppendorf tube. A total of 100 μl of the PCR reaction and 400 μl sterile water are loaded on the micron membrane. The tube is centrifuged for 15–20 min at 1000g. The membrane should be dry. The microcon device is put in a new eppendorf, 20 μl of sterile water added, briefly vortexed, and the microcon device placed upside down. The eppendorf is centrifuged for 1 min at 1000g to recover the DNA. The PCR amplification is checked by loading one tenth on a 1% agarose gel.

C. Synthesis of Antisense RNA

To 2.5 μl DNA (100–200 ng) 2.5 μl of the following mix is added: 1 μl transcription buffer (Promega, Cat. No. P118B), 0.5 μl DTT (Promega, Cat. No. P117B), 0.5 μl NTP-DIG-RNA (Roche, Cat. No. 1277073), 0.25 μl RNAsin inhibitor (Promega, Cat. No. N251X), and 0.25 μl RNA polymerase (T7 polymerase: Promega, Cat. No. P207B; T3 polymerase: Promega, Cat. No. P208C). The mixture is mixed and incubated for 2 h at 37 °C. Then, 2 μl RNase-free DNase I (Roche, Cat. No. 776785) and 18 μl sterile water are added and the mixture is incubated for 30 min at 37 °C. The reaction is stopped by adding 1 μl sterile 0.5 M EDTA and 9 μl sterile water. The RNA template is purified on a Sigmaspin Post Reaction Purification column (Sigma, Cat. No. S5059) as described previously.

III. Preparation of Embryos

Eggs are collected from single mating pairs about 1 h after laying. They are cleaned and unfertilized eggs are discarded. Embryos are allowed to develop in regular fish water until the end of gastrulation. For embryos older than 24 h, to prevent pigmentation, fish water is replaced at the end of gastrulation (10 hpf) by a 0.0045% solution of 1-phenyl-2-thiourea (Sigma, Cat. No. P-7629) in

$1\times$ Danieau's medium (58 mM NaCl, 0.7 mM KCl, 0.4 mM $MgSO_4$, 0.6 mM $Ca(NO_3)$, 2, 5 mM HEPES, pH 7.6).

Chorions are removed by pronase treatment (Sigma, Cat. No. P-6911) according to the online "Zebrafish Book" protocol (http://zfin.org/zf_info/zfbook/chapt4/4.1.html) prior to fixation in 4% paraformaldehyde. Alternatively, chorions can be removed after fixation by using a sharp forceps.

Embryos are fixed at the appropriate stage in 4% paraformaldehyde (PFA, Sigma, Cat. No. P-6148) in $1\times$ PBS overnight at $4\,^{\circ}$C. Paraformaldehyde powder is dissolved in $1\times$ PBS by heating on a hot plate with agitation using a magnet stirrer to $95\,^{\circ}$C (do no boil). Once the powder is completely dissolved, the solution is cooled on ice.

Fixed embryos are dehydrated in 100% methanol (MeOH) for 15 min at room temperature and then stored at $-20\,^{\circ}$C (for at least 2 h and up to several months) prior to proceeding with *in situ* hybridization experiments.

IV. Reagents and Buffers for *In Situ* Hybridization

- $10\times$ PBS (Dulbecco, Sigma, Cat. No. D-5652).
- MeOH.
- Tween 20 (Sigma, Cat. No. P-1379).
- Proteinase K (Boehringer, Cat. No. 1000 144).
- Anti-DIG antibody-alkaline phosphatase Fab fragment (Boehringer, Cat. No. 1 093 274).
- BSA fraction V, protease free (Sigma, Cat. No. A-3294).
- Formamide: high-purity grade (Sigma or Carlo Erba, Cat. No. 452286), deionized by adding and stirring slowly twice for 15 min each with 10 g/l Serdolit MB-3 (Serva, Cat. No. 40721). The solution is filtered to remove the resin and stored in the dark at $4\,^{\circ}$C.
- $20\times$ SSC.
- 5 mg/ml heparin (Sigma, Cat. No. H-3393).
- RNase-free tRNA (Sigma, Cat. No. R-7876): 50 mg/ml resuspended in water and extensively extracted several times in phenol/chloroform to remove protein.
- 1 M citric acid.
- Normal sheep serum (Jackson Immunresearch, Cat. No. 013-000-121).
- 1 M tris HCl, pH 9.5.
- 1 M $MgCl_2$.
- 5 M NaCl
- 50 mg/ml nitro blue tetrazolium [NBT; made from powder, Sigma, Cat. No. N-6876, NBT (50 mg) is dissolved in 0.7 ml anhydrous dimethylformamide and 0.3 ml H_2O]. Store in the dark at $-20\,^{\circ}$C.

- 50 mg/ml 5-bromo 4-chloro 3-indolyl phosphate [BCIP; made from powder, Sigma, Cat. No. B-8503 BCIP (50 mg) dissolved in 1 ml anhydrous dimethyl-formamide). Store in the dark at $-20\,^{\circ}$C.
- Embryo storage buffer: PBS, pH 5.5 (Na_2HPO_4 1.08 g/l, NaH_2PO_4 6.5 g/l, NaCl 8.0 g/l, KCl 0.2 g/l), 1 mM EDTA, 0.1% Tween 20.
- 0.5 M EDTA.
- 99% Pure glycerol (Sigma, Cat. No. G-6279).

V. *In Situ* Hybridization Protocol

This protocol is adapted from Thisse *et al.* (1993, 2001) and Thisse and Thisse (1998).

A. Day 1

Embryos (of the same developmental stage) are transferred into small baskets made of a metal or nylon mesh (mesh opening 100–150 μm) fused at the bottom of a plastic tube and placed in 24- or 6-well tissue culture plates. Baskets are made with 2-ml eppendorf tubes or 50-ml conical centrifuge tubes (Fig. 1A) cut with a cutter or a saw to produce a cylinder of plastic about 1.5- to 2-cm high, with a diameter of 1.2 cm (small baskets) or 3 cm (large baskets). A stainless steel mesh (for large baskets) or nylon mesh (for small baskets) is fused at the top of the tube as follows: on a hot plate, a small piece of aluminum foil is placed and then the metal or nylon mesh is put on the foil and the plastic tube pressed onto the mesh (Fig. 1B), until the fusion of the plastic glues the mesh to the tube. Once fused, they are rapidly removed from the hot plate. The aluminum foil, glued to the basket, cools down quickly and can be removed easily. Small baskets (convenient for treatment of up to 50 embryos) are usable in 24-well plates, large baskets made with 50-ml conical centrifuge tubes (for 500–1000 embryos) can be used in 6-well plates (Fig. 1C).

1. Rehydratation

Embryos stored in 100% MeOH are rehydrated by successive incubations (moving baskets from well to well) in the following solutions:

75% MeOH – 25% PBS for 5 min.
50% MeOH – 50% PBS for 5 min.
25% MeOH – 75% PBS for 5 min.
100% PBT (PBS/Tween 20 0.1%) four times for 5 min each.

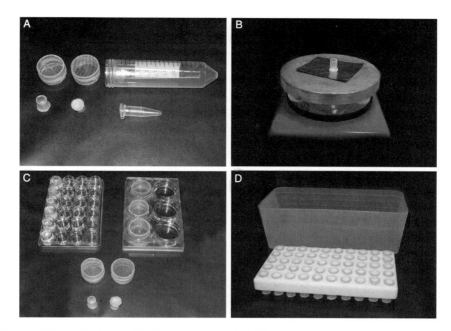

Fig. 1 Different devices used for large-scale *in situ* hybridization on whole-mount zebrafish embryos. (A) Small and large baskets are made from 2-ml or 50-ml plastic tubes cut with a saw or a cutter. (B) Fusion of metal or nylon mesh on the bottom of the plastic tube. (C) Incubation of the embryo (Day 1) are performed in multiwell plates (24-well plates for small baskets, 6-well plates for large baskets). (D) Washes from Day 2 to Day 3 are performed in small baskets made from 2-ml plastic tubes placed on a Styrofoam float in a plastic box.

2. Digestion with Proteinase K (10 μg/ml)

This step permeabilizes the embryos, permitting access of the RNA probe. The digestion time is dependent on the developmental stage. For blastula, gastrula, and somitogenesis stages (up to the 18-somite stage), 30 sec to 1 min is sufficient. For 24-h-old embryos, digestion is done for up to 10 min and for older embryos (36-h-old to 5-day-old embryos) for 20–30 min. Proteinase K digestion is stopped by incubation in 4% paraformaldehyde in 1× PBS for 20 min, followed by washes in 1× PBT 5 times for 5 min each.

3. Prehybridization

Embryos are transferred to 1.5-ml eppendorf tubes (up to 50 embryos per tube). At this step, embryos of different developmental stages can be pooled and treated together until the end of the *in situ* hybridization experiment. Prehybridization is performed by incubation in 700 μl of hybridization mix (HM) for 2–5 h at 70 °C in a waterbath. Prehybridized embryos can then be directly hybridized or stored in HM at −20 °C (up to several weeks). The HM is prepared as follows:

50% Formamide.
5× SSC.
0.1% Tween 20.
Citric acid to adjust HM to pH 6.0 (460 μl of 1 M citric acid for 50 ml of mix).
50 μg/ml heparin.
500 μg/ml tRNA.

4. Hybridization

The prehybridization mix is removed and discarded. It is replaced with 200 μl of HM containing about 100 ng of antisense DIG-labeled RNA probe and hybridized overnight in the eppendorf tube at 70 °C in a waterbath.

5. Preadsorbtion of Anti-DIG Antibody

In addition to the embryos used for the *in situ* hybridization, a batch of embryos is treated the same way, excluding the hybridization step, and used for the preadsorbtion of the anti-DIG antibody. 1000 embryos are used for 20 ml of anti-DIG antibody diluted 1:1000 in PBT − 2% sheep serum − 2 mg/ml BSA. Antibody is preadsorbed for several hours at room temperature under gentle agitation on a test tube rocker (Thermolyne, Vari-mix). Embryos used for pre-adsorbtion are removed and the preadsorbed antibody is stored at 4 °C until its use on Day 2.

B. Day 2

1. Washes

Embryos are removed from the eppendorf tube and placed (see Fig. 1D) in baskets made from 2-ml eppendorf tubes placed on a Styrofoam float (16 × 9 × 1.5 cm^3 with space for 50 small baskets) in a plastic box (21 × 10 × 7 cm^3) containing 200 ml of 100% HM wash solution at 70 °C. (HM used in washes does not contain tRNA and heparin.) Embryos stay in these baskets on the Styrofoam float until the staining step on Day 3. After a quick wash, the Styrofoam float carrying the 50 baskets is placed successively in another plastic box containing 200 ml of prewarmed wash solution and incubated at 70 °C in a shaking waterbath (with about 40 strokes/min). The successive steps and washing solutions are as follows:

15 min in 75% HM/25% 2× SSC at 70 °C.
15 min in 50% HM/50% 2× SSC at 70 °C.

15 min in 25% HM/75% 2× SSC at 70 °C.

15 min in 2× SSC at 70 °C.

These steps gradually facilitate the change from the HM to 2× SSC. Two washes of 30 min each are given in 0.2 × SSC. (These are high-stringency washes that remove nonspecifically hybridized probes.) Following the high-stringency washes, embryos are progressively moved from 0.2× SSC to 100% PBT by the following incubations (in 200 ml) at room temperature with slow agitation using an horizontal orbital shaker (about 40 rpm).

10 min in 75% 0.2× SSC/25% PBT.

10 min in 50% 0.2× SSC/50% PBT.

10 min in 25% 0.2× SSC/75% PBT.

10 min in PBT.

2. Incubation with Anti–DIG Antiserum

- Embryos are blocked for 3–4 h at room temperature in blocking buffer made in PBT containing 2% sheep serum and 2 mg/ml BSA.
- They are incubated in 200 ml of antibody solution diluted at 1:10,000 in blocking buffer overnight at 4 °C under slow agitation (30–40 rpm on the horizontal orbital shaker).

C. Day 3

1. Washes

The antiserum is removed and discarded. After a brief wash in PBT, it is washed extensively six times for 15 min in PBT at room temperature under slow agitation (30–40 rpm on the horizontal orbital shaker). After the last wash, and before moving into the staining buffer, embryos are dried by placing the Styrofoam float carrying the 50 baskets on an absorbing paper to remove remaining PBT (to avoid formation of a precipitate in the staining buffer). Embryos are then incubated at room temperature in the alkaline Tris buffer (100 mM Tris HCl, pH 9.5, 50 mM MgCl$_2$, 100 mM NaCl, 0.1% Tween 20) changed thrice at 5 min intervals.

2. Staining

Embryos are removed from the baskets and incubated in the staining solution at room temperature (in the dark) in a multiwell plate.

The staining solution (to keep out from the light) is as follows:

50 mg/ml NBT, 225 μl.

50 mg/ml BCIP, 175 μl.

Alkaline Tris buffer, to 50 ml.

The staining reaction is monitored regularly under a dissecting microscope with light from the top and with the plate on a white background.

When the signal is perceived as sufficient (and before apparition of background reaction time in a range of 15 min for genes strongly expressed, 1–1.5 h for most genes, and up to 5 h for genes that are weakly expressed), the staining reaction is stopped by transferring embryos into a 1.5-ml eppendorf tube, the staining solution is removed, and discarded and embryos are washed several times at room temperature with the stop solution, the composition of which is as follows:

PBS 1 × pH 5.5.
1 mM EDTA.
0.1% Tween 20.

Labeled embryos are stored in the stop solution (4 °C in the dark). Labeling stays unchanged for months under these conditions.

VI. Double *In Situ* Protocol

A. Preparation of Probes

Two different antisense RNA probes are used, one labeled with DIG 11-UTP and the second with fluorescein 12-UTP. The protocol for synthesis with fluorescein 12-UTP is identical to that used for the DIG 11-UTP described previously, except for the transcription mix. This fluorescein-labeled probe is kept in the dark as much as possible during the *in situ* hybridization steps. Incubate for 2 h at 37 °C 1 μg linearized DNA with the following:

- 2 μl 100 mM DTT.
- 1.3 μl NTP mix (16.4 μl ATP 100 mmol/l, 16.4 μl CTP 100 mmol/l, 16.4 μl GTP 100 mmol/l, 16.4 μl UTP 100 mmol/l, sterile water 34.4 μl).
- 0.7 μl fluorescein 12 UTP (Roche, Cat. No. 1.427.857).
- 1 μl RNasin inhibitor.
- 4 μl transcription buffer ×5.
- 1 μl T3 or T7 RNA polymerase.
- 6 μl sterile water.

B. *In Situ* Hybridization

1. Day 1

The preparation of embryos, prehybridization and preadsobtion of anti-DIG antibody and antifluorescein antibody (1:1000 dilution in PBT – 2% sheep serum – 2 mg/ml BSA) are performed as previously described. For the hybridization,

100 ng of both DIG- and fluorescein-labeled probes are added in 200 μl of HM and incubated overnight at 70 °C in a waterbath.

2. Day 2

Washes from HM until preincubation in PBT – 2% sheep serum – 2 mg/ml BSA are the same as those described for single *in situ* hybridization. Embryos are then incubated overnight in preadsorbed antifluorescein antibody (dilution 1:5000) with agitation at 4 °C in PBT – 2% sheep serum – 2 mg/ml BSA.

3. Day 3

The antiserum is removed, discarded, and washed with PBT (six washes of 15 min each, at room temperature) with gentle agitation (40 rpm). Embryos are then incubated in 0.1 M Tris HCl pH 8.2, 0.1% Tween 20 (three washes of 5 min each) and transferred to multiwell plates containing the Fast red staining solution. For the fast red solution, one tablet of Fast Red (Roche, 1.496.549) is dissolved for 2 ml of staining solution in 0.1 M Tris HCl, pH 8.2, 0.1% Tween 20 and filtered through a microfilter (3 μm).

Embryos are incubated in the staining solution at room temperature (covered with a box) and monitored regularly under a dissecting microscope. The reaction is stopped by removing the staining solution and embryos are washed three times for 15 min each in PBT.

Alkaline phosphatase activity carried by the antifluorescein antibody is inactivated by incubation in 0.1 M glycin HCl, pH 2.2, 0.1% Tween 20 at room temperature for 10 min. Embryos are washed four times for 5 min each in PBT and incubated for several hours in PBT – 2% sheep serum – 2 mg/ml BSA. Embryos are then incubated in preadsorbed anti-DIG antibody at a 1:10000 dilution in PBT – 2% sheep serum – 2 mg/ml BSA overnight with gentle agitation at 4 °C.

4. Day 4

The same protocol as described in Chapter V, Day 3, is followed.

C. Alternative Method: Enzyme-Labeled Fluorescence (ELF) Protocol (Molecular Probes, Cat. No. E 6604)

With this technology, alkaline phosphatase activity converts the ELF 97 phosphate compound to a brilliant-green fluorescent precipitate. The protocol used is the same as that described in Chapter V, from Day 1 to Day 3. However, after the PBT washes, instead of incubating the embryos in the alkaline Tris-buffer, embryos are washed for 5 min each in the ELF kit wash solution. Embryos are then incubated in the dark with 400 μl of the ELF reaction medium (as per

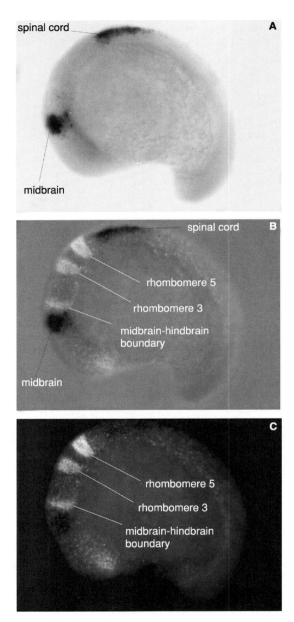

Fig. 2 Double *in situ* hybridization using NBT/BCIP and enzyme-labeled fluorescence (ELF) labeling methods. To identify the precise localization of the expression of EST CB313 (encoding a gene homologous to neuropeptide B), a double *in situ* hybridization was performed. Fluorescein-labeled CB313 probe was revealed by using NBT/BCIP reagents (Panels A and B). The digoxygenin-labeled probes Krox20 (a marker of rhombomere 3 and 5 of the rhombencephalon) and Pax2.1 (a marker of the midbrain–hindbrain boundary) were revealed by using the ELF kit (Panels B and C). This allowed to localize the expression of CB313 gene to the midbrain and anterior spinal cord.

manufacturer's instructions) from 30 min to 2 h. The alkaline phosphatase reaction is stopped by incubation in the ELF kit washing solution. Embryos are stored in PBS 1×, pH 5.5, 1 mM EDTA at 4 °C in the dark. The labeling is stable for several months under these conditions. An example of a double *in situ* hybridization with a combination of probes revealed with NBT/BCIP and the ELF kit is shown in Fig. 2.

VII. Recording Results

Labeled embryos are mounted in 100% glycerol. Because of the photosensitivity of the yolk cell, embryos at early developmental stages are first treated for 5 min in an acidic buffer (either PBS, pH < 3.5, or glycine buffer, pH 2.2). This treatment prevents photoreactivity of the yolk proteins, and even under intense light the yolk cell remains unstained. However, this acidic treatment affects embryo morphology. Therefore, although very convenient at early developmental stages (when only a few structures are formed and when the photolabeling of the yolk is a limiting factor for the observation), this acidic treatment should not be used for embryos older than the 15-somite stage. Embryos are observed in glycerol between the slide and coverslip (using bridges made of four coverslips of thickness 1.5 mm). Low-magnification pictures are taken with a Leica M420 Macroscope (which offers a large field of view and a long working distance and its vertical beam path provides for parallax-free imaging, resulting in high accuracy, top imaging, fidelity, and faithful photography) or with a microscope (Leica DM RA2) with a differential interference contrast (DIC), using a numeric camera (coolsnap CCD, Roper Scientific). Digitalized pictures are saved as TIFF files then adjusted for contrast, brightness, and color balance by using a Photoshop software, and stored as such or after conversion to the jpeg format to reduce the files size. For our large-scale *in situ* hybridization analysis, annotations are made by using the standardized anatomical dictionary. Pictures associated with text description and keywords are then deposited in the ZFIN database. Examples of such pictures are presented in Fig. 3.

VIII. Concluding Remarks

Over a 6-year period, we have analyzed more than 17,000 cDNAs and identified 4600 spatially restricted expression patterns. Because of redundancy (33% established by comparison with the genome sequence) this corresponds to about 3000 different genes. Descriptions of more than 1000 gene expression patterns have been released to the public through ZFIN (http://zfin.org, mirror sites in France at http://www.igbmc.u-strasbg.fr and in Japan at http://:www.grs.nig.ac.jp:6060) in the gene expression section (EST named CB*n*, *n* for the number of the clone). Users can find for these 1000 gene expressions 14,500 annotated pictures, key

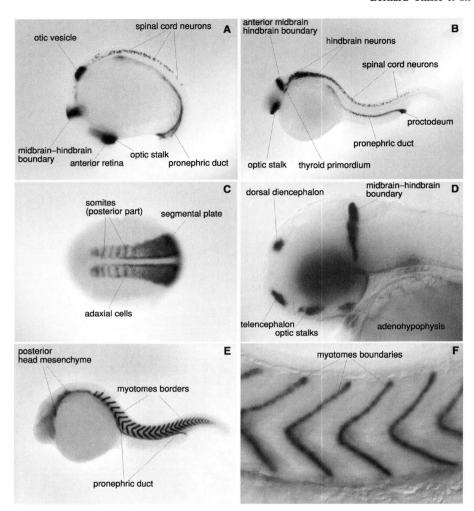

Fig. 3 Expression patterns of different genes as they appear (annotated pictures) in the Zebrafish Information Network (ZFIN) (http://zfin.org/). (A) CB378 (encoding Pax2.1) is expressed in optic stalk, anterior retina, midbrain–hindbrain boundary, otic vesicle, spinal cord neurons, and pronephric ducts at the 15–somite stage. (B) At 24 hpf, the transcripts of Pax2.1 are detected in optic stalk, thyroid primordium, anterior midbrain–hindbrain boundary, hindbrain and spinal cord neurons, pronephric ducts, and proctodeum. (C) CB641 (myf5) is expressed in the posterior part of the somites, in adaxial cells and segmental plate at the 7-somite stage. (D) High magnification of a 36-h-old embryo (DIC optics). CB110 (FGF8) is expressed in the adenohypophysis, the optic stalks, the telencephalon, the dorsal diencephalon, and at the midbrain–hindbrain boundary. (E) CB1045 (encoding a new protein) is expressed at 24 hpt in the myotome borders, posterior head mesenchyme, and pronephric ducts. (F) High magnification at the trunk level of the same embryo as that in (E), using the differential interference contrast (DIC) optics showing expression at the myotome boundaries.

words, and corresponding sequence analyses. The cDNA clones have been deposited at the Zebrafish International Research Center (ZIRC, Eugene, OR), which is in charge of the distribution of our clones.

Acknowledgments

We thank S. Geschier for maintenance of the fish and M. Fürthauer, A. Agathon, S. Obrecht-Pflumio, and B. Loppin for their contribution at different steps of the analysis. This work was supported by funds from the Institut National de la Santé et de la Recherche Médicale, the Centre National de la Recherche Scientifique, the Hôpital Universitaire de Strasbourg, and the National Institutes of Health (R01 RR15402).

References

Donovan, A., Brownlie, A., Dorschner, M. O., Zhou, Y., Pratt, S. J., Paw, B. H., Phillips, R. B., Thisse, C., Thisse, B., and Zon, L. I. (2002). The zebrafish mutant gene chardonnay (cdy) encodes divalent metal transporter 1 (DMT1). *Blood* **100**, 4655–4659.

Fürthauer, M., Thisse, C., and Thisse, B. (1997). A role for FGF-8 in the dorsoventral patterning of the zebrafish gastrula. *Development* **124**, 4253–4264.

Fürthauer, M., Reifers, F., Brand, M., Thisse, B., and Thisse, C. (2001). sprouty4 acts in vivo as a feedback-induced antagonist of FGF signaling in zebrafish. *Development* **128**, 2175–2186.

Fürthauer, M., Lin, W., Ang, S.-L., Thisse, B., and Thisse, C. (2002). Sef is a feedback-induced antagonist of Ras/MAPK-mediated FGF signalling. *Nat. Cell Biol.* **4**, 170–174.

Kikuchi, Y., Agathon, A., Alexander, J., Thisse, C., Waldron, S., Yelon, D., Thisse, B., and Stainier, D. Y. R. (2001). casanova encodes a novel Sox-related protein necessary and sufficient for early endoderm formation in zebrafish. *Genes Dev.* **15**, 1493–1505.

Schmid, B., Fürthauer, M., Connors, S. A., Trout, J., Thisse, B., Thisse, C., and Mullins, M. C. (2000). Equivalent genetic roles for bmp7/snailhouse and bmp2b/swirl in dorsoventral pattern formation. *Development* **127**, 957–967.

Thisse, C., Thisse, B., Schilling, T. F., and Postlethwait, J. H. (1993). Structure of the zebrafish snail1 gene and its expression in wild-type, spadetail and no tail mutant embryos. *Development* **119**, 1203–1215.

Thisse, C., and Thisse, B. (1998). High resolution whole-mount in situ hybridization. Available at ZFIN database at http://www.igbmc.u-strasbg.fr/zf-info/zbook/chapt9/9.82.html

Thisse, B., Pflumio, S., Fürthauer, M., Loppin, B., Heyer, V., Degrave, A., Woehl, R., Lux, A., Steffan, T., Charbonnier, X. Q., and Thisse, C. (2001). Expression of the zebrafish genome during embryogenesis. Available at ZFIN database at http://zfin.org/cgi-bin/webdriver?Mlval=aa-pubview2.apg&OID=ZDB-PUB-010810-1.

Tsang, M., Friesel, R., Kudoh, T., and Dawid, I. B. (2002). Identification of Sef, a novel modulator of FGF signalling. *Nat. Cell Biol.* **4**, 165–169.

CHAPTER 28

Design, Normalization, and Analysis of Spotted Microarray Data

F. B. Pichler,[*] M. A. Black,[†] L. C. Williams,[*] and D. R. Love[*]

[*]Molecular Genetics and Development Group
School of Biological Sciences
University of Auckland,
Auckland 1001, New Zealand

[†]Department of Statistics
University of Auckland,
Auckland 1001, New Zealand

I. Introduction

To date, the zebrafish has been considered an excellent model of vertebrate development and also as an emerging model of human disease. Although these considerations are legitimate, the repertoire of analytical tools that is generally used in the study of zebrafish appears limited. Importantly, the prospect of the complete genome sequence of this vertebrate species offers a means of viewing, as well as analyzing, the zebrafish in a novel way. This view can be simply stated as a four-dimensional gene expression program. This view does not seek primacy, but merely offers a different but nevertheless worthwhile perspective of a complex biological system.

The analysis of this gene expression program requires the development and application of appropriate tools. These tools can be found in large part by looking at what has been happening in the field of molecular genetics in the study of human disease and the analysis of other model organisms. Specifically, the development of microarrays has seen explosive growth, and it is here that we intend to discuss relevant aspects of microarray experiments that should find favor with respect to zebrafish-based studies. This discussion is not exhaustive, as many excellent reviews and books describe microarrays to which readers should refer (e.g., Bowtell and Sambrook, 2002; Holloway *et al.*, 2002).

In the context of zebrafish, microarray-based experiments can be divided into two different types. The first type involves the characterization of mutants compared with wild-type zebrafish at a given stage of development. The utility of such experiments is the ability to identify genes undergoing differential expression. The second type of experiment is a time course. These experiments can characterize coordinated changes in the transcription program during development as well as enable the monitoring of transient or regulatable effects of experimental manipulation such as chemical, physical, or heritable impacts.

II. Design of Microarray Experiments

Microarray experiments have the potential to incur considerable costs, particularly when using commercial arrays or gene chips. These experiments might also require a substantial investment in technical development, which applies to custom or cheaper spotted arrays. Prior to beginning microarray experiments, choices have to be made on which array approach to take. Such choices will be influenced by the scale and cost of experiments to be conducted, the specific questions being asked, and the available facilities. This section is divided into four parts. The first part provides an overview of different types of microarrays, the second concerns the design of custom arrays, the third discusses critical aspects of mRNA extraction and pooling, and the final part focuses on issues of experimental design.

A. Array Choice

There are essentially three different types of microarrays: cDNA arrays, oligo-nucleotide arrays, and Affymetrix gene chips. Each of these is described next, followed by a discussion of the advantages and disadvantages of large-scale and boutique arrays in the case of spotted cDNAs and oligonucleotides.

1. cDNA Arrays

cDNA or PCR product arrays consist of lengthy probes (100–1000+ nucleo-tides) spotted onto a solid support. This is the traditional microarray approach, originating with nylon arrays (e.g., Herwig *et al.*, 2001). The probes are long and hence bind robustly to their labeled targets and are relatively easy to use. The majority of available microarray protocols have been developed for cDNA arrays, and hence these arrays do not require lengthy periods of technical development. Their robustness, however, is also the Achilles' heel of this approach. Long probes tend to have problems with cross-hybridization to a variety of targets that are members of closely related genes. As such, cDNA arrays are being superceded by oligonucleotide arrays. Much of the microarray research in the zebrafish commu-nity to date has involved the use of cDNA arrays (e.g., Handley *et al.*, 2002; Lo *et al.*, 2003; Ton *et al.*, 2002).

2. Spotted Oligonucleotide Arrays

The probes for spotted oligonucleotide arrays are considerably shorter than those for cDNA arrays, typically 40- to 80-mer in length. A subset of spotted oligonucleotide arrays employs 10- to 26-mer probes and is used to detect single nucleotide polymorphisms (SNPs), but these are not discussed further here, except to state that this approach has already been demonstrated in zebrafish (Stickney *et al.*, 2002). Oligonucleotides are designed to be gene specific and are usually biased toward the 3′ end of a transcript as oligo d(T) primer driven cDNA synthesis can be used to synthesize the relevant target. Oligonucleotide probes can be spotted onto the slides, using printing pins to transfer the probes from a well to the side surface, or can be printed by using ink-jet technology either as entire probes or by *in situ* synthesis. At present, printing technology has improved to the point where it is possible to deposit more than 80,000 spots on a standard 1 in. × 3 in. glass slide. In the case of boutique arrays, in which the total number of genes being assayed is small, there is sufficient room on a glass slide to accommo-date several probes in order to target multiple domains within a gene. This approach allows the interrogation of alternative splicing events, thus improving the capture of biologically relevant information. The length of oligonucleotides influences the specificity of hybridization and hence affects the tolerance for mismatches between probe and target. This aspect can be an important factor if the design of probes is based on expressed sequence tags (ESTs) rather than fully characterized gene sequences. Oligonucleotides are more sensitive to hybridization temperature and washing stringency than cDNA or PCR products are and might require considerable technical development. At present, there are three commercially available oligonucleotide sets for zebrafish. MWG-biotech (www.mwg-biotech.com) offers a set of 14,067 oligonucleotides based on a combination of 1800 gene sequences and 12,768 open reading frames (ORFs) from the GenBank database of the National Center for Biotechnology Information (NCBI). Qiagen-Operon (oligos.qiagen.com) offers a more modest set of 3479 oligonucleotides consisting of 1206 well-annotated genes and 2273 ESTs designed from the Zebrafish Reference Sequence (RefSeq) and UniGene databases. Sigma-Genosys (www.sigma-genosys.com) offers a set of 16,399 oligonucleotides designed from a combination of publicly available mRNA and EST sequences from GenBank.

3. Affymetrix Gene Chips

Affymetrix (www.affymetrix.com) gene chips use multiple 25-mer probes that are synthesized directly on quartz wafers by using photolithography. The probe capacity of a standard-sized gene chip has increased to more than 1 million features and is continuing to increase through a steady decrease in feature size. The Affymetrix design approach uses combinations of 11–16 probes to target a single transcript. Each of these perfectly matched (PM) probes also has a complementary mismatch probe (MM), identical to the PM except for a single mismatch at the 13th oligonucleotide position. Unlike most spotted arrays, only a single labeled sample is hybridized to an Affymetrix gene chip. In the case of a single transcript, the extent of hybridization to all the relevant PM probes is assessed, minus the hybridization to the MM probes.

This approach allows an assessment to be made of the level of cross-hybridization to similar transcripts, the removal of background noise, and an estimation of the likelihood that the transcript is indeed present. In addition, this approach gives an indication of overall transcript abundance. The Affymetrix system is well characterized, and therefore technical development is generally not necessary. However, Affymetrix arrays can be considerably more costly than spotted arrays and do not lend themselves to customization. Affymetrix has developed a Zebrafish Genome Array gene chip that consists of probes to 14,900 transcripts compiled from the RefSeq, GenBank, NCBI Expressed Sequence Tags database (dbEST), and UniGene databases.

4. Large–Scale Arrays and Boutique Arrays

Although microarrays are most well known for their ability to interrogate simultaneously the entire transcriptome of an organism or tissue, small-scale boutique arrays can also be highly useful. The choice of whether to use probes targeting the largest possible number of genes or to use a small-scale focused array depends on the research question. There are distinct advantages to each approach. Large-scale arrays are useful for gene discovery and the assessment of perturbations for which the effect is unknown or poorly characterized. Large-scale arrays can have technical problems due to spatial bias, and subsequent data analysis can be a time-consuming process. Boutique arrays are most useful to interrogate defined subsets of well-characterized genes, such as gene families or biological pathways. In the case of boutique arrays, housekeeping genes or spiking controls are necessary to normalize the data, as small arrays violate many of the assumptions underlying several commonly used normalization and analysis procedures that have been developed for large-scale arrays (see Section III-C). Boutique arrays can also be more cost efficient because of a reduction in some reagent costs, but their main benefits are the ease with which slides can be made and the ease of analysis.

B. Array Design

1. Design of Oligonucleotide Probes

Standard probe design approaches aim to minimize secondary structure, standardize probe length and melting temperature (T_m), achieve a bias to the 3′ end of the transcript, and avoid repetitive or low-complexity regions. Depending on the slide surface chemistry, oligonucleotides can be modified to facilitate attachment. With intensive characterization of a gene, it is also possible to design oligonucleotides to span intronic regions and thus assess alternatively spliced transcripts. Long oligonucleotide probes (50- to 70-mer) can be considerably more sensitive than shorter probes. For example, Agilent Technologies (Fulmer-Smentek, 2003) reports that 60-mers are 5- to 8-fold more sensitive than 25-mers. However, the sensitivity of the oligonucleotide needs to be balanced with specificity, which tends to decrease with increasing length. When deciding on the optimal probe length, the quality of the target sequence also needs to be considered. When sequence reliability might be questionable, such as from EST databases, longer oligonucleotides might be desirable because of their relative insensitivity to 4- or 5-bp mismatches. Several different software packages are currently available for oligonucleotide probe design; however, most companies involved in oligonucleotide synthesis now also provide microarray-related probe design.

2. Design of Custom Arrays

In addition to designing one's own probes, it is often possible to obtain pre-designed probes for genes of interest. For example, although MWG-biotech offers a 14k zebrafish microarray oligonucleotide set, the company also provides a facility for selecting subsets of genes from its catalog for creating a boutique array. When designing a boutique custom array, it is important to include control probes to housekeeping genes, negative controls, and preferably spiking controls. Another factor to consider when designing boutique arrays is to maximize the number of features within as small an area as possible. By constraining the array to fit under a small 22×22 mm^2 coverslip, reagents costs can be reduced, spatial bias minimized, and the concentration of the target increased. In addition, multiple printing of each probe onto a slide will reduce the risk of loss of information caused by artifacts, overcome print tip problems, and avoid spatial bias issues. For example, we at present print each of our zebrafish probes 6–12 times onto a slide surface. It is also now possible to print small custom arrays (<1000 features) onto the bottom of 96-well plates, thus facilitating high-throughput microarray screening applications.

C. mRNA Extraction, Amplification, and Pooling

mRNA quantity and quality are critical factors in microarray experiments involving zebrafish. Young zebrafish embryos (<24 hpf) yield little RNA; therefore, consideration should be given to either amplify the RNA from a single or limited number of embryos or to extract RNA from a pool of many embryos. RNA amplification is expensive and is a linear as opposed to a logarithmic process, but allows RNA to be used from small amounts of tissue, such as single embryos or specific cells, or tissues harvested by laser microdissection. Unfortunately, amplification can run the risk of biasing a population of transcripts because of differential efficiencies using enzyme-based methods.

In microarray experiments, pooling refers to combining mRNA extracted from multiple organisms from the same treatment group. For example, in an experiment comparing gene expression between wild-type and mutant organisms, mRNA can be extracted from five organisms of each genotype and combined into two pooled mRNA samples, one for each genotype. The advantage of this approach is that it increases the amount of mRNA available for labeling and hybridization, which is particularly important for small organisms. It also has the effect of averaging out the variation in expression levels between individual organisms, or experimental units, of each genotype. This phenomenon can be considered either positively or negatively, depending on the nature of the experiment. The critical question is whether it is important to estimate the amount of intersubject variability within a treatment condition. If an estimation of this variability is not important, then pooled samples provide a suitable means of removing the effect of intersubject variability when assessing potentially differentially expressed genes. Kendziorski *et al.* (2003) provide a discussion of the statistical aspects of pooling in microarray experiments.

D. Experimental Design

The increasing popularity of microarray experimentation has led to a rediscovery of the fundamentals of experimental design in the statistical literature. Although there might exist a myriad of possible designs for any given experimental question, a number of simple designs are able to improve the quality of information achieved from microarray experiments.

By its very nature, microarray technology is prone to variability that needs to be understood and overcome by good design in order to extract meaningful data. Variability begins at the level of the slide. Most high-quality, commercially produced arrays have minimal intra- and interslide variability. The spots are often very consistent in morphology and spacing. Some slides arrive with CDs containing a spot location file specific to that array (e.g., Agilent piezoelectric printed slides) or are otherwise designed to automate spotfinding (e.g., Affymetrix). However, the expense of these slides tends to become a limiting factor for researchers. By contrast, cheaper slides that are often printed at academic facilities can be quite variable both at the intraslide level (feature shapes, pin-to-pin variation, and artifacts) and between batches of slides. Although cheaper than the commercial arrays, more technical replicates are needed to provide confidence in the data.

During the processing of microarray experiments, many additional factors can lead to technical variability. The most common source of problems arise during RNA extraction and handling and during hybridization and washing steps. These problems typically present themselves as high or uneven background, spatial bias, greatly mismatched channel intensities, or poor signal strength. In the case of competitive hybridizations applied to spotted arrays, an additional level of variability is caused by the potential for different degrees of binding of Cy3- and Cy5-labeled targets to a given probe. Several approaches can be used to account for technical variability. At the level of the array itself, the location of probes (particularly control probes) should be randomized over the array, and ideally each probe should be printed multiple times. At the level of the experiment, technical replication is required. For competitive hybridization, the minimum technical replicate is the dye switch, in which the same samples are simply labeled with the opposite dye and hybridized to a second array. In addition, simple replication of the experiment is advisable, although biological replication can be used as a substitute for technical replication, depending on the overall design of the experiment (see later).

The most simple two-treatment microarray experiment involves hybridizing two fluorescently labeled samples (one for each treatment) to an array and comparing the fluorescence intensities produced by laser scanning. Genes that exhibit differences in fluorescence intensity across the two conditions are then considered to have possibly undergone differential expression. Although this seems relatively straightforward, the second-hand nature of microarray data (in that transcript abundance at each spot is not directly observed and must be inferred by assessing fluorescence intensities) means that changes in fluorescence intensity are not necessarily a faithful indicator of differential expression. Even in such a simple setting, differences in the labeling efficiency and hybridization characteristics of the two fluorescent dyes, or spatial variation on the surface of the array, can make the detection of real differential expression challenging. As experiments become larger, for example in comparing more than two treatments, genotypes, or time-points, additional confounding factors begin to appear, all of which affect the accurate assessment of genes undergoing statistically significant differential expression.

1. Reference Designs

To facilitate comparisons between treatments on different arrays, many initial microarray experiments used a reference design (Kerr and Churchill, 2001) in which a mixture of a fluorescently labeled sample of interest and a differentially labeled reference sample were hybridized to a microarray. Additional samples relating to other treatments were then labeled and mixed with the differentially labeled reference sample for hybridization to subsequent arrays. An analysis of the expression levels across the two treatment conditions by using this approach required a comparison through the reference sample channel of each array. The advantage of this design is that each treatment is always compared to the same thing (the reference sample); the disadvantage is that half the channels in any experiment are taken up by the reference sample, which might not be of interest at all.

Two important attributes of the reference sample are that each gene expressed in the treatment sample is also represented in the reference sample and that the expression of the reference sample transcripts is completely consistent across each slide. A reference sample generated by combining transcripts from several treatment conditions ensures that every gene expressed in each treatment sample is also present in the reference sample. Provided there is a sufficient template, this approach is relatively straightforward. However, this sort of reference sample is useful for a given series of experiments only. If later expansion of the experiment is required, or if comparison of the results across many experiments is desired, then a universal reference sample can be considered. These reference samples can be generated from fragmented genomic DNA or from large quantities of RNA sourced from multiple tissues or cell cultures. [See Kim *et al.* (2002) for comparison of RNA and genomic DNA reference pools.] For some species, commercial reference samples are available to allow data comparison to be made between laboratories.

2. Alternative Designs

Although the reference design approach provides a flexible approach to microarray experimentation (particularly for experiments involving large numbers of treatments or if additional treatments are to be added at a later time), the fact that half the available data relates to the reference sample has been considered wasteful by some authors (Kerr and Churchill, 2001). An alternative to the reference design is to take a traditional statistical approach to experimental design that strives to achieve balance in terms of dyes (each treatment must be labeled by each dye the same number of times) and treatments (each treatment must appear in the experiment the same number of times as every other treatment), without the use of a reference sample. Through the appropriate pairing of differentially labeled treatment samples, designs can be created that use the same number of arrays as the reference design but for which the variance of comparisons of interest are greatly reduced, resulting in increased power to detect genes undergoing differential expression. One example of this approach is the loop design of Kerr and Churchill (2001), in which treatment pairs are arranged on arrays based on a cyclic pattern. Other more complex variations are also possible (Churchill, 2002). A disadvantage of this approach, except in the special circumstance that every treatment pair occurs together on an array at least once, is that the variance of comparisons of treatment pairs that do not occur together on an array is greater than the variance for treatment pairs sharing an array.

3. Technical and Biological Replication

Churchill (2002) provides an excellent review of the various issues encountered in designing microarray studies, the major theme of which involves recognizing the variance components present in the experiment. The three components that Churchill (2002) identifies are biological variation, technical variation, and measurement error. The first of these relates to the inherent variability that occurs between experimental units; in the case of zebrafish, it refers to the natural variation in transcript abundance for a given gene between different pooled mRNA samples. As an illustration of this point, consider extracting mRNA from 100 genetically identical zebrafish, all of which inhabit the same environment, and then creating two mRNA pools, each of size 50. If these pooled samples were labeled and then hybridized to a microarray, some level of variation in intensity levels would be observed for each gene. This difference does not involve an interesting change in gene expression, but rather reflects the natural variability of transcript abundance. The second of these variance components, technical variation, refers to the variability observed when a microarray experiment is repeated. This is accomplished by splitting each pooled mRNA sample in two and hybridizing them to two microarrays. Although there will often be good agreement between the intensities recorded on the two arrays, there will still be some differences, which simply reflect the variability of the experimental process.

The third variance component, measurement error, encompasses the familiar statistical concept of random error. Even when probes are replicated at multiple positions on an array, the fluorescence intensities recorded from each probe will vary because of natural variation in the hybridization process, which is essentially random. The use of replication at all levels of a microarray experiment is required if suitable estimates of variability are to be obtained. As Churchill (2002) notes, an experiment that does not contain biological replicates (e.g., multiple pools of distinct mRNA) is unable to estimate the presence of biological variation. The consequence of not addressing this design imperative is that any differentially expressed genes that might be detected in the experiment might simply be peculiar to that particular mRNA pool and not be reproduced with a second pool of mRNA.

III. Array Preprocessing

Prior to statistical analysis of a scanned slide, there needs to be several levels of normalization to account for the variability inherent in spotted microarray experiments. The degree of normalization required for a given slide is dependent on a multiplicity of factors, including the success of removing unbound probe, cross-hybridization, dye effects, background effects, spatial bias, and the presence of dust and other artifacts on the slide surface. Normalization attempts to balance the dye channels and account for noise that is unrelated to genuine differences between the two samples on the slide. Various normalization procedures can be applied to microarray data.

A. Background Correction

Most standard microarray scanners measure the average (mean or median) intensity of pixels within an area defined as the *spot feature* and also the median intensity of pixels from parts of the array outside the features, termed the *background*. This measurement of background intensity can be averaged over the entire array or taken from an area adjacent to the feature in question. The median background intensity is used to avoid high-intensity artifacts such as flecks of dust that have no bearing on the intensity of pixels within the feature, but would skew the measure of mean background intensity. For each channel, the background intensity can be subtracted from the raw feature intensity. This approach can correct for fluorescence artifacts that increase both background and feature intensity of one channel in a localized area of the array (Fig. 1A). In contrast,

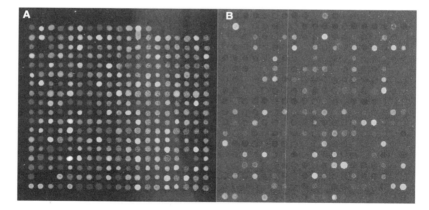

Fig. 1 Removal of background intensity. (A) A green swirl of hybridization solution overlays this block. Within the area of the swirl, both the background and feature pixels are affected. (B) A faint array resulting from washing at a high stringency reveals the natural fluorescence levels of the slide surface chemistry. The probes to which no target has bound have a lower fluorescence than that of the surrounding background. (See Color Insert.)

background removal can introduce variability into the data when slide surface chemistry of areas of the slide without spots is different from that of areas containing probe (Fig. 1B).

B. Visualization of Microarray Data

To determine the most appropriate normalization procedures for a microarray slide it is important to examine the slide, both at the level of the scanned image and by plotting the data under various transformations. An examination of the image itself will reveal many problems, such as abnormal background intensities and obvious spatial bias effects. However, plotting the data can reveal more subtle effects, particularly dye effects, and is useful to monitor the impact of subsequent normalization. A standard method used is the MA plot (Dudoit *et al.*, 2002) that plots the log2 of the fluorescence intensity ratio (M) against log10 of the product of the intensities (A); this plot is also referred to as an RI plot (Quackenbush, 2002). The MA plot identifies intensity-related biases, such as dye effects. Effects relating to the location of features on a slide can be detected by plotting the log2 of the fluorescence intensity ratio by the feature number or by plotting box plots of variability to detect pin effects. The Web site SNOMAD (http://pevsnerlab. kennedykrieger.org/snomadinput.html) includes a function that plots variation in channel intensities and overall differences in the average ratio across the slide. This plotting approach is used to detect spatial bias resulting from hybridization artifacts.

C. Global Normalization

In the case of hybridization of large arrays, an assumption is made that the majority of transcripts represented on the array will remain unchanged. If this assumption is valid, then it is possible to correct for different quantities of template between two samples or for different labeling efficiencies. For each channel, every feature is divided by the mean intensity of all the features. Table I shows a worked example indicating how the addition of three times as much of one sample than the other can affect the RNA ratio (Table I, actual RNA present in tissue and RNA used in microarray experiment) and how global mean normalization restores the original RNA ratio (Table I, global mean normalization). Although global mean normalization is extremely useful to correct dye intensity imbalances, it is not valid to use this method on boutique arrays.

D. Spike-Mix Normalization

Spike mixes consist of specific probes, typically of a different species from that to be studied on the array, corresponding to targets that are visually incorporated into samples at the cDNA synthesis step. Spike mixes have a variety of uses, including assessing the dynamic range and transcript detection limits, the calibrating ratios, and, particularly in the case of boutique arrays, balancing the dye channels. Spike

Table I
Global Mean Normalization

	Actual RNA present in tissue[a]			RNA used in microarray experiment[b]			Global mean normalization[c]		
Gene	RNA in treatment	RNA in Control	RNA ratio	3 × Treatment RNA on array	Control RNA on array	RNA ratio	Treatment RNA divided by mean	Control RNA divided by mean	RNA ratio
1	5000	1000	5:1	15,000	1000	15:1	2.5	0.5	5:1
2	2000	1000	2:1	6000	1000	6:1	1	0.5	2:1
3	1000	1000	1:1	3000	1000	3:1	0.5	0.5	1:1
4	1000	2000	1:2	3000	2000	1.5:1	0.5	1	1:2
5	1000	5000	1:5	3000	5000	1:1.5	0.5	2.5	1:5
				Mean intensity 6000	Mean intensity 2000				

[a]The relative transcript levels of five hypothetical genes in two samples designated *treatment* and *control*.
[b]In the microarray experiment, the researcher placed three times as much of the treatment than control sample on the array.
[c]The intensity of each gene is divided by the mean intensity level of all genes, resulting in restoration of the original RNA ratio levels.

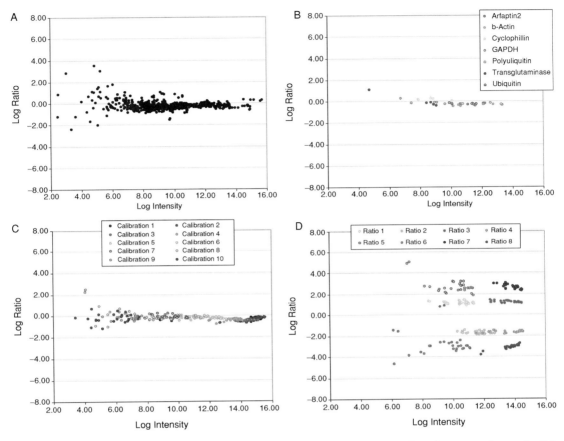

Fig. 2 Zebrafish spiking control genes and housekeeping genes. Data from one of our zebrafish apoptosis arrays is presented, partitioned into four components: genes of interest, housekeeping genes, calibration controls, and ratio controls. These data represent a pool of embryos that were harvested 6 h following treatment with 15 mJ of UV radiation, compared to embryos of a similar age (78 hpf) that were not exposed to UV radiation. Every probe was printed onto the array six times. The spiking controls were part of the Lucidea Universal Scorecard (Amersham). These data were achieved from one of our early slides printed onto polylysine-coated slides (ESCO), using a Stanford-type arrayer and Majer Precision pins. (A) MA plot of the genes of interest present on this array, prior to normalization using the spiking controls and housekeeping gene. (B) Seven housekeeping genes were used to cover a range of intensities and provide sufficient data to enable normalization of the genes of interest to a $\log_2$ ratio of zero. (C) Ten calibration spikes from the Lucidea Universal Scorecard were included in the array. The spiking concentration covers four orders of magnitude to enable an estimation of the limit of detection, saturation, and to allow for normalization of the genes of interest to a $\log_2$ ratio of zero. There are sufficient data points in these controls to enable visualization of nonlinear intensity-related biases in this array. (D) The Lucidea Universal Scorecard also includes eight ratio controls at both high and low intensities to determine the extent to which the ratios obtained varied from the expected values. This aspect is an often overlooked consideration, particularly if significance is being determined by a twofold cutoff. (See Color Insert.)

mixes can either be generated in-house, e.g., using PCR products, or purchased commercially. For our zebrafish boutique array, we have incorporated the Lucidea Universal Scorecard (Amersham) that comprises 23 probes to artificial genes (based on yeast intergenic regions) and their complementary targets (Fig. 2C, D). The scorecard consists of calibration controls, ratio controls, and negative controls. The calibration controls are useful for balancing the channels in boutique arrays.

E. Normalization Using Housekeeping Genes

Genes that are assumed to be present in both samples at unchanging ratios can be used as housekeeping genes to normalize the data. This approach is particularly important for boutique arrays for which many of the alternative normalization procedures cannot be used. An array should include housekeeping genes that cover a range of transcript abundances. For example, our boutique zebrafish array included arfaptin2, polyubiquitin, β-actin, ubiquitin, cyclophilin, transglutaminase, and GAPDH (Fig. 2B).

F. Smoothing Methods to Remove Experimental Effects

1. Global Loess

The most commonly used dyes for microarray experiments are Cy3 (indocarbocyanine) and Cy5 (indodicarbocyanine). These dyes have different emission profiles and slightly different labeling efficiencies. During scanning it is typical to try to balance the two dyes by altering the PMT setting of a scanner's lasers in order to make the majority of spots appear yellow. However, on plotting the ratios of each feature by the combined intensity of both channels for that feature (the MA plot), a nonlinear pattern in the distribution of the ratios is often seen (Fig. 3A). This nonlinear effect can be removed by using loess regression to calculate a local mean of the ratios of a proportion of the spots within a given intensity range. Both the range of intensity used and the proportion of outlier spots excluded from this analysis are customizable. The local mean ratio is then subtracted from each feature's ratio, resulting in the removal of the nonlinear dye effect (Fig. 3B).

2. Pin–Tip Loess

Global loess normalization can be extended to incorporate a separate smoother for each print tip, allowing the removal of pin-related artifacts. These generally occur as a result of differences in the physical characteristics of the pins and can lead to marked differences in the recorded intensities for the spots printed by different pins. Like the global approach, pin-tip loess makes the assumption that the majority of genes printed by each pin exhibit no differential expression and transforms the data to reflect this belief. In addition to removing effects relating directly to print tips, the pattern in which arrays are printed means that pin-tip loess also has the potential to act as a crude form of spatial normalization, applying

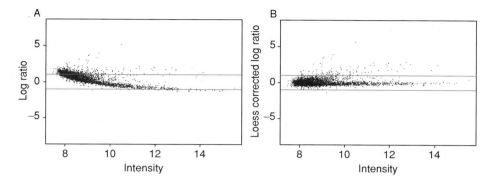

Fig. 3 Loess correction of mean feature intensity by channel. It is common for intensity-related biases to be present within microarray data. The effect of global loess smoothing on clearly biased data is shown here. The red lines demark an arbitrary fold significance level. (A) MA plot prior to correction. (B) MA plot of data after normalization. (See Color Insert.)

differing transformations at different spatial locations on the array (Fig. 4). More specialized approaches certainly exist, but pin-tip loess often provides an extremely effective method of normalization.

3. Local Mean Normalization Across the Array Surface

A relatively new method of array normalization is available online as part of the SNOMAD package (http://pevsnerlab.kennedykrieger.org/snomadinput.htm) and attempts to remove spatial bias occurring across the surface of the array. Spatial bias can occur for a variety of reasons, including printing effects (such as differences in print tips), failure to mix the hybridization solution adequately, desiccation at the edge of the coverslip, and coverslip irregularities. By using local mean normalization, the intensity of each element is divided by the local mean intensity determined by generating two loess curves, one for each axis of the slide. In this way, local mean normalization can correct for both spatial bias and background fluorescence artifacts that additionally affect feature intensity (e.g., Fig. 4).

IV. Analysis of Microarray Experiments

Once the intensity data in a microarray experiment have been normalized, it is possible to proceed with a statistical analysis. This logical progression from normalization to analysis is not mandatory, as some analysis methods incorporate normalization directly into the analysis process (e.g., the ANOVA approaches of Kerr *et al.*, 2000, and Wolfinger *et al.*, 2001). However, we have chosen to take this approach here as it has become relatively popular.

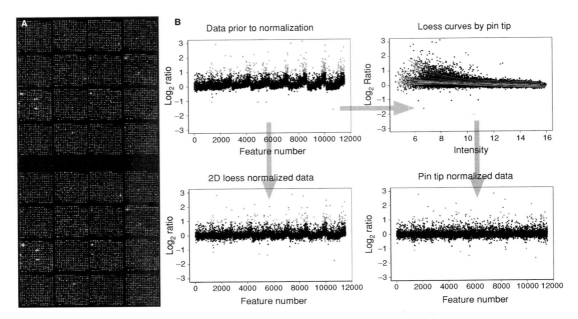

Fig. 4 Pin tip and local mean normalization across the array surface. Here we demonstrate the effects of normalizing to remove spatial bias. (A) This array is printed in two sets of 4 × 4 blocks and has a green smear (probably SDS) that begins on Block 8 of Set 1 and increases in severity in the rightmost blocks. (B) The effect of the green smear is most apparent when the $\log_2$ ratio of the fluorescent intensities is plotted against the feature number (in order from top left to bottom right). Pin tip normalization generates loess curves for each pin tip (each block) and the overall correction of this spatial effect is relatively good (bottom right graph). By contrast, local mean normalization across an array surface (2D loess) creates a form of topographical map using loess curves and attempts to flatten this map. Surprisingly, in this example, the 2D loess did not perform as well as the pin tip loess correction. (See Color Insert.)

The method of analysis that is chosen depends on the goal of the experiment. The most common goals are usually the identification of genes undergoing differential expression or the grouping of genes sharing similar expression profiles. In the former case, statistical hypothesis testing provides a natural method for the detection of genes undergoing statistically significant changes in expression level between pairs of treatment conditions. In the case of the latter, clustering methods are often used. Each of these approaches is discussed in detail next.

A. Hypothesis Testing

To identify genes undergoing differential expression, it is necessary to determine whether any changes in gene expression are at such a level that the observed fluctuations are unlikely to be due to chance. Standard hypothesis testing methods provide a means for undertaking this determination.

The generic form of a hypothesis test involves a simple comparison of expression levels, which can be stated as follows:

$$H_0 : \mu_{1g} - \mu_{2g} = 0$$

$$H_A : \mu_{1g} - \mu_{2g} \neq 0$$

where μ_{ig} denotes the mean log expression for gene g under condition i. By conducting the test on the log-intensity scale, the null hypothesis that the difference is equal to zero is equivalent to testing that the fold change in expression is equal to unity, i.e., no change in average expression level. To conduct the hypothesis test, a test statistic is required:

$$T = \frac{\bar{x}_{1g} - \bar{x}_{2g}}{\text{SE}}$$

where $\bar{x}_{ig}$ denotes the estimated mean log expression (normalized and background corrected if appropriate) of gene g under condition i, and SE represents a generic standard error estimate for the comparison, which depends on the variance structure of the data. The quantity $\bar{x}_{ig}$ can be calculated directly from the raw intensity data, which is not recommended, or more correctly from the normalized intensity data or the estimated parameters of a linear model.

In the context of the variance structure, the ANOVA approach of Kerr *et al.* (2000) can be used, but assumes a common variance for all genes on an array. In contrast, other approaches calculate standard errors based on the observed variability of the gene in question (Craig *et al.*, 2003; Dudoit *et al.*, 2002; Wolfinger *et al.*, 2001). If it is assumed that the variability of the intensity measurements is constant across the treatment conditions, then a pooled variance estimate can be formed for any gene; otherwise, variances are calculated under each condition. In the former case, the form of the test statistic is equivalent to a standard two-sample *t*-test with a pooled variance estimate, whereas the latter case is the standard two-sample test statistic (Welch's test) for the difference between two means. The more complex approach of Churchill (2002) can also be taken, with a calculated standard error term that incorporates multiple variance components. Other approaches such as the MIDAS tool, developed by the Institute for Genome Research (TIGR), use a sliding window approach across the MA plot to calculate intensity-dependent variance estimates. In all methods, however, the form of the test statistic is the same as the general form given previously, which is an estimate of the log differential expression divided by some estimate of the variability of the observed difference.

A variation on this approach has been proposed by Tusher *et al.* (2001), who suggest adding a constant to the denominator of the test statistic:

$$T = \frac{\bar{x}_{1g} - \bar{x}_{2g}}{\text{SE} + \alpha}$$

where α is a constant that minimizes the coefficient of variation to a relative difference statistic defined by the authors. This approach has the advantage of acting as a shrinkage estimator, reducing the effect of genes with very high or very low variances estimates.

1. Determining Statistical Significance

Once the form of the hypothesis to be tested has been chosen, statistical significance is determined by comparing the test statistic of each gene with the distribution of the test statistic under the null hypothesis. This distribution can be determined by using standard normal testing theory, in which the population distributions of $\bar{x}_{1g}$ and $\bar{x}_{2g}$ are assumed to be normal, with means of μ_{1g} and μ_{2g}, respectively. This approach allows the student's-t distribution to be used to generate a p value for each test.

Sometimes the assumption of normality might not be considered justified, (e.g. if there is extreme skewness present in the distribution of the residuals for each gene, under each treatment condition), in which case resampling-based methods can be used to achieve a distribution-free approach to hypothesis testing. Given the small number of observations generally available to make this assessment, however, it is often difficult to determine the appropriateness of the normality assumption. In practice, we have found that using resampling-based methods in the microarray setting tends to produce more liberal p values than do methods that rely on the normality assumption to produce p values (Craig *et al.*, 2003). This outcome tends to occur because the log of the intensity data tends to have "lighter tails" than those of the best-fitting normal distribution, which results in the resampling-based method producing lower variance estimates and thus lower p values. Depending on the setting, the conservative bias of normality-based methods might be an advantage as they provide greater protection against false positives. However, these methods have less power to detect nonnormally distributed genes undergoing significant differential expression.

2. Multiple Comparisons Procedures

Although the hypothesis testing approach described previously produces a p value for each gene, the selection of a significance threshold becomes important when dealing with the thousands of hypothesis tests involved in microarray experimentation. In the case of a single hypothesis test, choosing an α of 0.05 defines a procedure that will incorrectly reject the null hypothesis in 5% of cases. Such occurrences are also referred to as false positives, or type I errors. For a microarray experiment involving 10,000 genes, this corresponds to an average of 500 genes that would be incorrectly identified as differentially expressed.

As a means of overcoming this problem, multiple comparison procedures (MCPs) are used to provide a degree of control over the number of type I errors. Two of the most commonly controlled error rates for microarray experiments are the family-wise error rate (FWER) and the false discovery rate (FDR). Control of the FWER below some level α guarantees that the probability of at least one type I error (i.e., false positive) is less than or equal to α regardless of the number of tests being conducted or the proportion of true null hypotheses. In the microarray setting, FWER control is generally considered to be far too conservative, and FDR control is often presented as a favorable alternative (Craig *et al.*, 2003). Control of the FDR guarantees that the expected proportion of false discoveries (i.e., false positives, or type I errors) out of the total collection of rejected null hypotheses is less than or equal to α. That is, if 100 null hypotheses are rejected by using a FDR controlling method based on $\alpha = 0.1$, then, on average, 10 of those rejections are expected to actually be false positives. In terms of microarray experiments, this would be equivalent to having a list of 100 genes that are considered to be significantly differentially expressed, and knowing that, on average, 10 of those genes actually did not undergo differential expression. What the procedure does not tell is the identity of those 10 genes. Because the FDR is defined as an expectation, or long-run average, it is also possible that *none* of the genes reported as differentially expressed are false positives or that *all* the genes are false positives. Various methods can be used to control for these two error rates, most of them based on sequential examination of ordered p values. Dudoit *et al.* (2003) provide a thorough review of multiple comparison procedures for microarray experiments. Regardless of the specific details, the end result of this approach is a list of genes considered to have undergone statistically significant changes in expression at an appropriate level of confidence. This list can then be used as a basis for further experimentation or as a starting point for additional confirmatory analyses such as realtime PCR or Northern blotting.

B. Clustering

In timecourse experiments and various other experimental settings, the main focus of a microarray experiment is often to identify groups of genes that exhibit similar patterns of expression over time. These patterns, often called profiles, can be grouped by various statistical methods, which are generically referred to as clustering techniques. Two of the earliest examples of clustering microarray data were described in Chu *et al.* (1998) and Eisen *et al.* (1998). Both studies applied hierarchical clustering methods to a yeast developmental timecourse and were successful in identifying groups of genes exhibiting similar expression profiles. These groups were then confirmed to be functionally similar. The advantage of such an approach is that when genes of unknown function cluster with genes of known function, similarity of function can be inferred through similarity of expression profile. Although not a foolproof approach, it provides researchers with a useful tool for investigating gene function.

The general method for applying hierarchical clustering to timecourse microarray data is to form a log of expression ratios for each gene at each timepoint, relative to the reference sample. Thus, if there are 10 timepoints in the experiment, each gene will have 10 log ratios: time one relative to the reference, time two relative to the reference, etc. The next step is to calculate the distance between each pair of expression profiles, based on a suitable distance metric. Although many suitable metrics exist, the one proposed by Eisen *et al.* (1998) remains popular for microarrays. Calculation of pairwise distances results in the creation of an $n \times n$ distance matrix, where n is the number of genes represented in the experiment. Although this matrix contains all necessary information about expression profile similarity, it is not easy to infer relationships directly from these data. In practice, the most common tool for visualizing distance matrices is the dendrogram (e.g., Fig. 5), which generates a rooted tree based on the pairwise distances. The branch length between genes in the dendrogram indicates the degree of similarity between their expression profiles. Genes that have similar profiles are then grouped together as leaves of the tree, separated by short branches, while genes with dissimilar profiles are separated by long branches.

Since the original applications involving hierarchical clustering, a large number of alternative methods have been proposed, all having the goal of identifying groups of genes that share similar expression profiles. A review of these methods is beyond the scope of this chapter, and readers are therefore directed to the work of Quackenbush (2001).

C. Analytical Software

Although a number of commercial options are available for the analysis of microarray data, a large amount of freeware and open source software exists that is of extremely high quality. In particular, the Bioconductor (www.bioconductor. org) library for the R computing environment (Ihaka and Gentleman, 1996) provides a comprehensive range of open source tools for the analysis of microarray data, including methods covering the issues discussed here: normalization, detection of differentially expressing genes, and cluster analysis.

V. Discussion

This chapter has taken a relatively broad view of microarray analysis, supported in part by our own experiences in developing boutique zebrafish arrays. Our experiments have involved a close collaboration with statisticians, which should be considered mandatory for any biologist entering this challenging field of study. Of importance, however, is that a wealth of information is currently available to support the aspirations of those wishing to understand the gene expression program of zebrafish, whether in terms of developmental processes or in assessing transiently imposed or heritable impacts. Having obtained microarray data, however, a means for others to access it is required. An increasing trend within research communities is to house published microarray data into publicly available and searchable databases. These databases can be used to help fulfill MIAME compliance (Brazma *et al.*, 2001) requirements and are an extremely useful resource. Generic microarray databases are currently being developed and house both spotted array (cDNA/oligonucleotide) and Affymetrix data; for example,

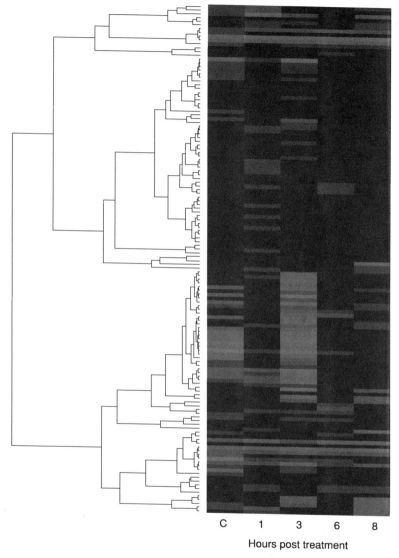

C   1   3   6   8

Hours post treatment

Fig. 5 Clustering. Clustering was used to examine the time course of expression of zebrafish genes encoding proteins comprising the apoptosis pathway following the exposure of 72 hpf zebrafish embryos to 15 mJ of UV radiation. Pools of 200 zebrafish were exposed to UV radiation and harvested at 1, 3, 6, and 8h following exposure. Transcripts from these embryos were labeled and hybridized against transcripts isolated from pooled embryos that were not exposed to UV radiation. "C" represents a control pool of 200 zebrafish that were not exposed to UV radiation. The dendrogram clusters genes by their similarity in expression profile. In this way, genes that ostensibly function in the same pathway can be determined. (See Color Insert.)

NCBI hosts the Gene Expression Omnibus (GEO; www.ncbi.nlm.nih.gov/geo). Users of model species that have their own information resource websites (such as ZFIN) have started to develop their own searchable microarray databases. This development allows more focused search options that are relevant to the given species (e.g., the *Arabidopsis* community has a Microarray Expression Viewer at www.arabidopsis.org) as opposed the more generic search options available through GEO. A similar database could be designed for zebrafish to enable search options relevant to this species. Such search options could include developmental stage, tissue type, and mutant characterization.

Use of microarray technology by the zebrafish community is in its infancy, and to date, initial applications have included the measurement of toxicity (Handley *et al.*, 2002), characterization of development (Ton *et al.*, 2002), mutation mapping (Stickney *et al.*, 2002), and drug discovery (Pichler *et al.*, 2003). We view the current tentative steps in developing zebrafish microarrays as a significant opportunity. Best practice developed by those studying human diseases and other model organisms can now be brought to bear for those working with zebrafish. We expect that the zebrafish can be more firmly placed in the field of comparative organism analysis by embracing the technologies used so successfully in other species. It should therefore be considered a given that the development and use of zebrafish microarrays offer a highly relevant technology platform to study vertebrate development and disease-based analyses, which will complement existing analytical approaches that have been used so far by the zebrafish community.

Acknowledgments

We acknowledge funding support of the University of Auckland Vice Chancellor's Development Fund, University of Auckland research Committee, Lottery Grams Board of New Zealand, and the Maurice and Phyllis Paykel Trust.

References

Bowtell, D., and Sambrook, J. (2002). "DNA microarrays: a molecular cloning manual." Cold Spring Harbor Laboratory Press, Cold Spring Harbor, NY.

Brazma, A., Hingamb, P., Quackenbush, J., Sherlock, G., Spellman, P, Stoeckert, C., Aach, J., Ansorge, W.,, Ball, C. A., Causton, H. C., Gaasterland, T., Glenisson, P., Holstege, F. C., Kim, I. F., Markowitz, V., Maltese, J. C., Parkinson, H., Robinson, A, Sarkans, U., Schulze-Kremer, S., Stewart, J., Taylor, R., Vilo, J., and Vingron, M. (2001). Minimum information about a microarray experiment (MIAME)—toward standards for microarray data. *Nat. Genet.* **29**, 365–371.

Chu, S., DeRisi, J., Eisen, M., Mulholland, J., Botstein, D., Brown, P. O., and Herskowitz, I. (1998). The transcriptional program of sporulation in budding yeast. *Science* **282**, 699–705.

Churchill, G. A. (2002). Fundamentals of experimental design for cDNA microarrays. *Natl. Genet. Suppl.* **32**, 490–495.

Craig, B. A., Black, M. A., and Doerge, R. W. (2003). Gene expression data: The technology and statistical analysis. *J. Agric. Biol. Environ. Stat.* **8**, 1–28.

Dudoit, S., Yang, Y. H., Callow, M. J., and Speed, T. P. (2002). Statistical methods for identifying differentially expressed genes in replicated cDNA microarray experiments. *Statistica Sinica* **12**, 111–139.

Dudoit, S., Shaffer, J. P., and Boldrick, J. C. (2003). Multiple hypothesis testing in microarray experiments. *Stat. Sci.* **18**, 71–103.

Eisen, M. B., Spellman, P. T., Brown, P. O., and Botstein, D. (1998). Cluster analysis and display of genome-wide expression patterns. *Proc. Natl. Acad. Sci. USA* **95**, 14863–14868.

Fulmer-Smentek, S. B. (2003). Performance comparison of Agilent's 60-mer and 25-mer *in situ* synthesized oligonucleotide microarrays. Publication No. 5988–5977 EN.Agilent Technologies, P90 Acto, CA, 12 pp. (http://www.chem.agilent.com/temp/radEB25E/00042209.pdf).

Handley, H., Grow, M., Fishman, M., and Stegeman, J. (2002). Generation of zebrafish cDNA microarrays for investigation of cardiovascular embryo toxicity by 2,3,7,8-tetrachlorodibenzo-*p*-dioxin. *Mar. Environ. Res.* **54**, 411–412.

Herwig, R., Aanstad, P., Clark, M., and Lehrach, H. (2001). Statistical evaluation of differential expression on cDNA nylon arrays with replicated experiments. *Nucleic Acids Res.* **29**, U1–U9.

Holloway, A. J., van Laar, R. K., Tothill, R. W., and Bowtell, D. D. L. (2002). Options available—from start to finish—for obtaining data from DNA microarrays II. *Nat. Genet. Suppl.* **32**, 481–489.

Ihaka, R., and Gentleman, R. (1996). R: A language for data analysis and graphics. *J. Comput. Graph. Stat.* **5**, 299–314.

Kendziorski, C. M., Zhang, Y., Lan, H., and Attie, A. D. (2003). The efficiency of pooling mRNA in microarray experiments. *Biostatistics* **4**, 465–477.

Kerr, M. K., and Churchill, G. A. (2001). Statistical design and the analysis of gene expression microarray data. *Genetical Research* **77**, 123–128.

Kerr, M. K., Martin, M., and Churchill, G. A. (2000). Analysis of variance for gene expression microarray data. *J. Computa. Biol.* **7**, 819–837.

Kim, H., Zhao, B., Snesrud, E. C., Haas, B. J., Toen, C. D., and Quackenbush, J. (2002). Use of RNA and genomic DNA references for inferred comparisons in DNA microarray analysis. *BioTechniques* **33**, 924–930.

Lo, J., Lee, S., Xu, M., Liu, F., Ruan, H., Eun, A., He, Y., Ma, W., Wang, W., Wen, Z., and Peng, J. (2003). 15,000 unique zebrafish EST clusters and their future use in microarray for profiling gene expression patterns during embryogenesis. *Genome Res.* **13**, 455–466.

Pichler, F. B., Laurensen, S., Williams, L. C., Copp, B., and Love, D. R. (2003). Chemical discovery and global gene expression analysis in zebrafish. *Nat. Biotechnol.* **21**, 879–883.

Quackenbush, J. (2001). Computational analysis of microarray data. *Nat. Rev. Genet.* **2**, 418–427.

Quackenbush, J. (2002). Microarray data normalization and transformation. *Nat. Genet. Suppl.* **32**, 496–501.

Stickney, H. L., Schmutz, J., Woods, I. G., Holtzer, C. C., Dickson, M. C., Kelly, P. D., Myers, R. M., and Talbot, W. S. (2002). Rapid mapping of zebrafish mutations with SNPs and oligonucleotide microarrays. *Genome Res.* **12**, 1929–1934.

Ton, C., Stamatiou, D., Dzau, V. J., and Liew, C.-C. (2002). Construction of a zebrafish cDNA microarray: Gene expression profiling of the zebrafish during development. *Biochem. Biophys. Res. Commun.* **296**, 1134–1142.

Tusher, V. G., Tibshirani, R., and Chu, G. (2001). Significance analysis of microarrays applied to the ionizing radiation response. *Proc. Natl. Acad. Sci. USA* **98**, 5116–5121.

Wolfinger, R. D., Gibson, G., Wolfinger, E. D., Bennett, L., Hamadeh, H., Bushel, P., Afshari, C., and Paules, R. S. (2001). Assessing gene significance from cDNA microarray expression data via mixed models. *J. Comput. Biol.* **8**, 625–637.

CHAPTER 29

Comparative Genomics, *cis*-Regulatory Elements, and Gene Duplication

Allan Force,[*] Cooduvalli Shashikant,[†] Peter Stadler,[‡] and Chris T. Amemiya[*]

[*]Molecular Genetics Program
Benaroya Research Institute
Seattle, Washington 98101

[†]College of Agricultural Sciences
The Pennsylvania State University
University Park, Pennsylvania 16802

[‡]Bioinformatics, Department of Computer Science
University of Leipzig
D-04103 Leipzig, Germany

I. Introduction

The completion of the genome sequence of *Danio rerio* is an unprecedented milestone for the zebrafish community. The genome sequence is greatly facilitating the identification of large numbers of mutant alleles generated in various mutagenesis screens and has largely transformed our logistical mindset from a positional cloning to a positional candidate strategy (Barrallo-Gimeno *et al.*, 2004; Collins, 1995; Malicki *et al.*, 2002). The genome sequence is also providing resources for high-throughput genomic analyses, including improved gene chips and other arrays (Lo *et al.*, 2003; Stickney *et al.*, 2002; Ton *et al.*, 2002). In this chapter, we discuss the component of the genome that is much less tangible than the coding sequences, but might be equally important in its physiological and ontogenetic functioning: the noncoding *cis*-regulatory component.

METHODS IN CELL BIOLOGY, VOL. 77

It was hypothesized as early as the 1960s and 1970s that for normal ontogeny to proceed, a tightly regulated program of genetic switches needs to be in place, both temporally and spatially (Britten and Davidson, 1969; Davidson, 2001; Davidson and Britten, 1973). Disruption of this process could lead to deviations in normal developmental pathways, the formation of homeotic mutations, and/or death. How this developmental blueprint was organized was not known; however, it was clear that genes played a major role and that changes to these gene networks could result in both disease in the short term and morphological change and speciation in the long term (Arnone and Davidson, 1997; Britten and Davidson, 1969; Davidson, 2001; Epstein *et al.*, 2004). It was not until the era of modern molecular biology that the components of such developmental pathways could be genetically dissected, albeit painstakingly (Myers *et al.*, 1985, 1986). Findings from these kinds of experiments corroborated the fundamental principles established earlier in simple prokaryotic systems, i.e., that genes are under the control of various regulatory units including promoter elements and distally located *cis* elements that serve to direct and/or abrogate functioning of these genes. These *cis*-regulatory elements (CREs) in the eukaryotic genome (enhancers, repressors) are now known to be surprisingly numerous and clearly outnumber the number of coding sequences in the genome. This realization has been borne out from the analysis of numerous metazoan genome sequences and has led to an intense interest in understanding the underlying biology of these elements. What is the logic of these CREs? How are they organized in the genome? How do they operate? Are they directly involved in human disease conditions? Can they be used as targets for pharmaceuticals? How has nature used CREs for adaptive purposes that drive morphological evolution? Can quantitative trait loci consist of CREs? And what happens to the regulatory component when genes become duplicated in the genome? These questions and more have culminated in a large effort by the National Human Genome Research Institute (NHGRI; www.genome.gov) to identify all the CREs and other genetically important elements in the human (vertebrate) genome, the so-called ENCODE project: ENCyclopedia Of DNA Elements (Collins *et al.*, 2003). This program consists of investigator-initiated projects that run the gamut from computational biology to high-throughput empirical approaches to identification of CREs by chromatin immunoprecipitation. However, a central component to all of these efforts is the employment of comparative genomics as an initial-pass method to delineate those regions that likely comprise CREs (Kim *et al.*, 2000; Pennacchio and Rubin, 2001, 2003; Santini *et al.*, 2003; Thomas *et al.*, 2003).

II. Comparative Genomics and Identification of *cis*-Regulatory Elements (CREs)

For the information embedded in a genome to be fully realized, the sequence needs to be compared with those of other taxa, that is, by procuring sequences from orthologous regions of different genomes for comparison and scrutiny.

Because of the extensive sequence divergences in the noncoding components of genomes, it is often very difficult to align such regions by using conventional local alignment tools such as BLASTN. Because of this demand, numerous tools have been developed to generate improved global alignments and graphically illustrate these results (Berezikov *et al.*, 2004; Bray *et al.*, 2003; Brudno *et al.*, 2003). The most popular visualization tools for this purpose are PIP (percent identity plot) and VISTA (visualizing global DNA sequence alignments of arbitrary length; Mayor *et al.*, 2000; Schwartz *et al.*, 2000). Figure 1 shows visualization of a global alignment of a portion of the HOX-A cluster by using both PIP and VISTA. In this demonstration, the genomic region from the 5′ end of the HOX cluster from the horn shark was used as the reference sequence in comparison to human and the duplicated HOX-Aa and -Ab clusters of the zebrafish. The two outputs (PIP and VISTA) show very similar results, except that the PIP plots use a series of dots to signify regions of high nucleotide identity, whereas VISTA plots use peaks. Nonetheless, the overall appearance and informational content of these graphs are essentially identical and both can be tweaked to specifications of the investigator.

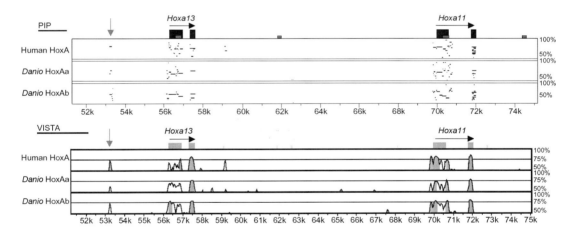

Fig. 1 Comparison of two commonly used methods for visualizing global alignments for the 5′ end of the HOX-A clusters of shark vs. that of humans, and the two duplicates of zebrafish. GenBank sequences used in the analyses were AF224262 (shark), AC004079 (human), AC107365 (zebrafish Aa), and AC107364 (zebrafish Ab). The *Hoxa13* and *-a11* genes are denoted by boxes. In the PIP plot, nucleotide identities are indicated by dots in contrast to the VISTA plot in which identities are indicated by peaks. In addition, the blue and red shading in the VISTA plot denote coding and noncoding sequence identities, respectively. Notable identities such as indicated by the red arrows and the region surrounding exon-2 of *Hoxa11* represent potential *cis*-regulatory elements (e.g., enhancers, suppressors, and micro-RNA sites). Not all pairwise comparisons are given in these outputs; only comparisons with the shark reference sequence are shown. Other conserved sequences might be present between given lineages, but to identify them one must swap reference sequences in the percent identity plot (PIP) and visualizing global DNA sequence alignment of arbitrary length (VISTA) analyses. (See Color Insert.)

The usefulness of such graphical tools for drawing inferences with respect to potential CREs is illustrated in Fig. 2. In this figure, PIP plots are given for two genes within an extended region being examined in a pilot ENCODE project; the plots were generated by using the human sequence as a reference and were ordered in increasing phylogenetic distance from the human sequence. The chimp shows almost identical levels of nucleotide identity across the entire region (including intronic regions), but the levels of identity are notably lower the farther away one gets from the human species. However, both genes are not evolving at the same rate: the CAV2 gene is diverging more slowly than the CAPZA2 gene. As would be predicted, for both genes the most distant species from humans (avian and fish) primarily show alignable regions within the exons only. Thus, to make meaningful comparisons (in which there is sufficient but not overwhelming phylogenetic signal), the choice of species for each genomic region being investigated might require consideration on a case-by-case basis. This principle, *phylogenetic shadowing*, necessitates strategic choice of taxa to maximize information content of the comparisons (Boffelli *et al.*, 2003).

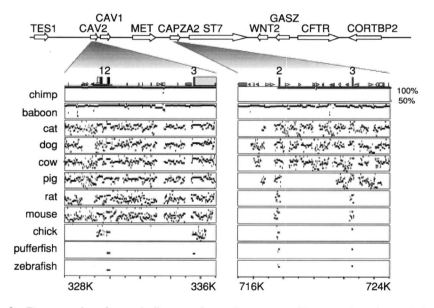

Fig. 2 Demonstration of genomic divergence in a region across a wide range of vertebrates. Pairwise comparisons across an extended region that encompasses the cystic fibrosis gene were made with a human reference sequence vs. those from 11 vertebrate taxa, primarily mammals. Partial PIP plots are shown for two gene regions, *CAV2* and *CAPZA2*. As would be predicted, the closer a species is phylogenetically to humans, the more nucleotide identity one observes, including intron sequences (e.g., chimp and baboon); sequences from more distant taxa show decreasing levels of identity. The degree of divergence between the two genes, however, is shown to vary. Adapted and modified from Thomas, T. W. *et al.* (2003). Comparative analyses of multi-species sequences from targeted genome regions. *Nature* **424**, 788–793, with permission. (See Color Insert.)

Although there is no question that we are far from understanding the underlying biology and evolution of CREs, the general comparative genomics strategy outlined previously appears to be highly effective as a first-pass method to identify potentially functional elements in the genome. Moreover, observation of overtly conserved *cis* regions across very wide phylogenetic distances strongly suggests that such sequences are functional and under purifying selection. An example of one such conserved region is shown upstream of *Hoxa13* in Fig. 1 (red arrow), a CRE shared by all vertebrate HOX-A clusters examined to date. An alignment of this region from five disparate vertebrate taxa (including two zebrafish duplicates) is shown in Fig. 3A. In this case, the strict conservation of the element across 800 million years of divergence (i.e., since sharks, bichirs, coelacanth, humans, and zebrafish last shared a common ancestor) implies that it might have a basal regulatory function. Although computational algorithms can be used to identify bona fide DNA-binding sites within these regions, the available reference databases (e.g., TRANSFAC) are largely incomplete, especially with regard to nonmammalian taxa. Thus, relatively tedious empirical methods are often necessary to interrogate these potential DNA-binding sites, such as electrophoretic gel shift assays (EMSAs), *in vitro* reporter expression assays (e.g., CAT and luciferase assays), and *in vivo* reporter expression assays (Hamada *et al.*, 1984; Nobrega *et al.*, 2003; Pennacchio and Rubin, 2001; Popperl and Featherstone, 1992; Shashikant and Ruddle, 1996; Trinklein *et al.*, 2004). An example of the last is given in Fig. 3B, in which the CRE upstream of zebrafish *hoxa13b* was used to drive expression of a basal promoter-*LacZ* construct in a mouse embryo. These results show that the element does exhibit enhancer-like activity by driving expression in limb structures as expected for posterior *Hox* genes. The mouse transgenic system is highly amenable to *in vivo* analysis of biological activity of exogenous promoter/enhancer elements by using reporter constructs (Anand *et al.*, 2003; Manzanares *et al.*, 2000).[1]

The methods described previously have proven useful as a first step in defining regulatory sequences in the eukaroytic genome. However, automated pipelines that allow much higher throughput are necessary to effectively analyze the massive amounts of sequencing information being brought online. Numerous bioinformatics groups are developing software for this purpose (Bigelow *et al.*, 2004; Blanchette and Tompa, 2003; Lenhard *et al.*, 2003; Markstein *et al.*, 2002; Sandelin and Wasserman, 2004). The CORG database catalogs conserved noncoding DNA sequences based on statistically significant local suboptimal alignments of the 15-kb regions upstream of the translation start sites of more than 10,000 pairs of orthologous genes (Dieterich *et al.*, 2003). A new computational tool has been developed by one of us (P.F.S) to be used for phylogenetic footprinting. The tracker program (Prohaska *et al.*, 2004b) was designed specifically to characterize

---

[1]Importantly, such transgenic reporter experiments can also be performed with zebrafish embryos; however, one must exercise caution in interpreting the results because of the notable levels of mosaicism in expression and integration in zebrafish transgenic experiments (Dickmeis *et al.*, 2004; Koster and Fraser, 2001; Manzanares *et al.*, 2000; Múller *et al.*, 2002; Stuart *et al.*, 1988; Westerfield *et al.*, 1992).

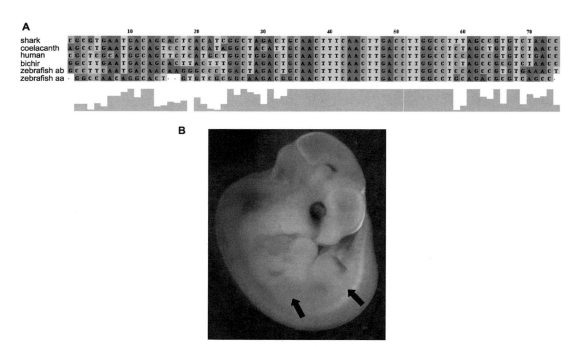

Fig. 3 Analysis of a putative CRE in the HOX-A cluster. (A) Nucleotide alignment of the conserved region upstream of *Hoxa13* in Fig. 1. (red arrow). This region is conserved in all vertebrate HOX-A clusters thus far identified. The alignment and histogram were generated by using the ClustalW program through the EMBL molecular biology server (http://www.ebi.ac.uk/clustalw/index.html). (B) Functional assay of the zebrafish *a13b* putative CRE. A 400-bp region encompassing the zebrafish element [second from bottom in (A)] was cloned into a *LacZ*-reporter construct that uses a mouse *Hsp68* minimal heat shock promoter (Shashikant *et al.*, 1995). The insert was released free of vector fragments and injected into a single-cell mouse embryo and examined (fixed and stained) at Day 10.5. The arrows denote limb buds, where expression is observed. Signals in the neural tube and head are unrelated to the enhancer activity and reflect cryptic activity of the *Hsp68* promoter and the effect of local enhancers present at the site of transgene integration, respectively. (See Color Insert.)

the noncoding DNA regions in the intergenic regions of extensive gene clusters in multiple species, in particular of the HOX clusters of vertebrate taxa. The program starts with pairwise BLAST searches and combines the resulting significant alignments into cliques. These are then processed to local multiple alignments of clusters of phylogenetic footprints (Fig. 4).

Comprehensive surveys of phylogenetic footprints provide a data basis for studying the evolution of regulatory patterns. A quantitative model of footprint loss in the wake of the teleost-specific HOX-cluster duplication indicates non-neutral evolution of CREs (Prohaska *et al.*, 2004b). This model is based on three assumptions: (1) if a gene is lost, the associated CREs will be lost too, (2) cross-regulatory interactions within the gene cluster might be lost, and (3)

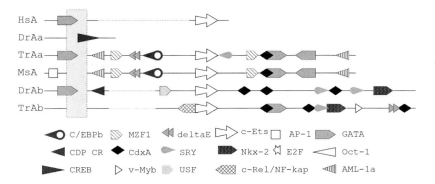

Fig. 4 Schematic overview of a phylogenetic footprint clique in the region between *Hoxa10* and *Hoxa9* of *Homo sapiens* (Hs) and teleosts (Dr, *Danio rerio*; Tr, *Takifugu rubripes*, tiger pufferfish; Ms, *Morone saxatilis*, striped bass). Individual transcription factor binding sites are tentatively identified here by using the TRANSFAC database (Heinemeyer *et al.*, 1998, 1999). The gray bar is a 15-nt sequence motif of unknown function that is absolutely conserved, with the exception of a single gap in the MsA sequence. Adapted from Prohaska, S. J. *et al.* (2004b). Surveying phylogenetic footprints in large gene clusters: Applications to Hox cluster duplications. *Mol. Phylogenet. Evol.* **31**, 581–604, with permission.

there is a loss of CREs because of stochastic resolution of genetic redundancy as described by the DDC model (see Section III).

The conserved noncoding DNA that is detected by large-scale phylogenetic footprinting contains a wealth of phylogenetic information, which we have used to study the duplication history of gnathostome HOX clusters (Chiu *et al.*, 2004; Prohaska and Stadler, 2004; Prohaska *et al.*, 2004a,b; e.g., Fig. 5). The recent discovery that the *Hoxb8* mRNA is cleaved under the direction of a microRNA, miR-196, which itself is located in the intergenic region upstream of *Hox-9* in many vertebrates (Yekta *et al.*, 2004), highlights that not all phylogenetic footprints are CREs. Apart from such noncoding RNAs, phylogenetic footprinting can also detect regulatory sequences that act at the mRNA level rather than on transcriptional regulation per se. The target for miR-196 in the 3′ UTR of the *Hoxb8* mRNA is a prominent example (Yekta *et al.*, 2004).

In addition to the high-throughput computational methods that will aid identification of potential CREs, complementary empirical approaches are also being developed to definitively identify DNA-binding sites within the CREs, most notably chromatin immunoprecipitation (Laganiere *et al.*, 2003; Weinmann, 2004). This method is used to identify en masse multiple targets of respective transcription binding factors as well as the activity state of chromatin (e.g., methylation or histone acetylation) and is being extensively employed in the ENCODE project.[2]

---

[2]Note that as daunting as it is to annotate the coding sequences of a genome, annotation of all the control elements of each respective gene will be a more onerous task and an adequate system to do so must be developed (Ashburner and Lewis, 2002; Harris *et al.*, 2004; Lewis *et al.*, 2000).

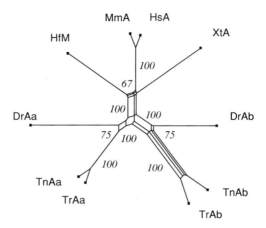

Fig. 5 Phylogenetic network (Bryant and Moulton, 2004) reconstructed from the sequences of the concatenated phylogenetic footprint cliques computed with the tracker program. From the noncoding DNA of respective HOX-A clusters we find strong support for the "duplication first" scenario, in which the duplication of the HOX clusters predates the divergence of the percomorph fishes (pufferfish) from the zebrafish lineage. Phylogenetic networks indicate the noise (uncertainty) by the width of the rectangles that replace the edges of the conventional trees. Numbers in italics give the bootstrap support of the individual splits (edges of the network). Species: *Homo sapiens* (Hs), *Mus musculus* (Mm), *Heterodontus francisci* (Hf; horn shark), *Xenopus tropicalis* (Xt; clawed frog), *Danio rerio* (Dr), *Takifugu rubripes* (Tr; tiger pufferfish), *Tetraodon nigroviridis* (Tn; green pufferfish).

III. Duplicated Genomes, Duplicated Genes

Based on genetic maps as well as analyses of whole-genome sequences, there is strong evidence that euteleosts (e.g., medaka, pufferfish, zebrafish) have undergone an independent genome duplication roughly 300 million years ago (Amores *et al.*, 1998, 2004; Cresko *et al.*, 2003; Naruse *et al.*, 2004; Postlethwait *et al.*, 1998, 2000; Taylor *et al.*, 2001a,b; Vandepoele *et al.*, 2004). Because many of these gene duplicates have been retained in zebrafish,[3] the genome sequence provides a very good opportunity to study the molecular evolution and fates of duplicated genes following a large-scale duplication event. Figure 6 illustrates a simple example of gross alterations in gene structure of duplicated genes in zebrafish. In this example, global alignments of the duplicated *pax6* genes in zebrafish were graphically plotted by using VISTA to show that both coding and noncoding sequences undergo notable divergence post gene duplication. This example is provided merely to illustrate that retained gene copies can undergo striking changes after

---

[3]In pufferfish, there is a substantially lower retention frequency of duplicated genes relative to that in zebrafish, perhaps related to its greatly reduced genome size (Taylor *et al.*, 2003; Vandepoele *et al.*, 2004). Loss of one copy of a gene duplicate is referred to as nonfunctionalization.

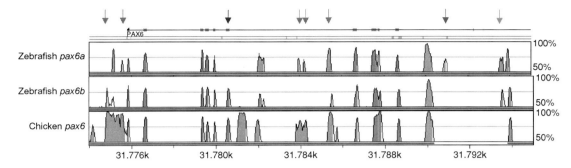

Fig. 6 VISTA analysis of *Pax6* gene duplicates of zebrafish. The two zebrafish *Pax6* duplicates, *Pax6a* and *pax6b* (AL929172 and BX000453, respectively), were compared with the human reference *Pax6* sequence (NT_009237); a plot of the chicken *Pax6* gene (AADN1060006) relative to that of humans is also shown for comparison. The blue arrow indicates a coding sequence loss in the *6a* duplicate and the red arrows indicate potential CREs that are overtly different between the *6a* and *6b* duplicates; the green arrow indicates a potential CRE that is shared between the two zebrafish duplicates but not present in humans or chicken; the purple arrow indicates a potential CRE that is shared between the human and zebrafish *6a* duplicate but not present in the *6b* duplicate or chicken. The clear differences in the *cis*-element composition between the two duplicates are highly suggestive of the two zebrafish *Pax6* genes having diverged in their functions; however, this inference awaits corroborative evidence from empirical experiments.

such a duplication event, and that such differences, particularly in potential CREs, can be readily detectable by routine global alignment and graphical analysis. Large-scale analysis of the zebrafish genome will ultimately reveal literally hundreds (if not thousands) of similar examples where duplicated genes have undergone marked divergence in their gene organizations. Of greater importance is the determination of the fate of these duplicates and how they have been retained over evolutionary time. For this discussion, we turn to population genetic modeling to address the theoretical and practical considerations regarding duplication of complex genes and their CREs.

Is 300 million years enough time for all the redundant gene duplicates to have been resolved? Under the classical model of gene duplication, a duplicate gene is assumed to be both redundant and sufficient for function in a single dose. Under this fitness model, called the double-recessive null model, all genotypes are considered to be equally fit except for those homozygous for null alleles at both duplicate loci. With a null mutation rate of 10^{-5}, an effective population size of 100,000, and a generation time of 1 year, the time to nonfunctionalization is approximately 1 million years, and for a generation time of 10 years the time to nonfunctionalization would be 10 million years (Watterson, 1983; Fig. 7). Therefore, under these parameters, the fates of all the redundant duplicates would be expected to be resolved. Even at an effective population size of 1 million, a null mutation rate of 10^{-7}, and a generation time of 1 year, the time to nonfunctionalization is approximately 20 million generations, and for a generation

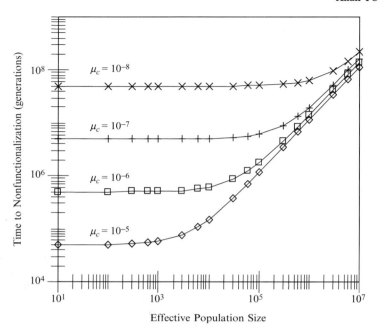

Fig. 7 The time to nonfunctionalization for a duplicate gene pair under the double-recessive null model. The analytical diffusion results of Watterson (1983) are plotted for four different null mutation rates ($\mu_c = 10^{-5}$, 10^{-6}, 10^{-7}, and 10^{-8}) and varying effective population sizes (also see Lynch and Force, 2000). In a practical sense, the results suggest that under the population genetic conditions for which teleost fish (zebrafish) would have generally evolved, all redundant duplicates should have either undergone nonfunctionalization or have been preserved by some mechanism well within the 300 million year estimate given for its whole genome duplication.

time of 10 years the time to nonfunctionalization would be 200 million years. Therefore, only if effective population sizes are very large or the null mutation rate is extremely low would the time to nonfunctionalization under the double-recessive null model exceed 200 million years. In a nutshell, these analyses suggest that all redundant duplicates conforming to the double-recessive fitness model should have either undergone nonfunctionalization or have been preserved by some mechanism.

There has been a renewed interest in the evolutionary fates of duplicate genes and their mechanisms of preservation (Clark, 1994; Force *et al.*, 1999; Hughes, 1994; Lynch and Force, 2000; Lynch *et al.*, 2001; Piatigorsky and Wistow, 1991; Sidow, 1996; Wagner, 1994, 1998; Wagner *et al.*, 2003; Walsh, 1995), which has resulted in the development of models that explicitly incorporate the complex, multifunctional organization of eukaryotic genes. For example, under the duplication–degeneration–complementation (DDC) model (Force *et al.*, 1999; Lynch and Force, 2000; Lynch *et al.*, 2001), genes are posited to contain independently mutable subfunctions that can be partitioned among descendent copies following a gene duplication event. A *gene subfunction* is defined as an independently mutable function of a gene that falls into a distinct complementation class and can correspond to regulatory and/or coding regions of genes (Force *et al.*, 1999). The partitioning of two ancestral subfunctions following gene duplication, each to a different daughter copy, is sufficient for duplicate preservation. A number of population-level mechanisms lead to duplicate-gene preservation by partitioning of gene subfunctions between gene duplicates, including adaptive-conflict neofunctionalization (Hughes, 1994; Piatigorsky and Wistow, 1991), subfunctionalization (Force *et al.*, 1999; Lynch and Force, 2000; Lynch *et al.*, 2001), dosage-mediated preservation (Force *et al.*, 1999), and epigenetic-mediated preservation (Adams *et al.*, 2003; Rodin and Riggs, 2003).

Although the double-recessive null model was originally applied to duplicate genes, it can also in principle be applied directly to duplicate gene subfunctions and for the purposes of this chapter to subfunctions that correspond to a set of regulatory elements that we refer to as *regulatory subfunctions*. Then, Fig. 7 can be interpreted as the time required for each individual regulatory subfunction to be resolved to one gene copy or the other following a genome duplication event. Nonfunctionalization of a regulatory subfunction with the null mutation rate of 10^{-7} and an effective population of 100,000 takes between 20 and 200 million years; therefore, the majority of subfunctions would be expected to have undergone nonfunctionalization of one copy or the other. Deviations from this prediction suggest various types of selection being involved in the maintenance of the duplicate gene subfunctions, such as specific dosage requirements or adaptive redundancy.

Several studies of duplicate genes in teleost fish seem to show partitioning of expression patterns, regulatory elements, and coding regions that are consistent with the DDC model, including *mitfa/mitfb* (Altschmied *et al.*, 2002; Lister *et al.*, 2001), *hoxb1a/hoxb1b* (McClintock *et al.*, 2001, 2002; Prince and Pickett, 2002), and *sox9a/9b* (Chiang *et al.*, 2001; Cresko *et al.*, 2003; Yan *et al.*, 2002). In the case of the *hoxb1a* and *hoxb1b* duplicates specific mutations in the ancestral regulatory elements have been identified that likely led to their silencing in a complementary manner: *hoxb1b* has mutations in an r4 enhancer element and *hoxb1a* does not have a 3′ RARE (retinoic acid receptor element) relative to the ancestral state seen in the mouse, whereas *hoxb1* has both a functional r4 element and a functional 3′ RARE element.

In the case of the r4 element, the conserved region is found in both *hoxb1* copies, but one of the copies has mutations in critical Hox/cofactor binding sites. It is important to note that these changes are relatively cryptic and would not be detectable in routine VISTA/PIP analysis. The caveat here is that although such broad brush techniques are extremely useful as initial predictors, they will not always detect regulatory subfunction partitioning events that have occurred, further emphasizing the need for empirical experimentation. The identification of regulatory subfunctions in duplicate genes from fish genomes will allow us to explore further the potential mechanisms responsible for duplicate-gene preservation and the forces acting on duplicate-gene subfunctions.

In summary, we have described computational tools for delineating conserved *cis* elements when comparing syntenic regions of different genomic sequences. Some of these methods result in relatively crude visualization of the conserved regions and can serve as first pass methods for examining the data. However, more computationally rigorous techniques (e.g., tracker) are being developed to better identify bona fide CREs. The predictive value of these computational methods, in general, will continue to improve as more knowledge is gained concerning the underlying logic of CREs. Zebrafish serves as a very good testbed for studying gene duplication and the retention of gene duplicates because of its available genome sequence and its tractability to experimental manipulation. Inferences regarding the role of regulatory subfunctions (e.g., CREs) in retention of specific gene duplicates as posited in the DDC model are, at this stage, difficult to assess by comparative genome data alone and require empirical experimentation.

Acknowledgments

We thank the members of our respective laboratories. Our work has been funded, in part, by grants from the National Institutes of Health (RR14085, HG02526-01), the National Science Foundation (IBN-0207870, IBN-0321461), the United States Department of Energy (DE-FG03-01ER63273) and DFG Bioinformatics Initiative BIZ 6/1-2 and FWF P#15893, and from the Pennsylvania State University.

References

Adams, K. L., Cronn, R., Percifield, R., and Wendel, J. F. (2003). Genes duplicated by polyploidy show unequal contributions to the transcriptome and organ-specific reciprocal silencing. *Proc. Natl. Acad. Sci. USA* **100,** 4649–4654.

Altschmied, J., Delfgaauw, J., Wilde, B., Duschl, J., Bouneau, L., Volff, J. N., and Schartl, M. (2002). Subfunctionalization of duplicate mitf genes associated with differential degeneration of alternative exons in fish. *Genetics* **161,** 259–267.

Amores, A., Force, A., Yan, Y. L., Joly, L., Amemiya, C., Fritz, A., Ho, R. K., Langeland, J., Prince, V., Wang, Y. L., Westerfield, M., Ekker, M., and Postlethwait, J. H. (1998). Zebrafish hox clusters and vertebrate genome evolution. *Science* **282,** 1711–1714.

Amores, A., Suzuki, T., Yan, Y. L., Pomeroy, J., Singer, A., Amemiya, C., and Postlethwait, J. H. (2004). Developmental roles of pufferfish Hox clusters and genome evolution in ray-fin fish. *Genome Res.* **14,** 1–10.

Anand, S., Wang, W. C., Powell, D. R., Bolanowski, S. A., Zhang, J., Ledje, C., Pawashe, A. B., Amemiya, C. T., and Shashikant, C. S. (2003). Divergence of Hoxc8 early enhancer parallels diverged axial morphologies between mammals and fishes. *Proc. Natl. Acad. Sci. USA* **100,** 15666–15669.

Arnone, M. I., and Davidson, E. H. (1997). The hardwiring of development: Organization and function of genomic regulatory systems. *Development* **124,** 1851–1864.

Ashburner, M., and Lewis, S. (2002). On ontologies for biologists: The Gene Ontology—untangling the web. *Novartis Found. Symp.* **247,** 66–80.

Barrallo-Gimeno, A., Holzschuh, J., Driever, W., and Knapik, E. W. (2004). Neural crest survival and differentiation in zebrafish depends on mont blanc/tfap2a gene function. *Development* **131,** 1463–1477.

Berezikov, E., Guryev, V., Plasterk, R. H., and Cuppen, E. (2004). CONREAL: Conserved regulatory elements anchored alignment algorithm for identification of transcription factor binding sites by phylogenetic footprinting. *Genome Res.* **14,** 170–178.

Bigelow, H. R., Wenick, A. S., Wong, A., and Hobert, O. (2004). CisOrtho: A program pipeline for genome-wide identification of transcription factor target genes using phylogenetic footprinting. *BMC. Bioinformatics* **5,** 27–46.

Blanchette, M., and Tompa, M. (2003). FootPrinter: A program designed for phylogenetic footprinting. *Nucleic Acids Res.* **31,** 3840–3842.

Boffelli, D., McAuliffe, J., Ovcharenko, D., Lewis, K. D., Ovcharenko, I., Pachter, L., and Rubin, E. M. (2003). Phylogenetic shadowing of primate sequences to find functional regions of the human genome. *Science* **299,** 1391–1394.

Bray, N., Dubchak, I., and Pachter, L. (2003). A VID: A global alignment program. *Genome Res.* **13,** 97–102.

Britten, R. J., and Davidson, E. H. (1969). Gene regulation for higher cells: A theory. *Science* **165,** 349–357.

Brudno, M., Do, C. B., Cooper, G. M., Kim, M. F., Davydov, E., Green, E. D., Sidow, A., and Batzoglou, S. (2003). LAGAN and Multi-LAGAN: Efficient tools for large-scale multiple alignment of genomic DNA. *Genome Res.* **13,** 721–731.

Bryant, D., and Moulton, V. (2004). Neighbor-net: An agglomerative method for the construction of phylogenetic networks. *Mol. Biol. Evol.* **21,** 255–265.

Chiang, E. F., Pai, C. I., Wyatt, M., Yan, Y. L., Postlethwait, J., and Chung, B. (2001). Two sox9 genes on duplicated zebrafish chromosomes: Expression of similar transcription activators in distinct sites. *Dev. Biol.* **231,** 149–163.

Chiu, C. H., Dewar, K., Wagner, G. P., Takahashi, K., Ruddle, F., Ledje, C., Bartsch, P., Scemama, J. L., Stellwag, E., Fried, C., Prohaska, S. J., Stadler, P. F., and Amemiya, C. T. (2004). Bichir HoxA cluster sequence reveals surprising trends in ray-finned fish genomic evolution. *Genome Res.* **14,** 11–17.

Clark, A. G. (1994). Invasion and maintenance of a gene duplication. *Proc. Natl. Acad. Sci. USA* **91,** 2950–2954.

Collins, F. S. (1995). Positional cloning moves from perditional to traditional. *Nat. Genet.* **9,** 347–350.

Collins, F. S., Green, E. D., Guttmacher, A. E., and Guyer, M. S. (2003). A vision for the future of genomics research. *Nature* **422,** 835–847.

Cresko, W. A., Yan, Y. L., Baltrus, D. A., Amores, A., Singer, A., Rodriguez-Mari, A., and Postlethwait, J. H. (2003). Genome duplication, subfunction partitioning, and lineage divergence: Sox9 in stickleback and zebrafish. *Dev. Dyn.* **228,** 480–489.

Davidson, E. H. (2001). "Genomic Regulatory Systems." Academic Press, San Diego, CA.

Davidson, E. H., and Britten, R. J. (1973). Organization, transcription, and regulation in the animal genome. *Q. Rev. Biol.* **48,** 565–613.

Dickmeis, T., Plessy, C., Rastegar, S., Aanstad, P., Herwig, R., Chalmel, F., Fischer, N., and Strahle, U. (2004). Expression profiling and comparative genomics identify a conserved regulatory region controlling midline expression in the zebrafish embryo. *Genome Res.* **14,** 228–238.

Dieterich, C., Wang, H., Rateitschak, K., Luz, H., and Vingron, M. (2003). CORG: A database for COmparative Regulatory Genomics. *Nucleic Acids Res.* **31,** 55–57.

Epstein, C. J., Erickson, R. P., and Wynshaw-Boris, A. (2004). "Inborn Errors of Development." Oxford University Press, New York.

Force, A., Lynch, M., Pickett, F. B., Amores, A., Yan, Y. L., and Postlethwait, J. (1999). Preservation of duplicate genes by complementary, degenerative mutations. *Genetics* **151,** 1531–1545.

Hamada, H., Seidman, M., Howard, B. H., and Gorman, C. M. (1984). Enhanced gene expression by the poly(dT-dG).poly(dC-dA) sequence. *Mol. Cell. Biol.* **4,** 2622–2630.

Harris, M. A., Clark, J., Ireland, A., Lomax, J., Ashburner, M., Foulger, R., Eilbeck, K., Lewis, S., Marshall, B., Mungall, C., Richter, J., Rubin, G. M., Blake, J. A., Bult, C., Dolan, M., Drabkin, H., Eppig, J. T., Hill, D. P., Ni, L., Ringwald, M., Balakrishnan, R., Cherry, J. M., Christie, K. R., Costanzo, M. C., Dwight, S. S., Engel, S., Fisk, D. G., Hirschman, J. E., Hong, E. L., Nash, R. S., Sethuraman, A., Theesfeld, C. L., Botstein, D., Dolinski, K., Feierbach, B., Berardini, T., Mundodi, S., Rhee, S. Y., Apweiler, R., Barrell, D., Camon, E., Dimmer, E., Lee, W., Chisholm, R., Gaudet, P., Kibbe, W., Kishore, R., Schwarz, E. M., Sternberg, P., Gwinn, M., Hannick, L., Wortman, J., Berriman, M., Wood, V., de la, C. N., Tonellato, P., Jaiswal, P., Seigfried, T., and White, R. (2004). The Gene Ontology (GO) database and informatics resource. *Nucleic Acids Res.* **32 Database issue,** D258–D261.

Heinemeyer, T., Chen, X., Karas, H., Kel, A. E., Kel, O. V., Liebich, I., Meinhardt, T., Reuter, I., Schacherer, F., and Wingender, E. (1999). Expanding the TRANSFAC database towards an expert system of regulatory molecular mechanisms. *Nucleic Acids Res.* **27,** 318–322.

Heinemeyer, T., Wingender, E., Reuter, I., Hermjakob, H., Kel, A. E., Kel, O. V., Ignatieva, E. V., Ananko, E. A., Podkolodnaya, O. A., Kolpakov, F. A., Podkolodny, N. L., and Kolchanov, N. A. (1998). Databases on transcriptional regulation: TRANSFAC, TRRD and COMPEL. *Nucleic Acids Res.* **26,** 362–367.

Hughes, A. L. (1994). The evolution of functionally novel proteins after gene duplication. *Proc. R. Soc. Lond. B. Biol. Sci.* **256,** 119–124.

Kim, C. B., Amemiya, C., Bailey, W., Kawasaki, K., Mezey, J., Miller, W., Minoshima, S., Shimizu, N., Wagner, G., and Ruddle, F. (2000). Hox cluster genomics in the horn shark, *Heterodontus francisci. Proc. Natl. Acad. Sci. USA* **97,** 1655–1660.

Koster, R. W., and Fraser, S. E. (2001). Tracing transgene expression in living zebrafish embryos. *Dev. Biol.* **233,** 329–346.

Laganiere, J., Deblois, G., and Giguere, V. (2003). Nuclear receptor target gene discovery using high-throughput chromatin immunoprecipitation. *Methods Enzymol.* **364,** 339–350.

Lenhard, B., Sandelin, A., Mendoza, L., Engstrom, P., Jareborg, N., and Wasserman, W. W. (2003). Identification of conserved regulatory elements by comparative genome analysis. *J. Biol.* **2,** 13.

Lewis, S., Ashburner, M., and Reese, M. G. (2000). Annotating eukaryote genomes. *Curr. Opin. Struct. Biol.* **10,** 349–354.

Lister, J. A., Close, J., and Raible, D. W. (2001). Duplicate mitf genes in zebrafish: Complementary expression and conservation of melanogenic potential. *Dev. Biol.* **237,** 333–344.

Lo, J., Lee, S., Xu, M., Liu, F., Ruan, H., Eun, A., He, Y., Ma, W., Wang, W., Wen, Z., and Peng, J. (2003). 15000 unique zebrafish EST clusters and their future use in microarray for profiling gene expression patterns during embryogenesis. *Genome Res.* **13,** 455–466.

Lynch, M., and Force, A. (2000). The probability of duplicate gene preservation by subfunctionalization. *Genetics* **154,** 459–473.

Lynch, M., O'Hely, M., Walsh, B., and Force, A. (2001). The probability of preservation of a newly arisen gene duplicate. *Genetics* **159,** 1789–1804.

Malicki, J. J., Pujic, Z., Thisse, C., Thisse, B., and Wei, X. (2002). Forward and reverse genetic approaches to the analysis of eye development in zebrafish. *Vision Res.* **42,** 527–533.

Manzanares, M., Wada, H., Itasaki, N., Trainor, P. A., Krumlauf, R., and Holland, P. W. (2000). Conservation and elaboration of Hox gene regulation during evolution of the vertebrate head. *Nature* **408,** 854–857.

Markstein, M., Markstein, P., Markstein, V., and Levine, M. S. (2002). Genome-wide analysis of clustered Dorsal binding sites identifies putative target genes in the *Drosophila* embryo. *Proc. Natl. Acad. Sci. USA* **99,** 763–768.

Mayor, C., Brudno, M., Schwartz, J. R., Poliakov, A., Rubin, E. M., Frazer, K. A., Pachter, L. S., and Dubchak, I. (2000). VISTA: Visualizing global DNA sequence alignments of arbitrary length. *Bioinformatics* **16,** 1046–1047.

McClintock, J. M., Carlson, R., Mann, D. M., and Prince, V. E. (2001). Consequences of Hox gene duplication in the vertebrates: an investigation of the zebrafish Hox paralogue group 1 genes. *Development* **128**, 2471–2484.

McClintock, J. M., Kheirbek, M. A., and Prince, V. E. (2002). Knockdown of duplicated zebrafish hoxb1 genes reveals distinct roles in hindbrain patterning and a novel mechanism of duplicate gene retention. *Development* **129**, 2339–2354.

Múller, F., Blader, P., and Strahle, U. (2002). Search for enhancers: Teleost models in comparative genomic and transgenic analysis of *cis* regulatory elements. *Bioessays* **24**, 564–572.

Myers, R. M., Lerman, L. S., and Maniatis, T. (1985). A general method for saturation mutagenesis of cloned DNA fragments. *Science* **229**, 242–247.

Myers, R. M., Tilly, K., and Maniatis, T. (1986). Fine structure genetic analysis of a beta-globin promoter. *Science* **232**, 613–618.

Naruse, K., Tanaka, M., Mita, K., Shima, A., Postlethwait, J., and Mitani, H. (2004). A medaka gene map: the trace of ancestral vertebrate proto-chromosomes revealed by comparative gene mapping. *Genome Res.* **14**, 820–828.

Nobrega, M. A., Ovcharenko, I., Afzal, V., and Rubin, E. M. (2003). Scanning human gene deserts for long-range enhancers. *Science* **302**, 413.

Pennacchio, L. A., and Rubin, E. M. (2001). Genomic strategies to identify mammalian regulatory sequences. *Nat. Rev. Genet.* **2**, 100–109.

Pennacchio, L. A., and Rubin, E. M. (2003). Comparative genomic tools and databases: Providing insights into the human genome. *J. Clin. Invest.* **111**, 1099–1106.

Piatigorsky, J., and Wistow, G. (1991). The recruitment of crystallins: New functions precede gene duplication. *Science* **252**, 1078–1079.

Popperl, H., and Featherstone, M. S. (1992). An autoregulatory element of the murine Hox-4.2 gene. *EMBO J.* **11**, 3673–3680.

Postlethwait, J. H., Woods, I. G., Ngo-Hazelett, P., Yan, Y. L., Kelly, P. D., Chu, F., Huang, H., Hill-Force, A., and Talbot, W. S. (2000). Zebrafish comparative genomics and the origins of vertebrate chromosomes. *Genome Res.* **10**, 1890–1902.

Postlethwait, J. H., Yan, Y. L., Gates, M. A., Horne, S., Amores, A., Brownlie, A., Donovan, A., Egan, E. S., Force, A., Gong, Z., Goutel, C., Fritz, A., Kelsh, R., Knapik, E., Liao, E., Paw, B., Ransom, D., Singer, A., Thomson, M., Abduljabbar, T. S., Yelick, P., Beier, D., Joly, J. S., Larhammar, D., and Rosa, F. (1998). Vertebrate genome evolution and the zebrafish gene map. *Nat. Genet.* **18**, 345–349.

Prince, V. E., and Pickett, F. B. (2002). Splitting pairs: The diverging fates of duplicated genes. *Nat. Rev. Genet.* **3**, 827–837.

Prohaska, S. J., Fried, C., Amemiya, C. T., Ruddle, F. H., Wagner, G. P., and Stadler, P. F. (2004a). The shark HoxN cluster is homologous to the human HoxD cluster. *J. Mol. Evol.* **58**, 212–217.

Prohaska, S. J., Fried, C., Flamm, C., Wagner, G. P., and Stadler, P. F. (2004b). Surveying phylogenetic footprints in large gene clusters: Applications to Hox cluster duplications. *Mol. Phylogenet. Evol.* **31**, 581–604.

Prohaska, S. J., and Stadler, P. F. (2004). The duplication of the *Hox* gene clusters in teleost fishes. *Theory Biosci.* (In Press).

Rodin, S. N., and Riggs, A. D. (2003). Epigenetic silencing may aid evolution by gene duplication. *J. Mol. Evol.* **56**, 718–729.

Sandelin, A., and Wasserman, W. W. (2004). Constrained binding site diversity within families of transcription factors enhances pattern discovery bioinformatics. *J. Mol. Biol.* **338**, 207–215.

Santini, S., Boore, J. L., and Meyer, A. (2003). Evolutionary conservation of regulatory elements in vertebrate Hox gene clusters. *Genome Res.* **13**, 1111–1122.

Schwartz, S., Zhang, Z., Frazer, K. A., Smit, A., Riemer, C., Bouck, J., Gibbs, R., Hardison, R., and Miller, W. (2000). PipMaker—a web server for aligning two genomic DNA sequences. *Genome Res.* **10**, 577–586.

Shashikant, C. S., Bieberich, C. J., Belting, H. G., Wang, J. C., Borbely, M. A., and Ruddle, F. H. (1995). Regulation of Hoxc-8 during mouse embryonic development: Identification and characterization of critical elements involved in early neural tube expression. *Development* **121,** 4339–4347.

Shashikant, C. S., and Ruddle, F. H. (1996). Combinations of closely situated cis-acting elements determine tissue-specific patterns and anterior extent of early Hoxc8 expression. *Proc. Natl. Acad. Sci. USA* **93,** 12364–12369.

Sidow, A. (1996). Gen(om)e duplications in the evolution of early vertebrates. *Curr. Opin. Genet. Dev.* **6,** 715–722.

Stickney, H. L., Schmutz, J., Woods, I. G., Holtzer, C. C., Dickson, M. C., Kelly, P. D., Myers, R. M., and Talbot, W. S. (2002). Rapid mapping of zebrafish mutations with SNPs and oligonucleotide microarrays. *Genome Res.* **12,** 1929–1934.

Stuart, G. W., McMurray, J. V., and Westerfield, M. (1988). Replication, integration and stable germline transmission of foreign sequences injected into early zebrafish embryos. *Development* **103,** 403–412.

Taylor, J. S., Braasch, I., Frickey, T., Meyer, A., and Van de, P. Y. (2003). Genome duplication, a trait shared by 22000 species of ray-finned fish. *Genome Res.* **13,** 382–390.

Taylor, J. S., Van de, P. Y., Braasch, I., and Meyer, A. (2001a). Comparative genomics provides evidence for an ancient genome duplication event in fish. *Philos. Trans. R. Soc. Lond. B. Biol. Sci.* **356,** 1661–1679.

Taylor, J. S., Van de, P. Y., and Meyer, A. (2001b). Genome duplication, divergent resolution and speciation. *Trends Genet.* **17,** 299–301.

Thomas, J. W., Touchman, J. W., Blakesley, R. W., Bouffard, G. G., Beckstrom-Sternberg, S. M., Margulies, E. H., Blanchette, M., Siepel, A. C., Thomas, P. J., McDowell, J. C., Maskeri, B., Hansen, N. F., Schwartz, M. S., Weber, R. J., Kent, W. J., Karolchik, D., Bruen, T. C., Bevan, R., Cutler, D. J., Schwartz, S., Elnitski, L., Idol, J. R., Prasad, A. B., Lee-Lin, S. Q., Maduro, V. V., Summers, T. J., Portnoy, M. E., Dietrich, N. L., Akhter, N., Ayele, K., Benjamin, B., Cariaga, K., Brinkley, C. P., Brooks, S. Y., Granite, S., Guan, X., Gupta, J., Haghighi, P., Ho, S. L., Huang, M. C., Karlins, E., Laric, P. L., Legaspi, R., Lim, M. J., Maduro, Q. L., Masiello, C. A., Mastrian, S. D., McCloskey, J. C., Pearson, R., Stantripop, S., Tiongson, E. E., Tran, J. T., Tsurgeon, C., Vogt, J. L., Walker, M. A., Wetherby, K. D., Wiggins, L. S., Young, A. C., Zhang, L. H., Osoegawa, K., Zhu, B., Zhao, B., Shu, C. L., de Jong, P. J., Lawrence, C. E., Smit, A. F., Chakravarti, A., Haussler, D., Green, P., Miller, W., and Green, E. D. (2003). Comparative analyses of multi-species sequences from targeted genomic regions. *Nature* **424,** 788–793.

Ton, C., Stamatiou, D., Dzau, V. J., and Liew, C. C. (2002). Construction of a zebrafish cDNA microarray: Gene expression profiling of the zebrafish during development. *Biochem. Biophys. Res. Commun.* **296,** 1134–1142.

Trinklein, N. D., Aldred, S. F., Hartman, S. J., Schroeder, D. I., Otillar, R. P., and Myers, R. M. (2004). An abundance of bidirectional promoters in the human genome. *Genome Res.* **14,** 62–66.

Vandepoele, K., De Vos, W., Taylor, J. S., Meyer, A., and Van de, P. Y. (2004). Major events in the genome evolution of vertebrates: Paranome age and size differ considerably between ray-finned fishes and land vertebrates. *Proc. Natl. Acad. Sci. USA* **101,** 1638–1643.

Wagner, A. (1994). Evolution of gene networks by gene duplications: A mathematical model and its implications on genome organization. *Proc. Natl. Acad. Sci. USA* **91,** 4387–4391.

Wagner, A. (1998). The fate of duplicated genes: Loss or new function? *Bioessays* **20,** 785–788.

Wagner, G. P., Amemiya, C., and Ruddle, F. (2003). Hox cluster duplications and the opportunity for evolutionary novelties. *Proc. Natl. Acad. Sci. USA* **100,** 14603–14606.

Walsh, J. B. (1995). How often do duplicated genes evolve new functions? *Genetics* **139,** 421–428.

Watterson, G. A. (1983). On the time for gene silencing at duplicate loci. *Genetics* **105,** 745–766.

Weinmann, A. S. (2004). Innovation: Novel ChIP-based strategies to uncover transcription factor target genes in the immune system. *Nat. Rev. Immunol.* **4,** 381–386.

Westerfield, M., Wegner, J., Jegalian, B. G., DeRobertis, E. M., and Puschel, A. W. (1992). Specific activation of mammalian Hox promoters in mosaic transgenic zebrafish. *Genes Dev.* **6**, 591–598.

Yan, Y. L., Miller, C. T., Nissen, R. M., Singer, A., Liu, D., Kirn, A., Draper, B., Willoughby, J., Morcos, P. A., Amsterdam, A., Chung, B. C., Westerfield, M., Haffter, P., Hopkins, N., Kimmel, C., Postlethwait, J. H., and Nissen, R. (2002). A zebrafish sox9 gene required for cartilage morphogenesis. *Development* **129**, 5065–5079.

Yekta, S., Shih, I. H., and Bartel, D. P. (2004). MicroRNA-directed cleavage of HOXB8 mRNA. *Science* **304**, 594–596.

PART V

Infrastructure

CHAPTER 30

Zebrafish Facilities for Small and Large Laboratories

Bill Trevarrow

Institute of Neuroscience
University of Oregon
Eugene, Oregon 97403

I. Introduction

Common principles underlie all sizes of zebrafish facilities. Each facility has different research goals and available space. Efficient functioning and avoidance of failures are primary concerns. Greater control over potential negative outside influences such as water supply, air supply, food, and new fish increases stability and security. This has to be balanced with considerations of space, equipment cost, labor, cost of maintenance, staff technical abilities, research usefulness, as well as efforts invested in a particular fish line. Understanding bacterial, chemical, and mechanical water processing; computers and databases; and zebrafish biology and research will be useful in keeping your water system functioning well. As facilities become larger, the accumulated investments increase and the repetitive tasks use up more labor, favoring increased automation, centralization, and security of operations. A small facility's staff might not have the time or expertise to deal with all these requirements. Instead, it might rely on institutional resources or outside contractors for particular tasks. Larger facilities tend to be more centralized and more robustly backed-up. Monitoring a larger but more centralized system is easier and protects against catastrophic failures and irreplaceable losses. Redundant systems and duplicate equipment increase the facility's stability, resilience to disruptions, and recovery from problems. Anticipating problems, determining how they can be rapidly diagnosed, and planning an efficient response ahead of time improve the facility's chances of long-term success.

Zebrafish program goals can change rapidly. Unfortunately, more tanks and more space are the most frequent desire. Tank densities can be increased with different rack designs, but it is ultimately space limited. An effective fish room packs in as many tanks as possible while still providing for other needs. Getting this balance right is difficult to design ahead of time. Building in flexibility permits later experience to shape the final relationships. As size increases, specialized functions are a more likely desire. A large facility is often better able to provide them. All facilities should keep records (written or computer) of which fish are in what tank and how they are related. As the numbers of tanks and numbers of lines increase, the burden of useful record keeping increases and is more easily handled with computer databases and barcode readers.

A. Size and Purpose Considerations

A zebrafish facility can range in size from a few independently filtered aquarium setups, through one or more self-contained racks, each equipped with a common filtration system, to larger facilities with many tank racks and large separate filtration units. As the operation increases in size, less labor efficient approaches become untenable. Larger facilities typically have a greater variety of different unique lines. These lines are an investment of time and careers and deserve protection, making interruptions and losses much more crucial. Water systems, filtration, monitoring, automation, and record keeping are affected by facility size. At the small end, a laboratory might have one or two hobbyist-style aquarium setups on a bench, perhaps for collecting embryos for *in situ* hybridizations. Each tank is maintained individually by hand and has its own filtration and lighting. A percentage of water would have to be changed periodically by hand. Monitoring of filter functioning would be by direct observation, probably daily. Records of facility conditions and fish genetics would probably be kept in a notebook. As more tanks are used, the labor demands of this hobbyist-style setup will overwhelm other aspects of the research project, requiring a new facility plan. By centralizing functions such as filtration and water exchange, less maintenance is required owing to the reduced number of units to be serviced (fewer filters to change, fewer moving parts). By reducing the number of machines and bodies of water to monitor, a centralized facility can be monitored at a very detailed level that would be prohibitively expensive with many independent units. On the other hand, such a centralized system is more susceptible to rapidly spreading diseases and catastrophic failures. This requires that it have both good biosecurity and a robust arrangement of back-up systems, based on redundant systems and duplicate parts.

Some tasks, such as water changes, can be designed away by automating processes. Something as simple as a yard sprinkler controller and valves can control water changes to several individual units. On the other hand, setting up automation of these bigger and more complex systems requires time, money, and more staff technical expertise.

B. Design Overview

A new facility design provides a rare opportunity to more cheaply install an extensive centralization friendly infrastructure, such as specialized plumbing and electrical equipment. When designing a new facility or reconsidering aspects of an existing one, it is helpful to break down the facility's functional goals (such as provide for a screen for ENU mutations) into subgoals (such as having healthy fish in breeding condition, a place to mutagenized them, and the means to hold sufficient numbers of individual fish through the screening). Each of these subgoals requires certain processes (such as temperature control and water filtration) that can be provided in different ways, some depending on scale. Such plans should be shaped by feedback at various stages to reveal problems as soon as possible. After the problems are corrected, the feedback cycle should be repeated. Guidelines for particular materials to use in rooms for fish (and other animals) are available (Astrofsky *et al.*, 2002; National Research Council, 1996).

II. Zebrafish Biology

Most people first encounter zebrafish in pet stores or home aquariums. Their hardiness, low maintenance, and ease of breeding make them good fish for beginners. These traits simplify maintaining large numbers of them without a lot of individual attention. Their small size and schooling also aids in keeping them at higher densities and in smaller boxes.

A. Different Sources of Fish

Zebrafish come from the Indian subcontinent. McClure (1998) described some of the North Indian environments where she found zebrafish. Several zebrafish lines have been established from pet store fish. The breeding populations of these pet store fish were probably established on fish farms in the 1930s. Since then, they have gone through many generations of selection. Several laboratory zebrafish lines have also been established from fish imported from India. Our knowledge of the conditions for keeping, breeding, and raising zebrafish comes from many years of hobbyist commercial breeder and laboratory zebrafish experience, as well as descriptions of their native habitat and research on other fish.

B. Materials and Testing for Toxins

A simple but sensitive bioassay for toxic materials has been developed at the Zebrafish International Resource Center (ZIRC; see Chapter 33). This test has been used to survey plastics, rubbers, and other materials contacting fish water. About one third of the plastics and rubbers killed larval fish in this test. Of particular interest were materials that caused fish death and were used in aquaculture, aquarium setups, and plumbing fittings, such as vinyl airline tubing and *n*-buna *o*-rings. Some of these materials might always be problematic; others might vary with lot or product. Each facility is encouraged to use this test on its own materials. Some toxins can be generated in extreme washing conditions (such as autoclave temperatures and high detergent pH) when some plastics break down. Crazing (many little cracks) or a frosted appearance of the clear plastic are visual clues to plastic breaking down. Such degraded materials should be replaced. If this happens often, a review of washing procedure is called for.

III. Fish Housing

Zebrafish can tolerate fairly wide extremes of environmental parameters (Table I). The water conditions can be maintained by hand or automatically with a water system. Fish containers (tanks) should be selected with several features in mind. Container size and desired population size will affect efficient space utilization and ease of use.

Fish in a fish facility live in a controlled microenvironment: the water in their tank. This is nested in a macroenvironment: the rest of the room containing the fish tanks. These macroenvironments are usually controlled by the building's heating ventilation and air conditioning (HVAC) system. The microenvironment is normally the direct responsibility of the facility. However, room air temperature, which is often used to control water temperature, is usually maintained institutionally.

A. Water Chemistry, Cycles, and Buffers

Good water quality is essential to the fish's welfare and should be periodically checked. Zebrafish water contains many dissolved solids and gases that interact with each other, the fish, and the atmosphere. Although zebrafish have a wide tolerance of water conditions, it is beneficial to keep changes small and gradual. To diagnose problems and make rational changes, many interactions among these chemicals have to be understood. Major water parameters, some of their interactions, and manipulations are briefly reviewed. These issues are addressed in much

Table I
Water Parameters

Parameter	Tolerated	Targeted
Chlorine (Cl_2)	Low	0 ppm
Gas saturation	Up to 102% saturation	Up to saturation
Copper (Cu^{2+})	1–10 ppb (or $\mu g/l$)	0 ppb (or $\mu g/l$)
Dissolved oxygen (DO_2)	5.0 to saturated	6.0 to saturated
pH	6.0–8.5	6.8–7.3
Buffering (alkalinity)	0–? ppm $Ca/MgCO_3$	60–120 ppm $Ca/MgCO_3$
Salinity	0–1.75 ppt	0.35–0.7 ppt
Ammonia (NH_4)	0.02–0.05 ppm	0 ppm
Nitrite (NO_2)	0–0.5 ppm?	0 ppm
Nitrate (NO_3)	200 ppm	<10 ppm
Water temperature	22–30+ °C	26–29 °C
	71.6–86+ °F	78–84 °F
Calcium (Ca^{2+})	10–160? ppm $Ca/MgCO_3$	60–120 ppm $Ca/MgCO_3$
Magnesium (Mg^{2+})	0–? ppm $Ca/MgCO_3$	60–120 ppm $Ca/MgCO_3$
Conductivity	0–1000+ μS	500–1000 μS

greater detail in many aquarium books and aquaculature texts (Moe, 1989; Spotte, 1992; Timmons *et al.*, 2001; Wheaton, 1977).

1. Dissolved Gases

Oxygen, chlorine, carbon dioxide, and total gases are important for zebrafish water systems.

a. Oxygen

Minimum levels of oxygen are required for fish survival and higher levels for them to thrive. To avoid any possible problems and support higher fish densities, dissolved oxygen (DO) levels should be kept near saturation. High fish densities, reduced water circulation, and high fish and bacterial metabolism decrease DO levels. In low oxygen conditions, fish will hover near the surface, where the oxygen levels are highest. DO levels can be conveniently measured by process instrumentation. Process probes can be left in the water for continuous monitoring. DO levels can be increased by aeration, using bubblers, wet–dry filters, venturi valves, or reactors under increased pressure.

b. Carbon Dioxide

Carbon dioxide (CO_2) is produced by fish and bacterial metabolism. At extreme levels, it can produce unconsciousness and then death. At lower levels, it can affect pH by interacting with the carbonate buffering system. Carbon dioxide can be removed by increasing aeration or agitating the water. CO_2 levels can be measured with test kits.

c. Total Gases

If the total dissolved gas concentration is higher than what the water can hold at equilibrium with the local atmosphere, then the water is supersaturated. Supersaturated water can occur in several ways and can rapidly cause widespread fish death. Water in pressurized plumbing can hold larger amounts of dissolved gases than at room pressure. Warming up cold water might cause it to become supersaturated because warm water can hold less dissolved gas. Together, the effects can be magnified. Alternatively, plumbing leaks on the suction side of a pump leak air into the pipe rather than water out. This air can be forced into the water at supersaturated levels by the high pressure and turbulent environment of the pump impeller.

Supersaturated water can be detected with a meter (a saturometer or a tensiometer) or by noticing bubbles rapidly forming on surfaces in water drawn from the water system. Conditions causing supersaturated water should be corrected as rapidly as possible. If the problem is a suction-side air leak, the connections should be fixed. Another possibility is that bubbles in some way might be drawn into the pump from the sump. If the problem is water temperatures and pressures in pipes that are not under your control, you can take steps to desaturate the water when it is available to you. An unpressurized aeration step rapidly drives dissolved gases to equilibrium with the atmosphere. From here, either the water will have to be pumped or gravity fed to the fish tanks. Supersaturation can also be controlled by reducing the flow to a tank until the incoming supersaturated water is diluted enough by the tank water and gas loss to the atmosphere to keep the tank from being supersaturated.

2. Salinity

Hardness, salinity, and buffering are related. Hardness is total dissolved solids, but there are frequently used subsets such as carbonate hardness or calcium and magnesium hardness that are closely related to buffering. Salinity is the nonorganic total dissolved solids, mostly ions. Buffering also involves ions, acting as pH buffers, mostly carbonates, sulfates, and phosphates. Most of the total dissolved solids are inorganic. About 80–100 ppm (mg/l) of calcium carbonate ($CaCO_3$) is considered good for zebrafish. Chloride (Cl^-), magnesium (Mg^{2+}), and sodium (Na^+) ions account for most of the remaining salinity. These can be mixed together or purchased in more complex mixes made for saltwater aquaria. When marine aquarium salts are used, Ca^{2+} can be added (Nüsslein-Volhard and Dahm, 2002). Alternatively, a simple mix of salts such as 14 kg of NaCl, 5 kg of $MgCl_2$, 0.8 kg of $CaCl_2$, 0.308 kg of KCl, and 35 mg of KI can be used. Salinity is measured by change in water density, optical properties, or conductivity. Conductivity measurements reflect the total ion charge carrying capacity of the water. They are influenced by the relative concentrations of different ions as well as the total ion concentration. Although not the choice for absolute measurements, conductivity works well for detecting relative changes in a controlled environment. Water density is measured with a hydrometer. Optical measurements can be made with a salinity refractometer. Refractometers are accurate but not easily monitored electronically.

Salinity can be manipulated by diluting with less salty water or adding salts or more salty water. Salt addition can be automated with a conductivity controller, a dosing pump, and a vat of high-concentration salts. Commercially available conductivity controllers can monitor conductivity levels and control pumps and valves to maintain salt levels. An independent shutoff relay, using a higher threshold, or a limited amount of salt available for pumping are typical safeties used to prevent oversalting. Zebrafish can tolerate zero salt, but can have problems at higher concentrations. Preferred concentrations seem to range from 0.35 to 0.7 ppt (parts per thousand, or g/l) of salts. To determine what conductivity value to use as a threshold, mix a small volume of the water source that would be used with the desired salt concentration. Then measure this solution with a conductivity meter to determine the set point for controlling the salt pump.

3. The Carbonate Buffering System

Carbonates comprise the major pH buffering system in freshwater aquariums. The system has equilibria with carbon dioxide (atmospheric and metabolic), several dissolved ionic forms of carbonate, and solid carbonates such as Mg^{2+} and Ca^{2+} carbonates. All these can cause changes in the buffering and pH. Fish and bacterial metabolism produces products that acidify the water, thereby lowering the pH. Carbonate buffering can be changed by adding acids, bases, and buffers, usually sodium bicarbonate or powdered aragonite ($Ca^{2+}/Mg^{2+}CO_3^{2-}$). Aragonite has a pK_a of approximately 8.3, and slowly dissolves and increases the pH until it approaches 8.3 (Moe, 1989). The metabolism of nitrifying bacteria is an additional influence on buffering capacity. They use 7.14 g of carbonate (CO_3) for each gram of ammonia nitrified as a carbon source (Timmons et al., 2001). Once a pH is established with a routine of water changes, salt, and buffer additions, it should be fairly stable. Adding buffer once a week will make up for water losses, changes, nitrifying bacterial metabolism, and loss to the atmosphere.

Buffering or alkalinity is often expressed in mg/l of $CaCO_3$, meq/l (1 meq/l = 50 mg/l $CaCO_3$), or German degrees of hardness (1 DH/l = 17.9 mg/l of $CaCO_3$). These measurements are usually taken with test kits.

4. Ammonia and Related Compounds

Nitrogen enters the water system as protein in the fish food. Ammonia, an excreted waste product of fish and bacterial protein metabolism, can be very toxic to fish, more so at pH higher than 8.0. Fortunately, *Nitrosomonas* bacteria oxidize ammonia to the less toxic nitrite (Moe, 1989; Timmons et al., 2001), and *Nitrospira* bacteria oxidize nitrite to nitrate (Burrell et al., 2001). Nitrate is much less toxic than nitrite. The nitrate then accumulates in the water system until it reaches an equilibrium with processes that remove nitrogen such as water exchanges, plant or algae growth and removal, chemical filtration, or by denitrifying bacteria that change nitrate into nitrogen gas. Denitrification (removing oxygen from nitrate and nitrite in anaerobic environments, releasing nitrogen gas) is a somewhat finicky process and not used in normal laboratory animal maintenance.

5. Carbon Cycle

Carbon, in the form of organic molecules, enters the water system as food or inorganically as carbonates. If the food is uneaten, it is either removed by particle filtration or eaten by autotrophic bacteria and broken down to CO_2 (mineralization). Eaten food is either digested by the fish, excreted as feces or as gametes, or becomes a corpse. These have the same fate as that of the food. Mineralization produces CO_2 that will equilibrate with atmospheric CO_2, the carbonate buffering system, and water loss.

6. Phosphorous

Phosphorous enters the water system in fish food and possibly as a component of tap water. Normally, dissolved phosphorous is reutilized by plants, but algae are the only available photosynthesizers in most zebrafish facilities. High phosphorous levels, combined with nitrates and sufficient illumination, provide good conditions for algal growth. Certain resin filtration media can remove phosphates from the water. High phosphorous is usually controlled by using low-phosphate food water sources.

B. Water System

1. Facility Water System Layout

As a whole, water can flow through the facility and tanks once and be disposed of (a flow-through system) or it can be reused after it is cleaned up (a recirculating system). Most facilities are mixes of these configurations. All recirculating systems require some water exchange and replacement of losses due to spillage and evaporation.

Water systems can supply fish tanks either continuously or discontinuously in flow-through or recirculating configurations. Flow-through water systems provide new, clean water to the tanks. The outflow from the tanks is disposed of. A recirculating water system processes the tank overflow water in a series of steps to remove undesirable and restore desirable compounds. The refreshed water is then returned (recirculated) to the tanks. A typical recirculating water system requires make-up water for evaporation and spillage, as well as 5–10% daily water changes. Water quality in tanks maintained by manual water changes can be stabilized by using small filters. In essence, these tanks become small, single-tank recirculating water systems.

Some facilities provide each tank with aeration. This is a redundant system for the plumbed supply of oxygenated water.

a. Water Sources

There are many possible sources and treatments for the new facility water. Potential problems are chlorine and chloramines, toxins and excessive amounts of materials that have to be removed, extreme temperature, and insufficient oxygenation. Minimum tolerable values depend on other water system parameters, but conservative values can usually be achieved.

Most academic research laboratories housing zebrafish use tap water. This can vary seasonally and when the weather, such as heavy rain, affects water quality. Typically, tap water contains chlorine (Cl_2) or sometimes chloramines (NH_2Cl, $NHCl_2$, or NCl_3) or ozone (O_3). One can determine how the water is treated by checking with the water supplier. There are many ways to remove these chemicals. Charcoal filtration, treatment with sodium thiosulfate, heavy or extended aeration, or processing with a reverse osmosis (RO) machine remove chlorine from the water (Timmons *et al.*, 2001; see Chapter 33). Chloramines will not diffuse out of the water, but can be removed with an RO machine. If only the chlorine component is removed with sodium thiosulfate, toxic ammonia levels can result. The ammonia can then be removed with ammonia-binding products such as zeolite. Unusual water sources such as spring water are used in some places instead of tap water. They might not have chlorine, but might require treatment for other problems. These sources should be tested both chemically and with zebrafish before committing to use them. Metal (such as copper and zinc) contamination can occur from pipes, valves, and other plumbing equipment, especially in new buildings. Levels at parts per billion (ppb) can cause problems (see Chapter 33). If it is unreasonable to replace the plumbing, then these metals can be removed with an RO machine.

2. Pumps

Pumps are rated by the volume of water moved [in gallons or liters per minute or hour (GPM, LPM, GPH, or LPH)] under different conditions of pressure and friction (in feet or meters of head or backpressure). The volumes pumped at different pressures define a flow vs. pressure curve for a pump. Approximate backpressure values can be calculated for a plumbing system, based on friction in the pipes and fittings. These values can be used to size pumps properly for particular operations. Common flow rates through tanks range from three to seven tank volume turnovers per hour.

Most pumps used in zebrafish facilities are centrifugal pumps that use a spinning impeller. They develop pressure by throwing the water to the outer edge of the impeller housing. Other pumps (usually for controlled additions of small amounts of chemicals) move water with pistons, diaphragms, or pressure on tubing. Some pumps can be used submerged. This is quieter but adds heat to the water. Pumps should be fish friendly in that they have nothing potentially toxic or corrodible in contact with the water.

C. Filters and Reactors

Fish water can be processed in many ways to make it better suited to the fish. These include removal of toxins, addition and removal of chemicals, or preventing the recirculation of infectious organisms.

1. Particle Filters

Particle filtration removes suspended particles from the circulation, such as fish feces, uneaten food, and dead fish. Filtered out materials remain in the water system, releasing chemicals as they decay, until the filter is cleaned. Some more expensive filters (bead filters, rotating drum filters, sand filters, and a few others) clean themselves, saving labor and more rapidly removing the filtered materials from the water. Most particle filters remove particles by sieving them out of the water flowing through particular pore sizes. Bead filters remove particles by their stickiness to the beads.

2. Biological Filters

Biological filtration uses metabolic processes to remove compounds from the water or to transform them into less toxic compounds. Recirculating water systems have biological filters to house nitrifying bacteria that process ammonia wastes. Biological filters also house autotrophic bacteria that eat dissolved organic material. More mature biological filters are assumed to have hundreds to thousands of different species of bacteria. This vast population might take quite a while to establish. The filters provide the bacteria with plastic or sand surfaces to grow on. The size of the bacterial populations and thereby their capacity for processing wastes are limited by surface area. Submerged media filters can be oxygen limited, and those with exposed media should not be. Plant filters have been used to remove nitrates and phosphates, but not in many zebrafish facilities.

The initial establishment of the biofilter can take several weeks if it is left to random colonization of the water system by airborne bacteria. Colonization can be speeded up by seeding bacterial-laden material from an established aquarium or biological filter. The major bacterial populations in such a seeded filter will grow rapidly if fed ammonium chloride and/or fish food. For each pound (or kilogram) of (50% protein) food eaten, fish, on average, excrete 20 g (or 44 g) of ammonia. Autotrophic bacteria eat uneaten fish food and produce ammonium. This and ammonium chloride feed the nitrifying bacteria. Nitrosomonas bacteria oxidize ammonia to produce nitrite. The nitrite feeds the *Nitrospira* bacteria that oxidize nitrite to nitrate. As these populations grow, the concentration of their nutrients should decrease. Ammonia, nitrite, and nitrate concentration changes can be followed through a series of test kit measurements. When the system is being fed the amounts of ammonia and/or food expected of the facility, when fully stocked, and the nitrogenous levels are acceptable, the biological filter will be ready for use. The drawback of this approach is that it may compromise good quarantine procedures. A good source of bacteria (many species) is desired for inoculation without introducing outside pathogens. Use of bacteria from a biological filter within the same facility (an internal source) avoids this problem. If you have no biological filter or want to avoid transferring bacteria from any water system, you can set up an unseeded biological filter months ahead of time and feed it lightly over an extended period. Eventually, it will be colonized and suitable for seeding the main filter. Commercial preparations of bacteria are also available for seeding biological filters. Many of these might not have *Nitrospira* bacteria to oxidize nitrites to nitrates. Once a filter is well colonized, the bacteria reside in a biofilm. This protects the bacteria and makes recolonization easier.

Autotrophic bacteria in a biological filter remove dissolved nutrients from the water, reducing nutrients for bacteria elsewhere in the water system. More rapidly growing colonies of autotrophic bacteria can outcompete more slowly growing nitrifying colonies for filter media space and bump them off. This reduces the filter's nitrification capacity. Because autotrophic bacteria can grow rapidly at times, it is good to have a large biological filter that can house both kinds of bacteria.

A mature biological filter is thought to have slow-growing bacteria specialized for extracting lower concentrations of nutrients. These more mature bacterial populations are thought to provide some probiotic protection by usurping available nutrients and inhibiting the establishment of the faster-growing opportunistically pathogenic bacteria. Bacteria in the water colonize the initially sterile gut of the larval zebrafish. The gut of older fish is continuously recolonized by the water system and biological filter bacterial populations.

3. Chemical Filters

Chemical filters remove chemicals from the water by interactions with the filter's media. Activated carbon is commonly used for dechlorinating, removing nonpolar organic molecules, and removing heavy metals. Activated carbon optimized for dechlorinating tap water might not efficiently remove organic molecules (Moe, 1989). Other media (e.g., zeolite, Amquel®) can remove ammonia and other nitrogen products. Water-softening filter resins can remove various ions from the water. A problem with using activated charcoal is determining when to change the media. When using charcoal to dechlorinate, filter functioning can be determined with a chlorine test. This can be used to determine when the media is becoming saturated. However, activated carbon is frequently used to remove unidentified toxins and organic molecules from the water. These unidentified or low-concentration contaminants are difficult to measure. This makes it difficult to determine when the media should be changed. Standard use of activated carbon can provide prophylactic protection against accidental introductions of toxins. Charcoal filtration could also remove chemicals that might leach from plastics.

4. Water Source Purification

Different water sources (e.g., tap water or well water) might have to be treated before they can be used for fish. This water conditioning can be provided institutionally (e.g., tap distilled, DI, or RO water) or by facility equipment. Facility supply requires a level of technical knowledge and labor necessary to operate and maintain the unit. Expenses include the machinery, backup machine, and periodic replacements of filters and membranes. Facility units have the advantage of being completely under your control, avoiding reliance on an outside organization. Institutional water systems might become contaminated with things growing in the long runs of pipes (Astrofsky et al., 2002).

RO is a process in which pressure forces water across a semipermeable membrane against a water concentration gradient. This separates purified water from many dissolved components that cannot cross the membrane. The delicate membranes are usually protected by particle filtration and activated charcoal. Affordable large-volume aquacultural-grade units can supply makeup water for most fish facilities.

Deionization removes ions with ion-binding resins. Units often include an activated carbon filter to remove organics. Deionized water should work well for a zebrafish facility, but alone it might not be able to remove all problematic compounds.

5. Water Sanitization

Ultraviolet (UV) sterilizers are used to disinfect the water before it is returned to the fish. This keeps bacterial levels low and prevents pathogens from one tank from being redistributed through the plumbing to other tanks. As the water passes through, the UV sterilizer irradiates the water, killing viruses and bacteria, mostly by damaging their DNA. Important parameters in sizing these units are water flow rate, power of UV emitted, water clarity, and the desired level of irradiation. The UV output from a sterilizer is computed in μWsec/cm^2 to reflect these factors. Because UV lamps vary, the watts powering a lamp are not necessarily reflected in the strength of its output in the germicidally significant wavelengths (Timmons *et al.*, 2001). Low-pressure mercury UV lamps most efficiently produce the germicidal UV (around 265 nm). UV output is also affected by lamp age. UV lamps can lose 40% of their output in 6 months. Therefore, the units should be sized to provide the desired UV dose over the planned age of the lamps. Lamps are usually replaced after 6 or 12 months. Doses required to kill particular infectious organisms can vary. Some are listed in reference tables (Creswell, 1993), but there are no data on many pathogenic species. UV killing is not absolute. Within a large population, a few bacteria survive a UV sterilizer as detectable colonies if enough water is sampled.

High doses of O_3 can be very effective at sterilizing water, but this approach requires a lot of equipment and space. This could include the ozone generator, possibly an oxygen source, a large-volume vessel to provide sufficient contact time with the water, followed by an ozone inactivation step to keep it from affecting the fish or biological filter. For these reasons, few, if any, zebrafish facilities sterilize water with O_3. O_3 is more commonly used to oxidize organics in the water to improve the water quality. Oxidation–reduction potential (ORP) controllers are used to provide feedback control to O_3 generators, with set points of 300–350 mV. More details can be found in Moe (1989) and Timmons *et al.* (2001).

D. Air Sources

If aeration is used in a facility, the air source will probably be either a high-pressure institutional airline or from one's own machines. Institutional high-pressure air sources are often supplied by compressors that tend to be oily and require filtering.

An aquacultural blower is a good choice for larger facilities. It makes large volumes of low-pressure oil-free air. The blower air source can be the building HVAC system or it can be plumbed from another location. The goal is to provide fresh uncontaminated air. Potential outside contaminants include automobile exhaust, insecticides, fume hood vents, and cigarette smoke. An outside source away from fumes is preferred to avoid HVAC system backup problems.

E. Monitoring, Alarms, and Redundancies

Good monitoring can identify problems before their effects become dramatic. To have a real effect on fish well being, a chain of events must be ready to go. Alerts have to be efficiently transmitted to the responsible people. Materials for effective responses should be handy, including duplicate pieces of equipment. The staff and equipment have to be able to respond rapidly.

To develop a good set of failure scenarios, first establish a good understanding of how your water system, building, and facility function. Examine each component and figure out what could break and what its consequences would be. Next, establish a monitoring program to ensure that indicators of problems are identified and communicated as soon as possible. Design and build in redundancies and backup systems and establish stores of essential parts. Train people to make sure that they know where things are and how to use them.

1. Monitoring and Logging

Standard maintenance tasks, water system changes, and problems should be logged in a notebook or on a checklist so that data are available for review. Monitoring machinery and water conditions can be done by hand. They should also be logged. Daily or weekly measurements of slowly changing water parameters (such as nitrites and nitrates) are usually sufficient. Many, but not all, monitoring tasks can be done better electronically, but require greater initial expenditures. Temperatures, motor and light function, light intensity, light timing, water on the floor (or other places), water flow and pressure in pipes, valve positions, water levels, pH, dissolved oxygen, conductivity, and ORP can be conveniently monitored electronically.

2. Electronic Monitoring

Electronic monitoring, alarm, and control systems can be assembled or purchased from several manufacturers. These systems can monitor a variety of different electronic inputs, store data, and send out programmed alarms to phones or pagers through an autodialer. Datalogged inputs can be useful in figuring out how a problem developed. Even when people are present, alerts provided by these systems can provide faster notification of a problem. At least one, preferably two, people should always be on call to respond to the pager, day and night, weekdays and weekends. Some monitoring systems can control outputs, such as solenoid valves and pumps, thus performing periodic tasks and taking programmed corrective or preventative actions.

3. Common Backup Systems and Redundancies

The most frequently backed-up systems in zebrafish facilities are the electrical system and water pumps. Alternative water sources and aeration in each tank as an oxygen supply in case of plumbing or pump failure are other backup systems. Zebrafish facility HVAC systems should be at least partially redundant. Temperature control and fresh unadulterated air for aeration are the main concerns. A redundant heat source (perhaps electric space heaters instead of stream) and a redundant way to cool the water (adding cold dechlorinated tap water) should be established.

4. Biological Monitoring

Periodic histopathological examinations of sentinel fish, fish from several different tanks, or sick fish can provide early warnings of diseases before they become a widespread problem. Sentinel fish are housed to receive the mixed unfiltered outflow from all of a water system's tanks. This is done by supplying the sentinel tank with water from the return sump before it is filtered. The fish can be examined histologically two or three times a year and replaced with juveniles or young adults. Identification of specific pathogens allows a much more directed treatment. There are not yet good methods to monitor the state of biological filters other than by the ammonia, nitrite, and nitrate levels.

F. Adult Tanks and Their Functions

1. Tanks

Fish containers (tanks or aquariums) are usually made of glass, acrylic, or polycarbonate. Polycarbonate tanks and lids can be cleaned and sterilized in many ways, including bleaching, or autoclaving, and tunnel washing. Glass tanks can crack when autoclaved or tunnel washed. Glass is heavier and breaks more easily, but is more resistant to scratching.

Slime in fish tanks is not necessarily a bad thing. Bacteria in the slime can provide some biological filtration for the tank; however, transparent walls are required to observe the fish. Inability to observe the fish is a good criterion for determining when to change out or clean tanks.

Zebrafish are good jumpers. Tank covers prevent them from jumping out, or worse, getting into the tank of a different strain. Unlike polycarbonate, flat acrylic covers warp because of absorbed water on the high-humidity side. Lids usually have holes for feeding, autofeeders, water, and air. There is often a compromise between a large hole for feeding and smaller holes to keep fish in.

Tank drains should be low maintenance, keep the fish in the tank, and allow the outflow of water and debris, preferably from the bottom. Ideally, they should not require cleaning until the tank is normally changed out.

2. Population Size

Different sizes of identifiable genetic populations are required for different research purposes. This and ease of use are major determining factors of useful tank sizes. Fish density is affected by factors such as water quality, food, feeding schedule, and age and size. Optimization of these factors can vary from facility to facility. Three sizes of populations are frequently encountered: individuals, families, and larger groups.

Families are a basic unit of genetic analysis, making them important in genetic screens and line maintenance. Families can vary in size from 10–20 to hundreds of individuals. Particular sizes depend on their intended use. Larger populations can be broken down into smaller ones, to be housed in several tanks. Other considerations favoring smaller-sized tanks are ergonomics, ease of use, utilizing space well, and ease of maintaining and cleaning tanks. Larger tanks are more difficult to move.

Frequently, one or two fish have to be kept so that they can be individually identified for later breeding. Without a good, long-term fish-labeling method, individual fish have to be housed separately or sometimes in pairs. Extended holding periods of days to months require filtration or water exchanges to provide good water quality. If long periods are required for securely establishing new genetic lines from screens or from a quarantine room, then fish in these tanks will do better on a water system. Although housing small numbers of fish in a large tank is not bad for the fish, it makes poor use of space. When the numbers of understocked tanks approach a rack's worth, a rack for smaller tanks becomes a space-efficient choice.

Larger populations of mixed stocks or mutagenized stocks are more efficiently housed in large tanks. These populations can range from a few hundred to thousands of individuals. Large populations can be housed in several large tanks (20–30 gallons) or in aquacultural vats or tubs.

3. Ergonomics

Ergonomics is concerned with making workstations better suited to people. This reduces discomfort and increases efficiency. Human ergonomics is a complex subject involving many factors, such as the weight being lifted, joint angles, repetitive twisting motions, force vectors, and how much a movement is repeated. Lighter tanks that can be removed easily should be less stressful on arms. Repetitive hand movements can be reduced by using foot-activated valves.

Common tank sizes are around 1 l (about 0.26 gallons, weighing 1 kg or 2.2 lb), or 3–4 l (0.79–1.05 gallons, weighing 3–4 kg or 6.6–8.8 lb), and larger tanks are

75.7–113.56 l (20–30 gallons, weighing 75.7–113.56 kg or 166.9–250.3 lb). The smaller sizes can be moved by hand when full; the larger ones can be difficult to move when empty.

G. Racks

In the academic environment, available space to put fish tanks into can often be a limiting factor. Racks provide a space-efficient way to keep many tanks in a given space. Their usefulness can be increased by including utilities for the tanks and adjacent work areas. These utilities can include incoming fish water, a water return for the tank outflow, air if the tanks are aerated, tank illumination to keep the fish in breeding condition, and possibly an illuminated temporary work space for netting and manipulating fish.

In general, racks support tanks on bars, shelves, water-collecting trays, or by suspending them. There are two ways to access fish in the tanks: reaching over the tank on a rack to net out the fish or removing a tank to net the fish out on a counter. Removable tanks have to be small enough for easy and ergonomic handling. They require no clearance above the tank for netting. This can result in more rows of tanks. In earthquake-prone areas, racks should be secured to the wall or floor to keep them from tipping over.

H. Breeding

1. Natural Crosses

The primary consideration in collecting eggs from breeding zebrafish is to keep the fish from eating the eggs after they are laid. Eggs can be collected in many ways from tanks of fish. They can be siphoned off the bottom of a tank after they are laid if the adults are prevented from eating them. Marbles, suspended meshes, or closely spaced rods can provide this protection (Westerfield, 1995). Alternatively, a collecting tray of some kind could be put in the tank and removed after the eggs are laid. More commonly, fish are bred in a small box with a mesh bottom through which the eggs can fall. The mesh bottom box, containing the fish, is suspended in a larger watertight box. This allows egg collections from pairs of fish, but requires more fish handling. Depending on the box size, one pair to six pairs of fish can be bred this way. Typically, these breeding boxes are about 1 l (0.26 gallons, 1 kg or 2.2 lb) in size.

2. Off Cycle

Zebrafish normally mate within 2 h of their perceived sunrise. Therefore, they can be induced to lay their eggs at different times of the actual day by shifting their lights on time. This is convenient for people working at certain developmental times. Changing their lights on time should be done without altering the 14 h of light and 10 h of dark required for good egg laying.

3. *In Vitro*

In Vitro fertilizations are performed by knocking out male and female fish, squeezing out the eggs and sperm, and combining them *in vitro* (Westerfield, 1995). This basic technique is used in many procedures for manipulating chromosome sets or generating haploid embryos or diploid embryos with no paternally derived chromosomes. Equipment for these techniques should be kept handy to the *in vitro* fertilization area.

a. Ultraviolet

Ultraviolet light is used to inactivate sperm DNA while leaving the sperm alive to fertilize the eggs. A Stratalinker® provides a well-regulated UV dose in a small closed container. The Stratalinker uses a UV sensor to determine when a dose has been given, thus compensating for an old dim lamp. This, combined with its ease of use, makes it the instrument of choice. We use the autolink setting.

b. Press

The early pressure treatment suppresses the second mitotic division to make partially homozygous embryos with no paternal chromosomes (Westerfield, 1995). Critical features for the equipment are a cylinder that has a bore large enough to receive the pressure transmitting egg holders and can maintain 8000 psi (or 39,056 kg/m^2). French presses are frequently used to apply the pressure, but other presses can also be used.

c. Heat–Shock Water Bath

A 41.4 °C heat shock is used to suppress the first mitotic division to make homozygous embryos. This is usually done by using an accurate waterbath temperature controller (Westerfield, 1995). This can be purchased as a single unit to mount on the waterbath unit or assembled from an industrial temperature controller, an aquarium heater, a thermocouple, and a power head.

d. Sperm Freezing

There are several methods for freezing sperm (Nüsslein-Volhard and Dahm, 2002; Ransom and Zon, 1999; Westerfield, 1995), and they continue to be refined. Sperm freezing requires a space wherein fish can be conveniently squeezed to obtain sperm, some counter space, a sperm-freezing apparatus, and a liquid nitrogen freezer for storing the sperm.

I. Nursery

Raising zebrafish from the embryos up to juveniles is considered more difficult than maintaining and breeding adults. High mortality rates (fertilized egg to juvenile) are not uncommon. An 80% survival rate is considered good. Rampant mold, bacterial infections, and *Coleps* can rapidly kill all the eggs in a container.

Poor water conditions can adversely affect embryonic development. Well-cleaned eggs, removal of dead eggs, copious food, water changes after feedings, and foods free of pathogens improve the well being of young fish.

1. Small Fish Housing

Maintaining young fish on a slow-exchange water system saves labor (water changes are automated) and provides better and more stable water quality. The main problem is keeping the fish and fish food in while letting the water flow easily out. The common solution is to use a mesh or screen sized to keep the small fish in and let the particulate wastes out. However, these meshes often become fouled with a mixture of uneaten food, bacteria, and biofilm, requiring frequent cleaning or replacement. This can be eliminated by using drains with a much larger mesh surface area. If the mesh is large enough, it will not require cleaning before the fish are due to be moved to a larger tank. Forming the mesh into a tube over the drain opening can provide about 10 times the mesh surface area and can still fit in a small container.

2. Live Foods

Depending on the strain of fish being maintained, details of their husbandry, and the desired survival rate, larval fish can be fed dried prepared foods or live foods. Live foods have several advantages, but raising them requires space and labor. If live foods are purchased, quarantine issues should be considered. Moving live foods are a more attractive prey item to the fish. They also have active enzymes that when eaten aid the immature larval digestive system in digesting food. Live foods can also retain water-soluble nutrients that would rapidly be lost from small food particles with high surface-to-volume ratios. Many live foods (filter feeding ones) can be enriched by feeding them various nutrients (algae, yeast, or fatty acid emulsions) just before they are fed to the fish. This makes them delivery vehicles for the added nutrient. The use of live foods for larval fish is widespread in commercial aquaculture.

J. Requirements of Standard Operations

Ideally, all materials used in contact with fish or the fish water should be used in one tank and then autoclaved, bleached, or treated in other ways to eliminate possible pathogens that might be transferred to other tanks. This requires large numbers of each item, enough storage space for them, and their ability to withstand the harsh cleaning conditions. Sterilizable nets can be difficult to find. Only a few (such as Aquatic Habitats[TM]) can take bleaching and many melt at autoclave temperatures.

K. Utilities and Infrastructure

Some building and room utilities and features are useful for zebrafish facilities. They include electricity, lights, HVAC, plumbing, network connections, storage space, and building security.

1. Electricity

Electricity is essential for operating a zebrafish facility. Without it, pumps and air blowers cease to function. In high-density tanks, this can lead to reduced oxygen levels and fish stress or death. It is therefore important to ensure its near-continuous delivery. Backup power is more important for certain pieces of equipment such as water pumps, air blowers, ventilation fans, freezers, and refrigerators. This is usually provided by a backup generator that automatically turns on when power is lost. Computer-based monitoring and alarming equipment and database computers should be provided with backup power from uninterruptible power sources (UPSs). If a computer is on a backed-up circuit powered by a generator, it will crash in the few seconds of power loss before the backup generator starts.

All circuits supplying power to rooms with significant amounts of water should be equipped with ground fault interruption (GFI) protection. These devices compare the current flow in the hot and neutral lines. They cut the power very rapidly (faster than a fuse or circuit beaker) if there is an imbalance between the two. This prevents electrocutions. Certain pieces of equipment do not work well with GFI circuits, such as large motors and refrigeration compressors. These should be wired separately.

2. Heating, Ventilation, and Air Conditioning (HVAC)

The HVAC system provides fresh air and controls the temperature. Animal facility HVAC systems are usually redundant to some degree, in case components fail. They should be provided with backup electrical power. Often, nonelectrical components can be backed up, such as the building's source of heat and cooling. Nonelectrical (steam) heat can be backed up by electrical space heaters in a fish room. Chilled water-cooling can be backed up with electrical air conditioners or by having a source of cool water (such as cold dechlorinated tap water) that could be added to the water system.

3. Lights

Lighting of zebrafish housing areas should be automatically turned on and off to provide the desired light cycle (14 h on and 10 h off). The light cycle is important in promoting reproductive activity of the fish. The circadian rhythm is much less disrupted by having the lights off for a while than having the lights on when it should be dark. It is therefore not so important that the lights have backup power. Timers with their own backup batteries do not require resetting. Many timer-controlled lights are wired with an override switch so that the lights can be turned on during dark periods and repairs can be performed. Unfortunately, such switches can be accidentally left on, resulting in constant light, which inhibits fish breeding. Detecting this can be difficult, because it would be unusual to go into the room when the lights are expected to be off. This can be prevented while preserving the functionality of the override switch by using a countdown timer (i.e., for a bathroom fan) instead. The timer should not have a hold setting. This limits overriding of the lighting controls to the longest setting on the timer.

Optimal light levels for breeding AB fish are in the range of 5–30 foot-candles of light at the water's surface. Illumination by inexpensive shop lights or 6000-K bulbs seems to make no difference to fish breeding.

4. Water and Process Plumbing

The water supply to the fish facility should have at least hot and cold water. Building RO, deionized, or tap-distilled water might also be available. Copper pipes can release copper into the water, affecting the fish's health (Chapter 33). In large facilities, having large distribution and return pipes for the water system installed professionally can save time and space.

5. Ethernet and Communications

Ethernet connections can be very useful. They allow a central database to be served out to computers all over the facility and the use of e-mail and Web browsers. Similarly, phone lines should be available and in large facilities can be the basis of an intercom system.

6. Storage

Storage is needed for many things, such as food (refrigerator and freezer), water system, and fish tank supplies (disposables, maintenance supplies, and replacement parts). A tool use and storage area is often very handy for repairs and fabrications. Some things have preferred storage methods. Clean tanks and equipment used with the fish should be stored in closed cabinets to keep them dust free. Some, possibly off site, long-term big-item storage will be useful if research directions change frequently.

7. Building Structure

Zebrafish facilities are best located in basements or in their own buildings, in case of floods. The floors should be sealed. Wall interiors can be prophylactically treated with boric acid to control insects. Rooms with large motors should be sound insulated. Larger facilities tend to have a greater separation of functions in different areas or rooms. It is also more likely that additional special functions will be desired.

L. Fish Record Keeping and Databasing

Accurate record keeping and labeling is important to prevent fish with different genetics from getting mixed up. As the number of tanks and genetic strains increase, this problem becomes more acute. Keeping such records is best done with computer databases. In addition to being able to perform searches, sorts, and make daily backups, databases also permit the programming of output for grant and Institutional Animal Care and Use Committee (IACUC) reports. FileMaker Pro is a commonly used, easily setup database.

1. Labeling and Barcoding

Labeling of tanks is also important for keeping track of fish. Tank labels can be either handmade or printed out from computer files. In either case, the labels should have waterproof print and adhere in wet environments. As tank and line numbers increase, standard printed labels become more appealing. Computer-printed labels always have the minimal required information, including barcodes if desired. Barcode readers can then be used in surveys and to update database records. This removes both the labor involved in writing and the associated human errors. Database records can be flagged for later database actions, such as sending out e-mail reminders. This will require that all the facility's fish users use their e-mail accounts.

IV. Quarantining

Fish coming into the fish facility from outside sources should be quarantined to prevent the introduction of new pathogens that they might carry. How this is done varies a lot with facility size. Small facilities might have a single isolated tank. Facilities with several tank racks might have a single rack dedicated to use as a quarantine rack. Large facilities often have separate rooms with separate water systems.

Quarantining fish often involves just holding fish for a few weeks to see whether they show disease symptoms. However, the standard for zebrafish facilities is to only let bleached eggs and embryos into the main fish facility. Thus, new fish do not enter the facility but are held in the quarantine room and bred to obtain eggs that enter after bleaching (Nüsslein-Volhard and Dahm, 2002; Westerfield, 1995). Ideally, the quarantine room is physically isolated, with a separate water system and its own set of equipment. Materials being removed from the quarantine room for use in the main fish room should be autoclaved or heavily bleached. Materials, water, and fish being brought into the quarantine room require no special treatment. One should be able to set up crosses, collect eggs, and raise baby fish in a well-equipped quarantine room.

Normally, a quarantine room uses a flow-through type of water system in which the water sees fish in a tank once and then goes down the drain. Any pathogens being shed by fish are not recirculated by water flow to other uninfected tanks.

The usual sequence of quarantining involves holding the parent fish for several months until their eggs have been bleached. The bleached eggs are then introduced into the main facility. Additional time is required to identify the F_1 mutation carriers in the main facility. After confirming that the mutation was successfully introduced into the facility, the quarantine room fish can be euthanized. Providing long-term housing for large numbers of different fish is important for efficient quarantining of large numbers of lines into a facility. Devices for bleaching many batches of eggs simultaneously can also save time.

V. Injections

Injections into zebrafish eggs and embryos are performed to transform the animal's genetics with DNA, to manipulate gene expression with RNA and morpholinos, and to trace cells through development with lineage tracers. The large size of the first few egg cells makes injecting them relatively easy with a dissecting scope, a micromanipulator, and a foot-switch-activated pressure injector apparatus. The space required is a vibration-free, 36-in.-wide, single-person counter space, with a pressurized air source (either filtered building air or a compressed gas cylinder with a regulator). Injecting smaller cells later in development requires a much more sophisticated setup.

VI. Mutagenesis Room

N-Nitroso-N-ethylurea (ENU) is at present the chemical mutagen of choice for zebrafish. Chemical mutagens in general are harmful molecules that should be used in a very controlled manner in a fume hood. Use of an apparatus for a well-controlled, hands-off addition and removal of the mutagen is both safer for the person involved and less stressful on the fish. After the ENU is washed out, the fish become hypersensitive to stimuli such as movements of people (especially sudden ones), sudden loud noises, and bumping by other fish. Controlling these stimuli and suppressing the startle responses of the mutagenized fish with 10 mg/l of MS-222 is important to their survival over the next several hours. The ENU can be removed and rinsed out into a waste storage and detoxification chamber by using valves not visible to the fish. Later, reactants can easily be added to break down the waste ENU to less toxic compounds.

These procedures require at least 3 ft of hood space and adjacent space for liquid and solid waste containers and for the mutagenesis and wash out solutions. A spectrophotometer (perhaps dedicated for ENU use) should be available for determining the ENU concentration, because the amounts in the ENU isopacs can be off by 50%. The ideal mutagenesis room would be close to where the fish are normally housed, sound insulated, have a sink, and room for a large bin to detoxify the mutagenizing apparatus.

VII. Treating Outflow

Some situations require the water outflow from part of a facility be specially treated.

A. Chlorinating Outflow

Some organisms such as infectious pathogens or invasive species should not be released into the local environment. To prevent this, the outflow is treated with chlorine for a long enough time (the contact period) to kill the organisms of concern. This can be done by running the questionable outflow through a long, narrow container that provides more than enough contact time (contact time = volume of contact chamber/rate of flow) for the chlorine to act. The chlorine (bleach) should be added at or before the beginning of the contact chamber and the final concentration determined at the end, ensuring that a minimum concentration has been maintained or exceeded throughout the treatment. This can be done with an expensive automated system or by dosing with a pool chlorine doser and measuring with a test kit or indicator strips with the proper resolution.

B. Charcoal-Treated Outflow

Some chemicals such as drugs or hormones might be present in the outflow from the tank. Certain chemicals should be removed from the water before it goes down the drain. Charcoal, or some other media known to absorb the chemicals, might be used by running a tube from the tank drain to a filter canister or bucket of charcoal so that the water flows through the media with a sufficient contact time for the absorption to occur. The outflow from the charcoal can then be sampled to determine whether the reduction in the chemical's concentration is sufficient for it to be put down the drain. The charcoal should be disposed of as chemical waste would be.

VIII. Disposing of the Dead

Dead zebrafish are often flushed down toilets or put down the drain through a garbage disposal. The frequent use and low perceived importance of this task can lead to the tragedy of garbage disposal abuse. This problem can be designed away by replacing the switch for the garbage disposal with a countdown timer without a hold setting. Installed this way, it cannot be left on very long. This timer can also be wired to simultaneously activate a solenoid valve to run water into the sink, protecting the garbage disposal from burning out and ensuring that the fish are washed down the drain. Animals exposed to toxins should be disposed of as if they were waste from that process.

IX. Disease Prevention

When many animals are held together closely, the possibility for rampantly spreading diseases is increased. There are several ways to counter this. Ultraviolet sterilizers can be used to kill circulating pathogens in the water system. Sentinel fish can provide an early warning of pathogens in the water system, and quarantining prevents the introduction of new pathogens. These have already been discussed.

A. Stress Reduction

Zebrafish are constantly exposed to bacteria in the water. Normally, their immune systems can fight off reasonable levels of bacteria, but stressed fish have less active immune systems (Astrofsky et al., 2002). Many things can contribute to stress, including handling and netting (Wedemeyer, 1996). Use of products that help restore a fish's protective slime coat, which can be rubbed off during handling, can help the fish recover. The fish's slime coat is its first line of defense against infection. Other stressors include crowding, loud noises, nondark vs. dark backgrounds, and levels of illumination. Some claim that sudden turning on of the lights in the morning stresses their fish. This can be avoided by ramping up the light intensity either gradually or in a series of steps.

Recommended light levels for the fish (5–30 foot-candles) are less than good task lighting required for detail work (sorting through fish and eggs, reading, and writing). Unless light levels are tested and adjusted, fish facilities can easily be brighter than the fish might like. Task lights can provide higher illumination levels in work areas. They should be turned off on the same schedule as the room lights are. Otherwise, left-on task lighting will disrupt the zebrafish's circadian rhythm and breeding. Similarly, computer screens that are left on all night can affect the breeding of nearby fish.

B. Nutrition

Zebrafish nutrition has not been heavily addressed (Stoskopf, 2002). Much of current practice is based on anecdotal knowledge or derived from studies of aquacultured species or other tropical fish. Increased diets can compensate to some degree for a more crowded or stressful environment. Breeding fish also require more food. General guidelines for laboratory fish feeding rates call for feeding 3–8% of the fish's body weight per day in dry food (AFS, AIFRB, and ASIH, 2004). All prepared foods should be stored in either a freezer or refrigerator to prevent their degradation and loss of nutritional value. A lot of food storage space is not required because most fish foods should not be stored for long periods.

C. Treatment

Treatments can be thought of as either eliminating the sick fish, isolating and treating them, or treating all the fish on a water system. Whenever drugs are used to treat disease, the treatment's effect on the biological filter should be considered. Many drugs can affect the bacteria in the biological filter. This can lead to big increases in ammonia levels, which can weaken or kill the fish.

X. Cleaning

Cleaning, sanitizing, and sterilizing can be done with sinks, large vats, washing machines, or larger tunnel washers. Bleach, strong detergents, and temperatures of 82.2 °C (180 °F) are typically used in these processes. Materials that are used frequently have to stand up to the cleaning cycle. In general, materials that can withstand autoclaving or several hours in strong bleach solutions (~1%) are acceptable. Sometimes products such as nets that can withstand this treatment are difficult to find. Most nets do not withstand bleaching. An Aquatic Habitats product is an exception. One might have to try many nets in order to find ones that can be autoclaved. When one is found, it is smart to buy lots of the type because the manufacturers might change the materials that they use, resulting in a net that could melt in the autoclave.

Nets and other items that go into tanks repeatedly should be cleaned after being used in a single tank. This requires large numbers of each item (and storage for them) and a rapid cleaning cycle. Crossing tanks should be cleaned after each use. In large facilities, hundreds of these have to be cleaned each day.

Plastic fish tanks can be switched out and cleaned either periodically or when it is difficult to observe the fish. They should also be cleaned when their fish are euthanized. Tanks are not washed as often as crossing tanks, but require more space in soaking bins. Clean equipment should be kept in closed, dust-free cabinets. Transportation patterns should be established to minimize the exposure of clean tanks to the dirty tanks that are collected and transported to the washroom.

XI. Human Safety

Use of GFI circuits to prevent electrocution and ergonomic issues have already been discussed.

A. Zoonotic Diseases

Zoonotic diseases are discussed in Chapter 33. Employees' understanding of these issues can be heightened by providing an information sheet on the potential zoonotic diseases they might be exposed to. Preventative measures include avoiding getting fish water on cut skin, using gloves, and having bactericidal soap and hydrogen peroxide available in case of cuts.

B. Chemical Exposures

Hazardous chemicals in zebrafish facilities can include mutagens such as ENU, acids and bases, and washing and sterilizing agents such as bleach and strong detergents. These chemicals should be properly stored. Material safety data sheets (MSDSs) should be available. Appropriate protective garb should be worn by people using them.

References

American Fisheries Society (AFS), American Institute of Fisheries Research Biologists (AIFRB), and American Society of Ichthyologists and Herpatologists (AISH) (2004). "Guidelines for the Use of Fishes in Research." American Fisheries Society, Bethesda, MD.

Astrofsky, K. M., Bullis, R. A., and Sagerström, C. G. (2002). Biology and management of the zebrafish. In "Laboratory Animal Medicine," 2nd ed., Chap. 19. Elsevier Science, Amsterdam.

Burrell, P. C., Phalen, C. M., and Hovanec, T. A. (2001). Identification of bacteria responsible for ammonia oxidation in freshwater aquaria. *Appl. Environ. Microbiol.* **57**(12), 5781–5800.

Creswell, R. L. (1993). "Aquaculture Desk Reference." Chapman and Hall, New York.

McClure, M. M. (1998). Development and evolution of pigmentation patterns in fishes of the genus *Danio* (Teleostei: Cyprinidae). Thesis. Cornell University, Ithaca, NY.

Moe, M. A. (1989). "The Marine Aquarium Reference, Systems and Invertebrates." Green Turtle Publications, Plantation, FL.

National Research Council (1996). "Guide for the Care and Use of Laboratory Animals." National Academy Press, Washington, D.C.

Nüsslein-Volhard, C., and Dahm, R. (2002). Zebrafish. Oxford University Press, New York.

Ranson, D. G., and Zon, L. I. (1999). Collection, storage, and use of zebrafish sperm. In "Methods in Cell Biology, Vol. 60, The Zebrafish, Genetics and Genomics." Academic Press, New York.

Spotte, S. (1992). "Captive Seawater Fishes, Science and Technology." Wiley-Interscience, New York, NY.

Stoskopf, M. K. (2002). Biology and health of laboratory fishes. In "Laboratory Animal Medicine," 2nd ed., Chap. 20. Elsevier Science, Amsterdam.

Timmons, M. B., Ebeling, J. W., Wheaton, F. W., Summerfelt, S. T., and Vinci, B. J. (2001). Recirculating aquaculture systems. Publication No. 01-002, Northwest Regional Aquaculture Center. Cayuga Aqua Ventures, Ithaca, NY.

Wedemeyer, G. (1996). "Physiology of Fish in Intensive Culture Systems." Chapman and Hall, New York.

Westerfield, M. (1995). "The Zebrafish Book." University of Oregon Press, Eugene, OR.

Wheaton, F. W. (1977). "Aquacultural Engineering." John Wiley & Sons, New York.

CHAPTER 31

A Nursery that Improves Zebrafish Fry Survival

Peter Cattin and Phil Crosier

Department of Molecular Medicine and Pathology
School of Medical Sciences
The University of Auckland
Auckland, New Zealand

I. Introduction

The need to maximize the survival of zebrafish embryos and fry is fundamental to all research that uses zebrafish as a model system. With the increasing number of mutants and transgenics being developed by zebrafish investigators, there is a need for a system that provides consistently high survival rates of young fry. It is widely recognized that a critical period in the life of a zebrafish is in the 2 weeks following absorption of the yolk. Two elements are essential to ensuring survival of zebrafish fry during this period: (1) to provide an adequate supply of nutritionally high quality feed of the appropriate particle size and (2) to ensure that the water quality in the fry tank is optimal. Excessive water flowing through a fry-rearing vessel and the use of air have the potential to physically harm fry or exhaust them in their attempts to swim against the current. A very moderate flow of good-quality water sufficient to remove uneaten feed and fecal waste without imposing any physical stress on the fry is required. There are various methods for growing fry, ranging from the use of containers of static water to customised tanks connected to some form of water recirculation.

This chapter briefly reviews existing methods used for raising zebrafish from the larval stage until 3–4 weeks of development and to describe a new system developed to ensure consistently high rates of survival during this developmental period. The need to design a new approach stemmed from the low rates of fry survival obtained by our group by using existing methods.

II. General Methods for Raising Embryos and Larvae

There are several detailed descriptions of methods for raising early larvae and fry (see Chapter 3, Brand *et al.*, 2002; Westerfield, 1993; *The Zebrafish Monitor* at http:/zfin.org/zf-info/monitor/mon.htm). In brief, following pairwise matings, sorted embryos are grown in static culture either in covered small beakers or in 90-mm petri dishes containing E3 embryo medium in an incubator at 28 °C. The numbers of embryos per vessel will vary from 25 to 50 per 100 ml of E3 in a beaker to ~50–60 embryos per 90-mm petri dish. Four days after fertilization, the larvae begin to be fed live infusoria. Investigators have their own preferences on the type of infusoria used for larval feeding. The organisms used include paramecium, *Tetrahymena*, and rotifers. Live cultures can be obtained from commercial suppliers or from other laboratories. Protocols for raising live infusoria are available (Brand *et al.*, 2002; Westerfield, 1993). The infusoria are washed, resuspended, and added directly to each dish of larvae. At around Day 5 to Day 8 of development, the fry are transferred to larger containers such as mouse cages containing system water. With the use of static conditions to grow larvae, there is a continual need to monitor water quality and replace one third to one half of the water each day. Another approach used for growing zebrafish fry is the use of baby tubes: plastic tubes with a nylon mesh attached at one end. Each baby tube is suspended in a tank and connected to the water inflow (Westerfield, 1993).

After about 3 weeks of development, the larvae are fed *Artemia* and methods for growing this feed are published. Brand *et al.* (2002) note that to ensure high survival rates and a homogeneously growing population, the larvae need to be fed within 48 h of swim bladder inflation, which occurs at 72 h postfertilization (Kimmel *et al.*, 1995). With feeding, there is a fine balance to be met between overfeeding, which negatively influences water quality with waste, and underfeeding larvae at this critical time of development and growth.

III. A Nursery that Improves Fry Survival

A. Background

Our group is investigating aspects of developmental hematopoiesis and angiogenesis by using a range of zebrafish methodologies, including *N*-ethyl-*N*-nitrosourea (ENU) genetic screens, morpholino knockdowns, and the development of transgenic lines. When we began working with zebrafish, we found inconsistent and often low fry survival rates ranging between 10% and 30%. The question of maximizing the survival of zebrafish fry is one that has troubled a number of laboratories. The systems described previously tend to use either static water, with methylene blue to reduce the possibility of fungal infection, or water containing methylene blue that is aerated or mechanically agitated. There are problems with these methods. The quality of the water quickly deteriorates as a result of the uneaten food and the fecal waste, necessitating either replacement of about one third to one half of the water daily or physically straining the fry from the water and transferring them to a clean container with fresh embryo medium. This process typically requires the use of some form of net to sieve the fry from the dirty water and to then transfer the fry to the new container, raising the possibilities of physical injury and/or promoting infection, resulting in early larval death.

We aimed to develop an approach that delivered consistently high fry survival rates and minimized stress to fry during the transition phase from endogenous to exogenous feeding. In designing this system, important factors considered were minimizing physical contact with the fry and providing consistently high quality water at a stable pH, temperature, and conductivity. We also wanted to devise a system that was labor saving and could be scaled up for large applications such as genetic screens. The fry nursery we have developed uses high-quality water with flow-through circulation and consistently provides 65–100% survival of the transgenic, mutant, and wild-type strains of zebrafish fry raised in our laboratory. The system developed is likely to have widespread utility for zebrafish investigators.

B. System Design and Operation

The system is called the MaxHatch^{TM} nursery. It uses flow-through water circulation and is shown photographically in Fig. 1 and schematically in Fig. 2. Embryos are hatched in 90-mm petri dishes in an incubator at 28 °C and transferred to the nursery at 5 days postfertilization (dpf). The nursery itself consists of 30×500 ml fry containers (fry pots) suspended in a water bath, with each fry pot readily accommodating up to 150 five-day-old fry. The nursery system illustrated can hold up to ~4500 fry.

Water is pumped from the reservoir through two filters (5 and 0.2 μm) to remove the solid waste, bacteria, spores, cysts, and other resting forms of pathogens to a UV sterilizer. The water is elevated from the UV sterilizer to a distribution manifold suspended over the water bath. A valve above each fry pot accurately controls water flow to each tank. Flow-through fry pots restrict the fry by means of a nylon screen across the base of the cylinder. The mesh of this screen is typically 500 μM. The fry pots convert from a flow-through container to a sealed container by the adjustment of a nut. This allows fry to be transferred between containers without the use of a strainer or net, thereby eliminating physical contact and manipulation.

Water returns from the water bath to the underbench reservoir. The reservoir contains the immersion heater and the media to support nitrifying bacteria.

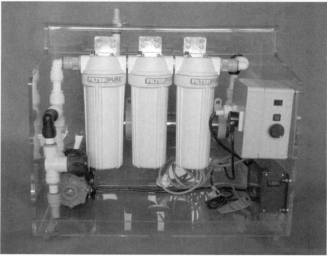

Fig. 1 The MaxHatch™ nursery consists of fry pots inserted into a holding manifold that is connected to the water purification and pump system. (See Color Insert.)

During feeding, the pump is switched off by an adjustable timer, enabling feeding to take place in no-flow conditions. This reduces waste of feed and water contamination while promoting higher levels of feed uptake. The pump is then turned back on after the predetermined feeding time has lapsed.

C. Our Experience

The nursery has been used routinely in our group for 3 years with successful raising of larvae occurring in a wide range of experimental situations. Direct survival comparisons have been made for a duration of up to 40 days with larvae

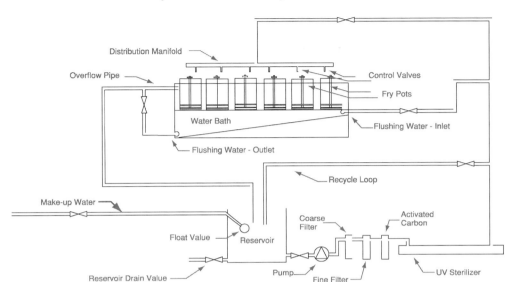

Fig. 2 Schematic outline of the MaxHatch nursery.

following injection of a number of transgenes (e.g., *gata1*-EGFP, *radar*-EGFP) and raising the embryos in either a static culture or the nursery. We have consistently found fry survival to be <10% in static culture compared with 65–100% in the nursery. We believe that the use of this system facilitates the successful generation of stable transgenic zebrafish lines. The system is used to raise larvae during genetic screens and in the ongoing maintenance of mutant lines resulting from such screens. It has potential to contribute to the successful raising of fry that are genetically compromised during the process of undertaking genetic screens.

D. Optimized Feeding Regimen for the Nursery

The stage at which fry transition from endogenous to exogenous feeding appears to be a critical time during zebrafish development and growth. If the water and feeding regimen are not adequate over the first 3 weeks of life, high mortalities of the fry will occur around 14 dpf, which is about 9 days after the yolk has been resorbed.

Assuming that the quality of the makeup water is appropriate, the nursery provides the necessary physical environment for coupling a feeding regimen that meets the nutritional needs of the fry. Key to determining the correct feed type is finding a feed of an appropriate size and nutritional value. We have found the nutrient profile and size of the Active Spheres Golden Pearls (www.brineshrimp direct.com) feed of 5–50 μm to be most suitable when feeding commences on day 5 postfertilization. Active Spheres Golden Pearls are fed twice daily in combination with a daily rotifer feeding (see later) until fry are of sufficient size for transfer to the adult fish system (which occurs between 21 and 28 dpf).

There are perceived benefits of providing live feed to young fry. The best live feed to offer younger fry are rotifers, *Brachionus plicatilis* in particular, which are approximately 250 μm in length. The single-cell microalgae *Nannochloropsis oculata* makes an excellent feed for rotifers. Although *B. plicatilis* and *N. oculata* are both marine species, they can be grown in fresh water. Our practice is to grow them in a brine solution consisting of 15 g of Instant Ocean per liter of distilled water.

Between 21 and 28 dpf, fry are gradually transitioned to the feeding regimen of adult fish. This involves introducing larger sized dry feed and live *Artemia* (initially newly hatched). Active Sphere Golden Pearls are initially alternated with, and after a week replaced by, a 50- to 150-μm-diameter high-protein powder (ZM100; www.zmsystems.demon.co.uk). The ZM100 feeds continue for the first 2 weeks after the fry are transferred from the nursery to the adult system. This ensures that smaller fry within the clutch obtain sufficient food to survive and grow. *Artemia* feeds are initially undertaken on a test basis between 25 and 28 dpf. A red color observed in the gut cavity of fry indicates the *Artemia* are being actively consumed. Daily feeds with live rotifers continue until fry are transferred out of the nursery.

Acknowledgments

We thank all members of our laboratory and the zebrafish community for helpful discussions and advice. Research in our laboratory is funded by grants from the Auckland Medical Research Foundation, Health Research Council of New Zealand, Lottery Health, Marsden Fund, and the Foundation for Research, Science & Technology (NERF).

References

Brand, M., Granato, M., and Nüsslein-Volhard, C. (2002). Keeping and raising zebrafish. *In* "Zebrafish: A Practical Approach" (C. Nüsslein-Volhard and R. Dahm, eds.), pp. 7–37. Oxford University Press, Oxford.

Kimmel, C. B., Ballard, W. W., Kimmel, S. R., Ullmann, B., and Schilling, T. F. (1995). Stages of embryonic development of the zebrafish. *Dev. Dyn.* **203,** 253–310.

Westerfield, M. (1993). "The Zebrafish Book," University of Oregon Press, Eugene. OR. Available at http:/zfin.org/zf_info/zbook/zfbk.html.

CHAPTER 32

Genetic Backgrounds, Standard Lines, and Husbandry of Zebrafish

Bill Trevarrow[*] and Barrie Robison[†]

[*]Institute of Neuroscience
University of Oregon,
Eugene, Oregon 97403

[†]Department of Biological Sciences
University of Idaho
Moscow, Idaho 83844

I. Introduction

Much has changed since the zebrafish was first advocated by Streisinger as a model system for developmental biology. Originally conceived as a vertebrate genetic model for developmental studies, the zebrafish now plays a significant and expanding role in many scientific disciplines, including biomedicine, neuroscience, genomics, toxicology, and evolutionary biology.

METHODS IN CELL BIOLOGY, VOL. 77

Table I
Zebrafish Standard Lines

Name (synonym)	Abbreviation	Normal maint. (# pairs, min. # eggs/clutch)	Squeeze well?	Clutch size	Natural cross well?	Map vs. what lines	Inbred?	Genetic bottleneck	Fish source	Labs or stock centers to get fish from	Disease suspectabilities	Comments	References
AB	AB	Round robin (90–100 fish)	Yes	100	Yes	SJD, WIK	No	15 Fish	Oregon pet store	ZIRC	Microsporidian ("skinny")	[a]	Streisinger et al., 1981
Tübingen (Tuebingen)	TU	Random pairs (20 pairs, 50–100 eggs)	No	150–300	Yes	WIK	No		German pet store	Mullins, TübSC, ZIRC	Mycobacteria (fish TB)	[b]	Mullins and Nüsslein-Volhard, 1993
SJD	SJD	inbred (1 pair)	Yes	100	No	AB, TU	≥90%	1–2 Haploid genomes	Derived from DAR	Johnson		[c]	Johnson et al., 1995
India	IND	50 Fish, by lineage	No	50	No	AB, TU	No	A few fish		Driever		[d]	
Wild type in Kalkutta (WIK)	WIK	Random pairs	No	150–200	Yes	AB, TU	No	2 Fish		TübSC, ZIRC		[e]	Rauch et al., 1997
tup Longfin	TL	Random pairs		100			?			TübSC		[f]	
Nadia	NA	Random pairs		100			No	51–71 Fish	Nadia region, 40 miles east of Kolkata	Robison			
AB/Tübingen (Tübingen/AB, Tüb/AB)	AB/TU	Random pairs (20 pairs)	Yes	100–200	Yes	WIK SJD?	?	4 Fish	Hybrid of AB and TU	Hopkins		[g]	
C32	C32	Inbred	Yes	>100		SJD	≥90%	1–2 Haploid genomes	Oregon pet store	Johnson		[h]	Streisinger et al., 1986
Ekk Will	EK	Random pairs	Yes	150	No		Not at first		Ekk Will, Gibsonton, FL	Grunwald		[i]	Stachel et al., 1993
Cologne	Koln						?			Campos-Ortega		[j]	
Hong Kong	HK	~100 Fish, by lineage	Yes	50	Yes		No		Hong Kong fish dealer	Driever			

Origin	Abbreviation	Breeding		Frozen sperm	Clonal genomes	Source	Status	Reference
Singapore	SING			?		Singapore fish dealer	Extinct?	?
Darjeeling	DAR	Inbred Random pairs	AB, TU	No		Collected in Darjeeling, India	Extinct?	[k]
C29	C29	Inbred	Yes	>100	Clonal 1–2 Haploid genomes	Oregon pet store	Extinct, 1991	Streisinger et al., 1986
Indonesia	INDO			?		Indonesian fish dealer	Extinct?	
Hong Kong/AB	HK/AB			?		Hybrid of HK and AB	Extinct?	Knapik et al., 1998
Hong Kong/Singapore	HK/SING			?		Hybrid of HK and SING	Extinct?	

Notes. ZIRC, Zebrafish International Resource Center; *TübSC*, Tübingen Stock Center.

[a] A widely used line, "good looking" embryos and haploids. There is no known frozen sperm from old stocks. Tends toward females. Created by hybridizing A and B lines. 1970s to 1990: breeding females selected based on haploid embryo morphology. By 1991, 71 generations from origin. AB rederivation from females only. Based on good haploid embryos 21 females selected from 180. Males derived from gynogenates; male line not continuous.

[b] A widely used line. No frozen sperm from old stocks. Fish from the Tubingen strain are being sequenced by the Sanger Institute.

[c] Derived from the Darjeeling by Stephen Johnson, used for mapping or mutageneses.

[d] India and DAR might be the same, but published dates of introduction differ: 1988 for DAR and 1990 for IND.

[e] Founders lethal free with 90% probability, based on family screening; derived from fish from the wild in India.

[f] selected for good layers.

[g] Sex ratio close to 50%. TU/AB carries alleles for a notochord and edemic embryo phenotypes in the background. Two of 25 lines selected based on a lack of recessive embryonic phenotypes from sib matings of screened families through 6 dpf.

[h] Crossed to SJD to introduce alleles for vigor. After nine generations of back crosses to C32, Johnson's version, C32bc9, was still about one third SJD by molecular markers rather than the predicted 0.5%. The most robust female is selected for reproduction each generation. Used for mapping vs. SJD.

[i] Mutations were not removed from the genetic background but not a lot of lethals in the background. This line was initiated in 1989 and is the source of the cDNA libraries that the Grunwald lab made.

[j] Background for several Campos-Ortega lab transgenics.

[k] Collected by Heiko Bleher in 1987 from 300 fish; DAR has been described as the same as India, but there are different published collection dates: 1988 for DAR (ZFIN) and 1990 for India; fast swimmer. No known non-SJD DARs are available.

Concurrent with the explosion in our knowledge of zebrafish biology has been a rapid increase in the number of wild-type lines used for study. Traditionally, a wild-type line has referred to a closed breeding population of fish that harbors no defined phenotypic mutations. Because some lines known to carry mutations are used as wild-type lines, *standard line* is a more widely useful term. Forward genetic mutagenesis and introgression of mutations onto known backgrounds have also produced many mutant lines, with genetic backgrounds similar to those of a specific wild-type line. As of publication, there are currently 19 wild-type lines listed on the Zebrafish Information Network (ZFIN; Table I). These lines delineate the genetic backgrounds currently used in most zebrafish laboratories.

The increasing number of zebrafish strains used for research raises a number of significant issues. Careful identification of zebrafish lines is critical in establishing the generality of research results. Each line of zebrafish is usually derived from a separate founder stock, and therefore each might harbor a unique genetic background. (See Table I for examples.)

Through epistatic gene action, the genetic background of a given line can have marked effects on experimental results. Differences among the backgrounds of different wild-type lines could conceivably result in a given mutation displaying different phenotypes, which could have a variety of effects. This is particularly important in developmental biology because of the extensive use of forward genetic mutagenesis screens. The potential variation in genetic background among lines highlights the need for a clear nomenclature system, allowing accurate identification of the genetic lines and backgrounds used in experiments.

Any standard line (e.g., the AB line) can be kept in many different laboratories worldwide, raising the issue of genetic divergence among sublines. In extreme cases, two laboratories performing the same experiments on their own version of AB fish would in fact be using genetically distinct lines, which can potentially confound experimental conclusions. Researchers can employ some form of genetic monitoring to mitigate this problem, but this is seldom done.

This chapter outlines the current issues surrounding the construction, maintenance, and use of genetically defined lines of zebrafish. We provide a summary of the potential pitfalls associated with the maintenance of standard lines of zebrafish, and where appropriate we provide an overview of the ways in which problems may be avoided.

II. Nomenclature and Definitions

The current zebrafish standard (wild-type) line nomenclature is summarized at http://zfin.org/zf_info/nomen.html. Standard strain names are nonitalicized with the first letter uppercase, and abbreviations are all uppercase and nonitalicized (see Table I). In its current incarnation, zebrafish standard line nomenclature does not address the full diversity of possible standard line types. Lines should have a unique identifier (name) that also indicates their relationships with other lines. As line pedigrees elaborate, these historically descriptive names will become more complex. Current mouse nomenclature (found at www.informatics.jax.org/mgihome/nomen/strains.shtml) addresses situations not yet formalized for zebrafish, such as inbred lines, substrain derivation, hybrid strains, and congenic lines that differ by single alleles. Zebrafish lines are at present named for their history, line originators, or sources of fish. Informal line nomenclature has also been used to track how many generations the line has been in a propagation scheme. (See Johnson's C32 description in ZFIN (http://zfin.org/cgi-bin/webdriver?MIval=aa-fishview apg&OID=ZDB-FISH-030501-1.))

Hybrid zebrafish lines have been dealt with implicitly (formal rule not stated, but examples shown at http://zfin.org/zf_info/nomen.html), with the two parental line abbreviations separated by a slash. There is no formalized relationship between order in the name and the sex of the fish in the founding cross. Unfortunately, there is no formalized way to distinguish between propagated hybrid lines and recurrently created genetic backgrounds derived from crossing two standard lines other than to state clearly the situation in the line description. Some zebrafish lines (SJD and C32; see Table I) are known to be highly inbred (Johnson and Zon, 1999); however, current zebrafish line nomenclature does not distinguish between inbred and noninbred lines. Mouse and medaka lines are considered inbred after 20 generations of matings between siblings, but no formal rule at present exists for zebrafish.

It is frequently desirable to introgress a particular mutation into a target background. The resultant line is considered congenic when the mutant locus is fully introgressed into the genetic background of the original strain. In mice, lines are considered congenic after 10 generations of backcrossing to the target strain. Although this might be expected to leave only a very small fraction of the original genetic background, genes within 10 cM are still likely to cosegregate with the allele of interest. Zebrafish mutants are often considered to be roughly introgressed into a new background after only three generations of crossing, leaving on average one eighth of the foreign genetic material remaining in the strain, but on average genes closer than 33 cM cosegregate with the target allele.

III. Goals for Line Use

The conventional wisdom is that standard lines are primarily used to provide genetic backgrounds for induction, manipulation, mapping, and identification of mutations. This is the main reason for the production of all major zebrafish wild-type lines. (See Table I and http://zfin.org for current information.) However, as zebrafish further expand their use as a genetic model organism, additional uses will arise in other fields such as toxicology and evolutionary biology. Even within the present zebrafish research community, there are often conflicting goals regarding the use of standard lines. For example, a line appropriate for mutagenesis might not excel for toxicological studies or genetic mapping.

Some lines (AB, TU, and AB/TU) are frequently used for mutagenic screens because embryonic lethals have been removed from the population. In addition, alleles that disrupt or interfere with embryonic phenotypes are often selected out, allowing a more productive screen for embryonic phenotypes. In some cases, screens require particular alleles for complementation, enhancer, or suppressor testing. In these situations, appropriate genetic elements have to be introgressed into the background of the standard line prior to mutagenesis.

Mapping strains (usually SJD and WIK) differ enough from AB and TU backgrounds to provide useful sets of polymorphic molecular markers. To map new mutations quickly, mapping compatibility with another line should be established (Pelegri and Schulte-Merker, 1999). SJD and WIK are most frequently used to map mutations on the AB and TU backgrounds. The highly inbred SJD (Johnson and Zon, 1999) tends to provide more consistently mappable markers, but many find the line more difficult to maintain. WIK is also considered to be molecularly distinct enough from both AB and TU to make it useful for mapping. TL is a line that has not been cleaned of background mutations, but is considered very robust and useful for easy maintenance of mutants.

IV. Breeding Strategies

The breeding strategies used to construct and maintain lines of experimental organisms depend on the experimental requirements of the investigator. In general, the goals of a breeding program can be organized into three broad categories: genetic uniformity, maintenance of genetic variation, and selection for desirable phenotypes. In this section, we describe general strategies that can be used to accomplish these goals when constructing standard lines of zebrafish.

A. Genetic Uniformity

It is often advantageous to establish a population that is genetically uniform in which all the members of the population are genetically very similar and in extreme cases genetically identical. The utility of this type of line lies in its uniformity. The line can be used by multiple laboratories without concern that variance in genetic background is confounding the results. Within laboratories, the line can be used in experiments requiring multiple treatments, minimizing the potential effect of genetic variance on the experiment.

Typically, the construction of genetically uniform lines relies on inbreeding, producing uniformity at the expense of genetic variation. This loss of genetic variation can cause a decline in the overall fitness of the line, commonly referred to as inbreeding depression. Some controversy surrounds the genetic causes of inbreeding depression (see Lynch and Walsh, 1998), but there is general agreement that it arises through dominance-related mechanisms.

Whatever the true genetic cause, inbreeding depression can have very significant consequences on husbandry. Associated declines in growth rate, breeding rate, fecundity, fertility, and other traits can have a fundamental impact on the viability of a line. Nevertheless, for some kinds of experiments, the advantages of repeatability and stability presented by genetically uniform lines outweigh potential declines in line fitness caused by inbreeding depression.

1. Sib Mating

For most organisms, the most straightforward method of producing a genetically uniform line is through the mating of close relatives such as full or half sibs. In this technique, siblings are mated together to produce an inbred family. Siblings from this family are subsequently mated together to form the next generation. This process is repeated until the desired level of homozygosity (and thus uniformity) is achieved. In pure sib mating, only one male and one female are selected to produce the next generation. By using basic quantitative genetic theory, the proportion of heterozygosity present in the population (relative to the level of heterozygosity in the source population) after t generations can be determined (Falconer and Mackay, 1996). Figure 1 shows these proportions plotted against a number of generations for a range of population sizes. In the case of full-sib mating ($N_e = 2$), 5.6% of the original heterozygosity remains in the population after 10 generations. Loss of heterozygosity means that the allele frequencies of the population are becoming fixed at more loci, and therefore the population is becoming increasingly geneticly uniform.

Given a generation time for the average zebrafish of 3 months, the production of a genetically uniform line through sib mating could take more than 3 years, depending on the degree of uniformity required by the investigator. In some species, the time taken to produce inbred lines can be even longer. Inbred medaka lines have been produced in 7 years (see http://biol1.bio.nagoya-u.ac.jp:8000/rw8-2.html). In mice, 20 generations are required for a line to be considered inbred, which would take more than 6 years in zebrafish. This might not be feasible for many research projects, particularly those for which a new line must be produced from a highly heterozygous founder population. Sib mating is therefore most useful for long-term research projects and for lines of fish that will be maintained for multiple users by a central agency such as the Zebrafish International Resource Center (ZIRC).

2. Clonal Lines

If production of a truly genetically uniform line is the ultimate goal, then sib mating is not the best breeding strategy. Total homozygosity is approached asymptotically under full-sib mating designs, and many generations are required to fully purge the line of residual genetic variation. As illustrated in Fig. 1, a point

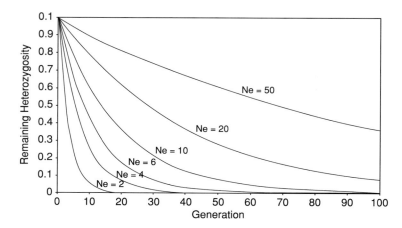

Fig. 1 The rate of losing heterozygosity decreases as the effective population size increases. However, large populations continue to lose heterozygosity at a slower rate.

of diminishing returns is approached, after which the gains in line uniformity (homozygosity) are negligible compared with the investment required.

As evidenced by the pioneering work of Streisinger and his colleagues, zebrafish are remarkably tolerant of chromosome set manipulation. This presents an effective short cut for producing genetically uniform lines. Streisinger's group showed that gynogenesis can be used to construct homozygous diploid fish in which only the maternal genome is passed to the offspring (Streisinger *et al.*, 1981). The technical details for producing gynogenetic diploid fish are described in Westerfield (1995) and have been further refined by Gestl *et al.* (1997).

The production of genetically uniform, totally homozygous lines of zebrafish is conceptually simple. In brief, sperm is collected from a number of males and the paternal genome is destroyed by exposure to UV radiation. Although the paternal genetic material is rendered nonfunctional, the sperm remain viable and will fertilize eggs squeezed from gravid females. After fertilization, development of the embryo begins by using only the haploid maternal genome. Left alone, the haploid embryos develop until approximately Day 4 and then die. However, the diploid state can be restored through the timely application of either a pressure or heat shock to inhibit the first cleavage (Westerfield, 1995). This results in a family of embryos that is genetically unique (as all the maternal gametes are unique in their genetic composition) but homozygous at most or all loci. These fish are commonly referred to as homozygous diploids.

Once the homozygous diploids reach maturity, females can then be used in a second round of gynogenesis. To the extent that the females are homozygous at all loci, all gametes produced by an individual female will be identical. Thus all gynogenetic offspring from a homozygous diploid female will be homozygous and genetically identical to each other and their mother, resulting in a clonal line of zebrafish.

Although the production of clonal lines in zebrafish is conceptually simple, a number of logistical difficulties are inherent to the technique. First, the treatments required (sperm irradiation and heat or pressure shock) cause very high levels of mortality, often approaching 100%. This drawback can be partially addressed by simply making more attempts at gynogenesis. Even with only 1% of the embryos surviving to adulthood, a moderately sized zebrafish facility can produce enough eggs for *in vitro* fertilizations to ensure that a reasonable number of homozygous diploids can be produced.

The second difficulty associated with the production of gynogenetic homozygous diploids, however, is more difficult to overcome. In some cases, the sex ratio of the homozygous diploids is highly biased toward males (Robison, personal observation). Propagation of a clonal line by gynogenesis is only possible if females are available. The sex determination mechanism of zebrafish is not known, and thus the cause of a male bias in the gynogenetic offspring of female zebrafish is difficult to determine. It is unclear even whether this sex ratio bias is dependent on the line used for gynogenesis or the subsequent rearing conditions. Androgenetic propagation of male homozygous diploids is technically possible in zebrafish (Corley-Smith *et al.*, 1999), but this technique has not been adequately explored in this species. Despite these difficulties, clonal lines of zebrafish have been developed in the past (Streisinger *et al.*, 1986).

Clonal lines have also been produced for many other fish species, including rainbow trout (Parsons and Thorgaard, 1985), carp (Bongers *et al.*, 1998), and tilapia (Hussain *et al.*, 1998). In salmonids, clonal lines have been effectively used to 'capture' interesting phenotypic variation found in wild and captive populations (Robison and Thorgaard, 2004; Robison *et al.*, 1999) in a stable and uniform genetic background. Clonal lines are also used for genetic mapping (Nichols *et al.*, 2003; Young *et al.*, 1998) and identification of quantitative trait loci (QTL) for a variety of economically, evolutionarily, or biomedically important traits (Robison *et al.*, 2001). Because the clones are genetically uniform, they have been distributed to numerous laboratories and serve as important reference strains for the construction of genomic resources such as bacterial artificial chromosome (BAC) libraries and cell lines (Ristow *et al.*, 1998).

B. Maintaining Genetic Variation

A common goal associated with the production of research lines is the maintenance of genetic variation. Typically, the purpose of maintaining genetic variation is to avoid the declines in fitness associated with inbreeding depression (see Section IV.A). Lines that harbor high levels of genetic variation are typically more robust than inbred lines, and thus breed more reliably, have higher fecundities, and higher survival rates from egg to adult. These characteristics are obviously desirable for researchers who require a near-constant supply of fish for use in experiments.

Genetic variation in laboratory lines is typically lost through genetic drift. Therefore, maintenance of genetic variation, regardless of organism, is centered around the concept of effective population size (N_e), which is the size of an idealized population that would undergo the same magnitude of genetic drift as the population under consideration. Effective population size is typically, but not always, lower than the census size of the population. This means that the magnitude of genetic drift is typically higher than what a simple count of the number of adults in the colony would indicate. Although a detailed discussion of the calculation of N_e is beyond the scope of this chapter, Falconer and Mackay (1996) provide an excellent discussion of the interplay between N_e and genetic drift. A number of factors typically seen during construction and maintenance of zebrafish lines can cause N_e to be lower than the census size (N), including an uneven sex ratio, uneven numbers in successive generations, variance among individuals in reproductive success, and overlapping generations. Each of these situations requires a different calculation for determining N_e. In some cases, such as an uneven sex ratio or temporal variance in population size, the calculations are quite straightforward (Falconer and Mackay, 1996). However, in other cases, such as that of overlapping generations, the calculation of N_e can be quite complex.

In general, when attempting to maintain high levels of genetic variation in a line, N_e should be as high as possible. Figure 1 shows the relative proportion of heterozygosity remaining in a line at varying effective population sizes over time. Two important points regarding the maintenance of genetic variation should be noted from this figure. First, at low N_e, genetic variation is lost very rapidly, the most extreme case being full-sib mating (described in the preceding Section IV.A.1) in which $N_e = 2$. Second, even with relatively large effective population sizes, genetic variation is still lost to drift, albeit at a much reduced rate.

In addition to losing genetic variation to drift, there are other ways in which lines that harbor high levels of genetic variation can evolve, thereby altering their genetic background. First, at high N_e selection is a much stronger evolutionary force than genetic drift. Genetically variable lines maintained at high effective population sizes are therefore subject to domestication selection. Despite knowing very little about the natural history of zebrafish, we can be reasonably certain that the zebrafish in the typical laboratory environment experience a very different set of selection pressures than those experienced by feral populations. In the laboratory, there is no predation, food is abundant and regularly presented, light levels and temperatures are strictly regulated, and water quality is carefully monitored. If the population was recently captured from a wild stock and sufficient genetic variation is present, adaptation of the line to the laboratory environment is simply unavoidable.

In many cases, domestication selection in populations of zebrafish is not undesirable (see Section IV.B.3). As the fish become better adapted to the laboratory environment, breeding rates, fecundities, and other important traits might evolve to more optimal levels. Nevertheless, one should not assume that maintaining a line of fish at high N_e will result in a genetically constant resource over time.

The concept of line evolution is central to another caveat associated with lines harboring high levels of genetic variation: subline evolution. Assume that a line of zebrafish is created that is very robust and breeds reliably. The line is maintained in large numbers to minimize inbreeding depression and genetic drift and is distributed to many laboratories worldwide. Each laboratory now has a subline of the original population, and these sublines are subject to different selection pressures and will experience different magnitudes and directions of genetic drift. After several generations, each laboratory will be working with genetically divergent lines, which can potentially confound experimental results. Despite these potential pitfalls, there are several techniques for maintaining genetic variation in populations of zebrafish, ensuring robust and readily breedable fish.

1. Increasing Effective Population Size

There are several ways that N_e can be increased relative to the census size, retarding the loss of genetic variation due to inbreeding or genetic drift. First, if a detailed pedigree is available, one can avoid matings between close relatives. This technique, however, offers only a transient reduction in the inbreeding coefficient (Falconer and Mackay, 1996). Another alternative is to equalize the family sizes among individual matings each generation (as has been done in the AB line). The potential pitfall to this approach is that if the family size is equalized, selection for fecundity is removed and deleterious alleles that reduce fecundity in females can accumulate in the population.

One example of a strategy to increase N_e and avoid inbreeding is the round robin mating technique (Westerfield, 1995), which uses relatively large numbers of males and females and partially fills a matrix of all the possible matings between these parents. Sperm from some of 60 or more males of different stocks or generations are pooled in several tubes. Unfertilized clutches from 30 or more females are collected, divided, and fertilized with different pools of sperm. The 15 best-looking embryos from each clutch are selected, if at least 13 of 15 (86.667%) of the clutch's embryos survive, develop swim bladders, and have a standard embryonic morphology. Embryos from good clutches are mixed to make the next generation of the standard line.

The round robin technique is commonly used to propagate the AB strain of zebrafish. Although the use of a relatively large breeding population compared to other methods reduces inbreeding, the round robin is a big labor investment, does not reveal all recessives, does not permit family-based screening, and allows random mating among sibs and parents. As discussed previously, equalizing the family sizes also removes any selection against small clutch size, potentially causing a long-term decline in fecundity.

2. Systematic Generation of Sublines

Ultimately, many researchers require a line of zebrafish that is genetically uniform yet highly robust and experimentally tractable. Because inbreeding depression typically compromises the fitness of inbred or clonal lines, and robust genetically variable lines are by definition not genetically uniform, it seems we are between a rock and a hard place. Given sufficient resources, however, there is a solution.

Starting with a population of zebrafish (or several populations), inbred or clonal sublines can be systematically constructed. Each of these lines will harbor only a fraction of the original genetic variation of the source population(s) and is also likely to be less fit. However, once the sublines are constructed, the criterion of genetic uniformity is then satisfied. When experimental animals are required, two inbred sublines can be crossed to produce an F1 line. Assuming appropriate levels of homozygosity in the sublines, the F1 line will be genetically uniform, yet because it is a hybrid of two genetically divergent lines the effects of inbreeding depression will be markedly reduced. The obvious downside to this approach is the cost, both in terms of the time required to produce all the sublines and the space and money required to house all the sublines.

3. Selection for Desirable Phenotypes

A common practice during the establishment of standard lines is selection for specific phenotypes. In zebrafish, this process can be roughly divided into two categories: screening and selection. Screening is commonly performed on qualitative traits such as developmental mutants. Lines can be screened to completely remove particular traits or to modify the frequencies of complex traits affected by multiple genes. The trick to screening is to be able to quickly and efficiently reveal all the alleles for removal. If the alleles have poor penetrance or are affected by multiple loci, true screening for a qualitative phenotype might not be appropriate.

Periodically, established lines can be intensively "cleaned up" with extra efforts to reveal and remove accumulated recessive mutations. This is most frequently done prior to a new mutagenesis. Background mutations with phenotypes similar to those sought make screening more difficult.

Complementation testing with tester fish rapidly identify all carriers in a line's broodstock of an unwanted mutation. Alternatively, general screens for carriers of alleles can be used to identify and remove multiple unidentified undesirable alleles. Walker (1999) scored mothers by screening clutches of haploid embryos for a stereotyped embryonic morphology. This clearly identified the mother's genotype for recessive alleles with high penetrance. Similar results could be obtained by using heat shock (HS) or early pressure (EP) to make gynogenotes or by using numerous sib matings to reveal a recessive gene carried by one of the parents.

Often the researcher wishes to select a trait that is quantitative in nature, such as growth rate or fecundity. In these cases, the process of selection incorporates standard quantitative genetic theory, which is well covered in introductory texts such as Falconer and Mackay (1996). Beaumont and Hoare (2003) have also described several ways to undertake artificial selection that are readily applicable to the construction of zebrafish standard lines. The most straightforward of these is mass selection, in which the best individuals from a mixed family population are selected for breeding. Mass selection is easy to perform and can be used over several generations. The response to selection can be easily predicted if some basic quantitative genetic parameters are known. At the most basic level, one can use the following breeder's equation to calculate the expected change in mean phenotype from one generation of mass selection:

$$R = h^2 S$$

The response of the trait to selection (R) depends on the selection differential (S), which is simply the difference between the mean phenotype of the population and the mean phenotype of the fish chosen for breeding. The response to selection also depends on the heritability (h^2) of the trait within the population, which is the proportion of phenotypic variance in the population that is due to the additive effects of genes. Although the math is quite simple, the experiments to calculate parameters such as heritability require a significant allocation of resources in terms of personnel and tank space. Several experimental designs for the calculation of heritability can be found in Lynch and Walsh (1998).

An alternative approach to standard artificial selection, family selection chooses whole families for breeding based on the family's mean value for the trait in question. Family selection is often more effective than mass selection when heritabilities are low. Early phenotypes (usually embryonic and early larval) are easily selected in families by using beakers or petri plates, but selection on adult traits requires a larger investment. During family selection, individual families must be housed separately, which can potentially occupy a large number of tanks. The selection of whole families and the exclusion of others also increase the variance in family size, which reduces effective population size and therefore increases the rate of inbreeding and genetic drift. Within-family selection can mitigate this somewhat by selecting the best individual from each family regardless of the family mean value. This approach is similar to mass selection, but with the additional burden of tracking individual families. In zebrafish, Pelegri (2002) reported successful selection for a more favorable sex ratio by selecting 5–10% of the clutches in a generation that is 50% female.

In another example of brief selection for zebrafish line modification, Pelegri and Schulte-Merker (1999) selected females amenable to gynogenetic techniques from at least 50 heat-shocked clutches of eggs. Because heat-shock gynogenesis causes the fish to pass through a homozygous state, there is also selection against recessive embryonic lethals. Although this technique can be used successfully, the small number of breeders selected imposes a substantial population bottleneck, reducing effective population size and increasing the rate of inbreeding. In addition, if the population chosen for this type of selection has a relatively high frequency of recessive deleterious alleles, survival rates will be very low (Pelegri, 2002).

There are few published accounts of selection on zebrafish lines in a traditional quantitative genetic sense. Although the experiments required for estimation of heritability and other quantitative genetic parameters in zebrafish are entirely feasible, they are not often done, and, if done, are rarely published. This is presumably because of the substantial effort required relative to the reward. (One is unlikely to get a grant renewed based on publication of heritabilities for breeding rate in zebrafish.) Nevertheless, even informal posting on ZFIN of results of selection during the construction and maintenance of standard lines will be useful to the general community.

In theory, molecular markers associated with QTL can be used to increase the efficiency of selection. Termed *marker-assisted selection* (*MAS*), this strategy first requires the identification of QTL for the trait of interest that is segregating in the population to undergo selection. Marker-assisted selection is relatively common in agriculture, but, as of publication, we are not aware of any QTL studies in zebrafish that would enable MAS.

V. Cryopreservation

Cryopreserved sperm are commonly used to restore lost genetic variation, usually in the form of specific mutants under the control of single genes. If line performance declines, frozen sperm is also an option for reintroducing lost genetic variation and genetic rejuvenation of the line. Sperm frozen immediately after a major line "clean up" is optimal for genetic reconstitution because it avoids the reintroduction of previously removed alleles while retaining diversity. However, the complexity of genetic backgrounds and the maternally derived oocyte genome prevent a straightforward and complete recovery of an entire genetic background.

Line-specific maternal effects (such as mitochondrial genes) cannot yet be stored as frozen material, as zebrafish eggs, or embryos, nuclei, and cytoplasm are not yet routinely cryopreserved. Fertilizing a nonline of interest egg with line of interest frozen sperm will make a nuclear–cytoplasmic hybrid of a hybrid nuclear genome and a nonline of interest mitochondrial genes. After inactivation, an oocyte's maternal nuclear genome androgenic diploids can be made with EP and frozen sperm (Corley-Smith *et al.*, 1999). This will give diploid, mostly homozygous, male-derived nuclear genome with a maternally derived mitochondrial genome. Recovering viable androgens will be more difficult if the line of interest carries deleterious or lethal alleles. Restorations will be more complete if the line has low diversity or genetic contributions from more unrelated males are used.

Frozen sperm can be a pathway for pathogen introductions and can become contaminated from other sperm samples if immersed in liquid nitrogen (Kirkwood and Colen, 2001). Eggs fertilized with sperm frozen during a disease outbreak or from a different facility should be quarantined as if they were from an external facility.

VI. Monitoring and Response

Genetic monitoring of standard lines has made use of morphological, behavioral, immunological, and general line performance traits. Molecular marker sets are now used for mapping genetic traits in zebrafish and have been used for genetic monitoring in other species (Sharp *et al.*, 2002). Monitoring lines with these sets should be useful in assaying the degree of inbreeding within a line (to optimize maintenance), determining the loss of genetic variation per generation (to reduce inbreeding), and identifying contamination by other lines of fish. Monitoring for contamination in this way assumes that line-specific alleles are not present elsewhere in the facility. If this assumption is not met, genetic monitoring might not identify all contamination events. Similar strategies can be used to measure the residual background when introgressing alleles into a new background and to speed the process up by choosing the statistical outliers with the least residual background (Sharp *et al.*, 2002).

Contamination of a breeding stock with other fish is usually considered the worst possible disruption to a genetic line. Fortunately, it is among the most preventable. Human error and escaped animals are the most frequent causes, and these should be preventable by design and training measures. Approaches to remove unwanted alleles include screening families every generation and keeping reproductively isolated and redundant copies of the line in parallel. Genetic monitoring of each population before breeding will determine whether either should be eliminated or propagated. If one is eliminated, the propagated group should then be split to reestablish the redundancy. Such redundancies for large breeding populations can be more efficiently handled by housing families separately and testing them before breeding. This approach combined with vigilant monitoring should eliminate most unwanted genetic events. If 10% of the families are eliminated and the line propagated with the remaining 90% of the families, inbreeding will increase slightly. To compensate for this reduced effective population size, the total number of families being maintained could be increased.

This approach is similar to that used in the round robin and described by Brand *et al.* (2002). However, these only keep families separate and screen through embryonic and early larval stages. The families are then combined losing the opportunity for later genetic monitoring of families. These methods do not detect recessive traits. Cryptic heterozygotes that are not completely revealed cannot be removed except when both parents are carriers, leaving a residual level of mutant frequencies. Maternally inherited recessive mutations can be revealed in haploids, gynogenotes, or F2 progeny of F1 families. F2 family screening will also test and transmit the paternal genome. To mix alleles with the rest of the line, the families will have to be interbred after scoring or interbred in alternate generations. Male gynogenotes raised to adults will carry only genes derived from these screened families, but they will not have been screened for male mating behavior and sperm production. Male mating behavior, important to "natural" cross success, will be selectively neutral if a line is maintained solely by squeezing.

VII. Distribution

Distribution of a line should provide the line's entire genetic background. This will be more complex for polymorphic than for inbred lines. The relative importance of distribution strategy depends on how the line will be used. For example, a stock center receiving a noninbred line would want as complete a sample of the line's genetic background as possible. Ideally, offspring from each fish in a breeding population would be transferred. Similarly, diverse stocks composed of offspring from several different parents should be distributed. The stock center's broodstock could be periodically refreshed from a source population (the lab maintaining the line) to reduce among line divergence.

VIII. Summary and Recommendations

In the long term, development of robust, fecund, and easy-to-raise genetically uniform zebrafish lines should be a goal of the research community. These lines can be derived from polymorphic strains with good performance characteristics, which will somewhat mitigate declines in fitness due to inbreeding depression. Development of genetically uniform standard lines will provide a framework for zebrafish research that is stable over time and repeatable among labs.

Shorter-term goals should include preservation of the current highly used polymorphic lines (AB and TU) and perhaps the establishment of additional polymorphic lines from new sources (wild catches or fish farms). To become more widely used, many lines might require removal of lethals (including embryonic lethals) and selection for good performance traits. Selection for line performance should take place under conditions in which that performance will be most critical.

Where possible, genetic monitoring should be implemented to track genetic changes within lines, identify potential contamination events, and monitor potential divergence among sublines. To be effective, monitoring must be coordinated with fish management so that problems can be quickly isolated and corrected. Maintaining and screening a line as several separately housed families instead of mixed family groups will provide an opportunity to eliminate the most drastic contamination events without having to completely replicate the line's entire breeding population. Using cryopreserved sperm is also a viable strategy for partially reconstituting a specific genetic background.

As use of the zebrafish becomes even more widespread across the scientific spectrum, standard lines will be adapted to new purposes. Adopting consistent community-wide standards for construction, maintenance, and nomenclature of standard lines of zebrafish should become a priority.

Acknowledgments

We acknowledge the many discussions with John Postlethwait, Charline Walker, Steve Johnson, David Grunwald, Joy Murphy, and many others.

References

Bongers, A. B. J., Sukkel, M., Gort, G., Komen, J., and Richter, C. J. J. (1998). Development and use of genetically uniform strains of common carp in experimental animal research. *Lab. Anim.* **32,** 349–363.

Beaumont, A. R., and Hoare, K. (2003). "Biotechnology and genetics in fisheries and aquaculture." Blackwell Science, Malden, MA.

Brand, M., Granato, M., and Nüsslein-Volhard, C. (2002). Keeping and raising zebrafish. *In* "Zebrafish: A Practical Approach, vol. 261, The Practical Approach Series," (C. Nüsslein-Volhard and R. Dahm, eds.), pp. 7–37. Oxford University Press, New York.

Corley-Smith, G. E., Broadhorst, B. P., Walker, C., and Postlethwait, J. H. (1999). Production of haploid and diploid androgenetic zebrafish (including methodology for delayed *in vitro* fertilization). *In* "Methods in cell Biology, vol. 60: The Zebrafish: Genetics and Genomics," (H. W. Detrich, III, M. Westerfield, and L. I. Zon, eds.). Academic Press, New York.

Falconer, D. S., and Mackay, T. F. C. (1996). *In* "Introduction to Quantitative Genetics," 4th ed., p. 464. Longman, Essex, England.

Gestl, E. E., Kauffman, E. J., Moore, J. L., and Cheng, K. C. (1997). New conditions for generation of gynogenetic half-tetrad embryos in the zebrafish. *J. Hered.* **88,** 76–79.

Hussain, M. G., Penman, D. J., and McAndrew, B. J. (1998). Production of heterozygous and homozygous clones in *Nile tilapia. Aquacult. Int.* **6,** 197–205.

Johnson, S. L., Africa, D., Horne, S., and Postlethwait, J. H. (1995). Half-tetrad analysis in zebrafish: Mapping the *ros* mutation and the centromere of linkage group I. *Genetics* **139,** 1727–1735.

Johnson, S. L., and Zon, L. I. (1999). Genetic backgrounds and some standard stocks and strains used in zebrafish developmental biology and genetics (Appendix 1). *In* "Methods in Cell Biology, vol. 60: The Zebrafish: Genetics and Genomics," (H. W. Detrich, III, M. Westerfield, and L. I. Zon, eds.). Academic Press, New York.

Kirkwood, J. K., and Colen, B. (2001). *In* "Disease Control Measures in Cryobanking the Genetic Resource, Wildlife Conservation for the Future," (P. F. Watson and W. V. Holt, eds.). Taylor & Francis, New York.

Knapik, E. W., Goodman, A., Ekker, M., Chevrette, M., Delgado, J., Neuhauss, S., Shimoda, N., Driever, W., Fishman, M. C., and Jacob, H. J. (1998). A microsatellite genetic linkage map for zebrafish (*Danio rerio*). *Nat. Genet.* **18,** 338–343.

Lynch, M., and Walsh, B. (1998). "Genetics and Analysis of Quantitative Traits." Sinauer Associates, Sunderland, MA.

Mullins, M. C., and Nüsslein-Volhard, C. (1993). Mutational approaches to studying embryonic pattern formation in the zebrafish. *Curr. Opin. Genet. Dev.* **3,** 648–654.

Nichols, K. M., Young, W. P., Danzmann, R. G., Robison, B. D., Rexroad, C., Noakes, M., Phillips, R. B., Bentzen, P., Spies, I., Knudsen, K., Allendorf, F. W., Cunningham, B. M., Brunelli, J., Zhang, H., Ristow, S., Drew, R., Brown, K. H., Wheeler, P. A., and Thorgaard, G. H. (2003). A consolidated linkage map for rainbow trout. *Anim. Genet.* **34,** 102–115.

Parsons, J. E., and Thorgaard, G. H. (1985). Production of androgenetic diploid rainbow trout. *J. Hered.* **76,** 177–181.

Pelegri, F. (2002). Mutagenesis. *In* "Zebrafish: A Practical Approach, vol. 261, the Practical Approach Series," (C. Nüsslein-Volhard and R. Dahm, eds.), pp. 145–175. Oxford University Press, New York.

Pelegri, F., and Schulte-Merker, S. (1999). A gynogenesis-based screen for maternal-effect genes in zebrafish, *Danio rerio. In* "Methods in Cell Biology, vol. 60: (The Zebrafish: Genetics and

Genomics," (H. W. Detrich, III, M. Westerfield, and L. I. Zon, eds.). pp. 1–20. Academic Press, New York.

Rauch, G. J., Grenato, M., and Haffter, P. (1997). A polymorphic zebrafish line for genetic mapping using SSLPs on high-percentage agarous gels. *TIGS-Technical Tips Online* **13,** 461.

Ristow, S. S., Grabowski, L. D., Ostberg, C., Robison, B. D., and Thorgaard, G. H. (1998). Development of long term cell lines from homozygous clones of rainbow trout. *J. Aquat. Anim. Hlth.* **10,** 75–82.

Robison, B. D., Sundin, K., Sikka, P., Wheeler, P. A., and Thorgaard, G. H. (2001). Composite interval mapping reveals a major locus influencing embryonic development rate in rainbow trout. *J. Hered.* **96,** 16–22.

Robison, B. D., and Thorgaard, G. H. (2004). The phenotypic relationship of a clonal line to its population of origin: Rapid embryonic development in an Alaskan population of rainbow trout. *Trans. Am. Fish. Soc.* **133,** 455–461.

Robison, B. D., Wheeler, P. A., and Thorgaard, G. H. (1999). Variation in development rate among clonal lines of rainbow trout. *Aquaculture* **173,** 131–141.

Sharp, J. J.., Sargent, E. E., and Schweitzer, P. A. (2002). genetic monitoring. *In* "Lab Animal Medicine" (J. G. Fox, L. C. Anderson, F. M. Loew, and F. W. Quimby, eds.). pp. 1117–1128. 2nd ed. Elsevier Science, Amsterdam.

Stachel, S. E., Grunwald, D. J., and Myers, P. Z. (1993). Lithium perturbation and goosecoid expression identify a dorsal specification pathway in the pregastrula zebrafish. *Development* **117,** 1261–1274.

Streisinger, G., Singer, F., Walker, C., Knauber, D., and Dower, N. (1986). Segregation analysis and gene-centromere distances in zebrafish. *Genetics* **112,** 311–319.

Streisinger, G., Walker, C., Dower, N., Knauber, D., and Singer, F. (1981). Production of clones of diploid homozygous zebrafish. *Nature* **291,** 293–296.

Walker, C. (1999). Haploid screens and gamma-ray mutagenesis. *In* "Methods in Cell Biology, vol. 60: The Zebrafish: Genetics and Genomics" (H. W. Detrich, III, M. Westerfield, and L. I. Zon, eds.). pp. 21–41. Academic Press, New York.

Westerfield, M. (1995). "The Zebrafish Book. A Guide for the Laboratory Use of Zebrafish (*Danio rerio*)," 3rd ed. University of Oregon Press, Eugene, OR.

Young, W. P., Wheeler, P. A., Coryell, V. H., Keim, P., and Thorgaard, G. H. (1998). A detailed linkage map of rainbow trout produced using doubled haploids. *Genetics* **148,** 839–850.

CHAPTER 33

Common Diseases of Laboratory Zebrafish

Jennifer L. Matthews

The Zebrafish International Resource Center
University of Oregon
Eugene, Oregon 97403

I. Introduction

This chapter is a review of diseases seen in laboratory zebrafish. It includes conditions caused by both infectious and noninfectious agents. This information is a summary of the current literature and actual diagnostic cases evaluated by the Zebrafish International Resource Center (ZIRC) Pathology Service.

II. Diagnostic Evaluation

You must know what you are treating to have an effective treatment plan. A vast array of chemicals and antibiotics is available for treating fish. Not all of these are used without risk to the fish or the filter system; thus, an accurate diagnosis is required before initiating any treatment. A diagnostic evaluation includes an assessment of the history, husbandry practices, water quality, and a clinical exam. Diagnostic procedures for zebrafish have been previously described in Astrofsky *et al.* (2002a) and Kent *et al.* (2002a). The ZIRC provides diagnostic pathology services to the research community (see http://zfin.org/zirc/health.php).

III. Common Diseases of Laboratory Zebrafish

A. Water Quality Related

1. Toxicities

The aquatic environment is susceptible to contamination by countless compounds and metabolic waste products. The toxicity of compounds dissolved in the aquatic environment can be affected by numerous interacting factors, including temperature, pH, hardness, alkalinity, other chemicals, and life stage of the fish. The toxic effects can vary with the nature and concentration of the pollutant. Clinical signs can range from acute mortalities to more subtle changes in behavior, respiration rate, or appearance. Other toxic effects might be even less apparent, such as immunosuppression or decreased fecundity.

a. Ammonia and Nitrite

The nitrogen cycle is the process responsible for converting toxic nitrogenous wastes to relatively nontoxic by-products and is carried out by a mixed population of bacteria present in the biological filter and in surface biofilms. Ammonia (NH_3) is the primary nitrogenous waste product of fish, and it also results from the bacterial breakdown of nitrogenous compounds in decaying matter within the system. In the nitrogen cycle, ammonia is oxidized first to nitrite (NO_2^-) and further to nitrate (NO_3^-). Both ammonia and nitrite are very toxic to fish. Nitrate is much less toxic, but it still should not be allowed to accumulate to high levels. In recirculating water systems, the majority of nitrate is removed by regular water changes.

Ammonia and nitrite toxicity can occur independently or together. Ammonia and nitrite toxicity are particularly problematic in newly established recirculating systems, in which insufficient numbers of bacteria are present in the biological filter to metabolize the amount of ammonia being produced. A similar problem can occur with biological filters that have been damaged by interruption of water supply, excessive cleaning, or antibacterial medications. Other common causes of ammonia and/or nitrite accumulation in zebrafish facilities include decomposition of excessive feed, overstocking, static water in crossing tanks or shipping bags, and insufficient rinsing of paramecium cultures prior to feeding.

Ammonia rapidly dissociates in solution to produce a pH- and temperature-dependent equilibrium of unionized ammonia (NH_3) and ammonium ions (NH_4^+; Emerson et al., 1975).

$$NH_3 + H_2O \Leftrightarrow NH_4^+ + OH^-$$

Unionized ammonia (NH_3) is much more toxic to fish than is ammonium (NH_4^+). Most aquarium test kits measure total ammonia nitrogen. To assess toxicity, the amount of unionized ammonia must be calculated based on the pH and temperature of the water sample. Table I lists the percentage of total ammonia nitrogen present in the unionized form at pH values and temperatures common to zebrafish systems.

Clinical signs of ammonia toxicity include hyperexcitability, anorexia, and death. Chronic exposure to sublethal levels of ammonia can cause hyperplasia and hypertrophy of gill epithelium, poor growth, and immunosuppression. Diagnosis is based on measurement of unionized ammonia in system or tank water. The lethal limit of unionized ammonia can be as low as 0.5 mg/l (ppm). Sublethal poisoning can occur at levels as low as 0.02–0.05 mg/l (ppm). As a general rule, unionized ammonia levels should be kept as close to 0 mg/l as possible. Treatment for ammonia toxicity should focus on correcting the initiating problem (e.g., decrease stocking density) and lowering levels of unionized ammonia with water changes, ammonia absorbing resins (e.g., zeolite) or binders (e.g., AmQuel®), reducing pH (shifts equilibrium toward less toxic ammonium), additional biological filtration, and reduced feeding.

Fish suffering from nitrite toxicity have tan- to brown-colored gills and show signs of hypoxia, such as gathering at water inlet and rapid respiration. Nitrite is actively transported across the gills, where it enters the bloodstream and oxidizes hemoglobin to methemaglobin. Methemaglobin does not efficiently transport

Table I

Percentage of Total Ammonia Present in Unionized Form Based on pH and Temperature

pH value	Temperature (°C)								
	22	23	24	25	26	27	x28	29	30
6.0	0.046	0.049	0.053	0.057	0.061	0.065	0.070	0.075	0.080
6.5	0.145	0.156	0.167	0.180	0.193	0.207	0.221	0.237	0.254
7.0	0.457	0.491	0.527	0.566	0.607	0.651	0.697	0.747	0.799
7.5	1.43	1.54	1.65	1.77	1.89	2.03	2.17	2.32	2.48
8.0	4.39	4.70	5.03	5.38	5.75	6.15	6.56	7.00	7.46
8.5	12.7	13.5	14.4	15.3	16.2	17.2	18.2	19.2	20.3
9.0	31.5	33.0	34.6	36.3	37.9	39.6	41.2	42.9	44.6

Adapted from Table 2 Emerson, K., et al. (1975) Aqueous ammonia equilibrium calculations: Effect of pH and temperature. J. Fish Res. Board Can. **32**, 2379–2383, with permission.

oxygen, and therefore tissues are deprived of oxygen. Diagnosis is based on the measurement of nitrite in system or tank water and some gross evidence of methemaglobinemia. Nitrite levels should be kept at <0.10 mg/l to avoid possible toxicity. As with ammonia toxicity, treatment consists of correcting the initiating cause and reducing levels of nitrite with water changes, biological filtration, and stoppage or reduction of feeding. Chloride ions can competitively inhibit nitrite absorption in the gill epithelium; therefore, adding salt can help alleviate clinical signs of toxicity. A general rule is to add a minimum of 3 mg/l NaCl for every 1.0 mg/l of measured nitrite. This is a relative small amount of salt, and most zebrafish systems are already maintained above this level of salinity.

b. Chlorine/Chloramine

Chlorine is commonly added to municipal water supplies as a disinfectant. The level of chlorine found in tap water samples can vary, but is typically in the range of 0.1–1.0 mg/l (Noga, 2000). Sodium hypochlorite (bleach) is also routinely used to disinfect zebrafish embryos and aquaculture equipment. Both tap water dechlorination failure and accidental exposure to chlorine used for disinfection are common sources of toxicity in zebrafish facilities.

In addition to chlorine, some municipal water facilities also add ammonia for disinfection. The ammonia reacts with the chlorine to form chloramines, a more chemically stable disinfectant. Chloramines are weaker germicides than chlorine, but their toxicity to fish is usually greater. The disinfectant used in a municipality can be determined by contacting the local water company.

Chlorine can be removed from municipal water sources by carbon filtration, vigorous aeration, or by neutralizing with sodium thiosulfate (7.4 ppm/ppm chlorine Wedemeyer, 1996). If chloramines are present, additional treatment will be necessary to remove the ammonia component. Filtering through ammonia-adsorbing resin (e.g., zeolite) or biological filtration can be used.

Both chlorine and chloramine toxicity cause acute necrosis of the gills. Clinical signs include respiratory distress and acute mortality. Chlorine toxicity is diagnosed by detecting elevated chlorine levels in the water. Chlorine should be at an undetectable level on testing with commercial test kits.

For emergency treatment of chlorine toxicity, fish should be immediately transferred to chlorine-free water. The chlorine can also be rapidly neutralized with sodium thiosulfate or a commercial dechlorinating agent. Aeration can also aid survival.

c. Metals

The most common metal toxicities that occur in zebrafish facilities are those associated with domestic water supplies. Copper, lead, galvanized iron (zinc-coated) and brass plumbing fixtures and pipes can leach metals (Olsson, 1998). Copper and zinc are probably the most commonly encountered heavy metal toxicities in research facilities and when present together can have a synergistic effect (Wedemeyer, 1996). The solubility and toxicity of metals are significantly influenced by other water quality parameters, including pH, temperature, alkalinity, and hardness. Both low pH and high temperature result in greater solubility. Hot water supply lines can contain significantly higher concentrations of dissolved metals than cold water pipes. Metal toxicity and solubility are also greater in soft, low-alkaline water. The amount of leaching that occurs is a function of contact time. The longer the water sits in a pipe, the higher the metal concentration. Metals can also be bound and inactivated by organic matter.

Clinical signs of metal toxicity in fish can be variable and nonspecific. Fish can show signs of lethargy, incoordination, osmoregulatory dysfunction, and immunosuppression. Copper has been shown to cause retarded sexual development, reduction in egg production, decreased larval survival, and teratogenic effects (Dave and Xiu, 1991; Olsson, 1998). Gill, liver, and kidney lesions might be noted on histopathology.

Diagnosis is based on the measurement of toxic metal levels in the water. If metal toxicity is suspected, it is best to send samples to a specialized laboratory for analysis so that it can be measured to the low end of the toxic range. Metal levels must be assessed together with water quality parameters, because metal toxicity is primarily due to dissolved ionic forms rather than the total concentration. Low levels of metals can be significant in systems with soft (<50 mg/l as $CaCO_3$), low-alkaline water. Parameters developed for rainbow trout on safe levels of copper as a function of hardness can serve as a useful guide (Alabaster and Lloyd, 1982). Safe copper levels increase with the level of hardness. [For example, at a hardness of 10 mg/l (as $CaCO_3$) the safe copper level is 1 μg/l (ppb), at hardness of 100 mg/l the safe copper level is 10 μg/l, and at a hardness of 300 mg/l the safe copper level jumps to 280 μg/l.]

Dave and Xiu (1991) investigated the effects of copper, mercury, lead, nickel, and cobalt on zebrafish embryos and larvae under standardized conditions (pH of 7.5–7.7 and hardness of 100 mg/l as $CaCO_3$). Exposures were started at the blastula stage (2–4 h postfertilization) and the effects of hatching and survival (for 16 days) were monitored. Safe levels or no-effect concentrations were determined from the dose relationships. The zero equivalent points for effect on hatching time were 0.05 μg Cu/l, 10 μg Hg/l, 20 μg Pb/l, 40 μg Ni/l, and 3840 μg Co/l, and those for effect on survival time were 0.25 μg Cu/l, 1.2 μg Hg/l, 30 μg Pb/l, 80 μg Ni/l, and 60 μg Co/l.

Treatment should focus on eliminating the source of leaching metal ions. Metal pipes and fittings should be switched out for polyvinyl chloride (PVC) or stainless steel. Metal chelators (EDTA) and ion-exchange resins are available that can bind toxic metals. Both of these also remove essential divalent cations (Ca^{2+}, Mg^{2+}) that can affect the toxicity of metals, and therefore these should be added back to the system water. Use of reverse osmosis to purify the incoming domestic water is an effective way to eliminate contamination from water supplies.

d. Rubber and Plastic Materials

Some common rubber and plastic materials found in and around aquaculture systems have been shown to be toxic to zebrafish embryos, including Buna rubber, Neoprene, ethylene proplylene diene monomer (EPDM), Nalgene 380 PVC (food and beverage grade) tubing, and clear vinyl airline tubing. These materials were tested on 3- to 4-day-old larval zebrafish, just posthatching, exposed to test material (approximately 10 cm^2 of surface area) in 200 ml of static fish water. Buna, Neoprene, and EPDM killed 100% of the larval fish within 24 h. Nalgene 380 PVC tubing and vinyl airline tubing killed up to 70% of the fish within a week. Long-term effects of sublethal exposures were not evaluated. Materials that were tested by using the same protocol and found not to kill larval zebrafish included Teflon®, silicon, Norprene®, Viton®, rigid PVC, and Kraton® (Austin Bailey, ZIRC, personal communication). It is recommended that all new materials to be used in fish systems be tested for toxicity prior to use.

2. Gas Bubble Disease

Under normal conditions, the partial pressures of dissolved gases in the water are in balance with the pressure exerted by these gases from the atmosphere. Supersaturation occurs when the partial pressures of the dissolved gases present in the water become greater than the atmospheric pressure. The blood and tissues of fish quickly reach equilibrium with the partial pressure of the dissolved gases present in their environment. Gas bubble disease occurs when fish absorb gases from supersaturated water, which then subsequently comes out of solution, forming gas bubbles in the circulation and tissues of the fish (Colt, 1986; Weitkamp and Katz, 1980).

In laboratory zebrafish systems, there are two primary causes of supersaturation. One is the rapid heating of water that is under pressure. This can occur in systems that use temperature-mixing valves. Because gas solubility decreases as the temperature rises, cold water can become supersaturated as it is warmed. The other common cause of supersaturation is a small leak on the suction side of centrifugal water pumps. Air can be drawn in rather than water coming out, making the leak difficult to detect. Within the pump, the air is forced into solution under pressure. Visible bubbles are often not present in the discharged water, which can further complicate detection.

When zebrafish are exposed to supersaturated water, they can show signs of exophthalmia (pop-eyed appearance), abdominal distension and hyperbuoyancy, gas bubbles in the skin especially around the eyes and head or general malaise. The bubbles under the skin can be quite visible to the naked eye (Fig. 1). The disease can result in areas of necrosis, secondary bacterial infections, and acute death. Diagnosis is based on the observation of gas emboli in capillaries of the gills or internal organs on wet mount exam or macroscopic gas bubbles in the eyes or skin. Measurement of dissolved gases requires a saturometer or tensiometer. Total gas pressures as low as 105% can be considered dangerous to zebrafish. Gas bubble disease is controlled by identifying the source and rectifying the problem.

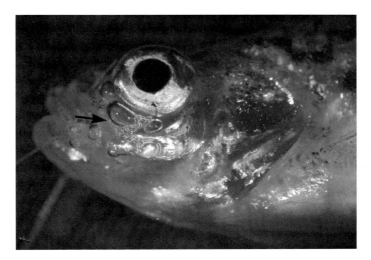

Fig. 1 Fish with gas bubble disease. Note the gas bubbles under the skin of the head and around the eyes (arrow).

B. Bacteria

1. Dermal and Systemic Infections

Almost all the bacterial pathogens affecting fish are found ubiquitously in the aquatic environment (Inglis, 1993). These opportunistic bacteria are capable of existing and multiplying outside of the fish host. Bacterial pathogenesis in fish involves both the virulence of the invading organism and the susceptibility of the fish host. The ability of bacteria to invade and infect fish most often requires the fish to be immunocompromised by some form of environmental or physical stress. Fish respond to stress with a series of physiological changes, resulting in the suppression of nonspecific defense mechanisms and an increased susceptibility to disease (reviewed in Reddy and Leatherland, 1998). Prevention and treatment of bacterial diseases often focus on improving hygiene and husbandry conditions to reduce stress.

Bacterial infections in zebrafish can present as localized skin infections, but more commonly the infections have systemic involvement. Clinical signs of bacterial infections include ulcerative or necrotic skin and fin lesions, superficial erythema (red coloration) and areas of hemorrhage, peritonitis, and generalized edema. Dropsy is a common term used to describe the edematous changes in the skin, causing protrusion of the scales (pine-cone appearance) and abdominal distension because of fluid accumulation (Fig. 2). Abdominal fluid associated with bacterial infections is often blood tinged.

Zebrafish are susceptible to a wide range of bacterial pathogens. Diagnosis by culture and antibiotic sensitivity testing has traditionally been under used and

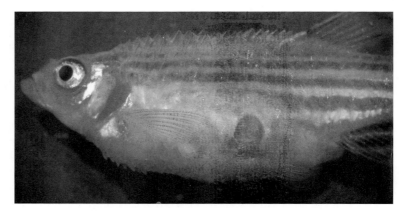

Fig. 2 Zebrafish with severe edema (dropsy), evident by the raised scales and swollen abdomen.

Fig. 3 Zebrafish with a bacterial skin infection. Note the frayed caudal fin/peduncle (arrow) and whitish appearance of the posterior region due to bacterial colonization.

therefore the responsible organism(s) is often never identified. Bacteria that can be found associated with skin and gill infections in fish include bacteria in the *Cytophaga, Flavobacterium*, and *Flexibacteria* group (Fig. 3). Skin infections can also be a sign of systemic disease. Bacteria known to cause systemic infections in freshwater fish include the motile aeromonads, *Aeromonas salmonicida, Pseudomonas* spp., *Edwardsiella* spp., *Streptococcus* spp., *Nocardia* spp., and *Mycobacterium* spp. (Inglis *et al.*, 1993; Stoskopf, 1993).

Pullium *et al.* (1999) reported an outbreak of motile aeromonad septicemia in a zebrafish facility characterized by lethargy, petechial (small spots) hemorrhages, skin ulceration, and high mortality. *Aeromonas hydrophilia, A. sobria*, and *Streptococcus* sp. were cultured from the fish. It was concluded that the infection was brought on by high nitrite levels and overcrowding.

Bacterial infections affecting the swimbladder (bacterial aerocystitis) have also been described in zebrafish (Kent *et al.*, 2002a). Kidney cultures from affected fish have been inconsistent, either yielding no growth or mixed infections (including *Aeromonas* spp. and *Pseudomonas* sp.). Histology typically reveals severe, chronic inflammation and bacteria within the swimbladder. The swimbladder of zebrafish is connected to the esophagus by a pneumatic duct, which is patent and functional. This duct can trap fine particulate feed or material and subsequently become colonized by bacteria in a susceptible host (Ferguson, 1989; Miyazaki *et al.*, 1984).

An experimental model of bacterial pathogenesis in zebrafish with *Streptococcus iniae* and *S. pyogenes* has been described (Neely *et al.*, 2002). Both *S. iniae* and *S. pyogenes* produced lethal infections following intramuscular and intraperitoneal injection in zebrafish. The authors were also able to induce infection by water exposure when fish were housed under conditions that mimicked aspects of a stressful environment.

Diagnosis of bacterial disease is based on clinical signs and culture of the organism. More important than identifying exactly what bacterial species you are dealing with are the antibiotic sensitivity results. This will allow appropriate antibiotic selection. Treatment should also focus on correction of water quality and husbandry conditions that might be causing undue stress.

2. Mycobacteriosis

Mycobacteriosis, often referred to as fish tuberculosis or fish TB, is a common disease of laboratory zebrafish as well as wild and captive fishes worldwide (Austin and Austin, 1999; Inglis *et al.*, 1993). Mycobacteria are nonmotile, weakly staining Gram-positive, pleomorphic rods that are acid fast. Many species of atypical (nontuberculosis) *Mycobacterium* are found ubiquitously in water and biofilms, and numerous species have been identified as pathogens of zebrafish (Astrofsky *et al.*, 2000; Hall-Stoodley and Lappin-Scott, 1998; Kent *et al.*, 2002a; Schulze-Robbecke *et al.*, 1992). The *Mycobacterium* species that have been identified in zebrafish facilities include *M. marinum, M. fortuitum, M. chelonae, M. abscessus, M. haemophilum,* and *M. septicum/peregrinum* (Astrofsky *et al.*, 2000; M. Kent, Oregon State University, personal communication).

Clinically, mycobacteriosis can manifest in a wide variety of signs, depending on the site and extent of infection. Signs include decreased fecundity, lethargy, anorexia/emaciation, skin ulceration, areas of hyperemia or hemorrhage, edema, abdominal distention, granulomatous nodules in internal organs, and mortality (Fig. 4). Deformities might occur with muscle and skeletal involvement. The severity of infections in zebrafish facilities can vary dramatically. Most common infections are chronic, with low-grade morbidity and mortality; however, severe outbreaks have occurred, causing acute disease and high mortality. As with most bacterial pathogens of fish, mycobacteria infections in zebrafish are most often opportunistic in nature. Poor water quality, high stress, or other type of husbandry failure commonly precede outbreaks. Zebrafish of certain genetic backgrounds

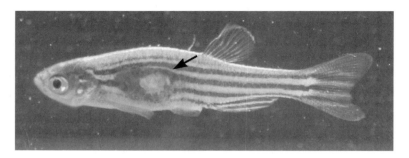

Fig. 4 Fish with skin ulceration (arrow) due to mycobacteriosis.

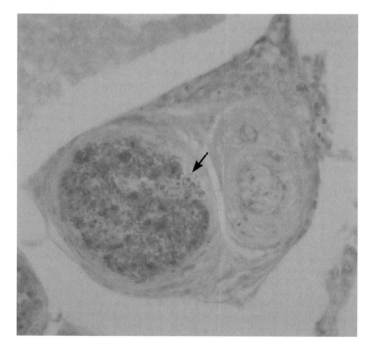

Fig. 5 Histological section showing typical mycobacteria granuloma with numerous acid-fast-positive (Ziehl–Neelsen) bacilli.

(e.g., Tu) appear to be more susceptible to mycobacteria infections. The virulence of a particular species or strain of *Mycobacterium* can also affect the severity of the disease.

Diagnosis is based on clinical signs, characteristic granulomatous inflammation, and the presence of acid-fast (Ziehl–Neelsen) staining bacteria in tissue sections or smears (Fig. 5). In severe cases, granulomas appearing as small, tan-colored nodules might be visible in visceral organs on gross internal examination or wet mount/squash preparations. Culture of the organism is considered definitive but can be difficult because of slow growth and special media requirements. PCR tests for the identification of mycobacteria infecting fish have been described (Colorni *et al.*, 1994; Talaat *et al.*, 1997). Such tests, although not commonly used for routine diagnostic purposes, can be useful in screening for subclinical infections or as a rapid diagnostic confirmation.

Mycobacteria infecting fish typically respond poorly to antimicrobial treatments. Control should be focused on the removal of infected fish, optimizing water quality and husbandry practices and the use of strict sanitation and quarantine procedures. In severe outbreaks with highly virulent stains of mycobacteria, control might require the eradication of infected stocks and subsequent disinfection of the system (Astrofsky *et al.*, 2000; Sanders and Swaim, 2001).

C. Protozoans

1. Microsporidiosis

Microsporidia are well-recognized pathogens of fish. A microsporidian infecting the central nervous system of zebrafish was first reported in 1980 by a group in France (de Kinkelin, 1980). The parasite has been more recently described as *Pseudoloma neurophilia* and is found commonly in zebrafish from both laboratory facilities and commercial suppliers (Matthews *et al.*, 2001).

Microsporidia are obligate intracellular parasites of eukaryotes with a complicated life cycle (Vávra and Larsson, 1999). The life cycle concludes with the production of an infectious and resistant spore. The formation of giant host cells filled with spores (a xenoma) is common for microsporidia species infecting fish. Spores have a characteristic posterior vacuole and polar tube apparatus, which function to transmit the spore contents and genetic material into the host cell. With most microsporidia, direct horizontal transmission occurs by ingestion of the infective spore; however, vertical transmission has also been demonstrated in some species of microsporidia (Bandi *et al.*, 2001; Dunn *et al.*, 2001).

The primary site of infection with *P. neurophilia* in zebrafish is the central nervous system (spinal cord, ventral nerve roots, and hindbrain) where xenomas are commonly found. Free spores or small xenomas can also be found occasionally in the ovary, skeletal muscle, and viscera. Mild to severe chronic myelitis (inflammation of the spinal cord) and myositis (inflammation of muscle) can be associated with infections. Inflammatory changes are most common when free spores are present. At present, only the spore can be readily identified in zebrafish and very little is known about the complete life cycle and mode(s) of infection of the parasite. The only confirmed host for *P. neurophilia* is zebrafish; however, we have detected a morphologically identical microsporidian in a neon tetra (*Paracheilodon innesi*).

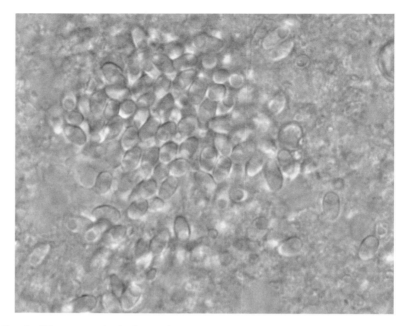

Fig. 6 Wet mount of spinal cord tissue showing microsporidian spores (*P. neurophilia*).

Clinical signs of microsporidiosis in zebrafish can include chronic wasting or emaciation, lethargy, spinal deformities, and dorsal darkening of the skin. Zebrafish can also be infected with *P. neurophilia* and show no abnormal clinical signs. It is often an incidental finding on routine histological exam. Severe infections are commonly associated with stressful husbandry conditions and immunosuppression.

Diagnosis can be made by finding characteristic spores on examination of dissected spinal cord tissue in a wet mount preparation (Fig. 6). Spores are approximately 3 μm × 5 μm, oval to pryriform in shape, and have a large posterior vacuole. Xenomas (up to 200 μm), spores, and associated inflammation can also be readily identified in histological sections (Fig. 7). Fungi-Fluor (Polysciences, Warrington, PA), a fluorescent stain, binds to chitin in the spore walls of microsporidia. This stain is excellent for demonstrating spores in either tissue smears or histological sections (Kent *et al.*, 2002a; Weber *et al.*, 1999).

At present, there is no treatment for microsporidiosis in zebrafish. Control should focus on optimizing husbandry conditions and removing all emaciated and moribund fish to prevent cannibalism. There have been several drug treatments (e.g., flubendazole, albendazole, fumagillin) tried in other species of fish with some success, but at this time they would be considered experimental in zebrafish (Shaw and Kent, 1999).

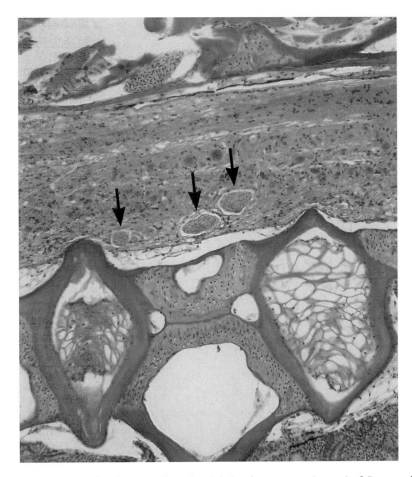

Fig. 7   Histological section (hematoxylin and eosin) showing xenomas (arrows) of *P. neurophilia* in spinal cord.

2. Velvet Disease

The causative agent of velvet disease in zebrafish is *Piscinoodinium pillulare*. The organism is a dinoflagellate that contains chlorophyll, which imparts a yellow-gold color to the parasite. The presence of this parasite in laboratory zebrafish colonies is less common now than it was previously. The parasite life cycle is direct and can be completed in 10–14 days under optimal conditions (23–26 °C). The parasitic stage, the trophont, is found on the skin and gills of infected fish, where it attaches and feeds on the host's epithelium. After feeding for several days, the trophont detaches from the host and becomes a tomont. The tomont undergoes a series of divisions to produce the motile and infective dinospores. The dinospores attach to a host, differentiate into a trophont, and continue the cycle.

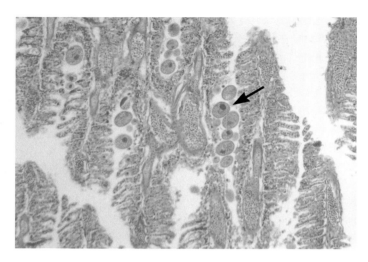

Fig. 8   Histological section (hematoxylin and eosin) of gills heavily infected with *Piscinoodinium* (arrow).

Piscinoodinium can be highly pathogenic. Gill epithelium reacts to the parasite with inflammation and hyperplasia, which can cause hypoxia. Damage to the fish epithelial surfaces also leads to osmoregulatory impairment and secondary bacterial infections. Infected fish show general signs of discomfort, flashing or rubbing, increased respiration, and decreased feeding. Diagnosis is made by demonstrating trophonts on wet mounts of skin or gill biopsies. Trophonts are almost round when mature, nonmobile, and can vary in size from 10 to 50 μm. In histological sections, trophonts appear as oval to round organisms on the gill and skin surface, with numerous cytoplasmic (often refractile) granules and a large nucleus (Fig. 8). On heavily infected fish, the skin can have a dusty appearance (velvet disease) when illuminated indirectly (i.e., with a flashlight) over a dark background.

Treatment has been successful with Atabrine™ (quinacrine dihydrochloride hydrate) added at 1 mg/l day for 3 days (Westerfield, 2000). Chloroquine hydrochloride (prolonged immersion at 10 mg/l) has also been used successfully to treat the marine counterpart, *Amyloodinium*, and is relatively nontoxic to fish (Noga, 2000). Strict quarantine practices are an important preventive measure. Infections can be easily avoided if only surface-sanitized (bleached) embryos are introduced into a system. There is also a much greater risk of finding this parasite in fish acquired from commercial sources such as pet suppliers.

3. *Ichthyophthirius multifiliis*

The ciliate *Ichthyophthirius multifiliis* is the causative agent of ich or white spot disease (reviewed in Dickerson and Dawe, 1995; Lom and Dyková, 1992). It is a common pathogen of freshwater tropical fish, but is a relatively uncommon pathogen of zebrafish. *Ichthyophthirius multifiliis* has been used as a parasitic disease model for immune studies in zebrafish (Clark, 2000). Considerable acquired immunity is present in fish that recover from infections. The life cycle of the parasite is direct and is similar to that of *Piscinoodinium* (velvet disease). The trophont form of *I. multifiliis* is found feeding under the epidermis of the skin and gills of the fish host. After feeding, it breaks through the epidermis and falls off the host and becomes an encysted tomont. The tomont adheres to an inanimate substrate and divides to produce large numbers of infective, motile theronts. Theronts swim to contact a fish host and then attach and penetrate the epithelium to continue the cycle. The life cycle is completed in 3–6 days at 25 °C.

Clinical signs of *I. multifiliis* include raised, white nodules (up to 1 mm in diameter) on the skin or gills, excess mucus, and an increased respiratory rate. The epithelial damage caused by the parasite can lead to osmoregulatory disturbances and predispose fish to secondary bacterial infections. Diagnosis is based on the detection of trophonts embedded within the epithelium of the skin or gills. On wet mount exam of skin scrapings or fin clip, trophonts are easily identified by the distinctive, often horseshoe-shaped, macronucleus and uniform motile cilia (Fig. 9). Trophonts can have a large variation in size (50 μm to 1 mm) and be pleomorphic in shape (oval to round). Histological sections reveal the parasite under the epithelium and associated with severe epithelial hyperplasia.

Infected fish should be immediately removed from the system and quarantined. Because the parasite is protected by penetration into the host epithelium, only the free-swimming tomont and theront stages are susceptible to chemical therapies. Drug treatments must be repeated several times to ensure that therapeutic levels are maintained while all the parasites pass through the unprotected stages. Formalin applied at 25 ppm (0.025 ml formalin/l) every other day for three treatments is typically effective. Different strains of zebrafish might respond differently to

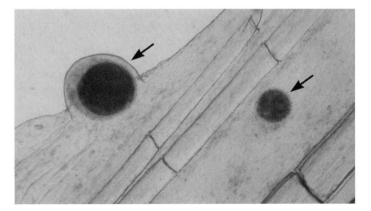

Fig. 9 Wet mount of fin biopsy showing trophonts (arrows) of *Ichthyophthirius multifiliis*. The larger trophont on the edge of the fin can be seen under the epithelium.

therapies; therefore, it is always advisable to do a test treatment on a few fish before applying the treatment to large numbers. Formalin can also have damaging effects on bacteria present in biological filters.

4. *Tetrahymena*

Tetrahymena spp. are free-living, ciliated protozoans commonly found in aquatic environments. Some species can be pathogenic to fish, especially when present in high numbers. Infections are often associated with poor environmental conditions, concurrent disease, or an immunocompromised fish host (Astrofsky *et al.*, 2002b). *Tetrahymena* spp. are ovoid to pear shaped, 30–60 μm × 50-100 μm in size, uniformly covered with cilia, and can have one to three posterior cilia (Fig. 10). In water, they move in a football-like spiraling motion. Reproduction is by binary fission and can result in rapid proliferation of the organism in favorable environmental conditions. Tetrahymena can invade skin, muscle, and internal organs, causing extensive necrotic changes. The infection most often results in mortality.

Astrofsky *et al.* (2002b) described a clinical outbreak of *Tetrahymena* spp. in a zebrafish facility, causing high mortalities of fry (30 day old). Excessive organic loads from fish waste and uneaten food as well as elevated nitrate levels were responsible for the rapid proliferation of *Tetrahymena* spp. within the fry tanks. Control of the outbreak was achieved by simply improving water quality and management practices.

Diagnosis of *Tetrahymena* spp. is based on finding characteristic ciliates in invasive lesions on wet mounts or histological sections. Penetration of ciliates into muscle and deep tissues is highly diagnostic (Noga, 2000). Formalin applied at 25 ppm (0.025 ml formalin/l) can be used to treat superficial infections (Noga, 2000;

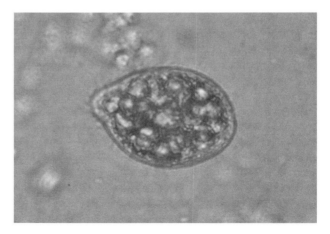

Fig. 10   Wet mount showing *Tetrahymena* sp.

Rothen *et al.*, 2002). Husbandry conditions and water quality should also be improved.

5. *Coleps*

Coleps are free-living protozoa commonly found in freshwater aquatic environments (Patterson, 1996). They have barrel-shaped bodies (50–80 μm) that are reinforced by calcareous plates (Fig. 11). They are scavengers that commonly feed on detritus but have a preference for tissues of dead or dying animals. Occasionally, *Coleps* are found to be predators of larval zebrafish (Mazanec and Trevarrow, 1998). *Coleps* can be inadvertently concentrated and fed to larval fish in contaminated paramecium cultures. Larval fish that have not yet formed a swim bladder and are resting on the bottom of the container are most susceptible to attack. Swarms of these protozoans can form and rapidly consume a larval fish.

Diagnosis is based on microscopically identifying the protozoan in larval zebrafish containers. Control has been accomplished by eliminating contamination in paramecium cultures by sterilizing starting culture media (Mazanec and Trevarrow, 1998). Increasing salinity in larval systems has also been suggested, but the *Coleps* can develop tolerance to the increased salinity over time.

D. Metazoans

1. Intestinal Capillarid Nematode

An intestinal nematode found infecting zebrafish has been identified as *Pseudocapillaria tomentosa*, a common nematode of cyprinid and other fishes (Kent *et al.*, 2002b). Infections in zebrafish have been associated with chronic wasting disease, decreased reproductive potential and growth rate, and intestinal neoplasms (Kent *et al.*, 2002b; Pack *et al.*, 1995). Capillarids are thin, transparent worms typically found within the lumen of the intestine and are locally tissue invasive. The eggs of parasitic nematodes have a distinctive oval shape with a cap- or plug-like structure at either end. Precise identification of capillarid nematodes to the species level requires careful examination of sexual organs of the male worm and is typically not necessary in a clinical setting.

The life cycle of capillarid nematodes often involves an invertebrate intermediate host; however, it has been shown that *P. tomentosa* can have a direct life cycle in zebrafish (Kent *et al.*, 2002b; Lomankin and Trofimeko, 1982). Oligochaete worms (e.g., *Tubifex tubifex*) have been shown to serve as paratenic host (transport host) for the parasite.

Clinically, infected zebrafish can be darker in color, emaciated, and lethargic. Infected fish can also appear normal or only show subtle abnormalities such as decreased fertility (Pack *et al.*, 1995). Diagnosis is made by finding adult worms in the intestinal tract of zebrafish on gross or histological exam. Worms can be identified grossly in infected fish by dissecting out fresh intestine and examining as a squash or wet mount preparation. The adult worms are motile, thin

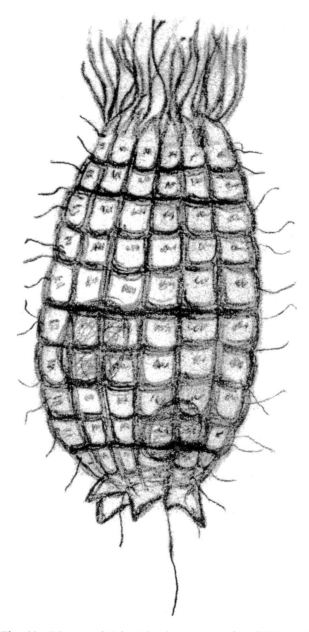

Fig. 11   Diagram of *Coleps* (drawing courtesy of April Mazanec).

(~50-μm), and 4–12 mm in length (Fig. 12). Characteristic ova (~30 μm × 60 μm) can be seen within gravid female worms or found free within the intestinal contents or feces. Histological sections reveal the worms within the lumen or wall of the intestine and can be associated with significant tissue reaction (Fig. 13). Infections are also associated with a higher incidence of intestinal neoplasms (Kent *et al.*, 2002b).

Nematode infections in zebrafish can be difficult to eliminate. As direct transmission occurs between fish, the infection can spread within a population if not controlled. Infections can be prevented with the use of strict quarantine procedures that allow only the introduction of surface sanitized (bleached) embryos.

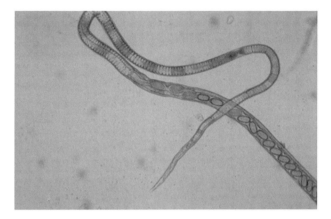

Fig. 12 Wet mount preparation of *Pseudocapillaria tomentosa*. Note the characteristic ova within the gravid female nematode (photograph courtesy of Michael Kent).

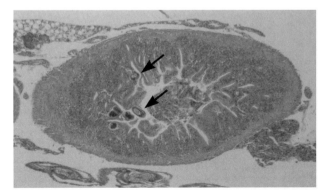

Fig. 13 Histological section (hematoxylin and eosin) of intestine infected with *Pseudocapillaria tomentosa*.

Oligochaete worms (*Tubifex*) can also carry the parasite and should thus be avoided as a food source. If fish are not highly valuable, the most effective treatment might be to cull the infected population and disinfect the system.

There are reports of anthelminthic drugs used to treat intestinal nematodes in fish, but these have not been extensively tested in zebrafish. Pack *et al.* (1995) reported that a mixture of trichlorofon and mebendazole in the form of Fluke-Tabs (Aquarium Products, Glen Burnie, MD) added to water (manufacturer recommended dose of 1 tablet/38 l) eliminated the infection in zebrafish. Treated fish gained weight, regained fertility, and when examined later the infection was not observed. No adverse effects of the treatment were reported; however, caution should be used with this treatment in zebrafish brood stock, because mebendazole is reported to be embryotoxic and teratogenic and trichlorfon is a neurotoxic organophosphate (Mashima and Lewbart, 2000). Levamisole has been reported to be ineffective against intestinal nematode infections (Pack *et al.*, 1995) and to cause sterility in zebrafish brood stock (Kent *et al.*, 2002b).

Oral fenbendazole (Panacur®) has been used to treat intestinal nematodes in fish (Noga, 2000). Fenbendazole is added to food at a concentration of 0.25% and fed for 3 days. Treatment should be repeated in 14–21 days. The drug can be mixed with commercial food enhanced with cod liver oil and bound with gelatin. Because fish are quick to refuse medicated feed, withholding food for a couple days prior to treatment can be beneficial. Zebrafish have been shown to ingest sufficient amounts of water to allow dosing with water-soluble oral medications in an immersion treatment (Jagadeeswaran and Sheehan, 1999). Fenbendazole in a prolonged immersion should be dosed at 2 mg/l and repeated once per week for 3 weeks (Noga, 2000). The ease of application makes this attractive; however, the effectiveness of this treatment has not yet been demonstrated.

2. Encysted Helminths

Unidentified helminths (cestodes or metacercaria of digenean trematodes) are occasionally seen encysted within various tissues of zebrafish. They have been observed in the liver, gills, mesenteries, vertebra, connective tissue around the eye, and in the visceral cavity. They are most commonly found in zebrafish that were wild caught or purchased from commercial (pond-raised) sources. Because these parasites have complex life cycles, requiring one or more intermediate host, they are typically dead-end infections in laboratory zebrafish. Snails used to control algae in zebrafish systems can conceivably continue the life cycle of these parasites and are thus not recommended.

3. Myxozoans

Myxozoan parasites are occasionally identified infecting zebrafish but have not been associated with significant pathological changes. Both celozoic (live in body cavities) and histocytic (live in tissues) species have been identified. Spores consistent with *Myxidium* spp. or *Zschokkella* spp. have been found within the lumen of the distal mesonephric duct. Pseudocysts of *Thelohanelus* spp. have been identified in the epithelium of the gill chamber. Most of the myxozoan infections identified in zebrafish have been in wild-caught fish or in zebrafish acquired from pet stores or commercial fish farms.

E. Water Molds

Aquatic fungi and fungus-like organisms (class Oomycetes) are important pathogens of both fish and eggs (Goven-Dixon, 1993). There are multiple genera of water molds found ubiquitously in the aquatic environment where they feed saprophytically on dead and decaying organic matter. They are opportunistic or secondary invaders usually associated with adverse environmental conditions or colonization of preexisting wounds. Factors that predispose infections include handling trauma, stagnant water, high densities, parasitic infections, and excess decaying feed or organic matter. Water molds typically produce a cottony growth on eggs or epithelial surfaces. They are often associated with body orifices such as the mouth, gill chamber, olfactory pits, and anus. Fungal infections in fish eggs most often begin with growth on unfertilized and nonviable eggs. Once a mold infection is established it can spread rapidly.

Saprolegnia spp. are among the most common water molds affecting fish. They are characterized by abundant vegetative growth in the form of aseptate hyphae. They produce specialized hyphae with long, cylindrical zoosporangia that release motile zoospores. Zoospores can germinate and grow into the skin of immunocompromised or injured fish, producing a superficial cotton-like growth.

Dykstra *et al.* (2001) reported a clinical outbreak of a water mold associated with high mortality of zebrafish fry (5–24 days posthatching). The fry developed dense growth of septate fungal hyphae from the mouth, operculum, and anal pore. The fungus *Lecythophora mutabilis* was isolated and cultured from the larval fish and environmental samples. Additional opportunistic bacterial and protozoal organisms were also identified within the lesions. The outbreak was attributed to environmental contamination and low calcium levels.

Diagnosis of water molds is based on observation of a cottony mass on eggs or the epithelium (skin and gills) of fish while in the water. Once removed from the water, the mass of fungal hyphae collapses and has a glistening, matted appearance. On wet mount examination, fungal hyphae can be observed. *Saprolegnia* spp. have aseptate hyphae of variable width ($\sim$7–20 μm), and occasionally zoosporangia can be identified (Fig. 14). Classification to the genus or species level requires the observation of asexual and sexual stages, respectively, and often involves special culture techniques. Presumptive diagnosis based on the identification of hyphae is sufficient for most clinical decisions. Fungal organisms can also be identified in histological sections and can be confirmed with special stains (e.g., Gomori's methenamine silver).

Fig. 14 Wet mount preparation of *Saprolegnia* sp. showing aseptate hyphae and cylindrical zoosporangia (arrow).

Treatment of water molds in fish is difficult, and therefore control should focus on identifying and correcting the predisposing factors. Fish pathogenic water molds are inhibited by high salinity ($\geq$3 g/l). Prolonged immersion in salt can also help counteract the osmotic stress associated with epithelial injury (Noga, 2000). Salt concentrations should be increased gradually and the fish monitored for signs of stress. Salt treatment should not be used on larval zebrafish. Concentrations as low as 1 g/l (ppt) have been shown to cause significant mortalities in yolk sac fry (2–3 days old) and free-swimming larval (5–7 day old) fish (Rothen *et al.*, 2002). Povidone iodine (1:10) can be used as a swab on focal lesions of valuable fish. The infected area can be carefully cleaned and debrided with a cotton swab soaked in disinfectant (Stoskopf, 1988).

Water molds can be prevented in eggs by maintaining optimal water quality in egg containers. Unfertilized and nonviable eggs should be immediately removed from egg dishes to prevent fungal colonization. Bleaching eggs (30–50 mg/l for 5 min) and transferring to sterile embryo media prevent most problems with water mold. Additional prophylactic treatment of eggs with methylene blue at 0.5–2 mg/l can reduce the incidence of bacterial and water mold infections in fish eggs (Herwig, 1979). This is particularly useful when embryos are being shipped and cannot be cleaned en route. Other treatments that have been used for water mold infections in fish eggs include formalin and hydrogen peroxide (Noga, 2000). For optimal embryo survival, it is recommend that prophylactic egg treatments be stopped 24 h prior to hatching.

F. Viral

Till the present time, no natural viral pathogens have been identified in zebrafish. Knowledge of fish viruses in tropical fish species, including zebrafish, is limited because diagnostics and research has traditionally been focused on the more commercially valuable cultured fish species. It is very likely that viral diseases affecting zebrafish will be identified in the future.

Under experimental conditions, zebrafish have been shown to be susceptible to several important fish viruses. LaPatra *et al.* (2000) demonstrated infection of whole zebrafish with viral supernatants of infectious hematopoietic necrosis virus (IHNV) and infectious pancreatic necrosis virus (IPNV). Sanders *et al.* (2003) also demonstrated that zebrafish are susceptible to infection by the spring viremia of carp virus (SVCV). Fish exposed to SVCV under conditions that mimic the natural route showed typical gross and histological lesions and up to 50% mortality.

G. Noninfectious and Idiopathic Conditions

1. Artemia Cyst Impaction

The shells and unhatched cyst of *Artemia* naupli are indigestible. They can lodge in the intestinal tract of zebrafish and lead to intestinal obstruction. To prevent this, care must be taken to separate the hatched naupli from the empty shells and unhatched cysts prior to feeding. Alternatively, *Artemia* naupli can be decapsulated prior to hatching. The decapsulation procedure removes the chitinous alveolar layer of the cyst shell. This increases the hatching percentage and produces a totally edible product regardless of hatch rate. The decapsulation procedure also disinfects the cysts (Hoff and Snell, 1999; Lavens and Sorgeloos, 1996).

2. Egg–Associated Inflammation and Fibroplasia

Chronic inflammation in the visceral cavity associated with degenerating eggs is a condition seen commonly in female zebrafish (Kent *et al.*, 2002a). The precise cause of the condition is unknown, but it appears to result from abnormal egg retention and absorption. Infectious agents are usually not found in the lesions. Occasionally, *Mycobacterium* spp. is observed within granulomas in these lesions, but these infections are probably not the primary cause. On histological examination, the condition appears as mild to severe chronic inflammation in the ovaries and can extend throughout the peritoneal cavity. Eggs of varying states (from intact to completely degenerate) are found within the lesion. Large rafts of amorphous eosinophilic debris are occasionally observed. Prominent fibroplasia often occurs and in some cases appears to lead to the formation of fibromas and fibrosarcomas.

3. Nephrocalcinosis

Nephrocalcinosis is a condition occasionally identified in laboratory-reared zebrafish (Kent *et al.*, 2002a). Nephrocalcinosis is caused by the formation of calcareous deposits in the kidney tubules or collecting ducts. It is similar to kidney stones in humans and animals. The mechanisms responsible for the development of this condition are not fully understood. A number of environmental and dietary factors have been associated with the disease in cultured fish species (Wedemeyer, 1996). Most consistently, high levels of dissolved carbon dioxide (>10–20 mg/l), with or without a dietary mediator, have been associated with the condition. The lesions must be severe before significant disease is observed. Most cases of nephrocalcinosis in zebrafish are diagnosed as an incidental finding on histopathology.

H. Spontaneous Neoplasm

Spontaneous neoplasm in laboratory zebrafish is relatively common and a much larger topic than can be sufficiently reviewed here. The incidences and histological types of neoplasia in zebrafish can vary widely and can be affected by the specific wild-type or mutant stain. Based on histological examination of retired brood stock from the Oregon zebrafish facilities and from diagnostic cases submitted to the ZIRC Pathology Service, some common neoplasms have been identified (Kent *et al.*, 2002a). The most common target tissues for spontaneous neoplasia are testis, gut, thyroid, liver, peripheral nerve, connective tissue, and the ultimobranchial gland. Less common target tissues include blood vessels, brain, gill, nasal epithelium, and the lymphomyeloid system. The incidence of spontaneous neoplasms in zebrafish appears to increase significantly with age (>1 year).

IV. Zoonosis

The overall incidence of transmission of disease-producing agents from aquarium fish to humans is very low. Humans typically contract diseases from aquarium fish through the ingestion of infected fish secretions and aquarium water or by contamination of lacerated or abraded skin. These can result in symptoms of gastroenteritis (nausea, vomiting, and diarrhea) or localized wound infections (Byers and Matthews, 2002; Harper, 2002; Nemetz and Shotts, 1993). The microorganisms of zoonotic concern are almost exclusively bacteria, many of which are part of the normal flora of the aquatic environment.

One important zoonotic disease relevant to laboratory zebrafish is cutaneous infections caused by atypical *Mycobacterium* spp. (Lehane and Rawlin, 2000; Lewis *et al.*, 2003; Nemetz and Shotts, 1993). The disease in humans is relatively uncommon. It is often referred to as fish tank granuloma or swimming pool granuloma. Humans are typically infected by contamination of lacerated or abraded skin with aquarium water or fish contact. A localized, cutaneous granulomatous nodule can form at the site of infection, most commonly on hands or fingers. The granulomas usually appear approximately 6–8 weeks after exposure to the organism. They initially present as reddish bumps (papules) that slowly enlarge into purplish nodules and nonhealing ulcers. The infection can spread to nearby lymph nodes. Individuals who have an immune-compromising medical condition or are taking medications that impair immune function (steroids, immunosuppressive drugs, or chemotherapy) are at a greater risk for disseminated forms of the disease. A physician should be consulted if suspect lesions are noted. The disease can be difficult to treat because of drug resistance. It is also possible for these species of *Mycobacterium* to cause some degree of positive reaction to the tuberculin skin test.

The single most effective measure to prevent zoonotic disease transmission is regular and thorough hand washing. Hands and arms should always be washed after handling fish and aquarium water. Never smoke, drink, or eat in fish rooms or before washing hands. Minor cuts and abrasions should be protected with sturdy, impervious gloves and cleansed immediately with antibacterial soap if exposure occurs. Any suspected zoonotic lesion or illness should be evaluated by a physician.

Acknowledgments

I thank Michael Kent for his expertise and discussions of fish pathology, Monte Westerfield for his support, and Bill Trevarrow for critical comments on this manuscript. The Zebrafish International Resource Center is supported by a grant from the NIH-NCRR (P40 RR12546).

References

Alabaster, J. S., and Lloyd, R. (1982). "Water Quality Criteria for Freshwater Fish," 2nd ed. Butterworth, Sydney.

Astrofsky, K. M., Harper, C. M., Rogers, A. B., and Fox, J. G. (2002a). Diagnostic techniques for clinical investigation of laboratory zebrafish. *Lab. Anim.* **31**, 41–45.

Astrofsky, K. M., Schech, J. M., Sheppard, B. J., Obenschain, C. A., Chin, A. M., Kacergis, M. C., Laver, E. R., Bartholomew, J. L., and Fox, J. G. (2002). High mortality due to *Tetrahymena* sp. infection in laboratory-maintained zebrafish (*Brachydanio rerio*). *Comp. Med.* **52**, 363–367.

Astrofsky, K. M., Schrenzel, M. D., Bullis, R. A., Smolowitz, R. M., and Fox, J. G. (2000). Diagnosis and management of atypical *Mycobacterium* spp. infections in established laboratory zebrafish (*Brachydanio rerio*) facilities. *Comp. Med.* **50**, 666–672.

Austin, B., and Austin, D. A. (1999). "Bacterial Fish Pathogens: Disease of Farmed and Wild Fish," 3rd ed. Praxis Publishing, Chichester, UK.

Bandi, C., Dunn, A. M., Hurst, G. D., and Rigaud, T. (2001). Inherited microorganisms, sex-specific virulence, and reproductive parasitism. *Trends Parasitol.* **17**, 88–94.

Byers, K. B., and Matthews, J. L. (2002). Use of zebrafish and zoonoses. *Appl. Biosaf.* **7**, 117–119.

Clark, T. G. (2000). A parasite model for the study of immunogenetics in zebrafish. *In* "International Conference on Zebrafish Development & Genetics Abstracts," April 26–30, 2000, p. 58. Cold Spring Harbor Laboratory, Cold Spring Harbor, New York.

Colorni, A., Ankaoua, M., Diamant, A., and Knibb, W. (1994). Detection of mycobacteriosis in fish using the polymerase chain reaction technique. *Bull. Eur. Ass. Fish Pathol.* **6**, 195–198.

Colt, J. (1986). Gas supersaturation: Impact on the design and operation of aquatic systems. *Aquacult. Eng.* **5**, 49–85.

Dave, G., and Xiu, R. (1991). Toxicity of mercury, copper, nickel, lead, and coblat to embryos and larvae of zebrafish, *Branchydanio rerio. Arch. Environ. Toxicol.* **21**, 126–134.

de Kinkelin, P. D. (1980). Occurrence of a microsporidian infection in zebra danio *Brachydanio rerio* (Hamilton-Buchanan). *J. Fish Dis.* **3**, 71–73.

Dickerson, H. W., and Dawe, D. L. (1995). *Ichthyophthirius multifillis* and *Cryptocaryon irritans* (Phylum Ciliophora). *In* "Fish Diseases and Disorders, Vol. 1: Protozoan and Metazoan Infections" (P. T. K. Woo, ed.), pp. 181–228. CAB International, Wallingford, U.K.

Dunn, A. M., Terry, R. S., and Smith, J. E. (2001). Transovarial transmission in the microsporidia. *Adv. Parasitol.* **48**, 57–100.

Dykstra, M. J., Astrofsky, K. M., Schrenzel, M. D., Fox, J. G., Bullis, R. A., Farrington, S., Sigler, L., Rinaldi, M. G., and McGinnis, M. R. (2001). High mortality in a large-scale zebrafish colony (*Brachydanio rerio* Hamilton & Buchanan, 1822) associated with *Lecythophora mutabilis* (van Beyma) W. Gam & McGinnis. *Comp. Med.* **51**, 361–368.

Emerson, K., Russo, R. C., Lund, R. E., and Thurston, R. V. (1975). Aqueous ammonia equilibrium calculations: Effect of pH and temperature. *J. Fish Res. Board Can.* **32**, 2379–2383.

Ferguson, H. W. (1989). "Systemic Pathology of Fish." Iowa State University Press, Ames, IA.

Goven-Dixon, B. A. (1993). Fungal and algal diseases of freshwater tropical fishes. *In* "Fish Medicine" (M. Stoskopf, ed.), pp. 563–572. W. B. Saunders, Philadelphia, PA.

Hall-Stoodley, L., and Lappin-Scott, H. (1998). Biofilm formation by the rapidly growing mycobacterium species *Mycobacterium fortuitum. FEMS Microbiol. Lett.* **168**, 77–84.

Harper, C. (2002). Zoonotic diseases acquired from fish. *Aquacult. Mag.* **28**, 55–58.

Herwig, N. (1979). "Handbook of Drugs and Chemical Used in the Treatment of Fish Diseases." Charles C. Thomas, Springfield, IL.

Hoff, F. H., and Snell, T. W. (1999). "Plankton Culture Manual." Florida Aqua Farms, Dade City, FL.

Inglis, V., Roberts, R. J., and Bromage, N. R. (1993). "Bacterial Diseases of Fish." Blackwell Science, Ames, IA.

Jagadeeswaran, P., and Sheehan, J. P. (1999). Analysis of blood coagulation in the zebrafish. *Blood Cells. Mol. Dis.* **25**, 239–249.

Kent, M. L., Spitsbergen, J. M., Matthews, J. M., Fournie, J. W., and Westerfield, M. (2002a). "Diseases of Zebrafish in Research Facilities." Online manual available at http://zfin.org/zirc/diseaseManual.php.

Kent, M. L., Bishop-Stewart, J. K., Matthews, J. L., and Spitsbergen, J. M. (2002b). *Pseudocapillaria tomentosa*, a nematode pathogen, and associated neoplasms of zebrafish (*Danio rerio*) kept in research colonies. *Comp. Med.* **52**, 354–358.

LaPatra, S. E., Barone, L., Jones, G. R., and Zon, L. I. (2000). Effects of infectious hematopoietic necrosis virus and infectious pancreatic necrosis virus infection on hematopoietic precursors of the zebrafish. *Blood Cells Mol. Dis.* **26**, 445–452.

Lavens, P., and Sorgeloos, P (1996). "Manual on the Production and Use of Live Food for Aquaculture." FAO Fisheries Technical Paper 361, Food and Agriculture Organization of the United Nations, Ghent, Belgium.

Lehane, L., and Rawlin, G. (2000). Topically acquired bacterial zoonoses from fish: A review. *Med. J. Aust.* **173**, 256–259.

Lewis, F. M. T., Marsh, B. J., and Fordham von Reyn, C. (2003). Fish tank exposure and cutaneous infections due to *Mycobacterium marinum*: Tuberculin skin testing, treatment and prevention. *Clin. Infect. Dis.* **37**, 390–397.

Lom, J., and Dyková, I. (1992). "Protozoan Parasites of Fishes." Elsevier, Amsterdam.

Lomankin, V. V., and Trofimenko, V. Y. (1982). Capillarids (Nematoda: Capillariidae) of freshwater fish fauna of the USSR. *Tr. Gelan.* **31**, 60–87.

Mashima, T. Y., and Lewbard, G. A. (2000). Pet fish formulary. *Vet. Clin. N. Am.: Exotic Anim. Pract.* **3**, 117–130.

Matthews, J. L., Brown, A. M. V., Larison, K., Bishop-Stewart, J. K., Rogers, P., and Kent, M. L. (2001). *Pseudoloma neurophilia* n.g., n.sp., a new genus and species of Microsporidia from the central nervous system of the zebrafish (*Danio rerio*). *J. Euk. Microbiol.* **48**, 229–235.

Mazanec, A. I., and Trevarrow, B. (1998). Coleps, scourge of the baby zebrafish. *Zebrafish Sci. Mont.* **5**. *Available at* http://zfin.org/zf_info/monitor/vol5.1/vol.5.1.html.

Miyazaki, T., Kubota, S., and Miyashita, T. (1984). A histopathological study of *Pseudomonas fluorescens* infection in tilapia. *Fish Pathol.* **19**, 161–166.

Neely, M. N., Pfeifer, J. D., and Caparon, M. (2002). Streptococcus-zebrafish model of bacterial pathogenesis. *Infect. Immun.* **70**, 3904–3914.

Nemetz, T. G., and Shotts, E. B. (1993). Zoonotic diseases. *In* "Fish Medicine" (M. Stoskopf, ed.), pp. 214–220. W. B. Saunders, Philadelphia, PA.

Noga, E. (2000). "Fish Disease: Diagnosis and Treatment." Mosby Electronic Publishing, St. Louis, IN.

Olsson, P. E. (1998). Disorders associated with heavy metal pollution. *In* "Fish Diseases and Disorders, Vol. 2: Noninfectious Disorders" (P. T. K. Woo, ed.), pp. 105–131. CAB International, Wallingford, UK.

Pack, M., Belak, J., Boggs, C., Fishman, M., and Driever, W. (1995). Intestinal capillariasis in zebrafish. *Zebrafish Sci. Mont.* **3**, 1–3.

Patterson, D. J. (1996). "Free-Living Freshwater Protozoa: A Colour Guide." John Wiley & Sons, New York.

Pullium, J. K., Dillehay, D. L., and Webb, S. (1999). High mortality in zebrafish (*Danio rerio*). *Contemp. Top Lab. Anim. Sci.* **38**, 80–83.

Reddy, P. K., and Leatherland, J. F. (1998). Stress physiology. *In* "Fish Diseases and Disorders, Vol. 1: Protozoan and Metazoan Infections" (P. T. K. Woo, ed.), pp. 279–301. CAB International, Wallingford, UK.

Rothen, D. E., Curtis, E. W., and Yanong, R. P. E. (2002). Tolerance of yolk sac and free-swimming fry of the zebra danio *Brachydanio rerio*, black tetra *Gymnicorymbus ternetzi*, buenos aires tetra *Hemigrammus caudovittatus*, and blue gourami *Trichogaster trichopterus* to therapeutic doses of formalin and sodium chloride. *J. Aquat. Anim. Hlth.* **14**, 204–208.

Sanders, G. E., Batts, W. N., and Winton, J. R. (2003). Susceptibility of zebrafish (*Danio rerio*) to a model pathogen, spring viremia of carp virus. *Comp. Med.* **53**, 514–521.

Sanders, G. E., and Swaim, L. E. (2001). Atypical piscine mycobacteriosis in Japanese medaka (*Oryzias latipes*). *Comp. Med.* **51**, 171–175.

Schulze-Robbecke, R., Janning, B., and Fischeder, R. (1992). Occurrence of mycobacteria in biofilm samples. *Tubercle Lung Dis.* **73**, 141–144.

Shaw, R. W., and Kent, M. L. (1999). Fish microsporidia. *In* "Microsporidia and Microsporidiosis" (M. Wittner and L. M. Weiss, eds.), pp. 418–446. American Society of Microbiology Press, Washington, D. C.

Stoskopf, M. (1988). Fish chemotherapeutics. *Vet. Clin. N. Am.: Sm. Anim. Pract.* **18**, 331–348.

Stoskopf, M. (1993). "Fish Medicine." W. B. Saunders, Philadelphia, PA.

Talaat, A. M., Reimschuessel, R., and Trucksis, M. (1997). Identification of mycobacteria infecting fish to the species level using polymerase chain reaction and restriction enzyme analysis. *Vet. Microbiol.* **58**, 229–237.

Vávra, J., and Larrson, J. I. R. (1999). Structure of the microsporidia. *In* "Microsporidia and Microsporidiosis" (M. Wittner and L. M. Weiss, eds.), pp. 7–84. American Society of Microbiology Press, Washington, D. C.

Weber, R., Schwartz, D. A., and Deplazes, P. (1999). Laboratory diagnosis of microsporidiosis. *In* "Microsporidia and Microsporidiosis" (M. Wittner and L. M. Weiss, eds.), pp. 315–362. *Am. Soc. Microbiol. Press*, Washington, D. C.

Wedemeyer, G. (1996). "Physiology of Fish in Intensive Aquaculture Systems." Chapman & Hall, New York.

Weitkamp, R. L., and Katz (1980). A review of dissolved gas supersaturation literature. *Trans. Am. Fish. Soc.* **109**, 659–702.

Westerfield, M. (2000). "The Zebrafish Book: A Guide for the Laboratory Use of Zebrafish," 4th ed. University of Oregon Press, Eugene, OR.

CHAPTER 34

Zebrafish Sperm Cryopreservation

Stéphane Berghmans, John P. Morris IV, John P. Kanki, and A. Thomas Look

Dana-Farber Cancer Institute
Department of Pediatric Oncology
Harvard Medical School
Boston, Massachusetts 02115

I. Introduction: Benefits of Zebrafish Sperm Cryopreservation

The small size and fecundity of the zebrafish make it amenable to large-scale genetic studies (see Chapters 1 to 11). However, limited animal facility space and the need to maintain mutant and stock lines simultaneously are often restricting factors to zebrafish research (Brand *et al.*, 2002; Harvey *et al.*, 1982; Ransom and Zon, 1999; Westerfield, 2000). Efficient sperm cryopreservation can help circumvent these constraints by reducing the number of live fish in a system while maintaining their reproductive capacity (Westerfield, 2000). Sperm cryopreservation also provides "genetic insurance" for recovering strains if living stocks are lost (Gwo *et al.*, 1999) and extends the functional reproductive lifetime of males as long as samples remain viable in storage (Vivieros *et al.*, 2000). Finally, the technique has been exploited in reverse-genetic mutagenesis approaches, in which a cryopreserved sperm library is used to recover heterozygote mutant fish of interest (Wienholds *et al.*, 2002; see also Chapters 4 and 5). Thus, an optimized zebrafish sperm cryopreservation protocol will not only increase the efficacy of this genetic screening method, but also benefit the zebrafish community as a whole.

METHODS IN CELL BIOLOGY, VOL. 77
Copyright 2004, Elsevier Inc. All rights reserved.
0091-679X/04 $35.00

A single zebrafish sperm freezing protocol had been published prior to the development of an alternative protocol in our laboratory. This technique yields sperm samples collected through abdominal massage, which are frozen by incubation on dry ice followed by liquid nitrogen immersion in a medium of 10% (v/v) methanol and 15% milk powder (w/v) diluted in Ginzburg Ringer's solution (Harvey *et al.*, 1982). This protocol has only been slightly modified in other laboratories since its original development (Brand *et al.*, 2002; Ransom and Zon, 1999; Westerfield, 2000). Despite Harvey *et al.* (1982) reporting an efficient recovery of embryos by using their protocol, it produced low, inconsistent fertilization yields in our hands. Therefore, we developed a new cryopreservation protocol to maximize zebrafish embryo recovery.

In this chapter, we first review the important elements that should be considered when developing a sperm cryopreservation protocol in zebrafish. We then present the development of our alternative freezing method and finally discuss further considerations and future directions for the continued optimization of zebrafish cryopreservation.

II. Critical Variables Affecting Sperm Cryopreservation

The goal of cryopreservation is to provide a safe chemical environment for reducing cellular temperatures below $-130\,^{\circ}\mathrm{C}$ in order to indefinitely store cells in an inactive state and to be able to thaw and recover these cells without compromising physiological function (Critser and Karow, 1997; Mazur, 1963, 1984). Spermatozoa present a unique challenge because they possess specialized cell structures, such as densely packed mitochondria and a long flagellum, that must be protected to maintain fertilization capacity (Critser and Karow, 1997). Cryopreservation of sperm from teleosts such as zebrafish presents the additional complication that they must not be prematurely activated by water exposure, which is needed postthaw, to trigger the changes in intracellular potassium concentration required for sperm motility and fertilization (Kime *et al.*, 2001; Takai and Morisawa, 1995). Therefore, an optimal zebrafish sperm cryopreservation technique must protect spermatozoa from freezing damage while maintaining spermatozoa inactivity. Freezing damage predominantly occurs between $-15\,^{\circ}\mathrm{C}$ and $-60\,^{\circ}\mathrm{C}$, a temperature range sperm is exposed to during both freezing and thawing. Freezing damage within this temperature range is determined by the relative rates at which extracellular and intracellular water are frozen as a new osmotic equilibrium is established. Both slow and rapid freezing rates can contribute to different types of freezing damage.

Slow freezing beyond $-15\,^{\circ}\mathrm{C}$ can damage cells through solution effects. When intracellular water is supercooled below $0\,^{\circ}\mathrm{C}$, it flows from the cell into the frozen extracellular environment, causing intracellular solutes to precipitate as their concentration increases (Mazur, 1963, 1984; Critser and Karow, 1997; Vivieros *et al.*, 2000). This intracellular precipitation leads to cellular dehydration, volume shrinkage, and pH change, all highly detrimental to the cell. In addition, increased solute concentration can disrupt interactions between lipids and proteins in the plasma membrane and cytoplasm, resulting in cell dysfunction after thawing (Critser and Karow, 1997). A slow cooling rate can also result in cell lysis due to hyperosmotic stress. This is caused by increased intracellular ionic concentration and the influx of extracellular water attempting to regain osmotic equilibrium. Lysis can also result from the irreversible reduction in cell volume caused by increased membrane phospholipid packing. When the freezing rate of cells is too rapid, it can lead to intracellular ice formation and deleterious structural damage to cells (Critser and Karow, 1997). In fish spermatozoa, freezing damage manifests structurally as extreme dehydration, swollen midpieces, ruptured mitochondria, disrupted flagellum, membrane destabilization, and DNA damage (Baulny et al., 1999; Cabrita et al., 2001, 2002; Dzuba and Kopieka, 2002; Legendre and Billard, 1980; Zilli et al., 2003). To counter the effects of osmotic stress and dehydration, freezing damage can be minimized by optimizing three critical variables: freezing rate, appropriate diluent, and cryoprotectant solution.

Several freezing rates have been tested in fish. In previous zebrafish cryopreservation protocols, sperm is frozen on dry ice prior to liquid nitrogen storage (Brand et al., 2002; Harvey et al., 1982). Similar freezing techniques that bring sperm below the eutectic point through incubation on a subzero material, followed by storage at $-196\,^{\circ}C$, have also been successful in other species (Lahnsteiner et al., 2000; Mounib, 1978; Table I). These techniques are particularly useful for laboratories without access to a controlled-rate freezer. Until now, controlled-rate freezing has never been tested in zebrafish, but it has enhanced postthaw sperm viability in rainbow trout, carp, and catfish (Cabrita et al., 2001, 2002; Magyary et al., 1996a,b; Rurangwa et al., 2000; Vivieros et al., 2000). The high variability in freezing rate studies across species makes it difficult to determine a master freezing technique. Thus, a titration of freezing rates by using a controlled-rate freezing unit is required to optimize the freezing rate for zebrafish sperm. Optimal thawing rates have also been investigated in other fish species (Lahnsteiner et al., 2000, 2002), and like freezing rates, they have been shown to be species specific and determined empirically. In established protocols, small volumes of frozen zebrafish sperm are thawed at room temperature. However, varying sperm sample volume and controlled freeze/thaw rates might prove more effective.

The diluent is the principal component of a cryoprotective medium in which sperm is collected because cryoprotectants alone, or at high concentrations, can be toxic to sperm (Magyary et al., 1996b). Optimal diluents must also mimic the physiological conditions found in teleost seminal fluid such that the stored sperm remains immotile until activated by water (Kime et al., 2001; Rana, 1995). Furthermore, a number of diluent additives, exhibiting a range of biochemical charactersitics, have been found to increase sperm viability. Dimethylacetamide (DMA), also a cryoprotectant, has been shown to increase the intracellular concentration of ATP in treated sperm and thus might potentially enhance sperm motility (Baulny et al., 1999). Methylxanthine phosphodiesterase inhibitors, such

Table I

Cryoprotectants, Freezing Technique, and Sperm:Medium Dilution in Various Fish Species

Species	Cryoprotectant	Freezing technique	Sperm:Medium dilution	Ref.
African catfish (*Clarius gariepinus*)	10% DMSO/10% egg yolk	Multistage controlled rate freezing	1:1	Rurangwa *et al.*, 2001
European catfish (*Silurius glanis*)	10% Methanol	Multistage controlled rate freezing	1:200	Vivieros *et al.*, 2000
	10/15% DMA; 10% egg yolk	Liquid nitrogen vapor incubation	1:3	Baulny *et al.*, 1999
Striped trumpeter (*Latris lineata*)	2.84 M DMSO/10% egg yolk	Liquid nitrogen vapor incubation	1:5	Ritar *et al.*, 2000
Salmon (*Salmo salar*)	12.5% DMSO	Dry ice/acetone incubation	1:3	Mounib, 1978
Formosan landlocked salmon (*Onchorhynchus maina Formosanus*)	10% DMSO	Dry ice incubation	1:5	Gwo, 1999
Northern pike (*Esox lucius*)	15% DMSO	Dry ice incubation	1:3	Babiak *et al.*, 1999
Bleak (*Chalcalburnus chlacordis*)	10% DMSO/0.5% glycin	Liquid nitrogen vapor incubation	1:7	Lahnsteiner *et al.*, 2000
Rainbow trout (*Salmo gairderi*)	10% DMSO/10% egg yolk	Dry ice incubation	1:3	Billard, 1983
Rainbow trout (*Oncorhyncus mykiss*)	7% DMSO/7.5 mg/ml DanPro670	Single-stage controlled-rate freezing	1:3	Cabrita *et al.*, 2001, 2002
Black porgy (*Acanthopagus schlegeli*)	5% Glucose/1.25% glycerol	Liquid nitrogen vapor incubation	1:1	Chao *et al.*, 1986
Walleye (*Stizostedion vitreum*)	7% DMSO/4 mg/ml BSA/7.5 mg/ml ProFam	Dry ice incubation	1:15	Bergeron *et al.*, 2002
Flounder (*Paralichthys olivaceus*)	12% Glycerol	Stepwise temperature decrease in liquid nitrogen vapor	1:2	Zhang *et al.*, 2003
Carp (*Cyprinus carpio* L.)	10% DMSO	Multistage controlled-rate freezing	1:9	Magyary *et al.*, 1996a,b
Zebrafish (*Danio rerio*)	10% Methanol/15% powdered milk	Dry ice incubation	1:5	Harvey *et al.*, 1982

Note: All freezing methods described are followed by subsequent immersion and storage in liquid nitrogen. DMSO, dimethylsulfoxide; DMA, dimethylacetamide; DanPro670, Profam, soybean protein extracts; BSA, bovine serum albumin.

as caffeine, 3-isobutyl-1-methylxanthine, and theophylline have also been proposed to increase sperm motility by increasing intracellular cAMP stores (Babiak *et al.*, 1999). These additives have not been specifically tested in zebrafish (Babiak *et al.*, 1999; Kopeika *et al.*, 2003). In addition to the diluent chemical composition, the dilution factor itself has also been shown to affect fish sperm viability (Table I). An excessively high sperm concentration might limit the oxygen necessary for sperm survival, and bubbling oxygen through sperm samples before freezing has been shown to improve carp sperm viability (Magyary *et al.*, 1996a). Thus, consistent sperm dilution conditions have not been established for zebrafish and other fish species (Rana, 1995; Table I).

Cryoprotectants are chemicals that prevent cellular freezing damage, and their use in cryoprotective media has been integral to maintaining sperm viability during the freeze/thaw process (Critser *et al.*, 1997; Karow *et al.*, 1969; Mazur, 1984). Cryoprotectants are classified by their cellular permeability, which defines their site of cellular protection (Critser *et al.*, 1997; Taylor *et al.*, 1974). Internal cryoprotectants vary in their cellular permeability and toxicity to sperm across species and function by regulating the intracellular environment during freezing (Critser *et al.*, 1997; Rana, 1995; Vivieros *et al.*, 2000). External cryoprotectants do not permeate the cell membrane and protect membrane integrity by stabilizing the extracellular environment (Critser *et al.*, 1997). However, the high osmotic potential for water leaving the cell when external cryoprotectants are used alone often negates their protective benefit by causing catastrophic cell shrinkage. Thus, external cryoprotectants are often used in conjunction with permeating ones to maximize freezing protection (Critser *et al.*, 1997; Karrow, 1969). Different cryoprotectant combinations have proven successful in various fish species (Table I). However, the high variability of effective cryoprotectants within and across species emphasizes the need to evaluate many cryoprotectant combinations for determining an optimal freezing medium in zebrafish. Efficient cryopreservation will be achieved when the cryoprotective medium minimizes toxicity and freezing damage in the context of optimal freeze/thaw and fertilization methods.

Finally, unlike mammals and some fish, zebrafish have no acrosomes and no known mechanisms to mediate sperm–egg chemotaxis (Kime *et al.*, 2001; Wolenski and Hark, 1987). Zebrafish sperm access the egg surface through a specialized opening in the chorion called the micropyle, probably providing a physical block to polyspermy (Wolenski *et al.*, 1987). The zebrafish micropyle indicates the requirement for sperm motility in order to access this structure for egg fusion (Kime *et al.*, 2001). Motility has been shown to correlate with fertilization capacity in zebrafish and other species (Chao *et al.*, 1986; Harvey *et al.*, 1982; Lahnsteiner *et al.*, 2000; Rurangwa *et al.*, 2000). Sperm frozen in media containing diluent that fails to completely inhibit motility display reduced fertilization capacity, indicating that sperm must be kept dormant before freezing to maximize postthaw fertility (Legendre *et al.*, 1980). Thus, motility is often used as a measure of postthaw sperm viability when developing cryopreservation protocols. The sperm:egg ratio is likely to be critical to fertilization rates and the volume of cryopreservation medium and fertilization solutions might also prove important for providing an optimal environment for the activated sperm to reach and fertilize all accessible eggs.

III. Zebrafish Sperm Cryopreservation with N,N–Dimethylacetamide

We have determined an efficient zebrafish sperm cryopreservation protocol that, in our hands, is significantly more effective than the current established zebrafish sperm cryopreservation technique (Fig. 1; Morris *et al.*, 2003). The freezing method presented here yields a fertilization rate of approximately 10% for each of four sperm samples obtained from a single male fish. To establish this cryopreservation procedure, we first determined which sperm collection method reliably provided the largest viable sperm samples and investigated the use of motility as an indicator of sperm quality. We then determined a sperm freezing medium that maximized sperm viability during freezing and thawing by comparing five cryoprotectants and five diluents. We also performed a preliminary evaluation of controlled-rate freezing and studied factors that might influence the *in vitro* fertilization of eggs by using frozen sperm.

The capacity to fertilize eggs is the best measure of sperm viability (Cabrita *et al.*, 2001a; Rurangwa *et al.*, 2000). However, fertilization assays can require very large numbers of available female fish and previous studies have suggested that motility assays can substitute for fertilization capacity in zebrafish and other species (Cabrita *et al.*, 2001a; Chao *et al.*, 1986; Harvey *et al.*, 1982; Lahnsteiner *et al.*, 2000; Rurangwa *et al.*, 2000). To determine whether motility was an accurate substitute for zebrafish sperm fertilization capacity, we investigated the correlation between sperm motility and the ability of sperm to fertilize eggs. The percent motility [(motile sperm/total sperm) × 100)] was used as the measure of sample motility such that a comparison between individual samples could be made, despite large standard deviations in total sperm count. Fertilization capacity was determined by calculating the percent fertilization produced by sperm samples used for *in vitro* fertilization assays. Eggs were fertilized at two different sperm dilutions (1:10 and 1:100) to eliminate the possibility of saturating eggs with viable sperm, which would have masked the effect of motility differences on fertilization capacity. We plotted the percent fertilization of each diluted sperm sample along with its corresponding percent motility and using linear regression analysis, and found that both the 1:10 and 1:100 dilution groups displayed a significant positive correlation between percent fertilization and percent motility ($R^2 = 0.82$, $P = 0.0056$; $R^2 = 0.81$, $P = 0.0044$; Fig. 2). Thus, we determined that sperm percent motility was a reasonable alternative measure of sperm viability and we used it as an assay for evaluating cryopreservation conditions.

Two methods are used to collect zebrafish sperm: abdominal massage and homogenization of testes removed by dissection (Brand *et al.*, 2002; Harvey

Protocol 1: Zebrafish Sperm Cryopreservation with 200 μl 10% DMA/BSMIS and
In Vitro **Fertilization with Frozen Samples**

Preparation

1. Prepare freezing solutions
   BSMIS: 75mM NaCl, 70 mM KCl, 2 mM CaCl$_2$, 1 mM MgSO$_4$ and 20 mM Tris pH 8. Store at 4 °C. For each male fish that sperm will be cryopreserved from, aliquot 20 μl of BSMIS into a microcentrifuge tube and place on ice.

   Freezing Medium Stock: 11.2% DMA in BSMIS. Vortex for 10 min to mix. Make DMA/BSMIS stock solution fresh before each round of cryopreservation.

2. Fill a large styrofoam box (at least 10 in. x 10 in. x 10 in.) with dry ice. For each male fish that sperm will be cryopreserved from, insert two 50-ml conical centrifuge tubes into the dry ice until just the cap is visible.

Sperm Collection and Freezing

1. Anesthetize male fish, dissect, and remove testes. Place testes in 20 μl chilled BSMIS.

2. Homogenize testes by grinding with a pestle.

3. Add 180 μl of 11.2% freezing medium stock and mix until uniformly cloudy. [Note: for samples frozen in 15% DMA (see results and discussion), 180 μl of a 16.75% DMA stock solultion is added to homogenized sperm in BSMIS to achieve a final DMA concentration of 15% DMA.]

4. Aliquot 50 μl of the sperm solution into four labeled cryotubes.

5. Insert two cryotubes each into 50-ml centrifuge tubes on dry ice.

6. Incubate for 30 min on dry ice.

7. Quickly remove samples from 50-ml centrifuge tubes and immerse directly in liquid nitrogen. Transfer samples to a liquid nitrogen freezer for long-term storage.

In Vitro Fertililzation

1. Preheat BSMIS to 37 °C.

2. Collect eggs from females in a petri dish. Pool egg clutches from at least three females (300–600 eggs).

3. Remove on 50-μl sample from liquid nitrogen and open cap. Add 1 ml of BSMIS at 37 °C and mix until sample is completely thawed. Apply thawed sperm to eggs.

4. Immediately add 1 ml of egg water.

5. After chorions separte from egg mass, fill petri dish with egg water and incubate embryos at 28.5 °C.

Fig. 1 Zebrafish sperm cryopreservation protocol with 200 μl 10% DMA/BSMIS and *in vitro* fertilization with frozen samples. From Morris J. P., IV., *et al.* (2003). Zebrafish sperm cryopreservation with *N,N*-dimethylacetamide. *Biotechniques* **35**, 958–968, with permission.

et al., 1982; Ransom and Zon, 1999; Westerfield, 2000). Because the ultimate goal of our project was to maximize the viability of cryopreserved sperm for fertilizing the largest possible number of eggs, it was important to choose the collection technique that most reliably provided large, viable sperm samples. To determine

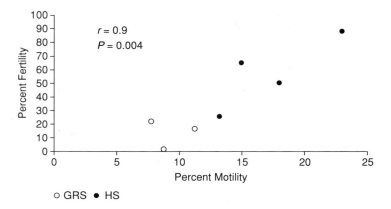

Fig. 2 Zebrafish sperm percent motility is correlated with sperm fertilization capacity. Sperm samples were incubated in either Ginzburg Ringers solution or Hank's solution. Sperm was diluted 1:10. From Morris J. P., IV., *et al.* (2003) Zebrafish sperm cryopreservation with *N,N*-dimethylacetamide. *Biotechniques* **35**, 958–968, with permission.

this, we recorded the average sperm motility as well as the total motile sperm collected by both methods. Although the samples collected by dissection yielded more sperm per sample than squeezing, there was no significant difference in percent motility or total number of motile sperm collected by these methods ($P = 0.046$ and $P = 0.58$, respectively). However, sperm was recovered from all dissected fish, whereas only half of the squeezed males yielded sperm samples. Because dissection resulted in sperm samples from all fish and was a less tedious procedure than abdominal massage, we conducted all further assays by using sperm collected from the dissection method.

Internal cryoprotectants covering a range of cellular permeabilities were tested for their toxicity and ability to provide sperm with the highest degree of freezing damage protection. We tested five cryoprotectants previously used for fish sperm cryopreservation: DMA, dimethylsulfoxide (DMSO), glycerol, ethylene glycol, and methanol. Each cryoagent was evaluated over a range of concentrations that have been used to cryopreserve sperm from various fish species [5%, 10%, 15%, and 20% (cryoprotectant volume/volume Hank's solution) (Westerfield, 2000)]. Cryoprotectants were first evaluated for toxicity by incubating fresh sperm at each concentration and assaying the percent motility of activated sperm at different time points. To evaluate the ability of cryoprotectants to prevent freezing damage, samples were frozen in the cryoprotectants at each concentration and their post-thaw percent motility examined. Our preliminary toxicity experiments indicated that 5% and 10% glycerol were the least chemically harmful over time. However, low toxicity did not correlate with the highest capacity for maintaining viability during cryopreservation. Sperm frozen in 10% and 15% DMA displayed the highest overall values of postthaw percent motility as well as the highest average postthaw percent motility across all cryoprotectant concentrations (Table II). Our experiments detected differences between postthaw percent motility due to cryoprotectant and concentration, statistically possessing 90% and 89% power, respectively. However, we were unable to determine a significant optimal combination ($P = 0.09$) at the 95% confidence level. Our analysis possessed significant power to detect differences and we observed that DMA samples exhibited the highest (1) overall postthaw percent motility at any concentration, (2) postthaw percent motility at three of the four concentrations tested, and (3) overall average postthaw percent motility. Therefore, we selected DMA (10% and 15%) as our cryoprotectant for further optimization of other aspects of the cryopreservation protocol.

An optimal diluent is one that both minimizes sperm toxicity and maintains sperm in an immotile state until it is activated for *in vitro* fertilization (Rana, 1995). We tested the following five diluents used in zebrafish or other fish species: BSMIS, Hank's solution, Ginzburg Ringer's solution, Mounib's solution, and Sperm solution (Table III). Testes were collected and homogenized in each of the five diluents, and sperm samples were observed over time to ensure that they remained inactive. Samples were then activated with water, and percent motility was recorded at different time intervals to determine diluent toxicity. All diluents maintained sperm in an inactive state, but statistical analysis of the percent motility over time failed to determine a diluent that gave significantly higher motility than all other diluents. Hank's solution and BSMIS afforded sperm the highest percent motility and average motility of all diluents tested. We decided to use BSMIS as our diluent, because the sperm displayed a low decrease in percent

Table II
Cryopreserved Sample Postthaw Motility

Cryoprotectant	Permeability coefficient ($P \times 10^{-5}$ cm/s)	Postthaw percent motility ($n = 3$)				Average postthaw percent, motility ($n = 12$)
		5%	10%	15%	20%	
DMA	14.7 ± 0.37	8.8 ± 1.2	11.6 ± 5.8	12.1 ± 4.9	7.2 ± 0.6	9.9 ± 3.9
Ethylene glycol	3.4 ± 0.1	7.7 ± 1.3	9.0 ± 6.8	7.4 ± 2.6	6.8 ± 0.4	7.6 ± 3.3
Methanol	11.4 ± 0.4	3.1 ± 2.4	8.9 ± 4.5	9.9 ± 2.2	9.3 ± 2.2	7.5 ± 3.7
DMSO	1.30 ± 0.1	8.0 ± 2.1	6.0 ± 1.0	7.6 ± 2.2	6.8 ± 0.8	7.1 ± 1.6
Glycerol	0.58 ± 0.04	5.7 ± 0.6	7.4 ± 0.2	8.4 ± 1.8	5.3 ± 0.4	6.7 ± 1.5

Note: Cryoprotectant permeability coefficients are from Naccache and Sha'afi (1973) and reported in units of P (permeability coefficient) $\times 10^{-5}$ cm/s. Postthaw percent motility is presented as mean $\pm$ standard deviation for sperm frozen in medium consisting of each cryoprotectant diluted to the listed concentration in HS; $n = 3$ for all concentrations. Average postthaw percent motility represents sperm frozen with all four concentrations of each cryoprotectant ($n = 12$). From Morris, J. P., IV., *et al.* (2003). Zebrafish sperm cryopreservation with *N,N*-dimethylacetamide. *Biotechniques* **35,** 958–968, with permission.

Table III

Composition of Diluents Evaluated During Development of Cryopreservation Protocol by Utilizing the DMA/BSMIS Medium

Diluent	Components	Ref.
Ginzburg Ringer's	0.11 M NaCl, 4 mM KCl, 2.7 mM $CaCl_2$, 11 ddH_2O, and 77 mM $NaHCO_3$	Westerfield, 2000
Sperm solution	80 mM KCl, 45 mM sodium acetate, 0.4 mM $CaCl_2$, 0.2 mM $MgCl_2$, and 10 mM HEPES, pH 7.7	Corley-Smith *et al.*, 1999
Mounib's solution	25 mM sucrose, 6.5 mM reduced glutathione, and 100 mM potassium bicarbonate	Billiard, 1983
Hank's solution	0.137 mM NaCl, 5.4 mM KCl, 0.25 mM Na_2 HPO_4, 0.44 mM KH_2PO_4, 1.3 mM $CaCl_2$, 1.0 mM $MgSO_4$, and 4.2 mM $NaHCO_3$	Westerfield, 2000
BSMIS	75 mM NaCl, 70 mM KCl, 2 mM $CaCl_2$, 1 mM $MgSO_4$, and 20 mM Tris, pH 8.0	Lahnsteiner *et al.*, 2000

motility over time as well as a trend toward low toxicity. Without a statistically significant result, this experiment must be seen as preliminary and BSMIS cannot be considered the optimal diluent for zebrafish sperm cryopreservation without further analysis.

Each cell type has a specific optimal freezing rate at which freezing damage will be minimized (Critser and Karow, 1997). To determine the optimal freezing rate for zebrafish sperm, samples were frozen at different rates and their prefreeze and postthaw motility were compared. Four freezing rates were evaluated: $-10\,°C/$ min, $-40\,°C/min$, $-60\,°C/min$, and $-80\,°C/min$. Sperm frozen at $-60\,°C/min$ displayed the highest overall postthaw percent motility (Table IV). In pairwise comparisons, sperm frozen at $-60\,°C/min$ possessed significantly higher motility than sperm frozen at $-80\,°C/min$ and $-40\,°C/min$ ($P = 0.002$ and $P = 0.02$, respectively). All other pairwise comparisons were not significant, although a $-60\,°C/min$ freezing rate displayed higher postthaw percent motility than the $-10\,°C/min$ freezing rate (Table IV). More importantly, sperm frozen at $-60\,°C/min$ did not display significantly higher percent motility than sperm frozen by incubation on dry ice followed by liquid nitrogen storage. For its relative ease of use, we therefore decided to rely on dry ice incubation in our cryopreservation protocol.

High concentrations of sperm in frozen samples have been reported to decrease fertilization capacity (Rana, 1995). Therefore, we investigated the fertilization capacity of sperm samples frozen in relatively higher volumes of medium. Sperm was frozen in both 10% and 15% DMA in BSMIS, but diluted by a factor of 10 compared to the volume of our preliminary assays. This 10-fold increase in freezing medium volume had the advantage of providing four separate 50-μl

Table IV
Pre- and Postthaw Motility of Zebrafish Sperm Frozen by Using Fixed Controlled Rates

Freezing rate	Average prefreeze motility ($n = 6$)	Average postthaw motility ($n = 6$)	Average change in percent motility ($n = 6$)
Incubation on dry ice	20.77 ± 5.41	12.04 ± 3.78	8.73 ± 9.00
$-10\,°C/min$	15.78 ± 1.46	12.04 ± 1.07	3.46 ± 1.74
$-40\,°C/min$	18.64 ± 5.02	11.02 ± 1.52	7.62 ± 5.40
$-60\,°C/min$	19.86 ± 6.94	13.58 ± 6.94	6.94 ± 1.80
$-80\,°C/min$	16.36 ± 2.39	9.02 ± 2.04	7.54 ± 1.93

Table V
In Vitro **Fertilization with Cryopreserved Sperm Samples**

Protocol ($N = 8$)	Freezing medium	Clutch size	Recovered embryos	Percent fertilization
Capillary/20μl	10% Methanol and 15% powdered milk/GRS	442.1 ± 172.8	1.0 ± 1.2	0.2 ± 0.2
Capillary/20μl	10% DMA/HS	169.6 ± 124.6	27.9 ± 32.3	14.0 ± 10.1
Capillary/20μl	15% DMA/HS	312.1 ± 142.7	26.0 ± 15.9	10.7 ± 12.2
Protocol 1/200μl	10% DMA/BSMIS	478.4 ± 192.2	38.6 ± 10.6	10.2 ± 4.1
Protocol 1/200μl	15% DMA/BSMIS	436.6 ± 147.0	47.6 ± 42.4	9.3 ± 5.5

Note: *In vitro* fertilization with sperm frozen by using the standard freezing protocol derived from Harvey *et al.* (1982) is designated *capillary/20 μl* and was performed with the entire 20 μl sample collected from one male. Fertilization with sperm frozen by using Protocol 1, is designated *protocol 1/200 μl*, and was performed with one 50-μl aliquot of the 200-μl total sample frozen from individual males. $N = 8$ for all trials. Data presented as mean ± standard deviation. From Morris, J. P., IV., *et al.* (2003). Zebrafish sperm cryopreservation with N,N-dimethylacetamide. *Biotechniques* **35,** 958–968, with permission.

cryopreserved sperm samples for each male. Fertilization with diluted aliquots of sperm frozen in 10% DMA/BSMIS medium resulted in a fertility rate of $10.2 \pm 4.1\%$, whereas diluted samples frozen in 15% DMA/BSMIS medium gave a fertility rate of $9.3 \pm 5.5\%$ (Table V). When compared with undiluted samples, percent fertilization, clutch size, and the number of embryos recovered were not significantly different ($P = 0.99$, 0.63, and 0.57, respectively). We used the 1:10 sperm dilution in our final protocol as it yielded fertilization rates comparable to those of undiluted samples and increased the effectiveness of embryo recovery because four frozen aliquots, rather than one, could be obtained from each male fish. In addition, we observed a greater variance in the number of embryos recovered from fertilizations performed with aliquots frozen with 15% DMA vs. those frozen with 10% DMA ($P < 0.001$), and we therefore chose 10% DMA diluted in BSMIS as our freezing solution.

During our studies, we observed the most efficient embryo recovery when using large clutches for *in vitro* fertilization. Our protocol calls for a target clutch size of 300–600 healthy eggs, often pooled from as many as three females. In our experiments, we observed that clutch size was positively correlated with fertilization rate when samples frozen at high 10% and 15% DMA media volumes were used for *in vitro* fertilization ($r = 0.82$ and 0.98, respectively). Also, the number of recovered embryos was positively correlated with clutch size when samples frozen by the Harvey-derived protocol with 10% and 15% DMA were used for *in vitro* fertilization ($r = 0.81$ and 0.98, respectively). Thus, although clutch size does not necessarily increase postthaw fertilization capacity, we emphasize it as a method for maximizing the effective recovery of fertilized embryos from a single aliquot of frozen sperm.

Finally, we investigated the length of time for which archived frozen stocks can be stored while maintaining fertilization capacity. We cryopreserved large numbers of sperm aliquots by using the described protocol and performed *in vitro* fertilization following different periods of storage in liquid nitrogen. Thus far, our experiments have shown that sperm samples retain their fertilization capacity for over 12 months, and we plan on continuing their evaluation over longer periods of time.

IV. Future Directions

We are confident that this work has established a more efficient zebrafish sperm cryopreservation protocol that should have broad utility within the zebrafish community. Our method provides the conditions for sperm frozen from any healthy zebrafish male to consistently recover offspring through *in vitro* fertilization. In addition to its higher fertilization rate, this protocol yields four cryopreserved sperm samples per individual. This increases the potential recovery rate fourfold and can extend the effective period of time for recovering embryos from a given zebrafish line, helping preserve its genetic diversity. Although the ability to consistently recover embryos from frozen sperm samples represents a significant improvement, we view our work as a preliminary evaluation of cryopreservation conditions in zebrafish and a starting point to improve additional elements of cryopreservation to increase its efficiency in the future. Therefore, we encourage others in the community to continue cryopreservation studies to further optimize this protocol.

Many factors might ultimately increase the efficiency of zebrafish sperm cryopreservation, and very few have been systematically investigated. Each of the variables we addressed in this protocol can be examined in more detail. Cryopreservation can be improved by a more thorough testing of optimal freeze/thaw rates. Our preliminary results showed that sperm frozen at $-60\,°C/min$ produces significantly lower freezing damage than sperm frozen at $-80\,°C/min$ and $-40\,°C/min$. Other freezing rates, including multiple stages of freezing might be more effective. Controlled-rate thawing can also lead to improved fertilization capacity.

The testing of additional types of both diluents and cryoprotectants can be explored further and can represent a major step toward cryopreservation optimization. Furthermore, additional combinations of diluents with cryoprotectants can prove to be more successful than those reported here. However, these experiments are likely to represent a relatively major research effort. Internal cryoprotectants, other than the five presented here, have been used successfully in other fish species (Rana, 1995; Table I). Combining different internal cryoprotectants together and their pairing with external cryoprotectants can lead to increased cryoprotectant media efficacy. Furthermore, several additional phospholipids, peptide, and sugar-based external cryoprotectants have also been examined for their ability to protect sperm from freezing damage in other fish species (Babiak et al., 1999; Cabrita et al., 2001a; Lahnsteiner et al., 2000; Table I). Because so many different potential internal and external cryoprotectants are available, careful consideration of their chemical and physiological characteristics will be required to avoid evaluating a prohibitively large number of different combinatorial conditions. To help investigating cryoprotectant and diluent effects, more accurate and efficient methods to determine sperm viability can be employed, such as computer-aided sperm analysis (CASA) or viability-staining followed by flow cytometric analysis (Cabrita et al., 2001b; Kime et al., 2001; Pena et al., 1999; Rurangwa et al., 2000). Improving the assays for sperm motility and viability can make assays on cryoprotectant media technically more feasible.

Other aspects of cryopreservation can be optimized by advancing current technical methods or improving aspects of sperm production and the fertilization process. For example, developing a more consistent and efficient abdominal massage collection method as an alternative to testes dissection might be preferred because of the continued survival of the male fish for future use and their relatively higher numbers of mature sperm. Although higher in number, most of the sperm isolated by testes dissection are immature. Thus, an increase in mature sperm yield will be particularly beneficial, which might be possible by promoting spermatogenesis through the use of spermatogenesis-inducing hormones. This technique has been used on other fish, and such a hormone has recently been isolated in zebrafish (Magyary et al., 1996a; Todo et al., 2000). Fertilization rates can benefit from increasing the oxygen supply to the cryopreservation process, which has been postulated to affect subsequent sperm viability (Magyary et al., 1996a). Lastly, the use of chemical additives that metabolically enhance motility after sperm activation can improve their effectiveness in zebrafish in vitro fertilization.

The major challenge facing any investigator trying to develop a better sperm cryopreservation protocol is the sheer number of variables to test. Consequently, the obstacle we encountered was the large number of both male and female fish required to determine statistically significant differences in these assays. Thus, if additional zebrafish sperm cryopreservation studies are to be completed in the future, we propose a concerted zebrafish community effort to help decrease the burden on single laboratory groups. We are aware that other zebrafish cryopreservation protocols have been developed in the community, and several of these protocols were made available to us. Unfortunately, given limited resources for these studies, we were unable to methodically test these other protocols and could not include them in this chapter. We are in contact with the ZFIN Web site to make our protocol available, and we suggest that other laboratories use this community Web site to share worthwhile advances contributing to the improvement of zebrafish sperm cryopreservation.

References

Babiak, I., Glowgowski, J., Luczynski, M. J., Luczynski, M., and Demanianowicz, W. (1999). The effect of egg yolk, low density lipoproteins, methylxanthines and fertilization diluent on cryopreservation efficiency of northern pike (*Esox lucius*) spermatozoa. *Theriogenology* **52**, 473–479.

Baulny, B. O., Labbe, C., and Maisse, G. (1999). Membrane integrity, mitochondrial activity, ATP content, and motility of the European catfish (*Silurus glanis*) testicular spermatozoa after freezing with different cryoprotectants. *Cryobiology* **39**, 177–184.

Bergeron, A., Vandenberg, G., Proulx, D., and Bailey, J. L. (2002). Comparison of extenders, dilution ratios and theophylline addition on the function of cryopreserved walleye semen. *Theriogenology* **57**, 1061–1071.

Bilard, R. (1983). Ultrastructure of trout spermatozoa: Changes after dilution and deep-freezing. *Cell Tissue Res.* **228**, 205–218.

Brand, M., Granato, M., and Nusslein-Volhard, C. (2002). Keeping and raising zebrafish. *In* "Zebrafish: A Practical Approach" (C. Nusslein-Volhard and R. Dahm, eds.), pp. 7–37. Oxford University Press, New York.

Cabrita, E., Anel, L., and Herraez, M. P. (2001a). Effect of external cryoprotectants as membrane stabilizers on cryopreserved rainbow trout sperm. *Theriogenology* **56**, 623–635.

Cabrita, E., Martinez, F., Alvarez, M., and Herraez, M. P. (2001b). The use of flow cytometry to assess membrane stability in fresh and cryopreserved trout spermatozoa. *Cryo Lett.* **22**, 263–272.

Chao, N., Chao, W., Liu, K., and Liao, I. (1986). The biological properties of black porgy (*Acanthopargus* schlegel) sperm and its cryopreservation. *Proc. Natl. Sci. Coun. B. ROC* **10**, 145–149.

Corley-Smith, G. E., Brandhorst, B. P., Walker, C., and Postlethwait, J. H. (1999). Production of haploid and diploid androgenetic zebrafish (including methodology for delayed *in vitro* fertilization). *Methods Cell Biol.* **59**, 45–60.

Critser, J. K., and Karow, A. M. (1997). "Reproductive Tissue Banking: Scientific Principles." Academic Press, San Diego.

Dzuba, B. B., and Kopeika, E. F. (2002). Relationship between the changes in cellular volume of fish spermatozoa and their cryoresistance. *Cryo Lett.* **23**, 353–360.

Gwo, J. C., Ohta, H., Okuzawa, K., and Wu, H. C. (1999). Cryopreservation of sperm from the endangered formosan landlocked salmon (*Oncorhynchus masou formosanus*). *Theriogenology* **51**, 569–582.

Harvey, B., Norman Kelley, R., and Ashwood-Smith, M. J. (1982). Cryopreservation of zebrafish spermatozoa using methanol. *Can. J. Zool.* **60**, 1867–1870.

Karrow, A. M. (1969). Cryoprotectants—a new class of drugs. *J. Pharm. Pharmacol.* **1**, 209–223.

Kime, D. E., Van Look, K. J., McAllister, B. G., Huyskens, G., Rurangwa, E., and Ollevier, F. (2001). Computer-assisted sperm analysis (CASA) as a tool for monitoring sperm quality in fish. *Comp. Biochem. Physiol C. Toxicol. Pharmacol.* **130**, 425–433.

Kopeika, J., Kopeika, E., Zhang, T., Rawson, D. M., and Holt, W. V. (2003). Detrimental effects of cryopreservation of loach (*Misgurnus fossilis*) sperm on subsequent embryo development are reversed by incubating fertilised eggs in caffeine. *Cryobiology* **46**, 43–52.

Lahnsteiner, F., Berger, B., Horvath, A., Urbanyi, B., and Weismann, T. (2000). Cryopreservation of spermatozoa in cyprinid fishes. *Theriogenology* **54**, 1477–1498.

Lahnsteiner, F., Mansour, N., and Weismann, T. (2002). The cryopreservation of spermatozoa of the burbot, *Lota lota* (Gadidae, Teleostei). *Cryobiology* **45**, 195–203.

Legendre, M., and Billard, R. (1980). Cryopreservation of rainbow trout sperm by deep-freezing. *Reprod. Nutr. Dev.* **20**, 1859–1868.

Magyary, I., Urbanyi, B., and Horvath, L. (1996a). Cryopreservation of common carp (*Cyprinus carpio L.*) sperm I. The importance of oxygen supply. *J. Appl. Ichthyol.* **12**, 113–115.

Magyary, I., Urbanyi, B., and Horvath, L. (1996b). Cryopreservation of common carp (*Cyprinus carpio L.*) sperm II. Optimal conditions for fertilization. *J. Appl. Ichthyol.* **12**, 117–119.

Mazur, P. (1963). Kinetics of water loss from cells at subzero temperatures and the likelihood of intracellular freezing. *J. Gen. Physiol.* **47**, 347–369.

Mazur, P. (1984). Freezing of living cells: mechanisms and implications. *Am. J. Physiol.* **247**, C125–C142.

Morris, J. P., IV., Berghmans, S., Zahrieh, D., Neuberg, D. S., Kanki, J. P., and Look, A. T. (2003). Zebrafish sperm cryopreservation with *N,N*-dimethylacetamide. *Biotechniques* **35**, 956–968.

Mounib, M. S. (1978). Cryogenic preservation of fish and mammalian spermatozoa. *J. Reprod. Fert.* **228**, 205–218.

Naccache, P., and Sha'afi, R. I. (1973). Patterns of nonelectrolyte permeability in human red blood cell membrane. *J. Gen. Physiol.* **62**, 714–736.

Pena, A., Johannisson, A., and Linde-Forsberg, C. (1999). Post-thaw evaluation of dog spermatozoa using new triple fluorescent staining and flow cytometry. *Theriogenology* **52**, 965–980.

Rana, K. J. (1995). Cryopreservation of fish spermatozoa. *In* "Methods in Molecular Biology," vol. 38, pp. 151–165. Academic Press, San Diego.

Ransom, D. G., and Zon, L. I. (1999). Collection, storage, and use of zebrafish sperm. *In* "Methods in Cell Biology" (H. W. Detrich, M. Westerfield, and L. I. Zon, eds.), vol. 60, pp. 365–372. Academic Press, San Diego.

Ritar, A. J., and Campet, M. (2000). Sperm survival during short-term storage and after cryopreservation of semen from striped trumpeter (*Latris lineata*). *Theriogenology* **54**, 467–480.

Rurangwa, E., Volckaert, F. A. M., Huyskens, G., Kime, D. E., and Ollevier, F. (2001). Quality control of refrigerated and cryopreserved semen using computer-assisted sperm analysis (CASA), viable staining and standardized fertilization in African catfish (*Clarias gariepinus*). *Theriogenology* **55**, 751–769.

Takai, H., and Morisawa, M. (1995). Change in intracellular K+ concentration caused by external osmolality change regulates sperm motility of marine and freshwater teleosts. *J. Cell Sci.* **108**, 1175–1181.

Taylor, R., Adams, G. D., Boardman, C. F., and Wallis, R. G. (1974). Cryoprotection—permeant vs. nonpermeant additives. *Cryobiology* **11**, 430–438.

Todo, T., Ikeuchi, T., Kobayashi, T., Kajiura-Kobayashi, H., Suzuki, K., Yoshikuni, M., Yamauchi, K., and Nagahama, Y. (2000). Characterization of a testicular 17alpha, 20beta-dihydroxy-4-pregnen-3-one (a spermiation-inducing steroid in fish) receptor from a teleost, Japanese eel (*Anguilla japonica*). *FEBS Lett.* **465**, 12–17.

Vivieros, A. T. M., So, N., and Komen, J. (2000). Sperm cryopreservation of African catfish. *Clarius gariepinus*: cryoprotectants, freezing rates, and sperm:egg dilution ratio. *Theriogenology* **54**, 1395–1408.

Westerfield, M. (2000). "The Zebrafish Book: A Guide for the Laboratory Use of Zebrafish (*Danio rerio*)." University of Oregon Press, Eugene, OR.

Wienholds, E., Schulte-Merker, S., Walderich, B., and Plasterk, R. H. (2002). Target-selected inactivation of the zebrafish rag1 gene. *Science* **297**, 99–102.

Wolenski, J. S., and Hart, N. H. (1987). Scanning electron microscope studies of sperm incorporation into the zebrafish (*Brachydanio*) egg. *J. Exp. Zool.* **243**, 259–273.

Zhang, Y. Z., Zhang, S. C., Liu, X. Z., Xu, Y. Y., Wang, C. L., Sawant, M. S., Li, J., and Chen, S. L. (2003). Cryopreservation of flounder (*Paralichthys olivaceus*) sperm with a practical methodology. *Theriogenology* **60**, 989–996.

Zilli, L., Schiavone, R., Zonno, V., Storelli, C., and Vilella, S. (2003). Evaluation of DNA damage in *Dicentrarchus labrax* sperm following cryopreservation. *Cryobiology* **47**, 227–235.

INDEX

VOLUMES IN SERIES

Founding Series Editor
DAVID M. PRESCOTT

Volume 1 (1964)
Methods in Cell Physiology
Edited by David M. Prescott

Volume 2 (1966)
Methods in Cell Physiology
Edited by David M. Prescott

Volume 3 (1968)
Methods in Cell Physiology
Edited by David M. Prescott

Volume 4 (1970)
Methods in Cell Physiology
Edited by David M. Prescott

Volume 5 (1972)
Methods in Cell Physiology
Edited by David M. Prescott

Volume 6 (1973)
Methods in Cell Physiology
Edited by David M. Prescott

Volume 7 (1973)
Methods in Cell Biology
Edited by David M. Prescott

Volume 8 (1974)
Methods in Cell Biology
Edited by David M. Prescott

Volume 9 (1975)
Methods in Cell Biology
Edited by David M. Prescott

Advisory Board Chairman
KEITH R. PORTER

Series Editor
LESLIE WILSON

Series Editors
LESLIE WILSON AND PAUL MATSUDAIRA

Volume 38 (1993)
Cell Biological Applications of Confocal Microscopy
Edited by Brian Matsumoto

Volume 39 (1993)
Motility Assays for Motor Proteins
Edited by Jonathan M. Scholey

Volume 40 (1994)
A Practical Guide to the Study of Calcium in Living Cells
Edited by Richard Nuccitelli

Volume 41 (1994)
Flow Cytometry, Second Edition, Part A
Edited by Zbigniew Darzynkiewicz, J. Paul Robinson, and Harry A. Crissman

Volume 42 (1994)
Flow Cytometry, Second Edition, Part B
Edited by Zbigniew Darzynkiewicz, J. Paul Robinson, and Harry A. Crissman

Volume 43 (1994)
Protein Expression in Animal Cells
Edited by Michael G. Roth

Volume 44 (1994)
***Drosophila melanogaster:* Practical Uses in Cell and Molecular Biology**
Edited by Lawrence S. B. Goldstein and Eric A. Fyrberg

Volume 45 (1994)
Microbes as Tools for Cell Biology
Edited by David G. Russell

Volume 46 (1995) (in preparation)
Cell Death
Edited by Lawrence M. Schwartz and Barbara A. Osborne

Volume 47 (1995)
Cilia and Flagella
Edited by William Dentler and George Witman

Volume 48 (1995)
***Caenorhabditis elegans:* Modern Biological Analysis of an Organism**
Edited by Henry F. Epstein and Diane C. Shakes

Volume 49 (1995)
Methods in Plant Cell Biology, Part A
Edited by David W. Galbraith, Hans J. Bohnert, and Don P. Bourque

Volume 50 (1995)
Methods in Plant Cell Biology, Part B
Edited by David W. Galbraith, Don P. Bourque, and Hans J. Bohnert

Volume 51 (1996)
Methods in Avian Embryology
Edited by Marianne Bronner-Fraser

Volume 52 (1997)
Methods in Muscle Biology
Edited by Charles P. Emerson, Jr. and H. Lee Sweeney

Volume 53 (1997)
Nuclear Structure and Function
Edited by Miguel Berrios

Volume 54 (1997)
Cumulative Index

Volume 55 (1997)
Laser Tweezers in Cell Biology
Edited by Michael P. Sheez

Volume 56 (1998)
Video Microscopy
Edited by Greenfield Sluder and David E. Wolf

Volume 57 (1998)
Animal Cell Culture Methods
Edited by Jennie P. Mather and David Barnes

Volume 58 (1998)
Green Fluorescent Protein
Edited by Kevin F. Sullivan and Steve A. Kay

Volume 59 (1998)
The Zebrafish: Biology
Edited by H. William Detrich III, Monte Westerfield, and Leonard I. Zon

Volume 60 (1998)
The Zebrafish: Genetics and Genomics
Edited by H. William Detrich III, Monte Westerfield, and Leonard I. Zon

Volume 61 (1998)
Mitosis and Meiosis
Edited by Conly L. Rieder

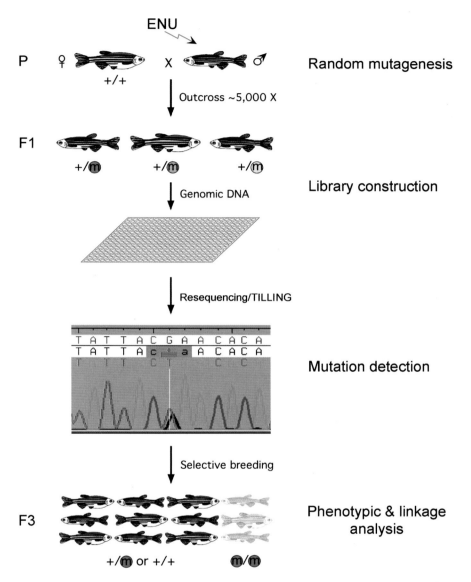

Chapter 4, Fig. 1 Target-selected mutagenesis in zebrafish. Approximately 100 adult male zebrafish are randomly mutagenized with *N*-Ethyl-*N*-nitrosourea (ENU) and outcrossed against wild-type females. A library of approximately 5000 F1 animals is constructed, in principle having independent mutations. Genomic DNA of these F1 animals is isolated and arrayed in 384-well PCR plates, suitable for robotic handling. The DNA is screened for mutations in target genes by resequencing or TILLING. Animals with interesting mutations are recovered from the library [reidentified from a pool of living F1 fish or recovered by *in vitro fertilization* (IVF) with frozen sperm] and outcrossed against wild-type fish or incrossed with other mutants. Finally, the mutations are homozygosed and animals are analyzed for phenotypes and linkage to the mutation.

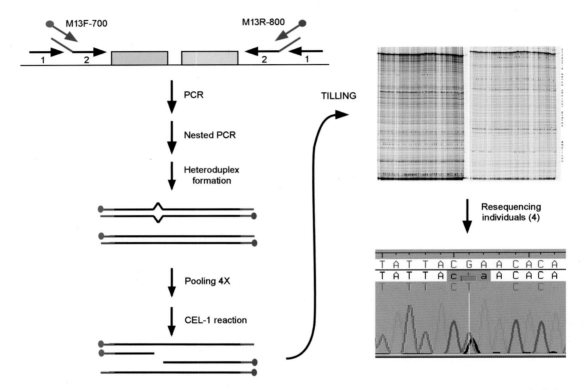

Chapter 4, Fig. 3 Mutation detection by TILLING. See text for detailed description of all the steps involved.

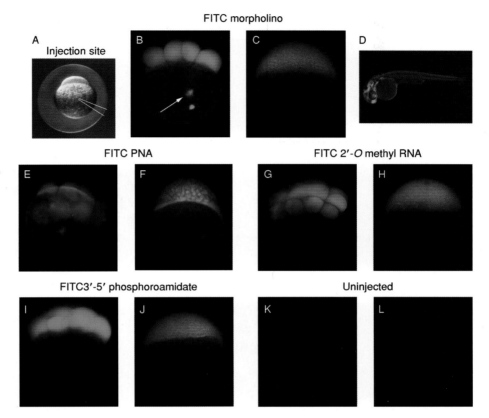

Chapter 7, Fig. 1 Uniform distribution of fluoresccin isothiocyanate (FITC)-modified oligonu-cleotides in zebrafish embryos after microinjection. FITC-labeled modified oligonucleotides of various chemistries were injected at the one- to two-cell stage as described (Nasevicius and Ekker, 2000). (A–D) Morpholino (MO)-injected embryos. (E, F) Peptide nucleic acid (PNA)-injected embryos. (G, H) 2′-*O* methyl RNA-injected embryos. (I, J) 3–5′ Phosphoroamidate oligonucleotide-injected embryos. (K, L) Uninjected embryos. Fluorescence was assayed by using a modified FITC filter set. Fluorescence indicates the injected oligonucleotide localization. The compounds are completely translocated to blastomeres as early as the eight-cell stage (<1 h after injection, Panels B, E, G, and I compared with Panel K). Later in development oligonucleotides remain uniformly distributed among blastomeres (midblastula, Panels C, F, H, and J compared with Panel L) and at 30 h of development (panel D). From Nasevicius and Ekker, (2001). The zebrafish as a novel system for functional genomics and therapeutic development applications. *Curr. Opin. Mol. Ther.* **3**, 224–228, with permission.

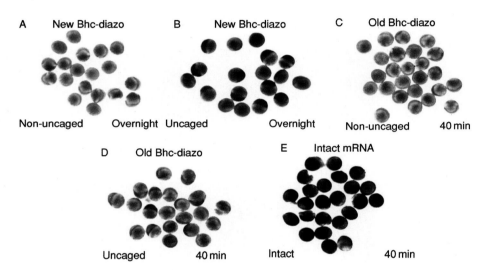

Chapter 9, Fig. 4 Comparison of the efficiency of mRNA caging with fresh and old Bhc-diazo and by injection of caged mRNA dissolved in DMSO. The embryos were injected with β-galactosidase mRNA caged with fresh (A, B) or old (C, D) Bhc-diazo, uncaged at 3 hpf, and stained for β-galactosidase activity at 6 hpf. The embryos injected with intact β-galactosidase mRNA were fixed at 6 hpf and stained for β-galactosidase activity for comparison (E).

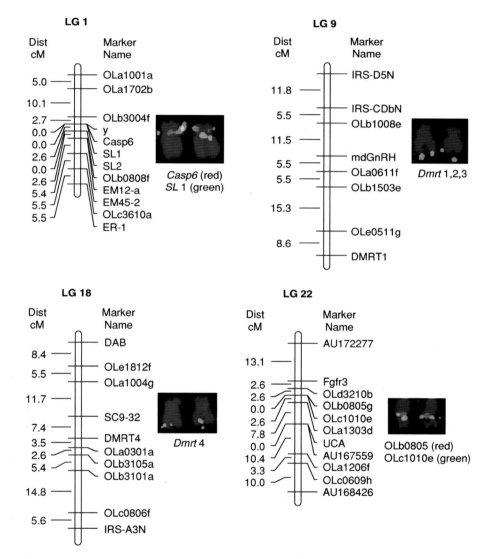

Chapter 10, Fig. 1 Preliminary mapping of genomic clones to four different linkage groups in medaka.

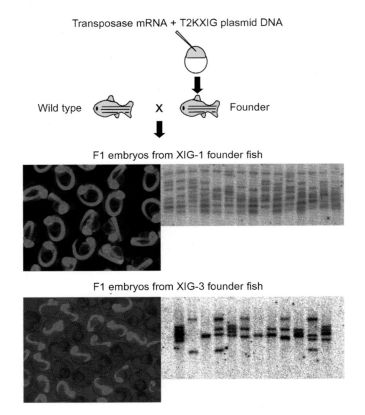

Transposase mRNA + T2KXIG plasmid DNA

Wild type   X   Founder

F1 embryos from XIG-1 founder fish

F1 embryos from XIG-3 founder fish

Chapter 11, Fig. 3   Transgenesis by using T2KXIG. Fish injected with the transposase mRNA and the plasmid DNA containing the transposon vector was mated with noninjected wild-type fish. The GFP expression in F1 embryos from the XIG-1 and the XIG-3 founder fish and Southern blot hybridization analysis of the F1 fish by using the GFP probe are shown.

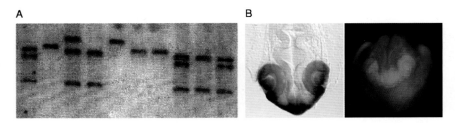

A   B

Chapter 11, Fig. 4   Transgenesis by using T2Ksix3.2G. (A) Southern blot hybridization analysis by using the GFP probe of F1 fish from the founder fish injected with the transposase mRNA and the plasmid DNA containing T2Ksix3.2G. (B) Expression of the endogenous *six3.2* mRNA revealed by whole mount *in situ* hybridization (left) and GFP expression in the F2 embryo carrying a single T2Ksix3.2G insertion (right).

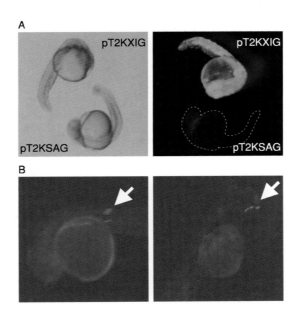

Chapter 11, Fig. 5 Transient GFP expression by using T2KXIG and T2KSAG. (A) GFP expression in embryos injected with the T2KXIG plasmid DNA and the transposase mRNA or the T2KSAG plasmid DNA and the transposase mRNA. (B) GFP expression in some muscle cells at Day 1 (left) or some neurons at Day 2 (right) in embryos injected with the T2KSAG plasmid DNA and the transposase mRNA.

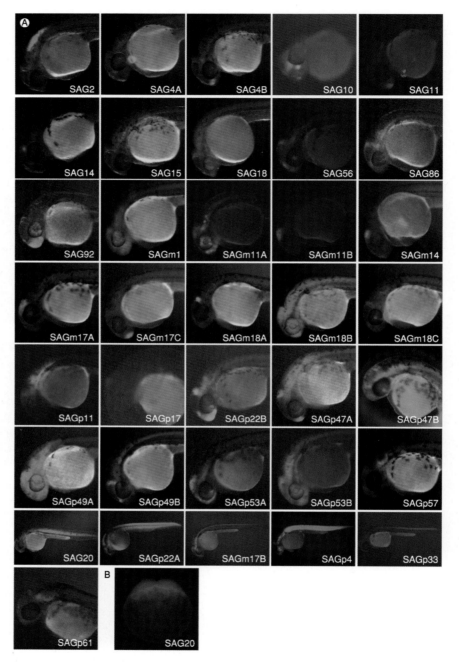

Chapter 11, Fig. 6 Unique GFP expression patterns identified in embryos carrying gene trap insertions. (A) Unique GFP expression patterns in embryos carrying T2KSAG insertions at 30–36 hpf. The patterns are named SAGXX. (B) Maternal GFP expression in an SAG20 embryo at the two-cell stage.

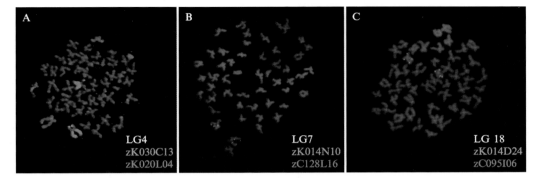

Chapter 13, Fig. 1 Representative two-color FISH images. Two-color FISH images for (**A**) the near-telomeric short-arm probe (red) and long-arm heterochromatin probe (green) of linkage group 4, (**B**) the near-centromeric (green) and near-telomeric long-arm probe (red) of linkage group 7, (C) the near-telomeric short-arm probe (red) and near-telomeric long-arm probe (green) for linkage group 18.

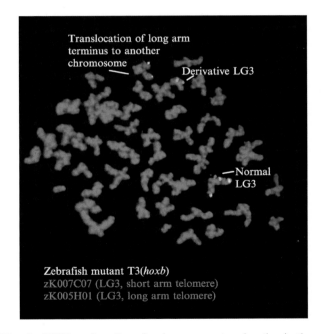

Chapter 13, Fig. 3 FISH confirmation of a chromosome translocation in the zebrafish mutant, T3(*hoxb*). A two-colored FISH experiment shows syntenic hybridization of a near-telomeric short-arm probe and a near-telomeric long-arm probe on one normal LG3 chromosome. The nonsyntenic hybridization of these same probes to two different chromosomes in this zebrafish mutant is consistent with a chromosomal translocation involving one LG3 chromosome and another nonhomologous chromosome.

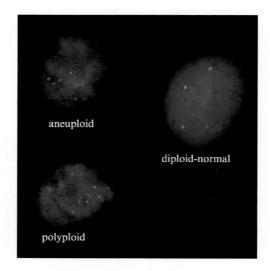

Chapter 13, Fig. 4 Use of BAC probes to assess genomic instability. Normal hybridization patterns would show two copies of each BAC probe in a given nucleus (diploid-normal). A polyploid cell would be expected to have the same number of signals for each BAC probe, provided there are three or more signals for each probe in a given nucleus. The polyploid cell shown has four green and four red signals, consistent with a tetraploid cellular content. Aneuploid cells would be expected to have a different number of signals for each BAC probe in a given nucleus. The aneuploid cell shown here has two copies of the green-labeled BAC probe and four copes of the red-labeled probe.

Four sets of RH mapping templates are arrayed in a 384-well PCR plate on a Robbins Hydra 96 Robot, dried at 55°C, and then stored at –20°C.

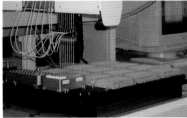

In 90 min, PCR reactions of 24 markers can be assembled in duplicates in 12 384-well PCR plates on the chilled deck of *GENESIS RSP 150* robot (TECAN).

In 2.5 h, PCR reactions of twelve 384-well plates can be completed using six GenAmp 9700 PCR machines (Applied BiosystemsInc.) (Total time = 4 h).

In 1 h, PCR reactions of each 384-well PCR plates (two-marker worth PCR reactions) can be analyzed on a 2% agarose gel in 1XTBE (total time 5 h).

Gel images are acquired and RH scores generated by using a Windows-based image processing software developed in the lab.

Chapter 15, Fig. 1 A high-throughput system for radiation hybrid panel mapping.

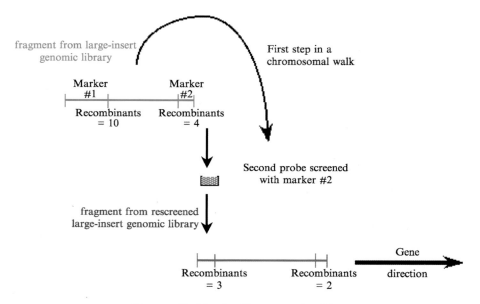

Chapter 17, Fig. 3 Chromosomal walking.

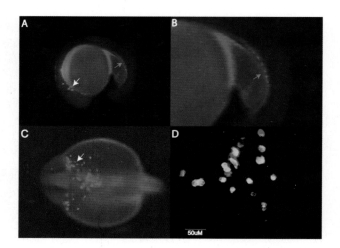

Chapter 18, Fig. 2 Fluorescent images of *TG(pu.1:EGFP)^{df5}* embryos at 22 hpf. (A) Lateral view of GFP+ cells. (B) Magnified lateral view of the posterior tail in (A). (C) Dorsal view of the anterior head region. White arrow, *pu.1*-expressing myeloid cells migrating from the anterolateral mesoderm over the yolk; red arrow, *pu.1* cells in the ICM. (D) Confocal image of two-color coexpression assays on cells migrating over the anterior yolk at 22 hpf using a *pu.1* RNA probe (red) and an antibody to EGFP (green). Yellow indicates coexpression. From Hsu, K. *et al.* (in press). The pu.2 promoter drives myeloid gene expression in zebrafish. *Blood*, with permission.

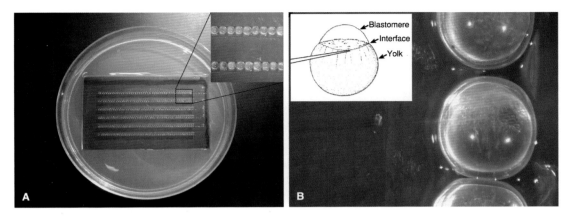

Chapter 19, Fig. 2 Microinjection process for the generation of transgenic fish, using the *Sleeping Beauty* transposon. (A) Agarose microinjection tray is loaded with one-cell-stage zebrafish embryos. (B) Transposon DNA/transposase RNA injection solutions are injected at the yolk–blastomere interface in one-cell zebrafish embryos. In this example, a tracer dye is included to visualize the solution transfer.

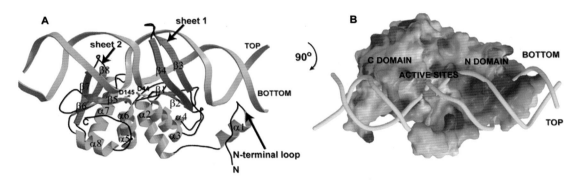

Chapter 21, Fig. 3 Structure of the I-SceI protein–DNA complex. (A) The secondary structure elements of the protein interacting with DNA (cyan ribbon) have been labeled as sheet 1 and sheet 2 (major groove contacts) and the N-terminal loop (minor groove contacts). α-Helices are depicted in green and β-strands in magenta. The catalytic aspartate residues are represented as ball-and-stick models. (B) GRASP potential surface area of the protein. The DNA is shown as a cyan ribbon. Positive charges are more abundant in the N domain. This figure was obtained by rotating the model shown in (A) by 90° about a horizontal axis, from Moure, C. M. *et al.* (2003). The crystal structure of the gene targeting homing endonuclease I-SceI reveals the origins of its target site specificity. Reprinted from *J. Mol. Biol.* **334,** 685–695, with permission.

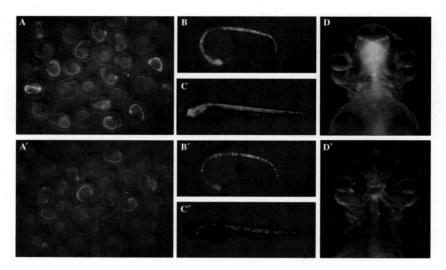

Chapter 21, Fig. 6 G0 expression of *shh*-GFP in zebrafish on microinjection with and without I-SceI (also see Table II). Circular vector containing an expression cassette [green fluorescent protein (GFP) driven by the *zfshh* promoter] flanked by I-SceI recognition sites (15 ng/μl) was injected into the cytoplasm of one-cell-stage zebrafish embryos with (A–D) or without (A′–D′) I-SceI meganuclease (0.3 U/μl). (A, A′) Overview of GFP-expressing embryos at 24 npf. Total numbers of GFP-expressing embryos increased on coinjection of meganuclease (A) compared to injection of DNA alone (A′). (B, B′) Representative samples of whole embryos exhibiting uniform promoter-dependent GFP expression at 24 hpf. Coinjection (B) of meganuclease yielded more transgene-expressing cells within the primary expression domain (notochord) than injection without I-SceI (B′), resulting in less mosaic GFP expression. (C, C′) The same representative samples as in B and B′ are shown at 48 hpf. The meganuclease coinjected embryos retained uniform GFP expression (C), but expression of GFP was greatly reduced in embryos injected without I-SceI (C′). (D, D′) A close-up of the head region of embryos at 48 hpf showing primary and secondary SHH expression domains. Overall GFP expression levels in meganuclease-injected embryos were still strong, and secondary domains (retinal ganglion cells and amacrine cells) also showed uniform expression of GFP (D). GFP expression levels in conventionally injected embryos were poor and highly mosaic in primary and secondary domains of SHH expression (D′).

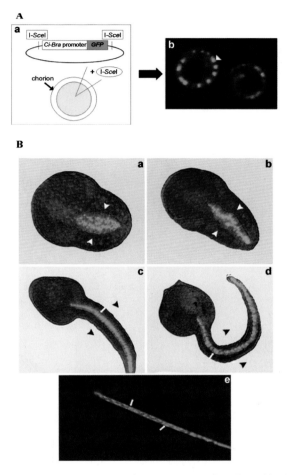

Chapter 21, Fig. 7 Generation of transgenic *Ciona savignyi* by I-SceI. (A, a) Circular transgene containing flanking I-SceI recognition sites was coinjected with meganuclease. (b) G0 late tailbud stage embryos. GFP expression in the notochord is evident. Arrowhead points to a GFP-positive notochord cell. (B) GFP-expressing embryos and larva derived from a transgenic founder. (a, b) Early tailbud stage embryos. Notochord cells are converging toward the midline (white arrowheads). (c) Mid-tailbud stage embryo. Mediolateral intercalation is complete and the notochord now consists of a single row of GFP-positive cells (white arrow). (d) Late tailbud stage embryo. Note the discontinous GFP signal due to formation of vacuoles in the notochord cells (white arrow). (e) GFP-positive swimming larva. Note the position of GFP-labeled cell nuclei (white arrows). From Deschet, K. *et al.* (2003). Generation of Ci-Brachyury-GFP stable transgenic lines in the ascidran *Ciona savignyi. Genesis* **35,** 248–259. Reprinted by permission of Wiley-Liss, Inc., a subsidiary of John Wiley & Sons, Inc. with permission.

Chapter 24, Fig. 3 Compact discs (CDs) and digital video discs (DVDs) serve as efficient vehicles for collecting visual data into a common visualization database.

Chapter 24, Fig. 4 "Zebrafish: The Living Laboratory." This CD visualization database was compiled from contributions from more than 10 laboratories. As a portable compendium, CDs and DVDs provide easily accessible repositories for large quantities of visualizations and other visual data.

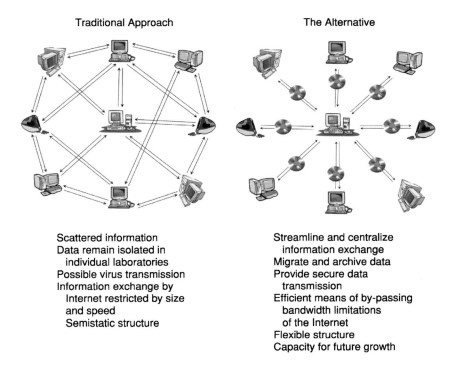

Traditional Approach

The Alternative

Scattered information
Data remain isolated in
    individual laboratories
Possible virus transmission
Information exchange by
    Internet restricted by size
    and speed
Semistatic structure

Streamline and centralize
    information exchange
Migrate and archive data
Provide secure data
    transmission
Efficient means of by-passing
    bandwidth limitations
    of the Internet
Flexible structure
Capacity for future growth

Chapter 24, Fig. 5   The DVD Exchange Project represents a virtual Intranet for mass visual data exchange.

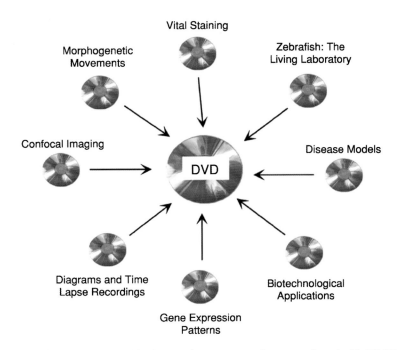

Vital Staining

Morphogenetic
Movements

Zebrafish: The
Living Laboratory

Confocal Imaging

DVD

Disease Models

Diagrams and Time
Lapse Recordings

Gene Expression
Patterns

Biotechnological
Applications

Chapter 24, Fig. 6   Because of its increased storage capacity, approximately 10-GB DVDs will soon become a major archival medium for visual information.

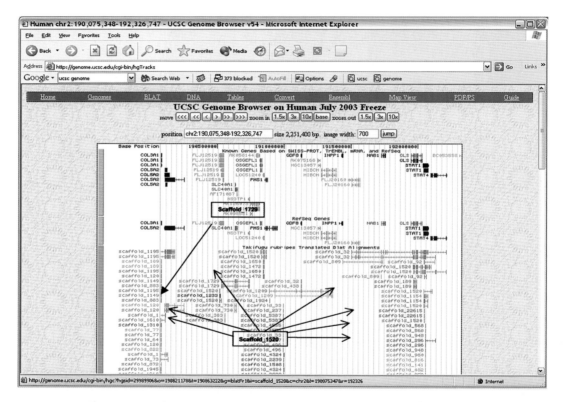

Chapter 25, Fig. 4 Comparative genomic functions available in the UCSC human genome browser. (A) Option selection buttons for controlling information displayed in the browser window. (B) A browser window showing selected information for human genome nucleotide bases, human known genes, and comparative genomic matches between humans and the fugu fish.

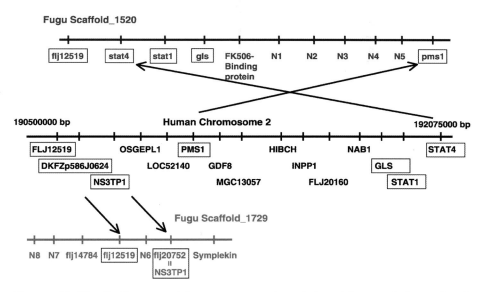

Chapter 25, Fig. 7 Summary of the synteny comparison of the *weh* locus equivalent regions in humans and fugu. The synteny between the FLJ12519-GLS-STAT1 interval and fugu scaffolds does not appear to be intact, but parts of this synteny interval do.

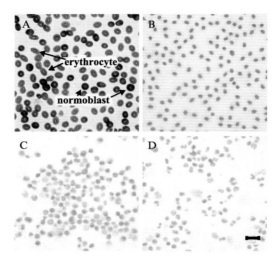

Chapter 26, Fig. 3   *In situ* hybridization of α-globin cRNA to blood cells from *N. coriiceps* and *C. aceratus*. Blood smears from *N. coriiceps* (A, B) and *C. aceratus* (C, D) were hybridized either to antisense (A, C) or sense (B, D) DIG-labeled α-globin RDA transcripts. Following application of alkaline-phosphatase-conjugated anti-DIG Fab fragments, the bound RNAs were detected as a brownish-black reaction product resulting from enzymatic action on the color reagents NBT and BCIP. After stopping the reaction, cells were counterstained with methylene green. Bar 25 μm.

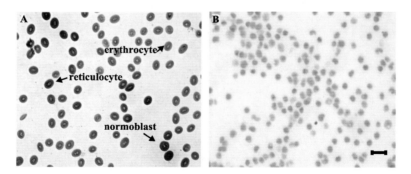

Chapter 26, Fig. 5   *In situ* hybridization of RDA 197 antisense cRNA to blood cells. (A) *N. coriiceps*. (B) *C. aceratus*. Sense controls did not produce a signal. Bar 25 μm.

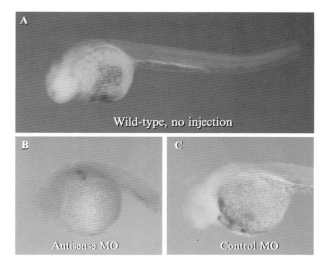

Chapter 26, Fig. 6 Suppression of red blood cell formation in zebrafish embryos by antisense morpholino oligonucleotides targeted to the *bty* mRNA. (A–C) Hemoglobin detection by *o*-dianisidine. (A) Uninjected wild-type embryo. The circulation stains reddish brown when reacted with *o*-dianisidine, indicating the presence of hemoglobin-expressing red cells. (B) Antisense MO Zebb302. Note the nearly complete absence of red blood cells. (C) Control MO Zebb302b. Red blood cells were present at near wild-type levels. Embryos were age matched (32 hpf) and micrographed in 70% glycerol/PBS, using a Nikon dissecting microscope with digital imaging system.

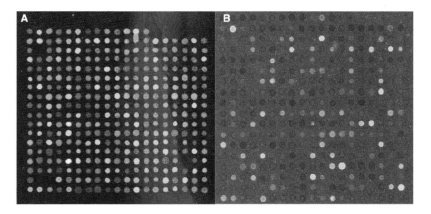

Chapter 28, Fig. 1 Removal of background intensity. (A) A green swirl of hybridization solution overlays this block. Within the area of the swirl, both the background and feature pixels are affected. (B) A faint array resulting from washing at a high stringency reveals the natural fluorescence levels of the slide surface chemistry. The probes to which no target has bound have a lower fluorescence than that of the surrounding background.

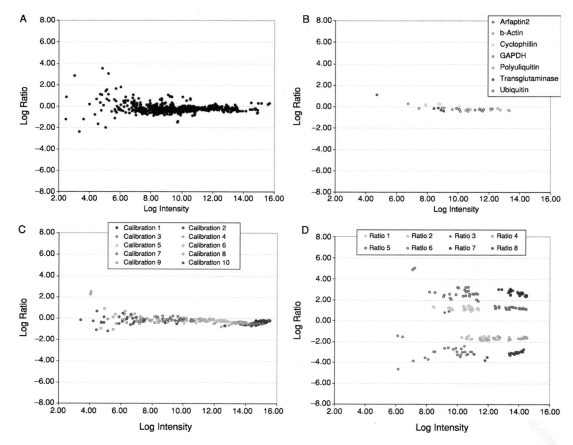

Chapter 28, Fig. 2 Zebrafish spiking control genes and housekeeping genes. Data from one of our zebrafish apoptosis arrays is presented, partitioned into four components: genes of interest, housekeeping genes, calibration controls, and ratio controls. These data represent a pool of embryos that were harvested 6 h following treatment with 15 mJ of UV radiation, compared to embryos of a similar age (78 hpf) that were not exposed to UV radiation. Every probe was printed onto the array six times. The spiking controls were part of the Lucidea Universal Scorecard (Amersham). These data were achieved from one of our early slides printed onto polylysine-coated slides (ESCO), using a Stanford-type arrayer and Majer Precision pins. (A) MA plot of the genes of interest present on this array, prior to normalization using the spiking controls and housekeeping gene. (B) Seven housekeeping genes were used to cover a range of intensities and provide sufficient data to enable normalization of the genes of interest to a $\log_2$ ratio of zero. (C) Ten calibration spikes from the Lucidea Universal Scorecard were included in the array. The spiking concentration covers four orders of magnitude to enable an estimation of the limit of detection, saturation, and to allow for normalization of the genes of interest to a $\log_2$ ratio of zero. There are sufficient data points in these controls to enable visualization of nonlinear intensity-related biases in this array. (D) The Lucidea Universal Scorecard also includes eight ratio controls at both high and low intensities to determine the extent to which the ratios obtained varied from the expected values. This aspect is an often overlooked consideration, particularly if significance is being determined by a twofold cutoff.

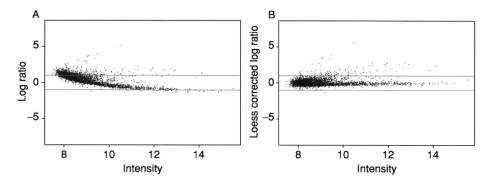

Chapter 28, Fig. 3 Loess correction of mean feature intensity by channel. It is common for intensity-related biases to be present within microarray data. The effect of global loess smoothing on clearly biased data is shown here. The red lines demark an arbitrary fold significance level. (A) MA plot prior to correction. (B) MA plot of data after normalization.

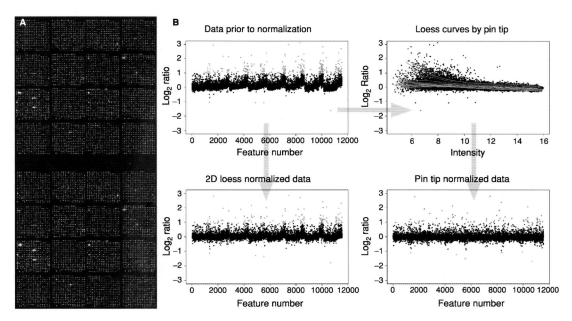

Chapter 28, Fig. 4 Pin tip and local mean normalization across the array surface. Here we demonstrate the effects of normalizing to remove spatial bias. (A) This array is printed in two sets of 4×4 blocks and has a green smear (probably SDS) that begins on Block 8 of Set 1 and increases in severity in the right-most blocks. (B) The effect of the green smear is most apparent when the $\log_2$ ratio of the fluorescent intensities is plotted against the feature number (in order from top left to bottom right). Pin tip normalization generates loess curves for each pin tip (each block) and the overall correction of this spatial effect is relatively good (bottom right graph). By contrast, local mean normalization across an array surface (2D loess) creates a form of topographical map using loess curves and attempts to flatten this map. Surprisingly, in this example, the 2D loess did not perform as well as the pin tip loess correction.

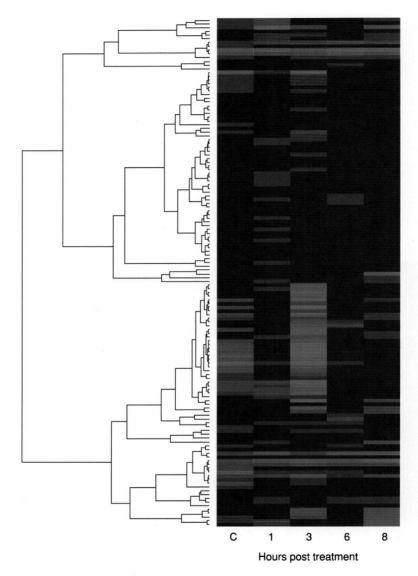

Chapter 28, Fig. 5 Clustering. Clustering was used to examine the time course of expression of zebrafish genes encoding proteins comprising the apoptosis pathway following the exposure of 72 hpf zebrafish embryos to 15 mJ of UV radiation. Pools of 200 zebrafish were exposed to UV radiation and harvested at 1, 3, 6, and 8 h following exposure. Transcripts from these embryos were labeled and hybridized against transcripts isolated from pooled embryos that were not exposed to UV radiation. "C" represents a control pool of 200 zebrafish that were not exposed to UV radiation. The dendrogram clusters genes by their similarity in expression profile. In this way, genes that ostensibly function in the same pathway can be determined.

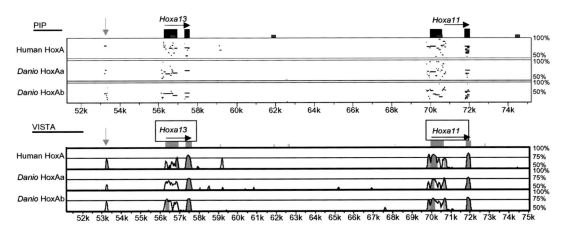

Chapter 29, Fig. 1 Comparison of two commonly used methods for visualizing global alignments for the 5′ end of the HOX-A clusters of shark vs. that of humans, and the two duplicates of zebrafish. GenBank sequences used in the analyses were AF224262 (shark), AC004079 (human), AC107365 (zebrafish Aa), and AC107364 (zebrafish Ab). The *Hoxa13* and *-a11* genes are denoted by boxes. In the PIP plot, nucleotide identities are indicated by dots in contrast to the VISTA plot in which identities are indicated by peaks. In addition, the blue and red shading in the VISTA plot denote coding and noncoding sequence identities, respectively. Notable identities such as indicated by the red arrows and the region surrounding exon-2 of *Hoxa11* represent potential *cis*-regulatory elements (e.g., enhancers, suppressors, and micro-RNA sites). Not all pairwise comparisons are given in these outputs; only comparisons with the shark reference sequence are shown. Other conserved sequences might be present between given lineages, but to identify them one must swap reference sequences in the percent identity plot (PIP) and visualizing global DNA sequence alignment of arbitrary length (VISTA) analyses.

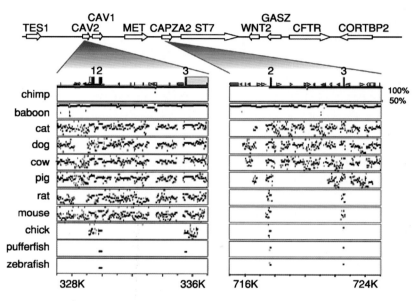

Chapter 29, Fig. 2 Demonstration of genomic divergence in a region across a wide range of vertebrates. Pairwise comparisons across an extended region that encompasses the cystic fibrosis gene were made with a human reference sequence vs. those from 11 vertebrate taxa, primarily mammals. Partial PIP plots are shown for two gene regions, *CAV2* and *CAPZA2*. As would be predicted, the closer a species is phylogenetically to humans, the more nucleotide identity one observes, including intron sequences (e.g., chimp and baboon); sequences from more distant taxa show decreasing levels of identity. The degree of divergence between the two genes, however, is shown to vary. Adapted and modified from Thomas, T. W. *et al.* (2003). Comparative analyses of multi-species sequences from targeted genome regions. *Nature* **424**, 788–793, with permission.